W9-DHW-504

Common organic solvents

Name	Boiling point (°C)	Density (g·mL^{-1})	Flammable	Miscible with H$_2$O
Acetic acid	118	1.049	no	yes
Acetone (2-propanone)	56.5	0.791	yes	yes
Chloroform*	61	1.484	no	no
Cyclohexane	80.7	0.778	yes	yes
Dichloromethane	40	1.325	no	no
Dimethylformamide (DMF)	153	0.944	yes	yes
Ethanol (95% aq. azeotrope)	78.2	0.816	yes	yes
Ethanol (anhydrous)	78.5	0.789	yes	yes
Ether (diethyl)	34.6	0.713	yes	no
Ethyl acetate	77	0.902	yes	slightly
Heptane	98.4	0.684	yes	no
Hexane	69	0.660	yes	no
Methanol	64.7	0.792	yes	yes
Pentane	36.1	0.626	yes	no
2-Propanol (Isopropyl alcohol)	82.5	0.785	yes	yes
Tetrahydrofuran (THF)	66	0.889	yes	yes
Toluene	110.6	0.866	yes	no

* Suspect carcinogen.

Selected data on common acid, base and salt solutions

Compound	Molarity	Density (g·mL^{-1})	% by weight
Acetic acid (glacial)	17	1.05	100
Ammonia (concentrated)	15.3	0.90	28.4
Hydrochloric acid (concentrated)	12	1.18	37
Nitric acid (concentrated)	16	1.42	71
Phosphoric acid (concentrated)	14.7	1.70	85
Sodium hydroxide	6	1.22	20
Sodium bicarbonate (saturated)	0.74	1.04	6
Sodium chloride (saturated)	5.3	1.19	26
Sulfuric acid (concentrated)	18	1.84	96.5

EXPERIMENTAL

ORGANIC

CHEMISTRY

A Balanced Approach:

Macroscale and Microscale

EXPERIMENTAL ORGANIC CHEMISTRY

A Balanced Approach:

Macroscale and Microscale

JERRY R. MOHRIG
Carleton College

CHRISTINA NORING HAMMOND
Vassar College

TERENCE C. MORRILL
Rochester Institute of Technology

DOUGLAS C. NECKERS
Bowling Green State University

W. H. FREEMAN AND COMPANY

New York

Acquisitions Editor: Michelle Russel Julet

Development Editor: Randi Rossignol

Marketing Manager: John Britch

Project Editor: Penelope Hull

Text and Cover Designer: Blake Logan

Cover Photographer: Joshua Sheldon

Cover photographic manipulation: Patricia McDermond

Illustrator: Fine Line Illustration

Production Coordinator: Sheila Anderson

Compositor: Progressive Information Technologies

Manufacturer: Quebecor/KP

All experiments contained herein have been performed several times by students in college laboratories under supervision of the authors. If performed with the materials and equipment specified in this text, the authors believe the experiments to be a safe and valuable educational experience. However, all duplication or performance of these experiments is conducted at one's own risk. The authors and publisher hereby disclaim any liability for any loss or damage claimed to have resulted from or related in any way to the experiments, regardless of the form of action.

Library of Congress Cataloging-in-Publication Data
Experimental organic chemistry : a balanced approach, macroscale and
    microscale / Jerry R. Mohrig . . . [et al.].

    p. cm.
    Includes bibliographical references (p.   –   ) and index.
    ISBN 0-7167-2818-4
    1. Chemistry, Organic—Laboratory manuals. I. Mohrig, Jerry R.
QD261.E96 1998
547′.0078—DC21

                    97-9409

Printed in the United States of America

First printing, 1997

W. H. Freeman and Company
41 Madison Avenue, New York, NY 10010
Houndmills, Basingstoke RG21 6XS, England

Contents

M   = macroscale
       procedure

m   = microscale
       procedure

🔒⊸   = discovery
       based

 = CD ROM

vi   Contents

Contents

Part 3 Organic Qualitative Analysis   497

Part 4   Spectrometric Methods   **577**

Part 5   Techniques of the Organic Laboratory   **687**

Preface

This book builds on the success of the three editions of *Laboratory Experiments in Organic Chemistry* by Jerry Mohrig and Douglas Neckers, bringing it into the twenty-first century with modern chemistry and thoughtful pedagogy. Chemistry is an experimental science. It is in the laboratory that students learn its process. Our primary objective was to write a laboratory text that involves students in the process. Using the laboratory only to have students confirm what they have seen in the classroom wastes a precious opportunity. Although we include a number of standard synthesis experiments, we also provide the chance for students to test ideas and address experimental questions in an atmosphere where they can succeed, making the organic chemistry laboratory rewarding and enjoyable for both students and their teachers.

(Mm) A Balanced Approach

We provide a wide variety of experiments and projects, including a balanced approach of macroscale and microscale, thereby allowing a great deal of flexibility in planning a laboratory program. The **mix of macroscale and microscale experiments** allows students to start with macroscale work to gain confidence before progressing to microscale. Many experiments are written with both macroscale and microscale procedures. Our macroscale experiments use the smallest reagent quantities that lead to successful results. The scale is tailored to fit the conditions of the reaction and the methods used to characterize the product. Our microscale experiments use 150–250 mg of starting material, which provides significant savings in chemical and waste disposal costs, yet ensures that students almost always have the satisfaction of a successful outcome.

An Investigative Approach

The book has many **discovery-based experiments** that involve students in active learning. They take students beyond the "cookbook" approach to address questions whose answers come from their experimental data. Which isomer was formed? What does the product ratio tell you about the reaction mechanism? What is the identity of the product? Answering questions requires mastery of good laboratory technique, learned in the process of scientific investigation. The experiments are carefully designed to reward student efforts with success.

We have included a wide range of **multiweek projects** in the book. Unlike most other organic chemistry lab texts, there are projects designed for the introductory level as well as more sophisticated projects for the second-semester laboratory. Indeed, it is possible to plan an entire laboratory program around the project approach. Multiweek projects allow students to become much more engaged with their laboratory work than is often possible with a set of separate 3- to 4-hour experiments. Certainly chemists do not practice their science in short segments. Research is always a project approach. The flexible use of time in multiweek projects allows students more opportunity to engage in an active learning environment. Projects also provide effective opportunities for teamwork.

A Practical Approach

In addition to offering students many attractive choices and options that we hope will catch their interest, **the experiments work.** For the past three years, they have been class tested at Vassar College, Carleton College, and Bowling Green State University. We have used this student and faculty input to revise and improve them.

Our experiments teach the fundamental methods of **modern laboratory technique,** including an appreciation of organic synthesis, bioorganic chemistry, and the analysis and characterization of organic compounds. There is extensive use of NMR and IR, analyzing reactions by TLC and GLC, Wittig methodology, and phase-transfer catalysis. Although 300-MHz FTNMR spectra are standard throughout the book, it is also possible to do most experiments and projects successfully using 60-MHz NMR.

In any laboratory, **safety** and **waste disposal** are important issues. They have been considered in the development of every experiment. Each experiment includes information on the safe handling of chemicals. How to handle reaction by-products and waste materials is also provided at the end of every procedure, with specific directions relating to cleaning up and neutralizing aqueous solutions.

Organization

The book is divided into five parts. The table of contents shows that **Part 1, Experiments,** is comprehensive enough to permit a complete course using either macroscale experiments or microscale techniques. **Part 2, Projects,** offers students the opportunity to probe areas of organic chemistry that would be impossible to explore in a single lab period in an environment that fosters active learning. **Part 3, Organic Qualitative Analysis,** is based on a flexible combination of modern spectrometry and classical chemical methods. Modern spectrometry has revolutionized structure determination in organic chemistry, and **Part 4, Spectrometric Methods,** discusses them in detail. In **Part 5,** the major **Techniques of the Organic Laboratory** are discussed comprehensively, and they are extensively cross referenced in all experimental procedures. We use square brackets—[see Technique 3.2]—to make it easy for the student to find the details of a specific technique when they encounter it in an experiment.

Supplements

An *Instructor's Manual* is available. It gives approximate times required for completion of the experiments, the amounts of reagents, equipment, and supplies needed, good ways to dispense reagents, and answers to the questions that appear at the end of each experiment. It also includes sample syllabi that provide options for structuring an organic chemistry laboratory program.

A *CD-ROM* by Melanie Cooper and Timothy S. Kerns, which contains virtual reality movies of selected techniques, is packaged inside the back cover of the book. These techniques are identified in the book by the icon 📀. We suggest that the appropriate movie be viewed as part of the prelab preparation for using a new technique.

Acknowledgments

We hope that teachers and students of organic chemistry find our text useful in their laboratory work. We would be pleased to hear from users of the book regarding impressions of the experiments and projects and ways in which the book might be improved. Please write to us in care of the Chemistry Acquisitions Editor at W. H. Freeman and Company, 41 Madison Avenue, New York, NY 10010, or e-mail us at chemistry@whfreeman.com.

We have benefitted greatly from the insights and thoughtful critiques of the following reviewers:

Steven R. Angle, University of California, Riverside
Gottfried Brieger, Oakland University
William D. Closson, deceased, State University of New York at Albany
Thomas A. Evans, Denison University
Thomas W. Flechtner, Cleveland State University
Rosemary Fowler, Cottey College
Jacquelyn Gervay, University of Arizona
Gordon W. Gribble, Dartmouth College
Jhong K. Kim, University of California, Irvine
Larry L. Lambing, Missouri Western State College
Joseph W. LeFevre, State University of New York, Oswego
Kirk P. Manfredi, University of Northern Iowa
John N. Marx, Texas Tech University
Robert Minard, Pennsylvania State University

Roger K. Murray, University of Delaware
Quentin R. Petersen, Central Michigan University
Paul Schueler, Raritan Valley Community College
Sara Selfe, University of Washington
Stansilaw Skonieczny, University of Toronto
Ronald Starkey, University of Wisconsin—Green Bay
Julie Tan, Cumberland College
Jeffrey P. Ward, Georgetown University
Michael A. Walters, Dartmouth College
James J. Worman, Rochester Institute of Technology

We are indebted to the late David Todd, who provided many helpful suggestions during the development of the experiments, reviewed the manuscript, and read the entire book in galley proofs with careful attention to detail.

We greatly appreciate the comments and suggestions made by our colleagues who class-tested the experiments and projects in their laboratory sections. We thank Curt Beck, Nicholas Benfaremo, Brian LeBlanc, Anthony Arnold, Christopher Smart, and Sarjit Kaur at Vassar College, David Alberg and Charles Carlin at Carleton College, and Kurt Deshayes at Bowling Green State University.

We are grateful to Edith Stout and Melanie Levesque for recording spectral data and to Joseph LeFevre for consultation on NMR spectra. Matthew Finacure of Pennsylvania State University checked the safety information and cleanup protocols for accuracy. We also acknowledge the contribution of the following Vassar students who assisted in the development and testing of experiments: Rebecca Ballard, Jon Ihle, Ansu Gebeh, Stephan Muhlebach, Jessica Boxhill, Michael Gaskin, Sathya Theodore, and David Michalak.

Special thanks go to the Vassar students who posed for the cover and part opening photographs. On the cover, left to right, they are Samuel Rutherford, Nina Pascuzzi, David Michalak, and Sathya Theodore.

We wish to thank Deborah Allen, Michelle Russel Julet, and Randi Rossignol, our editors at W. H. Freeman and Company, for their vision and direction in keeping the project going forward. We also thank Maria Epes, Art Director, and Blake Logan, Designer, for the elegant book design; Penelope Hull, Project Editor, for her masterly orchestration of the production stages; and Jodi Simpson for her careful copy editing;.

TCM gratefully acknowledges support for a sabbatical year from RIT and Dartmouth College. Last, we could not have completed such a large task without the unending patience and support of our spouses, Jean Mohrig, William Hammond, and Deanne Morrill, and our families.

GETTING ACQUAINTED WITH THE LABORATORY

I.1 Experimental Organic Chemistry

Chemistry is an experimental science. The laboratory is where you learn its processes. It is much easier to understand concepts when a laboratory experience ties in with the theories discussed in the classroom. As you do your laboratory work, you will have numerous opportunities to design experiments and test out your ideas. Sometimes your task will be determining which products form in a chemical reaction; other times you will investigate the best ways to carry out reactions.

To learn experimental organic chemistry, you need to master an array of techniques for handling and analyzing organic compounds and reactions. We feel that these techniques are important enough to deserve an entire section devoted just to them. The same is true of spectrometry, particularly nuclear magnetic resonance and infrared spectrometry, which have almost revolutionized the practice of chemistry in the last few decades. The separate sections on techniques and spectrometry will make them more accessible to you in the wide range of experiments found in the book. To make it easy for you to use these sections when you need them, we include within the experiments cross references that will look like this: [see Technique 3.2].

I.2 Laboratory Safety

As you begin your study of experimental organic chemistry, you need to learn how to work safely with organic chemicals. Many organic compounds are flammable or toxic. Some are volatile and can easily get into the air that you breathe; others are absorbed through the skin. Despite the hazards, these compounds can be handled with a minimum of risk if you are adequately informed about the hazards and necessary safe-handling practices, and if you use common sense while you are in the laboratory. Safety information may be found in a variety of sources that are discussed in Technique 1.6. We have consulted these sources in developing the experiments in this book and have provided you with a summary of the most important chemical safety information about each compound used in an experiment. You will find this safety information at the beginning of each experimental procedure in the following format:

┌───┐
SAFETY INFORMATION

Salicylic acid is toxic and an irritant to skin, eyes, mucous membranes, and the upper respiratory tract. Avoid breathing salicylic acid dust; and do not allow it to contact skin, eyes, and clothing.

Acetic anhydride is toxic, corrosive, and a lachrymator (causes tears). Use it only in a hood, wear gloves, and avoid contact with skin, eyes, and clothing.
└───┘

In addition to acquainting yourself with safe-handling practices for the chemicals that you will be using, you need a basic understanding of the general principles of laboratory safety. These are discussed in detail in Technique 1, a chapter that we urge you to read before you begin work in the laboratory. At the first meeting of your lab class, local safety issues will be discussed. These may include policies on safety glasses and goggles, the location of safety showers, eye wash stations, gloves, and procedures to be followed in emergency situations.

I.3 Cleanup Following an Experiment

After the completion of any experimental procedure, there will usually be some by-products, unused reagents, or other materials remaining. Concern for the environment and laws protecting the environment mean that these substances cannot simply be poured down the sink or thrown into the wastebasket. Therefore, we have provided a brief section entitled *Cleanup* at the end of each experimental procedure. The information given therein tells you what to do with each by-product remaining after you have completed the experimental procedure. The following example shows the format used in this book for cleanup information.

Cleanup: Neutralize the aqueous filtrate with solid sodium carbonate before washing the solution down the sink or placing it in the container for aqueous inorganic waste. Place the methanol solution from the recrystallization in the waste container designated for flammable (organic) liquids.

The information provided in the cleanup protocol is intended as a general guideline. Local environmental laws will always take precedence over the procedures given here. Consult your instructor about any changes in the cleanup procedure for your own laboratory.

I.4
Desk Equipment

You will find a variety of equipment in your laboratory desk, some of which will be familiar to you from earlier lab experiences. A typical student desk contains an assortment of beakers, Erlenmeyer flasks, filter flasks, thermometers, heating devices, clamps, test tubes, and a variety of other items. Your desk will probably have many, if not all, of the equipment items shown in Figure I.1.

In addition, you will have a set of carefully constructed and polished glassware with ground glass joints, called **standard taper**

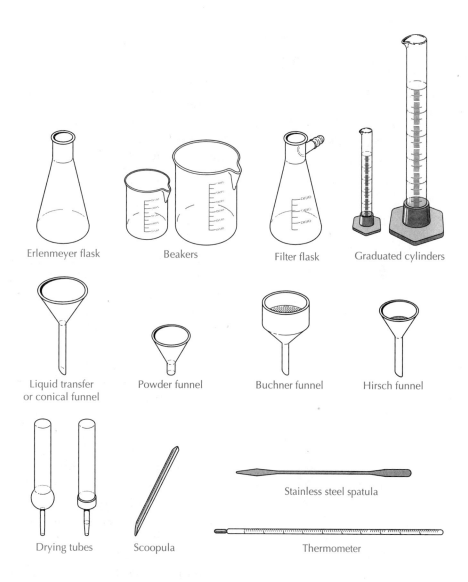

FIGURE I.1 Typical equipment in a student desk.

glassware, which is designed to fit together snugly in a variety of configurations. The use and care of standard taper glassware is discussed in Technique 2. Figure 2.2 of Technique 2 shows the pieces of this glassware commonly found in a student laboratory desk.

As you check over your desk equipment at the beginning of the term, examine each piece of glassware carefully for cracks or chips. Glassware with spherical surfaces, such as round-bottomed flasks, can develop small star-shaped cracks. Heating cracked glassware can result in leakage, thereby ruining your experiment and possibly causing a serious spill or fire. Replace any damaged glassware.

1.5
Introduction to Macroscale and microscale Laboratory Work

You will find that the experiments in this book are designed for two different quantities of chemical reagents and are labeled either macroscale or microscale. These terms refer both to the amounts of reagents and to the size of the glassware used. If your laboratory program includes microscale experiments, your desk may also contain microscale glassware that looks similar to the larger standard taper glassware, only of a much smaller size. This kind of glassware is illustrated in Figure 2.5 of Technique 2.

Traditionally, laboratories in organic chemistry courses have been equipped with glassware suitable for quantities of reagents in the 2–20 g range. These are designated macroscale experiments in this book. The quantities of chemicals that we suggest are the smallest that can be used successfully in macroscale glassware for a given experiment. The more recent introduction of microscale glassware and techniques into undergraduate organic chemistry laboratories has brought about a significant decrease in the amounts of chemicals used. In this book microscale experiments usually involve 150–250 mg of starting materials and appropriately scaled-down glassware.

Both microscale and macroscale procedures have advantages and disadvantages. The smaller quantities of chemicals used in microscale experiments can significantly decrease their economic and environmental costs and increase their safety. These are substantial advantages. However, when a synthesis requires multiple steps, like those in the project section, macroscale procedures are usually used to allow success in achieving a reasonable quantity of the final product. Even if every step in a synthetic sequence produces a 70% yield, a three-step synthesis would give only a 34% overall yield. If only 200 mg of starting material were used, there might not be enough final product to isolate and analyze. Microscale procedures can also be more difficult to use in the introductory organic chemistry laboratory, especially at the beginning. The manipulation of microscale quantities of liquids, particularly in distillation and extraction, can be more demanding than macroscopic distillation or extraction with a separatory funnel.

The experiments in this book have been designed to provide you with experience in both macroscale and microscale techniques. Many experiments offer a choice of scale. Some are only presented in a microscale version when, for example, only milligram quantities are needed for the analysis of the product or a starting reagent is particularly expensive or hazardous. When a macroscale approach will be more effective, it is the one used. In other words, the choice will depend on the scale that best suits the specific experiment.

I.6
Using Handbooks

You will need to use handbooks from time to time to look up physical constants, such as melting and boiling points, densities, and other useful facts. Three sources are particularly helpful.

The *Aldrich Catalog of Fine Chemicals* is published annually by the Aldrich Chemical Company of Milwaukee, Wisconsin. It lists over 37,000 organic and inorganic compounds and includes the chemical structure for each one, a brief summary of its physical properties, references on IR, UV, and NMR spectra, and safety information. There are also references to Beilstein's *Handbuch der Organischen Chemie* and to *Reagents for Organic Synthesis* by Fieser and Fieser (see Appendix B for more information). Figure I.2 shows a page from the current *Aldrich Catalog*.

The *CRC Handbook of Chemistry and Physics* is a commonly used handbook that is also published annually. The 77th Edition (1996–1997) contains a wealth of information, including extensive tables of physical properties and solubilities, as well as structural formulas, for 12,000 organic and 2400 inorganic compounds. Newer editions of the *CRC Handbook* have an arrangement of topics different from that of earlier editions. In the 77th Edition, organic compounds appear in Section 3 and inorganic compounds in Section 4. To use these tables successfully, you must pay close attention to how compounds are listed, because substituted derivatives of compounds are found under the listing of the parent compound. Figure I.3 shows a page from the 77th Edition of the *CRC Handbook of Chemistry and Physics*, which illustrates how the compound 1-bromobutane is listed under the parent compound, butane.

■ **Dimethylbu** ■

$

18,310-5   **3,3-Dimethyl-1-butanol,** 99% [624-95-3] $(CH_3)_3CCH_2CH_2OH$ FW 102.18 mp -60°.............   **1g**   **8.70**
★   bp 143° n_D^{20} 1.4140 d 0.844 Fp 118°F(47°C) Beil. 1(3),1677 FT-NMR 1(1),172A   **10g**   **38.90**
  FT-IR 1(1),115D SI 14,A,9 Safety 2,1359B R&S 1(1),125F   **50g**   **137.10**

13,682-4   **3,3-Dimethyl-2-butanol,** 98% [464-07-3] (pinacolyl alcohol) $(CH_3)_3CCH(OH)CH_3$.........   **25g**   **30.20**
★   FW 102.18 mp 4.8° bp 119-121° n_D^{20} 1.4150 d 0.812 Fp 84°F(28°C) [α]^{25} 0° (neat)   **100g**   **84.20**
  Beil. 1,412 FT-NMR 1(1),184B FT-IR 1(1),122D SI 15,C,9 Safety 2,1359C R&S 1(1),133E
  RTECS# EL2276000 FLAMMABLE LIQUID

  3,3-Dimethyl-2-butanone, see P4,560-5, Pinacolone page 1184

19,040-5   **2,3-Dimethyl-1-butene,** 97% [563-78-0] $(CH_3)_2CHC(CH_3)=CH_2$ FW 84.16 mp -158°.......   **5g**   **19.35**
★   bp 56° n_D^{20} 1.3890 d 0.680 Fp -1°F(-18°C) Beil. 1(3),816 FT-NMR 1(1),31B FT-IR 1(1),22A   **25g**   **62.10**
  SI 3,D,5 Safety 2,1359D R&S 1(1),23I FLAMMABLE LIQUID IRRITANT

22,015-9   **2,3-Dimethyl-2-butene,** 99 + % [563-79-1] $(CH_3)_2C=C(CH_3)_2$ FW 84.16 mp -75°...........   **10mL**   **34.15**
★   bp 73° n_D^{20} 1.4120 d 0.708 Fp 2°F(-16°C) Beil. 1,218 FT-NMR 1(1),35B FT-IR 1(1),25A   **25mL**   **52.40**
  SI 3,B,8 Safety 2,1360A R&S 1(1),25H FLAMMABLE LIQUID IRRITANT   **100mL**   **140.80**
  Substrate in photoinduced molecular transformations involving 2-hydroxy-1,4-
  naphthoquinones. J. Org. Chem. 1993, 58, 4614.

12,925-9   **2,3-Dimethyl-2-butene,** 98% [563-79-1] $(CH_3)_2C=C(CH_3)_2$...   **10mL**   **18.10**
★   **25mL**   **36.15**
    100mL   **109.15**
    1L   **521.15**

30,403-4   **2,3-Dimethyl-2-butene,** 1.0M solution in tetrahydrofuran [563-79-1]...........................   **100mL**   **14.65**
★ $(CH_3)_2C=C(CH_3)_2$ FW 84.16 d 0.857 Fp -5°F(-20°C) FT-NMR 1(1),25A FT-IR 1(1),25A   **800mL**   **35.30**
  SI 3,B,8 Safety 2,1360B R&S 1(1),25H FLAMMABLE LIQUID IRRITANT
  Used for the preparation of thexylborane. See 22,079-5, Thexylborane
  Preparation Kit.
  (Packaged under nitrogen in Sure/Seal™ bottles)

11,905-9   **3,3-Dimethyl-1-butene,** 95% [558-37-2] (neohexene) $(CH_3)_3CCH=CH_2$ FW 84.16........   **50mL**   **10.05**
★   mp -115° bp 41° n_D^{20} 1.3760 d 0.653 Fp -20°F(-28°C) Beil. 1,217 FT-NMR 1(1),29B   **250mL**   **41.85**
  FT-IR 1(1),25B SI 3,C,4 Safety 2,1360C R&S 1(1),23B FLAMMABLE LIQUID IRRITANT
  Stabilized with 50-150 ppm BHT

36,952-7   **N,N-Dimethylbutylamine,** 99% [927-62-8] $CH_3(CH_2)_3N(CH_3)_2$ FW 101.19....................   **250mL**   **13.50**
★   bp 93.3°/750mm n_D^{20} 1.3980 d 0.721 Fp 25°F(-3°C) Beil. 4,1,371 FT-NMR 1(1),483B   **1L**   **37.45**
  SI 44,A,3 R&S 1(1),321D FLAMMABLE LIQUID CORROSIVE

12,641-1   **1,3-Dimethylbutylamine,** 98% [108-09-8] $(CH_3)_2CHCH_2CH(CH_3)NH_2$ FW 101.19...........   **5g**   **22.70**
★   bp 108-110° n_D^{20} 1.4085 d 0.717 Fp 55°F(12°C) Beil. 4,191 FT-NMR 1(1),460B   **25g**   **77.80**
  FT-IR 1(1),288A SI 42,A,1 Safety 2,1360D R&S 1(1),307D RTECS# EO4460000
  FLAMMABLE LIQUID TOXIC

18,311-3   **3,3-Dimethylbutylamine,** 98% [15673-00-4] $(CH_3)_3CCH_2CH_2NH_2$ FW 101.19..................   **250mg**   **12.65**
  bp 114-116° n_D^{20} 1.4135 d 0.752 Fp 42°F(5°C) FT-NMR 1(1),460C FT-IR 1(1),287D   **1g**   **37.35**
  SI 42,B,1 Safety 2,1361A R&S 1(1),307E FLAMMABLE LIQUID CORROSIVE

24,439-2   **3,3-Dimethyl-1-butyne,** 98% [917-92-0] (tert-butylacetylene) $(CH_3)_3CC\equiv CH$.............   **5g**   **35.00**
  FW 82.15 mp -78° bp 37-38° n_D^{20} 1.3740 d 0.667 Fp < -30°F(-34°C) Beil. 1,256   **25g**   **116.65**
  FT-NMR 1(3),504C FT-IR 1(2),933B SI 425,A,4 Safety 2,1361B R&S 1(2),2697N
  FLAMMABLE LIQUID IRRITANT

35,990-4   **3,3-Dimethylbutyraldehyde,** 95% [2987-16-8] $(CH_3)_3CCH_2CHO$ FW 100.16....................   **1mL**   **24.50**
  bp 104-106° n_D^{20} 1.3970 d 0.798 Fp 51°F(10°C) Beil. 1,3,2843 FT-NMR 1(1),729B   **5mL**   **81.75**
  SI 73,B,3 R&S 1(1),509K FLAMMABLE LIQUID IRRITANT

D15,260-9   **2,2-Dimethylbutyric acid,** 96% [595-37-9] $C_2H_5C(CH_3)_2CO_2H$ FW 116.16...................   **5mL**   **15.60**
  bp 94-96°/5mm n_D^{20} 1.4154 d 0.928 Fp 175°F(79°C) Beil. 2,335 FT-NMR 1(1),761A   **100mL**   **17.30**
  FT-IR 1(3),579A SI 76,A,3 R&S 1(1),539A IRRITANT   **500mL**   **57.60**

  3,3-Dimethylbutyric acid, see B8840-3, tert-Butylacetic acid page 264

  3,3-Dimethylbutyryl chloride, see B8880-2, tert-Butylacetyl chloride page 264

D15,280-3   **Dimethylcarbamyl chloride,** 98% [79-44-7] $(CH_3)_2NCOCl$ FW 107.54 mp -33°.............   **5g**   **10.45**
★   bp 167-168°/775mm n_D^{20} 1.4530 d 1.168 Fp 155°F(68°C) Beil. 4,73 FT-NMR 1(1),1217A   **100g**   **10.85**
  FT-IR 1(1),743C SI 133,A,6 Safety 2,1361C R&S 1(1),871G RTECS# FD4200000   **500g**   **35.10**
  CANCER SUSPECT AGENT CORROSIVE

41,539-1   **1-(N,N-Dimethylcarbamoyl)-4-(2-sulfoethyl)pyridinium hydroxide, inner salt**........   **10g**   **12.50**
  [136997-71-2] FW 258.30 mp 185°(dec.) SI 404,C,2 IRRITANT   **50g**   **41.35**

41,539-1

FOR LABORATORY SUPPLIES SEE THE TECHWARE SECTION   ■ **579** ■

FIGURE 1.2 Page from the 1996–1997 *Aldrich Catalog.* Listings provide a summary of the physical properties for each compound. (Reprinted with permission from Aldrich Chemical Co., Inc., Milwaukee, WI.)

PHYSICAL CONSTANTS OF ORGANIC COMPOUNDS (continued)

No.	Name / Synonym	Mol. Form. / Mol. Wt.	CAS RN / mp/°C	Merck No. / bp/°C	Beil. Ref. / den/g cm^{-3}	Solubility / n_D
3202	Butanamide, 2,4-dihydroxy-N-(3-hydroxypropyl)-3,3-dimethyl-, (R)- / Dexpanthenol	$C_9H_{19}NO_4$ / 205.25	81-13-0 / dec	2924	4-04-00-01652 / 1.20^{20}	H$_2$O 4; EtOH 4; eth 2; MeOH 4 / 1.497^{20}
3203	Butanamide, N,N-dimethyl-	$C_6H_{13}NO$ / 115.18	760-79-2 / -40	186; 125^{100}	4-04-00-00185 / 0.9064^{25}	ace 4; bz 4; eth 4; EtOH 4 / 1.4391^{25}
3204	Butanamide, N,N'-(3,3'-dimethyl[1,1'-biphenyl]-4,4'-diyl)bis[3-oxo- / N,N'-Bis(acetoacetyl)-3,3'-dimethylbenzidine	$C_{22}H_{24}N_2O_4$ / 380.44	91-96-3 / 212		3-13-00-00490	DMSO 2
3205	Butanamide, N-(4-hydroxyphenyl)- / 4'-Hydroxybutyranilide	$C_{10}H_{13}NO_2$ / 179.22	101-91-7 / 139.5	4745	4-13-00-01109	H$_2$O 4; EtOH 4
3206	Butanamide, 3-methyl- / Isovaleramide	$C_5H_{11}NO$ / 101.15	541-46-8 / 137	5119 / 226	4-02-00-00902	H$_2$O 3; EtOH 3; eth 3; peth 4
3207	Butanamide, 3-oxo-N-phenyl- / Acetoacetanilide	$C_{10}H_{11}NO_2$ / 177.20	102-01-2 / 86	50	4-12-00-00955	H$_2$O 2; EtOH 3; eth 3; bz 3
3208	Butanamide, N-phenyl-	$C_{10}H_{13}NO$ / 163.22	1129-50-6 / 97	189^{15}	4-12-00-00387 / 1.134^{25}	H$_2$O 1; EtOH 4; eth 4; chl 2
3209	1-Butanamine / Butylamine	$C_4H_{11}N$ / 73.14	109-73-9 / -49.1	1543 / 77.0	4-04-00-00540 / 0.7414^{20}	H$_2$O 5; EtOH 3; eth 3 / 1.4031^{20}
3210	2-Butanamine, (±)- / sec-Butylamine, (±)-	$C_4H_{11}N$ / 73.14	33966-50-6 / <-72	1544 / 63.5	4-04-00-03618 / 0.7246^{20}	H$_2$O 3; EtOH 5; eth 5; ace 4 / 1.3932^{20}
3211	1-Butanamine, N-butyl- / Dibutylamine	$C_8H_{19}N$ / 129.25	111-92-2 / 62	3019 / 159.6	4-04-00-00550 / 0.7670^{20}	H$_2$O 3; EtOH 4; eth 4; ace 3 / 1.4177^{20}
3212	1-Butanamine, N-butyl-N-nitroso- / Dibutylnitrosamine	$C_8H_{18}N_2O$ / 158.24	924-16-3	105^8	4-04-00-03389	
3213	1-Butanamine, N,N-dibutyl- / Tributylamine	$C_{12}H_{27}N$ / 185.35	102-82-9 / -70	9530 / 216.5	4-04-00-00554 / 0.7770^{20}	H$_2$O 2; EtOH 4; eth 4; ace 3 / 1.4299^{20}
3214	1-Butanamine, 4,4-diethoxy-	$C_8H_{19}NO_2$ / 161.24	6346-09-4	196	4-04-00-01928 / 0.933^{25}	/ 1.4275^{20}
3215	1-Butanamine, N,N-dimethyl-	$C_6H_{15}N$ / 101.19	927-62-8	95	4-04-00-00546 / 0.7206^{20}	H$_2$O 5; EtOH 5; eth 5; ace 5 / 1.3970^{20}
3216	2-Butanamine, 2,3-dimethyl-	$C_6H_{15}N$ / 101.19	4358-75-2 / -70.5	104.5	4-04-00-00733 / 0.7683^0	/ 1.4096^{17}
3217	2-Butanamine, 3,3-dimethyl-	$C_6H_{15}N$ / 101.19	3850-30-4 / -20	102	4-04-00-00730 / 0.7668^{20}	H$_2$O 4 / 1.4105^{25}
3218	1-Butanamine, N-ethyl- / Butylethylamine	$C_6H_{15}N$ / 101.19	13360-63-9	107.5	4-04-00-00547 / 0.7398^{20}	EtOH 5; eth 5; ace 5; bz 5 / 1.4040^{20}
3219	2-Butanamine, N-ethyl-, (±)- / sec-Butyl ethyl amine (DL)	$C_6H_{15}N$ / 101.19	116724-10-8 / -104.3	98	2-04-00-00636 / 0.7358^{20}	ace 4; bz 4; eth 4; EtOH 4
3220	1-Butanamine, 3-methyl- / Isoamylamine	$C_5H_{13}N$ / 87.16	107-85-7	4994 / 96	4-04-00-00696 / 0.7505^{20}	H$_2$O 5; EtOH 5; eth 5; ace 3 / 1.4083^{20}
3221	1-Butanamine, N-methyl- / Butylmethylamine	$C_5H_{13}N$ / 87.16	110-68-9	91	4-13-00-00546 / 0.7637^{15}	
3222	2-Butanamine, 2-methyl-	$C_5H_{13}N$ / 87.16	594-39-8 / -105	77	4-04-00-00694 / 0.731^{25}	H$_2$O 4; ace 4; eth 4; EtOH 4 / 1.3954^{25}
3223	2-Butanamine, 3-methyl-	$C_5H_{13}N$ / 87.16	598-74-3 / -50	85.5	2-04-00-00644 / 0.7574^{19}	H$_2$O 4; EtOH 3 / 1.4096^{18}
3224	1-Butanamine, 2-methyl-N,N-bis(2-methylbutyl)-	$C_{15}H_{33}N$ / 227.43	620-43-9	232	0-04-00-00179 / 0.9^{13}	ace 4; bz 4; eth 4; EtOH 4 / 1.4330^{20}
3225	1-Butanamine, 3-methyl-N,N-bis(3-methylbutyl)-	$C_{15}H_{33}N$ / 227.43	645-41-0	235	4-04-00-00700 / 0.7848^{20}	H$_2$O 1; EtOH 4; eth 5; bz 5 / 1.4331^{20}
3226	1-Butanamine, 3-methyl-N-(3-methylbutyl)- / Diisoamylamine	$C_{10}H_{23}N$ / 157.30	544-00-3 / -44	3178 / 188	4-04-00-00699 / 0.7672^{21}	H$_2$O 1; EtOH 5; eth 5 / 1.4235^{20}
3227	2-Butanamine, N-(1-methylpropyl)-	$C_8H_{19}N$ / 129.25	626-23-3	135	4-04-00-00620 / 0.7534^{20}	H$_2$O 4; EtOH 3; os 3 / 1.4162^{20}
3228	1-Butanamine, 1,1,2,2,3,3,4,4-nonafluoro-N,N-bis(nonafluorobutyl)-	$C_{12}F_{27}N$ / 671.10	311-89-7	178	4-02-00-00819 / 1.884^{25}	ace 3 / 1.291^{25}
3229	Butane	C_4H_{10} / 58.12	106-97-8 / -138.2	1507 / -0.5	4-01-00-00236 / *0.573^{25}	H$_2$O 3; EtOH 4; eth 4; chl 4 / 1.3326^{20}
3230	Butane, 2,2-bis(ethylsulfonyl)- / Sulfonethylmethane	$C_8H_{18}O_4S_2$ / 242.36	76-20-0 / 76	8937 / dec	3-01-00-02790 / 1.199^{85}	chl 3
3231	Butane, 1-bromo- / Butyl bromide	C_4H_9Br / 137.02	109-65-9 / -112.4	1553 / 101.6	4-01-00-00258 / 1.2758^{20}	H$_2$O 1; EtOH 5; eth 5; ace 5 / 1.4401^{20}
3232	Butane, 2-bromo-, (±)- / (±)-sec-Butyl bromide	C_4H_9Br / 137.02	5787-31-5 / -111.9	1554 / 91.2	4-01-00-00261 / 1.2585^{20}	ace 4; eth 4; chl 4 / 1.4366^{20}
3233	Butane, 1-bromo-4-chloro-	C_4H_8BrCl / 171.46	6940-78-9	175; 63^{10}	4-01-00-00264 / 1.4886^{20}	H$_2$O 1; EtOH 3; eth 3; ctc 2 / 1.4885^{20}
3234	Butane, 2-bromo-1-chloro- / 2-Bromo-1-chlorobutane	C_4H_8BrCl / 171.46	79504-01-1	146.5	4-01-00-00264 / 1.468^{20}	bz 4; eth 4; EtOH 4; chl 4 / 1.4880^{20}
3235	Butane, 1-bromo-3,3-dimethyl-	$C_6H_{13}Br$ / 165.07	1647-23-0	138	3-01-00-00409 / 1.1556^{20}	eth 4; EtOH 4; chl 4 / 1.4440^{20}
3236	Butane, 2-bromo-2,3-dimethyl- / 2-Bromo-2,3-dimethylbutane	$C_6H_{13}Br$ / 165.07	594-52-5 / 24.5	133; 87^{180}	4-01-00-00374 / 1.1772^{10}	eth 4; chl 4 / 1.4517
3237	Butane, 1-bromo-4-fluoro- / 1-Bromo-4-fluorobutane	C_4H_8BrF / 155.01	462-72-6	135	4-01-00-00263	eth 4; EtOH 4 / 1.4370^{25}
3238	Butane, 1-bromo-2-methyl-, (S)- / d-Amyl bromide	$C_5H_{11}Br$ / 151.05	534-00-9	636 / 121.6	4-01-00-00327 / 1.2234^{20}	H$_2$O 1; EtOH 3; eth 3; chl 4 / 1.4451^{20}
3239	Butane, 1-bromo-3-methyl- / Isoamyl bromide	$C_5H_{11}Br$ / 151.05	107-82-4 / -112	4996 / 120.4	4-01-00-00328 / 1.2071^{20}	H$_2$O 1; EtOH 3; eth 3; ctc 2 / 1.4420^{20}

3-90

FIGURE I.3 Page from the 77th Edition of the *CRC Handbook of Chemistry and Physics.* (Reprinted with permission from CRC Press Inc., Boca Raton, FL.)

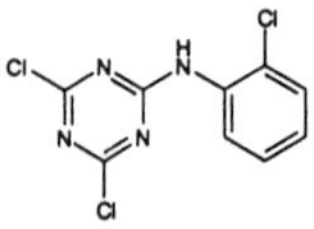

White to tan crystals, mp 159-160°. Insol in water. Soly at 30° (g/100 ml): toluene, 5; xylene, 4; acetone, 10. Subject to hydrolysis; not compatible with oils and alkaline materials. LD$_{50}$ orally in rats: > 5000 mg/kg, Mobay Technical Information Sheet, Jan. 1979.

USE: Fungicide.

695. Anileridine. *1-[2-(4-Aminophenyl)ethyl]-4-phenyl-4-piperidinecarboxylic acid ethyl ester; 1-(p-aminophenethyl)-4-phenylisonipecotic acid ethyl ester; ethyl 1-(4-aminophenethyl)-4-phenylisonipecotate; N-[β-(p-aminophenyl)ethyl]-4-phenyl-4-carbethoxypiperidine; N-β-(p-aminophenyl)-ethylnormeperidine; Leritine; Nipecotan; Alidine; Apodol.* C$_{22}$H$_{28}$N$_2$O$_2$; mol wt 352.48. C 74.97%, H 8.01%, N 7.95%, O 9.08%. Synthesis: Weijlard *et al., J. Am. Chem. Soc.* **78**, 2342 (1956); U.S. pat. 2,966,490 (1960 to Merck & Co.).

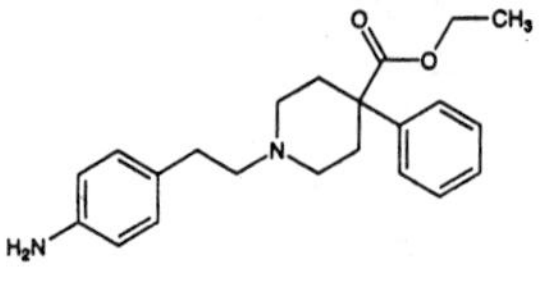

mp 83°.

Dihydrochloride, C$_{22}$H$_{28}$N$_2$O$_2$.2HCl, crystals from methanol + ether, mp 280-287° (dec). Freely soluble in water, methanol. Solubility in ethanol: 8 mg/g. pH of aq solns 2.0 to 2.5. Solns are stable at pH 3.5 and below. At pH 4 and higher the insol free base is precipitated. uv max (pH 7 in 90% methanol contg phosphate buffer): 235, 289 nm (A$_{1 cm}^{1\%}$ 293, 34.5). Distribution coefficient (water, pH 3.6/n-butanol): 0.9.

Note: This is a controlled substance (opiate) listed in the U.S. Code of Federal Regulations, Title 21 Part 1308.12 (1995).

THERAP CAT: Analgesic (narcotic).

696. Aniline. *Benzenamine; aniline oil; phenylamine; aminobenzene; aminophen; kyanol.* C$_6$H$_7$N; mol wt 93.13. C 77.38%, H 7.58%, N 15.04%. First obtained in 1826 by Unverdorben from dry distillation of indigo. Runge found it in coal tar in 1834. Fritzsche, in 1841, prepared it from indigo and potash and gave it the name aniline. Manuf from nitrobenzene or chlorobenzene: Faith, Keyes & Clark's *Industrial Chemicals*, F. A. Lowenheim, M. K. Moran, Eds. (Wiley-Interscience, New York, 4th ed., 1975) pp 109-116. Procedures: A. I. Vogel, *Practical Organic Chemistry* (Longmans, London, 3rd ed., 1959) p 564; Gattermann-Wieland, *Praxis des organischen Chemikers* (de Gruyter, Berlin, 40th ed., 1961) p 148. Brochure "*Aniline*" by Allied Chemical's National Aniline Division (New York, 1964) 109 pp, gives reactions and uses of aniline (877 references). Toxicity study: K. H. Jacobson, *Toxicol. Appl. Pharmacol.* **22**, 153 (1972).

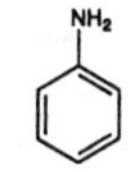

Oily liquid; colorless when freshly distilled, darkens on exposure to air and light. *Poisonous!* Characteristic odor and burning taste; combustible; volatile with steam. d$_4^{20}$ 1.022. bp 184-186°. Solidif −6°. Flash pt, closed cup: 169°F (76°C). n$_D^{20}$ 1.5863. pKb 9.30. pH of 0.2 molar aq

soln 8.1. One gram dissolves in 28.6 ml water, 15.7 ml boil. water; misc with alcohol, benzene, chloroform, and most other organic solvents. Combines with acids to form salts. It dissolves alkali or alkaline earth metals with evolution of hydrogen and formation of anilides, e.g., C$_6$H$_5$NHNa. *Keep well closed and protected from light. Incompat.* Oxidizers, albumin, solns of Fe, Zn, Al, acids, and alkalies. LD$_{50}$ orally in rats: 0.44 g/kg (Jacobson).

Hydrobromide, C$_6$H$_7$N.HBr, white to slightly reddish, crystalline powder, mp 286°. Darkens in air and light. Sol in water, alc. *Protect from light.*

Hydrochloride, C$_6$H$_7$N.HCl, crystals, mp 198°. d 1.222. Darkens in air and light. Sol in about 1 part water; freely sol in alc. *Protect from light.*

Hydrofluoride, C$_6$H$_7$N.HF, crystalline powder. Turns gray on standing. Freely sol in water; slightly sol in cold, freely in hot alc.

Nitrate, C$_6$H$_7$N.HNO$_3$, crystals, dec about 190°. d 1.36. Discolors in air and light. Sol in water, alc. *Protect from light.*

Hemisulfate, C$_6$H$_7$N.½H$_2$SO$_4$, crystalline powder. d 1.38. Darkens on exposure to air and light. One gram dissolves in about 15 ml water; slightly sol in alc. Practically insol in ether. *Protect from light.*

Acetate, C$_6$H$_5$NH$_2$.HOOCCH$_3$. Prepd from aniline and acetic acid: Vignon, Evieux, *Bull. Soc. Chim. France* [4] **3**, 1012 (1908). Colorless liquid. d 1.070-1.072. Darkens with age; gradually converted to acetanilide on standing. Misc with water, alc.

Oxalate, C$_6$H$_5$NH$_2$.HOOCCOOH.H$_2$NC$_6$H$_5$. Prepd from aniline and oxalic acid in alc soln: Hofmann, *Ann.* **47**, 37 (1843). Triclinic rods from water, mp 174-175°. Readily sol in water; sparingly sol in abs alc. Practically insol in ether.

Caution: Intoxication may occur from inhalation, ingestion, or cutaneous absorption. *Acute Toxicity:* cyanosis, methemoglobinemia, vertigo, headache, mental confusion. *Chronic Toxicity:* anemia, anorexia, wt loss, cutaneous lesions, *Clinical Toxicology of Commercial Products*, R. E. Gosselin *et al.,* Eds. (Williams & Wilkins, Baltimore, 4th ed., 1976) Section III, pp 29-35.

USE: Manuf dyes, medicinals, resins, varnishes, perfumes, shoe blacks; vulcanizing rubber; as solvent. Hydrochloride used in manuf of intermediates, aniline black and other dyes, in dyeing fabrics or wood black.

697. Aniline Mustard. *N,N-Bis(2-chloroethyl)benzenamine; N,N-bis(2-chloroethyl)aniline; phenylbis[2-chloroethylamine]; β,β'-dichlorodiethylaniline; Lymphochin; Lymphocin; Lymphoquin.* C$_{10}$H$_{13}$Cl$_2$N; mol wt 218.13. C 55.06%, H 6.01%, Cl 32.51%, N 6.42%. Prepd by the action of phosphorus pentachloride on *N,N-*bis-[2-hydroxyethyl]-aniline (phenyldiethanolamine): Robinson, Watt, *J. Chem. Soc.* **1934**, 1538; Korshak, Strepikheev, *J. Gen. Chem. USSR* **14**, 312 (1944).

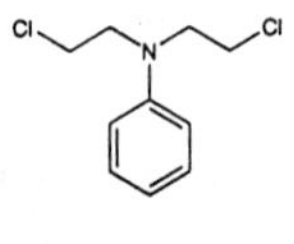

Stout prisms from methanol, mp 45°. bp$_{14}$ 164°; bp$_{0.5}$ 110°. Sol in hot methanol, ethanol. Very slightly sol in ether.

Hydrochloride, C$_{10}$H$_{14}$Cl$_3$N, crystals. *Vesicant.* Freely sol in water. Sol in alcohol.

USE: In cancer research.

698. Anilinephthalein. *3,3-Bis(4-aminophenyl)-1(3H)-isobenzofuranone; 3,3-bis(p-aminophenyl)phthalide.* C$_{20}$H$_{16}$N$_2$O$_2$; mol wt 316.36. C 75.93%, H 5.10%, N 8.85%, O 10.11%. Prepn: Hubacher, *J. Am. Chem. Soc.* **73**, 5885 (1951).

Consult the Name Index before using this section.   **Page 111**

FIGURE I.4 Page from the 12th Edition of *The Merck Index.* (Reprinted with permission from Merck and Co., Inc., Whitehouse Station, NJ.)

The Merck Index, currently in its 12th Edition, has over 10,000 entries of organic compounds, giving physical properties and solubilities as well as many references to the compounds' syntheses, safety information, and uses. *The Merck Index* is particularly comprehensive for organic compounds of medical and pharmaceutical importance. Figure I.4 shows a page from the 12th Edition of *The Merck Index.*

Appendix B contains a more thorough description of the literature of organic chemistry, detailing additional sources such as textbooks, reference books, chemistry journals, abstracts, and indexes.

I.7
Your Laboratory Notebook

A few general comments are in order about the laboratory notebook that will be the primary record of your experimental work. First, although many college bookstores sell books that are specifically designed as lab notebooks, it is often sufficient to use any notebook with a strong binding. Spiral and three-ring binders are inappropriate for lab notebooks. All your entries must be made directly in your laboratory notebook in ink. The use of scraps of paper for any records is unacceptable because these are easily lost; the practice will probably be strictly forbidden in your lab. The notebook should begin with a table of contents (leave a few pages for this); the following pages should be numbered sequentially.

Some flexibility in format and style may be allowed, but proper records of your experimental results must answer certain questions. *When* did you do the work? *What* are you trying to accomplish in the experiment? *How* did you do the experiment? *What* did you observe? *How* do you explain your observations? Your notebook must be written with accuracy and completeness. It must be organized and legible but does not need to be a work of art.

A lab record needs to be written in three steps: pre-lab, during lab, and post-lab. It should contain the following categories for each experiment you do.

To Be Done before You Come to the Laboratory

This notebook entry is designed to help you prepare for an experiment in an effective and safe fashion. It includes the date and title of the experiment, the balanced chemical reaction, and a statement of purpose; a table of reagents and solvents; the way you will calculate your percent yield; an outline of the procedure to be used; and answers to any pre-lab questions.

Title: Use a title that clearly identifies what you are doing in this experiment. Most experiments are syntheses. Exceptions include special analytical techniques or projects.
Date(s): Use the date when an experiment is actually carried out. In some research labs, where patent issues are important, a witnessed signature of the date is required.

Balanced Chemical Reaction: Write a chemical equation that shows the overall process. Any details of reaction mechanisms go into the summary.

Purpose Statement: Write a brief statement of purpose for the synthesis or analysis, with a few words on major analytical or conceptual approaches.

Table: Include all reagents and solvents. The table normally lists molecular weights, the number of moles and grams of reagents used, densities of the liquids used, boiling points of compounds that are liquids at room temperature and melting points of all organic solids, and pertinent hazard warnings. In some cases, special properties such as specific rotation or index of refraction are also listed.

Method of Yield Calculation: Outline the computation to be used in a synthesis experiment, including calculation of the theoretical yield.

Procedure and Pre-Lab Questions: Write an experimental outline in sufficient detail so that the experiment could be done without reference to your lab textbook, along with the answers to any assigned pre-lab questions.

To Be Done during the Laboratory Session(s)

Recording observations during the experiment is a crucial part of your lab notebook. If your observations are not complete, you will be unable to interpret the results of your experiment; once you have left the laboratory, it is difficult, if not impossible, to reconstruct them.

Observations: Observations must be recorded in your notebook in ink while you are doing an experiment. The actual quantities of all reagents must be recorded as they are used, as well as the amounts of crude and purified products that you obtained. Mention which measurements (temperature, time, melting point, etc.) and spectra are taken. Because organic chemistry is primarily an experimental science, your observations are crucial to your success. Things that seem insignificant may be important in understanding and explaining your results later. Some typical laboratory observations might be

A white precipitate appeared, which dissolved when the sulfuric acid was added; the solution turned cloudy when it was cooled to 10°C; 10 mL more solvent were required to completely dissolve the yellow solid; the reaction was heated at 50°C for 25 min with a warm-water bath; the NMR sample was prepared with (state amount of compound used), using $CDCl_3$ added to a height of 52 mm in the NMR tube; a polar GC column was used at a temperature of 137°C; the infrared spectrum was run on a mineral oil mull; a short puff of white smoke appeared when the sodium hydroxide was added.

The recording of your observations may be done in a variety of ways. They may be written on right-hand pages across from the corresponding section of the experimental outline. It is a good idea to cross-index your observations to specific steps in the procedure that you wrote out as part of your pre-lab preparation. Your instructor will likely provide specific advice on how you should record your observations during the laboratory.

To Be Done after the Experimental Work Has Been Completed

In this section of your notebook you evaluate and interpret your experimental results. Entries include calculation of the percent yield, interpretation of physical and spectral data, a summary of your conclusions, and answers to any assigned post-lab questions.

Percent Yield: Write out your calculation of the percent of the theoretical yield that you actually obtained; base your calculation on the limiting reagent in the synthesis experiment.

The single most important criterion of success in a chemical synthesis is how much of the desired product is produced. To be sure, the purity of the product is also crucial; but if a synthesis route produces only small amounts of the needed product, it is not much good. Reactions on the pages of textbooks are often far more difficult to carry out in good yield than the books suggest. The best yields of products come when experienced chemists carry out a reaction a second or third time.

When you report on a synthetic step, the measure of your success is the percent yield, a parameter defined as the amount of material obtained divided by the theoretical yield (amount expected) multiplied by 100. You calculate the theoretical yield from the balanced chemical equation and the limiting reagent, assuming 100% conversion of the starting materials to product(s):

$$\% \text{ yield} = \frac{\text{actual yield of product}}{\text{theoretical yield}} \times 100$$

Conclusions and Summary: Include a succinct discussion of your results. It is important that you write up a summary of your work as soon as possible after finishing an experiment, drawing conclusions consistent with your observations and results. These remarks must include a thorough interpretation of your NMR and IR spectra and other analytical results, such as TLC and GC analyses. Properly labeled spectra and chromatograms should be firmly stapled in your notebook. Answers to any assigned post-lab questions should be included. You should also cite any journal articles, books, and other reference sources that you have used.

p a r t

1

Experiments

Part 1 contains a wide variety of experiments covering most of the topics included in an organic chemistry course. The experiments provide a number of choices in planning a laboratory program. Many of the experiments have both macroscale and microscale procedures. For several topics there is more than one experiment from which to choose. Other experiments use a "discovery-based" format in which the answer to a question, such as "Which isomer is formed?" or "What does the product ratio tell you about the mechanism?" comes from the experimental data. Answering these questions requires mastery of good laboratory technique, learned in the context of scientific investigation. Optional experiments allow further exploration of a particular reaction.

While conducting these experiments you will develop an appreciation of organic synthesis and learn the fundamental methods of modern laboratory technique and the analysis and characterization of organic compounds. We have used a square-bracket notation—[see Technique 7.3]—to make it easy for you to find the details of a specific technique or spectrometric method when you encounter it in an experimental procedure.

The experiments have been tested by our students and are carefully designed to reward your efforts in the laboratory with success. We hope that you will find the experiments interesting and enjoyable as you learn the process of organic chemistry.

EXTRACTION OF CAFFEINE FROM TEA

Learn some of the basic techniques of organic chemistry—extraction, filtration, evaporation of a solvent and drying methods—in the context of working with a chemical known to all, caffeine.

Pure caffeine is a white, tasteless substance that makes up as much as 5% of the weight of tea leaves. By structure, caffeine is closely related to the purine bases, guanine and adenine, found in deoxyribonucleic acids (DNA).

Purine

Caffeine   Guanine   Adenine

A number of plants contain caffeine, and its use as a stimulant predates written history. The origins of tea and coffee are lost in legend. In addition to being in tea leaves and coffee beans, caffeine is a natural constituent of kola nuts and cocoa beans. Cola soft drinks contain 14–25 mg of caffeine per 100 mL (3.6 oz), and a sweet chocolate bar weighing 20 g (0.7 oz) contains about 15 mg of caffeine. "Stay awake" preparations such as No Doz have caffeine as a main active ingredient.

   The caffeine content of tea leaves depends on the variety and where they were grown; most tea has 3–5% by weight. Coffee beans contain only about 2% caffeine by weight, yet a cup of coffee has about 3.5 times as much caffeine as does a cup of tea. How can this be? Coffee is usually boiled in its brewing or else ground extremely fine; tea leaves are simply steeped in hot water for a few minutes. Furthermore, more ground coffee than tea is used to brew one cup of beverage. A cup of tea contains about 25 mg of caffeine.

The biological action of caffeine includes cardiac and respiratory stimulation, and it has a diuretic effect as well. Tea also contains a trace of the alkaloid theophylline, which is similar in structure to caffeine; it stimulates muscle action and relaxes the coronary artery. Theophylline also has veterinary applications as a diuretic and a cardiac stimulant.

Theophylline

Obtaining pure caffeine from tea requires a method for separating caffeine from the other substances found in tea leaves. Cellulose, the primary leaf component, poses no problem, because it is virtually insoluble in water. However, a large class of weakly acidic molecules called tannins also dissolve in the hot water used to dissolve the caffeine from tea leaves. Tannins are colored compounds having molecular weights between 500 and 3000 and phenolic groups that make them acidic. If calcium carbonate, a base, is added to tea water, calcium salts of these acids form in the tea solution. The caffeine can then be separated from the alkaline tea solution by a process of extraction [see Technique 4] using dichloromethane, an organic solvent in which caffeine readily dissolves. The calcium salts of the tannins remain dissolved in the aqueous solution. Flavinoid pigments and chlorophylls also contribute to the color of a tea solution. Although chlorophylls have some solubility in dichloromethane, the other pigments do not. Thus, the dichloromethane extraction of a basic tea solution removes nearly pure caffeine, which has a slight green color from the chlorophyll impurity.

After the extraction procedure, the organic solution of dichloromethane and caffeine is dried with an anhydrous inorganic salt [see Technique 4.6]. Crude caffeine is recovered as a solid residue by evaporation of the dichloromethane.

The solubility of caffeine in water at 20°C is 2.2 g per 100 mL, so there is no problem in keeping it in water solution while you filter off the spent tea leaves and calcium salts. Caffeine is far more soluble in dichloromethane: 10.2 g per 100 mL at 20°C. So this extraction takes advantage of a distribution coefficient (K) of 4.6 [see Technique 4.1 for a discussion of distribution coefficients]. To conserve dichloromethane and time, we will settle for two 15-mL extractions of the aqueous tea solution. This method does not extract all the caffeine but yields more than enough for the purification step. The 10 g of tea that you boil with water should contain at least 300 mg of caffeine. You will be able to recover 10–30% of this amount.

A comment about filtering the boiled tea solution should be made before you begin. If it is filtered when it is too hot, messy bubbling occurs in the filtrate and some solution may be lost. Yet if it is filtered when it is too cool, the gelatinous material that separates on cooling will clog the pores of the filter paper. Fast, nonretentive filter papers such as Schleicher and Schuell (S&S) No. 410 and Whatman No. 54 work well.

Macroscale Procedure

Techniques   Vacuum Filtration Apparatus: Technique 5, Figure 5.4

Extraction: Techniques 4.2, 4.3, and 4.4

Drying Agents: Techniques 4.6 and 4.7

Solvent Removal: Technique 4.8

SAFETY INFORMATION

Dichloromethane is toxic, an irritant, absorbed through the skin, and harmful if swallowed or inhaled. Use it in a well-ventilated hood. Wear gloves and wash your hands thoroughly after handling it.

Solid **caffeine** is toxic and an irritant. Avoid contact with skin, eyes, and clothing.

Before you begin your laboratory work, be sure to read Technique 4, Extraction and Drying Agents. It tells you how to do an extraction and explains why it works.

Place approximately 10 g of tea leaves in a 400-mL weighed (tared) beaker; record the mass of the tea leaves. If you use tea bags, four bags should contain about 10 g of tea; remove the tea leaves from the bags and place the tea in the beaker. Add 4.8 g of calcium carbonate and pour 100 mL of water over the tea. Boil the mixture gently on a hot plate for 15 min, stirring every minute or two with a stirring rod.

Let the tea mixture cool to about 55°C, then filter it, using vacuum filtration [Technique 5, Figure 5.4], through S&S No. 410 or Whatman No. 54 filter paper. Pour the tea mixture into the Buchner funnel in two portions. If the filter paper clogs while the first

Although you will often want to shake a separatory funnel vigorously to bring about efficient mixing of the two layers, in this extraction frustrating emulsions (a milky looking mixture that does not separate easily into the two phases) result from such a procedure. Swirling the two layers together rather than shaking should prevent or minimize the formation of an emulsion.

The amount of caffeine transferred from the aqueous layer to the dichloromethane layer depends on the amount of contact between the two phases, so you need to swirl the funnel fast enough to produce a good mixing of the phases without forming an appreciable amount of emulsion.

This process is called washing the organic layer.

If the amount of emulsion is more than a thin film, set up a vacuum filtration, using a 4.5-cm Buchner funnel containing a silicone-treated filter paper such as Whatman 1PS. Drain the lower layer and any emulsion layer directly from the separatory funnel into the Buchner funnel with the vacuum turned on. Pour the remaining water layer out of the separatory funnel, before carefully pouring the filtrate into the separatory funnel (be sure to use a conical funnel in the top of

portion is filtering, replace it with a fresh piece before filtering the remainder of the tea mixture.

Cool the filtered solution to 15–20°C by adding a few ice chips. Set up a 125-mL separatory funnel as shown in Technique 4, Figure 4.3 and pour the cooled tea solution into the separatory funnel (be sure the stopcock is closed). Add 15 mL of dichloromethane to the funnel. Stopper the separatory funnel, hold the stopper firmly in place with your index finger, and invert the funnel. Open the stopcock to vent the vapors. Rotate the inverted funnel for 2–3 min, so that the two layers swirl together many times, opening the stopcock frequently to vent the funnel.

Allow the layers to separate and then drain the dichloromethane layer into a 50-mL Erlenmeyer flask. If a small emulsion layer is present at the interface between the organic and aqueous phases, add it to the Erlenmeyer flask. Cork the Erlenmeyer flask to prevent evaporation of the dichloromethane. Add 15 mL of fresh dichloromethane to the separatory funnel (still containing the tea solution) and repeat the extraction process. Again, allow the layers to separate and drain the dichloromethane layer, including any emulsion layer, into the Erlenmeyer flask containing the dichloromethane solution from the first extraction. Pour the tea solution out of the top of the separatory funnel into a beaker.

Rinse the separatory funnel with water before pouring the combined dichloromethane solutions into the funnel; add about 20 mL of water. Stopper the funnel, invert and rock it gently to mix the two layers. Some emulsion layer may be present at this point. If only a thin layer of emulsion exists at the interface between the aqueous phase and the dichloromethane solution, push a small piece of glass wool to the bottom of the dichloromethane layer with a large stirring rod. The glass wool will break the membranes of the emulsion. Drain the lower dichloromethane layer slowly into a clean, dry 50-mL Erlenmeyer flask.

Add anhydrous magnesium sulfate to the dichloromethane solution [see Technique 4.6]. Cork the flask and allow the mixture to stand for at least 10 min, swirling the flask occasionally.

Weigh (tare) a dry 50-mL Erlenmeyer flask on a balance that measures to 0.001 g. Place a fluted filter paper in a dry conical funnel and filter the drying agent from the dichloromethane solution *the separatory funnel). Slowly drain the lower organic phase into a clean, dry 50-mL Erlenmeyer flask.*

[see Technique 4.7a], collecting the filtrate in the tared 50-mL Erlenmeyer flask. Rinse the magnesium sulfate remaining in the flask with approximately 2 mL of dichloromethane and also pour this rinse through the funnel. Add a boiling stick or boiling chip to the flask containing the dichloromethane solution so that it boils without bumping [see Technique 3.1].

Evaporate the dichloromethane on a steam bath or water bath heated on a hot plate in a hood. Alternatively, the dichloromethane may be removed by evaporation, using a stream of nitrogen, in a hood, or with a rotary evaporator [see Technique 4.8]. Continue the evaporation until a dry greenish residue of crude caffeine forms on the bottom of the flask. Weigh the flask and determine the mass of crude caffeine. Calculate the percent recovery. Cork the flask and store it in your laboratory drawer for purification and analysis in Experiment 2.

Cleanup: Place the tea leaves in the nonhazardous solid waste container. Wash the tea solution remaining from the initial extractions and the water remaining in the 125-mL separatory funnel down the sink. Allow the flask containing the magnesium sulfate drying agent to dry in a hood before putting the spent drying agent in the inorganic waste container.

microscale Procedure

Techniques Vacuum Filtration Apparatus: Technique 5, Figure 5.4
Extraction: Techniques 4.2, 4.3, and 4.4
Drying Agents: Techniques 4.6 and 4.7
Solvent Removal: Technique 4.8

--- SAFETY INFORMATION ---

Dichloromethane is toxic, an irritant, absorbed through the skin, and harmful if swallowed or inhaled. Use it in a well-ventilated hood. Wear gloves and wash your hands thoroughly after handling it.

Solid **caffeine** is toxic and an irritant. Avoid contact with skin, eyes, and clothing.

Pour the contents of two tea bags into a tared (weighed) 150-mL beaker; weigh and record the mass of the tea. Add 50 mL of water and 2.4 g of calcium carbonate to the beaker containing the tea. Boil the mixture gently on a hot plate for 10 min.

Let the tea mixture cool to about 55°C. Using a 2-in. Buchner funnel and a 125-mL filter flask, filter the mixture, using vacuum filtration [Technique 5, Figure 5.4], through S&S No. 410 or Whatman No. 54 filter paper.

Pour 12 mL of water into a clean 150-mL beaker and mark the outside of the beaker at the 12-mL level. Discard the water before pouring the tea solution into the beaker. Boil the solution on a hot plate until the liquid level reaches the 12-mL mark on the beaker. Cool the solution briefly on the bench top.

Read Technique 4, Extractions, before beginning the extractions. It tells you how to do an extraction and explains why it works.

Transfer the tea solution to a 15-mL screw-capped centrifuge tube, then cool the tube in a water bath containing a few ice chips for 2–3 min. Add 2 mL of dichloromethane to the centrifuge tube, cap the tube tightly, and shake the mixture gently by repeatedly inverting the tube. Allow the layers to separate. If an emulsion layer (milky) forms between the lower organic phase and the upper aqueous tea solution, spin the tube containing the mixture in a centrifuge for approximately 1 min.

SAFETY PRECAUTION

Balance the centrifuge by placing another centrifuge tube of equal mass in the hole opposite the one containing the sample tube.

Prepare a Pasteur filter pipet [see Technique 2.4]. Press the air from the rubber bulb and insert the pipet to the bottom of the centrifuge tube [see Technique 4.5a]. Partially release the pressure on the bulb and draw the lower organic layer into the pipet until the interface between the two layers is exactly at the bottom of the centrifuge tube. Maintain a steady pressure on the rubber bulb while transferring the pipet to a second 15-mL centrifuge tube. Cap the second tube to prevent evaporation of the dichloromethane. The tea solution remains in the first centrifuge tube.

Repeat the extraction process two more times, using a new 2-mL portion of dichloromethane each time. After each extraction, transfer the lower dichloromethane layer to the second centrifuge tube that contains the dichloromethane/caffeine solution from the previous extraction.

Add anhydrous magnesium sulfate to the combined dichloromethane/caffeine solution [see Technique 4.6]. Cap the tube and allow the mixture to stand for at least 10 min, swirling the tube occasionally.

Prepare a microfunnel from a Pasteur pipet and cotton as shown in Technique 4.6, Figure 4.15b. Be sure that the cotton is packed tightly in the pipet. Clamp the microfunnel so that the tip is inserted about halfway into a tared (weighed) 25-mL Erlenmeyer flask. Transfer the dried dichloromethane/caffeine solution to the microfunnel, using a new Pasteur filter pipet. Rinse the magnesium sulfate remaining in the centrifuge tube with about 0.5 mL of dichloromethane and transfer this rinse to the microfunnel.

Evaporate the dichloromethane on a steam bath in a hood. Alternatively, the dichloromethane may be removed by evaporation, using a stream of nitrogen, in a hood. Continue the evaporation until a dry greenish residue of crude caffeine forms on the bottom of the flask. Weigh the flask and determine the mass of crude caffeine. Calculate the percent recovery. Cork the flask and store it in your laboratory drawer for purification and analysis in Experiment 2.

Cleanup: Place the tea leaves in the nonhazardous-solid waste container. Wash the tea solution remaining from the initial extractions and the water remaining in the centrifuge tube down the sink. Allow the centrifuge tube containing the magnesium sulfate drying agent to lie on its side in a hood until the residual dichloromethane evaporates before putting the spent drying agent in the inorganic waste container.

Questions

1. Why is the tea boiled with water in this experiment?
2. Why is the aqueous tea solution cooled to 15–20°C before the dichloromethane is added?

3. Why does the addition of salt (NaCl) to the aqueous layer sometimes help to break up an emulsion that forms in an extraction?
4. The distribution coefficient for caffeine in dichloromethane and water is 4.6. Assume that your 100-mL tea solution contained 0.30 g of caffeine. If you had only extracted with one 15-mL portion of dichloromethane, how much caffeine would have been left in the water solution? How much would be left in the water after the second 15-mL dichloromethane extraction? How much caffeine would be left in the water solution if only one extraction with 30 mL of dichloromethane were performed?
5. Why is less caffeine actually isolated than is suggested by the calculation in Question 4?

Experiment 2

PURIFICATION AND THIN-LAYER CHROMATOGRAPHIC ANALYSIS OF CAFFEINE

Continue the study of caffeine that you isolated in Experiment 1. Use sublimation to separate chlorophyll from caffeine and thin-layer chromatography to assess the purity of the caffeine.

The extraction of caffeine from tea yields a remarkably pure product, considering the complex composition of tea leaves. The isolated caffeine is not completely pure, however. The greenish color immediately suggests the presence of an impurity, because caffeine is white. The major impurity is the chlorophyll present in the leaves.

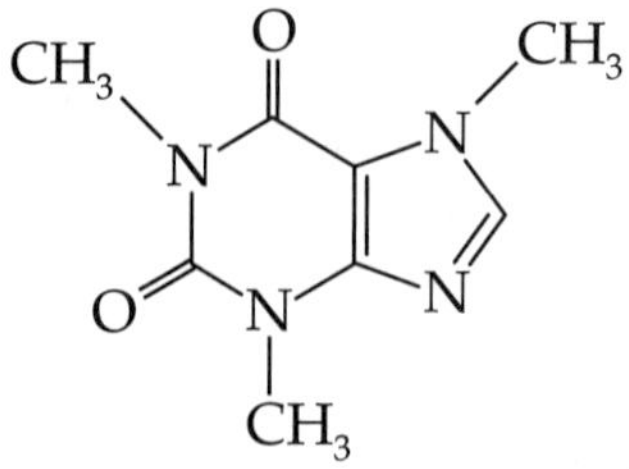

Caffeine

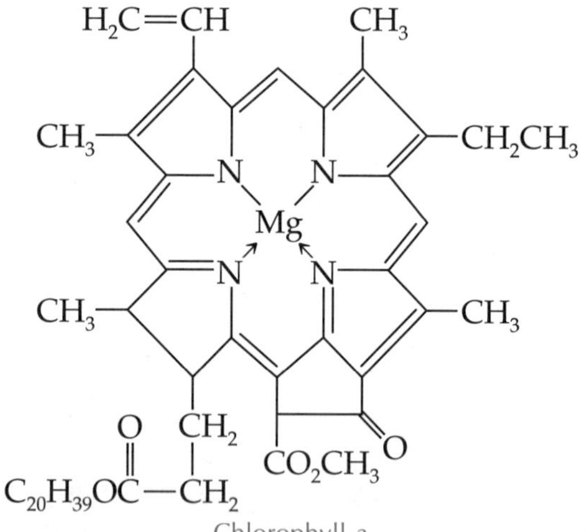

Chlorophyll a

Caffeine and chlorophyll can be separated by recrystallization, but in this experiment, you will use a simpler technique, vacuum sublimation [see Technique 8]. When a substance sublimes, it goes directly from the solid to the gas phase without ever being a liquid. This method works well for purifying an organic compound if these conditions are satisfied: the compound vaporizes without melting or decomposing, and the vapor recondenses to the solid, and the impurities present do not also sublime.

microscale Procedure

Techniques   Sand Bath for Heating: Technique 3.2
Sublimation: Technique 8
Thin-Layer Chromatography: Technique 10

Sublimation of Crude Caffeine

Begin heating a sand bath to 170–180°C [see Technique 3.2]. Place about 30 mg of your crude caffeine from Experiment 1 in the bottom of a 25-mL filter flask. Put the inner test tube or 15-mL centrifuge tube firmly in place, using a filter adapter to make a tight connection with the side-arm test tube [see Technique 8, Figure 8.1b]. Half fill the test tube with cold tap water (ice is not necessary for this sublimation).

> *SAFETY PRECAUTION*
>
> The inner test tube must fit tightly in the filter adapter, or the difference in pressure between the atmosphere and the vacuum inside the apparatus may push the test tube forcibly against the bottom of the filter flask, shattering both.

Clamp the flask securely in the sand bath and attach the side arm to a water aspirator or the vacuum line with a guard flask or trap bottle between them. Use thick-walled tubing. Turn on the vacuum source.

Adjust the level of the filter flask so that sand covers a few millimeters of the flask wall. Be sure that the temperature of the sand bath does not rise above 180–185°C, because a higher temperature will cause the crude caffeine to char or melt. As you heat the sample, you will notice white caffeine migrating first to the surface of the crude caffeine, then to the walls of the flask and the outside of the cold inner tube.

When the residue in the bottom of the flask has become dark green and no more caffeine appears to be collecting on the inner tube and the flask walls, the sublimation is complete. Carefully raise the sublimation apparatus from the sand bath and slowly let air back into the system by gently removing the rubber tubing from the water aspirator nipple before turning off the aspirator. If you are using a vacuum line, turn off the vacuum and slowly remove the rubber tubing from the side arm. Carefully remove the inner test tube from the flask. Use a spatula to scrape the pure caffeine onto a piece of tared (known mass) weighing paper and weigh your purified product. Then store the caffeine in a small test tube. The sublimed caffeine should be much lighter in color than the crude material.

Thin-Layer Chromatography of Caffeine

Be sure to read Technique 10, Thin-Layer Chromatography, before undertaking this part of the experiment. It explains the theory and techniques of TLC.

— **SAFETY INFORMATION** ——

Chloroform is highly toxic and a suspected carcinogen. Avoid inhalation or skin contact. Wear gloves, dispense it only in a hood.

Ethyl acetate is toxic, an irritant, and very flammable. Avoid inhalation or skin contact. Keep the developing chamber closed except when inserting or removing a thin-layer plate.

Never look directly into an **ultraviolet lamp.** Radiation in this region of the spectrum can cause eye damage.

Plastic-backed or aluminum-backed, precoated silica gel plates with fluorescent indicator and 2.5×6.7 cm dimensions work well for the chromatographic separation of caffeine.*

Weigh about 8 mg of your remaining crude caffeine and dissolve it in 0.50 mL of chloroform in a small test tube. Cork the test tube when the solution is not being used.

Dissolve 6 to 8 mg of your sublimed caffeine in 0.50 mL of chloroform in a second test tube. Label the two tubes. Again, cork this solution when it is not being used.

---

*These plates are available from many suppliers. In developing the experiment, we used Eastman Chromatogram Silica Gel sheets with fluorescent indicator.

Prepare a developing chamber, using a wide-mouth, capped bottle [see Technique 10, Figure 10.5]. A piece of 7- or 9-cm filter paper makes a good wick to ensure solvent saturation in the developing chamber's atmosphere. Use ethyl acetate as the developing solvent.

Ethyl acetate, a solvent of intermediate polarity, will separate caffeine from the highly polar chlorophyll molecules.

Add 4–5 mL of ethyl acetate to the developing chamber, cap it, and shake it vigorously. After the filter paper and the atmosphere have been saturated, the solvent should be about 2 to 3 mm deep in the bottom of the chamber. If you have a greater depth, remove some of the solvent with a Pasteur pipet and save it in a corked test tube. If you do not have enough solvent, add another 1 or 2 mL of ethyl acetate to the developing chamber.

Prepare several micropipets according to the instructions in Technique 10.3 or obtain micropipets from the supply in the laboratory.

Make a mark about 2 mm long with a pencil at the long edge of a 2.5×6.7 cm silica gel thin-layer plate approximately 1 cm from the bottom edge. Dip a clean micropipet into the known solution of caffeine available in the laboratory and apply one spot of the solution at about one-third the width of the thin-layer plate. With another micropipet, apply one spot of your crude caffeine solution at about two-thirds the width of the thin-layer plate [see Technique 10, Figure 10.4].

Don't disturb the developing chamber during the chromatography.

Develop the chromatogram by placing the thin-layer plate in the ethyl acetate developing chamber with a pair of tweezers. Keep the chamber capped at all times. Allow the solvent to rise up the plate until it is about 1.0 cm from the top.

Alternatively, an iodine chamber may be used to visualize the spots on the developed chromatogram [see Technique 10.5]. The iodine visualization method shows a small amount of an impurity having a higher R_f value than caffeine.

Remove the plate from the chamber with a pair of tweezers and *immediately mark the solvent front* with a pencil line before the solvent begins to evaporate. After marking the solvent front position, let the solvent evaporate in a hood for 2–3 min.

If your spots turn out to be very large and show tailing, you can spot less sample on a different plate by diluting your caffeine solution with a few more drops of chloroform and by using a micropipet with a thinner tip.

Visualize the results of your chromatogram by shining ultraviolet radiation (254 nm) on the plate in a darkened room or dark box [see Technique 10, Figure 10.7]. You will notice dark spots wherever the caffeine is present on the plate. Outline each spot with a pencil while the plate is under the ultraviolet lamp. When you have the thin-layer plate in ordinary light, you can see the small, pale green spot of chlorophyll; circle it with a pencil as well. Calculate the R_f values for caffeine and chlorophyll under your

If the spots are very small or very faint, you can spot more sample by applying repetitive spots to the same position; let the solvent evaporate before overspotting the first spot.

Avoid holding the micropipet against the thin-layer plate for a prolonged time; this practice will only produce a larger diffuse spot, not one containing a more concentrated sample.

experimental conditions [see Technique 10.1 and Figure 10.2]. Be sure to record in your notebook all the data needed to reproduce your R_f values; draw diagrams of your chromatograms. Also estimate the size of the spots and note whether there was any tailing.

Prepare a second thin-layer plate with spots of the known caffeine solution and your sublimed (purified) caffeine solution. If a chlorophyll solution is available, prepare another thin-layer plate with spots of chlorophyll and your crude caffeine. Repeat the development and visualization procedures and again calculate the R_f values. Replenish the ethyl acetate in the developing chamber if the level falls below 2 mm.

Optional Experiment: Evaluating TLC Parameters

You will probably use TLC analyses a number of times during your study of organic chemistry, so you might want to see how variable conditions change your results. Calculate the R_f value for caffeine (and chlorophyll), using one of the following experimental conditions (or more, if you have time). Remember that you can apply two or more separate samples on each plate, and do not forget to mark the starting point and solvent front on each thin-layer plate.

1. Apply more of the crude caffeine sample to the thin-layer plate. Use repetitive spotting, and try using three times as much as you used before. In a separate application, try six times as much.

2. Use a more polar solvent to develop your chromatogram. Make up a solution containing 1 part ethanol to 9 parts ethyl acetate by volume. Five milliliters of solution should be enough, so combine 0.5 mL of ethanol and 4.5 mL of ethyl acetate and mix them well. Remove the ethyl acetate and the filter paper wick from your developing bottle and dry the bottle. Add a dry filter paper and 3–4 mL of the 1:9 ethanol/ethyl acetate solvent. Shake the capped bottle to ensure saturation. Do the rest of the chromatography in the usual way. Calculate the R_f value.

3. Try the thin-layer chromatography of caffeine, using a developing solvent of 1:1 solution of ethanol/ethyl acetate. Calculate the R_f value.

4. Try the TLC of caffeine, using hexane as the developer. Calculate the R_f value.

Cleanup: Pour your chloroform solutions of crude and purified caffeine into the container for halogenated organic waste. Pour any remaining developing solvents—ethyl acetate, ethyl acetate/ethanol, or hexane—into the container for flammable or organic waste. Place the solvent-saturated filter paper and the chromatograms in a waste container for solids in a hood.

Questions

1. Why does caffeine have a larger R_f value than chlorophyll?
2. Why can there be no breaks in the thin-layer surface?
3. Two compounds have the same R_f (0.87) under identical conditions. Does this show that they have identical structures? Explain.

SYNTHESIS OF ETHANOL BY FERMENTATION OF SUCROSE

Use the age-old biochemical process of fermentation to produce ethanol, which you will purify by distillation.

$$C_{12}H_{22}O_{11} \xrightarrow[\text{invertase}]{H_2O} C_6H_{12}O_6 + C_6H_{12}O_6 \xrightarrow{\text{zymase}} 4CH_3CH_2OH + 4CO_2$$

Sucrose	Glucose	Fructose	Ethanol
MW 342.3			MW 46.1

Sucrose, a disaccharide of molecular formula $C_{12}H_{22}O_{11}$, is hydrolyzed into the monosaccharides fructose and glucose by the enzyme invertase, which is present in yeast. The fructose and glucose then undergo a series of reactions catalyzed by zymase, the name for a collection of enzymes that are found in yeast and catalyze the individual steps in converting these monosaccharides to ethanol and carbon dioxide. The fermentation solution also contains a mixture of potassium phosphate, calcium phosphate, magnesium sulfate, and ammonium tartrate. Over 100 years ago, Pasteur discovered that these salts increase yeast growth and promote the fermentation process. Later research determined that these salts promote the formation of phosphates of fructose and glucose, one of the steps leading from the monosaccharides to ethanol and carbon dioxide.

The ethanol formed in the reaction serves to inhibit the fermentation process by killing the yeast cells, so fermentation stops when the alcohol content of the solution approaches 12% by volume. Death of the yeast cells accounts for the approximately 12% alcohol content of most wines and beer. More concentrated alcohol solutions are obtained by fractionally distilling the aqueous alcohol solution produced by fermentation [see Technique 7.4]. Fractional distillation still does not yield 100% pure ethanol, because the azeotropic mixture of 95% ethanol and 5% water (by weight) boils at 78.1°C, whereas 100% ethanol boils at 78.4°C [see Technique 7.5].

You may notice that your ethanol solution obtained after the fractional distillation has a yeastlike odor somewhat different from that of the ethanol found in the laboratory. In addition to ethanol, the fermentation process produces small amounts of other compounds such as acetaldehyde, 1-propanol, 2-propanol (isopropyl alcohol), 2-methyl-1-butanol, and 3-methyl-1-butanol, all of which contribute to the odor of the biosynthesized ethanol. For industrial alcohol, where chemical purity is essential, these compounds are removed by treatment of the distilled alcohol with activated charcoal.

All lab operations following the fermentation can be completed within one laboratory period. The action of the yeast on sucrose, however, does not occur quickly, so the reactants need to be mixed together a week before the ethanol is to be isolated.

*MMM*Macroscale Procedure

Techniques   Boiling Points: Technique 7.1
Simple Distillation: Technique 7.3
Fractional Distillation: Technique 7.4
Azeotropic Distillation: Technique 7.5

— SAFETY INFORMATION —

Celite is a lung irritant. Avoid breathing the dust when handling this material.

Note to instructor: Prepare Pasteur's salts solution by dissolving 2.0 g of potassium phosphate, 0.20 g of calcium phosphate, 0.20 g of magnesium sulfate and 10.0 g of ammonium tartrate in 860 mL water.

Place 40 g of sucrose in a 500-mL round-bottomed flask. Add 200 mL of water and 3.0 g of dry yeast. Stir until the sugar dissolves and no yeast granules remain. Then add 35 mL of Pasteur's salts solution and stir to mix. Close the flask with a one-hole rubber stopper fitted with a piece of bent glass tubing, as shown in Figure 3.1. Fill an 18 × 150 mm test tube halfway with saturated

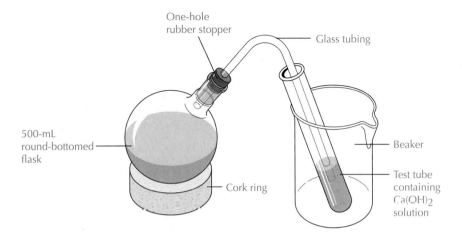

FIGURE 3.1 Apparatus for fermentation of sucrose.

The presence of oxygen in the fermenting mixture causes oxidation of ethanol to acetic acid (vinegar).

aqueous calcium hydroxide (limewater) and submerge the other end of the glass tubing about 1 cm below the surface of the solution. Limewater serves to exclude atmospheric oxygen from the fermentation mixture and simultaneously prevents a pressure increase from the CO_2 formed by the reaction. Store the mixture at room temperature for 1 week to allow complete fermentation. (Bubbling ceases when the reaction is complete.)

Celite is a trade name for diatomaceous earth, a powdered inert material made from the shells of diatoms (a type of oceanic phytoplankton). The tiny particles of yeast cell debris clog the pores of filter paper. The Celite catches the cell debris before it reaches the filter paper, thereby allowing rapid filtration of the solution.

When the reaction is complete, add 10 g of Celite filter aid to the fermentation mixture and stir to wet the Celite. Set up a vacuum filtration apparatus, using a 500-mL filter flask [see Figure 5.4, Technique 5.3]. Wet the filter paper with water and turn on the vacuum source. Pour the reaction mixture slowly onto the filter paper. Rinse the flask with a few milliliters of water and also pour this rinse over the filter cake. Save the filtrate for the next step.

Simple Distillation

Wash the 500-mL round-bottomed flask and pour the filtrate into it. Add two boiling chips. Assemble the apparatus for simple distillation, as shown in Technique 7.3, Figure 7.7, substituting a 100-mL graduated cylinder for the receiving flask. Heat the ethanol solution to boiling and adjust the rate of heating so that the distillate flows at a rate of about 1 drop per second. Record the temperature as each 2-mL portion of distillate is collected. Turn off the heat and remove the heating mantle from the round-bottomed flask after 50 mL of distillate have been collected.

Note well the position of the thermometer bulb shown in the insert of Technique 7, Figure 7.7.

Table 3.1. Density, percentage by weight, and percentage by volume of ethanol in H$_2$O at 20°C

Density	% by weight	% by volume	Density	% by weight	% by volume
0.9893	5.0	6.2	0.8557	75.0	81.3
0.9819	10.0	12.4	0.8436	80.0	85.5
0.9752	15.0	18.5	0.8310	85.0	89.5
0.9687	20.0	24.5	0.8180	90.0	93.3
0.9617	25.0	30.4	0.8153	91.0	94.0
0.9539	30.0	36.2	0.8125	92.0	94.7
0.9450	35.0	41.8	0.8098	93.0	95.4
0.9352	40.0	47.3	0.8070	94.0	96.1
0.9248	45.0	52.7	0.8042	95.0	96.8
0.9139	50.0	57.8	0.8013	96.0	97.5
0.9027	55.0	62.8	0.7984	97.0	98.1
0.8911	60.0	67.7	0.7954	98.0	98.8
0.8795	65.0	72.4	0.7923	99.0	99.4
0.8676	70.0	76.9	0.7893	100.0	100.0

Tare (weigh) a dry, corked 10-mL Erlenmeyer flask on a balance that weighs to 0.001 g. Carefully pipet 10.00 mL of the distillate into the tared flask and then weigh the flask again. Use this information to determine the density of your ethanol solution. From the density value, you can then determine the ethanol content, using Table 3.1.

Fractional Distillation   Pour all the distillate (including the 10.0 mL used for the density determination) into a 100-mL round-bottomed flask, add two boiling chips and assemble the fractional distillation apparatus shown in Technique 7.4, Figure 7.12, substituting a 10-mL graduated cylinder for the receiving flask. The fractionating column is packed with stainless steel sponge from scouring pads available in any supermarket.

Heat the solution to boiling, then moderate the rate of heating so that the vapor (visible by noting the position of the condensate ring) in the fractionating column ascends slowly, thereby giving the vapor and condensate time to equilibrate. During the course of the distillation, gradually increase the rate of heating so that vapor continues to reach the condenser. A distillation rate of 1 drop every 2–3 s is usually satisfactory. You must not allow the distillation to stop until you have collected the amount of distillate desired. A smooth continuous progression through the fractions described in the next paragraph, with ever-increasing temperature changes, indicates that the procedure is being done correctly. Record the temperature as each milliliter of distillate is collected.

Note well the position of the thermometer bulb shown in the inset of Figure 7.7 in Technique 7.

Label two 50-mL Erlenmeyer flasks "fraction 1" and "fraction 2." Collect fraction 1 from 77–80°C, fraction 2 from 80–96°C, and fraction 3 above 96°C. If the fractionating column is working well and the rate of distillation is satisfactory, the observed boiling point should remain steady while the ethanol/water azeotrope distills (fraction 1). When 10 mL of distillate have been collected, quickly pour the distillate into the flask labeled "fraction 1." Continue adding 10-mL portions to this flask until the temperature rises above 80°C. At this point, quickly add whatever amount is in the cylinder to fraction 1 and begin collecting fraction 2. If the column is efficient, fraction 2 will be a small volume collected during a rapid rise in temperature to 95°C or slightly higher. When the temperature begins to stabilize again, pour the distillate into the flask labeled "fraction 2." Stop the distillation after 4–6 mL of fraction 3 have been collected.

Weigh a dry 10-mL Erlenmeyer flask, pipet 10.00 mL of fraction 1 into the flask, and weigh it again. Calculate the density of fraction 1 and determine the % ethanol by weight and volume from Table 3.1. From the total volume of fraction 1, its density, and the weight % of ethanol, calculate the percent yield of ethanol from the fermentation reaction. Submit your product (all of fraction 1) to your instructor in a container labeled with your name, the name of the product, the boiling-point range, the volume, the density, and the date.

Cleanup: Place the Celite in the nonhazardous-solid waste container. The residues in the boiling flasks from both the simple and fractional distillations may be washed down the sink. Fractions 2 and 3 may also be discarded down the sink.

Treatment of Data Plot the data for both the simple distillation and the fractional distillation on the same set of coordinates, using the volume of distillate as the abscissa (*x*-axis) and the temperature as the ordinate (*y*-axis).

Questions

1. After the fermentation, what is the precipitate in the $Ca(OH)_2$ solution? Write the balanced equation for this reaction.
2. Why is it impossible for 100% pure ethanol to be obtained from fractional distillation of the fermentation mixture?
3. What is the composition of fraction 3?
4. Does your graph of the simple and fractional distillations show any difference in the shapes of the two curves? What does this difference, if any, indicate about the efficiency of fractional distillation relative to that of simple distillation for separating the azeotrope of ethanol and water from water?

Experiment 4

SYNTHESIS OF SALICYLIC ACID

Starting with oil of wintergreen, produce salicylic acid by a base-catalyzed hydrolysis reaction. Learn how to carry out a reaction at an elevated temperature and purify your product by recrystallization.

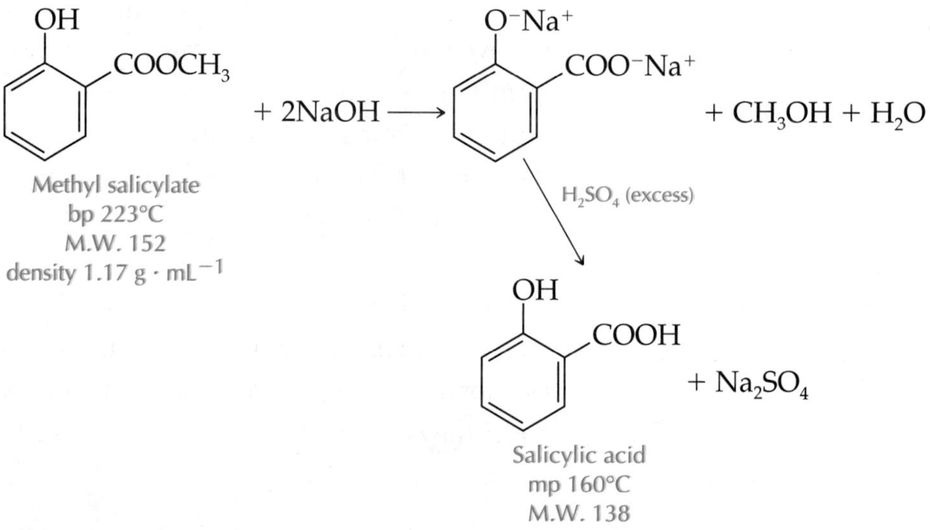

Methyl salicylate
bp 223°C
M.W. 152
density 1.17 g · mL^{-1}

+ 2NaOH ⟶

H_2SO_4 (excess)

+ CH_3OH + H_2O

Salicylic acid
mp 160°C
M.W. 138

+ Na_2SO_4

Derivatives of salicylic acid are familiar and important compounds. Among them are methyl salicylate (oil of wintergreen) and acetylsalicylic acid, aspirin. While you explore the chemistry of these compounds, you will be learning techniques for purifying and analyzing organic solids in this and the next experiment. The techniques of recrystallization and melting-point determination are highlighted in this experiment.

Methyl salicylate   Salicylic acid   Acetylsalicylic acid (aspirin)

Methyl salicylate has a fragrant, minty smell that has made it a favorite flavoring in candies. It is also the major constituent of oil of wintergreen, making up over 90% of the oil from the wintergreen plant. However, most of the methyl salicylate used in foods is made synthetically, a cheaper process than its extraction from wintergreen leaves and sweet birch bark.

Salicylic acid is a white, crystalline compound. It is commonly used in ointments and plasters for the removal of warts from the skin.

Methyl salicylate has two functional groups that show themselves in this experiment; one is the ester moiety and the other is the phenol group.

Esters are derivatives of carboxylic acids and can easily be converted into them by base-promoted *hydrolysis.* Hydrolysis is literally the splitting of a substrate (here the ester) by water, and this reaction is very effectively promoted by NaOH. Base-promoted hydrolysis of esters is also called saponification.

Let's examine the steps in the hydrolysis process. Phenols react with the strong base NaOH to form sodium salts, as do carboxylic acids. The latter, however, are over 100,000 times stronger acids than are most phenols. So when methyl salicylate hydrolyzes in the presence of sodium hydroxide, reactions 1 through 3 occur. The phenol is converted to its conjugate base—a phenoxide ion (1); the ester is hydrolyzed (2), and the carboxylic acid product is converted to its conjugate base (3).

$$\text{(structure: salicylate methyl ester with OH)} + OH^- \rightleftharpoons \text{(structure: with } O^-) + H_2O \qquad (1)$$

$$\text{(structure: } O^-, COOCH_3) + OH^- \rightleftharpoons \text{(structure: } O^-, COOH) + CH_3O^- \qquad (2)$$

$$\text{(structure: } O^-, COOH) + CH_3O^- \rightleftharpoons \text{(structure: } O^-, COO^-) + CH_3OH \qquad (3)$$

Addition of a strong acid after hydrolysis leads to protonation of the salicylate salt, and salicylic acid is produced. The salicylic acid is insoluble in cold water. The methanol that is also formed is soluble in water and is not recovered. In overall terms, we have

$$\text{(structure: OH, COOCH}_3) \xrightarrow[\text{H}_2\text{O}]{\text{NaOH}} \text{(structure: ONa, COONa)} + CH_3OH + H_2O \xrightarrow{\text{H}_2\text{SO}_4} \text{(structure: OH, COOH)} + Na_2SO_4$$

As you would expect, the hydrolysis of esters goes faster at high temperature. Even with the concentrated sodium hydroxide solution that we suggest for this experiment, the hydrolysis would take a long time at room temperature. Therefore, the reaction mixture is heated at the boiling point of the solution for a period of time, using a process called refluxing. Boiling causes the solvent vapor to move up the reflux condenser, where the cooled surface condenses the vapor. The reliquified solvent then drops back into the reaction flask.

The process of refluxing is discussed in Technique 3.3.

The final step in the synthesis is the purification of the crude salicylic acid by recrystallization [see Technique 5]. Water can be a good recrystallization solvent for polar organic compounds such as salicylic acid and will be used in this experiment. The information given in Table 4.1 indicates that water fulfills the solubility criteria for a good recrystallization solvent because salicylic acid is soluble near the boiling point of water and relatively insoluble at the temperature of an ice-water bath (0–10°C).

Once you have obtained a recrystallized product, assess its purity by determining the melting point [see Technique 6]. The

Table 4.1. **Solubility of salicylic acid in water**

Temperature, °C	Amount of salicylic acid per 100 mL water, g
0	0.10
10	0.13
25	0.23
50	0.63
75	1.8
90	3.7

presence of even a small quantity of impurity usually depresses a compound's melting point a few degrees and causes melting over a range of several degrees. The melting point of salicylic acid has been reported in the chemical literature as 159 or 160°C. Because you will do only one recrystallization of your salicylic acid, you will likely find a melting range of two or three degrees, with the upper limit of the range slightly below 159°C.

WWMacroscale Procedure

Techniques   Macroscale Reflux: Technique 3.3
Boiling Stones: Technique 3.1
Macroscale Recrystallization: Technique 5.3
Melting Points: Technique 6.3

— SAFETY INFORMATION —

Sodium hydroxide is corrosive and causes burns. Wear gloves and avoid contact with skin, eyes, and clothing. Notify the instructor if any solid NaOH is spilled.

Sulfuric acid solutions are corrosive and cause burns. Avoid contact with skin, eyes, and clothing.

Methyl salicylate is toxic and an irritant. Wear gloves and avoid contact with skin, eyes, and clothing.

Solid NaOH is hygroscopic and rapidly absorbs water from the atmosphere. Keep the reagent bottle tightly closed.

Clamp a 100-mL round-bottomed flask to a ring stand or vertical support. Place 4.6 g (0.12 mol) of sodium hydroxide and 25 mL of water in the flask; stir the mixture until the solid dissolves. Add

ALWAYS grease ground glass joints [see Technique 2.1].

2.0 mL (2.3 g, 0.015 mol) of methyl salicylate. A white solid will quickly form (see Question 3). Attach a water-cooled reflux condenser to the round-bottomed flask [see Technique 3.3].

Add one or two boiling stones to the reaction mixture to prevent bumping of the solution when it is heated and place a heating mantle under the flask [see Technique 3.2]. Heat the reaction mixture at boiling for 15 min. The solid that forms initially will dissolve as the mixture is warmed.

After the reflux period, remove the heating mantle and let the mixture cool to room temperature. Placing a beaker of tap water under the flask speeds the cooling process.

Carefully add 3 M sulfuric acid solution in approximately 3-mL increments until a heavy white precipitate of salicylic acid forms and *remains* when the mixture is well stirred. You will need approximately 15–20 mL of the sulfuric acid solution.

After you have added just enough 3 M sulfuric acid to give a heavy white precipitate, add 2 mL more acid to ensure complete precipitation of the salicylic acid. Cool the mixture in an ice-water bath to about 5°C. Collect the precipitated crude product by vacuum filtration, using a Buchner funnel [see Technique 5, Figure 5.4].

Recrystallization of Salicylic Acid

Read Technique 5, Recrystallization, carefully before doing this part of the experiment.

Recrystallize the crude salicylic acid by the following procedure. Heat about 60 mL of water in a 125-mL Erlenmeyer flask on a hot plate until the water almost boils. Place the crude salicylic acid in another 125-mL Erlenmeyer flask and add a boiling stone or wooden boiling stick. Carefully pour approximately 20 mL of hot water over the salicylic acid and heat the flask containing the salicylic acid to boiling.

> **SAFETY PRECAUTION**
>
> Use a pair of flask tongs to hold any hot flask.

Continue adding approximately 5-mL portions of hot water until the solid has completely dissolved, allowing a little time after each addition for the dissolution process to occur. When dissolu-

Remember that the correct amount of solvent is just a little over the minimum that will dissolve the crystals when the recrystallization solution is boiling.

tion is complete, add 8–10 mL of excess hot solvent. Estimate the total amount of solvent that you used and record this in your notebook (see Question 5).

Set the flask on the bench top until extensive crystallization has occurred throughout the solution and it has cooled nearly to room temperature. Then place the flask in an ice-water bath for 10 min. Collect the crystals, using vacuum filtration. When the filtration is complete, disconnect the vacuum source and pour a few milliliters of *ice-cold* water over the crystals to dissolve any impurities coating the crystals from residual crystallization solution. Turn on the vacuum again to remove the solvent.

Pull air through the crystals for a few minutes to facilitate drying. Water does not evaporate very quickly, so the crystals should be kept open to the atmosphere on a piece of filter paper or on a watch glass at least overnight to complete the drying process. You can also dry salicylic acid more quickly by heating it in an oven at 90–110°C for 10–15 min.

Weigh your salicylic acid product when it is thoroughly dry. Calculate your percent yield.

Melting Point

Read Technique 6, Melting Points and Melting Ranges, before doing this part of the experiment.

Carefully follow the directions in Technique 6.3 for determining the melting point of your dried salicylic acid. This compound can sublime with prolonged heating; nevertheless, using a seal capillary tube should not be necessary. Reference tables give the melting point of salicylic acid as 159 or 160°C. Be sure to record the melting range that you observe; you will likely find a melting range of two or three degrees.

Your instructor may ask you to obtain an IR spectrum of your salicylic acid [see Spectrometric Method 1]. If so, compare your spectrum with that shown in Figure 4.1 on page 38.

Submit your product, properly labeled, to your instructor, unless you are instructed to use it for Experiment 5.

Cleanup: Neutralize any excess acid in the filtrate from the reaction mixture with sodium carbonate before washing the solution down the sink or placing it in the container for aqueous inorganic waste. Dilute the filtrate from the recrystallization with water and wash it down the sink.

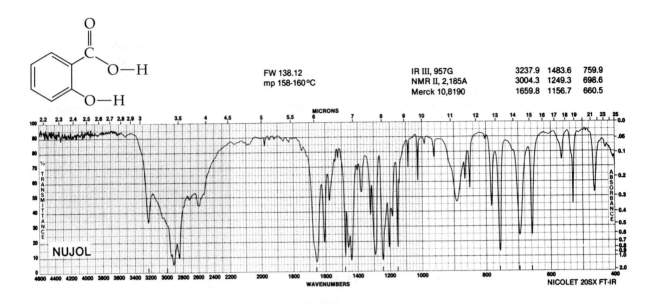

FW 138.12
mp 158-160 °C

IR III, 957G
NMR II, 2,185A
Merck 10,8190

3237.9 1483.6 759.9
3004.3 1249.3 698.6
1659.8 1156.7 660.5

FIGURE 4.1 IR spectrum of salicylic acid. (Provided by Aldrich Chemical Company, Inc., Milwaukee, WI.)

mmmicroscale Procedure

Techniques   Microscale Reflux: Technique 3.3
Boiling Stones: Technique 3.1
Microscale Recrystallization: Technique 5.5a
Melting Points: Technique 6.3

SAFETY INFORMATION

Sodium hydroxide solutions are corrosive and cause burns. Wear gloves and avoid contact with skin, eyes, and clothing. Notify the instructor if any solid NaOH is spilled.

Sulfuric acid solutions are corrosive and cause burns. Avoid contact with skin, eyes, and clothing.

Methyl salicylate is toxic and an irritant. Wear gloves and avoid contact with skin, eyes, and clothing.

Solid NaOH is hygroscopic and rapidly absorbs water from the atmosphere. Keep the reagent bottle tightly closed.

Pour 3.5 mL of water into a 10-mL round-bottomed flask. Add 0.48 g (0.012 mol) of sodium hydroxide, being careful not to get any solid on the inside of the ground glass joint. Swirl the flask gently until the solid dissolves. Add 0.20 mL (0.23 g, 0.0015 mol) of methyl salicylate, measured with a graduated pipet, to the

NaOH solution. A white solid will quickly form (see Question 3). Add a boiling stone. Attach a water-cooled reflux condenser [see Technique 2, Figure 2.6b and Technique 3, Figure 3.4b] after you have applied a light coating of grease to the lower ground glass joint of the condenser. Gently tighten the screw cap to hold the joint together.

Heat an aluminum heating block to 120–130°C on a hot plate. Place the reflux apparatus in the flask depression of the aluminum block and heat the mixture at reflux for 15 min. The white solid that formed initially will dissolve as the mixture is warmed.

After the reflux period, lift the apparatus off the aluminum block and cool the mixture to room temperature. Placing a small beaker of tap water under the flask speeds the cooling process.

Remove the condenser and set the flask in a 50-mL beaker. Carefully add 3 M sulfuric acid solution in approximately 0.5-mL increments until a heavy white precipitate of salicylic acid forms and *remains* when the mixture is well stirred; then add another 0.5 mL to ensure complete precipitation of the salicylic acid. Two to three milliliters of acid will need to be added. Cool the flask containing the salicylic acid in a 50-mL beaker containing an ice-water mixture. Collect the product by vacuum filtration, using a Hirsch funnel, as shown in Technique 5, Figure 5.7.

Recrystallization of Salicylic Acid

Read Technique 5, Recrystallization, before doing this part of the experiment.

Remember that the correct amount of solvent is just a little over the minimum *that will dissolve the crystals when the recrystallization solution is boiling.*

Transfer the crude product from the Hirsch funnel to a glassine weighing paper, using a spatula. Roll the paper into a funnel or crease it, and transfer the salicylic acid crystals to a 10-mL Erlenmeyer flask. Add 2 mL of water to the flask containing the crystals, put a boiling stick in the flask, and heat it to boiling on a hot plate. Add water in 0.5-mL increments until the solid dissolves in the boiling solvent; then add an additional 0.5 mL of water. Estimate the total amount of solvent that you used and record this in your notebook (see Question 5).

— **SAFETY PRECAUTION** —

Use a pair of flask tongs to hold any hot flask.

Set the flask on the bench top until extensive crystallization has occurred throughout the solution and it has cooled nearly to room temperature. Then place the flask in an ice-water bath for

5 min. Collect the crystals, using vacuum filtration and a Hirsch funnel. When the filtration is complete, disconnect the vacuum source and pour approximately 1 mL of *ice-cold* water over the crystals to dissolve any impurities coating the crystals from residual crystallization solution. Turn on the vacuum again to remove the solvent.

Pull air through the crystals for a few minutes to facilitate drying. Water does not evaporate very quickly, so the crystals should be kept open to the atmosphere on a piece of filter paper or on a watch glass at least overnight to complete the drying process. You can also dry salicylic acid more quickly by heating it in an oven at 90–110°C for 10–15 min.

Weigh your salicylic acid product when it is thoroughly dry. Calculate your percent yield.

Melting Point   Determine the melting point of your salicylic acid, as described in the Macroscale Procedure.

Cleanup: Neutralize any excess acid in the filtrate from the reaction mixture with sodium carbonate before washing the solution down the sink or placing it in the container for aqueous inorganic waste. Dilute the filtrate from the recrystallization with water and wash it down the sink.

Questions

1. Describe the factors that could cause you to obtain an incorrect melting point.
2. Suppose the material that you are recrystallizing fails to precipitate out of the cold solvent. What would you do to recover the material from the solution?
3. What is the white solid that quickly forms when the methyl salicylate is added to the sodium hydroxide solution?
4. Suppose that you obtain 10 g of salicylic acid from the hydrolysis of 13 mL of methyl salicylate. Calculate your percent yield.
5. (a) Use Table 4.1 and the total estimated volume of solvent that you used for the recrystallization to calculate the amount of salicylic acid that would remain dissolved in the cold recrystallization solvent if you cooled it to 10°C before filtration. (b) How much would this loss lower your percent yield?

Experiment 5
SYNTHESIS OF ASPIRIN

Beginning with salicylic acid, make aspirin, purify it, and determine its purity by a melting-point determination.

Salicylic acid	Acetic anhydride	Acetylsalicylic acid	Acetic acid
M.W. 138	M.W. 102	(Aspirin)	
mp 160°C	bp 139°C	M.W. 180	
	density 1.08 g/mL	mp 135–136°C	

Few synthetic organic compounds have enjoyed such widespread medicinal use as aspirin. Over 30 million pounds of it are consumed each year in the United States alone, as a first line of defense against the discomforts of colds, minor pains, headaches, and arthritis. Even though extracts of willow leaves and bark have been used for centuries for their pain-relieving (analgesic) and fever-reducing (antipyretic) properties, only in the late 1800s was the active ingredient of willow and poplar bark discovered to be salicylic acid. This substance, it was found, could be produced cheaply and in large amounts, but its use had severe limitations because of its acidic properties. Membranes lining the stomach and passages leading to it are irritated by the acid. The side effects of salicylic acid use were often worse than the original discomfort. The breakthrough came in 1893 when a German chemist, Felix Hofmann, synthesized the acetyl derivative of salicylic acid; it proved to have the same kind of medicinal properties without the high degree of irritation to mucous membranes.

Salicylic acid

Acetylsalicylic acid (aspirin)

Aspirin tablets also include a binder such as starch to hold the tablet in a stable shape. The usual 5-grain (an apothecary unit of mass) aspirin tablet contains 0.325 g of acetylsalicylic acid. Acetaminophen, sold as Tylenol, Datril, or Anacin 3, is another widely used analgesic. A newer analgesic, ibuprofen (brand names include Advil and Nuprin), became available over the counter in the United States during the 1980s, although it was available earlier as the prescription drug Motrin.

Acetaminophen   Ibuprofen

The contents of the stomach are acidic, and most of the ingested aspirin passes through the stomach unchanged. However, under the alkaline conditions in the intestines, aspirin forms sodium acetylsalicylate, which is absorbed through the intestinal wall. Few people suffer serious toxic effects from using aspirin, although some people are allergic to it. People suffering from ulcers may find their condition made worse by the use of aspirin. Aspirin has been cited as a possible contributing factor in Reye's syndrome, which can lead to death. In addition, aspirin seems to interfere with blood clotting. This anticoagulant property of aspirin limits its use for patients anticipating surgery but renders it effective in reducing heart attacks and strokes by preventing or hindering the formation of blood clots. Excess amounts, however, can cause pain, fever, or inflammation. Aspirin reduces the activity of the enzyme prostaglandin synthetase, thereby inhibiting prostaglandin synthesis. Acetaminophen and ibuprofen also work by reducing the level of prostaglandins in the body.

Aspirin is prepared by acetylating salicylic acid in a process called *esterification.* Esterification is the reaction of a carboxyl (—COOH) group, and an —OH group to form a carboxylate ester group. In this case the source of the —OH group is the phenolic —OH attached to the ring of the salicylic acid. The acetyl group comes from acetic anhydride, and the reaction is catalyzed by phosphoric acid:

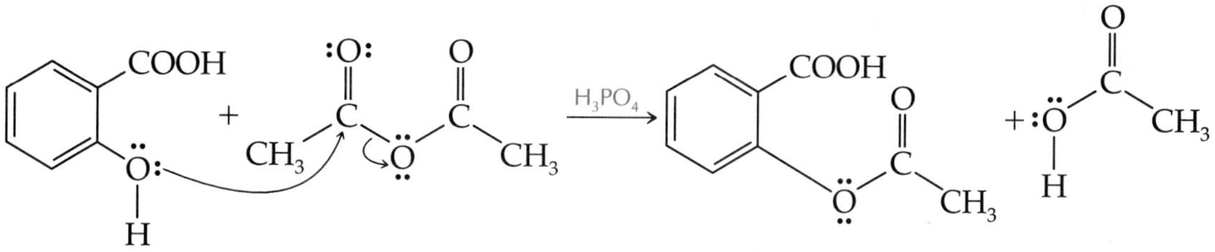

| Salicylic acid | Acetic anhydride | Acetylsalicylic acid (aspirin) | Acetic acid |

You will use recrystallization in a mixed solvent of ethanol and water to purify the crude aspirin. You will then assess the purity of your product by a melting-point determination.

ⵡⵡⵡMacroscale Procedure

Techniques Mixed Solvent Recrystallization: Technique 5.2a

Recrystallization: Technique 5.3

Melting Points: Technique 6.3

SAFETY INFORMATION

Salicylic acid is toxic and an irritant to skin, eyes, mucous membranes, and the upper respiratory tract. Avoid breathing the dust. Avoid contact with skin, eyes, and clothing.

Acetic anhydride is toxic, corrosive, and a lachrymator (causes tears). Wear gloves, use it in a hood, and avoid contact with skin, eyes, and clothing.

Concentrated **phosphoric acid** (85%) is corrosive and causes burns. Avoid contact with skin, eyes, and clothing.

Weigh 1.0 g (7.2 mmol) of the salicylic acid that you synthesized in Experiment 4 or from the supply available in the lab; your salicylic acid from Experiment 4 must be *thoroughly* dry before using it to make aspirin. Place the salicylic acid in a 50-mL Erlenmeyer flask. Under a hood, add 2.0 mL (21 mmol) of acetic anhydride measured with a graduated pipet. Add 5 drops of 85% phosphoric acid and mix the chemicals well by rotating the flask. Loosely stopper the flask with a cork before leaving the hood. The mixture may become warm from the exothermic reaction; allow it to stand for about 10 min. To complete the reaction, loosen the cork and heat the flask for 5 min in a 45–50°C water bath. Hot tap water may be used to make the bath.

The freshly broken surface of the scratched glass provides the nuclei for the formation of aspirin crystals.

The odor of acetic acid will no longer be apparent if you have washed the crystals enough.

Chill the mixture in an ice-water bath and scratch the inside wall of the flask with a stirring rod until a semicrystalline paste forms. Add 10 mL of cold water and 6 g of ice. Stir the mixture to break up the pasty solid.

When the ice has melted, collect the crystals by vacuum filtration on a small Buchner funnel [see Technique 5.3, Figure 5.4]. Wash the remaining crystals out of the flask, using 2–3 mL of *ice-cold* water, and rinse the crystals on the funnel with several 1- to 2-mL portions of *ice-cold* water. Press the product with a large cork to remove as much water as possible. Set aside about 10 mg for later determination of the melting range.

Purify the crude aspirin by recrystallization from a mixed solvent [see Technique 5.2a]. Stir your crude aspirin with 1.5 mL of ethanol in a 50-mL Erlenmeyer flask. If the crystals do not dissolve at room temperature, warm the mixture briefly on a steam bath or on a hot plate set at the lowest setting.

Probably no filtration is necessary at this stage; however, if the ethanol solution has any insoluble particles in it, you should filter it through a small fluted filter paper [see Technique 5.3a]. If you carry out this filtration step, you will have to rinse the filter paper carefully with another milliliter of ethanol to recover all the

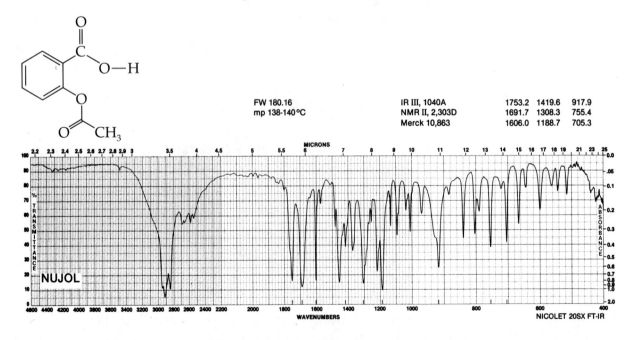

FW 180.16
mp 138-140°C

IR III, 1040A
NMR II, 2,303D
Merck 10,863

1753.2	1419.6	917.9
1691.7	1308.3	755.4
1606.0	1188.7	705.3

NUJOL

NICOLET 20SX FT-IR

FIGURE 5.1 IR spectrum of acetylsalicylic acid. (Provided by Aldrich Chemical Company, Inc., Milwaukee, WI.)

aspirin. If some of the aspirin has crystallized during the filtration, reheat the filtrate until the crystals dissolve.

Pour 10 mL of warm water (55–60°C) into the ethanol/aspirin solution and let the solution cool at room temperature for 10–15 min. Then cool it in an ice-water bath for 5 min to complete crystallization. Collect the product by vacuum filtration. Wash the product with 1 mL of *ice-cold* water and remove as much liquid as possible through suction. Allow the crystals to dry thoroughly before weighing them. Calculate your percent yield.

The melting range is a good way to assess the purity of aspirin. Take the melting ranges for both your crude aspirin and your recrystallized aspirin samples [see Technique 6.3]. What do these data tell you about the purity of your product? At the discretion of your instructor, determine the IR spectrum of your aspirin [see Spectrometric Method 1]. Compare it to the spectrum shown in Figure 5.1.

Cleanup: The filtrate from the crude aspirin should be neutralized with solid sodium bicarbonate before it is poured down the sink or placed in the container for aqueous inorganic waste. (**Caution:** **Foaming.**) The filtrate from the recrystallization can be poured down the sink.

⸺**m***icroscale*
Procedure

Techniques   Microscale Filtration: Technique 5.5a
Mixed Solvent for Recrystallization: Technique 5.2a
Microscale Recrystallization in Craig Tube: Technique 5.5b
Melting Points: Technique 6.3

⸺ *SAFETY INFORMATION* ⸺

Salicylic acid is toxic and an irritant to skin, eyes, mucous membranes, and the upper respiratory tract. Avoid breathing the dust. Avoid contact with skin, eyes, and clothing.

Acetic anhydride is toxic, corrosive, and a lachrymator (causes tears). Wear gloves, use only in a hood, and avoid contact with skin, eyes, and clothing.

Concentrated **phosphoric acid** (85%) is corrosive and causes burns. Avoid contact with skin, eyes, and clothing.

Using weighing paper, weigh a sample in the range of 220–230 mg (record the exact mass), rather than trying to weigh out exactly 225 mg.

Place 225 mg $\pm$ 5 mg of salicylic acid in a 13 $\times$ 100 mm test tube. Working under a hood, add 0.50 mL (5.3 mmol) of acetic anhydride (measured with a graduated pipet) and 1 drop of 85% phosphoric acid (measured with a Pasteur pipet). Cork the test tube before leaving the hood. Loosen the cork, place the test tube in a beaker of water at 45–50°C; hot tap water may be used to make the water bath. Shake the tube gently until the salicylic acid dissolves. Continue heating for 15 min. Remove the test tube from the water bath and add 1.4 mL of water, measured with a graduated pipet. Cool the solution at room temperature until crystallization begins, then cool in an ice-water bath until crystallization is complete. It may be necessary to scratch the inside of the test tube to initiate crystallization.

The freshly broken surface of the scratched glass provides nuclei for the formation of aspirin crystals.

Collect the crude product by vacuum filtration on a Hirsch funnel [see Technique 5.5a] and wash the crystals three times with 0.5-mL portions of distilled water, using a calibrated plastic pipet to measure and deliver the water. With the vacuum turned on, draw air through the Hirsch funnel for 5 min. Then dry the crude product by pressing it between pieces of fine-grained filter paper.

The odor of acetic acid will no longer be apparent if you have washed the crystals enough.

Purify the crude aspirin by recrystallization from a mixed solvent [see Technique 5.2a], using a Craig tube [see Technique 5.5b]. Weigh the crude aspirin on tared weighing paper and set aside about 10 mg for a melting-point determination. Transfer the crystals to a Craig tube by rolling the weighing paper into a funnel. The following solvent quantities are based on 140–160 mg of crude aspirin; if your crude product weighs more or less than this range, change the quantities of solvent proportionally. Add a boiling stick and 0.40 mL of ethanol, measured with a graduated pipet, to the Craig tube. Place the Craig tube in an aluminum block heated to 90°C on a hot plate [see Technique 3.2b]. When the crystals have dissolved, add 1.0 mL (graduated pipet) of water and warm the solution to dissolve any crystals that form.

If no crystals have formed after 15 min, add a seed crystal from the crude aspirin you have saved. Do this by picking up one or two tiny crystals on the tip of a microspatula and touching the spatula briefly to the top of the solution.

Place the plug in the Craig tube and set the apparatus in a 25-mL Erlenmeyer flask to cool slowly to room temperature. When crystal growth has ceased, cool the tube in an ice-water bath for 3–4 min. Centrifuge [see Technique 5.5b] to remove the solvent and save a few crystals for determining a melting point.

Determine the yield by weighing the recrystallized aspirin, and calculate the percent yield. Determine the melting range of the crude aspirin and the recrystallized aspirin [see Technique 6.3]. Has the purity been improved by the recrystallization? At the discretion of your instructor, determine the IR spectrum [see Spectrometric Method 1]. Compare it with the spectrum shown in Figure 5.1.

Cleanup: The filtrate from the crude aspirin should be neutralized with 5% sodium bicarbonate solution before it is poured down the sink or placed in the container for aqueous inorganic waste. (Caution: **Foaming.**) The ethanol/water solution remaining in the centrifuge tube may be poured down the sink.

Optional Experiment

Read Technique 10, Thin-Layer Chromatography, before carrying out this procedure.

Most painkillers contain at least one of the two most popular analgesic agents: aspirin or acetaminophen. In addition, caffeine is often added as a stimulant. Sometimes an antihistamine or sedative is included; for example, Excedrin PM contains the sedative methapyrilene hydrochloride.

If the label on a pill bottle were lost, how could a hospital laboratory tell what the bottle contained? This may seem like an academic question until you think about the possibility of drug overdose.

Chemists are frequently called upon to identify unknown substances. An identification may be needed for medical or legal reasons or because a chemical reaction leads to an unexpected product. The identifications can be based on complete physical properties, chemical reactivities, or spectroscopic characteristics. Of course, melting points can be helpful, but they are difficult to use with mixtures, and most medicinal preparations contain mixtures. For fast qualitative drug analysis, thin-layer chromatography (TLC) is hard to beat. The theory and practice of TLC is discussed in detail in Technique 10. Samples of acetaminophen, ibuprofen, aspirin, and caffeine will be available in the laboratory for your use as standards.

Procedure. We recommend ethyl acetate as the developing solvent and 2.5 × 6.7 cm plastic-backed, precoated silica gel thin-layer plates with fluorescent indicator. Dilute solutions (1–2%) work well for sample application. You will find that aspirin tails considerably on the thin-layer plate when ethyl acetate is used as the developing solvent, but this behavior causes no serious problem in the analysis. You can calculate the R_f value by using the center of the oblong aspirin spot.

After running thin-layer chromatograms for the three standards, you should pick one or two analgesic products and determine the analgesic it contains. Suggested products to test include Anacin, Advil, Excedrin Extra Strength, and Tylenol.

To analyze the analgesic, add a spatula-sized portion of a crushed tablet to 6 mL of a 50:50 chloroform/ethanol solution and gently heat the mixture on a steam bath in the hood for 2–3 min. The binder in the tablet (starch) and the inorganic buffers will not dissolve, but the analgesics and other medicinal agents will go into solution. You should heat the mixture just enough to dissolve the active ingredients of the tablet but not enough to boil away any solvent. After the solid settles, you can analyze the solution by TLC, again using ethyl acetate as the developing solvent and UV radiation for visualization. How many components are indicated in the tablet by your TLC analysis? What are they? Calculate the R_f values [see Technique 10.1a] and compare them with the R_f values for the known analgesic solutions.

Cleanup: Pour the ethyl acetate left in the developing jar into the container for flammable (organic) waste. Pour the remaining solutions of analgesic tablets into the container for halogenated-waste. The thin-layer plates should be placed in the container for nonhazardous solid waste.

Questions

1. Assuming that 1 g of aspirin dissolves in 450 mL of water at 10°C, how much aspirin would be lost in the 16 mL of water (macroscale preparation) or 1.4 mL of water (microscale preparation) added to the reaction mixture if the mixture were at 10°C during the filtration?
2. How much difference would the amount lost in the water (Question 1) make in your percent yield?
3. What is the purpose of adding the concentrated phosphoric acid to the reaction mixture in the synthesis of aspirin?
4. Suggest a possible composition for buffered aspirin. What benefits might it have?

RADICAL CHLORINATION REACTIONS

Study a reaction—the free-radical halogenation of alkanes— that you have probably already read about in your textbook. Use gas chromatography to determine the possible isomers produced in a chlorination reaction.

$$CH_3CH_2CH_2CH_2Cl \xrightarrow[\substack{HCl \\ H_2O}]{NaOCl} \text{mixture of dichlorobutanes}$$

1-Chlorobutane
bp 78.5°C
$d = 0.886 \text{ g} \cdot \text{mL}^{-1}$

$$\underset{\substack{| \\ CH_3}}{CH_3\overset{\substack{CH_3 \quad CH_3 \\ | \qquad |}}{C}CH_2CHCH_3} \xrightarrow[\substack{HCl \\ H_2O}]{NaOCl} \text{mixture of monochlorooctanes}$$

2,2,4-Trimethylpentane
bp 99.2°C
$d = 0.692 \text{ g} \cdot \text{mL}^{-1}$

Chlorine is a pale green gas that is highly toxic and hard to handle. In fact, Cl_2 was the first poisonous gas used in modern warfare, during World War I in the trenches of western Europe.

The reaction conditions used here produce Cl_2, in situ, where it reacts with alkanes by a free-radical substitution pathway. The Cl_2 results from treating sodium hypochlorite, NaOCl (household bleach), with HCl by the following reactions:

$$NaOCl + HCl \longrightarrow HOCl + NaCl$$

$$HOCl + HCl \longrightarrow Cl_2 + H_2O$$

Alkanes are relatively unreactive compounds and halogenation is one of the few reactions that they undergo. For example, the following equation represents the chlorination of methane to produce chloromethane:

$$CH_4 + Cl_2 \xrightarrow{\Delta \text{ or } h\nu} CH_3Cl + HCl$$

The reaction occurs in a series of steps, the first called initiation, in which light or heat induces the dissociation of a chlorine molecule into two free chlorine atoms.

$$Cl_2 \xrightarrow{\Delta \text{ or } h\nu} 2Cl\cdot$$

These free chlorine atoms, called radicals, have an unpaired, or odd, electron. They promote chlorination by abstraction of hydrogen from the organic starting material, methane in this example.

$$\dot{C}l + C\overset{\frown}{H}_4 \longrightarrow CH_3\cdot + HCl$$
$$CH_3\overset{\frown}{\cdot + \dot{C}}l_2 \longrightarrow CH_3Cl + Cl\cdot$$

These two steps are called the propagation steps; we find that the chlorine atoms necessary for the first step are generated in the second step. This allows the propagation sequence to occur over and over again. Ultimately, the reaction undergoes several termination steps that provide no radical products:

$$2Cl\cdot \longrightarrow Cl_2$$

$$R\cdot + Cl\cdot \longrightarrow RCl$$

$$2R\cdot \longrightarrow R{-}R$$

Chlorination reactions can lead to mixtures of products, like those you will observe in this experiment. For example, monochlorination of propane gives two products:

$$CH_3CH_2CH_3 + Cl_2 \longrightarrow CH_3CH_2CH_2Cl + CH_3\underset{\underset{Cl}{|}}{C}HCH_3$$

Considering only positional isomers (and ignoring cis/trans isomers and other stereoisomers), methylcyclopentane gives rise to four monochlorination products:

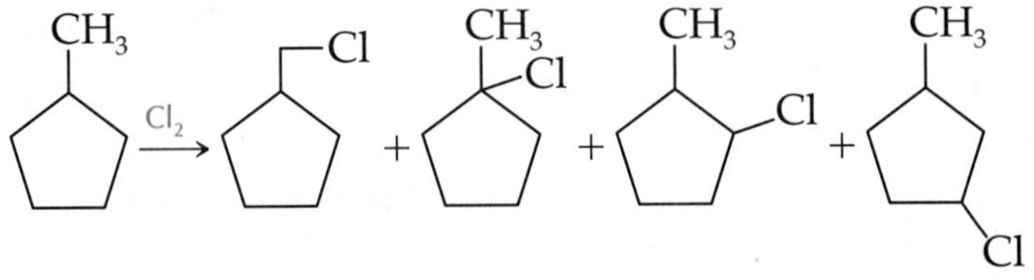

Two major factors control the preference for isomer formation during radical halogenation reactions, namely, a statistical factor and the relative stability of the carbon radical intermediates. For example, if the statistical factor is the only one directing the chlorination of 2-methylpropane (isobutane), then the predominate product would be the primary chloride, isobutyl chloride because there are nine methyl hydrogens and only one methine (CH)

hydrogen. In fact, the reaction produces a mixture of primary and tertiary chlorides:

$$(CH_3)_2CHCH_3 + Cl_2 \longrightarrow (CH_3)_2CHCH_2Cl + (CH_3)_3CCl$$

| 2-Methylpropane (isobutane) | 2-Methyl-1-chloropropane (isobutyl chloride) a primary chloride | 2-Methyl-2-chloropropane (*tert*-butyl chloride) a tertiary chloride |

Because tertiary radicals have more stability than secondary ones, which in turn show more stability than primary radicals, we would expect the chlorination of 2-methylpropane to give a large proportion of tertiary product (2-methyl-2-chloropropane). This is indeed the case.

In analyzing the course of halogenations, both statistical and radical stability factors must be considered. In fact, using both often leads to excellent product ratio predictions.

The radical chlorination of 1-chlorobutane and 2,2,4-trimethylpentane studied in this experiment both produce a mixture of products. The number and identity of the products will be determined by gas chromatography [see Technique 11].

mmmicroscale Procedure

Techniques Microscale Extraction, Technique 4.5 and 4.5d
Pasteur Filter Pipets, Technique 2.4
Gas Chromatography, Technique 11

--- **SAFETY INFORMATION** ---

1-Chlorobutane is harmful if inhaled, ingested, or absorbed through the skin. Wear gloves and use the reagent in a hood, if possible.

2,2,4-Trimethylpentane is flammable and a skin irritant. Use the reagent in a hood, if possible.

Sodium hypochlorite solution, when acidified, emits chlorine gas that is toxic as well as an eye and respiratory irritant. Use the reagent in a hood, if possible.

Freshly purchased household bleach such as Clorox or a supermarket brand works well.

You will carry out the following radical chlorination procedure on 1-chlorobutane and 2,2,4-trimethylpentane. As part of your pre-lab preparation, draw the structures for the possible monochlorination products in each reaction and give the IUPAC name of each.

Label the bottom of a 5-mL conical vial for each reaction. Place 1.0 mL of the compound being tested in the vial. Add 1.0 mL of 5% sodium hypochlorite solution. Working in a hood, add 0.5 mL of 3 M hydrochloric acid to the vial and *immediately* close it with a septum and screw cap. Shake the vial until the yellow color of the chlorine has moved from the aqueous layer to the organic layer (about 30 s). Repeat this procedure with the other organic compound as well.

Place the vials at a distance of 5–8 cm from an unfrosted light bulb and irradiate them until the yellow color of the chlorine disappears from the organic layer (usually within 1 to 2 min). Shake each vial occasionally during the irradiation period.

SAFETY PRECAUTION

The unfrosted light bulb is very bright; do not look directly at it.

Set the conical vial in a small beaker so it does not tip over.

Working with one vial at a time, carry out the following procedure on each. Add 100 mg of anhydrous sodium carbonate in several portions; wait for the foaming to cease before making the next addition. When the addition is complete and foaming ceases, cap the vial and shake it vigorously. Prepare a Pasteur filter pipet [see Technique 2.4]. Remove the lower aqueous layer with the Pasteur filter pipet [see Technique 4.5d]. Add 100 mg of anhydrous calcium chloride and allow the solution to dry for at least 10 min.

The halogenation has not proceeded far enough to appreciably change the density from that of the starting material.

Prepare two more Pasteur filter pipets while you are waiting for the solutions to dry. Using one of these filter pipets, transfer one of the dried solutions to a labeled sample vial or small test tube fitted with a cork. Repeat this procedure for the other solution, using the other filter pipet. Analyze the solutions on a gas chromatograph that has a nonpolar column [see Technique 11]. Consult your instructor about specific operating instructions and sample size for the gas chromatographs in your laboratory. Record the operating parameters, including column temperature, injector temperature, detector temperature, helium flow rate, chart speed, and the voltage or attenuation for the recorder. See Question 1 for the elution order of the dichlorobutanes from a nonpolar column.

The following order of elution from a nonpolar column has been reported by Russell and Haffley (Reference 3) for the monochlorination of 2,2,4-trimethylpentane:

1. 2-chloro-2,4,4-trimethylpentane
2. 3-chloro-2,2,4-trimethylpentane
3. 1-chloro-2,4,4-trimethylpentane
4. 1-chloro-2,2,4-trimethylpentane.

For each reaction, use the chromatogram to determine the percentage of each product [see Technique 11.6].

Cleanup: The aqueous phase separated from the reaction mixture may be washed down the sink or placed in the container for aqueous inorganic waste. The calcium chloride should be placed in the container for hazardous solid waste because it is coated with halogenated hydrocarbons. After performing the GC analysis, the product mixture should be poured into the halogenated-waste container.

Optional Experiments

Carry out the preceding procedure using 3-methylpentane and 2-chlorobutane as the substrates. Both of these compounds give diastereomers for one of the products that have boiling points different enough to be partially separated on the gas chromatograph.

References

1. Gilow, H. M. *J. Chem. Educ.* **1991,** *68,* A122–A124.
2. Reeves, P. C. *J. Chem. Educ.* **1971,** *48,* 636–637.
3. Russell, G. A.; Haffley, P. G. *J. Org. Chem.* **1966,** *31,* 1869–1871.

Questions

1. A handbook will reveal the boiling points of the four dichlorobutanes formed by the chlorination of 1-chlorobutane. The

dichlorobutanes elute in the following order: 1,1-dichlorobutane, 1,2-dichlorobutane, 1,3-dichlorobutane, and 1,4-dichlorobutane. Use the boiling points to explain the GC elution order.

2. Calculate the proportions of isomeric products that would be expected from each reaction carried out in this experiment, if only statistical factors determined the course of the reaction. Compare your experimental results with these proportions.

3. Using your experimental results for each compound tested, calculate the relative reactivity (rate) for abstraction of each type of hydrogen by a chlorine radical.

4. From the study of radical chlorination reactions using many hydrocarbons, chemists have found that the relative reactivity for 1°, 2°, and 3° hydrogen atoms is 1:3.8:5.0. Compare your experimental values for relative reactivity to these and suggest a reason for any possible differences.

5. 1-Chlorobutane has two types of 2° hydrogens relative to the position of the chlorine atom in the molecule. Did your results show an equal amount of chlorination at each 2° site? Give a possible explanation for any difference observed in the amount of chlorination at each site.

Experiment 7
ISOLATION OF ESSENTIAL OILS AND THE THEORY OF ODOR

Investigate the relationship between smell and three-dimensional chemical structure, starting with either caraway seeds or orange peels. Optical activity of the compounds will be an important technique in their analysis.

The first organic compounds to be identified came from plant and animal tissues. Although most organic compounds used today are synthesized from a variety of raw materials, living organisms continue to provide some industrial chemicals and a number of new drugs.

Essential oils make up one group of organic compounds still obtained from plants. Essential oils are often characterized by very distinctive odors, and these odors are dependent to a certain degree on the stereochemistry found in the particular molecule. As we shall see in the following experiments, essential oils of relatively low molecular weights can be obtained easily by steam distillation.

Many essential oils belong to the class of compounds called terpenes, compounds whose structures have been studied for many years and whose carbon skeletons are composed of five carbon units, called isoprene units. For example, citronellol *(oil of geranium)* is composed of two isoprene units.

2-Methyl-2-butene (isoprene)   Isoprene unit   Citronellol (a monoterpene)

Terpenes are classified as monoterpenes (contain 10 carbons), sesquiterpenes (C_{15}), diterpenes (C_{20}), and so on. Vitamin A is a diterpene:

Vitamin A (a diterpene)

Theories to explain the sense of smell have been the subject of scientific analysis for many years. In fact, the Roman atomist Lucretius suggested that a substance could give off "atoms" of vapor of a given type and the odor perceived for that substance would depend on the nature of pores in the nose amenable to specific "atoms" of vapor. This idea was extended by R. W. Moncrief in 1949. He suggested that nasal pores had shapes and sizes such that they would act as receptor sites and that these sites would fit all molecular structures corresponding to a given primary odor.

J. E. Amoore continued to build the theory of primary odors. In the 1950s he postulated seven primary odors: camphoraceous, ethereal, floral, musky, pepperminty, pungent, and putrid. He concluded that there was a general shape and size for all molecules in a primary odor category and that all structures in that odor group would fit into a certain cavity. Amoore recognized that the picture was more complex, because he also concluded that charge distribution on the molecule and correspondingly in the receptor site is assumed to be important for both the putrid and the pungent class.

The next level in odor theory involved stereochemistry. Specifically, the biological response of smell can be induced by one enantiomer (because of a proper fit in a receptor, which is composed of optically active proteins), whereas the other enantiomer often shows no response (because of a lack of fit), or it may cause some very different physiological response (because of a fit with another enantiomeric receptor site).

Current theories of smell suggest that the situation may be much more complex and that there may be many different kinds of odor receptors. We still have much to learn about what controls our sense of smell and why different people smell the same compounds with distinctly different results.

In Experiment 9.1 you will isolate the stereoisomer of carvone found in caraway oil and analyze the optical properties of this isomer and its enantiomer from spearmint oil. You will also prepare derivatives of each. It is worthy of note that these two enantiomers have very different odors. In Experiment 9.2 you will isolate one enantiomer of limonene from orange peels and analyze the optical properties of this compound.

7.1

Isolation of (S)-(1)-Carvone from Caraway Seeds

Extract the essential oil from caraway seeds and analyze it by GC and polarimetry.

(S)-(+)-Carvone
bp 230°C
density 0.965 g · mL^{-1}
$[\alpha]_D^{20}$ +61.2°
MW 150.2
Major component of caraway
and dill seed oils

(R)-(−)-Carvone
bp 230°C
density 0.965 g · mL^{-1}
$[\alpha]_D^{20}$ −62.5°
MW 150.2
Major component of spearmint oil

It is remarkable that the distinctly different smells of caraway seeds and spearmint leaves come from two isomers that differ only in their chirality. The major compound in both caraway oil and spearmint oil is carvone, an unsaturated cyclic ketone with the formula $C_{10}H_{14}O$. One oil contains the (+) isomer and the other the (−) isomer. Current theories on smell hold that because odor receptors of the nose are chiral, the chiral carvone isomers, with their mirror image shapes, fit into quite different receptor sites. Curiously, it has been reported that about 1 person out of 10 cannot tell the difference between the odors of caraway and spearmint.

As far as we know, chiral (or "asymmetric") compounds in nature exist only in living tissue or in matter that was once a part of living tissue. Chirality plays a major role in the mechanisms of biochemical recognition. Yet it is still a mystery why caraway plants, *Carum carvi*, produce (S)-(+)-carvone and spearmint plants produce its mirror image (R)-(−)-carvone. Other plants such as gingergrass produce racemic carvone.

Even more curious is the fact that the α-pinenes taken from different pine trees in the same grove can have opposite optical activities. Nature goes even one step further; some botanically indistinguishable plants that grow in different countries can carry out complete metabolic sequences of mirror image reactions. How such differences developed is still unknown.

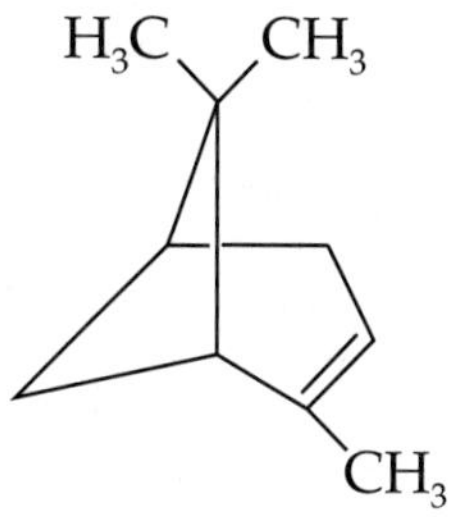

α-Pinene

Caraway seed oil can be isolated by the steam distillation of ground caraway seeds. The compound that is responsible for its characteristic odor, (S)-(+)-carvone, constitutes about 70–80% of the oil. The other major component is limonene, a terpene also found in spearmint oil and the oils of oranges and lemons.

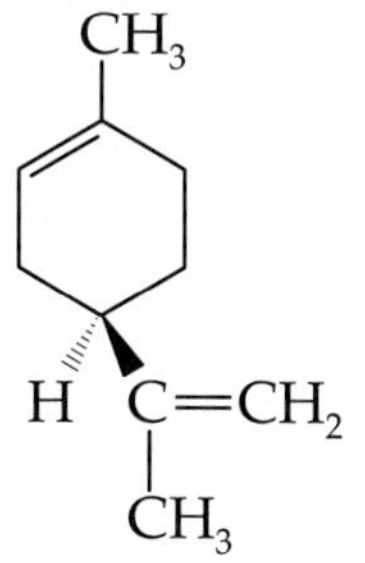

(R)-(+)-Limonene

Caraway seeds are inexpensive when purchased in bulk and will be your source of (+)-carvone. Although carvone boils at 230°C, far above the boiling point of water, it can easily be steam distilled from ground seeds because it is largely insoluble in water. Extraction of the distillate with dichloromethane, followed by drying and evaporation, produces caraway seed oil.

In this experiment you will investigate the differences between the carvone from spearmint and that from caraway oil by using optical activity measurements, gas-liquid chromatography, and your sense of smell. While the steam distillation of the caraway oil is bubbling along merrily, you can proceed with your analysis of (−)-carvone. (−)-Carvone is commercially available, and you will be comparing this enantiomer with the (+)-carvone you isolate from caraway seeds.

Because it is far easier to work with small amounts of solids than of liquids, it is useful to make a solid derivative of your rather small amount of (*S*)-(+)-carvone. Carvone is a ketone, and 2,4-dinitrophenylhydrazones are the most common derivatives of such carbonyl compounds. However, their deeply colored solutions make them impossible to use for the polarimetric studies, because they do not transmit enough light. The solid derivative of choice here is the semicarbazone, which forms white crystals and colorless solutions.

| Carvone | Semicarbazide hydrochloride mp 175°C MW 111.5 | Carvone β-semicarbazone mp 141–142°C MW 207.2 |

There are two possible diastereomers for the semicarbazones of (−)-carvone and (+)-carvone. They result from the restricted rotation about the carbon-nitrogen double bond.

The α isomers of (−)- and (+)-carvone semicarbazone melt at 162–163°C; the β isomers melt at 141–142°C. The β-semicarbazone forms under our experimental conditions.

The preparation of carvone's semicarbazone is doubly useful here because it lets us separate (+)-carvone from the other constituents of caraway seed oil. The other major component, limonene, has no carbonyl group, so it will not form a semicarbazone and will be washed away in the filtrate from the recrystallization.

Macroscale Procedure

Techniques   Steam Distillation: Technique 7.6
Extraction: Technique 4.2
Melting Points: Techniques 6.3 and 6.4
Gas-Liquid Chromatography: Technique 11
Polarimetry: Technique 13

— **SAFETY INFORMATION** —

Dichloromethane is toxic, an irritant, absorbed through the skin, and harmful if swallowed or inhaled. Use it in a hood, if possible, and wash your hands thoroughly after handling. Be sure to do the evaporation process in a hood.

(S)-(+)-carvone is highly toxic and absorbed through the skin. Avoid contact with skin, eyes, and clothing.

Methanol is toxic and flammable. Pour it only in a hood.

Steam Distillation

A bakery or a bakery supplier is the best source of fresh caraway seeds. Seeds purchased in a supermarket may have spent months on the shelf and still contain carvone but tend to have little or no limonene remaining in them.

No additional water is needed for this steam distillation; replace the dropping funnel with a ⊤ glass stopper.

Weigh 25 g of fresh caraway seeds and grind them in an electric blender. Put the ground seeds into a 500-mL round-bottomed flask and add 250 mL of water. Set up a distillation apparatus as shown in Technique 7.6, Figure 7.15, but *without* the separatory (dropping) funnel and round-bottomed receiving flask. Use a 100-mL graduated cylinder for the receiver and close the second neck of the Claisen adapter with a glass stopper.

Boil the mixture vigorously, using a heating mantle as the heat source, but be careful not to let any solid material bump over into the condenser. Collect about 80 mL of distillate. Faster distillation seems to make for a smoother, less bumpy process, so do not decrease the rate of heating once the distillation begins. Distill the mixture as rapidly as the cooling capacity of the condenser will permit. The distillation should take approximately 30–40 min. During this time, prepare the semicarbazone of (R)-(−)-carvone, spearmint oil (see p. 61).

Isolation of Caraway Oil

Review Technique 4.2, Macroscale Extractions, before doing this part of the experiment.

The salt helps to minimize emulsions during the extractions by making the organic layer less soluble in the water layer.

Be sure to vent the funnel frequently during the extractions.

Pour your distillate into a 125-mL separatory funnel. Add 5 g of sodium chloride and shake the mixture to dissolve the salt. Then add a few pieces of ice to make the solution distinctly cool before adding any volatile dichloromethane.

Obtain 20 mL of dichloromethane. Place a conical funnel in the top of the separatory funnel. Hold the condenser and the vacuum adapter used to collect the distillate above the separatory funnel and rinse them with a few milliliters of the dichloromethane; let the dichloromethane drain into the separatory funnel. Use the remaining dichloromethane to rinse the 100-mL cylinder and also add that solution to the separatory funnel. This rinse recovers all the caraway oil clinging to the glass surfaces.

Extract the carvone from the aqueous layer by inverting the funnel and shaking it back and forth gently [see Technique 4.2]. Repeat the gentle shaking and venting for about 2 min. Allow the layers to separate before draining the bottom (organic) layer into a dry 50-mL Erlenmeyer flask. Carvone is quite soluble in dichloromethane; any organic membranes that may form at the interface should be left with the aqueous layer.

Repeat the extraction of the aqueous layer, using 15 mL of fresh dichloromethane. Combine the two dichloromethane extracts and dry the solution with anhydrous magnesium sulfate for at least 10 min.

Weigh a clean, dry 50-mL Erlenmeyer flask on an analytical balance. Filter half of the dichloromethane solution through a conical funnel fitted with a fluted filter paper [see Technique 5.3a, Figure 5.5] into the tared Erlenmeyer flask. Add a boiling stick and evaporate the dichloromethane on a steam bath or hot plate in a hood. Cool the flask briefly and filter the remaining dichloromethane solution into it. Continue heating the flask until the solvent has completely evaporated. Alternatively, the dichloromethane can be removed with a rotary evaporator [see Technique 4.8].

Cool the flask and carefully wipe the outside dry with a tissue; then weigh it on an analytical balance. Heat the flask several additional minutes, cool, and weigh again. The two masses should agree within 0.05 g. If they do not, repeat the heating and weighing procedures. Record the final mass of your caraway oil to the nearest milligram. The residue in the flask is caraway seed oil. Carefully compare the smell to that of (*R*)-(−)-carvone.

Gas Chromatographic Analysis

Review Technique 11, Gas-Liquid Chromatography, before you do this part of the experiment.

Use both the nonpolar and polar columns at 170–200°C to find the retention time for a known sample of (−)-carvone on each column under your conditions. If you are using a capillary column chromatograph, prepare a solution containing 0.5 mL of ether and 1 drop of the compound being tested. After you have analyzed (*R*)-(−)-carvone, analyze your caraway seed oil, using the same nonpolar and then polar GC columns at the same temperature and flow rate. Also chromatograph samples of known (*S*)-(+)-carvone, and limonene, using identical GC conditions.

Calculate the retention times of (−)- and (+)-carvone and limonene. Also calculate the percentages of (+)-carvone and limonene in caraway seed oil by computing the areas under their respective peaks if you are using a nonintegrating recorder (or peak heights, if the peaks are very narrow) [see Technique 11.6].

Derivatives: Preparation of Semicarbazones

Sodium acetate provides the proper pH for rapid carbonyl addition-dehydration reactions.

Weigh 0.38–0.39 g of (*R*)-(−)-carvone (spearmint oil) into a 25-mL Erlenmeyer flask and add 4.0 mL of 95% ethanol. Dissolve 0.40 g of semicarbazide hydrochloride and 0.40 g of anhydrous sodium acetate *or* 0.64 g of sodium acetate trihydrate in 2.0 mL of water in a 13 × 100 mm test tube. Pour the resulting solution into the flask containing the carvone solution and add a boiling stone or stick. Warm the mixture on a steam bath set for a gentle steam flow for 15 min.

Add 2.5 mL of water to the warm solution; then set the reaction mixture aside to cool slowly to room temperature. Under these conditions, crystallization may take 30–45 min; slow crystallization gives nearly pure crystals that probably will not need to be recrystallized for the optical activity studies. After crystallization is complete, cool the solution for 5 min in an ice-water bath.

Collect the crystals by vacuum filtration, using a small Buchner funnel, and wash them on the funnel with a few milliliters of cold water. Allow the solid to dry for several hours or overnight before taking the melting point [see Technique 6.3]. If it is necessary to recrystallize the semicarbazone, you may do so from an ethanol/water mixture [see Technique 5.2a].

Prepare the semicarbazone of the carvone in your caraway seed oil in the Erlenmeyer flask containing the oil, using all of your sample except what you need for the GC analysis. Add

If possible, make this derivative on the same day that you isolate the caraway seed oil.

4.0 mL of ethanol to your caraway seed oil. Prepare the semicarbazide solution (as directed above) in a 13 × 100 mm test tube and pour it into the ethanol solution of caraway seed oil in the Erlenmeyer flask. Follow the procedure given earlier for heating and crystallization. Adjust the proportions of all reagents if you have less than 0.30 g of caraway seed oil. Again, determine the melting point after the solid dries.

Also grind together a small amount of approximately equal portions of the carvone semicarbazones from each source and take a melting point of the mixture [see Technique 6.4].

Polarimetry

Read Technique 13, Polarimetry, before doing your measurements of optical activity. Use a 5.0% or 10.0% sucrose solution in water as a standard, to become acquainted with polarimetric measurements. The specific rotation of sucrose, $[\alpha]_D^{20}$, equals + 66.4° (H_2O). Compare your calculated specific rotation with the accepted value.

dm = decimeter

If you are using 2-dm polarimetry tubes, use two 25-mL volumetric flasks and prepare a 25-mL solution of each of your *dry* carvone semicarbazones in anhydrous methanol. If you are using 1-dm periscope polarimeter tubes, prepare each solution in a 10-mL volumetric flask. The concentration should be 1.50% if you have enough material (0.15 g/10 mL of solution). If you do not have enough for this concentration, use what you have, saving enough for the melting points. Weigh the dry carvone semicarbazone samples to the nearest milligram. After dissolving the semicarbazone, stopper the flask and shake it a number of times to ensure a completely homogeneous solution.

If you see any undissolved particles such as paper fibers or pieces of dust in the solution, filter the solution by gravity through a small plug of glass wool or a small fluted filter paper, using a short-stemmed funnel. If you are using a periscope polarimeter tube, filter the solution directly into the tube; if you are using a straight polarimeter tube, filter the solution into a 25- or 50-mL Erlenmeyer flask; then fill the polarimeter tube. Keep the solutions tightly stoppered, except during transfer, to avoid evaporation of the solvent.

Compare the specific rotations of the carvone semicarbazones from caraway seed oil and from (*R*)-(−)-carvone. Are they mirror images?

Cleanup: Filter the caraway seed residue from the aqueous liquid remaining in the distillation flask. (Do not put caraway seeds down the sink.) Wash the aqueous filtrate down the sink; dispose of the seed residue as food garbage. Allow any residual dichloromethane to evaporate from the magnesium sulfate drying agent in a hood before placing it in the container for inorganic waste or the container for solid hazardous waste. Pour the ether solutions used for GC analysis, the filtrate from the semicarbazone preparations, and the methanol solutions from your polarimetric measurements into the container for flammable (organic) waste.

References

1. Garin, D. L. *J. Chem. Educ.* **1976,** *53,* 105.
2. Glidewell, C. *J. Chem. Educ.* **1991,** *68,* 267–269.
3. Murov, S. L.; Pickering, M. *J. Chem. Educ.* **1973,** *50,* 74.

Questions

1. Critically evaluate your evidence on whether the carvone isolated from caraway seed oil is the mirror image of (−)-carvone.
2. What caused the melting point of the mixture of (+)- and (−)-carvone semicarbazones to be higher than the melting point of either pure compound?
3. Propose a method that you could use to isolate the limonene in caraway seed oil.

7.2
Isolation of (*R*)-(+)-Limonene from Orange Peels

Extract the essential oil from orange peels and analyze it by polarimetry.

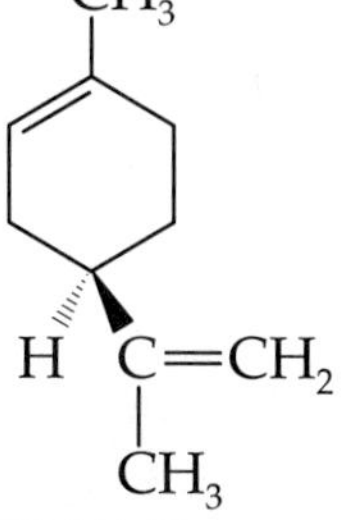

(*R*)-(+)-Limonene
bp 175.5–176°C
density 0.8402 g · mL^{-1}
[α] +123.8°
MW 136.2
Major component of orange peels

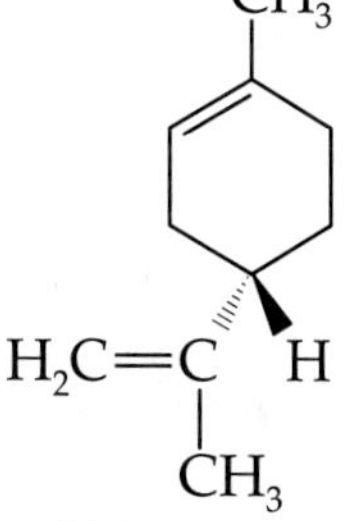

(*S*)-(−)-Limonene
bp 175.5–176.5°C
density 0.8407 g · mL^{-1}
[α] −101.3°
MW 136.2
Minor component of caraway seeds

Limonene is a terpene whose stereochemistry is dependent on its source. Orange peels provide limonene that is virtually 100%

(*R*)-(+)-limonene, whereas pine needles provide essentially 100% (*S*)-(−)-limonene. The biosynthetic pathway to limonene is quite complex, but we do know how different plants produce one enantiomer or the other.

The key intermediate in the formation of limonene is neryl pyrophosphate, which is formed biosynthetically from acetyl coenzyme A after a long series of steps. During cyclization, the stereocenter that differentiates the (+) and (−) forms of limonene is formed. The active site of the enzyme that promotes the cyclization of neryl pyrophosphate in pine needles must be different from that in oranges.

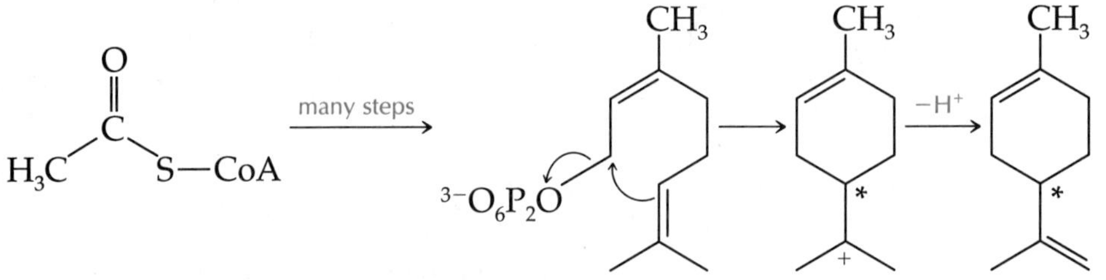

Neryl pyrophosphate   Limonene

∿∿∿Macroscale Procedure

Techniques   Steam Distillation: Technique 7.6
Extraction: Technique 4.2
Polarimetry: Technique 13

--- SAFETY INFORMATION ---

Dichloromethane is toxic, an irritant, absorbed through the skin, and harmful if swallowed or inhaled. Use it in a hood, if possible, and wash your hands thoroughly after handling.

(+)-Limonene is an irritant and absorbed through the skin.

Steam Distillation

You will need the peels from two oranges for this experiment. Your instructor will specify whether you are to provide the oranges or whether they will be available in the laboratory.

Peel the oranges just before grinding the peels to prevent loss of the volatile limonene.

Remove most of the white pulp from the orange peels with a knife or spatula before grinding the peels in a blender with 200–250 mL of water to make a slurry that can be easily poured into a 500-mL round-bottomed flask. Add 4 drops of an antifoaming agent to the flask.

Assemble the steam distillation apparatus shown in Technique 7.6, Figure 7.15 but *without* the separatory (dropping) funnel. Use a 50-mL round-bottomed flask as the receiver. Before attaching the receiving flask, pour 35 mL of water into it and mark the level on the outside of the flask; then pour out the water (the flask does not have to be dried).

Boil the mixture at a moderately rapid rate, using a heating mantle as the heat source, but be careful not to let any solid material bump over into the condenser. Collect 35 mL of distillate.

Isolation of Limonene

Review Technique 4.2, Macroscale Extractions, before doing this part of the experiment.

Pour your distillate into a 125-mL separatory funnel and add 2 g of sodium chloride. Stopper the funnel and shake it to dissolve the salt. Then add a few pieces of ice to make the solution distinctly cool before adding any volatile dichloromethane.

The NaCl helps to minimize emulsions during the extractions by making the organic layer less soluble in the aqueous layer.

Obtain 15 mL of dichloromethane. Place a conical funnel in the top of the separatory funnel. Rinse the condenser and vacuum adapter used to collect the distillate with a few milliliters of the dichloromethane by holding them above the funnel and letting the dichloromethane drain into the separatory funnel. Use the remaining dichloromethane to rinse the 50-mL round-bottomed flask and also add that solution to the separatory funnel. This rinse recovers any limonene clinging to the glass surfaces.

Extract the limonene from the aqueous layer by inverting the funnel and shaking it back and forth gently [see Technique 4.2]. Be sure to vent the funnel frequently. Repeat the gentle shaking and venting for about 2 min. Allow the layers to separate before removing the bottom organic layer into a dry 50-mL Erlenmeyer flask. Limonene is quite soluble in dichloromethane; any organic membranes that may form at the interface should be left with the aqueous layer.

Repeat the extraction of the aqueous layer, using another 10 mL of dichloromethane. Combine the two dichloromethane extracts and dry the solution with anhydrous calcium sulfate for at least 10 min.

Weigh a dry 50-mL Erlenmeyer flask on an analytical balance to the nearest milligram. Filter the dichloromethane through fluted filter paper into the tared flask [see Technique 4.7]. Carry out the evaporation of dichloromethane in a hood. Put a boiling stick into the flask and evaporate the dichloromethane on a steam bath, using a gentle flow of steam. (Alternatively, the evaporation process may be done with a stream of nitrogen and gentle warming of the flask in a beaker of hot tap water or with a rotary evaporator [see Technique 4.8].) Cool the flask and carefully wipe the outside dry with a tissue before weighing it on an analytical balance. Heat the flask several additional minutes, cool, and weigh it again. The two masses should agree within 0.05 g. If they do not, repeat the heating and weighing procedures until this agreement is reached. Record the final mass of limonene to the nearest milligram. Describe its appearance and carefully note the odor **(do NOT breathe the vapors of limonene).**

Polarimetry

Read Technique 13, Polarimetry, before doing your measurements of optical activity. Use a 5.0% or 10.0% sucrose solution in water as a standard, to become acquainted with polarimetric measurements. The specific rotation of sucrose, $[\alpha]_D^{20}$, equals $+66.4°$ (H_2O). Compare your calculated specific rotation with the accepted value.

1 dm = 1 decimeter

The following directions for preparation of the polarimetry solution are for 1-dm periscope polarimeter tubes. Consult your instructor if your laboratory is equipped with another type or size of polarimeter tubes.

Obtain 10 mL of 95% ethanol in a clean, dry graduated cylinder. Dissolve your limonene in 3 mL of ethanol (use a Pasteur pipet to transfer the ethanol) and quantitatively transfer the resulting solution to a 10-mL volumetric flask as follows: Set the volumetric flask in a small beaker so that it will not tip, and place a very small funnel in the neck of the volumetric flask. Carefully pour the limonene solution into the volumetric flask. Rinse the Erlenmeyer flask three times with approximately 1-mL portions of ethanol and add these rinses to the volumetric flask. Rinse the funnel with 1 mL of ethanol and remove it from the volumetric flask. Fill the volumetric flask to the calibration mark, using a Pasteur pipet; stopper the flask and invert it several times, until the contents are thoroughly mixed.

Calibrate the polarimeter, using 95% ethanol as the reference solvent.

If the solution is cloudy, filter it directly into a polarimeter tube, using a small funnel and fluted filter paper. If the solution is clear, it can be poured directly into a polarimeter tube. Determine and record the observed rotation for your limonene solution; also record the temperature [see Technique 13.4]. Calculate the specific rotation $[\alpha]_D$ of limonene and its enantiomeric excess (optical purity) [see Technique 13.5].

Cleanup: Filter the orange peel residue from the aqueous liquid remaining in the distillation flask. (Do not put orange peels down the sink.) Wash the aqueous filtrate down the sink; dispose of the orange peel residue as food garbage. Allow any residual dichloromethane to evaporate from the calcium sulfate drying agent in a hood before placing it in the container for inorganic waste or the container for solid hazardous waste. Pour the ethanol solution used for your polarimetric measurements into the container for flammable (organic) waste.

Reference

1. Glidewell, C. *J. Chem. Educ.* **1991,** *68,* 267–269.

Questions

1. How would the presence of residual dichloromethane affect the observed rotation of limonene? Explain.
2. The optical rotations of enantiomers are, in principle, equal in magnitude, but opposite in sign. The rotations that we have provided have opposite signs, but the magnitudes are unequal. Explain this inequality, keeping in mind that the same explanation probably applies to why the densities of the enantiomers are not equal, although—again, in principle—they should be.
3. The conjugated diene, α-terpinene, is formed in some of the same biosynthetic pathways that give rise to limonene. Make simple mechanistic proposals for how α-terpinene may be formed and state why stereochemistry is no longer an issue in the structure of α-terpinene.

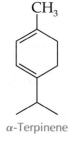

CH$_3$

α-Terpinene

Experiment 8

NUCLEOPHILIC SUBSTITUTION REACTIONS

Investigate the relationship of structure and reactivity in substitution reactions, a topic that you already have or soon will be studying in the classroom. The techniques encountered range from simple qualitative tests to NMR analysis, to rate measurements, to organic synthesis.

One of the most-studied and well-established mechanisms of organic chemistry is that for nucleophilic substitution. In this reaction a *nucleophile* (literally, a nucleus-seeking reagent) is used to displace a group from an organic substrate:

$$R—G \quad + \quad Nu^- \quad \longrightarrow R—Nu + G^-$$
Substrate   Nucleophile
(G = leaving group)

The nucleophile need not be negatively charged, but it must have a negatively polarized atom, such as the oxygen of an alcohol, that acts as the nucleophilic atom. A typical example of a nucleophilic substitution reaction is the displacement of chloride ion from 2-chloropropane (isopropyl chloride) by methoxide ion, provided by sodium methoxide:

$$\begin{array}{ccc} H_3C & H \\ & C & + \ Na^+ \ ^-OCH_3 \longrightarrow \\ H_3C & Cl \end{array} \quad \begin{array}{cc} H_3C & OCH_3 \\ & C & + Na^+Cl^- \\ H_3C & H \end{array}$$
2-Chloropropane   Sodium methoxide

There are two limiting mechanisms for nucleophilic substitution: a direct displacement (S_N2) process and a carbocation (S_N1) process. (In the earlier literature, carbocations are called carbonium ions.) The notation S_N2 means that a reaction is a substitution (S) reaction induced by a nucleophile (N) and has a molecularity of two in the rate-determining step. Similarly, S_N1 means that a reaction is a substitution reaction with a molecularity of one. Molecularity is the number of species involved in the transition state for a step in the reaction mechanism.

The direct displacement (S_N2) process, favored by primary substrates, involves backside attack by a nucleophile that is directly displacing the leaving group. Effective nucleophiles include hydroxide ion and other anionic species with highly concentrated negative charges. Good leaving groups include halide ions (especially iodide ion) that have readily polarizable bonds to carbon. The bond polarization leads to the formation of a negative charge on the leaving group (in the case of a halide ion):

$$HO^- + CH_3I \longrightarrow [\overset{\delta-}{HO}----CH_3----\overset{\delta-}{I}]^{\ddagger} \longrightarrow HOCH_3 + I^-$$

An interesting general aspect of nucleophilic substitution is the alternative pathways provided by the two limiting mechanisms. For example, tertiary halides, substrates that are very poor candidates for an S_N2 reaction, are able to undergo substitution by another process, the S_N1 mechanism. This process takes advantage of the fact that, comparatively speaking, tertiary carbocations are quite stable. Thus the hydrolysis of *tert*-butyl bromide (2-bromo-2-methylpropane) takes place by a process in which the transition state in the first step is unimolecular:

$$(CH_3)_3CBr \rightleftharpoons (CH_3)_3C^+ + Br^-$$

$$(CH_3)_3C^+ + H_2O \rightleftharpoons (CH_3)_3COH_2^+ \xrightarrow{-H^+} (CH_3)_3COH$$

A variety of methods has been used to investigate these mechanisms, including kinetics, stereochemistry, and variations in the nature of the substrate. Kinetic investigations reveal how the rate expression is affected by the reactant concentrations. A rate law is a simple differential equation that indicates how the concentrations of the reactants affect the rate of reaction. The rate law can be given in terms of the rate of formation of product, RNu, over time. For the model reaction at the beginning of this experiment, we obtain the following rate law as an experimental result:

$$\text{rate} = k(RG)(Nu^-)$$

in which the proportionality constant, k, is called the specific rate constant of the reaction. The exponents of the concentration terms are referred to as the order for these reagents (here, each is first order). Overall, the reaction is second order.

Stereochemical investigations also reveal mechanistic information. The direct displacement (S_N2) reaction gives rise to inversion of configuration at the carbon atom bearing the leaving group; thus we can be more explicit about this transition state:

Inversion

We call this structural change inversion because the position of the —OH group in the alcohol product is exactly opposite that of the iodine group in the reactant.

The transition state allows us to appreciate the effect caused by changes in the substrate. Specifically, any substrate alteration that increases the steric hindrance about the reactive center makes that transition state species more difficult to form and slows the reaction. For example, increases in the bulk of one or more of the R groups in this transition state slows the rate of reaction:

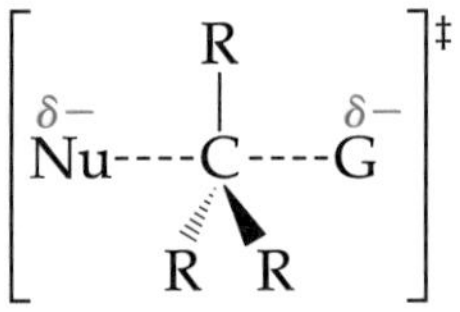

$$\left[\begin{array}{c} R \\ | \\ \overset{\delta-}{Nu}\text{----}C\text{----}\overset{\delta-}{G} \\ \diagup \quad \diagdown \\ R \quad R \end{array} \right]^{\ddagger}$$

Thus we find that S_N2 reactivity has the following dependence on substrate type:

$$\text{methyl} > \text{primary} > \text{secondary} > \text{tertiary}$$

and this dependence on the identity of the alkyl group:

$$\text{methyl} > \text{ethyl} > \text{isobutyl (2-methylpropyl)} >>> \text{neopentyl (2,2-dimethylpropyl)}$$

Stereochemical studies of S_N1 processes also reveal details consistent with the carbocation mechanism. Substitution proceeding by a carbocation results in the formation of racemic mixtures when the leaving group is attached to a chiral center because the nucleophile can attack with equal probability from either side:

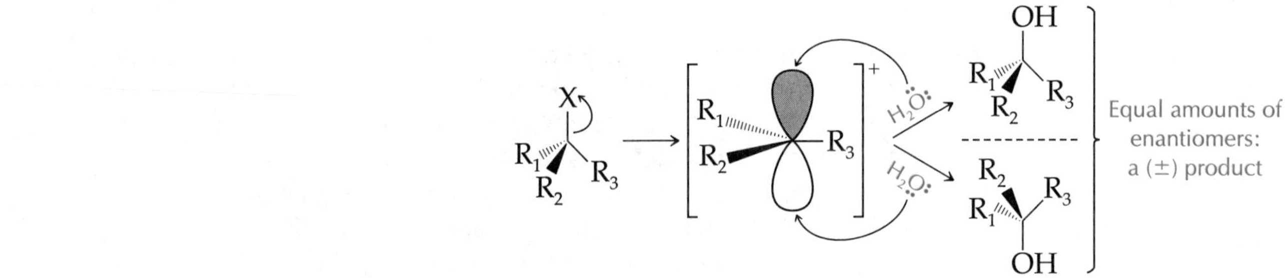

Moreover, ease of S_N1 reactions also follows a substrate dependence pattern that is consistent with carbocation formation. Thus,

$$\text{tertiary} > \text{secondary} > \text{primary} > \text{methyl}$$

This pattern is probably due to at least two factors; one is the following order of carbocation stability:

$$\text{tertiary} > \text{secondary} > \text{primary} > \text{methyl}$$

Any structural feature leading to carbocation stabilization (for example, allylic or benzylic stabilization) enhances the rate of the S_N1 substitution.

The other factor that enhances S_N1 substitution rates is steric. Specifically, as carbocations are formed, the carbon center undergoing reaction changes from sp^3 to sp^2 hybridization. This change results in an increase in bond angle (from $109°$ to $120°$) that relieves strain about the reactive center and enhances the rate. Therefore, the more highly branched centers, such as that found in a tertiary center, provide the greatest source of strain relief.

In Experiments 8.1 and 8.2, you will investigate various nucleophilic substitution reactions. In Experiment 8.1 you will use qualitative chemical tests to differentiate between S_N2 and S_N1 processes. In Experiment 8.2 you will carry out a typical synthesis illustrating direct displacement (an S_N2 process) by preparing an alkyl halide from an unknown alcohol; then you will identify the product by its boiling point and NMR spectrum.

8.1

S_N1/S_N2 Reactivity of Alkyl Halides

Study structure-reactivity relationships by using S_N1 and S_N2 chemistry.

$$S_N1: \quad RX + AgNO_3 \xrightarrow{CH_3CH_2OH} ROCH_2CH_3 + AgX \text{ (ppt)} + HNO_3$$
$$X = Cl, Br, I$$

$$S_N2: \quad RX + NaI \xrightarrow{acetone} RI + NaX \text{ (ppt)}$$
$$X = Cl, Br$$

In Experiment 8.1 you will use two reagents to test the ability of substrates to undergo S_N1 and S_N2 reactions. The "S_N1" reagent is silver nitrate in ethanol solution. The silver ion provided by this reagent coordinates with the halide ion in the organic substrate and enhances the ability of the carbon-halogen bond to polarize, thus readily producing halide ion and a carbocation (when carbocations can form):

$$AgNO_3 \longrightarrow Ag^+ + NO_3^-$$

$$RX + Ag^+ \longrightarrow \overset{\delta+ \quad \delta- \quad \delta+}{R\text{----}X\text{----}Ag} \longrightarrow R^+ + XAg \text{ (ppt)}$$

$$R^+ + Nu^- \longrightarrow RNu$$

Compounds that undergo rapid reaction with ethanolic silver nitrate solutions are assumed to be substrates that readily produce carbocations. Ease of carbocation production is based on increased stabilization afforded either by extensive branching (for example, a tertiary C) or by resonance (as in allylic or benzylic substrates):

Cation stability: $(CH_3)_3C^+ > (CH_3)_2CH^+ > CH_3CH_2^+ > CH_3^+$

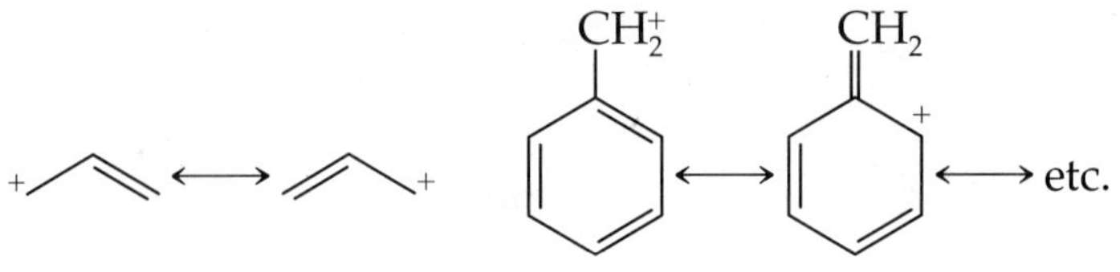

Allylic stabilization   Benzylic stabilization

Benzylic and allylic substrates readily undergo reaction with ethanolic silver nitrate because ionization is driven by the formation of a carbocation stabilized by the resonance processes shown above.

What may seem like similar substrates—halobenzenes and vinyl halides—are unreactive toward ethanolic silver nitrate because of the conjugation of the orbitals (and the electron pair) on the halogen with the aromatic or alkene carbon π system and the instability of the necessary carbocations. The cations are unstable because the carbon that would bear the positive charge is sp^2, rather than sp^3, hybridized. Moreover, bonds to sp^2 hybrids are stronger.

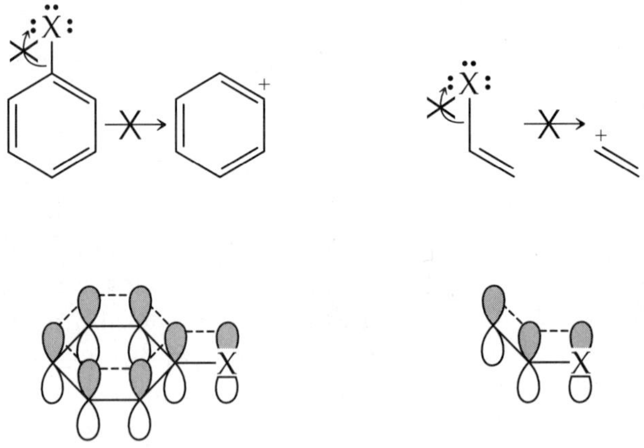

Halobenzene   Vinyl halide

The "S_N2" reagent is sodium iodide in acetone. The iodide ion provided by this reagent can attack the back side of a carbon bearing a halide, readily displacing chloride or bromide ion, which precipitates as the insoluble sodium salt:

$$RX + NaI \longrightarrow [I\text{----}R\text{----}X]^\ddagger \longrightarrow I\!-\!R + NaX$$

The ability of this S_N2 reaction to occur is based on the facts that the highly polarizable iodide ion in acetone is an excellent nucleophile, and that the NaCl or NaBr formed is less soluble in acetone than NaI is. Precipitation of the sodium halide product is a driving force for the reaction. This reaction is expected to follow an S_N2 reactivity sequence in which, for example, primary halides are more reactive than secondary, which are in turn more reactive than tertiary:

$$CH_3X > RCH_2X > R_2CHX > R_3CX$$

An interesting aspect of the NaI/acetone reaction is the fact that benzylic and allylic substrates also readily undergo this S_N2 reaction. It is driven by the overlap of the benzylic and allylic *p*-orbitals with the orbital of the iodide ion:

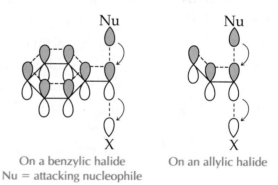

On a benzylic halide   On an allylic halide
Nu = attacking nucleophile

The NaI/acetone reagent is unreactive toward conjugated substrates such as halobenzenes and vinyl halides.

microscale Procedure

Do the tests with 15% NaI in acetone first because these tests require dry test tubes. Label 10 × 75 mm test tubes and fit each with a cork. The following compounds will be tested with each reagent unless otherwise noted:

1-Chlorobutane
1-Bromobutane
1-Iodobutane (for $AgNO_3$/ethanol test only)
Bromocyclopentane
Bromocyclohexane
2-Chlorobutane (*sec*-butyl chloride)
2-Bromobutane (*sec*-butyl bromide)
2-Chloro-2-methylpropane (*tert*-butyl chloride)
2-Bromo-2-methylpropane (*tert*-butyl bromide)

S_N2 Reaction Conditions (15% NaI in Acetone)

Label your eight test tubes before obtaining any reagents. Place 3 drops, measured with a Pasteur pipet, of each halide (omit 1-iodobutane for this reaction) in its own dry test tube. Immediately close each tube with a cork and keep the tubes stoppered except while adding the test reagent. Obtain 10 mL of 15% NaI/acetone solution in a small flask.

Using a graduated pipet, add 1.0 mL of NaI solution to the first test tube; recork the tube, record the time, and shake the tube to ensure complete mixing. Continue to monitor the reaction and record the time required for a precipitate to form. In the meantime, continue adding the test reagent to the other tubes. If no precipitate forms after 5 min at room temperature, arrange the cork loosely in the tube and heat it in a warm-water (50°C) bath. Check the reaction periodically and record the time that any turbidity (cloudiness) or precipitate appears. Consider the alkyl halide unreactive under the test conditions if no change occurs after 15 min of heating. Periodically inspect any tube in which a precipitate has not formed for signs of changes.

Cleanup: Pour the contents of all test tubes in the container for halogenated organic waste. Wash the test tubes thoroughly before using them for the S_N1 reactions.

S_N1 Reaction Conditions (1% AgNO₃ in Ethanol)

Label nine clean (but not necessarily dry) test tubes. Use 3 drops of each halide, one halide per test tube. Obtain 10 mL of 1% silver nitrate/ethanol solution in a clean small flask. Using a graduated pipet add 1.0-mL portions of silver nitrate to each test tube in the same way you did in the S_N2 tests. Record the time of addition, cork the tubes, and shake to mix thoroughly. Note the time for the first signs of turbidity or for the first distinct precipitate formation. Any test tube that has not shown signs of reactivity after 5 min at room temperature should be heated in a warm-water (50°C) bath for 15 min. Loosen the cork before heating and check periodically for precipitation.

Cleanup: Pour the contents of all test tubes in the container for halogenated organic waste. Pour any remaining silver nitrate solution in the container for inorganic waste.

Optional Experiment

> ── **SAFETY INFORMATION** ──────
>
> The compounds marked with an asterisk are lachrymators (cause tears) and should be used only in a hood.

Test the following compounds with both 15% NaI in acetone and 1% AgNO$_3$ in ethanol, using the same procedures employed in Experiment 8.1.

Lachrymators. Use in a hood.

1-Chloro-2-butene (crotyl chloride)*
2-Chloro-2-butene*
1-Chloro-2-methylpropane
1-Chloroadamantane
Chlorobenzene
Benzyl chloride*

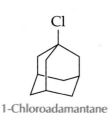

1-Chloroadamantane

Place these compounds in the lists of reactivity for each reagent prepared from the data collected in the first part of this experiment. Explain the reactivity observed.

Treatment of Data

1. List the alkyl halides in order of decreasing reactivity toward each of the reaction conditions. Briefly discuss why this order of reactivity was observed.
2. Order the reactivity of the primary halides to each reagent. Briefly explain.

3. Order the reactivity of the secondary halides to each reagent. Briefly explain.
4. Order the reactivity of the tertiary halides to each reagent. Briefly explain.
5. Did you observe any difference in the reactivity of bromocyclopentane and bromocyclohexane to each reagent? Briefly explain.

8.2

Preparation and Identification of Alkyl Bromides Synthesized from Unknown Alcohols

 Synthesize an alkyl bromide and identify it on the basis of its boiling point and NMR spectrum.

$$R—CH_2OH + NaBr + H_2SO_4 \longrightarrow R—CH_2Br + H_2O + NaHSO_4$$

The conversion of alcohols to organic halides is an important first step in a variety of synthetic routes. In this experiment you will use a mixture of a strong mineral acid (sulfuric acid) and sodium bromide to convert a primary alcohol to the corresponding bromoalkane. The role of the acid is to produce a better leaving group by protonating the alcohol. Sodium bromide provides the bromide ion, which is sufficiently nucleophilic to directly displace the protonated hydroxyl group in this S_N2 reaction:

$$R\ddot{O}H + H^+ \longrightarrow R—\ddot{O}H_2^+ \xrightarrow{Br^-} [\overset{\delta-}{Br}\text{----}R\text{----}\overset{\delta+}{\ddot{O}H_2}]^{\ddagger} \longrightarrow Br—R + \ddot{O}H_2$$
$$\qquad\qquad\qquad\qquad\qquad\qquad S_N2 \text{ transition state} \qquad\qquad \text{Water}$$

If we examine the transition state for the nucleophilic displacement process in more detail, we find that there are effectively five bonds to the central carbon. This cluttered state is achievable with primary alcohols because two of the three groups originally bonded to C-1 are very small hydrogen atoms:

$$CH_3CH_2CH_2CH_2OH_2^+ \xrightarrow{Br^-} \left[\overset{CH_2CH_2CH_3}{\underset{\overset{|}{\underset{H\quad H}{}}}{\overset{\delta-}{Br}\text{-----}\text{-----}\overset{\delta+}{OH_2}}} \right]^{\ddagger} \longrightarrow BrCH_2CH_2CH_2CH_3 + H_2O$$
$$\qquad\qquad\qquad\qquad\qquad\qquad \text{Pentavalent} \atop \text{transition state}$$

In fact, partly for these steric reasons, if the two protons of ethanol, a primary alcohol, are replaced by alkyl groups in the original structure, the displacement will proceed by the alternate S_N1 pathway:

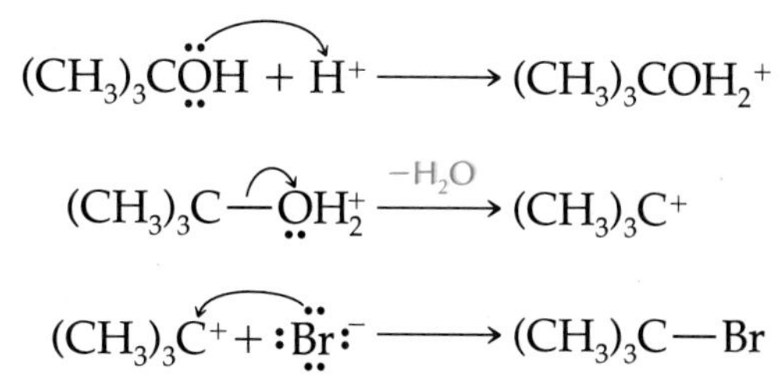

$$(CH_3)_3C\ddot{O}H + \overset{..}{H}{}^+ \longrightarrow (CH_3)_3COH_2{}^+$$

$$(CH_3)_3C \overset{\frown}{-} \overset{..}{O}H_2^+ \xrightarrow{-H_2O} (CH_3)_3C^+$$

$$(CH_3)_3\overset{\frown}{C}{}^+ + :\overset{..}{\underset{..}{Br}}:^- \longrightarrow (CH_3)_3C-Br$$

You will characterize the R group in the bromoalkane product (RCH_2Br), and thus the R group in the original alcohol, by determining its boiling point and also by interpreting its ^1H NMR spectrum.

Macroscale Procedure

Techniques   Reflux: Technique 3.3
Simple Distillation: Technique 7.3
Microscale Extractions: Technique 4.5b
Short-Path Distillation: Technique 7.3a
NMR Spectrometry: Spectrometric Method 5.2

— **SAFETY INFORMATION** —

The **primary alcohols** used in this experiment are flammable and irritants to skin and eyes. Avoid contact with skin, eyes, and clothing.

The **alkyl halides** synthesized are harmful if inhaled, ingested, or absorbed through the skin.

Sodium bromide is a skin irritant.

Sulfuric acid solution (8.7 m) is corrosive and causes severe burns.

Concentrated hydrochloric acid is corrosive, causes burns, and emits HCl vapors. Use it in a hood.

Pour 25 mL of 8.7 M sulfuric acid into a 50-mL Erlenmeyer flask and place the flask in an ice-water bath to cool.

Weigh your vial of unknown alcohol, then pour all the liquid *except a few drops* into a 100-mL round-bottomed flask and weigh the vial again. Save the vial and the remaining alcohol for NMR analysis.

Obtain 10.4 g of sodium bromide and transfer it to the round-bottomed flask, using a powder funnel so that salt granules do not stick to the neck of the flask. Attach a water-cooled condenser positioned for refluxing [see Technique 3.3]. Pour the chilled sulfuric acid solution slowly (over a period of 15–20 s) down the condenser while the flask is gently swirled.

Loosen the clamp holding the flask while you are swirling it.

After retightening the clamp holding the flask, place an electric heating mantle under it and heat the mixture at a moderately vigorous boil (the two phases should mix well) for 45 min. Adjust the rate of heating so that the vapors condense in the *lower third* of the condenser; otherwise, some of the organic materials may be lost. Consult your instructor about performing the NMR analysis of your unknown alcohol during this time.

The distillate contains water, your alkyl halide, and any alcohol that failed to react. To separate the alkyl bromide, we will take advantage of three facts: (1) the alkyl bromide is insoluble in water, (2) it is insoluble in concentrated hydrochloric acid, whereas any remaining alcohol dissolves in the acid, and (3) its density is greater than that of either water or concentrated HCl.

At the end of the reflux period, lower the heating mantle and cool the reaction mixture for 10 min. Add 20 mL of water and assemble the apparatus for a simple distillation [see Technique 7.3, Figure 7.7]. Use a 50-mL Erlenmeyer flask for the receiver. Distill until the condensate no longer contains water-insoluble droplets (about 15–25 mL of condensate). To test for completeness of the distillation, collect 1 or 2 drops of distillate in a clean test tube, add 0.5 mL of water, and look for any insoluble droplets. The distillation is complete when no droplets are visible. The final temperature of the distillation will range from 100–115°C.

The boiling point gradually rises during the azeotropic distillation of the alkyl bromide and water. The ternary azeotrope of HBr, H_2SO_4, and H_2O boils at 115°C.

Extractions [see Technique 4.5b]

— SAFETY PRECAUTION —

Wear gloves while doing these extractions and work in a hood while using concentrated hydrochloric acid, if possible.

In all of the following separations, the organic phase is the lower layer.

Transfer the distillate to a 15-mL centrifuge tube fitted with a tight cap. Add 3 mL of water, cap the tube and shake it, then allow the layers to separate. Record the volume of the lower phase. Remove the upper layer with a Pasteur filter pipet and transfer it to a 100-mL beaker [see Technique 4.5b, Method A].

Chill 8 mL of concentrated hydrochloric acid in an ice-water bath. Add about half of the cold HCl to the crude product, swirl the centrifuge tube briefly in the ice-water bath, then cap the tube tightly and shake to thoroughly mix the layers. If the phases do not separate cleanly, spin the tube in a centrifuge for 1 min. Record the volume of the lower phase. Carefully remove the upper HCl layer (all the aqueous layers can be combined in the same beaker). Wash the organic phase with the rest of the cold HCl. Record the volume of the lower (organic) layer after the phases separate (see Questions 2 and 3). Remove the upper (acid) layer.

Wash the alkyl halide with 5 mL of ice-cold water, allow the phases to separate, and remove the upper aqueous phase. Repeat the washing procedure with 3 mL of 5% sodium bicarbonate. **(Caution: Foaming may occur.)** Allow the layers to separate and transfer the lower organic phase to a clean, dry centrifuge tube [see Technique 4.5b, Method B].

Add anhydrous calcium chloride pellets to the product in several small portions until a few pellets do not immediately clump together or form a milky liquid. Cap the tube and shake it. Allow the mixture to stand for 5 min. Using a clean Pasteur pipet, transfer the product to a dry 10-mL Erlenmeyer flask and add small portions of anhydrous calcium chloride pellets until several move freely in the liquid. Cork the flask and allow it to stand at least 15 min.

Final Distillation

Place a dry 25-mL round-bottomed flask in a 100-mL beaker and determine the combined mass. You will use this flask as the receiving flask.

Using a Pasteur pipet, carefully transfer the crude product to a clean, dry 25- or 50-mL round-bottomed flask. Add a boiling stone. Assemble the apparatus for short-path distillation as shown in Technique 7.3a, Figure 7.8, using the tared 25-mL round-bottomed flask as the receiver. Hold the apparatus together by placing Keck clips on both ends of the vacuum adapter. Submerge the receiving flask in a beaker of cold tap water. Record the boiling range from the temperature at which the first drop falls into the

Do not distill to dryness. receiver to the highest temperature noted. Weigh the receiving flask and contents to determine your yield.

NMR Analysis

Consult your instructor about the correct sample preparation for the NMR spectrometer available in your laboratory. A general procedure used with a highfield (300 MHz) spectrometer involves placing 1 Pasteur-pipet drop of the compound being analyzed in an NMR tube and adding 0.7 mL of deuterated chloroform. Cap the tube and invert it several times to mix the solution thoroughly. Determine the ^{1}H NMR spectrum for both your unknown alcohol and the alkyl bromide synthesized from it. Analyze your spectra, using the chemical shifts of the proton signals, as well as the splitting pattern and the integration ratio to support your assignment of all peaks.

Identification of Your Alkyl Bromide

Table 8.1 lists the molecular weights and boiling points of the alkyl bromides that could be synthesized in this experiment. Use your observed boiling point and the ^{1}H NMR spectra to identify your alkyl bromide and the unknown alcohol from which you synthesized it.

After you have identified the alcohol and the alkyl bromide, calculate the theoretical yield of alkyl bromide on the basis of the mass of alcohol that you used in the synthesis. Calculate your percent yield. Submit your product to your instructor in a properly labeled vial.

Cleanup: Carefully pour the sulfuric acid solution remaining in the round-bottomed flask into a large beaker that contains about 200 mL of water. Rinse the flask with water and add the rinse water to the beaker. Also pour all the aqueous phases from the extractions into this beaker. Add sodium carbonate in small portions with stirring **(Caution: Foaming!)** until the acid is neutralized. The solution can then be washed down the sink or placed in the container for aqueous inorganic waste. Place the calcium chloride in the con-

Table 8.1. **Boiling points of alkyl bromides**

Alkyl bromide	Molecular weight	Boiling point, °C
1-Bromopropane	123	71
1-Bromobutane	137	101
1-Bromo-2-methylpropane	137	90–92
1-Bromopentane	151	130
1-Bromo-3-methylbutane	151	120–121
1-Bromohexane	165	154–158
1-Bromo-4-methylpentane	165	146–147

tainer for hazardous solid waste or the container for inorganic solid waste. Pour the residue remaining in the distillation flask after the final distillation into the container for halogenated waste.

Questions

1. What is the upper layer that forms in the reaction flask during the reflux period? Why does it separate from the aqueous layer?
2. In the extraction steps, why was the crude product washed with concentrated hydrochloric acid?
3. You observed a decrease in the volume of the crude organic product during the extractions with HCl. Why?
4. Write the equation for the reaction referred to in Question 1.
5. Why was the product washed with aqueous sodium bicarbonate?
6. Treatment of 1-butanol with phosphoric acid (H_3PO_4) mixed with sodium chloride should result in the formation of 1-chlorobutane (butyl chloride). Write a mechanism for this reaction.
7. Treatment of *tert*-butyl alcohol (2-methyl-2-propanol) with HCl leads to a reasonable yield of *tert*-butyl chloride (2-chloro-2-propane). Write a mechanism for this reaction.
8. The IR spectrum of 1-bromobutane is shown in Figure 8.1. (a) What causes the peaks in the $3000-2800 \text{ cm}^{-1}$ region? (b) Where do you expect to find C—Br stretching? (c) What would be the most obvious indication of unreacted 1-butanol in the IR spectrum?

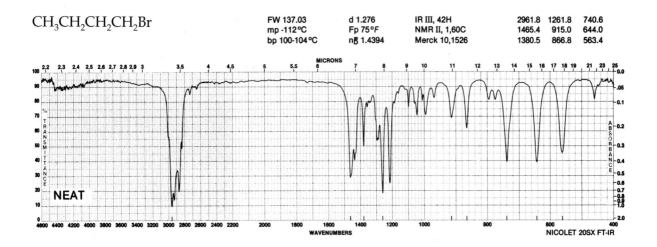

$CH_3CH_2CH_2CH_2Br$

FW 137.03	d 1.276	IR III, 42H	2961.8	1261.8	740.6
mp -112°C	Fp 75°F	NMR II, 1,60C	1465.4	915.0	644.0
bp 100-104°C	n₫ 1.4394	Merck 10,1526	1380.5	866.8	563.4

NEAT

FIGURE 8.1 IR spectrum of 1-bromobutane. (Provided by Aldrich Chemical Company, Inc., Milwaukee, WI.)

9. With an eye on the process used here for preparing 1-bromo-butane, comment on whether each of the following reactions might lead to the same bromide:
   a. 1-butanol + NaBr
   b. 1-butanol + NaBr/H_3PO_4
   c. 1-butanol + KBr/H_2SO_4
   d. 1-butanol + NaBr/HCl
   e. 1-butanol + NaBr/acetic acid

Experiment 9

ADDITION REACTIONS OF ALKENES

Combine theory and experiment to compare ionic and radical addition products, isomers formed in a dimerization reaction, and the stereochemistry of ionic addition. Use gas chromatography, spectrometry, and melting-point determination to analyze the reaction products.

A common reaction of alkenes is that of 1,2-addition, where the C=C is broken and an atom or group of atoms adds to each carbon. A wide range of reagents (AB) can add by different mechanisms to the double bond of alkenes:

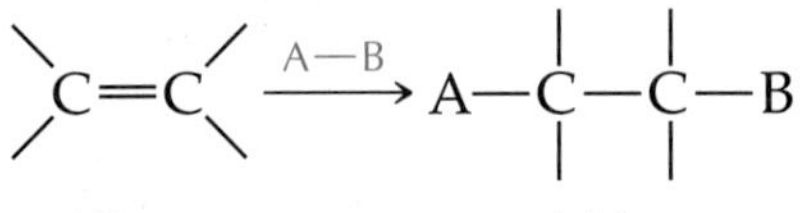

$$\overset{}{\underset{}{>}}C=C\overset{}{\underset{}{<}} \xrightarrow{\text{A—B}} A-\overset{|}{\underset{|}{C}}-\overset{|}{\underset{|}{C}}-B$$

Alkene   Adduct

One of the more common addition processes is ionic. The majority of ionic addition reactions of simple alkenes involves initial attack of the double bond by the electrophilic portion (E^+) of the ionic reagent (ENu):

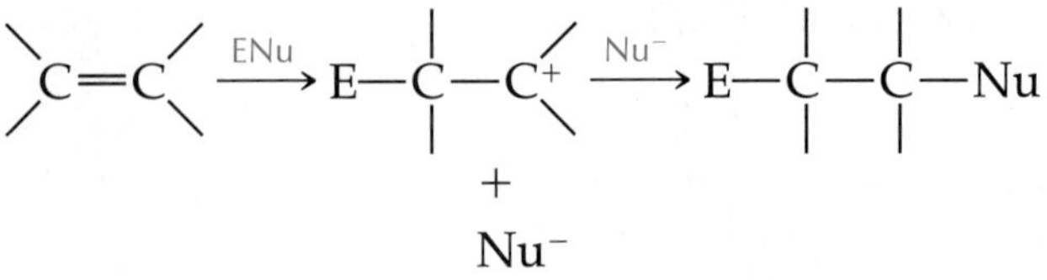

$$\overset{}{\underset{}{>}}C=C\overset{}{\underset{}{<}} \xrightarrow{\text{ENu}} E-\overset{|}{\underset{|}{C}}-\overset{}{\underset{}{C^+}}\overset{}{\underset{}{<}} \xrightarrow{\text{Nu}^-} E-\overset{|}{\underset{|}{C}}-\overset{|}{\underset{|}{C}}-Nu$$
$$+$$
$$Nu^-$$

Ionic additions to double bonds are initiated by electrophilic attack, because the alkene π-bond is electron rich and thus is electrostatically vulnerable to attack by positive reagents, such as electrophiles. In general, the more stable the carbocation intermediate is, the faster it forms.

$$\xrightarrow{E^+} \overset{+}{C}-C-E \longrightarrow \text{product}$$

π-system

A wide range of electrophiles can initiate such addition reactions, including the proton (H^+) and halonium ions (X^+) contributed by Br_2, Cl_2, and I_2. We will be examining proton-initiated reactions in Experiments 9.1 and 9.2, and bromine addition in Experiment 9.3.

9.1
Free-Radical versus Ionic Addition of Hydrobromic Acid to Alkenes

Compare the controlling factors and contrast the results in free-radical and ionic additions to alkenes.

Understanding the addition of HBr to alkenes requires an appreciation and comparison of two types of mechanisms, ionic additions and radical additions:

Ionic addition

$$\underset{}{\overset{}{C}}{=}\underset{}{\overset{}{C}} \xrightarrow{\text{HBr}} H-C-\overset{+}{C} \xrightarrow{\text{Br}^-} H-C-C-Br$$

Carbocation

Radical addition

$$\underset{}{\overset{}{C}}{=}\underset{}{\overset{}{C}} \xrightarrow[\text{ROO·}]{\text{HBr}} Br-C-C· \xrightarrow{\text{HBr}} Br-C-C-H$$

Carbon radical

Ionic addition of hydrohalic acids (HX) to double bonds can be pictured as an initial proton attack on an alkene π-system to form a carbocation, followed by entrapment of that carbocation by halide ion to produce an organic halogen compound:

$$\xrightarrow{} \overset{+}{C}-C-H \xrightarrow{X^-} X-C-C-H + Br·$$

H^+

π-system   Carbocation

Thus, the fact that HCl adds to propene to produce only 2-chloropropane is readily explained by the formation of the more stable secondary carbocation. The alternative primary carbocation, $CH_3CH_2CH_2^+$, is far too unstable to be formed. Therefore, when the chloride ion traps the positively charged intermediate, only the secondary, 2-chloropropane product is formed:

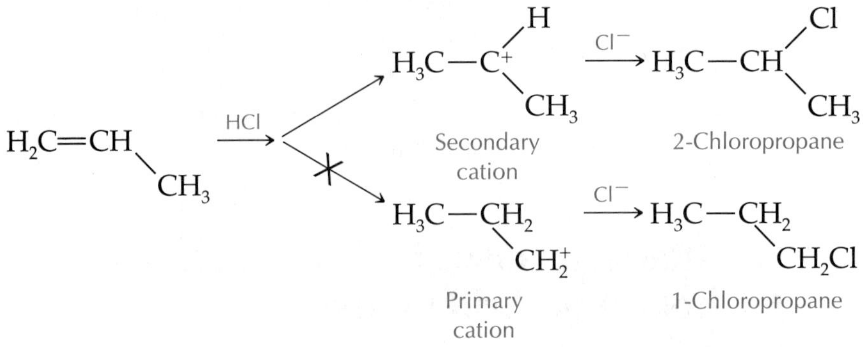

When HBr is added to an alkene, a free-radical chain reaction can compete with the ionic reaction described earlier. This process is identifiable during the addition of HBr to propene by the fact that 1-bromopropane (rather than the 2-bromo isomer) is obtained:

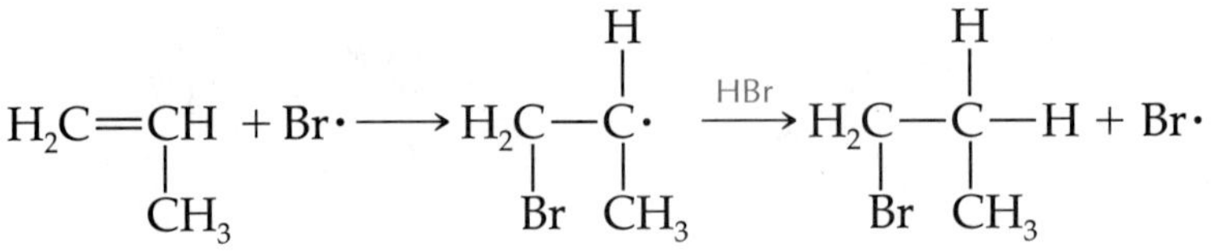

The bromine radical bonds to the terminal carbon of propene because this structure produces the more highly branched and more stable radical intermediate.

The addition of HX to an unsymmetrical alkene occurs in a Markovnikov fashion when the HX reagent adds a proton to the alkene carbon that already has more attached hydrogens. Thus, Markovnikov addition of HBr to propene gives 2-bromopropane. If, on the other hand, 1-bromopropane is formed, the reaction is said to have occurred in an anti-Markovnikov fashion.

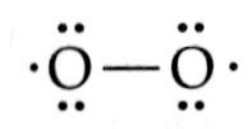

Peroxides can be used to induce anti-Markovnikov addition. In fact, the oxygen in the air can, by its paramagnetic character, form alkyl peroxides or hydroperoxide from hydrocarbons or HBr:

Vladimir V. Markovnikov, 1838–1904, was a Russian chemist at Odessa and later at Moscow University who carried out early studies of such additions.

Paramagnetic oxygen molecule

$$O_2 + RH \longrightarrow R\cdot + H-O-O\cdot$$

$$O_2 + R\cdot \longrightarrow R-O-O\cdot$$

Prior to 1933, chemists did not understand why addition of HBr to alkenes often occurred with inconsistent results. Sometimes Markovnikov addition was the major process, and sometimes anti-Markovnikov addition predominated. The pioneering work of M. S. Kharasch and F. W. Mayo lead to the discovery that, unless the reagents for this reaction were scrupulously purified, HBr addition could indeed yield mixed results. Small amounts of dissolved oxygen can give rise to peroxides as just described, and these would lead to anti-Markovnikov addition. Highly purified reagents and the presence of free-radical inhibitors give rise to largely Markovnikov addition. Thus, if ROO· is our general radical-initiation species, the mechanism of anti-Markovnikov addition is the following chain reaction:

Initiation

$$R—O—O· + HBr \longrightarrow Br· + ROOH$$

Propagation

$$H_2C{=}CH \ + \ Br· \longrightarrow H_2C—\overset{\overset{\displaystyle H}{|}}{\underset{\underset{\displaystyle CH_3}{|}}{C}}·$$
$$\underset{CH_3}{|} \qquad\qquad\qquad \underset{Br}{|}$$

Termination

$$H_2C—\overset{\overset{\displaystyle H}{|}}{C}· \ + \ HBr \longrightarrow H_2C—\overset{\overset{\displaystyle H}{|}}{C}—H + Br·$$
$$\underset{Br}{|}\ \underset{CH_3}{|} \qquad\qquad\qquad \underset{Br}{|}\ \underset{CH_3}{|}$$

$$R· + R· \longrightarrow R—R$$

In Experiment 9.1, the addition of HBr to an alkene is carried out in three ways. The first and second reactions use a mixture of HBr in acetic acid to carry out the addition; the first reaction uses 1-hexene as the substrate, and the second uses 2-methyl-2-butene. In the first reaction, we can assume that the ratio of 2-bromohexane to 1-bromohexane corresponds to the ratio of the Markovnikov to anti-Markovnikov processes and thus, in turn, to the ratio of ionic to radical addition. In like fashion, the 2-methyl-2-butene reaction allows the use of relative amounts of Markovnikov versus *anti*-Markovnikov product as a measure of ionic versus free-radical processes. The third reaction involves very different conditions, but again we can use the relative amounts of Markovnikov versus *anti*-Markovnikov product as a measure of ionic versus free-radical processes.

More recently, newly developed reagents use heterogeneous conditions to ensure ionic addition to an alkene. You will use one of these methods in the third reaction. Specifically, when an alkene is dissolved in dichloromethane and treated with HX (or a hydrohalic acid source) while in contact with alumina or silica gel, an efficient ionic addition takes place. In this experiment, you will treat 1-hexene in dichloromethane with oxalyl bromide in the presence of alumina. The protons of the HBr that eventually add to the alkene are obtained from the hydroxylated surface of the alumina, and the bromide ions are obtained from oxalyl bromide:

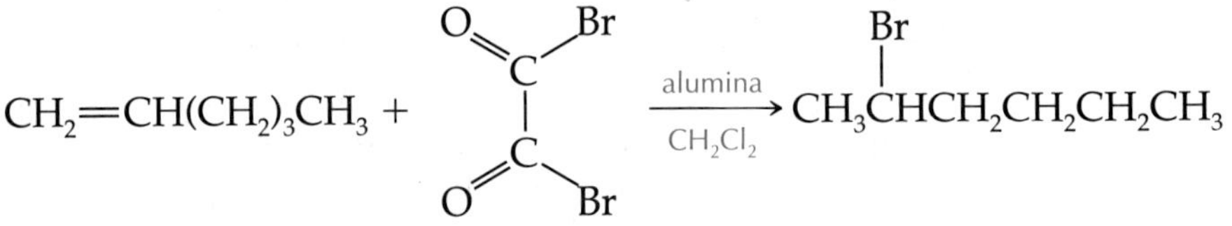

$$CH_2{=}CH(CH_2)_3CH_3 \ + \quad \xrightarrow[\text{CH}_2\text{Cl}_2]{\text{alumina}} \quad CH_3CHCH_2CH_2CH_2CH_3$$

| 1-Hexene | Oxalyl bromide | 2-Bromohexane |

An important aspect of the alumina is its pretreatment at 120°C. This process removes all the water loosely associated with the alumina surface but allows the water that is covalently bonded to the alumina to remain as a monolayer on the surface:

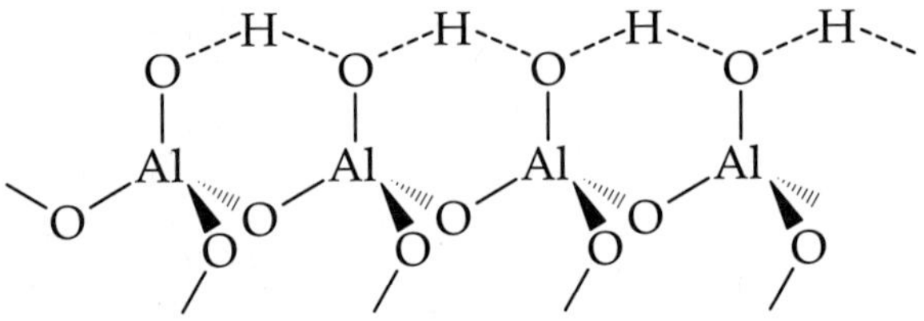

It is this surface that promotes ionization of the HX that adds to the alkene. In this way the alumina promotes ionic addition to the double bond.

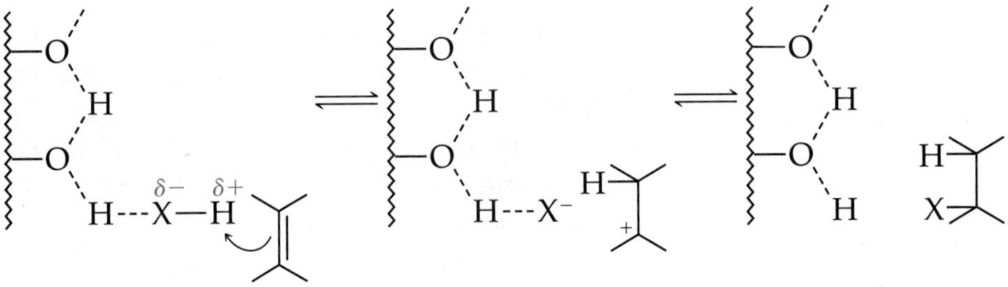

mmmicroscale
Procedure

Techniques Microscale Extractions: Techniques 4.5d, 4.5a, and 4.5b, Method B
Pasteur Filter Pipet: Technique 2.4
Gas Chromatography: Technique 11

Prelaboratory Assignment:
Name the possible products of
HBr addition to 1-hexene and
2-methyl-2-butene. Locate the
boiling points of these
products in a handbook such
as the CRC Handbook *or*
Lange's Handbook of
Chemistry *(see*
Introduction I.6).

SAFETY INFORMATION

Wear gloves while conducting all of these experiments.

1-Hexene and **2-methyl-2-butene** are very flammable and are skin irritants. Avoid contact with skin, eyes, and clothing. Use them in a well-ventilated area.

30% Hydrogen bromide in acetic acid and **oxalyl bromide** are toxic and corrosive. Avoid breathing the vapors. Avoid contact with skin, eyes, and clothing. Use these reagents in a hood and wear gloves.

Diethyl ether is very flammable. Use it in a hood and keep it away from flames or electrical heating devices.

Alumina is an eye and respiratory irritant. Avoid breathing any fine particles while weighing it.

Experiment 1

Measure 0.35 mL of 1-hexene into a 5-mL conical vial. Working in a hood, add 1.0 mL of 30% HBr in acetic acid (5 M) to the vial. Cap the vial with the Teflon (dull) side of the septum down and shake it frequently for 10 min. Occasionally loosen the cap to release any buildup of pressure.

After the reaction period, allow the phases to separate and remove the lower acetic acid layer with a Pasteur filter pipet [Pasteur filter pipet: Technique 2.4; microscale extraction: Technique 4.5d]. Place the acid layer in a 150-mL beaker containing about 50 mL of water. Add 2.0 mL of ether and 2.0 mL of water to the organic phase remaining in the conical vial. Cap the vial and shake it to mix the phases. Remove the lower aqueous layer, adding it to the beaker containing the previously removed acid layer. Wash the ether layer with 2.0 mL of 5% sodium bicarbonate solution. Remove the lower aqueous phase. Add anhydrous calcium chloride pellets to the remaining ether solution, cap the vial, and allow the solution to dry for 10–15 min.

Set the conical vial in a small
beaker so it does not tip over.

SAFETY PRECAUTION

There may be pressure inside the vial from carbon dioxide while washing with Na_2HCO_3, open the cap slowly to relieve the pressure.

Prepare another Pasteur filter pipet. Transfer the ether/product solution to a clean test tube. If you are analyzing the product mixture on a capillary column gas chromatograph, simply inject 1 μL of the ether solution into the chromatograph. If your laboratory is equipped with packed column chromatographs, you will probably need to evaporate some of the ether with a stream of nitrogen or on a steam bath before injecting 1 μL into the chromatograph. Consult your instructor about specific sample preparation techniques and sample size for your instruments. Use a nonpolar column such as OV-1 or SE-30, with a column temperature of 60–70°C.

Experiment 2

Repeat the procedure for Experiment 1, substituting 2-methyl-2-butene for 1-hexene and using a 3-mL conical vial for the reaction vessel.

The first step of the extraction procedure differs from that of Experiment 1 because the organic (product) phase moves from the top layer to the bottom layer as the reaction proceeds. Remove the lower (organic) layer with a Pasteur filter pipet and transfer it to a clean 5-mL conical vial [see Technique 4.5a]. Add 2.0 mL of water and 2.0 mL of ether to the organic phase. Continue as directed in the procedure for Experiment 1.

Experiment 3

─── *SAFETY PRECAUTION* ────────

Conduct this experiment in a hood until you have transferred the water/dichloromethane mixture to a centrifuge tube.

The alumina must be heated at 120°C for 48 h before using it in this experiment.

Weigh 110–120 mg of 1-hexene by adding it dropwise from a Pasteur pipet to a dry 25-mL Erlenmeyer flask. Cork the flask while you obtain the other reagents. Add 5 mL of dichloromethane, 2.5 g of alumina (Al_2O_3), and a magnetic stirring bar to the flask. Prepare an ice-water bath in a small beaker or crystallizing dish; set the ice bath on a magnetic stirrer and clamp the flask in the bath. Stir the mixture for 5 min to cool its contents. Wearing gloves, take a corked 10 × 75 mm test tube to the reagent-dispensing hood and place 1.0 mL of 2 M oxalyl bromide in CH_2Cl_2 in the test tube; immediately recork the test tube. Using a Pasteur pipet, add the oxalyl bromide solution dropwise over a period of 1 min to the stirred reaction mixture. Continue stirring the mixture for an additional 10 min.

Gravity filter the reaction mixture, using a small funnel and fluted filter paper; collect the filtrate in a 25-mL Erlenmeyer flask [see Technique 4.7, Figure 4.13]. Retrieve the magnetic stirring bar and wash it. Add 5 mL of ice-cold water to the filtrate, place the stirring bar in the flask containing the filtrate, and stir the mixture until bubbling ceases (about 5 min). Pour the mixture into a 15-mL centrifuge tube with a tight cap.

Transfer the lower organic phase to another centrifuge tube, using a Pasteur filter pipet [see Technique 4.5b, Method B]. Add 2.0 mL of 5% sodium hydroxide solution to the centrifuge tube containing the dichloromethane solution (organic phase). Cap the tube and shake it to mix the layers. If the phases do not separate cleanly, spin the tube for 1 min in a centrifuge. Transfer the lower organic phase to a clean 10-mL Erlenmeyer flask and dry the solution with anhydrous calcium chloride pellets for 10–15 min.

Separate the drying agent and analyze the product solution as directed in Experiment 1. Utilize the same gas chromatograph and instrument parameters you used for Experiment 1.

Identification of the Products

Because you are using a nonpolar column on the gas chromatograph, the products usually elute in order of increasing boiling point. Having known samples of the products available will allow you to verify the identity of the products by using the peak enhancement method [see Technique 11.6]. After you have taken a gas chromatogram of the product of Experiment 1, add 1 drop of 1-bromohexane to the ether (product) solution. Inject a 1-μL sample of this "enhanced" solution into the same chromatograph (at the same parameters) already used for Experiment 1. The peak whose height has increased relative to the others corresponds to the known compound that you added.

In the same way, add 1 drop of 2-bromo-2-methylbutane to the ether/product solution from Experiment 2 after you have analyzed it by GC. Carry out the chromatographic determination as described in the previous paragraph.

Compare the retention times of the peaks found in Experiment 3 with those for Experiment 1. What product(s) forms in Experiment 3?

Cleanup: Combine all the aqueous solutions from the extractions in the beaker containing the acetic acid / HBr extracts. Neutralize the acid by adding solid sodium carbonate in small portions. **(Caution: Foaming.)** Wash the neutralized solution down the sink or pour it into the container for aqueous inorganic waste. Pour the product solutions into the container for halogenated organic waste. Place the calcium chloride drying agent and alumina in the container for hazardous solid waste or in the container for inorganic waste.

Treatment of Data

1. Compare the results of Experiments 1 and 3. Explain any differences.

2. Compare the results of Experiments 1 and 2. Does the difference in structure of the substrates affect the results? Explain.

References

1. Brown, T. M.; Dronsfield, A. T.; Ellis, R. *J. Chem. Educ.* **1990**, *67*, 518.
2. Kropp, P. J.; Daus, K. A.; Crawford, R.; Tubergen, M. W.; Kepler, K. D.; Craig, S. L.; Wilson, V. P. *J. Am. Chem. Soc.* **1990**, *112*, 7433–7434.

Questions

1. In Experiment 3, what would you expect the other organic product to be?
2. Hydrogen chloride adds ionically to alkenes and, in fact, the competing free-radical reactions described for HBr additions do not occur for HCl additions. In each case, predict the course of the reaction of HCl with the following hydrocarbons by writing the structure of the product: (a) styrene (phenylethene); (b) isobutylene (2-methylpropene); (c) ethylene (ethene); (d) α-methylstyrene (1-methyl-1-phenylethene).
3. Predict the course of the reaction of HBr with the substrates listed in Question 2 by providing the structure of the organic product. Assume that a free-radical process applies in all cases.

4. Silica gel may be used to promote the ionic addition of hydrogen halides in much the same fashion as alumina does. Structures A and B are different representations of a silica gel surface. Surface A contains two hydroxyl groups on each silicon atom (geminal —OH groups) that lead to the hydrogen bonding shown; surface B does not yield such hydrogen bonding. (a) Give a simple explanation of why structure A allows hydrogen bonding and B does not. (b) The hydrogen-bonded hydroxyl groups of A (pK_a 5–7) are much more acidic than the hydroxyl groups of B (pK_a 9.5). The acidic groups are thus sufficiently reactive to promote the addition of hydrogen halides to alkenes (in other words, a structure with halide bonded directly to Si, analogous to that described earlier for alumina incorporating Al-halide bonds, is not necessary). Outline a mechanism for the addition of a general hydrogen halide (HX) to alkene, using silica gel.

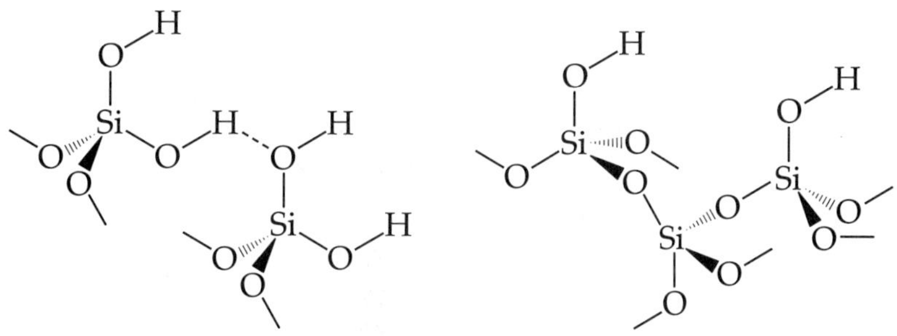

A   B

9.2
Dimerization of 2-Methylpropene (Isobutylene)

Study the dimerization of an alkene and the isomeric products that are formed.

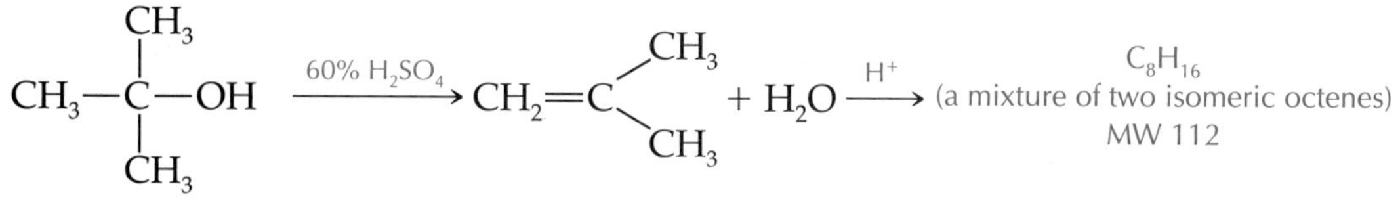

2-Methyl-2-propanol
(*tert*-butyl alcohol)
MW 74
bp 83°C
density 0.775 g · mL^{-1}

2-Methylpropene
(isobutylene)

C_8H_{16}
(a mixture of two isomeric octenes)
MW 112

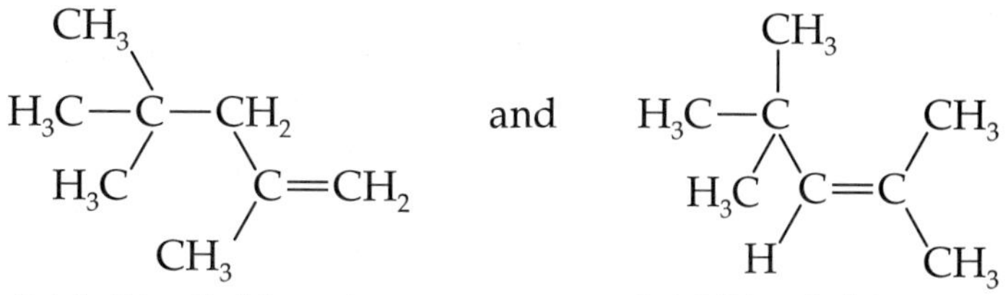

2,4,4,-Trimethyl-1-pentene
bp 101.4°C

and

2,4,4-Trimethyl-2-pentene
bp 104.9°C

The dimerization reaction observed in this experiment follows the dehydration of a tertiary alcohol precursor. In this case, 2-methyl-2-propanol undergoes dehydration to 2-methylpropene. This alkene then dimerizes in the presence of a catalytic amount of sulfuric acid:

$$(CH_3)_3COH + H_2SO_4 \longrightarrow (CH_3)_2C{=}CH_2 + H_2O$$

$$2\ (CH_3)_2C{=}CH_2 + H_2SO_4 \longrightarrow$$

$$CH_2{=}\underset{\underset{CH_3}{|}}{C}CH_2C(CH_3)_3 + (CH_3)_2C{=}CHC(CH_3)_3$$

This dimerization takes advantage of the fact that the tertiary alcohol readily forms the 2-methyl-2-propyl (*tert*-butyl) cation, which in turn deprotonates to form 2-methyl-2-propene (isobutylene):

$$(CH_3)_3COH \xrightarrow{\ H^+\ } (CH_3)_3COH_2^+ \xrightarrow{\ -H_2O\ } (CH_3)_3C^+ \xrightarrow{\ -H^+\ } (CH_3)_2C{=}CH_2$$

2-Methyl-2-propanol   2-Methylpropene
(*tert*-butyl alcohol)   (isobutylene)

The carbocation (*tert*-butyl cation) is an excellent electrophile and attacks the π-system of a second isobutylene molecule, a reaction resulting in formation of a C_8 carbocation:

$$(CH_3)_3C^+ + (CH_3)_2C{=}CH_2 \longrightarrow (CH_3)_2C^+{-}CH_2{-}C(CH_3)_3$$

This carbocation can deprotonate to form C_8 alkenes in two ways, by loss of a methyl proton or by loss of a methylene proton:

$$(CH_3)_2\overset{+}{C}{-}CH_2{-}C(CH_3)_3 \xrightarrow{\ -H^+\ } \underset{\underset{CH_2}{\|}}{CH_3C}{-}CH_2{-}C(CH_3)_3 + (CH_3)_2C{=}CH{-}C(CH_3)_3$$

and

2,4,4-Trimethyl-1-pentene   2,4,4-Trimethyl-2-pentene

If thermodynamic factors control the course of the deprotonation, loss of the methylene proton to form the more substituted (and thus more stable) alkene 2,2,4-trimethyl-2-pentene would take place. Alternatively, the other isomer, 2,4,4-trimethyl-1-pentene, is the favored product when it is the more rapidly formed product (even though it is less stable). This mechanism is rationalized by the fact that the proton is abstracted by a weak base, such as water, and thus the approach of the base to the C_8 cation is sterically more favorable when the water approaches the terminal methyl rather than the internal methylene group (also there is a statistical factor of $9:2$ favoring the terminal methyls):

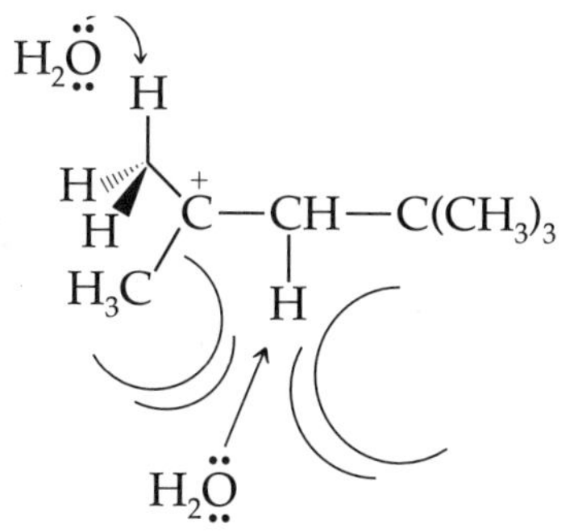

Actually, the reaction conditions produce a mixture of the two possible products. You will use gas chromatography to determine their relative ratio. You will also carry out two qualitative tests on the product for the presence of unsaturated C—C bonds, namely, the oxidation of the double bond to a diol by potassium permanganate and the addition of bromine to the double bond.

Macroscale Procedure

Techniques   Reflux: Technique 3.3
Extraction: Technique 4.2
Short-Path Distillation: Technique 7.3a
Drying Agents: Technique 4.6
Gas Chromatography: Technique 11
NMR Spectrometry: Spectrometric Method 2

— *SAFETY INFORMATION* —

Sulfuric acid is corrosive and causes severe burns. Notify the instructor if any acid is spilled.

2-Methyl-2-propanol (*tert*-butyl alcohol) causes skin and eye irritation. Avoid contact with skin, eyes, or clothing.

The product mixture is a skin irritant and harmful if inhaled or swallowed. Avoid contact with skin, eyes, or clothing.

If too much heat is applied during reflux charring occurs in the reaction mixture, 2-methylpropene (isobutylene), a gas, is driven out of the reaction, and the yield is reduced.

Place 32 mL of 9 M sulfuric acid in a 100-mL round-bottomed flask. Slowly add 15 mL of 2-methyl-2-propanol (*tert*-butyl alcohol). Swirl the flask briefly to mix the two liquids. Immediately attach a water-cooled condenser and heat the reaction at a *gentle* reflux for 30 min [see Technique 3.3].

Cool the reaction flask to room temperature and transfer its contents to a separatory funnel. Carefully drain the lower acid layer into a labeled 125-mL Erlenmeyer flask [see Technique 4.2]. Wash the hydrocarbon layer twice with 10 mL of water, and finally with 10 mL of 5% sodium bicarbonate solution. Separate the lower aqueous layer and pour the organic layer into a 25-mL clean Erlenmeyer flask. Add anhydrous calcium chloride to the product and allow it to dry for 10–15 min.

Weigh a corked 25-mL round-bottomed flask in a small beaker or flask holder. Filter the product through a funnel containing a small plug of cotton into a 50-mL round-bottomed flask [see Technique 4.7, Figure 4.14]. Assemble the apparatus for short-path distillation, using the tared 25-mL round-bottomed flask as the receiver [see Technique 7.3a]. Position the receiving flask in a beaker of ice water. Collect the product fraction from 98°C to 106°C. Do not allow the distilling flask to boil dry. Wipe the outside of the receiving flask with a towel, recork the flask, and weigh it in the same beaker or holder used previously. Calculate the percent yield and perform the following tests on the product mixture.

Do not allow the distilling flask to boil dry.

Bromine Addition   Put 3 drops of the product in a small test tube and add 1.0 mL of dichloromethane. Make dropwise additions of a 5% bromine in dichloromethane solution, shaking the test tube after each addition. Record your observations in your notebook. Repeat the test, using 2-methyl-2-propanol. A positive test for unsaturation is the disappearance of the red-brown bromine color, which usually happens immediately upon addition of the bromine solution to an alkene.

Oxidation with Potassium Permanganate

Place 3 drops of the product and 2.0 mL of water in a small test tube. Add 1 drop of 2% potassium permanganate solution. Cork the test tube and shake it thoroughly. Record your observations in your notebook. Repeat the test, using 2-methyl-2-propanol. The disappearance of the purple permanganate color and the formation of a brown precipitate (MnO_2) indicate a positive test for unsaturation.

Gas Chromatographic Analysis of Products [see Technique 11]

If your laboratory is equipped with capillary column chromatographs, prepare a sample for analysis by dissolving 2 drops of your product mixture in 0.5 mL of ether; inject 1 μL of the ether solution into the GC. Consult your instructor about sample preparation and injection volume for a packed column GC. Use a nonpolar column such as SE-30 or OV-101 with the column temperature at 55–70°C. The products elute in order of increasing boiling point. You may wish to verify this order of elution by using the peak enhancement method [see Technique 11.6].

NMR Analysis of Products [see Spectrometric Method 2]

Prepare an NMR sample of your product mixture as directed by your instructor.

The area under the integration line for each peak is proportional to the number of molecules giving that signal. Quantification of the product mixture is best done from the ratio of the peak areas (or peak heights, if sharp) for the two peaks produced by the methyl protons of the *tert*-butyl groups on each product isomer.

Identify the protons associated with each set of peaks in your NMR spectrum. There are three types of protons in the octenes formed in the reaction under study—aliphatic, allylic and vinylic—and each type gives signals in a distinct region of the spectrum.

— **SAFETY PRECAUTION** —

Remember that concentrated acid solutions are *always added to water.*

Cleanup: Carefully pour the acid solution remaining from the first extraction into a 1000-mL beaker containing about 400 mL of water. Add the aqueous phases from the other extraction steps to this beaker as well; then add solid sodium carbonate in small portions until the acidic solution is completely neutralized. (**Caution: Foaming.**) Wash the neutralized solution down the sink or pour it into the container for aqueous inorganic waste. Place the calcium chloride pellets in the container for nonhazardous solid or inorganic waste. Pour the residue remaining in the distillation flask into the container for flammable (organic) waste and rinse the flask with a few milliliters of acetone; also pour the acetone rinse into the container for flammable waste.

Alkenes have an unpleasant musty odor. The acetone rinse keeps the odor out of the laboratory when the flask is washed at the sink.

mmicroscale Procedure

Techniques   Microscale Reflux: Technique 3.3
Microscale Extraction: Technique 4.5d
Microscale Distillation: Technique 7.3b
Gas Chromatography: Technique 11
NMR Spectrometry: Spectrometric Method 2

SAFETY INFORMATION

Sulfuric acid is corrosive and causes severe burns. Notify the instructor if any acid is spilled.

2-Methyl-2-propanol causes skin and eye irritation. Wear gloves and avoid contact with skin, eyes, or clothing.

The product mixture is a skin irritant and harmful if inhaled or swallowed. Avoid contact with skin, eyes, or clothing.

Before measuring the reagents, place an aluminum heating block on a hot plate/stirrer and begin heating the block to 120°C.

Set the 10-mL round-bottomed flask in a small beaker before putting any reagents into it.

Place 4.3 mL of 9 M sulfuric acid in a 10-mL round-bottomed flask. Add 2.5 mL 2-methyl-2-propanol to the flask. Put a one-half inch magnetic stirring bar in the flask and fit the water-jacketed condenser in the flask with an O-ring and a screw cap (do *not* grease the ground glass joint). Set the flask in the aluminum block and secure the apparatus with a clamp fastened to the condenser [see Technique 3.3, Figure 3.5b].

Begin stirring the reaction mixture at a moderate speed and heat the reaction under *gentle* reflux for 20 min. Maintain the temperature of the aluminum block at 115–125°C. Cool the flask to room temperature before removing the condenser. A small beaker of tap water hastens the cooling process.

Wear gloves while performing the following extraction procedure. Using a Pasteur pipet (no cotton in the tip), carefully transfer

Conical vials tip over easily; place the vial in a 30- or 50-mL beaker whenever you set it down.

approximately half of the cooled reaction mixture to a 5-mL conical vial. Remove the lower acid layer with the same Pasteur pipet, placing the acid in a small labeled beaker or flask [see Technique 7.4d]. Transfer the rest of the reaction mixture to the conical vial that still contains the organic phase. Remove the lower acid layer with the Pasteur pipet and add it to the acid solution separated previously.

Prepare a Pasteur filter pipet [see Technique 2.4]. Wash the organic layer twice with 1-mL portions of water, then once with 1 mL of 5% sodium carbonate. For each wash, cap the vial, shake it, then remove the lower aqueous layer with the Pasteur filter pipet. Combine the aqueous phases in a small flask or test tube. Dry the organic phase with a few pellets of anhydrous calcium chloride; cap the vial during the 10-min drying period.

Set the vial in a small beaker.

Begin heating an aluminum block on a hot plate/stirrer at the lowest heat setting while you transfer the dried product mixture to a 3-mL conical vial, using a clean Pasteur filter pipet. Put a spin vane into the vial, attach the Hickman still to the vial with an O-ring and cap, and then attach an air-cooled condenser at the top of the still. Set the vial in the aluminum block and fasten a clamp around the Hickman still [see Technique 7.3b]. Start the magnetic stirrer.

Slowly increase the rate of heating until boiling begins. Then maintain the heating block at 125–135°C during the distillation. The distillate will condense on the inside of the Hickman stillhead and run into the well. When the well fills with condensate, open the side port and remove the liquid with a Pasteur pipet, or use a syringe and needle to withdraw the product through the septum in the side port. Put the product into a tared (weighed) screw-capped vial. Stop the distillation before the vial reaches dryness. Determine the mass of your product.

Product Analysis

Carry out the alkene tests and the GC and NMR analyses on your product as described in the Macroscale procedure.

Cleanup: Carefully pour the acid solution remaining from the first extraction into a 150-mL beaker containing about 40 mL of water. Add the aqueous phases from the other extraction steps to this beaker, then add solid sodium carbonate in small portions until the acidic solution is completely neutralized. Wash the neutralized solution down the sink or pour it into the container for aqueous inorganic waste. **(Caution: Foaming.)** Place the calcium chloride pellets in the container for nonhazardous solid or inorganic waste. Pour the residue remaining in the distillation vial into the container for flammable (organic) waste and rinse the vial with a few drops of acetone; also pour the acetone rinse into the container for flammable waste.

Alkenes have an unpleasant musty odor. The acetone rinse keeps the odor out of the laboratory when the flask is washed at the sink.

References

1. Allen, M.; Joyner, C.; Kubler, P. G.; Wilcox, P. *J. Chem. Educ.* **1976,** *53,* 175.
2. Tremelling, M. J.; Hammond, C. N. *J. Chem. Educ.* **1982,** *59,* 697.

Questions

1. From your knowledge of alkene stability, predict which isomeric octene should be formed in the larger amount. Do your experimental results support this? Explain why or why not.
2. Write balanced equations for any positive reactions observed in the bromine addition test and the permanganate test.
3. If you analyzed your product mixture by both GC and NMR, compare your results from the two methods.
4. Treatment of 1,1-diphenylethanol with acid produces two compounds: compound X of molecular formula $C_{14}H_{12}$ and compound Y of molecular formula $C_{28}H_{24}$. Predict the structures of these two compounds by using likely mechanisms.

9.3

The Stereochemistry of Bromine Addition to *trans*-Cinnamic Acid

Discover how the geometry of an alkene and a stereoselective reaction process produce a particular product stereoisomer.

trans-Cinnamic acid
(*E*-isomer)
MW 148
mp 133°C

+ Br$_2$ $\xrightarrow{CH_2Cl_2}$

2,3-Dibromo-3-phenylpropanoic acid
MW 308
Which racemic mixture is formed?
enantiomers (2*S*,3*S*) and (2*R*,3*R*)
mp 93.5–95°C
enantiomers (2*S*,3*R*) and (2*R*,3*S*)
mp 202–204°C

The addition of molecular bromine to the double bond of an alkene is a reaction whose stereochemical and mechanistic details are well established. Diatomic molecular bromine acts as source of electrophilic bromine, and the addition of the electrophilic Br^+ species to the π-system of the double bond produces a bridged bromonium ion. This bridged ion is in turn nucleophilically captured by bromide ion to produce dibromide in an *anti* fashion:

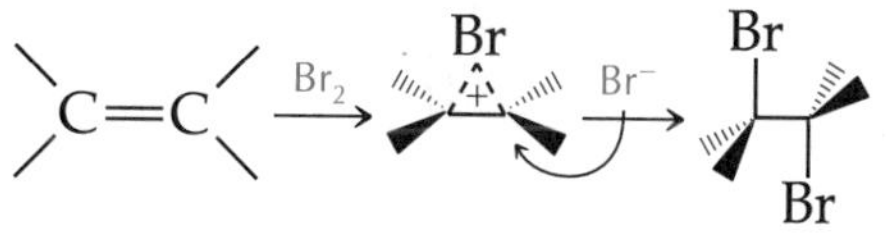

Support for an *anti* process comes from the fact that bromine adds to cyclopentene to form the *trans*-dibromide:

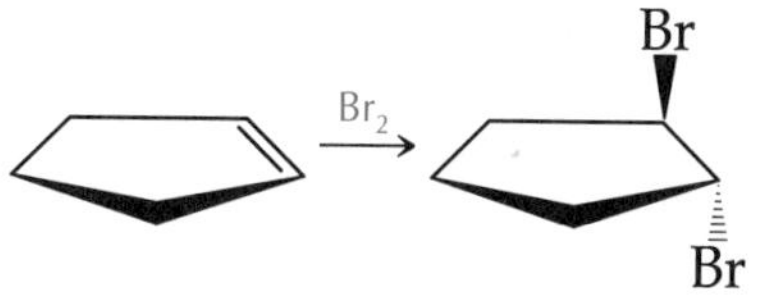

The addition of bromine to *trans*-cinnamic acid gives a dibromide with two stereocenters (chiral centers) of different types. Thus, there are four stereoisomers possible and these are shown here in their Fischer projections:

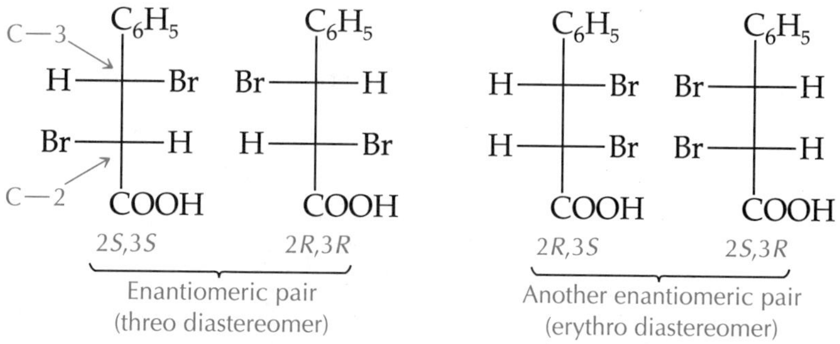

	Enantiomeric pair (threo diastereomer)		Another enantiomeric pair (erythro diastereomer)
2S,3S	2R,3R	2R,3S	2S,3R

2,3-Dibromo-3-phenylpropanoic acids

In this experiment, you will determine the melting point of the dibromide product and use this result to draw a conclusion as to which enantiomeric pair has formed. You can then use the stereochemistry of the products to deduce the mechanism of the bromination reaction by which these products were formed.

Procedure

Techniques   Reflux: Technique 3.3
Mixed Solvent Recrystallization: Technique 5.2a

SAFETY INFORMATION

Dichloromethane is toxic, an irritant, absorbed through the skin, and harmful if swallowed or inhaled. Use it in a hood and wash your hands thoroughly after handling it.

Bromine is very corrosive and causes serious burns. Its vapors are toxic and irritating to the eyes, mucous membranes, and respiratory tract. A solution of bromine also emits bromine vapor and should be used only in a well-ventilated hood. Wear gloves impermeable to Br_2 while measuring and transferring the Br_2 solution.

trans-**Cinnamic acid** is a mild irritant. Avoid skin contact and wash your hands after handling it.

Place 0.60 g of *trans*-cinnamic acid in a 25-mL round-bottomed flask. Add 3.5 mL of dichloromethane and 2.0 mL of 10% bromine in dichloromethane solution. Put a boiling stone in the flask and attach the water-cooled condenser [see Technique 3.3]. Clamp the flask in a beaker of water whose temperature is 45–50°C. Reflux the reaction mixture *gently* for 30 min. The product will begin to precipitate as the reaction proceeds. If the bromine color disappears during the reflux period, add 10% bromine in dichloromethane solution dropwise through the top of the condenser until a light orange color persists.

Cool the reaction flask to room temperature, then cool it further in an ice-water bath for 10 min, to ensure complete crystallization of the product. Collect the crude product by vacuum filtration in a Buchner funnel. Wash the crystals three times with 2.0-mL portions of cold dichloromethane by disconnecting the vacuum, pouring the solvent over the crystals, and then restarting the vacuum.

Transfer the crystals to a 50-mL Erlenmeyer flask and add 2.0 mL of ethanol. Heat until boiling on a steam bath or hot plate. If the crystals do not all dissolve, continue adding ethanol in

0.5-mL increments, boiling briefly after each addition, until the crystals do dissolve. Add a volume of water equal to the total amount of ethanol used and warm the mixture until any crystals that formed when the water was added have dissolved. Cool the flask to room temperature before placing it in an ice-water bath. Recover the crystals by vacuum filtration. Allow the crystals to dry overnight before determining the melting point and mass. Calculate the percent yield.

From the melting point of your product, decide whether the addition of bromine followed a *syn* or *anti* mechanism (or both). Propose a mechanism that explains your results.

Cleanup: Pour the filtrates from the reaction mixture and from the recrystallization in the container for halogenated waste.

microscale Procedure

Techniques   Microscale Reflux: Technique 3.3

Mixed Solvent Recrystallization:
Technique 5.2a

Microscale Filtration: Technique 5.5,
Figure 5.7

— SAFETY INFORMATION —

Dichloromethane is toxic, an irritant, absorbed through the skin, and harmful if swallowed or inhaled. Wear gloves, use it only in a hood, and wash your hands thoroughly after handling it.

Bromine is very corrosive and causes serious burns. Its vapors are toxic and irritating to the eyes, mucous membranes, and respiratory tract. A solution of bromine also emits bromine vapor and should be used only in a well-ventilated hood. Wear gloves impermeable to Br_2 while measuring and transferring the Br_2 solution.

trans-**Cinnamic acid** is a mild irritant. Avoid skin contact and wash your hands after handling it.

Place 100 mg of *trans*-cinnamic acid in a 3-mL reaction vial. Add 0.7 mL of dichloromethane and 0.35 mL of 10% bromine in dichloromethane solution. Put a boiling stone in the vial and fit it with the water-cooled condenser [see Technique 3, Figure 3.5b]. Clamp the apparatus so that the lower portion of the vial is submerged in a beaker of water whose temperature is 45–50°C. Reflux the reaction mixture gently for 20 min. The product will begin to precipitate as the reaction proceeds. If the bromine color disappears during the reflux period, add 10% bromine in dichloromethane solution dropwise through the top of the condenser until a light orange color persists.

Cool the reaction vial in an ice-water bath for 5 min to ensure complete crystallization of the product. Collect the crude product by vacuum filtration in a Hirsch funnel [see Technique 5.5, Figure 5.7]. Wash the crystals three times with 0.5-mL portions of cold dichloromethane.

Transfer the crystals to a 10-mL Erlenmeyer flask and add 0.4 mL of ethanol. Heat until boiling on a steam bath or hot plate. If the crystals do not all dissolve, continue adding ethanol in 0.1-mL increments, boiling briefly after each addition, until the crystals do dissolve. Add a volume of water equal to the total amount of ethanol used and warm the mixture until any crystals that formed when the water was added have dissolved. Cool the flask to room temperature before placing it in an ice-water bath. Recover the crystals by vacuum filtration on a Hirsch funnel. Allow the crystals to dry overnight before determining the melting point and mass. Calculate the percent yield.

From the melting point of your product, decide whether the addition of bromine followed a *syn* or *anti* mechanism (or both). Propose a mechanism that explains your results.

Cleanup: Pour the filtrates from the reaction mixture and the recrystallization into the container for halogenated waste.

Questions

1. Why is it necessary to maintain excess bromine in the reaction mixture? How can you tell that an excess of bromine is present?
2. When ionic bromination of cyclopentene is carried out, the product is *trans*-1,2-dibromocyclopentane. Gas chromatographic analysis with a heated inlet can result in the elimination of halogen from the organic dihalide product to regenerate the starting alkene. Thus, the GC properties of both the starting material and

the adduct are of interest. Predict the relative retention times of cyclopentene and *trans*-1,2-dibromocyclopentane. (*Hint:* Look up the physical properties of these two compounds in a handbook.)

3. Treatment of Z-2-butene with bromine in dichloromethane results in an organic dibromide product that, when treated with zinc, regenerates the original Z-2-butene. Bromination of E-2-butene gives a different dibromide product; when this product is treated with zinc, the original E-2-butene is regenerated. What assumption must be made about the stereochemistry of the zinc-promoted elimination?

4. Write the structure of the dibromide product(s) that you think will be formed as the result of the treatment of Z-2-pentene with bromine in dichloromethane. Also draw the structure of the dibromide product(s) predicted when E-2-pentene is treated with bromine in dichloromethane.

Experiment 10

DEHYDROHALOGENATION OF 2-BROMOHEPTANE: VARIATION OF PROMOTING BASE

Investigate an elimination reaction in order to study the effect of reaction conditions on product composition. Change the size of the base and determine by gas chromatography how the ratio of heptene isomers varies.

$$CH_3(CH_2)_4CHCH_3 \xrightarrow[\text{alcohol}]{\text{alkoxide ion}}$$

$$|$$
$$Br$$

2-Bromoheptane
bp 165–167°C
density 1.128 g·mL^{-1}
MW 179.10

CH$_3$(CH$_2$)$_3$CH$_2$ H   CH$_3$(CH$_2$)$_2$CH$_2$ H   CH$_3$(CH$_2$)$_2$CH$_2$ CH$_3$

$$\text{C=C} \quad + \quad \text{C=C} \quad + \quad \text{C=C}$$

H   H   H   CH$_3$ H   H

1-Heptene
bp 93.6°C

E-2-Heptene
bp 98.0°C
(*trans*)

Z-2-Heptene
bp 98.5°C
(*cis*)

alkoxide = methoxide or *tert*-butoxide
alcohol = methanol or *tert*-butyl alcohol (2-methyl-2-propanol)

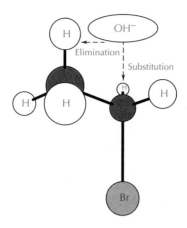

FIGURE 10.1 Substitution versus elimination.

The chemical reaction of a strong base and an alkyl bromide can lead to an alkene:

$$CH_3CH_2Br + NaOH \longrightarrow CH_2{=}CH_2 + NaBr + H_2O$$
$$\quad\;\beta \quad\;\; \alpha$$

In this reaction, the base abstracts one hydrogen (called a β-hydrogen) from carbon-2 of bromoethane and, at the same time, the bromide ion leaves carbon-1 resulting in the formation of ethene. By using reaction conditions that enhance the basic rather than the nucleophilic character of sodium hydroxide, we can influence the competition between the elimination reaction and nucleophilic substitution. This competition is illustrated in Figure 10.1. When there is greater steric hindrance to displacement of bromide (substitution) than there is to β-hydrogen abstraction (elimination), elimination becomes the favored reaction.

The E2 mechanism describes the course of many elimination reactions. The reaction is bimolecular in the sense that the transition state involves two molecular species: a base and the organic substrate. In this process (Figure 10.2), an *anti*-periplanar arrangement is adopted between the base-induced proton abstraction and the halide ion that is being lost. The transition state illus-

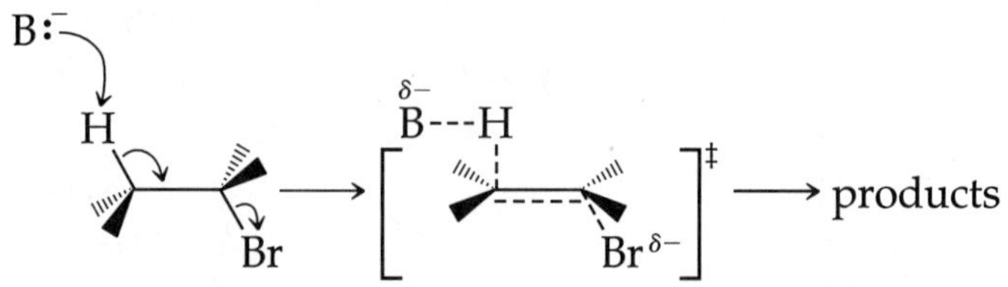

FIGURE 10.2 E-2 Elimination.

trated in Figure 10.2 also suggests that considerable double bond character develops in the transition state. Hence, it is not surprising that double bond stability as well as steric effects contribute to the course of this reaction. Indeed, we find that in many eliminations the more stable alkene is the favored product. The following order of alkene stability has been experimentally determined:

$$R_2C{=}CR_2 > R_2C{=}CHR > \textit{trans-}RCH{=}CHR > R_2C{=}CH_2 >$$
$$\textit{cis-}RCH{=}CHR > RCH{=}CH_2 > H_2C{=}CH_2$$

In the reactions carried out in this experiment, you need to consider both the size of the base and the stability of the alkene as factors that may influence the product composition. You will use sodium methoxide and potassium *tert*-butoxide as bases and determine the product composition from each reaction by gas chromatography.

$$CH_3O^-Na^+$$

Sodium methoxide

$$\begin{array}{c} CH_3 \\ | \\ H_3C{-}\overset{\displaystyle |}{\underset{\displaystyle |}{C}}{-}O^-K^+ \\ | \\ CH_3 \end{array}$$

Potassium *tert*-butoxide
(potassium 2-methyl-2-propoxide)

**microscale
Procedure**

Techniques Pasteur Filter Pipet: Technique 2.4

Microscale Reflux / Anhydrous Conditions: Technique 3.3a

Microscale Extractions: Technique 4.5d

Gas Chromatography: Technique 11

— *SAFETY INFORMATION* —

2-Bromoheptane is flammable and an irritant. Wear gloves. Avoid contact with skin, eyes, and clothing.

Sodium methoxide in methanol solution is flammable, corrosive, and moisture sensitive. **Potassium *tert*-butoxide** is corrosive and moisture sensitive. Avoid contact with skin, eyes, and clothing. Keep the containers tightly closed and store both reagents in desiccators.

Methanol is flammable and toxic. **2-Methyl-2-propanol** (*tert*-butyl alcohol) is flammable and an irritant.

Pentane is extremely flammable.

Your instructor may decide that you will work with a laboratory partner. If so, each of you will do one of the two reactions and share your results with the other person.

Sodium Methoxide Reaction

Prepare a microscale drying tube containing anhydrous calcium chloride [see Technique 3.3, Figure 3.6b]. Pipet 1.65 mL of a 25% (by weight) solution ($d = 0.945 \text{ g} \cdot \text{mL}^{-1}$) of sodium methoxide in methanol into a 5-mL conical vial. Add 0.35 mL of dry methanol and 0.22 mL of 2-bromoheptane. Put a magnetic spin vane into the vial. Fit the vial with a water-cooled condenser and place the drying tube containing calcium chloride in the top of the condenser. Begin stirring and heat the reaction for 30 min under reflux in a hot-water (75°C) bath.

Potassium tert-Butoxide Reaction

Prepare a microscale drying tube containing anhydrous calcium chloride [see Technique 3.3, Figure 3.6b]. Begin heating an aluminum block on a hot plate/stirrer to 95°C. Put a magnetic spin vane into a 5-mL conical vial. Pipet 2.0 mL of dry 2-methyl-2-propanol (*tert*-butyl alcohol) into the vial. Weigh 0.825 g of potassium *tert*-butoxide and immediately add it to the vial containing the alcohol. Add 0.22 mL of 2-bromoheptane. Fit the vial with a water-cooled condenser and place the drying tube containing calcium chloride in the top of the condenser. Begin stirring and heat the reaction under reflux in the aluminum block for 30 min.

Isolation of Product Mixture for Both Reactions

At the end of the reflux period, cool the reaction to room temperature. Remove the spin vane with a pair of tweezers. Add 1.3 mL of pentane and 1.0 mL of water to the reaction vial. Cap the vial tightly and shake the mixture gently, but thoroughly, until the white solid dissolves. A few additional drops of water may be necessary to completely dissolve the solid. Remove the lower aqueous phase with a Pasteur filter pipet [see Technique 2.4] and transfer it to a 50-mL Erlenmeyer flask, leaving the organic phase in the conical vial [see Technique 4.5d]. Repeat the above extraction procedure with two 1.0-mL portions of water, removing the lower aqueous phase before adding the next portion of water; all the aqueous phases should be combined in the Erlenmeyer flask. After removing the last portion of water, add 200 mg of anhydrous sodium sulfate to the pentane solution remaining in the vial. Cap the vial and allow the mixture to dry for at least 10 min. Transfer the dried solution to a clean dry test tube, using a new Pasteur filter pipet.

Gas Chromatographic Analysis of Product Mixture

Analyze the product mixture, using a gas chromatograph equipped with a nonpolar column such as SE-30 or OV-101 [see Technique 11]. If the product solution is too concentrated for capillary gas chromatography, prepare a GC solution in a small test tube by using 0.5 mL of pentane and 5 drops of the product mixture. The products elute in order of increasing boiling point. Calculate the percent of each heptene isomer in your product mixture [see Technique 11.6].

Cleanup: Determine the pH of the combined aqueous solution from the extractions, using pH test paper. Add 5% hydrochloric acid dropwise until the pH is about 7; the neutralized solution may be washed down the sink or poured into the container for aqueous inorganic waste. The pentane solution from the GC analysis should be poured into the container for halogenated waste because it contains some unreacted 2-bromoheptane.

References

1. Sayed, Y.; Ahlmark, C. A.; Martin, N. H. *J. Chem. Educ.* **1989,** *66,* 174–175.
2. Leone, S. A.; Davis, J. D. *J. Chem. Educ.* **1992,** *69,* A175–A176.

Questions

1. Compare the results of the two reactions and discuss how the use of different bases has affected the product composition.
2. Which factor predominates in each reaction: the size of the base or the stability of the alkenes formed? Explain.
3. Because alcohol protons are reasonably acidic and easily undergo exchange, it is important that the base be matched to the appropriate solvent. Explain this, telling why it would be virtually impossible to understand what factors in Question 2 predominate in a reaction carried out in methanol solvent and using potassium *tert*-butoxide as the base.
4. Explain why it is common to use lithium diisopropyl amide (LDA) as a base in elimination reactions, especially when it is desirable to avoid substitution products.
5. Simple internal alkenes are usually found to be somewhat more stable when trans (the *E* form) rather than cis (the *Z* form). Explain why.
6. In contrast to Question 5, cyclohexene normally occurs as a *cis*-cycloalkene and is impossible to isolate in a *trans*-form. Explain.

Experiment 11

DEHYDRATION OF ALCOHOLS

Demonstrate acid catalysis in the loss of water from alcohols and study the alkene mixtures that are formed.

Elimination reactions are one of the four fundamental classes of chemical reactions, the other three being substitution, addition, and rearrangement. Most of the important elimination reactions are 1,2-eliminations, where an atom or group of atoms is removed from each of two adjacent carbon atoms. Carbon-oxygen and carbon-nitrogen double bonds, as well as carbon-carbon and carbon-nitrogen triple bonds, can also be formed by the same kind of process.

$$-\overset{\mid}{\underset{\underset{X}{\mid}}{C}}-\overset{\mid}{\underset{\underset{Y}{\mid}}{C}}- \longrightarrow \overset{\diagdown}{\diagup}C=C\overset{\diagup}{\diagdown} + X-Y$$

An elimination requires a leaving group, X, that departs *with* its bonding electrons. The second group, lost from the adjacent carbon atom, is often hydrogen, which goes *without* its bonding electrons. In other words, it is lost as a proton. Reactions involving loss of HX are of primary importance for the synthesis of alkenes. If the HX lost is water, then the elimination reaction is called a dehydration.

When an alcohol is heated in the presence of a strong acid, the major product is an alkene or a mixture of alkenes:

$$-\overset{\mid}{\underset{\mid}{C}}-\overset{\mid}{\underset{\mid}{C}}- \xrightarrow[\text{heat}]{\text{H}^+} \overset{\diagdown}{\diagup}C=C\overset{\diagup}{\diagdown} + H_2O$$

The acid is essential because it converts the very poor leaving group, —OH, into a reasonably good one, water, by protonation of the —OH group. Because alcohols are weak bases, strong acids such as sulfuric acid or phosphoric acid are required to protonate them:

$$H-\overset{\mid}{\underset{\underset{OH}{\mid}}{C}}-\overset{\mid}{\underset{\mid}{C}}- + H_3PO_4 \rightleftharpoons H-\overset{\mid}{\underset{\underset{\overset{+}{O}-H}{\underset{\mid}{H}}}{C}}-\overset{\mid}{\underset{\mid}{C}}- + H_2PO_4^-$$

When secondary or tertiary alcohols are used, protonation and heating provide the driving forces for the loss of water to form a positively charged carbon species called a carbocation:

$$H-\underset{\underset{H}{\overset{|}{\underset{+O-H}{}}}}{\overset{|}{C}}-\overset{|}{\underset{|}{C}}- \rightleftharpoons H-\overset{|}{\underset{|}{C}}-C^{+} \diagup_{\diagdown} + H_2O$$

Carbocation

The rate of formation of a carbocation by expulsion of a water molecule depends heavily on the stability of the carbocation being formed. Secondary and especially tertiary carbocations are normally stable enough to be important reaction intermediates. Carbocation stability increases markedly as the number of alkyl or aryl substituents attached to the carbon atom bearing the positive charge increases. These substituents allow the charge to be dispersed throughout a larger space and therefore stabilize the carbocation.

In all but the most highly acidic environments, even tertiary carbocations are too unstable to endure long. They react with nucleophiles to give substitution products, through what is called the S_N1 pathway. They also react with bases to give elimination products, through the *E1 pathway*. The term E1 simply means unimolecular elimination. The unimolecular nature of the reaction pertains to the nature of the slowest or rate-determining step. For the acid-catalyzed dehydration of secondary and tertiary alcohols, the slow step is the expulsion of a water molecule from the protonated alcohol. This reaction involves only one reactant; hence, it is unimolecular.

We have already said that reaction of a carbocation with a base gives an alkene:

$$H-\overset{|}{\underset{|}{C}}-C^{+}\diagup_{\diagdown} + :B \rightleftharpoons \diagdown_{\diagup}C=C\diagup_{\diagdown} + HB^{+}$$

Yet, what bases can there be in a strongly acidic sulfuric or phosphoric acid solution? Surely any base present in this environment cannot be very basic. Fortunately, a strong base is not necessary to pull a proton away from a carbocation, because carbocations are very strong acids. Weak bases such as water, bisulfate ion (HSO_4^-), or dihydrogen phosphate ion ($H_2PO_4^-$) are potent enough to abstract a proton from a carbocation:

$$H-\overset{|}{\underset{\underset{\displaystyle H_2\ddot{O}:}{}}{C}}-\overset{/}{C+} \rightleftharpoons \overset{\backslash}{\underset{/}{C}}=\overset{/}{\underset{\backslash}{C}} + H_3O^+$$

The entire pathway for the E1 elimination of water from an alcohol in the presence of strong acid is

$$-\overset{|}{\underset{\underset{H}{|}}{C}}-\overset{|}{\underset{\underset{OH}{|}}{C}}- \underset{\text{fast}}{\overset{H^+}{\rightleftharpoons}} -\overset{|}{\underset{\underset{H}{|}}{C}}-\overset{|}{\underset{\underset{\underset{H}{|}}{\overset{+}{O}-H}}{C}}- \underset{\text{slow}}{\overset{-H_2O}{\rightleftharpoons}} -\overset{|}{\underset{\underset{H}{|}}{C}}-\overset{/}{C+} \underset{\text{fast}}{\overset{-H^+}{\rightleftharpoons}} \overset{\backslash}{\underset{/}{C}}=\overset{/}{\underset{\backslash}{C}}$$

Sulfuric acid and phosphoric acid are chosen over hydrochloric acid and hydrobromic acid as the catalysts for dehydration of alcohols in part because the conjugate bases of sulfuric and phosphoric acids are poor nucleophiles and do not lead to the formation of large amounts of substitution products. Naturally, when a reaction is being used for the synthesis of an alkene, we want as high a yield from the elimination reaction as possible. Because Cl^- and Br^- are good nucleophiles, using HCl or HBr could allow the carbocation to be trapped by these halides, thereby leading to undesirable alkyl halide products.

All the reaction steps in the E1 mechanism are reversible. Thus, the mechanism for dehydration of alcohols is just the reverse of the mechanism for hydration of alkenes. This reversibility also means that alkenes formed under these conditions can revert to alcohols unless proper experimental conditions are used.

To drive the elimination to completion, that is, to achieve 100% conversion, the alkene is distilled from the reaction mixture as it is formed. This strategy allows the various equilibria to be shifted continually in favor of the alkene. The constant removal of a product as it is formed is an important technique for obtaining high yields of products from equilibrium reactions.

11.1
Synthesis of Cyclohexene by Acid-Catalyzed Dehydration of Cyclohexanol

Carry out the synthesis of an alkene; then isolate, purify, and analyze it.

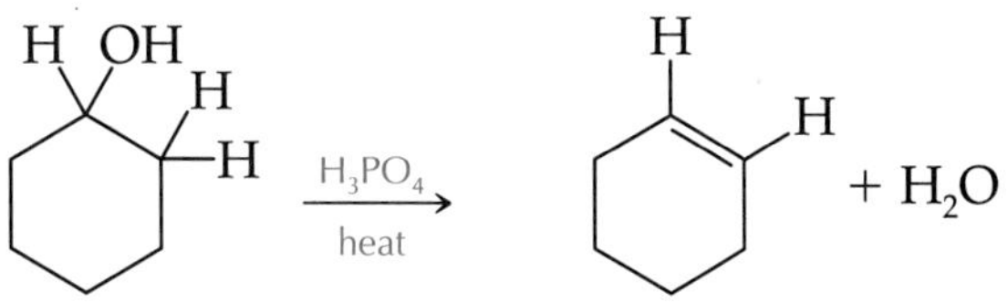

Cyclohexanol
bp 161°C
MW 100.2
density 0.962 g · mL^{-1}

Cyclohexene
bp 83°C
MW 82.1
density 0.810 g · mL^{-1}

Phosphoric acid catalyzes the 1,2-elimination of water from cyclohexanol by an E1 pathway. Because of the symmetry of this alcohol, only one alkene can be formed when a proton is abstracted from either carbon atom adjacent to the positively charged carbon atom of the cyclohexyl cation:

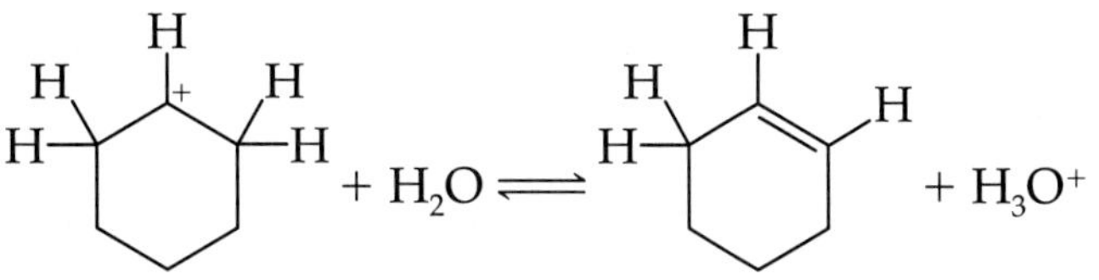

The reaction itself is carried out in a distillation apparatus. The equilibrium is shifted toward complete conversion of cyclohexanol to cyclohexene by distillation of the cyclohexene and water as they are formed.

Phosphoric acid, rather than sulfuric acid, is the suggested acidic catalyst in this reaction for two reasons. The first involves safety. If care is not taken in completely mixing the sulfuric acid and cyclohexanol before heating the mixture, irritating and toxic sulfur dioxide fumes may be evolved into the laboratory atmosphere. This undesired fuming can also occur near the end of the reaction if the flask is heated for too long. The sulfur dioxide results from oxidation-reduction reactions in which hot sulfuric acid acts as an oxidizing agent.

The second reason for choosing phosphoric acid is that less polymerization occurs when it is used. Black tarry polymers can result when carbocations undergo reactions with alkenes to form a long-chain molecule. Under the proper conditions, a cyclohexyl cation can react with the electron-rich cyclohexene to form a new carbocation, which in turn can react with another molecule of cyclohexene. Along the way toward polymer formation, a carbocation may lose a proton, but the resulting alkene can be reprotonated under these equilibrium conditions. Eventually polymers with molecular weights in the thousands may form:

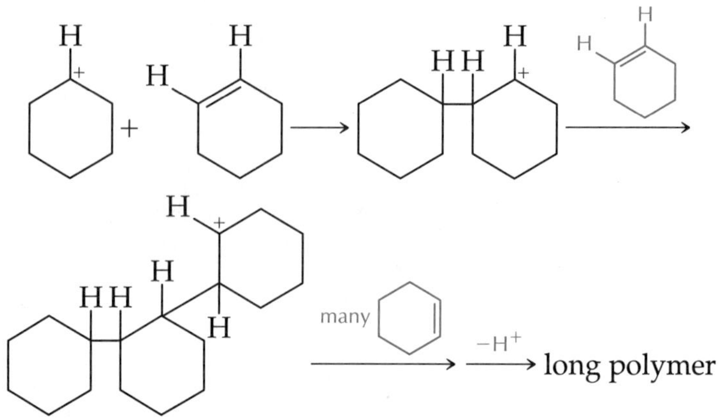

The use of phosphoric acid in this experiment does have one disadvantage relative to sulfuric acid catalysis. The dehydration reaction is slower when phosphoric acid is the catalyst, and the yield of cyclohexene is apt to be a bit lower. Nevertheless, we think that the safety factors favor using phosphoric acid for this dehydration.

Cyclohexyl cations also may react with a molecule of cyclohexanol, forming the conjugate acid of dicyclohexyl ether. Again, this reaction is reversible under the experimental conditions:

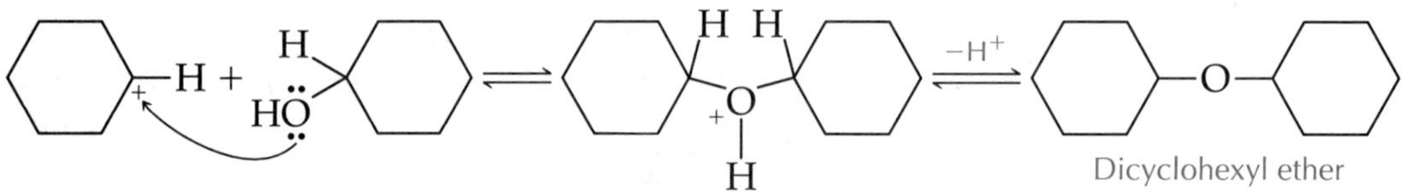

Cyclohexene bp 83°C
Cyclohexanol bp 161°C
Dicyclohexyl ether bp 243°C

If the distillation temperature is controlled, no dicyclohexyl ether or cyclohexanol should distill from the reaction mixture into the receiving flask because their boiling points are considerably higher than that of cyclohexene.

Two simple identification tests can quickly distinguish alkenes from other classes of compounds. Almost all alkenes react quickly and smoothly with a dilute solution of bromine in dichloromethane to form dibromoalkanes:

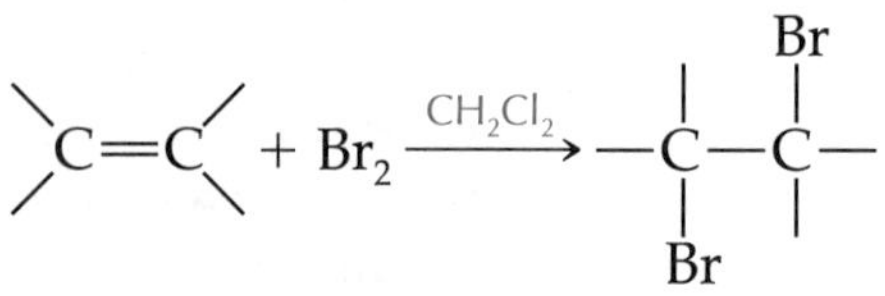

The molecular bromine is consumed and its characteristic red-brown color disappears.

The second simple qualitative test for alkenes involves their rapid oxidation by a solution of potassium permanganate. The purple color of the permanganate solution disappears within 2 or 3 min. In its place, a brown precipitate of manganese dioxide appears:

$$3 \,\diagup\!\!\!C\!\!=\!\!C\diagdown\!\! + 2\,MnO_4^- + 4\,H_2O \longrightarrow 3 -\underset{OH}{\overset{|}{C}}-\underset{OH}{\overset{|}{C}}- + 2\,MnO_2 + 2\,OH^-$$

Because alcohols may also be oxidized by potassium permanganate, both a positive bromine addition test and a positive permanganate oxidation test are necessary for a compound to be characterized as an alkene.

〰〰**Macroscale** **Procedure**	**Techniques** Simple Distillation: Technique 7.3 Drying Agents: Technique 4.6 Short-Path Distillation: Technique 7.3a Gas Chromatography: Technique 11 IR Spectrometry: Spectrometric Method 1

─ **SAFETY INFORMATION** ─

Cyclohexanol is an irritant. Avoid contact with skin, eyes, and clothing.

Concentrated (85%) phosphoric acid is very irritating to the skin and mucous membranes. Wear gloves. If you spill any phosphoric acid on your skin, wash it off immediately with copious amounts of water.

Remember that the position of the thermometer bulb is crucial for obtaining an accurate measure of the temperature of the distilling vapor.

Pour 18 mL (17.3 g, 0.17 mol) of cyclohexanol and 5.0 mL of 85% phosphoric acid into a 100-mL round-bottomed flask. Mix the contents of the flask thoroughly by swirling it and add two boiling stones. Assemble a simple distillation apparatus above the flask [see Technique 7.3]. Use a 25-mL round-bottomed flask as the receiving flask. Immerse the receiving flask up to its neck in an ice-water bath to minimize the escape of cyclohexene vapors into the laboratory.

Heat the reaction mixture with a heating mantle to distill the products. The temperature of the distilling vapor must not exceed 100–102°C. The liquid that collects in the receiving flask is a mixture of cyclohexene and water. Stop the distillation when only a few milliliters of residue remain in the distilling flask.

The solubility of NaCl is 35 g per 100 mL of water at room temperature.

Add enough solid sodium chloride to the distillate in the receiving flask to saturate the water layer; the salt minimizes the solubility of the organic product in the aqueous layer. Decant the distillate into a separatory funnel, leaving any undissolved sodium chloride in the receiving flask. Add 10 mL of 0.5 M sodium carbonate solution and shake the separatory funnel gently [see Technique 4.2]. Shake the funnel until no more gas is evolved when the funnel is vented.

The sodium carbonate will neutralize any acid that may have distilled with the product.

SAFETY PRECAUTION

Be sure to vent the separatory funnel often, because carbon dioxide evolution will cause a pressure buildup.

Draw off the lower aqueous layer. If it is still acidic to pH test paper or litmus paper, add another 10 mL of 0.5 M sodium carbonate solution to the upper organic layer, shake the separatory funnel gently, and draw off the bottom aqueous layer. Check a drop of the aqueous extract with pH paper; you have finished the neutralization of all the acid when the test paper indicates a basic solution, which is due to the presence of excess sodium carbonate.

Pour the crude cyclohexene out of the top of the separatory funnel into a dry 25-mL Erlenmeyer flask and add anhydrous magnesium sulfate to dry the product [see Technique 4.6]. Allow the product to stand over the drying agent for 10 min.

If any of the glassware is wet, the distilled cyclohexene will be cloudy, as the result of a water emulsion.

Place a small plug of cotton in a conical funnel and decant the product through the funnel into a dry 50-mL round-bottomed flask [see Technique 4.7]. Add a boiling stone and assemble the apparatus for a short-path distillation [see Technique 7.3a]. Be sure that you have washed and dried the stillhead, thermometer adapter, and vacuum adapter, because they may be wet from your first distillation. Use a weighed 25-mL or 50-mL round-bottomed flask submerged up to its neck in an ice-water bath as the receiver.

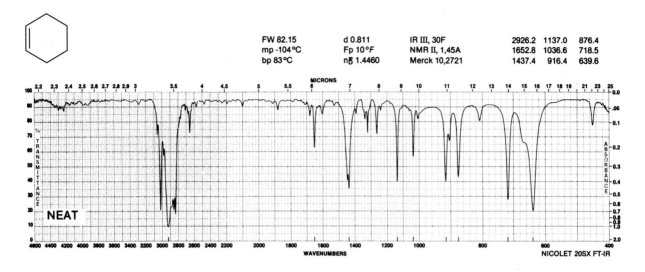

FW 82.15	d 0.811	IR III, 30F	2926.2 1137.0 876.4
mp -104 °C	Fp 10°F	NMR II, 1,45A	1652.8 1036.6 718.5
bp 83°C	n꜀ 1.4460	Merck 10,2721	1437.4 916.4 639.6

FIGURE 11.1 IR spectrum of cyclohexene. (Provided by Aldrich Chemical Company, Inc., Milwaukee, WI.)

Stop the distillation before the boiling flask reaches dryness.

Collect the fraction that boils between 79°C and 84°C in the weighed flask. Determine the mass of your product and calculate the percent yield.

Check the purity of your product by gas chromatography [see Technique 11]. Use a nonpolar column heated to about 60°C. For a capillary GC, inject a 1-μL sample of a solution prepared from 0.5 mL of ether and 1 drop of your product. Consult your instructor about sample preparation for a packed column gas chromatograph. Your instructor may also suggest that you determine the IR spectrum of your product. Then compare your spectrum to that shown in Figure 11.1.

Perform the qualitative tests for alkenes (potassium permanganate oxidation and bromine addition) described in Experiment 9.2 (p. 94) on your cyclohexene product and on cyclohexane or hexane as a reference in separate test tubes. Record your results and explain them. After performing these tests, submit your remaining product to your instructor.

Cleanup: Add about 30 mL of *cold* water to the residue remaining in the reaction flask, then add solid sodium carbonate in *small* portions until the phosphoric acid is neutralized. **(Caution: Foaming will occur when Na$_2$CO$_3$ is added.)** Wash the solution down a sink *in a hood* or pour it into the container for

The acetone rinse will minimize the unpleasant musty odor of cyclohexene in the laboratory.

aqueous inorganic waste. Pour the residue remaining in the boiling flask after the final distillation into the container for flammable (organic) waste that is located in a hood. While working in the hood, rinse the flask with a few milliliters of acetone and pour this rinse into the container for flammable waste. Place the spent drying agent in the container for nonhazardous solid waste or for inorganic waste. Pour the solutions remaining from the bromine test into the container for halogenated organic waste and the solutions from the potassium permanganate test into the container for inorganic waste.

Questions

1. What side reactions compete with the acid-catalyzed dehydration reaction?
2. Rather than continuing the distillation to dryness, why is it stopped when a residue remains in the flask?
3. Why is the distillation temperature kept below 102°C in the dehydration of cyclohexanol?
4. Why must any acid be neutralized with sodium carbonate before the final distillation of cyclohexene?
5. What product would result from the addition of a concentrated permanganate solution to cyclohexene and prolonged heating?

11.2 Dehydration of 2-Methylcyclohexanol

Determine by GC the composition of isomeric alkenes that form in the acid-catalyzed dehydration of an alcohol.

$$ \text{2-Methylcyclohexanol} \xrightarrow{\text{H}_3\text{PO}_4} \text{1-Methylcyclohexene} + \text{3-Methylcyclohexene} + \text{Methylenecyclohexane} + \text{H}_2\text{O} $$

2-Methylcyclohexanol
(mixture of cis/trans isomers)
bp 165–166°C
MW 114
density 0.930 g · mL^{-1}

1-Methylcyclohexene
bp 110°C
MW 96

3-Methylcyclohexene
bp 104°C
MW 96

Methylenecyclohexane
bp 102–103°C
MW 96

Product ratio will be determined.

The dehydration of 2-methylcyclohexanol is more challenging than the same reaction described for cyclohexanol in Experiment 11.1. First, the alcohol starting material is normally a 50:50 mixture of cis and trans isomers:

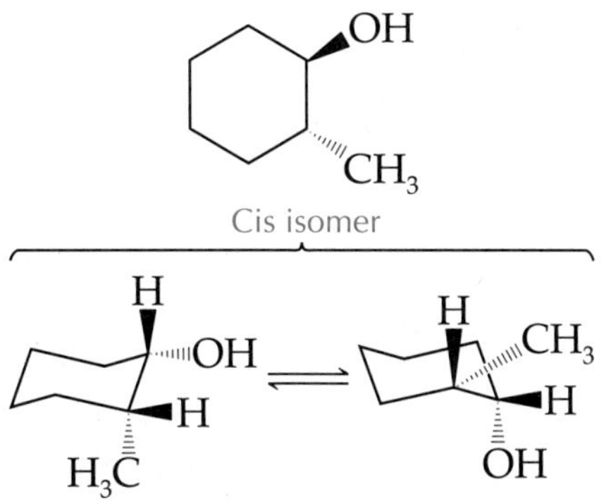

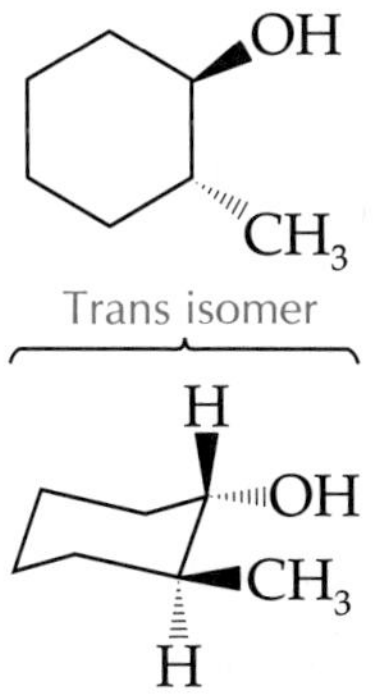

Cis isomer

Trans isomer

This mixture of isomers can lead to variations in product ratios because the two alcohols undergo dehydration and rearrangement at different rates (the cis isomer is known to be about 30 times more reactive toward dehydration), but ratio variations can be minimized if we use conditions that ensure complete dehydration of both alcohol isomers. Second, the product mixture contains three alkenes that are formed in various ratios that depend on reaction conditions. Finally, quantitative separation and characterization of the three alkene products can be demanding, because their structures are similar.

Protonation of the alcohol followed by loss of the improved leaving group (H_2O) gives rise to a secondary carbocation that can lose a proton to form either 3-methylcyclohexene or 1-methylcyclohexene (E1 process).

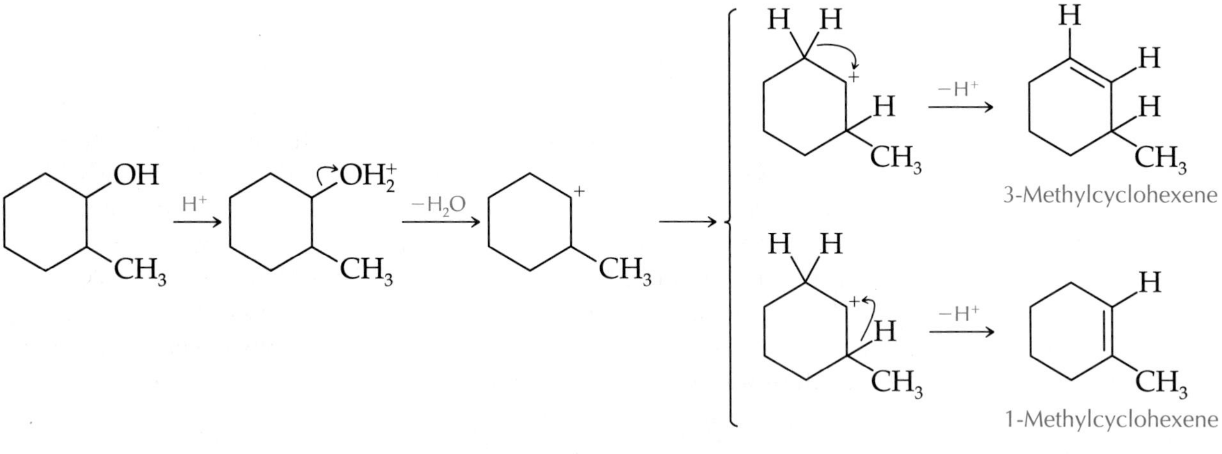

3-Methylcyclohexene

1-Methylcyclohexene

We expect 1-methylcyclohexene to be more stable than 3-methylcyclohexene, because the double bond of the former has more alkyl groups directly attached to it. Under conditions of thermodynamic control, the more stable 1-methyl isomer should predominate. Kinetic control will give rise to the isomer that is formed more rapidly. In other words, the transition state of lower energy leads to the predominant product. It is often found that the products of kinetic control and thermodynamic control are different.

The protonated alcohol can also rearrange directly to a tertiary carbocation as a molecule of water is lost. Alternatively, the secondary carbocation that is formed by loss of water can undergo a hydride shift to form the 1-methylcarbocation. Loss of a proton from this cation can lead to either 1-methylcyclohexene or methylenecyclohexane:

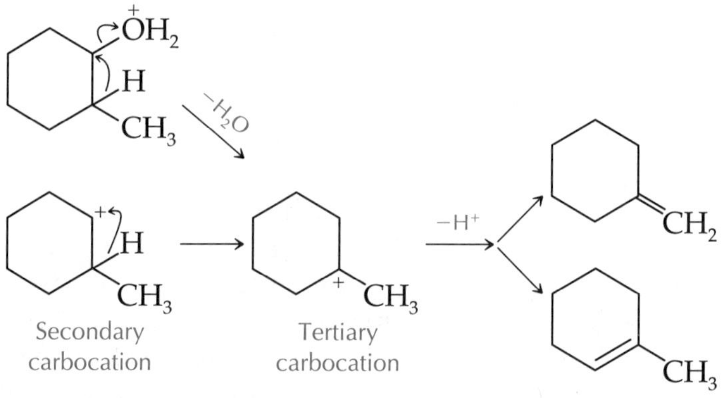

Such rearrangements are common for carbocations, and in this case we clearly have a strong driving force, because the tertiary carbocation is substantially more stable than its secondary carbocation precursor:

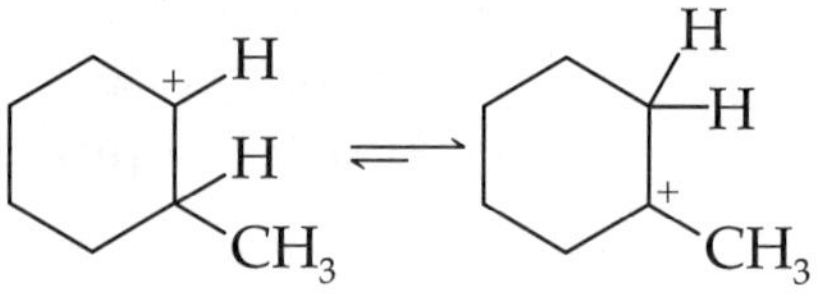

A simple distillation will not readily separate all the products. In fact, even if a fractional distillation process using a packed column is used, we may only separate 1-methylcyclohexene from the other two isomers, because the latter pair of compounds have very close boiling points. Even gas chromatography may be a challenge. GC using packed (1/4–3/4 in.) columns that separate on a boiling-point basis will probably only separate 1-methylcyclohexene from the other two possible alkenes. GC using packed columns intended for separations based on polarity differences will likely be no more successful. Capillary GC works better, and we have recommended conditions in the experimental section.

microscale Procedure

Techniques   Microscale Distillation: Technique 7.3b
Microscale Extraction: Technique 4.5d
Drying Agents: Technique 4.6
Gas Chromatography: Technique 11

---SAFETY INFORMATION---

2-Methylcyclohexanol is harmful if inhaled and may cause skin or eye irritation. Avoid contact with skin, eyes, and clothing.

Concentrated (85%) phosphoric acid is very irritating to the skin and mucous membranes. Wear gloves. If you spill any phosphoric acid on your skin, wash it off immediately with copious amounts of water.

Begin heating an aluminum block to 175°C. Measure 2.5 mL of 2-methylcyclohexanol with a graduated pipet and transfer it to a 5-mL conical vial. Add 0.4 mL of 85% phosphoric acid. Put a spin vane in the vial and fit a Hickman stillhead above the vial with a screw cap and an O-ring [see Technique 2.3]. In the same manner, fit a water-cooled condenser above the stillhead. Set the vial in the heated aluminum block and clamp the entire apparatus as shown in Technique 7, Figure 7.10. Wrap the vial and screw cap below the cup of the stillhead with several layers of glass wool, then place two auxiliary aluminum blocks around the vial. Turn on the magnetic stirrer.

---SAFETY PRECAUTION---

Use tongs or large tweezers to handle the glass wool so that you avoid burns from the hot aluminum block and irritation from the tiny glass fibers.

When the upper portion of the Hickman still becomes coated with a film of water droplets, slowly increase the rate of heating so that the block temperature rises to 200–210°C. As more vapor

*Do not interrupt the heating
while removing condensate.*

condenses in the stillhead, droplets will run into the cup. When the liquid level in the cup reaches the half-full point, open the port and withdraw the condensate with a Pasteur pipet [see Technique 7.3b]. Transfer the condensate to a 5-mL conical vial and cap the vial tightly. Close the cap on the stillhead port as quickly as possible. Alternatively, remove the distillate with a syringe and needle inserted through the septum. Continue collecting and removing distillate until the reaction is complete. Move the auxiliary aluminum blocks with tongs to see if the contents of the vial are still boiling. The phosphoric acid in the reaction vial does not distill and becomes yellow near the end of the distillation.

Add 1.0 mL of water to the vial containing the distillate, cap the vial tightly, and shake it to mix the layers. Allow the layers to separate and remove the aqueous layer with a Pasteur filter pipet [see Technique 2.4 for the filter pipet and Technique 4.5d for microscale extraction]. Repeat this washing process with 1.0 mL of 5% sodium carbonate solution. After removing the aqueous sodium carbonate layer, add 6–10 pellets of anhydrous calcium chloride to the product layer remaining in the conical vial. Cap the vial and allow the product to dry for 10–15 min.

Prepare another Pasteur filter pipet. Weigh a small sample vial and transfer the dried product to this vial with the Pasteur filter pipet. Weigh the product and calculate the percent yield. Carry out the product analysis described in the next section, then submit any remaining product to your instructor.

Product Analysis

Determine the ratio of products by gas chromatographic analysis on a nonpolar column heated to 50–55°C [see Technique 11]. For a capillary GC, prepare a solution containing 1 drop of your product in 1.0 mL of ether. Inject a 1-μL sample of this solution into the GC. Consult your instructor about the appropriate sample dilution and sample volume if you are using a packed column gas chromatograph. The methylcyclohexenes elute in order of increasing boiling point. If known samples of the products are available in your laboratory, confirm the identification of each product by the peak enhancement method [see Technique 11.6].

Perform the bromine addition and the potassium permanganate oxidation tests described in Experiment 9.2 (p. 94) on your product. Do these tests indicate the presence of unsaturated carbon-carbon bonds? Also perform these tests on 2-methylcyclo-hexanol. Compare the results with those that you obtained from your product.

Cleanup: Pour the phosphoric acid remaining in the reaction vial and the aqueous solutions saved from the extractions into a beaker containing about 50 mL of water; neutralize this solution with solid sodium carbonate before washing it down the sink or pouring it into the container for aqueous inorganic waste. Place the spent drying agent in the container for solid inorganic waste. Pour the ether solution from the GC analysis into the container for flammable (organic) waste and the solutions from the bromine addition tests into the container for halogenated waste. The solutions from the potassium permanganate oxidation test may be diluted with water and washed down the sink or placed in the container for inorganic waste, as directed by your instructor.

Optional Experiments

Feigenbaum (Ref. 2) reports that 1-methylcyclohexene largely free of other isomers can be synthesized from 1-methylcyclohexanol by using the procedure described in this experiment. Synthesize 1-methylcyclohexene from 1-methylcyclohexanol by the same procedure used in this experiment. Confirm its identity by IR analysis and its purity by GC analysis. Compare your IR spectrum to that shown in Figure 11.2.

In the dehydration of 2-methylcyclohexanol, Todd (Ref. 3) found that the ratio of 1-methylcyclohexene to 3-methylcyclohex-ene changed during the course of the distillation. Carry out the dehydration, using 15 mL of 2-methylcyclohexanol and 3 mL of 85% phosphoric acid in a 50-mL round-bottomed flask (remember to add 2 boiling stones). Assemble the apparatus for simple distil-lation [see Technique 7.3], with the addition of a Claisen adapter between the flask and the stillhead; close the second opening in the Claisen adapter with a glass stopper. Use 15-mL centrifuge tubes with tight-fitting caps as receiving vessels for the distillate. Collect the distillate in two fractions, approximately 3–4 mL as the first fraction and the rest as the second fraction. Follow the wash-ing procedure used in this experiment for each fraction, working

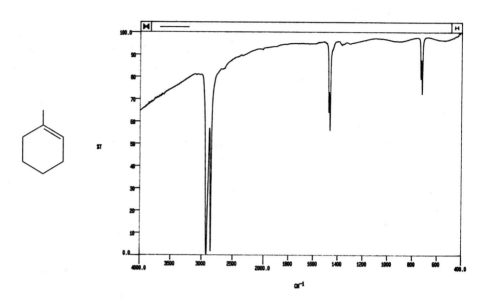

FIGURE 11.2 IR spectrum of 1-methylcyclohexene.

with each fraction in the 15-mL centrifuge tube in which it was collected [see Technique 4.5d]. Wash the product with 3 mL of water; then wash it with 3 mL of 5% sodium carbonate solution. Dry each product fraction with anhydrous calcium chloride for 10–15 min. Analyze each fraction by gas chromatography.

References

1. Taber, R. L.; Champion, W. C. *J. Chem. Educ.* **1967,** *44,* 620.
2. Feigenbaum, A. *J. Chem. Educ.* **1987,** *64,* 273.
3. Todd, D. *J. Chem. Educ.* **1994,** *71,* 440.

Questions

1. Because acid-catalyzed dehydration is an equilibrium situation, how was the reaction forced to completion?
2. Why would a poor yield of alkenes be realized if the dehydration were carried out by using hydrochloric acid?
3. Ignoring rearrangements, outline a mechanistic pathway showing the alkene products possible from 3-methylcyclohexanol.

4. Write a mechanism that would allow for the formation of methylenecyclohexane from 3-methylcyclohexanol.

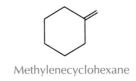

Methylenecyclohexane

5. In some cases, ethers arise when alcohols are treated with acid. Write a mechanism to illustrate this.

Experiment 12
ETHER SYNTHESES

Synthesize an ether, using either S_N2 or S_N1 chemistry; purify it by distillation; and characterize the structure of the product by NMR spectrometry.

Ethers are important organic compounds, especially as solvents. Diethyl ether (also called ethyl ether, or simply ether) is the most commonly used solvent for organic extractions because it is relatively inert, has a low boiling point, and is relatively insoluble in water. Thus, water and ether form a useful pair of immiscible solvents for extractions [see discussion in Technique 4.1]. Tetrahydrofuran (THF) and 1,4-dioxane are also frequently used as solvents in organic reactions because they are relatively inert and can solvate a large variety of organic molecules, thus enhancing their dissolution and reactivity. Moreover, they are water-soluble; thus, water/THF or water/dioxane solvent mixtures can be used as homogeneous reaction solvents to accommodate both nonpolar and polar solutes.

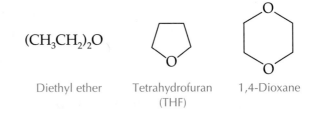

$(CH_3CH_2)_2O$ Diethyl ether   Tetrahydrofuran (THF)   1,4-Dioxane

The synthesis of symmetrical ethers from alcohols occurs in a straightforward manner. For example, when ethanol is treated with sulfuric acid at 140°C, diethyl ether is formed. If we extend this approach of acid-catalyzed synthesis from symmetrical ethers to the synthesis of unsymmetrical or "mixed" ethers, we should anticipate a problem, because more than one product is possible:

$$ROH + R'OH \xrightarrow{H_2SO_4} ROR' + ROR + R'OR'$$

There are, however, methods for preparing unsymmetrical ethers. The Williamson ether synthesis used in Experiment 12.1 demonstrates a very important synthetic route to ethers by the nucleophilic substitution (S_N2) reaction of an alkyl halide and an alkoxide. Experiment 12.2 uses S_N1 chemistry with acid catalysis to produce just one unsymmetrical ether, because a very stable carbocation intermediate directs the course of the reaction.

12.1 Williamson Ether Synthesis of 1-Ethoxybutane

Synthesize and purify an unsymmetrical ether.

$$CH_3CH_2CH_2CH_2 \!-\! Br + CH_3CH_2 \!-\! O^-Na^+ \xrightarrow{CH_3CH_2OH}$$

1-Bromobutane	Sodium ethoxide
MW 137	MW 68
bp 101.3°C	
density 1.27 g·mL^{-1}	

$$CH_3CH_2CH_2CH_2 \!-\! O \!-\! CH_2CH_3 + NaBr$$

1-Ethoxybutane
(butyl ethyl ether)
MW 102
bp 92.3°C

In the Williamson or S_N2 ether synthesis, one part of the ether is obtained from an alkyl halide and the other part from an alcohol. Therefore, the choice of the two types of starting materials determines the structure of the ether formed:

$$ROH + Na \longrightarrow RO^-Na^+ + \tfrac{1}{2} H_2$$

$$RO^-Na^+ + R'X \longrightarrow ROR' + Na^+X^-$$

$$X = Cl, Br$$

An alcohol can be converted to an alkoxide by treatment of the alcohol with a reactive metal such as sodium. This oxidation-reduction reaction produces hydrogen gas as a side product. One advantage in using an alkoxide prepared in this fashion is that the hydrogen gas readily escapes from the reaction mixture and does not interfere with the synthesis reaction. You will not have to prepare the alkoxide, however, because it is commercially available.

When an unhindered alkoxide combines with a primary alkyl halide, an ether is produced by an S_N2 reaction, the principles of which have already been discussed in Experiment 8. You will treat 1-bromobutane with sodium ethoxide, a strongly nucleophilic but sterically unencumbered alkoxide. Thus, the desired displacement reaction is achieved with a minimum of competing elimination, leading to a good yield of 1-ethoxybutane.

Macroscale Procedure

Techniques   Reflux: Technique 3.3

Extraction: Technique 4.2

Short-Path Distillation: Technique 7.3a

SAFETY INFORMATION

1-Bromobutane is harmful if inhaled, ingested, or absorbed through the skin.

Sodium ethoxide powder is corrosive; if you spill any, clean it up immediately. An ethanolic solution of sodium ethoxide is toxic and flammable. Wear gloves. Both the solid and the ethanolic solution rapidly undergo reaction with moisture in the atmosphere; keep the bottle tightly capped.

Diethyl ether is extremely volatile and flammable.

Pour 41 mL of a 21% (weight) solution (0.868 g·mL^{-1}) of sodium ethoxide in ethanol into a 100-mL round-bottomed flask. (Alternatively, you may use 7.5 g of sodium ethoxide and 40 mL of anhydrous ethanol.) Add 9.0 mL of 1-bromobutane and a boiling stone. Fit the flask with a water-cooled condenser and heat the reaction mixture under reflux for 45 min [see Technique 3.3], using a hot-water (90°C) bath as the heat source. A white solid will precipitate as the reaction proceeds (see Question 1).

Diethyl ether bp 35°C

Cool the reaction mixture to room temperature or lower, add 50 mL of water and 10 mL of diethyl ether. Stir the contents of the flask until the solid dissolves, and then transfer the solution to a separatory funnel. Rinse the round-bottomed flask with 10 mL of diethyl ether and add this rinse to the separatory funnel. Shake the funnel to extract the product from the aqueous solution into the diethyl ether layer [see Technique 4.2]. Drain the lower aqueous solution into a labeled Erlenmeyer flask, then pour the upper organic layer into another labeled Erlenmeyer flask. Return the aqueous solution to the separatory funnel and extract it with 10 mL of fresh diethyl ether. Separate the lower aqueous phase. Combine the two ether solutions in the separatory funnel. Wash the combined ether solution twice, using 30 mL of water each time, and once with 20 mL of saturated sodium chloride, removing each aqueous solution before adding the next one. After you separate the NaCl solution from the ether layer, pour the ether layer into a clean 50-mL Erlenmeyer flask. Dry the ether solution with anhydrous calcium chloride for at least 10 min.

Filter the ether solution through a conical funnel that contains a small cotton plug into a 50-mL round-bottomed flask [see Technique 4.7]. Add a boiling stone. Assemble the apparatus for short-path distillation, using a dry 100-mL round-bottomed flask as the receiving flask for the first fraction [see Technique 7.3a]. Weigh a 25-mL round-bottomed flask to use as the receiving flask for the product fraction. Use a steam bath as the heat source for the distillation of diethyl ether and submerge the 100-mL receiving flask in a beaker of ice and water up to the neck of the flask. When the distillation of diethyl ether slows to an occasional drop, change to a

See Technique 4.3 for a discussion of using NaCl for "salting out."

Be sure to cork the flask while the solution is drying so that ether vapors do not escape into the laboratory.

$CH_3OCH_2CH_2CH_2CH_3$

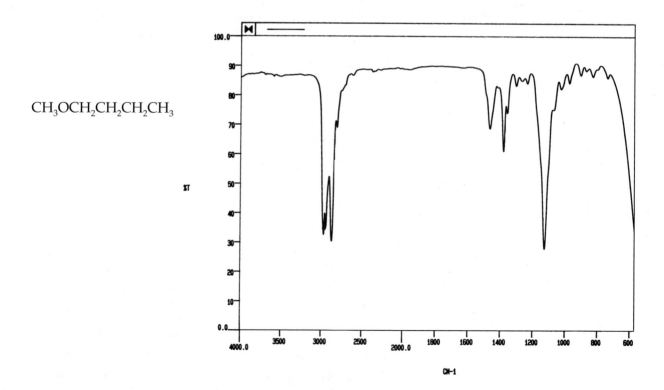

FIGURE 12.1 IR spectrum of 1-ethoxybutane.

heating mantle as the heat source, using a low to moderate rate of heating. Continue the distillation until the temperature reaches 88°C, then quickly replace the 100-mL flask with the tared 25-mL round-bottomed flask to collect the product fraction that boils from 88°C to 92°C.

SAFETY PRECAUTION

Ethers can form peroxides, which are explosive, if the distilling flask reaches dryness during a distillation.

Stop the distillation by removing the heating mantle **before the boiling flask reaches dryness.** Determine the weight of 1-ethoxybutane and calculate the percent yield from your synthesis. Your instructor may ask you to determine the IR or the NMR spectrum of your product.

$$CH_3CH_2OCH_2CH_2CH_2CH_3$$

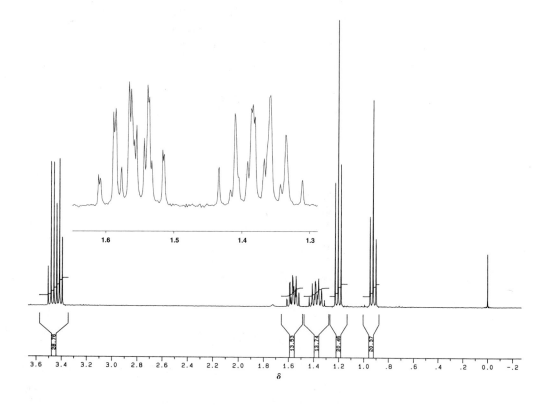

FIGURE 12.2 NMR spectrum (300 MHz) of 1-ethoxybutane.

Cleanup: Combine all of the aqueous solutions from the extractions. Determine the pH with pH test paper and neutralize the solution with 6 M hydrochloric acid added with a dropper. Wash the neutralized solution down the sink or pour it into the container for aqueous inorganic waste. Place the used calcium chloride pellets in the container for nonhazardous solid or inorganic waste. Pour the diethyl ether fraction into the container labeled "Recovered Ether" or the container for flammable (organic) waste, as directed by your instructor. Pour the residue remaining in the distillation flask into the container for flammable (organic) waste.

Questions

1. What is the solid that forms in the reaction flask during the reflux period and later dissolves when water is added?
2. Write an equation for the reaction that occurs between water and the excess sodium ethoxide remaining in the reaction mixture after the synthesis is complete.
3. Predict the products of the following reactions:
   (a) 1-iodobutane + sodium methoxide
   (b) iodomethane + sodium butoxide (NaO(CH$_2$)$_3$CH$_3$)
   (c) isobutyl bromide + sodium ethoxide
   (d)   CH$_2$Br

4. Suggest one or two Williamson syntheses, beginning with an alkyl halide and an alkoxide, for each of the following ethers. If one of the pair is clearly superior, state which and why.
   (a) dibutyl ether
   (b) 1-methoxypentane (methyl pentyl ether)
   (c) isopropyl methyl ether (2-methoxypropane)
   (d)   CH$_2$OCH$_2$CH$_3$

5. Treatment of *tert*-butyl chloride with potassium *tert*-butoxide leads to no significant amounts of di(*tert*-butyl) ether. What would you expect to be formed and why?

12.2
Synthesis and Analysis of a Propyl Benzhydryl Ether

Synthesize an ether by using acid catalysis and characterize the product by using NMR.

Benzhydrol
mp 69°C
MW 184.2

1-Propanol
(propyl alcohol)
bp 97°C
MW 60.1
density 0.804 g · mL^{-1}

1-Propyl benzhydryl ether

or

2-Propyl benzhydryl ether

In this reaction you will treat a mixture of benzhydrol (diphenyl-methanol) and 1-propanol with sulfuric acid. This mixture produces primarily one benzhydryl alkyl ether. The question is which alkyl group, a propyl or an isopropyl group, is part of the ether product? When the alcohol mixture is treated with sulfuric acid, both alcohols can have their —OH groups protonated, but only benzhydrol readily undergoes formation of a carbocation, because the benzhydryl cation is highly stabilized by resonance:

Benzhydrol
(diphenylmethanol)

This carbocation can then undergo reaction with an alcohol to form the ether product.

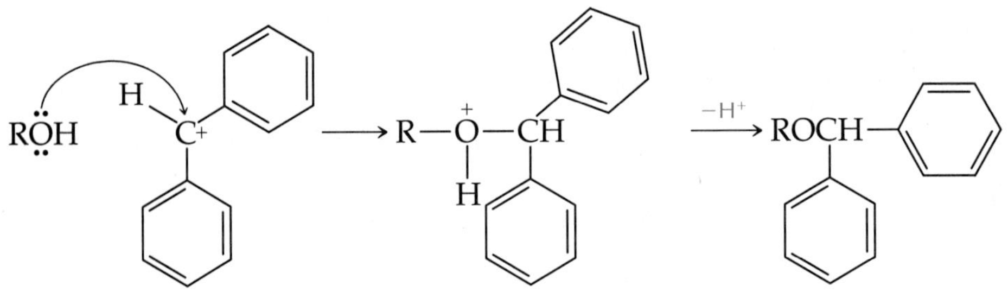

Thus, the mechanism for a synthesis completed in this way can be summarized as follows:

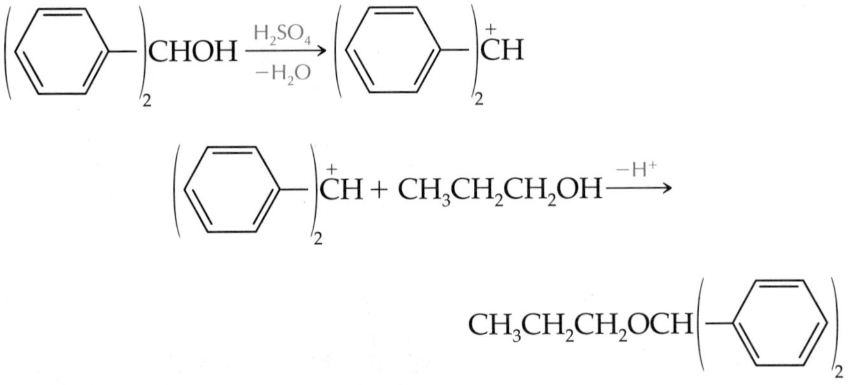

Another route to an ether product could involve formation of an isopropyl cation. If this cation were trapped by benzhydrol, a rearranged ether product would be obtained:

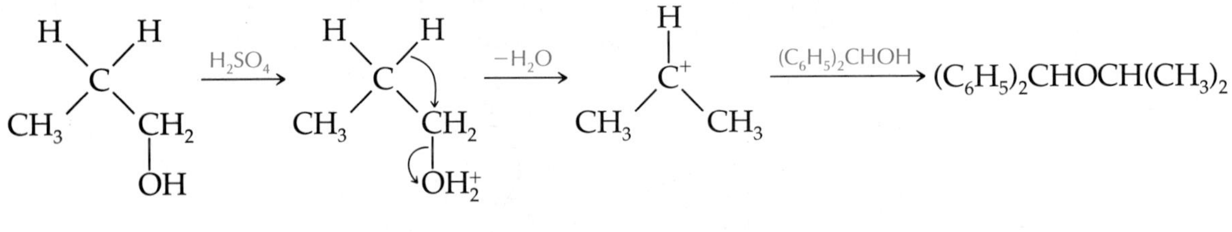

Still another possibility is that under these reaction conditions 1-propanol rearranges to 2-propanol.

The identity of the ether product will be determined by ^{1}H NMR. You should focus especially upon the integration and splitting pattern for the alkyl NMR signals. This pattern will reveal which product forms and thus which mechanism applies.

ʍʍMacroscale
Procedure

Techniques   Reflux: Technique 3.3
Extraction: Technique 4.2
Vacuum Distillation: Technique 7.7
Gas Chromatography: Technique 11
NMR Spectrometry: Spectrometric Method 2

SAFETY INFORMATION

Benzhydrol is an irritant. Avoid contact with skin, eyes, and clothing.

Sulfuric acid is corrosive and causes severe burns. Wear gloves while handling it.

1-Propanol is an irritant and flammable. Avoid contact with skin, eyes, and clothing.

Heptane is flammable.

Dissolve 7.4 g of benzhydrol in 24 mL of 1-propanol in a 100-mL round-bottomed flask. While swirling the flask, add 0.8 mL of concentrated sulfuric acid (98% H_2SO_4). Attach a reflux condenser and heat the reaction mixture at reflux for 25 min. Remove the heating mantle and cool the flask to below room temperature with an ice-water bath.

Transfer the reaction mixture to a separatory funnel. Add 25 mL of water and extract the aqueous mixture with 30 mL of heptane [see Technique 4.2]. Separate the aqueous phase and wash the organic layer twice with 15-mL portions of water, and once with 15 mL of 0.5 M sodium carbonate solution **(Caution: Vent frequently.),** removing each aqueous solution before adding the next one. Check the pH of the carbonate extract with litmus paper to make sure that it is basic. If it is not, extract the heptane solution again with another 15 mL of 0.5 M Na_2CO_3 solution.

Excess H_2SO_4 must be removed from the organic layer by neutralization with Na_2CO_3. If H_2SO_4 remains in the organic phase, your product will decompose into a brown tar during the distillation.

Dry the heptane solution with anhydrous magnesium sulfate. Filter the dried solution into a dry 100-mL round-bottomed flask, using fluted filter paper [see Technique 4.7, Figure 4.13]. Set up vacuum distillation equipment as shown in Technique 7.7, Figure 7.17 if you are using small pieces of wood splints in place of boiling stones, or as shown in Figure 7.18 if you are using a capillary bubbler. Use a 50-mL round-bottomed flask placed in an ice-water bath as the receiving flask for the distillation of heptane. Also determine the mass of your 25-mL round-bottomed flask before you start the distillation; this will be the receiving flask for the product fraction.

Do not use wood splints or boiling stones if you use a capillary bubbler.

Distill the heptane at atmospheric pressure, using a low setting on the transformer controlling the heating mantle. When the heptane has distilled, increase the rate of heating and continue the distillation at atmospheric pressure until the distilling temperature reaches 100–125°C, or until no vapor reaches the thermometer bulb. This practice ensures that any remaining 1-propanol or any di-*n*-propyl ether that may have formed in the reaction will distill over and not contaminate your final product. The product is moderately pure at this point. A vacuum distillation will lead to a product of high purity.

If your laboratory is equipped for carrying out a vacuum distillation, proceed as described here. If you are not carrying out a vacuum distillation on the product, proceed to the analysis of the product.

The product is a high-boiling liquid that should be distilled under vacuum to avoid decomposition from excessive heating. At a pressure of 52 mmHg, the propyl benzhydryl ether distills at 200–205°C; at a pressure of 20 mmHg, it distills at 170–180°C. To get ready for the vacuum distillation, attach a piece of thick-walled tubing to a guard flask that is connected to the water-tap aspirator [see Technique 7.7].

Turn on the water-aspirator pressure full force and distill the propyl benzhydryl ether under vacuum. If your aspirator system will not maintain a pressure of 50 mmHg or lower, you will be unable to collect your product by using a vacuum distillation.

GC and NMR Analysis of the Product

Check the purity of your product by gas chromatography [see Technique 11]. Operating parameters for a packed column gas chromatograph are a silicone column at 190–196°C, a carrier gas flow rate of 25 mL·min^{-1}, and 1-μL sample injection. For a capillary column gas chromatograph equipped with a silicone (nonpolar) column, use a column temperature of 190°C, an injector temperature of 220°C, and 1 μL of an ether solution containing one drop of the product in 0.5 mL of diethyl ether. The order of elution will be benzhydrol, 2-propyl benzhydryl ether, then 1-propyl benzhydryl ether. You can verify the benzhydrol peak by the peak enhancement method [see Technique 11.6].

Weigh your product and calculate the percent yield. An NMR spectrum of your product is the only simple way to prove its structure. If an NMR spectrometer is available in your laboratory, obtain a spectrum of your product. Prepare the NMR sample as discussed in Spectrometric Method 2. Compare your NMR spectrum to that shown in Figure 12.3.

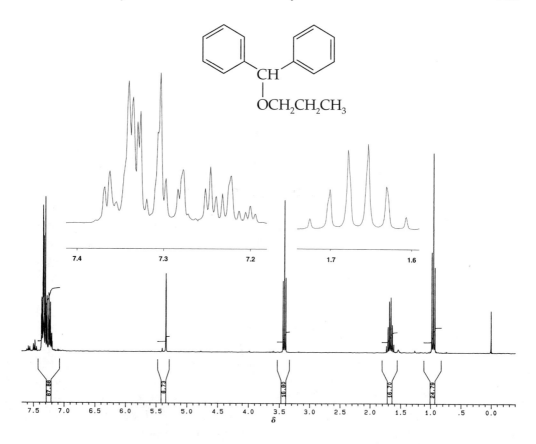

FIGURE 12.3 NMR spectrum (300 MHz) of 1-propyl benzhydryl ether.

Questions

1. Why does lowering the pressure in a distillation apparatus lower the boiling point of the compound to be distilled?

2. Explain how the structure of 2-propyl benzhydryl ether can be determined from its NMR spectrum.

3. John Sophomore used only 10 mL of 1-propanol instead of 24 mL. He isolated, in addition to some of the expected product, a white solid with mp 110–111°C, which gave only two signals in its NMR spectrum. A signal was present at $\approx \delta\,7.2$ (relative area of 10), and a singlet was present at $\delta\,5.3$ (relative area of 1). Explain what the white solid was and how it was formed.

4. Treatment of a *tert*-butyl alcohol/ethanol mixture with sulfuric acid leads to *tert*-butyl ethyl ether in reasonable yield, and a little diethyl ether or di-*tert*-butyl ether. Explain, along with a mechanism.

5. Treatment of an isobutylene (2-methylpropene)/ethanol mixture with sulfuric acid leads to *tert*-butyl ethyl ether in reasonable yield. Explain, along with a mechanism.

6. Millions of pounds of *tert*-butyl methyl ether (MTB) are manufactured each year as an antiknock additive for gasoline. Suggest a possible method for its commercial synthesis.

GRIGNARD SYNTHESES

Investigate how organometallic Grignard reagents can be used to make a variety of carbon-carbon bonds.

Organometallic compounds are versatile intermediates in the synthesis of alcohols, carboxylic acids, alkanes, and ketones, and their reactions form the basis of some of the most useful methods in synthetic organic chemistry. They readily attack the carbonyl double bonds of aldehydes, ketones, esters, acyl halides, and carbon dioxide. The use of organometallic reagents can produce the synthesis of highly specific carbon-carbon bonds in excellent yields.

Among the most important organometallic reagents are the alkyl- and arylmagnesium halides, which are almost universally called Grignard reagents after the French chemist Victor Grignard, who first realized their tremendous potential in organic synthesis. Their importance in the synthesis of carbon-carbon bonds was recognized immediately after the report of their discovery in 1901. Grignard received the 1912 Nobel prize in chemistry for applications of this reagent to organic synthesis.

Grignard reagents, like all organometallic compounds, are substances containing carbon-metal bonds. Because metals are electropositive elements, carbon-metal bonds have a high degree of ionic character, with a good deal of negative charge on the carbon atom. This ionic character gives organometallic compounds a high degree of carbon nucleophilicity:

$$\overset{\delta-}{R}-\overset{\delta+}{Mg}-\overset{\delta-}{X}$$

Grignard reagents undergo a wide variety of nucleophilic addition reactions with compounds containing polar C=O bonds. The resulting carbon-carbon bond formation yields larger and more complex molecules; and because a variety of different organic (R or Ar) groups can be introduced into organic structures, a wide array of organic compounds can be produced. For example, the reaction of Grignard reagents with aldehydes produces primary and secondary alcohols:

$$R-Br + Mg \longrightarrow RMgBr \xrightarrow{HCHO} \xrightarrow[H^+]{H_2O} RCH_2OH$$
(a primary alcohol)

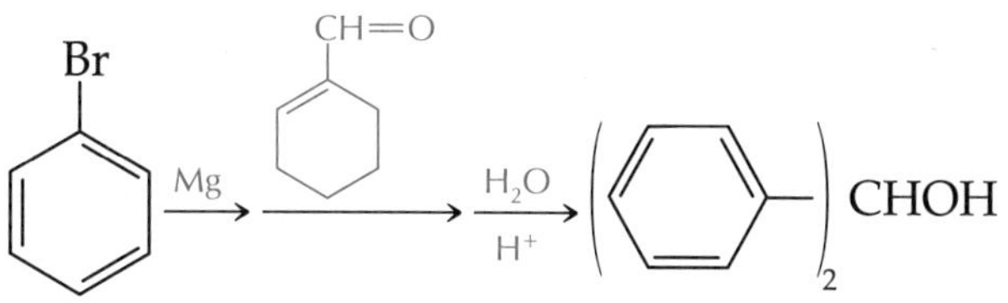

Benzhydrol
(a secondary alcohol)

and their reaction with ketones gives tertiary alcohols:

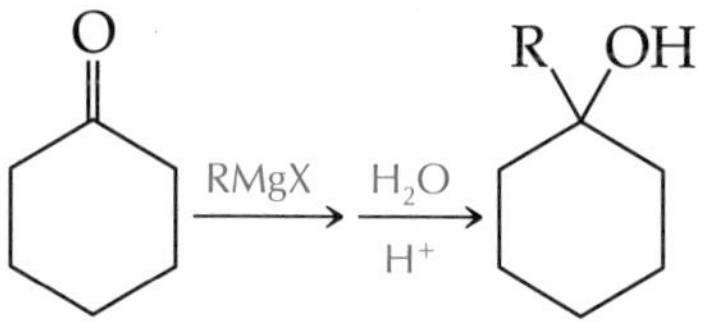

The mechanism of the Grignard reaction with aldehydes and ketones is actually quite complex, but it can easily be rationalized as a simple nucleophilic addition reaction:

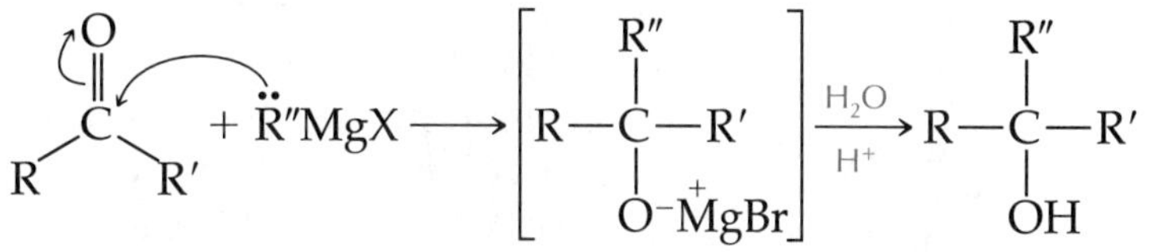

The hydrolysis step is important in a Grignard synthesis. It is common to use an aqueous mineral acid, such as sulfuric or hydrochloric acid, to expedite hydrolysis. Not only does this cause the reaction to go more readily, but Mg(II) is converted from the much less manageable hydroxide or alkoxide salts to water-soluble sulfates or chlorides. For preparing labile products, such as tertiary alcohols, the weaker acid ammonium chloride is an excellent alternative. Strong acids, such as sulfuric acid, may cause tertiary alcohols to dehydrate.

Synthesis of Grignard Reagents

The synthesis of a Grignard reagent requires the reaction of an alkyl or aryl halide with magnesium metal in the presence of an ether solvent:

$$R—X + Mg \longrightarrow R—Mg—X$$

Relative reactivity: R—I > R—Br > R—Cl

Alkyl and aryl bromides are the common starting materials for Grignard syntheses; the reaction of alkyl and aryl chlorides with magnesium is often rather slow to react and alkyl iodides are generally expensive.

Formation of a Grignard reagent takes place in a heterogeneous reaction at the surface of the solid magnesium metal, and the surface area and reactivity of the magnesium are crucial factors in the rate of the reaction. It is thought that the alkyl or aryl halide reacts with the surface of the metal to produce a carbon-free radical and a magnesium-halogen bond. The free radical R·, then reacts with the ·MgX to give the Grignard reagent, RMgX.

Grinding a few of the magnesium turnings with a mortar and pestle promotes the formation of the Grignard reagent by exposing an unoxidized metallic surface and providing a larger reactive surface area. For an alkyl halide, this procedure will usually be all that is necessary to initiate the reaction quickly; and, in many instances, breaking only one magnesium turning suffices.

When an aryl halide is used, grinding a few magnesium turnings and adding a small iodine crystal promote the heterogeneous reaction at the surface of the magnesium. There is some question about iodine's exact function; it may react with the metal surface to provide a more reactive interface or it may activate the aryl halide. Some of the color changes that one sees are due to the presence of iodine.

The proper selection of solvent is crucial in carrying out a reaction involving a Grignard reagent. Diethyl ether is the most frequently used solvent because it is inexpensive and promotes good yields. The yields of Grignard reagents are highest when a large amount of ether is present and when pure, finely divided magnesium metal is used.

The magnesium atom in a Grignard reagent has a coordination number of four. The alkyl magnesium halide already has two covalent bonds to magnesium. The other two sites can be occupied by ether molecules. Both electrons in the magnesium-ether bond are provided by the oxygen atom of the ether molecule:

$$H_5C_2 - \overset{\cdot\cdot}{\underset{\cdot\cdot}{O}} - C_2H_5$$
$$R - \overset{\cdot\cdot}{\underset{\cdot\cdot}{Mg}} - X$$
$$H_5C_2 - \underset{\cdot\cdot}{O} - C_2H_5$$

These complexes are quite soluble in ether. In the absence of the solvent, the reaction of magnesium and the alkyl halide takes place rapidly but soon stops because the surface of the metal becomes coated with the organomagnesium halide. In the presence of a solvent, the surface of the metal is kept clean and the reaction proceeds until all the limiting reagent is consumed. Although we usually write the reagent as simply RMgX, the actual structure of the compound in ether solution is often dimeric.

Di-*n*-butyl ether and tetrahydrofuran are occasionally used as solvents for Grignard reagents that form with particular difficulty. These solvents have a higher boiling point than diethyl ether. At a temperature above 0°C, most Grignard reagents exist in rapid equilibrium with the dialkyl- or diarylmagnesium and the magnesium halide:

$$R_2Mg + MgX_2 \xrightleftharpoons{K_{eq}} 2\,RMgX$$

The RMgX structure that Grignard originally proposed for this reagent is normally favored in the equilibrium. For phenylmagnesium bromide, K_{eq} has a value of 55 when diethyl ether is the solvent.

Because the reactivity of Grignard reagents as nucleophiles is due in large part to the strongly polarized carbon-metal bond of the reagent, it is only natural that they are also strong bases. They react quickly and easily with acids, even relatively weak acids such as water, alcohols, amines, and alkynes. Although this reaction is sometimes synthetically useful for the reduction of a carbon-halogen bond, it is more often a nuisance because it destroys the Grignard reagent.

Anhydrous Reaction Conditions

It is critical to maintain completely dry conditions throughout the formation and reaction of the Grignard reagent.

Be sure to disassemble and remove the stopcock of your separatory funnel during the drying period so that no water remains trapped in the bore.

The presence of water or other acids inhibits the initiation of the reaction and destroys the organometallic reagent once it forms. All glassware and reagents must be thoroughly dry before beginning a Grignard experiment. Oven-drying of the glassware is essential when the laboratory atmosphere is humid. When the humidity in the laboratory is low, as it is during the winter heating season, air-drying the glassware overnight will usually be sufficient for macroscale preparations. The glassware for microscale reactions must always be dried in an oven just prior to beginning the reaction because even trace amounts of moisture become significant at this scale.

Commercially available anhydrous ether, alkyl halides, and aryl halides are sufficiently pure for most Grignard reactions. Keep the ether container tightly closed except when actually pouring the reagent, and do not let your ether stand in an open container, because water from the air will dissolve into it.

Drying tubes containing anhydrous calcium chloride or other desiccants should be placed at the top of the reflux condenser and the dropping funnel during the preparation of the Grignard reagent and its subsequent reaction with a carbonyl compound. The drying tubes prevent atmospheric moisture from entering the reaction system. The apparatus for carrying out a macroscale

Grignard reaction is shown in Technique 3, Figure 3.7 and in Figure 3.8 for a microscale Grignard synthesis.

Side Reactions   Other reactions that may interfere with Grignard syntheses include Grignard coupling and reaction with oxygen:

$$2\,RX + Mg \longrightarrow R\!-\!R + MgX_2$$

$$RMgX + O_2 \longrightarrow ROOMgX \xrightarrow{\;RMgX\;} 2\,ROMgX$$

The coupling reaction takes place at the active surface of the solid magnesium when two free radicals react to form a stable hydrocarbon dimer. This side reaction is favored by a high concentration of the alkyl halide, where the possibility of two free radicals forming near each other at the metal surface is more likely.

Combination of Grignard reagents with oxygen is a side reaction when the syntheses are run in the presence of air. The low-boiling-point diethyl ether (bp 35°C) helps to exclude air near the surface of the reaction solution because of its great volatility. Organometallic compounds are often handled under a nitrogen or argon atmosphere to avoid reactions with oxygen.

13.1
Synthesis of Phenylmagnesium Bromide and Its Use as a Synthetic Intermediate

Synthesize a Grignard reagent.

Bromobenzene
bp 156°C
MW 157
density 1.50 g · mL^{-1}

Phenylmagnesium bromide
(an intermediate to be used immediately
in Experiments 13.1a, 13.1b, or 13.1c)

Treatment of an ether solution of bromobenzene with magnesium metal results in the formation of the Grignard reagent phenylmagnesium bromide. Insertion of the magnesium between carbon and bromine greatly enhances the reactivity of the aromatic structure and reverses the polarity of the bond between the carbon and the bromine functional group. In the organometallic compound, the carbon has a negative charge that makes it a strong nucleophile:

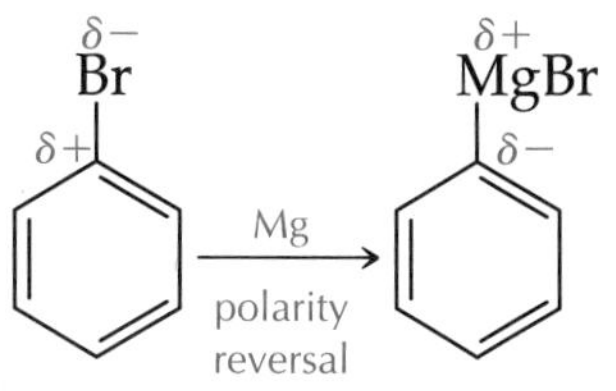

Usually a Grignard reagent is used immediately after synthesis, as you will do in this experiment. You have the option of combining phenylmagnesium bromide with benzaldehyde to yield the secondary alcohol diphenylmethanol (benzhydrol), with methyl benzoate to give the tertiary alcohol triphenylmethanol, or with carbon dioxide (dry ice) to form benzoic acid:

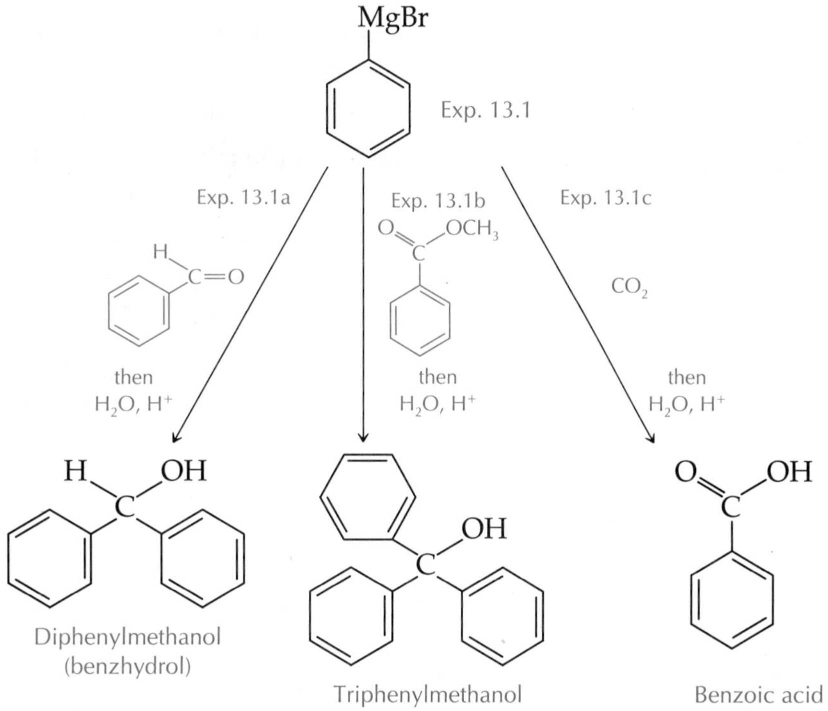

Diphenylmethanol
(benzhydrol)

Triphenylmethanol

Benzoic acid

Macroscale Procedure

Technique Anhydrous Reaction Conditions: Technique 3.3a

--- SAFETY INFORMATION ---

Diethyl ether, commonly called ether, is extremely flammable. Be certain that no flames are used in the laboratory and that hot electrical devices are not in the vicinity where ether is being used. Work in a hood, if possible.

Bromobenzene is an irritant. Wear gloves. Avoid contact with skin, eyes, and clothing.

Make sure that all your glassware and reagents are thoroughly dry before beginning the experiment. Refer to p. 137 in the introduction to Experiment 13 concerning the anhydrous conditions required for a Grignard reaction.

Assemble the apparatus shown in Technique 3.3a, Figure 3.7a or 3.7b, using a 250-mL round-bottomed flask, a water-jacketed condenser, and a 125-mL separatory funnel. For a one-necked flask, use a Claisen adapter to make two openings; for a three-necked flask, close the third opening with a glass stopper. Prepare two drying tubes containing anhydrous calcium chloride. Fit these to the top of the condenser and the separatory funnel with thermometer adapters or one-holed rubber stoppers.

Obtain 2.0 g (0.082 mol) of magnesium turnings, a *small* iodine crystal, and 15 mL of anhydrous ether. Open the round-bottomed flask and add the magnesium and the ether to the flask, then add the iodine crystal. With a dry stirring rod, gently break one or two magnesium pieces to expose a fresh surface. Close the flask.

> ### — SAFETY PRECAUTION —
>
> Be careful not to press so hard on a magnesium turning that you crack the round-bottomed flask.

Prepare a solution of bromobenzene in ether by dissolving 8.6 mL of bromobenzene (12.9 g, 0.082 mol) in 20 mL of anhydrous ether, and pour it into the separatory (dropping) funnel (with the stopcock closed). Add 4–5 mL of this solution to the reaction flask and swirl the flask to mix the reagents thoroughly.

Place the flask in a warm-water bath (45–50°C) and swirl the flask from time to time. The ether will begin to boil. The brownish solution should soon turn colorless, and a small amount of cloudy precipitate may appear. Formation of the organometallic reagent is exothermic. Periodically remove the water bath and note whether the reaction has started. This process will be indicated by refluxing of the ether, which results from the heat produced by the reaction.

When the reaction is well under way, remove the warm water bath, and add 20 mL of anhydrous ether through the condenser to dilute the solution. Then add the rest of the bromobenzene/ether

The precipitate is probably some magnesium hydroxide, which results from any moisture in the flask or the reagents.

solution from the dropping funnel at a speed that maintains a moderate reflux rate. Have a cold-water bath available to *briefly* cool the reaction in case the condensation ring in the condenser is more than halfway up the condenser. Do not cool the reaction so much that it stops. When all the bromobenzene has been added and the reaction stops refluxing spontaneously, heat the reaction in the warm-water bath (45–50°C) for 20 min. Cool the reaction flask in a cold-water bath before proceeding immediately to 13.1a for the macroscale preparation of benzhydrol, or to 13.1b for the macroscale preparation of triphenylmethanol.

microscale Procedure

Technique   Anhydrous Reaction Conditions: Technique 3.3a

— SAFETY INFORMATION —

Diethyl ether, commonly called ether, is extremely flammable. Be certain that no flames are used in the laboratory and that hot electrical devices are not in the vicinity where ether is being used. Work in a hood, if possible.

Bromobenzene is an irritant. Avoid contact with skin, eyes, and clothing.

Prepare a microscale drying tube, using two small pieces of cotton and anhydrous calcium chloride, as shown in Figure 3.8, Technique 3.3a. Place the drying tube, a 5-mL conical vial, a water-jacketed condenser, Claisen adapter, and two small screw-capped vials (without the caps) in a 250-mL beaker and dry the glassware in a 110°C oven for at least 30 min. You also need a rubber septum or a screw cap and a plastic septum for the Claisen adapter and a 1-mL syringe; be sure that these items are dry.

While the glassware is cooling, sand or scrape the oxide coating from a 6- to 7-cm piece of magnesium ribbon and cut the cleaned ribbon into 1-mm pieces. Weigh 51 mg (2.2 mmol) of magnesium, using tweezers or forceps to pick up the metal pieces. Place the magnesium pieces in the dried conical vial; add one small crystal of iodine and a magnetic spin vane to the vial.

Assemble the apparatus, shown in Technique 3.3a, Figure 3.8, as quickly as possible.

Obtain 0.30 mL of anhydrous diethyl ether in the 1-mL syringe. Inject the ether through the septum on the Claisen adapter into the conical vial.

Prepare a solution consisting of 0.21 mL of bromobenzene (314 mg, 2.0 mmol) and 0.50 mL of anhydrous ether in one of the oven-dried vials. Draw this solution into the same syringe used previously and insert the needle through the septum. Add another 0.20 mL of ether to the vial as a rinse, cap the vial, and set it aside.

Begin stirring the reaction mixture and inject about 0.1 mL of the bromobenzene/ether solution into the reaction vial. It may be necessary to warm the conical vial in a beaker of warm water (45–50°C) for a few minutes to initiate the reaction. A cloudiness in the reaction solution and the disappearance of the yellow iodine color indicate that the reaction has begun; remove the warm-water bath. The heat of reaction will cause the ether to reflux. Once the reaction has begun, add the remaining bromobenzene solution dropwise from the syringe over a period of 5–7 min. After the addition is complete, draw the ether rinse solution remaining in the vial into the syringe and inject it into the reaction mixture in one portion. Heat the reaction vial in a warm-water bath (45–50°C) for 15 min, then cool the mixture to room temperature. Proceed immediately to 13.1a for the microscale preparation of benzhydrol or to 13.1c for the preparation of benzoic acid.

13.1a Synthesis of Diphenylmethanol (Benzhydrol)

Use a Grignard reagent to prepare a secondary alcohol; purify and characterize the product.

Phenylmagnesium bromide

Benzaldehyde
bp 179°C
MW 106
density 1.046 g · mL^{-1}

H_2O, H^+

Diphenylmethanol
(benzhydrol)
mp 69°C
MW 184.2

Benzaldehyde, a reagent in this synthesis, is used in the manufacture of dyes and perfumes. As the artificial essential oil of almond, benzaldehyde is used as a flavoring; it occurs naturally in bitter almonds. The product of this Grignard synthesis, benzhydrol, finds use as a reagent in organic syntheses.

Treatment of benzaldehyde with phenylmagnesium bromide gives, after cold acid hydrolysis, benzhydrol. This reaction takes advantage of the high nucleophilicity of the phenyl portion of the Grignard reagent and the electrophilicity of the carbonyl carbon atom.

Benzaldehyde   Benzhydrol
                                                                            (diphenylmethanol)

Macroscale Procedure

Techniques   Extraction: Technique 4.2
                 Recrystallization: Technique 5.3

--- SAFETY INFORMATION ---

Benzaldehyde is highly toxic. Wear gloves while using it.

Hexane is extremely flammable. Heat it only on a steam bath.

Ether is also extremely flammable. The extractions, distillation of ether, and addition of the crude benzhydrol to hot water should be done in a hood. Use a steam bath as the heat source for the distillation.

Reaction of the Grignard reagent and benzaldehyde is quite vigorous and produces large amounts of a white precipitate.

Place a solution made from 8.0 mL (8.4 g 0.079 mol) of benzaldehyde and 10 mL of anhydrous ether in the separatory funnel and begin dropwise addition of the solution to the phenylmagnesium bromide solution. After the addition is complete, reflux the mixture for 15 min, using a warm-water bath (45–50°C). Cool 40 mL of 2.7 M sulfuric acid in an ice-water bath during this time.

Cool the reaction flask in an ice-water bath. Remove the condenser and separatory funnel.

SAFETY PRECAUTION

Carry out the hydrolysis step and the extractions in a hood to minimize ether fumes in the laboratory.

This is a convenient stopping point. Cork the flask and store it as directed by your instructor.

Carefully add small portions of the *cold* sulfuric acid solution to the reaction flask over a 20-min period. Swirl the flask frequently during the addition of the acid. The precipitate will decompose and you should have a yellowish ether solution and a colorless aqueous phase.

Transfer the mixture to a separatory funnel and separate the lower aqueous phase into a labeled Erlenmeyer flask. Pour the upper ether layer into another labeled Erlenmeyer flask. Keep both layers. Return the aqueous phase to the separatory funnel. Extract the aqueous phase twice with 30-mL portions of ether; after each extraction, add the ether layer to the flask containing the ether layer from the first separation [see Technique 4.2]. Dry the combined ether extract with anhydrous potassium carbonate for 10–15 min.

The ether for the extractions does not need to be anhydrous.

Filter the solution through a conical funnel containing a small cotton plug into a dry 250-mL round-bottomed flask [see Technique 4.7]. Set up a simple distillation apparatus, using a steam bath as the heat source, and distill the ether from the crude benzhydrol [see Technique 7.3]. Alternatively the ether can be removed with a rotary evaporator [see Technique 4.8].

Hot water vaporizes any remaining traces of ether.

This is a convenient stopping point.

Pour the residue from the distillation into 120 mL of hot water (60–70°C) contained in a 250-mL beaker. Cool the water mixture by adding ice. Solid benzhydrol will precipitate. Collect the solid by vacuum filtration and allow it to dry completely before doing the recrystallization.

Save a few crystals of the crude benzhydrol for a melting point before you recrystallize the rest of the crude product from hexane in a 250-mL Erlenmeyer flask, using a steam bath as the heat source [see Technique 5.3]. After crystallization is well advanced, cool the solution in an ice-water bath, and collect the product by vacuum filtration. Using the vacuum line, pull air over the crystals

If time permits, use TLC to check the product for the presence of biphenyl, a possible side product [see Technique 10].

to dry them. Determine the melting point of both the crude and the purified product, weigh your product, and calculate the percent yield. Obtain an IR spectrum or an NMR spectrum of your purified product as directed by your instructor. Submit your product to your instructor.

Cleanup: Neutralize the aqueous solution remaining from the extractions with solid sodium carbonate before washing it down the sink or pouring it into the container for aqueous inorganic waste. Place the spent drying agent in the container for nonhazardous waste or inorganic waste. Place the recovered ether from the distillation in a container for recovered ether or in the container for flammable (organic) waste, as directed by your instructor. Place the hexane filtrate in the container for flammable (organic) waste.

microscale Procedure

Techniques   Extraction: Technique 4.5c

Microscale Recrystallization: Technique 5.5b

— **SAFETY INFORMATION** —

Benzaldehyde is highly toxic. Wear gloves while using it.

Hexane is extremely flammable. Heat it only on a steam bath.

Ether is also extremely flammable.

Prepare a solution of 138 mg (1.3 mmol) of benzaldehyde and 0.40 mL of anhydrous ether in a tared, oven-dried vial. Immediately draw this solution into the same syringe used in the preparation of phenylmagnesium bromide. Insert the syringe needle through the septum in the Claisen adapter. Rinse the vial with 0.20 mL of ether and set it aside. Add the benzaldehyde solution in small portions to the stirred phenylmagnesium bromide solution over a period of 2–3 min. Draw the rinse solution into the syringe and inject it in one portion into the reaction mixture. Continue to stir the reaction mixture for 5 min or until it reaches room temperature.

Remove the Claisen adapter assembly and place the reaction vial in a small beaker of ice water. Add 5–6 drops of water to hydrolyze the alkoxide magnesium salt and stir the reaction mixture for 5 min. Add 2.0 mL of ice-cold 1 M sulfuric acid dropwise. Continue stirring until all remaining magnesium particles dissolve and both phases become clear solutions. Remove the spin vane with a pair of tweezers. Rinse the spin vane with a few drops of ether while holding it above the conical vial. If most of the ether has evaporated, add an additional 1.0 mL of ether. Remove the lower aqueous phase with a Pasteur filter pipet [see Technique 2.4 and Technique 4.5c]. Dry the ether solution with 200 mg of anhydrous potassium carbonate for 10 min.

Additional drying agent may be used, if necessary.

Transfer the dried ether solution to a tared Craig tube in 0.5-mL portions, using a clean Pasteur filter pipet. After each transfer, evaporate the ether in a hood, using a gentle stream of nitrogen. Rinse the drying agent with 0.5 mL of ether and transfer the rinse to the Craig tube; completely evaporate all of the ether. Weigh the Craig tube containing the dried product to determine the yield of crude product.

Recrystallize the crude product in a Craig tube, using a minimum volume of warm hexane. Put about 2 mL of hexane in a small test tube with a boiling stick and heat the test tube in a warm-water bath (55–60°C), using a steam bath as the heat source. Add a boiling stick to the Craig tube, place the Craig tube in the same water bath, and add warm hexane dropwise until the crystals just dissolve. Remove the Craig tube from the water bath, remove the boiling stick and insert the plug, and set the tube in a 25-mL Erlenmeyer flask to cool slowly [see Technique 5.5b]. Once crystallization is complete at room temperature, cool the Craig tube in an ice bath for 5 min before removing the solvent by centrifugation. Allow the recrystallized benzhydrol to dry completely before determining the mass, the percent yield, and the melting point. Determine the IR or NMR spectrum of your product as directed by your instructor.

Cleanup: Neutralize the aqueous solution remaining from the extractions with solid sodium carbonate before washing it down the sink or pouring it into the container for aqueous inorganic

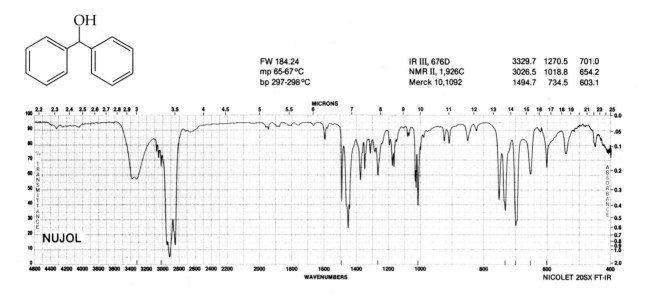

FW 184.24
mp 65-67°C
bp 297-298°C

IR III, 676D
NMR II, 1,926C
Merck 10,1092

3329.7	1270.5	701.0
3026.5	1018.8	654.2
1494.7	734.5	603.1

FIGURE 13.1 IR spectrum of benzhydrol. (Provided by Aldrich Chemical Company, Inc., Milwaukee, WI.)

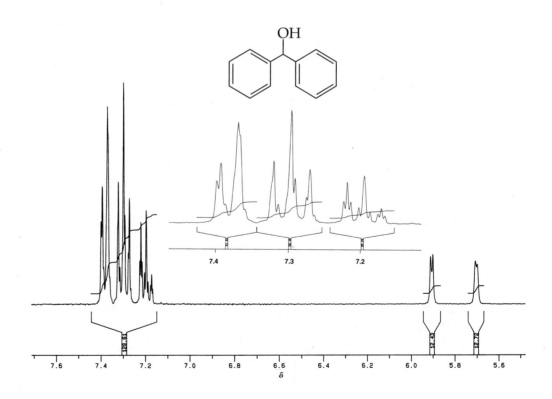

FIGURE 13.2 NMR spectrum (300 MHz) of benzhydrol.

waste. Place the spent drying agent in the container for nonhazardous waste or inorganic waste. Place the hexane filtrate remaining in the centrifuge tube into the container for flammable (organic) waste.

(13.1b) Synthesis of Triphenylmethanol

Demonstrate how two moles of a Grignard reagent convert an ester to a tertiary alcohol; purify and characterize the product.

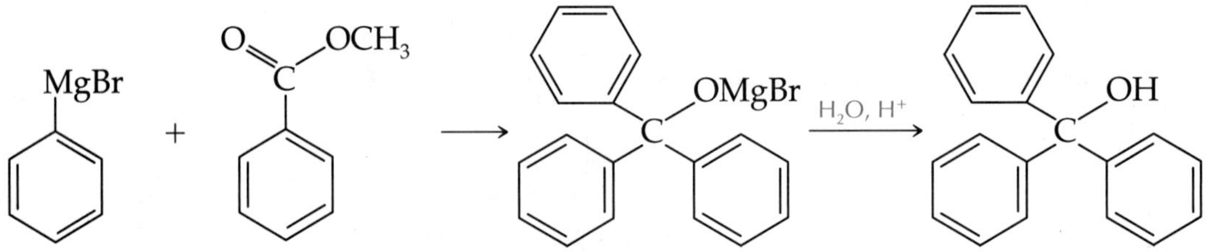

Phenylmagnesium bromide

Methyl benzoate
bp 199.5°C
MW 136
density 1.09 g · mL^{-1}

Triphenylmethanol
(triphenylcarbonal)
mp 162.5°C
MW 260.3

Treatment of methyl benzoate with excess phenylmagnesium bromide gives, after acid hydrolysis, triphenylmethanol (triphenylcarbinol).

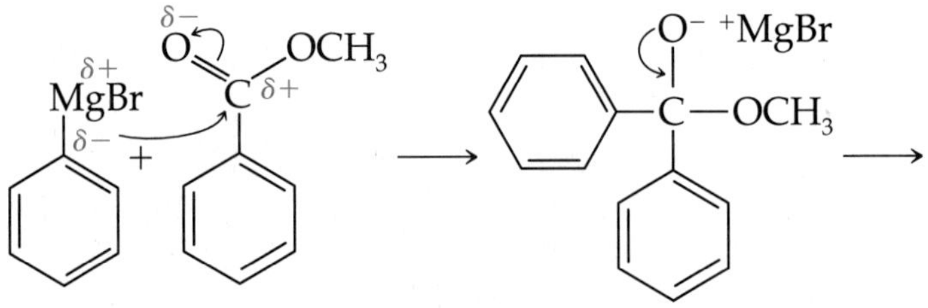

Methyl benzoate

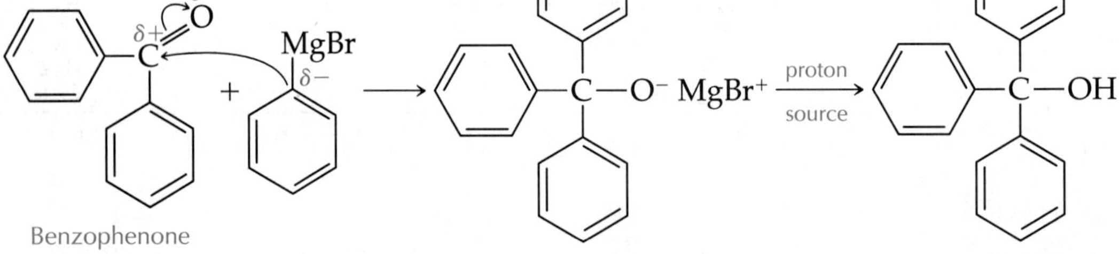

Benzophenone
(a ketone)

Triphenylmethanol

This reaction takes advantage of the high nucleophilicity of the phenyl portion of the Grignard reagent and the electrophilicity of the carbonyl carbon atom of the ester. The first equivalent of Grignard reagent displaces the ethoxy group from the ester, converting it to the ketone benzophenone. This ketone cannot be isolated under the conditions of the reaction, because it is quickly converted to the magnesium salt of the tertiary alcohol product by a second equivalent of Grignard reagent.

You will use a mixed solvent recrystallization to separate the major by-product, biphenyl, from triphenylmethanol. This method makes use of the fact that biphenyl is quite soluble in hexane whereas the desired product, triphenylmethanol, is not. By dissolving the triphenylmethanol in a minimum amount of dichloromethane, a solvent in which it is very soluble, then adding an excess of hexane, the triphenylmethanol will crystallize from the solution while the relatively soluble biphenyl remains dissolved in the solution.

Macroscale Procedure

Techniques   Extraction: Technique 4.2

Recrystallization: Technique 5.3

— *SAFETY INFORMATION* —

Carry out the extractions, distillation, and recrystallization in a hood. Use a steam bath as the heat source for the distillation and recrystallization.

Methyl benzoate is an irritant. Avoid contact with skin, eyes, and clothing.

Hexane is extremely flammable. Heat it only on a steam bath.

Ether is also extremely flammable.

Dichloromethane is toxic, an irritant, absorbed through the skin, and harmful if inhaled. Use it only in a hood and wear gloves while doing the recrystallization and filtration.

Cool the flask containing the phenylmagnesium bromide briefly with an ice bath. Place a solution made from 4.5 mL (4.9 g, 0.036 mol) of methyl benzoate and 10 mL of anhydrous ether in the separatory funnel and begin dropwise addition of the solution to the phenylmagnesium bromide solution. Reaction of phenylmagnesium bromide and methyl benzoate is quite vigorous and produces large amounts of a white precipitate. After the addition is complete, reflux the mixture for 15 min, using a water bath at 45–50°C.

Cool the reaction flask in an ice-water bath. Remove the condenser and separatory funnel. Place 50 mL of 1 M sulfuric acid and about 50 g of ice in a 250-mL Erlenmeyer flask. Set aside about 10 mL of this cold aqueous acid solution in a graduated cylinder before pouring the cooled reaction mixture into the flask containing the acid/ice mixture. Rinse the reaction flask with the reserved acid solution; you may also need to add about 10 mL of ether to dissolve all the material remaining in the flask. Add this rinse to the Erlenmeyer flask containing the reaction mixture and acid. Stir the mixture until the salts and bits of unreacted magnesium dissolve. It may be necessary to add additional ether to maintain a volume of approximately 50 mL.

Ether added during the hydrolysis or for the extractions does not have to be anhydrous.

When the hydrolysis of the salts is complete, the milkiness in the ether layer will disappear. Transfer the mixture to a separatory funnel and remove the lower aqueous phase [see Technique 4.2]. Wash the ether layer with 20 mL of water, then with 20 mL of saturated sodium bicarbonate solution. **(Caution: Foaming; vent the funnel frequently.)** Lastly, wash the ether layer with 25 mL of saturated sodium chloride; remove the lower NaCl solution and pour the organic phase into a clean 125-mL Erlenmeyer flask. Dry the ether/product solution with anhydrous potassium carbonate for 10–15 min.

This is a convenient stopping point. Cork the flask tightly and store it as directed by your instructor.

Filter the dried solution through a conical funnel containing a small cotton plug into a dry 250-mL round-bottomed flask [see Technique 4.7]. Add 30 mL of hexane to the flask. Set up a simple distillation apparatus, using a steam bath as the heat source, and slowly distill the ether/hexane solvent until crystals of triphenylmethanol begin to form. At this point, stop the distillation, remove the distilling head and condenser, and cool the flask to room temperature. Complete the crystallization process in an ice-water bath before collecting the product by vacuum filtration. Use the vacuum line to pull air over the crystals to dry them. You can recover a second crop of product by partially evaporating the filtrate on a steam bath in a hood. Set aside a few milligrams of crude product for a melting-point determination.

Recrystallize the crude triphenylmethanol, using a mixed solvent of dichloromethane and hexane [see Technique 5.2]. Obtain 20 mL of dichloromethane in a graduated cylinder. Place the crude product in a 125-mL Erlenmeyer flask, add a boiling stick, and dissolve the crystals in the smallest possible amount of boiling dichloromethane, using a steam bath as the heat source. Note the amount of dichloromethane that was used to dissolve the product and obtain four times that volume of hexane. Add the hexane to the hot dichloromethane solution. If crystals start to form, warm the mixture briefly to dissolve them, then set the flask on the desk top until it cools to room temperature. Cool the flask in an ice-water bath and collect the purified product by vacuum filtration.

Determine the melting point of both the crude and the purified product, weigh your product, and calculate the percent yield. Obtain an IR spectrum or an NMR spectrum of your purified product as directed by your instructor. Submit your product to your instructor.

Cleanup: Neutralize the aqueous solution remaining from the extractions with solid sodium carbonate before washing it down the sink or pouring it into the container for aqueous inorganic waste. Place the spent drying agent in the container for non-hazardous waste or inorganic waste. Place the recovered ether/hexane from the distillation and the filtrate from the crude product in the container for flammable (organic) waste. Place the dichloromethane/hexane filtrate from the purified product in the container for halogenated waste.

(*13.1c*) Synthesis of Benzoic Acid

Investigate the addition of phenylmagnesium bromide to carbon dioxide to produce benzoic acid.

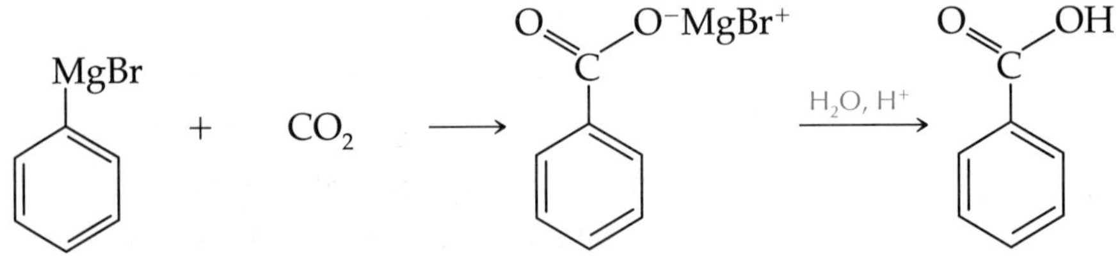

Phenylmagnesium   Carbon dioxide   Benzoic acid
bromide   (dry ice)   mp 122.4°C
                     mp −78.5°C   MW 122
                      MW 44

Treatment of excess dry ice (CO_2) with phenylmagnesium bromide gives, after acid hydrolysis, benzoic acid. This reaction takes advantage of the high nucleophilicity of the phenyl portion of the Grignard reagent and the electrophilicity of the carbon aom of CO_2. The Grignard reagent attacks the carbon of CO_2 to form a carboxylate salt that is readily converted to benzoic acid by acidification with hydrochloric acid. The use of dry ice as the source of carbon dioxide helps regulate the reaction, because the extremely low temperature of this solid moderates the usual high exothermicity of Grignard additions.

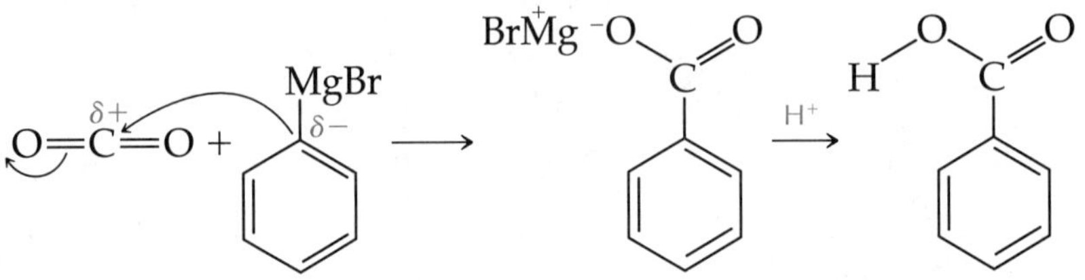

Benzoic acid

microscale
Procedure

Techniques   Extraction: Technique 4.5d
                 Microscale Recrystallization: Technique 5.5b

─ SAFETY INFORMATION ─

Handle solid CO_2, dry ice, with a towel or wear heavy cloth gloves. Contact with the skin can cause frostbite.

Diethyl ether is extremely flammable.

Aqueous hydrochloric acid solutions are a skin irritant. Wash your hands thoroughly if any acid spills on your hands. Aqueous sodium hydroxide solutions are corrosive and cause burns. Solutions as dilute as 2.5 M can cause severe eye injury.

Any moisture on the dry ice will lower your yield of benzoic acid.

Obtain 1.0 mL of anhydrous ether in a test tube fitted with a cork. Obtain a small piece of dry ice (solid CO_2), wipe the condensed frost off the surface with a dry towel, and place the piece in a dry 30-mL beaker. Open the conical vial containing the phenylmagnesium bromide prepared in Section 13.1, quickly draw the solution into a dry Pasteur pipet, and drip the solution over the dry ice. Using the same Pasteur pipet, rinse the conical vial with the ether in the test tube and transfer the rinse to the beaker containing the dry ice. Stir the contents of the beaker occasionally until the excess dry ice sublimes.

Add 2.0 mL of 3 M hydrochloric acid and stir the reaction mixture to hydrolyze the magnesium salt. Transfer the mixture to a 5-mL conical vial, using a Pasteur pipet. Rinse the beaker with 0.5 mL of ether and add this rinse to the conical vial. Cap the vial and shake it thoroughly. This procedure should produce two clear layers. It may be necessary to add a few more drops of ether or hydrochloric acid. Remove the lower aqueous layer with a Pasteur filter pipet [see Technique 4.5d]. Wash the ether phase with 1.0 mL of water; remove the water layer to the container that already holds the previous aqueous phase and set the container aside, leaving the ether layer in the conical vial.

Extract the ether layer with 0.7 mL of 3 M sodium hydroxide solution by capping the vial and shaking the mixture thoroughly. Remove the lower aqueous phase to a 30-mL beaker. Repeat the extraction procedure with 0.5 mL of 3 M NaOH, then with 0.5 mL of water, each time transferring the lower aqueous phase to the 30-mL beaker containing the first sodium hydroxide extract. Set aside the ether layer remaining in the conical vial; it contains some biphenyl, a reaction by-product.

Warm the beaker containing the sodium hydroxide extracts for several minutes on a steam bath in a hood to expel any dissolved ether. Cool the solution to room temperature and add 6 M hydrochloric acid dropwise until the solution is acidic to pHydrion paper (pH 2–3), and a thick white precipitate of benzoic acid forms. Chill the mixture in an ice-water bath before collecting the benzoic acid by vacuum filtration on a Hirsch funnel [see Technique 5.5a]. Wash the crystals twice with 1-mL portions of ice-cold water. Allow the benzoic acid to dry thoroughly before determining the melting point and percent yield. Determine the IR or NMR spectrum as directed by your instructor.

Cleanup: Combine the aqueous extracts and the aqueous filtrate; neutralize the solution with sodium carbonate before washing it down the sink or pouring it into the container for aqueous inorganic waste. Pour the ether remaining in the conical vial into the container for flammable (organic) waste.

References

1. Grignard, V. (translated by P. R. Jones and E. Southwick) *J. Chem. Educ.* **1970,** *47,* 290–299.
2. Ashby, E. C.; Laemmle, J.; Newmann, H. M. *Accts. Chem. Res.* **1974,** *7,* 272–280.

Questions

1. What is the white precipitate that forms when benzaldehyde is added to the solution of phenylmagnesium bromide?
2. Which of the following will undergo reaction with a Grignard reagent? What will the products be if a reaction occurs?

$$\text{(a) } (CH_3)_2CHC\overset{\displaystyle O}{\overset{\displaystyle \|}{H}} \qquad \text{(d) } CH_3\overset{\displaystyle O}{\overset{\displaystyle \|}{C}}OCH_2CH_3$$

$$\text{(b) } CH_3\overset{\displaystyle O}{\overset{\displaystyle \|}{C}}CH_2CH_3 \qquad \text{(e) } CH_3CH_2CH_2CH_2CH_3$$

(c) CH_3NH_2 (f) CH_3CH_2Cl

3. Write an equation showing how benzoic acid would react with phenylmagnesium bromide.
4. Suggest organometallic syntheses of the following materials.

(a) 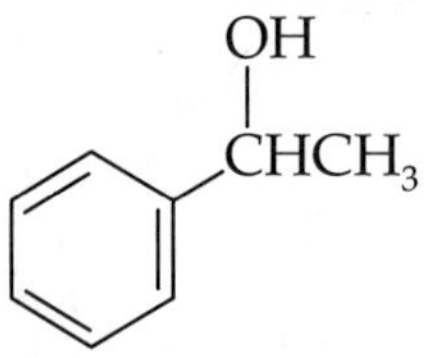 [phenyl]–CHCH₃ with OH

(d) 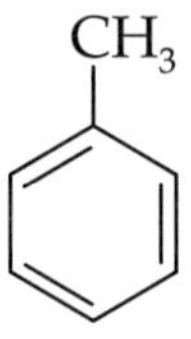 [phenyl]–CH₃

(b) CH₃CHCH₃ with OH

(c) [phenyl]–CH₂COOH

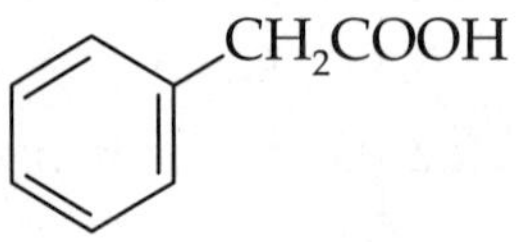

(e) 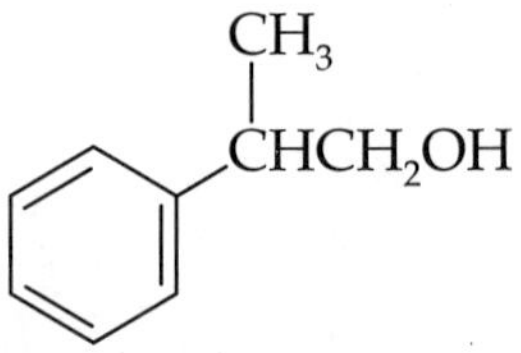 [phenyl]–CHCH₂OH with CH₃

13.2 Grignard Synthesis of a Triphenylmethane Dye: Crystal Violet or Malachite Green

Use Grignard reactions to synthesize dyes and test the dyes on various fabrics.

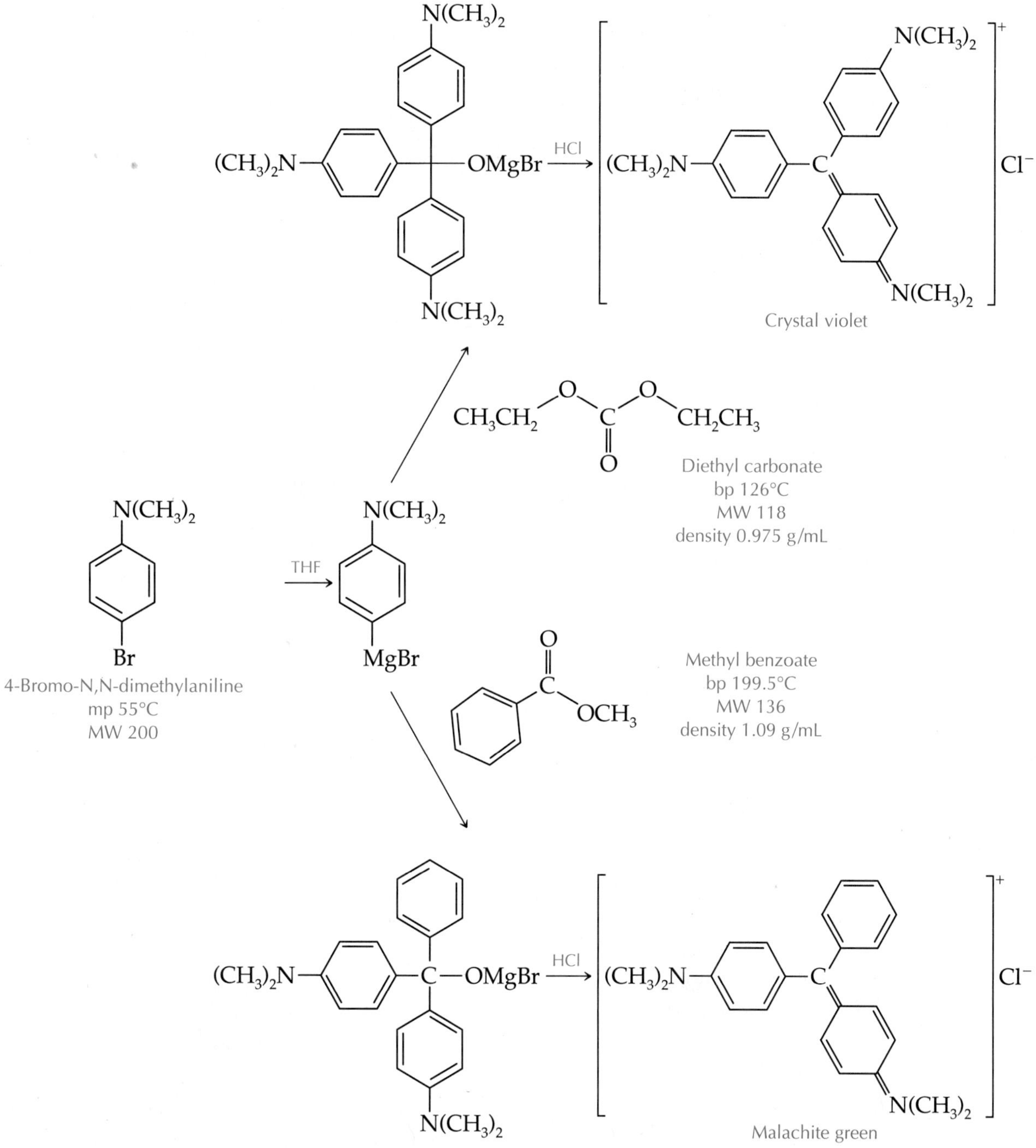

N(CH3)2
(CH3)2N
OMgBr
HCl
Cl−
Crystal violet
CH3CH2
O
C
CH2CH3
Diethyl carbonate
bp 126°C
MW 118
density 0.975 g/mL
Br
4-Bromo-N,N-dimethylaniline
mp 55°C
MW 200
THF
MgBr
OCH3
Methyl benzoate
bp 199.5°C
MW 136
density 1.09 g/mL
Malachite green

Crystal violet and malachite green are two highly colored compounds that can be made by multiple Grignard reactions. Crystal violet is also known as gentian violet. This compound is highly soluble in polar solvents (including water) and has the interesting property that although it is a deep purple in solution, it is a green solid. Crystal violet has been used as a dye, a stain, and an ink and has antiseptic properties as well.

Malachite green is also known as aniline green, China green, and Victoria green. This compound is also highly soluble in polar solvents such as water, and although in solid form it is green, in solution it is blue-green. It has been used as a dye, a stain, and an antiseptic.

Crystal violet is formed by the Grignard reaction between *p*-dimethylaminophenylmagnesium bromide and diethyl carbonate.

Diethyl carbonate, an ester of carbonic acid, undergoes two nucleophilic displacements to form a diaryl ketone. This ketone cannot be isolated and rapidly undergoes addition of a third mole of the arylmagnesium compound to form a magnesium salt, which when acidified gives the tertiary alcohol. This, in turn, in the presence of HCl loses a molecule of water and forms a highly stable colored carbocation.

An important aspect of both malachite green and crystal violet is the fact that they are highly conjugated. The conjugation is responsible for the colors of these dyes. The positive charge is extensively delocalized due to the overlap of the orbitals on the cationic center with the conjugated π-system of the benzene rings and even the *para*-dimethylamino groups.

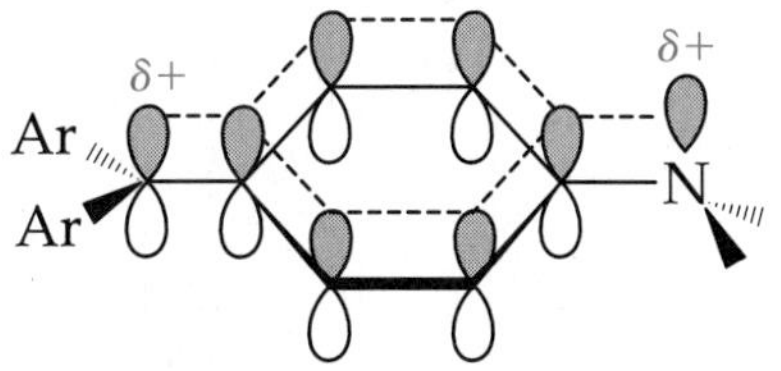

You may choose which compound you wish to synthesize. Both syntheses start with the preparation of the Grignard reagent from 4-bromo-N,N-dimethylaniline. Crystal violet is formed when three moles of the Grignard reagent add to one mole of the ester, diethyl carbonate. Malachite green is synthesized in a similar fashion by adding two moles of the Grignard reagent to one mole of the ester, methyl benzoate.

⁄⁄⁄⁄⁄Macroscale Procedure

Technique   Anhydrous Reaction Conditions: Technique 3.3a.

— *SAFETY INFORMATION* —

4-Bromo-N,N-dimethylaniline is toxic and an irritant. Wear gloves while weighing it.

Tetrahydrofuran (THF) is very flammable.

Diethyl carbonate and methyl benzoate are irritants. Avoid contact with skin, eyes and clothing.

If possible, conduct the experiment in a hood.

(continued on next page)

The Grignard reagent is water sensitive. All glassware used for this experiment must be dry. Once the glassware is rinsed with THF, the reagents need to be added quickly and the apparatus assembled to minimize exposure to atmospheric moisture.

Prepare a drying tube filled with anhydrous calcium chloride [see Technique 3.3a]. Obtain the following reagents before setting up the reaction: 2.5 g of 4-bromo-N,N-dimethylaniline, 0.40 g Mg turnings, 30 mL of dry tetrahydrofuran (THF), and 2 or 3 small crystals of iodine.

Keep the reagent bottle of THF and your flask containing THF tightly closed. THF rapidly absorbs water from the atmosphere.

Rinse a 50-mL round-bottomed flask with 2–3 mL of THF, pour 2–3 mL of THF through the condenser, and discard these rinses in the flammable waste container. Clamp the round-bottomed flask in place, and add the bromoaniline, the remaining 22–23 mL of THF, the Mg turnings, and the I_2 crystals. With a clean, dry stirring rod, gently crack one or two Mg turnings to expose a fresh surface, which helps initiate the reaction.

— *SAFETY PRECAUTION* —

Be careful not to press so hard on a magnesium turning that you crack the round-bottomed flask.

Fit the condenser to the flask and swirl the mixture. Adjust the water so that it flows gently through the condenser, and place the drying tube at the top of the condenser. Heat the reaction with a 70–75°C water bath. Maintain a gentle reflux for 30 minutes; swirl the flask every 5 minutes during the heating period. The initial dark color fades and is replaced by the greyish solution typical of

Grignard reagents. Cool the reaction flask in a beaker of tap water until it reaches room temperature.

Choose the dye you wish to make.

For malachite green: Weigh 0.21 g of methyl benzoate into a small vial (set in a 30 mL beaker so it will not tip over). Add 1.0 mL of THF to the vial. **For crystal violet:** Weigh 0.30 g of diethyl carbonate into a small vial (set in a 30 mL beaker so that it will not tip over). Add 1.0 mL of THF to the vial.

Remove the condenser. Using a Pasteur pipet, add the ester solution drop by drop, swirling after each drop. After the addition is complete, replace the condenser and heat the reaction mixture to reflux for 5 minutes. Swirl the flask occasionally while heating it. Cool the flask to room temperature.

Dyeing Test Samples

Put on gloves before starting the dye synthesis. Pour the reaction mixture into a 150-mL beaker. Slowly add 5 mL of 5% HCl solution to the beaker with stirring; some bubbling will occur as the residual magnesium reacts with the acid.

Test fabric strips 10 cm wide, containing a variety of fibers, are available from TestFabrics, Inc., 200 Blackford Avenue, P.O. Box 420, Middlesex, NJ, 08846 (phone: 908-469-6446).

A variety of fabric samples will be available. Dip each sample in the dye solution using a pair of forceps and leave it at least 1 minute. Remove the sample with forceps, rinse it with tap water, and blot it dry. The dye solution is very concentrated, so intense color should be produced on the fabrics, depending on how well the particular type of fiber accepts the dye. Record the types of fabrics you tested and describe any variations in intensity observed for different fibers. Compare your results with those of another student who synthesized the alternative dye.

Allow the samples to dry and attach them to your report. Keep them in a small plastic bag or wrapped in paper because the dye may rub off on anything the samples touch.

UV-Visible Spectrum of the Dye

Touch the tip of a stirring rod to the dye solution and then dip it in a beaker containing 50 mL of distilled water. Stir to mix thoroughly. Fill a sample cuvette, and determine the UV-visible spectrum using distilled water as the reference solvent. Determine the wavelength(s) of maximum absorbance.

Cleanup: The dye solution should be poured in the waste container labeled "Dye Solution." The aqueous solution used for the UV-VIS spectrum is very dilute and can be flushed down the drain. Dye stains on glassware can be removed with a few milliliters of 6 M HCl then washing with water. Neutralize the acid washings with sodium carbonate before pouring them down the sink or pouring them into the container for aqueous inorganic waste.

Reference

1. Taber, D. F.; Meagley, R. P.; Supplee, D. *J. Chem. Educ.* **1996,** *73,* 259–260.

Questions

1. Why does it require three moles of the Grignard reagent for each mole of ester in the synthesis of crystal violet, whereas only two moles of the Grignard reagent are required for each mole of ester in the synthesis of malachite green?
2. What precipitate formed when the ester was added to the solution of your Grignard reagent?
3. What advantage(s) does tetrahydrofuran (THF) have over diethyl ether as a solvent in this experiment?
4. What is the correlation of the color of your dye solution with the UV-VIS spectrum of the solution?

Experiment 14
PHASE TRANSFER CATALYSIS

Discover how quaternary ammonium salts can be used to catalyze a variety of organic reactions.

Chemical reactions tend to occur most readily if the reagents and solvents form a homogeneous reaction medium. This condition of homogeneity is often difficult to attain, however, when the system involves nonpolar organic substrates and polar inorganic reagents. We usually find that the inorganic reagent is soluble in a highly polar solvent like water but not in many organic solvents, whereas the organic compound is soluble in the nonpolar organic solvents but not in water.

One approach to achieving a homogeneous reaction medium has been to use a mixture of two miscible solvents of quite different polarities, such as the organic solvent tetrahydrofuran (THF) and water. This method requires that the solvents readily dissolve in each other, a situation that is certainly the case for THF and water but that does not occur for many common organic solvents.

Another problem with the use of a mixed solvent is its environmental cost. Often it is very expensive to recover organic solvents that are miscible with water, so they form waste that may lead to pollution. How much simpler and safer if we could just use water and an immiscible organic solvent.

The development of phase transfer catalysis (PTC) over the last three decades has provided a very useful alternative to mixed solvents for reactions involving organic and inorganic reagents. *Phase transfer catalysts* are reagents that undergo reactions with inorganic salts, forming new compounds that are soluble to an appreciable degree in both polar and nonpolar solvents. These new compounds can move from one liquid phase to another, an ability that allows useful organic reagents to be moved from one liquid phase to another.

The quaternary ammonium halides, *tetra-N-butylammonium bromide* and trioctylmethylammonium chloride (abbreviated Q^+X^-) represent common types of phase transfer catalysts:

$$(CH_3CH_2CH_2CH_2)_4N^+Br^- \qquad [CH_3(CH_2)_7]_3NCH_3^+Cl^-$$
$$Q^+Br^- \qquad\qquad\qquad Q^+Cl^-$$

These "quat" halides are two of many quaternary ammonium halides that display useful PTC properties. Aliquat 336, Q^+Cl^-, is the phase transfer catalyst used in Experiments 14.1 and 14.2. In

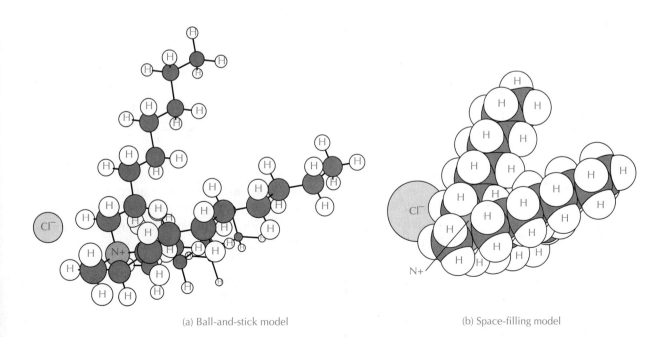

(a) Ball-and-stick model   (b) Space-filling model

FIGURE 14.1 Representations of Aliquat 336, trioctylmethylammonium chloride. The only unlabeled atoms are carbons.

both these compounds, the nitrogen atom is tetravalent and bears a formal positive charge for the same reason as the ammonium ion (NH_4^+) does. In addition, the compounds have four nonpolar alkyl groups. Thus, a quaternary compound of this type has both ionic and nonpolar character, properties that allow it to dissolve in both polar and nonpolar solvents (Figure 14.1).

14.1

Williamson Ether Synthesis of Bis-4-Chlorobenzyl Ether Using Phase Transfer Catalysis

Use a phase transfer catalyst, which combines polar and nonpolar characteristics, in the S_N2 synthesis of an ether under microscale conditions.

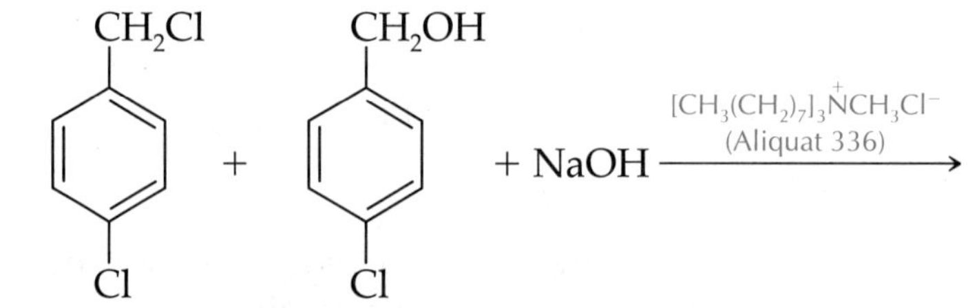

4-Chlorobenzyl
chloride
mp 30°C
MW 161

4-Chlorobenzyl
alcohol
mp 72°C
MW 142.6

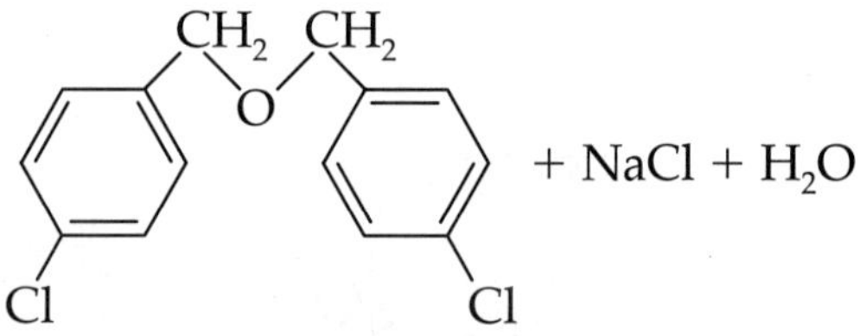

Bis-4-chlorobenzyl ether
mp 54°C
MW 267

The preparation of ethers from alcohols or from derivatives of alcohols has already been described in Experiment 12. Here you will use a phase transfer catalyst, trioctylmethylammonium chloride (Aliquat 336), to increase the efficiency of an ether synthesis:

$$[CH_3(CH_2)_7]_3\overset{+}{N}CH_3\ Cl^-$$

Aliquat 336
(tricaprylmethylammonium chloride
or methyltrioctylammonium chloride)

A discussion of how quaternary ammonium salts are used as phase transfer catalysts is given in the introduction to Experiment

The solvent appears in parentheses in these equations.

14. In this synthesis, Aliquat 336 acts as a transfer agent to move the basic catalyst (OH⁻) from the aqueous layer to the dichloromethane (organic) phase. The following sequence of reactions occurs:

1. Ion exchange in the aqueous layer:

$$R_3\overset{+}{N}CH_3\,Cl^-\,(H_2O) + Na^+\,OH^-\,(H_2O) \rightleftharpoons R_3\overset{+}{N}CH_3\,OH^-\,(H_2O) + Na^+Cl^-\,(H_2O)$$

2. Phase transfer at the interface:

$$R_3\overset{+}{N}CH_3\,OH^-\,(H_2O) + CH_2Cl_2 \rightleftharpoons R_3\overset{+}{N}CH_3\,OH^-\,(CH_2Cl_2) + H_2O$$

3. Formation of the conjugate base of the benzyl alcohol:

$$R_3\overset{+}{N}CH_3\,OH^-\,(CH_2Cl_2) + 4\text{-}ClC_6H_4CH_2OH \rightleftharpoons 4\text{-}ClC_6H_4CH_2O^-\,R_3\overset{+}{N}CH_3 + H_2O$$

4. The Williamson (S$_N$2) reaction in the CH$_2$Cl$_2$ layer:

$$4\text{-}ClC_6H_4CH_2O^-\,R_3\overset{+}{N}CH_3 + 4\text{-}ClC_6H_4CH_2Cl \longrightarrow 4\text{-}ClC_6H_4CH_2OCH_2C_6H_4Cl + R_3\overset{+}{N}CH_3\,Cl^-$$

5. Catalyst recycling:

$$R_3\overset{+}{N}CH_3\,Cl^-\,(CH_2Cl_2) + H_2O \rightleftharpoons R_3\overset{+}{N}CH_3\,Cl^-\,(H_2O) + CH_2Cl_2$$

Steps 1–5 repeat continuously until all the benzyl alcohol has been consumed.

microscale Procedure

Techniques Microscale Extractions: Technique 4.5b, Method B

Microscale Recrystallization: Technique 5.5

Pasteur Filter Pipets: Technique 2.4

--- SAFETY INFORMATION ---

Wear gloves and, if possible, conduct the entire experiment in a hood.

4-Chlorobenzyl chloride is corrosive, a severe skin irritant, and a lachrymator. Wear gloves. Avoid contact with skin, eyes, and clothing, and avoid inhaling its vapors. If possible, weigh the amount needed in a hood. Wear gloves while using this compound.

(continued on next page)

> ── *SAFETY INFORMATION (continued)* ────
>
> **4-Chlorobenzyl alcohol** may be a skin and eye irritant. Avoid contact with skin, eyes, and clothing.
>
> Aqueous **sodium hydroxide** solutions are corrosive and cause burns. Solutions as dilute as 9% (2.5 M) can cause severe eye injury. You will be using a 30% solution in this experiment.
>
> **Dichloromethane** is toxic, an irritant, absorbed through the skin, and harmful if inhaled. Use it only in a hood. Wash your hands thoroughly after handling it.

Remember to set the conical vial in a small beaker so that it does not overturn while you add the reagents.

Place an aluminum heating block on a hot plate and heat the block to 120°C. Put 1 drop of trioctylmethylammonium chloride (Aliquat 336) in a 5-mL conical vial. Add 200 mg of 4-chlorobenzyl chloride, 170 mg of 4-chlorobenzyl alcohol, and 0.6 mL of 30% sodium hydroxide. Put a boiling chip in the vial and attach a water-cooled reflux condenser [see Technique 3.3]. Heat the reaction mixture under reflux for 45 min. At the end of the reflux period, cool the reaction vial to room temperature.

Add 1.5 mL of water and 2 mL of dichloromethane to the reaction mixture. Cap the vial and shake it thoroughly to mix the two phases. Use a Pasteur filter pipet [see Technique 2.4] to transfer the lower organic phase to a clean 5-mL conical vial [see Technique 4.5b, Method B]. Wash the organic phase with 0.6 mL water, then 0.6 mL 10% hydrochloric acid, and last with 0.6 mL half-saturated (3 M) sodium chloride solution. For each wash, cap the vial, shake it to mix the contents, and transfer the organic phase to another conical vial before adding the next aqueous reagent. After the last wash, transfer the dichloromethane solution to a clean, dry 13 × 100 mm test tube and add 200 mg of anhydrous sodium sulfate. Cork the test tube and allow the solution to dry for 10 min.

A Craig tube is too small to hold all the product solution. Therefore, the solvent will be evaporated in two steps using half of the solution each time.

Using a clean Pasteur filter pipet, transfer *half* of the dichloromethane solution to the tube of the Craig apparatus. In a hood, remove the dichloromethane with a stream of nitrogen. Then transfer the remaining dichloromethane solution to the Craig tube and again remove the solvent with a stream of nitrogen.

Recrystallize the crude product in a minimum volume of methanol in the Craig tube [see Technique 5.5b]. Start by adding 0.5 mL of methanol and a boiling stick to the Craig tube. Place the tube in an aluminum block heated to 70°C. Continue adding methanol in dropwise increments until the crude product dissolves completely. Put the plug into the tube and set the apparatus in a 25-mL Erlenmeyer flask to cool slowly. When crystal growth has ceased, cool the tube in an ice-water bath for 5 to 10 min. Centrifuge the Craig apparatus to remove the supernatant liquid from the crystals.

Allow the product to dry before determining the mass and the melting point. Calculate your percent yield. Take an NMR or IR spectrum of your product as directed by your instructor.

Cleanup: Combine all the aqueous solutions, check the pH with test paper, and add either 10% hydrochloric acid or solid sodium bicarbonate until the solution is neutral to the pH test paper

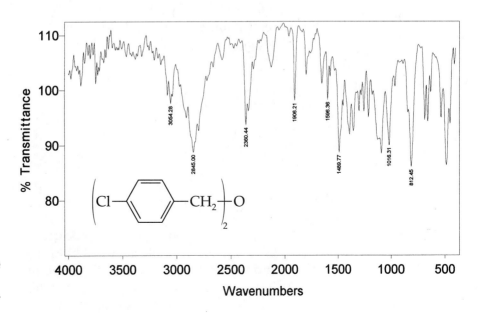

Spectrum run on a sample dispensed on KBr using the diffuse reflectance technique on ATI Mattson Genesis Mattson Series Spectrometer.

FIGURE 14.2 IR spectrum of bis-(4-chlorobenzyl) ether.

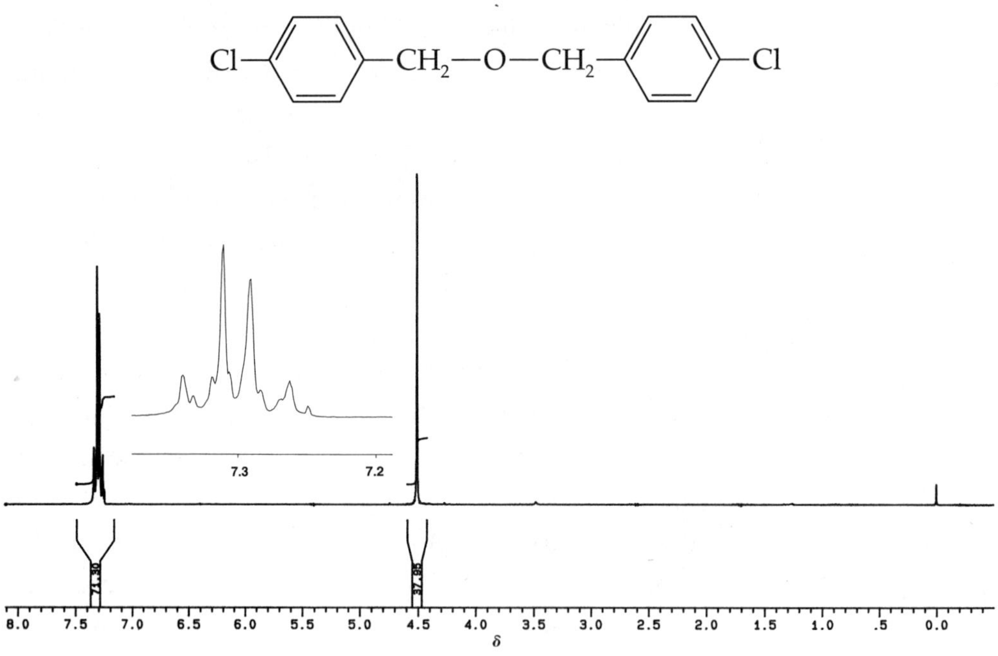

FIGURE 14.3 ^{1}H NMR spectrum (300 mHz) of bis-4-chlorobenzyl ether.

before pouring the solution down the sink or into the container for aqueous inorganic waste. The supernatant liquid from the recrystallization should be poured into the container for halogenated waste.

Reference

1. Hammond, C. N.; Tremelling, M. J. *J. Chem. Educ.* **1987,** *64,* 440–441.

Questions

1. What would you expect the product of a Williamson synthesis to be if you used 3-nitrobenzyl chloride and *m*-ethylbenzyl alcohol and the procedure described in Experiment 14.1?
2. Propose a possible side product for this reaction, keeping in mind that 4-chlorobenzyl chloride could be vulnerable to attack by any other nucleophiles available in this reaction system.

Synthesis of Butyl Benzoate Using Phase Transfer Catalysis

Use a phase transfer catalyst, which combines polar and nonpolar characteristics, in an S_N2 reaction to synthesize an ester, which you will characterize by IR spectroscopy.

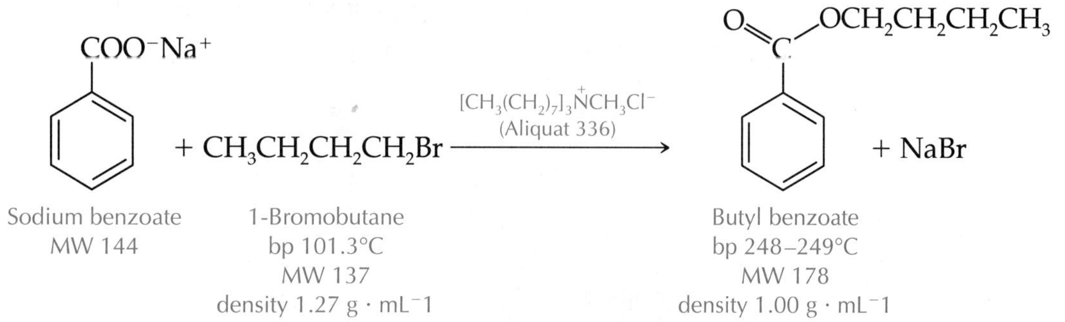

Sodium benzoate	1-Bromobutane	Butyl benzoate
MW 144	bp 101.3°C	bp 248–249°C
	MW 137	MW 178
	density 1.27 g · mL⁻¹	density 1.00 g · mL⁻¹

A large variety of compounds can be prepared by using nucleophilic substitution reactions. A detailed introduction to these reactions is found in Experiment 8, where we describe the specifics of the S_N2 mechanism. A quick review of only those details related to this synthesis is presented here to illustrate the mechanism of ester formation. The nucleophilic species is the benzoate anion, and the desired chemical reaction is the displacement of bromide (a good leaving group):

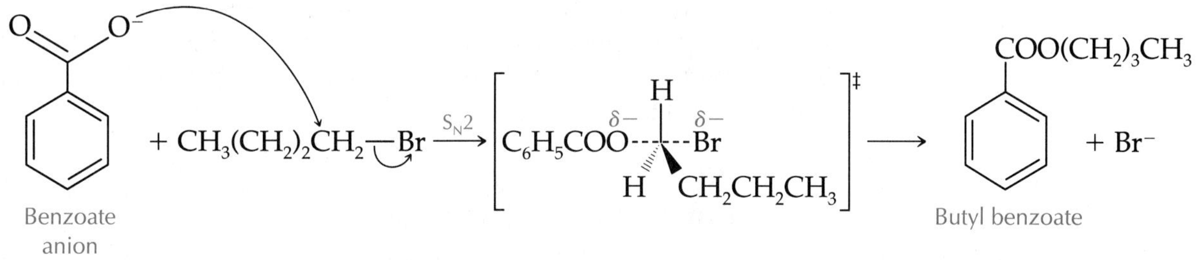

Because we are using two immiscible phases (water and 1-bromobutane), it is important to keep in mind that the ionic salt, sodium benzoate, greatly prefers to be in the aqueous phase and not in the 1-bromobutane layer. Therefore, a phase transfer catalyst is useful because it transports benzoate ion to the organic layer, where an efficient substitution reaction can occur. You will notice that in this experiment no solvents are used other than water and 1-bromobutane, one of the reactants. This tactic eliminates any organic waste materials and is especially environmentally friendly. The phase transfer catalyst employed in this experiment is Aliquat 336, a quaternary ammonium salt. Figure 14.1 (p. 161) shows that this phase transfer catalyst has a very large proportion of nonpolar character.

A discussion of phase transfer catalysts appears in the introduction to Experiment 14. The phase transfer catalyst serves to carry the benzoate anion into the organic layer. Because the benzoate anion exists in large excess relative to chloride anion, we rely on the favorable position of the equilibria in the following reactions:

1. Ion exchange in the water layer:

$$[CH_3(CH_2)_7]_3\overset{+}{N}CH_3\,Cl^-\,(H_2O) + C_6H_5COO^-\,(H_2O) + Na^+ \rightleftharpoons$$

$$[CH_3(CH_2)_7]_3\overset{+}{N}CH_3\,C_6H_5COO^-\,(H_2O) + Na^+\,(H_2O) + Cl^-\,(H_2O)$$

2. Phase transfer from water to the organic layer:

$$[CH_3(CH_2)_7]_3\overset{+}{N}CH_3\,C_6H_5COO^-\,(H_2O) \rightleftharpoons [CH_3(CH_2)_7]_3\overset{+}{N}CH_3\,C_6H_5COO^-\,(org)$$

3. Direct displacement in the organic layer:

$$C_6H_5COO^-\,(org) + CH_3(CH_2)_2CH_2Br \longrightarrow C_6H_5COOCH_2(CH_2)_2CH_3 + Br^-\,(org)$$

Butyl benzoate

4. Phase transfer (catalyst recycling) to the water layer:

$$[CH_3(CH_2)_7]_3\overset{+}{N}CH_3\,Br^-\,(org) \rightleftharpoons [CH_3(CH_2)_7]_3\overset{+}{N}CH_3\,(H_2O) + Br^-\,(H_2O)$$

Steps 1–4 are repeated many times as vigorous stirring facilitates the exchanges between layers. You will use IR spectroscopy [see Spectrometric Method 1] to differentiate the product from the starting alkyl bromide.

Macroscale Procedure

Techniques   Reflux: Technique 3.3
Extraction: Technique 4.2
IR Spectroscopy: Spectrometric Method 1

―― *SAFETY INFORMATION* ――――

1-Bromobutane is a skin irritant and harmful if inhaled, ingested, or absorbed through the skin. Wear gloves and use it in a hood, if possible.

Place 2.0 mL of 1-bromobutane, 3.0 g of sodium benzoate, 5.0 mL of water, 4 drops of Aliquat 336, and a boiling stone in a 50-mL round-bottomed flask. Fit a reflux condenser above the flask and heat the reaction mixture under reflux for 1 h [see Technique 3.3].

Cool the flask in a beaker of room-temperature water. If solid forms in the cooled reaction mixture, add 2–3 mL of water, stopper the flask, and shake the mixture until the solid dissolves. Transfer the cooled contents of the flask to a separatory funnel. Rinse the flask with 15 mL of dichloromethane and add the dichloromethane rinse to the separatory funnel. Add 10 mL of water to the funnel and gently shake it to mix the layers [see Technique 4.2]. Drain the lower organic phase into a clean labeled Erlenmeyer flask; pour the aqueous phase out the top of the funnel into another Erlenmeyer flask. Return the organic phase to the separatory funnel and wash it with 5 mL of 15% NaCl solution. Drain the organic phase into a clean dry Erlenmeyer flask and add a small amount of anhydrous sodium sulfate as the drying agent [see Technique 4.6]. Allow the solution to dry for at least 15 min.

Weigh a dry 25-mL Erlenmeyer flask to the nearest milligram. Prepare a Pasteur filter pipet [see Technique 2.4] and use this pipet to transfer the solution from the drying agent to the 25-mL flask. Working in a hood, remove the dichloromethane with a stream of nitrogen while warming the flask in a beaker of water at 50°C. When the volume is reduced to about 2 mL, dry the outside of the flask and weigh it. Blow nitrogen over the solution for another minute and then weigh the flask again. Continue this process until two weighings agree within 0.025 g.

Determine the IR spectrum of your product. Liquids need to be completely dry before taking an IR spectrum. If your product appears to have any water droplets in it after the evaporation of the dichloromethane, add a few grains of anhydrous sodium sulfate and wait 10 min before determining the IR spectrum. Use a clean Pasteur filter pipet to remove the product from the drying agent. Compare your product spectrum to the IR spectrum of 1-bromobutane shown in Figure 14.4 and determine whether there is evidence of starting material in the product. Analyze the IR

CH₃CH₂CH₂CH₂Br

FW 137.03	d 1.276	IR III, 42H	2961.8	1261.8	740.6
mp -112°C	Fp 75°F	NMR II, 1,60C	1465.4	915.0	644.0
bp 100-104°C	nᵦ 1.4394	Merck 10,1526	1380.5	866.8	563.4

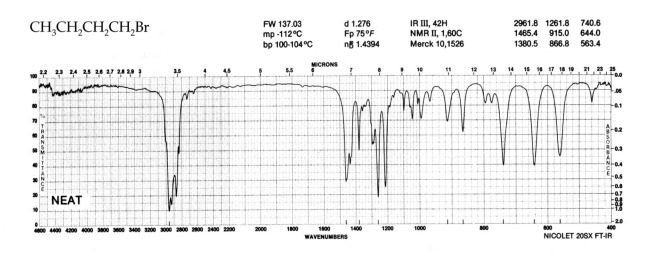

FIGURE 14.4 IR spectrum of 1-bromobutane. (Provided by Aldrich Chemical Company, Inc., Milwaukee, WI.)

$$\text{O}$$
$$\overset{\text{O}}{\overset{\|}{\text{C}}}-\text{OCH}_2\text{CH}_2\text{CH}_2\text{CH}_3$$

FW 178.23	Fp 223°F	3075.0	1456.0	1106.8
bp 249°C	nᵦ 1.4983	2969.7	1269.8	1025.7
d 1.010		1739.3	1178.3	709.3

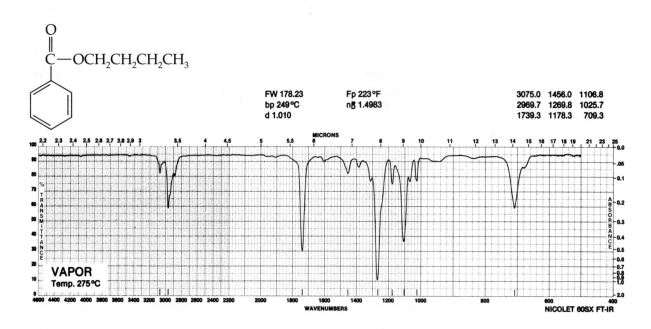

FIGURE 14.5 IR spectrum of butyl benzoate. (Provided by Aldrich Chemical Company, Inc., Milwaukee, WI.)

spectrum of your product and assign its major absorption bands [see Spectrometric Method 1]. The IR spectrum of butyl benzoate is shown in Figure 14.5.

Cleanup: The aqueous solutions from the extractions may be washed down the sink or poured into the container for aqueous inorganic waste. Place the used sodium sulfate in the container for solid waste.

Questions

1. How is the phase transfer catalyst removed from the product?
2. What purpose does washing with 15% (half-saturated) NaCl solution serve?
3. There was a slight excess of sodium benzoate in the reaction. How is this removed from the product?
4. What peak in the IR spectrum suggests the presence of water in the product? Why? What should be done to remove this water?
5. Using the theory of solvent effects on nucleophilic substitution reactions, suggest reasons for the enhanced reactivity of nucleophiles in the presence of phase transfer catalysts.
6. Look (in this book or in a class text) at other examples of syntheses that use PTC reagents and briefly explain why they are successful.

Experiment 15
OXIDATION REACTIONS

Investigate the oxidation of secondary alcohols to ketones, using household bleach, or the oxidation of a primary alcohol to an aldehyde, using pyridinium chlorochromate.

The concept of oxidation-reduction (redox) reactions is usually introduced by examining electron transfer. That is, you learn that the loss of electrons is oxidation and the gain of electrons is reduction ("LEO the lion says GER"). Except for simple ionic reactions, however, these definitions can be difficult to apply. A more general approach uses the concept of the oxidation states of the elements involved in the reaction. Thus, in the following reaction (which is not balanced), Br_2 is reduced and HSO_3^- is oxidized:

$$Br_2 + HSO_3^- \longrightarrow 2\,Br^- + SO_4^{2-}$$

Oxidation states: $Br = 0$ $S = +4$ $Br = -1$ $S = +6$

Another approach to oxidation and reduction, commonly used in organic chemistry, looks for the number of hydrogen atoms in the organic compounds. A simple example is the hydrogenation of alkenes to form alkanes; this reaction illustrates a type of reduction because the organic product has more hydrogen atoms than the reactant does:

$$CH_2{=}CH_2 \xrightleftharpoons[Pt]{H_2/Pt} CH_3{-}CH_3$$
$$(C_2H_4) \hspace{4cm} (C_2H_6)$$

The reverse reaction (dehydrogenation of an alkane to form an alkene) can also be done and is an example of oxidation. A similar approach is to look for the number of oxygen atoms in the structure of each organic compound. The one with more oxygen atoms is more highly oxidized. It is important to be able to scan organic reactions and deduce whether the chemical change is oxidation, reduction, or neither. In the following sequence, the oxidation state of each organic compound increases as we proceed from left to right:

Increasing state of oxidation $\longrightarrow$

$R{-}CH_3$	RCH_2OH	$R(C{=}O)H$	$R(C{=}O)OH$
Alkane	Alcohol	Aldehyde	Carboxylic acid

Every oxidation-reduction (redox) reaction has both a reducing reagent and an oxidizing reagent. In organic synthesis, you will often find that when an organic compound is being oxidized, an inorganic compound is the oxidizing agent. Likewise, in the reduction of an organic compound, an inorganic compound is often the reducing agent.

In this experiment we deal only with oxidations where alcohols with at least one α-hydrogen are oxidized to the corresponding carbonyl compounds:

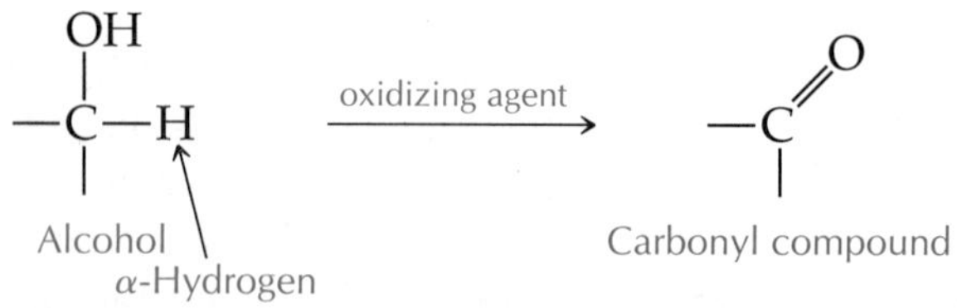

Many alcohols can be oxidized in this way; primary alcohols can be oxidized to aldehydes and secondary alcohols to ketones. Tertiary alcohols, because they bear no α-hydrogens, do not undergo oxidation under these conditions.

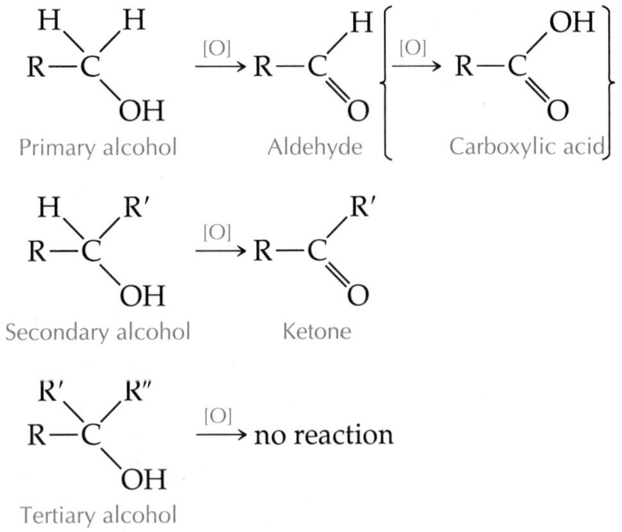

Primary alcohol → Aldehyde → [Carboxylic acid]

Secondary alcohol → Ketone

Tertiary alcohol → no reaction

[O] = oxidizing conditions

Extensive work has been done using chromium(VI) reagents to induce such oxidations. They have probably been the most popular oxidizing agents used in organic synthesis over the last 50 years. A problem associated with the oxidation of primary alcohols to aldehydes with Cr(VI), however, is the potential for further oxidation of an aldehyde to a carboxylic acid. Studies have shown that aldehydes are oxidized to carboxylic acids when these reactions are carried out under aqueous conditions; in such solvent mixtures, aldehydes are hydrated to form *gem*-diols, and these in turn form readily oxidizable chromate esters:

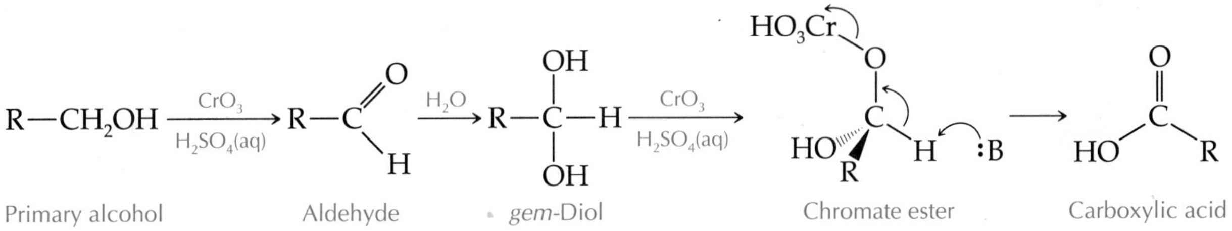

Primary alcohol   Aldehyde   *gem*-Diol   Chromate ester   Carboxylic acid

It should therefore not be surprising to learn that controlled oxidation of alcohols to aldehydes requires anhydrous conditions. One method uses chromium trioxide in pyridine. Thus, 1-decanol can be converted to decanal:

$$CH_3(CH_2)_8CH_2OH \xrightarrow[\text{pyridine}]{CrO_3} CH_3(CH_2)_8\overset{\displaystyle O}{\overset{\displaystyle \|}{C}}H$$
63–66%

A problem with the simple chromium trioxide/pyridine mixture is its potentially pyrophoric nature. The order of mixing of

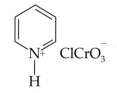

Pyridinium chlorochromate
(PCC)
MW 215.56

the reagents is crucial. A safer and superior method is oxidation with pyridinium chlorochromate (PCC); this reaction is also carried out under anhydrous conditions. In Experiment 15.3 you will see how cinnamyl alcohol can be converted to cinnamaldehyde, using PCC.

Although chromium trioxide is still a popular oxidizing agent, it has two major drawbacks. Chromium compounds are toxic, and they pose a hazard to the environment. Therefore, they must be handled and disposed of carefully, which requires added expense. By and large, chromium trioxide has been replaced in the undergraduate organic chemistry laboratory.

An excellent alternative is sodium hypochlorite (NaOCl), which is the active component of household bleach. This reagent has low toxicity and also gives products that are nontoxic. Thus, in Experiment 15.1 you will use NaOCl to oxidize cyclohexanol to cyclohexanone, and in Experiment 15.2 a variety of secondary alcohols are oxidized to the corresponding ketone, using the same reagent.

15.1
Oxidation of Cyclohexanol Using Sodium Hypochlorite

Investigate the oxidation of cyclohexanol in an environmentally friendly experiment.

Cyclohexanol $+ NaOCl \xrightarrow{CH_3COOH}$ Cyclohexanone $+ H_2O + NaCl$

Cyclohexanol
bp 161°C
density 0.962 g · mL^{-1}
MW 100.16

Cyclohexanone
bp 155.6°C
density 0.947 g · mL^{-1}
MW 98.14

In this experiment you will use common household bleach (for example, Clorox) as the oxidizing agent. Most bleaches are aqueous sodium hypochlorite, NaOCl(aq), prepared by adding chlorine to aqueous sodium hydroxide:

$$Cl_2 + 2\,NaOH \rightleftharpoons NaOCl + NaCl + H_2O$$

When sodium hypochlorite is added to acetic acid, the following acid-base reaction occurs:

$$NaOCl + CH_3COOH \rightleftharpoons HOCl + CH_3COO^-Na^+$$

NaOCl, Cl_2, and hypochlorous acid (HOCl) are all sources of positive chlorine (Cl^+), which has two fewer electrons than does chloride anion, Cl^-. As you might expect from the deficiency of electrons in the valence shell of Cl^+, positive chlorine is more reactive than chloride ion, especially in electrophilic processes. Although evidence for the discrete existence of Cl^+ in aqueous solution has never been found, it is apparent that a key step in reactions of positive chlorine reagents is the transfer of the Cl^+ entity from Cl_2 or HOCl to the substrate to form a new compound. It is reasonable to expect that the first step in the oxidation of cyclohexanol is exchange of positive chlorine with the hydroxyl proton and subsequent elimination of HCl from the resulting alkyl hypochlorite to form the ketone cyclohexanone:

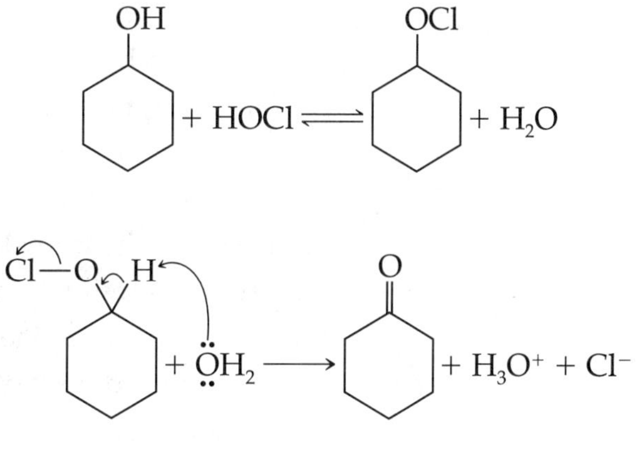

In the first reaction, Cl^+ is transferred to the substrate; and in the second, Cl^- is lost. The change is a reduction by two electrons. Cyclohexanol provides the two electrons and is thereby oxidized to cyclohexanone. Hypochlorite (or positive chlorine) oxidations provide a distinct advantage over traditional Cr(VI) reagents because the toxicity of chromium reagents poses very significant handling and waste disposal problems. The salt formed in hypochlorite oxidations can be disposed of down the sink.

Procedure

Techniques   Steam Distillation: Technique 7.6
Gas Chromatography: Technique 11
IR Spectrometry: Spectrometric Method 1

SAFETY INFORMATION

Cyclohexanol is an irritant. Avoid contact with skin, eyes, and clothing.

Glacial acetic acid is a dehydrating agent, an irritant, and causes burns. Dispense it in a hood and avoid contact with skin, eyes, and clothing.

Sodium hypochlorite solution emits chlorine gas, which is a respiratory and eye irritant. Dispense it in a hood.

Clorox or a supermarket brand of household bleach works well; use only newly opened bleach.

Pour 16 mL (0.15 mol) of cyclohexanol and 8.0 mL of glacial acetic acid into a three-necked 500-mL round-bottomed flask. At a hood, pour 115 mL of 5.25% (0.74 M) sodium hypochlorite (bleach) solution into a 125-mL dropping funnel. (Your separatory funnel may be used as a dropping funnel.) Stopper the dropping funnel before carrying it to your work station. Insert the dropping funnel in a neck of the flask, then put a small piece of paper between the stopper and neck of the funnel so that there is no buildup of vacuum and the solution can drip smoothly. Place a condenser in the middle neck and a thermometer inserted through the thermometer adapter in the third neck of the flask. Prepare an ice bath for cooling the flask if the reaction becomes too warm.

Add approximately one-fourth of the bleach solution to the reaction flask, then loosen the clamp holding the flask and swirl the flask to mix the reagents. Add the rest of the sodium hypochlorite solution over a period of 15 min. Adjust the rate of addition so that the temperature remains between 40°C and 45°C during the addition. Cool the reaction briefly with the ice bath if the temperature exceeds 45°C; however, keeping the reaction below 40°C for any period of time may result in a lowered yield.

When the addition of sodium hypochlorite is complete, remove the dropping funnel, and temporarily stopper the reaction flask. Pour another 115 mL of sodium hypochlorite solution into the funnel and again add this portion of bleach to the reaction flask over a period of 15 min, taking care that the temperature again remains between 40°C and 45°C during the addition. After the sodium hypochlorite addition is complete, remove the dropping funnel and stopper the flask opening. Allow the reaction mixture to stand at room temperature for 20 min with occasional swirling of the flask.

The yellow color of chlorine should no longer disappear near the end of the addition period.

Test the reaction mixture for excess hypochlorite by placing a drop of the reaction solution on a piece of wet starch-iodide indicator paper. The appearance of a blue-black color from the formation of the triiodide-starch complex on the indicator paper signifies the presence of excess hypochlorite. Add 2.0 mL of saturated sodium bisulfite solution to the reaction flask and swirl the flask to mix the contents thoroughly. Again test the reaction solution with starch-iodide paper. If necessary, continue adding bisulfite solution and testing with starch-iodide paper until excess oxidant is removed.

If the bleach is too concentrated, colorless iodate may form on the indicator paper.

Add 2.0 mL of thymol blue indicator solution to the mixture in the reaction flask. Place a conical funnel in one neck of the flask and add 6 M sodium hydroxide solution over 3 min until the indicator changes to blue (30–40 mL will be needed). Swirl the flask during the addition of NaOH.

The blue color of thymol blue appears at about pH 9.

Rinse the thermometer, condenser, and dropping funnel in a sink with water, and rinse again with sodium bisulfite solution before using them for the steam distillation.

You will remove the organic product from the reaction mixture by steam distillation [see Technique 7.6]. Set up the distillation apparatus as shown in Figure 7.15, omitting the dropping funnel, and closing the second neck of the Claisen adapter plus the other two necks of the round-bottomed flask with glass stoppers; instead of the Claisen adapter, a short column can be used.

The dropping funnel can be omitted, because the reaction mixture already contains a substantial amount of water and it is unnecessary to add more.

Collect 70–80 mL of distillate. Add 10 g of solid sodium chloride to the distillate to decrease the solubility of cyclohexanone in the aqueous phase. Stir the mixture until most of the NaCl dissolves. Decant the liquid from undissolved NaCl into a separatory funnel. Allow the phases to separate and remove the lower aqueous layer. Pour the upper product layer into a clean 50-mL Erlenmeyer flask and dry it with anhydrous magnesium sulfate or potassium carbonate for 20 min [see Technique 4.6].

The dried product is fairly pure at this point. If time permits, you can distill the final product, using a simple distillation apparatus [see Technique 7.3] and collecting the fraction boiling from 150–156°C. To separate the product from the drying agent, decant the product through a conical funnel containing a small piece of cotton into a tared vial or small Erlenmeyer flask, or into a 50-mL round-bottomed flask if you will be doing a final distillation [see Technique 4.7]. Weigh your product and calculate the percent yield.

Analyze the product by gas chromatography [see Technique 11] or by IR spectrometry [see Spectrometric Method 1], as directed by your instructor. You can also determine the refractive index [see Technique 9] and prepare the solid 2,4-dinitrophenyl-hydrazone described in the following optional experiment. Submit your remaining product to your instructor.

Cleanup: Adjust the pH of the solution remaining in the reaction flask to 7 with dilute hydrochloric acid before washing the solution down the sink or pouring it into the container for aqueous inorganic waste. Place the drying agent in the container for non-hazardous solid waste or the inorganic waste.

Optional Experiment

The synthesis of a solid derivative from your product provides another check on its identity. The melting point of the derivative can be compared with the melting point of known compounds. Ketones form solid derivatives upon treatment with 2,4-dinitro-phenylhydrazine. Follow the procedure in Organic Qualitative Analysis 6.2 (p. 555) for the preparation of the 2,4-dinitrophenyl-hydrazone of your cyclohexanone. Allow the crystals to dry and determine the melting point. The reported melting point for the 2,4-dinitrophenylhydrazone of cyclohexanone is 162°C.

The 2,4-Dinitrophenylhydrazone can be recrystallized from ethanol [see Technique 5.2]

Cleanup: Pour both the filtrate from the reaction mixture and the filtrate from the recrystallization into the container for flammable or organic waste.

References

1. Mohrig, J. R.; Neinhuis, D. M.; Linck, C. F.; Van Zoeren, C.; Fox, B. G.; Mahaffy, P. G. *J. Chem. Educ.* **1985**, *62*, 519–521.
2. Perkins, R. A.; Chau, F. *J. Chem. Educ.* **1982**, *59*, 981.

Questions

1. Describe what your IR spectrum would show if any cyclohexanol remained in the product.
2. What was the purpose of adding sodium hydroxide to the reaction mixture before the steam distillation? Write an equation for the reaction that occurred when the base was added.
3. Balance the equation for the redox reaction that occurs between bisulfite (HSO_3^-) and hypochlorite (OCl^-) and yields sulfate and chloride ions.
4. Mohrig and coworkers (Ref. 1) suggest that 2-propanol, rather than sodium bisulfite ($NaHSO_3$), can be used to destroy the excess sodium hypochlorite. What organic compound would be formed and why should it not be a contamination problem in this synthesis?
5. Predict the product that will be obtained if *trans*-2-methylcyclohexanol is oxidized with NaOCl. What will the product be if the *cis*-isomer is oxidized?

15.2

General Method for Oxidizing Secondary Alcohols to Ketones

Design and carry out an experiment for the oxidation of a secondary alcohol with household bleach in the presence of acetic acid.

$$R-\underset{\underset{R'}{|}}{\overset{\overset{H}{|}}{C}}-OH + NaOCl \xrightarrow{CH_3COOH} \underset{R}{\overset{O}{\underset{}{\|}}}\underset{R'}{C} + H_2O + NaCl$$

2° alcohol   Ketone

In this experiment you will use NaOCl in a water/acetic acid mixture to oxidize a secondary alcohol (RCHOHR') of your choice to a ketone [R(C=O)R']. The mechanism of the reaction is identical to that described in Experiment 15.1. It is your responsibility to analyze the product to confirm that indeed it is the compound that you expected. You can do this by measuring a variety of physical properties, including boiling point and refractive index. An especially important method is the use of IR spectrometry. All these ketone products will yield a carbonyl stretching band in the IR spectrum near 1715 cm^{-1}. If the oxidation has gone to completion, the strong, broad O—H stretching band of the secondary alcohol in the 3550–3200 cm^{-1} region will also have disappeared. The rest of the IR spectrum will yield a fingerprint that should allow straightforward identification of the ketone; in addition, ^1H NMR could also be used [see Spectrometric Method 2].

Procedure

Techniques   Extraction: Technique 4.2
Simple Distillation: Technique 7.3
Short-Path Distillation: Technique 7.3a
IR Spectrometry: Spectrometric Method 1

SAFETY INFORMATION

Glacial acetic acid is corrosive and causes burns; the vapor is extremely irritating to mucous membranes and the upper respiratory tract. Wear gloves, dispense it only in a hood. Avoid contact with skin, eyes, and clothing.

Sodium hypochlorite solution emits chlorine gas, which is a respiratory and eye irritant. Dispense it in a hood.

All secondary alcohols should be considered moderate skin irritants. Avoid contact with skin, eyes, and clothing.

Options for the 2° Alcohol

You will need to calculate the number of moles for your particular alcohol and ascertain from a handbook the boiling point and refractive index of the ketone that is produced by its oxidation.

Select one of the following secondary alcohols (or your instructor may assign one) as your starting reagent. Use 7.0 g of the alcohol.

2-pentanol	2-heptanol
3-hexanol	cyclohexanol
cyclopentanol	2-hexanol
3-pentanol	3-heptanol

Alternatively, you may use the secondary alcohol that you prepared in the Grignard reaction (Project 2.2) as the starting reagent for this synthesis. Determine the number of moles from the mass of your alcohol.

Quantities of Other Reagents

Household bleach solution supplies the sodium hypochlorite for the oxidation. The amount of sodium hypochlorite (NaOCl) in recently purchased household bleach is 5.25% by weight, or 0.74 M. You need to have a 10% molar excess of sodium hypochlorite relative to the secondary alcohol. Calculate how many milliliters of sodium hypochlorite solution you will need.

Acetic acid is the other reagent. Use 0.5 mL of acetic acid for each 0.010 mol of alcohol.

Oxidation

Select an Erlenmeyer flask that will be filled no more than one-third to one-half full with your reaction mixture. Put the secondary alcohol and the requisite amount of acetic acid in the flask together with a magnetic stirring bar. Set a separatory funnel above the flask to serve as a dropping funnel and place the calculated amount of sodium hypochlorite in the funnel. Begin stirring and add NaOCl solution at a rate that maintains the temperature of the reaction between 40°C and 50°C. If the temperature exceeds 50°C, cool the flask briefly in a beaker of ice water; the reaction temperature, however, should not be less than 40°C, or the product yield will be low. The addition of the NaOCl solution should take 15–20 min. Continue stirring another 20 min after the addition of NaOCl is complete.

Turn off the stirring motor before you put the thermometer into the reaction mixture to check the temperature.

Test the reaction mixture for excess hypochlorite by placing a drop of the reaction solution on a piece of wet starch-iodide indicator paper. The appearance of a blue-black color from the formation of the triiodide-starch complex on the indicator paper signifies the presence of excess oxidant. Add 2.0 mL of saturated sodium bisulfite solution to the reaction flask and swirl the flask to mix the contents thoroughly. Again test the reaction solution with starch-iodide paper. If necessary, continue adding 2.0-mL portions of bisulfite solution and testing with starch-iodide paper until the excess hypochlorite is gone.

If the bleach is too concentrated, colorless iodate may form on the paper.

Transfer the reaction mixture to a separatory funnel. Rinse the reaction flask with 25 mL of dichloromethane and add this rinse to the separatory funnel. Shake the separatory funnel gently to extract the product into the dichloromethane. Allow the phases to separate and remove the lower organic phase. Repeat the extraction of the upper aqueous phase with another 25-mL portion of dichloromethane. Combine the two dichloromethane extracts; set the aqueous phase aside for treatment later.

In these extractions, it is necessary to remove both phases from the separatory funnel, then return the organic phase to the funnel for the next step.

Return the combined dichloromethane solution to the separatory funnel. Wash the dichloromethane solution twice with 20-mL portions of saturated sodium bicarbonate. **(Caution: Foaming occurs.)** Transfer the dichloromethane solution to a clean flask and add anhydrous potassium carbonate; cork the flask and allow the solution to dry for 15 min.

Set up a simple distillation apparatus on a steam bath or a water bath on a hot plate [see Technique 7.3]. Filter the dried

Alternatively, the dichloromethane can be removed on a rotary evaporator [see Technique 4.8].

product solution through a small fluted filter paper into a 100-mL round-bottomed flask, rinse the drying agent remaining in the Erlenmeyer flask with a few milliliters of dichloromethane, and also pour the rinse solution through the filter paper. Add a boiling stone to the flask. Distill the dichloromethane fraction.

If the volume of product remaining in the distillation flask appears to be at least 4 or more milliliters, transfer the liquid to a 25-mL round-bottomed flask, using a Pasteur pipet; add a boiling stone and set up a short-path distillation, collecting the fraction at the boiling point of your particular product in a tared round-bottomed flask [see Technique 7.3a]. Determine the mass and percent yield of your ketone.

SAFETY PRECAUTION

Stop the short-path distillation before the distilling flask reaches dryness.

If the volume of product remaining in the distillation flask is too small for a final distillation, disconnect the stillhead from the boiling flask and heat the product for several more minutes in a hood to remove the last traces of dichloromethane. Cool the flask to room temperature. Transfer the product to a tared vial with a Pasteur pipet; determine the mass and percent yield of your ketone.

Determine the purity of your product by gas chromatographic analysis [see Technique 11]. Determine the IR spectrum [see Spectrometric Method 1] and refractive index [see Technique 9] of your product.

Cleanup: Combine the aqueous extracts and neutralize any remaining acetic acid with solid sodium carbonate (**Caution: Foaming.**) before washing the solution down the sink or pouring it into the container for aqueous inorganic waste. Put the distilled dichloromethane in the container for recovered dichloromethane or the container for halogenated waste, as directed by your instructor. Place the drying agent in the container for nonhazardous solid waste or the container for inorganic waste.

References

1. Mohrig, J. R.; Neinhuis, D. M.; Linck, C. F.; Van Zoeren, C.; Fox, B. G.; Mahaffy, P. G. *J. Chem. Educ.* **1985,** *62,* 519–521.

2. Perkins, R. A.; Chau, F. *J. Chem. Educ.* **1982,** *59,* 981.

Questions

1. What is the purpose of washing the dichloromethane solution with sodium bicarbonate? Write an equation for the reaction.

2. Compare your experimental refractive index to that reported in a handbook for your product.

3. Describe what your IR spectrum would show if any starting alcohol remained in the product.

4. Mohrig and coworkers (Ref. 1) suggest that 2-propanol, rather than sodium bisulfite ($NaHSO_3$), can be used to destroy the excess sodium hypochlorite. What organic compound would be formed and why should it not be a contamination problem in your synthesis?

5. Predict the product that will be obtained if *trans*-2-methylcyclohexanol is oxidized with NaOCl. What will the product be if the *cis*-isomer is oxidized?

15.3

Oxidation of Cinnamyl Alcohol Using Pyridinium Chlorochromate

Oxidize a primary alcohol to an aldehyde under anhydrous conditions, using TLC to monitor the reaction and column chromatography to purify the product.

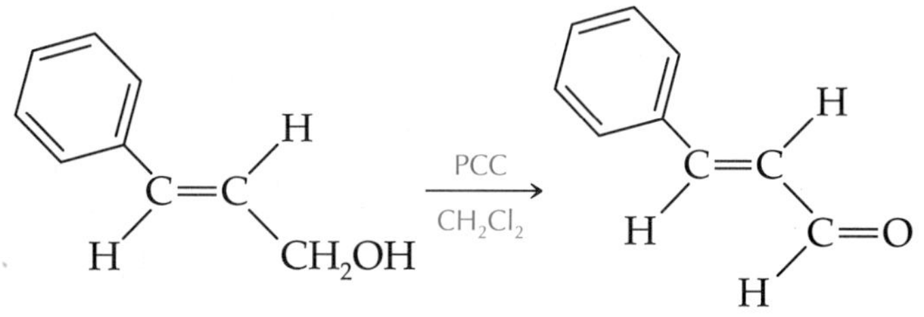

trans-Cinnamyl alcohol
(*E*)-3-phenyl-2-propen-1-ol
mp 33°C
MW 134.18

trans-Cinnamaldehyde
(*E*)-3-phenyl-2-propenal
bp 252°C
MW 132.16

Cinnamaldehyde is a major component of cinnamon oil, obtained from the bark of different trees in the genus *Cinnamomum*. Esters of cinnamyl alcohol occur naturally in Peruvian balsam, cinnamon leaves, and hyacinth oil. Both cinnamyl alcohol and cinnamaldehyde find application in the flavor and fragrance industry.

Chromium(VI) salts have a long history of use for oxidation of primary and secondary alcohols. As described in the introduction to Experiment 15, the controlled oxidation of primary alcohols to aldehydes requires anhydrous conditions so that the aldehyde is not oxidized to the carboxylic acid.

In 1975 Corey and Suggs (Ref. 1) discovered that pyridinium chlorochromate (PCC) was soluble in dichloromethane and other organic solvents; thus chromium (VI) oxidations could be carried out under anhydrous conditions. Pyridinium chlorochromate is prepared by treating pyridine with chromium trioxide in the presence of HCl:

$$\text{Pyridine} + CrO_3 + HCl \longrightarrow \text{PCC} \quad ClCrO_3^-$$

In this experiment you will oxidize cinnamyl alcohol to cinnamaldehyde, using a PCC/sodium acetate/molecular sieve mixture as the oxidant and carrying out the reaction in dichloromethane solution. The course of the reaction will be monitored by thin-layer chromatography, and purification of the cinnamaldehyde will be done by column chromatography.

microscale Procedure

Techniques   Thin-Layer Chromatography: Technique 10
Column Chromatography: Technique 12
NMR Spectrometry: Spectrometric Method 2

SAFETY INFORMATION

Cinnamyl alcohol is an irritant. Avoid contact with skin, eyes, and clothing.

Pyridinium chlorochromate is a suspected carcinogen. Wear gloves while handling it and avoid breathing the dust.

Dichloromethane is toxic, an irritant, absorbed through the skin, and harmful if inhaled. Use it only in a hood and wash your hands thoroughly after handling it.

Diethyl ether and **pentane** are extremely volatile and flammable.

Silica gel is a lung irritant. Avoid breathing the dust.

Following the Reaction by Thin-Layer Chromatography

A standard solution of 2% cinnamyl alcohol in dichloromethane will be available as a reference for monitoring the reaction by thin-layer chromatography [see Technique 10]. Use silica gel on glass or aluminum thin-layer plates and 30:70 ethyl acetate/pentane as the developing solvent. Visualize the chromatograms by dipping the TLC plate in *p*-anisaldehyde reagent solution and then heating the TLC plate on a hot plate adjusted to a medium-heat setting until the color appears [see Technique 10.5].

Alternatively, plastic-backed TLC plates can be used with iodine visualization [see Technique 10.5].

Oxidation and Chromatography

Place 1.8 g of the pyridinium chlorochromate/sodium acetate/4A molecular sieve mixture (1:1:1 by wt) and 10 mL of dichloromethane in a dry 50-mL Erlenmeyer flask. Suspend the oxidant mixture with magnetic stirring and add 250 mg of cinnamyl alcohol. Continue stirring the reaction mixture and monitor the course of the reaction by TLC at approximately 20-min intervals. When TLC indicates that the reaction is complete, add 2.0 g of Florisil (a filter aid), stir vigorously, add 10 mL of ether, and stir 5 min longer. Vacuum filter the mixture, using a few milliliters of ether to rinse the Erlenmeyer flask; pour the rinse over the solid in the Buchner funnel. Transfer the filtrate to a 50-mL round-bottomed flask, add 1.0 g of silica gel (70–230 mesh column chromatography grade), and evaporate the solvent in a hood, using a gentle stream of nitrogen. Warm the flask in a beaker of water at 40–45°C during the evaporation.

Silica gel may spatter out of the flask during the evaporation if the flow of nitrogen is too vigorous.

Alternatively, the solvent can be removed on a rotary evaporator [see Technique 4.8].

We will use a dry-packed chromatographic column [Technique 12.8]. Push a small amount of glass wool to the bottom of a 19 × 200 (or 300) mm chromatography column, using a long stirring rod. Clamp the column in a vertical position and add enough sand to cover the glass wool. Obtain 7.0 g of silica gel (70–230 mesh for column chromatography) and pour it slowly into the column, tapping the column to settle the particles. The top of the silica gel must be level. Add the silica gel/product mixture to the top of the column and tap the column to again level the silica gel. Cover the silica gel with a thin layer of sand.

It is critical that the silica gel surface not be disturbed.

Open the stopcock at the bottom of the chromatography column and slowly pour 60 mL of 20:80 (v/v) ethyl acetate/pentane solution (the eluting solvent) onto the column, using a small funnel to direct the flow against the inner wall of the column so that

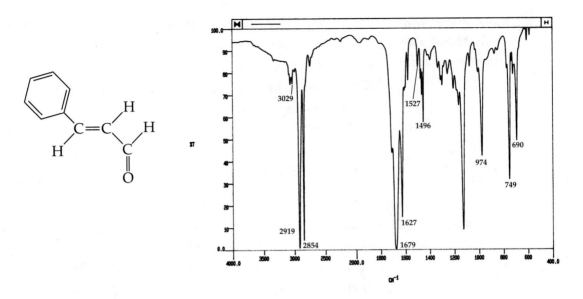

FIGURE 15.1 IR spectrum of cinnamaldehyde.

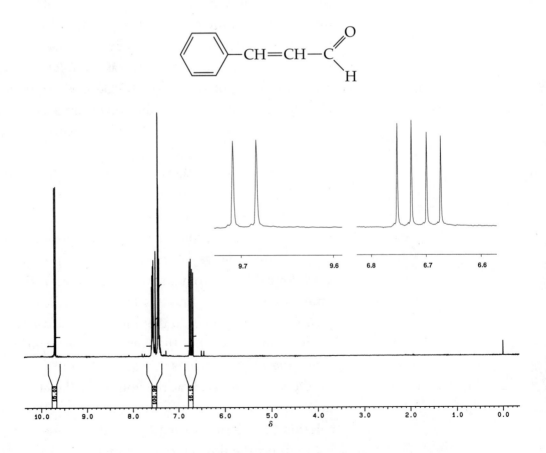

FIGURE 15.2 ^{1}H NMR (300 MHz) spectrum of cinnamaldehyde.

the silica gel surface is not disturbed. Collect 10-mL fractions of eluent. Check each fraction by TLC. Combine the fractions containing pure cinnamaldehyde in a tared round-bottomed flask and evaporate most of the solvent on a steam bath in a hood, using a boiling stick to prevent bumping. Then, using a stream of nitrogen and working in a hood, continue the evaporation of solvent until a constant weight is reached. Determine the percent yield. Obtain an IR or NMR spectrum of the product as directed by your instructor.

Alternatively, the solvent may be removed on a rotary evaporator [see Technique 4.8].

Cleanup: Place the mixture of Florisil and pyridinium chlorochromate in the container for chromium (or hazardous inorganic) waste. Pour the developing solvent from the TLC chamber and any fractions from the column chromatography that did not contain cinnamaldehyde into the container for flammable (organic) waste.

References

1. Corey, E. J.; Suggs, J. W. *Tetrahedron Lett.* **1975,** 2647–2650.
2. Taber, D. F.; Wang, Y.; Liehr, S. *J. Chem. Educ.* **1996,** *73,* 1042–1043.
3. Herscovici, J.; Antonakis, K. *J. Chem. Soc. Chem. Comm.* **1980,** 561.

Questions

1. Compare this oxidation procedure with that using other aqueous Cr(VI) oxidation reagents. What are the advantages of the reagent used in this experiment?
2. Why does cinnamaldehyde have a higher R_f than cinnamyl alcohol in TLC?
3. Predict the product that will be obtained if *cis*-4-*tert*-butylcyclohexanol is oxidized (a) with NaOCl; (b) with CrO_3/H_2SO_4; (c) with $Na_2Cr_2O_7/H_2SO_4$; (d) with PCC.
4. Propanal (bp 50°C) is much more volatile than either 1-propanol (bp 97°C) or propanoic acid (bp 140°C). (a) Explain. (b) Make use of these volatility differences to devise a method for preparing propanal that minimizes oxidation of the aldehyde to carboxylic acid. (*Hint:* The setup of the reaction system can allow you to take advantage of the boiling-point differences.)

DIELS-ALDER REACTIONS

Discover how Diels-Alder reactions can be used to synthesize cyclic compounds.

The Diels-Alder reaction has great utility in the chemical synthesis of ring compounds, from simple cyclohexanes to complex steroids. Curiously, this efficient synthetic method doesn't seem to be used at all for biochemical syntheses in nature. Otto Diels and Kurt Alder received the Nobel prize in 1950 for its discovery.

The Diels-Alder reaction is a cycloaddition reaction in which two molecules add together to form a cyclic compound. Probably the simplest example is the reaction of ethene and 1,3-butadiene to form cyclohexene:

| 1,3-Butadiene (the diene) | Ethene (the dienophile) | Six-electron transition state | Cyclohexene product |

As this example implies, the Diels-Alder reaction involves the rearrangement of six π-electrons in a cyclic array of atoms. It is an example of a concerted pericyclic reaction in which the p-orbitals of all six carbon atoms are overlapping simultaneously in the transition state. Four of the π-electrons come from the conjugated diene. The other two electrons come from the alkene, which is often called the dienophile because it loves to react with the diene. Overall, it is an example of a [4 + 2] cycloaddition reaction. The transition state for the Diels-Alder reaction has been described as having aromatic character, because it has the same number of π-electrons in extended cyclic overlap as benzene does. It should not be surprising that Diels-Alder transition states are quite stable. Often only gentle heating of a mixture of the diene and dienophile is required for almost quantitative yields of cyclic products. In some cases, no heat is required.

Although simple in concept, the Diels-Alder reaction of ethene and 1,3-butadiene proceeds slowly and with a modest yield of cyclohexene. The reaction is far faster when the carbon-carbon double bond of the dienophile is electron deficient. Electron-withdrawing groups on the dienophile speed up the reaction significantly. Electron-donating groups on the diene also help.

In molecular orbital (MO) notation, the Diels-Alder reaction proceeds as shown here:

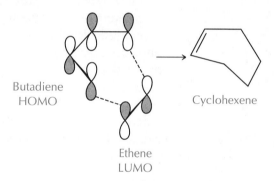

1,3-Butadiene

Ethene   Cyclohexene

With cycloaddition reactions, it is common to be concerned with the overlap between the atomic orbital components of the HOMO (Highest Occupied Molecular Orbital) of one reactant (for example, 1,3-butadiene) with the LUMO (Lowest Unoccupied Molecular Orbital) of the other reactant. The in-phase overlap (denoted by the same shading in the orbital lobes) results in relatively easy bond formation and ready production of the cyclic product.

Butadiene
HOMO   Cyclohexene

Ethene
LUMO

Diels-Alder reactions can be modified to allow preparation of a variety of different compounds by substituting various groups onto the diene and dienophile. In addition, the dienophile can also be a carbon-oxygen double bond or a carbon-carbon triple bond. Diels-Alder reactions have been of great importance in preparing useful compounds in chemical research laboratories and in pharmaceutical and polymer industrial laboratories.

16.1
Synthesis of Dimethyl 3,4,5,6-Tetraphenylphthalate from 2,3,4,5-Tetraphenylcyclopentadienone and Dimethyl Acetylenedicarboxylate

Synthesize a substituted aromatic ring by a Diels-Alder reaction. Characterize the product by melting point and NMR spectrometry.

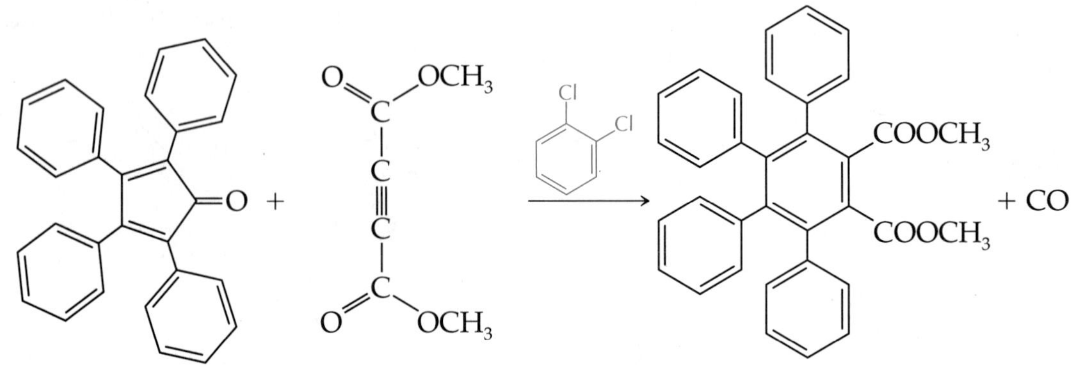

2,3,4,5-Tetraphenylcyclo-
pentadienone
mp 219°C
MW 384.4

Dimethyl
acetylenedicarboxylate
bp ~300°C
MW 142.1
density 1.156 g · mL^{-1}

Dimethyl
3,4,5,6-tetraphenylphthalate
mp 258°C
MW 474.5

In this Diels-Alder reaction, the dienophile is a substituted alkyne, or acetylene. The dienophile undergoes ready reaction with 2,3,4,5-tetraphenylcyclopentadienone (the diene) to initially form a bridged bicyclic compound. The term *bridged* refers here to the one carbon bridge linking the two ends of the boatlike cyclohexadiene skeleton. The internal C—C—C bond angle of this bridge is far less than the 120° angle desired by the sp^2 hybridization for such a carbonyl carbon, and the compound undergoes a rapid reaction to extrude carbon monoxide and form the hexasubstituted benzene compound shown:

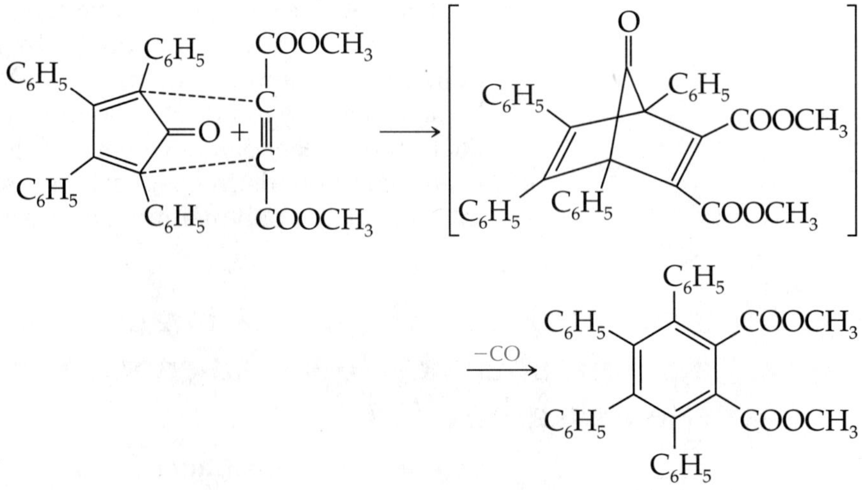

Final product

In this experiment, a useful characteristic of the diene is its ring structure because this ring locks the two double bonds in the so-called *s-cis* conformation, which is a better orientation for a Diels-Alder reaction than is the *s-trans* conformation:

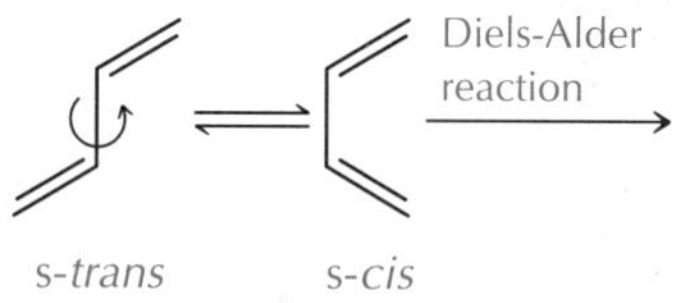

Another favorable aspect of this reaction is the presence of the ester groups on the ends of the alkyne triple bond. It is well known that electron-withdrawing groups such as —COOR or other carbonyl or cyano (—C≡N) groups attached to the carbon-carbon double bond of the dienophile will increase the rate of the Diels-Alder reaction.

1,2-Dichlorobenzene is the solvent for the reaction. It has a convenient boiling point (bp 180°C) to carry out the Diels-Alder reaction at reflux.

Macroscale Procedure

Technique   NMR Spectrometry: Spectrometric Method 2

— SAFETY INFORMATION —

Carry out this synthesis in a hood.

Dimethyl acetylenedicarboxylate is corrosive and a lachrymator (causes tears). Wear gloves while handling it.

1,2-Dichlorobenzene is toxic and an irritant.

This procedure is based on 0.50 g of tetraphenylcyclopentadienone. If you are using the tetraphenylcyclopentadienone that you prepared in Experiment 22, you need to proportionally adjust all reagent quantities if you use an amount of ketone other than 0.50 g.

Heat an aluminum block on a hot plate to 180°C. Weigh 0.50 g of tetraphenylcyclopentadienone and place it in a dry 25 × 150 mm test tube. Add 0.25 mL of dimethyl acetylenedicarboxylate measured with a graduated 1-mL pipet and 2.5 mL of 1,2-dichlorobenzene. Clamp the test tube in the aluminum block, set a pair of aluminum auxiliary (half) blocks around the tube [see Technique 3, Figure 3.2b] and insert a thermometer in the test tube. Allow the solution to reflux gently; the ring of condensate should be just above the top of the half blocks. Continue heating at reflux for 5–10 min until no further color change occurs.

The product is colorless to light tan, and the purple color of tetraphenylcyclopentadienone disappears as the reaction proceeds.

Cool the solution to 80°C and add 3.5 mL ethanol; crystals should begin to form. Stir to mix and allow crystallization to continue until the test tube is at room temperature. Cool the test tube in an ice-water bath for about 5 min and collect the crystals by vacuum filtration. Rinse any remaining crystals from the test tube with a few drops of ice-cold methanol. If you are going to immediately prepare a sample for NMR analysis, allow the crystals to dry by drawing air over them for 15 min. Determine the mass of your dried product and its melting point. Prepare a sample for NMR analysis as directed by your instructor.

Cleanup: Pour the filtrate into the container for halogenated waste.

microscale Procedure

Technique   NMR Spectrometry: Spectrometric Method 2

> **SAFETY INFORMATION**
>
> Carry out this synthesis in a hood.
>
> **Dimethyl acetylenedicarboxylate** is corrosive and a lachrymator (causes tears). Wear gloves while handling it.
>
> **1,2-Dichlorobenzene** is toxic and an irritant.

This procedure is based on 0.100 g of tetraphenylcyclopentadienone. If you are using the tetraphenylcyclopentadienone prepared in Experiment 22, you need to proportionally adjust all reagent quantities if you are using more or less than 0.100 g.

Heat an aluminum block on a hot plate to 180°C. Weigh 0.100 g of tetraphenylcyclopentadienone and place it in a dry

5-mL conical vial containing a magnetic spin vane. Add 50 μL of dimethyl acetylenedicarboxylate measured with an automatic pipettor and 0.5 mL of 1,2-dichlorobenzene measured with a 1-mL graduated pipet. Fit an air condenser to the conical vial and set it in the aluminum block. Clamp the air condenser with a microclamp and allow the solution to reflux gently. Continue heating the mixture at reflux until no further color change occurs (5–10 min).

The product is colorless to light tan, and the purple color of tetraphenylcyclopentadienone disappears as the reaction proceeds.

Cool the solution for 5 min and add 0.70 mL ethanol; crystals should begin to form. Stir to mix, and allow crystallization to continue until the vial is at room temperature. Cool the vial in an ice-water bath for approximately 5 min, then collect the crystals by vacuum filtration on a Hirsch funnel. Rinse any remaining crystals

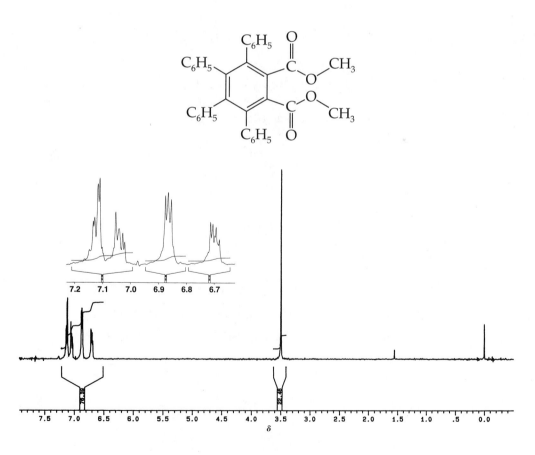

FIGURE 16.1 ^{1}H NMR (300 MHz) spectrum of dimethyl 3,4,5,6-tetraphenylphthalate.

from the vial with a few drops of ice-cold methanol. If you are going to immediately prepare the sample for NMR analysis, allow the crystals to dry by drawing air over them for 15 min. Determine the mass of your product and its melting point. Prepare a sample for NMR analysis as directed by your instructor.

Cleanup: Pour the filtrate into the container for halogenated waste.

Questions

1. A driving force for this reaction is the formation of carbon monoxide, a stable compound. How else can stability be used to rationalize the ease of the reaction?
2. One of the following two compounds much more readily takes the part of the diene in the Diels-Alder reaction. Which one and why?

3. Ethene (ethylene) is actually a fairly poor dienophile. Order the following dienophiles on the basis of their relative reactivity toward a common diene.

16.2

Synthesis of 4-Cyclohexene-*cis*-1,2-Dicarboxylic Acid from Butadiene Sulfone and Maleic Anhydride

Synthesize a cyclic compound by a Diels-Alder reaction.

Butadiene sulfone
(2,5-dihydrothiophene-1,1-dioxide)
(3-sulfolene)
mp 65–66°C
MW 118.2

1,3-Butadiene
bp −4°C
MW 54.1

| 1,3-Butadiene | Maleic anhydride mp 52°C MW 98.1 | 4-Cyclohexene-*cis*-1,2-dicarboxylic anhydride mp 105–106°C MW 152 | 4-Cyclohexene-*cis*-1,2-dicarboxylic acid mp 165°C MW 70 |

This experiment demonstrates the use of 1,3-butadienesulfone (or 3-sulfolene) as an in situ source of 1,3-butadiene. Because butadiene is a gas, it presents some handling difficulties. We can, however, use the cyclic sulfone to produce 1,3-butadiene while all the reactants are dissolved in a solvent, xylene in this experiment:

Butadiene sulfone   Butadiene

While still in the reaction mixture, 1,3-butadiene rapidly reacts with maleic anhydride:

cis   *cis*

Two electron-withdrawing groups on the alkene double bond make maleic anhydride an excellent dienophile. Xylene (dimethylbenzene) is a convenient inert solvent, because the reaction proceeds readily at its reflux temperature (bp 139°C). The initial Diels-Alder product is an anhydride, which can easily be isolated. You will, however, hydrolyze the anhydride directly to the dicarboxylic acid.

The geometry of the two carbonyl groups is cis in both the anhydride and the dicarboxylic acid products, because the configuration of the two carbonyl groups is *cis* in the maleic anhydride starting material. The Diels-Alder reaction is a concerted *syn* cycloaddition process that occurs stereospecifically with retention of configuration. The *cis* geometry is retained throughout the reaction.

Macroscale Procedure

Techniques   Drying tube used as a gas trap: Technique 3.3
IR Spectrometry: Spectrometric Method 1

— SAFETY INFORMATION —

Conduct the experiment in a hood, if possible.

Butadiene sulfone (3-sulfolene) is an irritant. Wear gloves and avoid contact with skin, eyes, and clothing. This compound emits toxic, corrosive sulfur dioxide when it is heated. Be sure that the gas trap is positioned before you begin heating the reaction mixture.

Maleic anhydride is toxic and corrosive. Avoid breathing the dust and avoid contact with skin, eyes, and clothing.

Xylene is flammable.

Sodium hydroxide is corrosive and causes burns. Wear gloves and avoid contact with skin, eyes, and clothing. Clean up any spilled pellets immediately. Keep the jar tightly closed.

Solid sodium hydroxide absorbs moisture from the atmosphere; spilled pellets can form droplets of concentrated NaOH solution rapidly.

Synthesis of the Diels-Alder Adduct

Prepare a gas trap, using a drying tube. Place a small piece of cotton in the bottom of the tube, fill it approximately three-fourths full with 20–40 mesh sodium hydroxide pellets, and put a piece of cotton on top of the NaOH pellets.

Combine 2.0 g of butadiene sulfone, 1.2 g of finely ground maleic anhydride, 0.80 mL of xylene (a mixture of isomers is all right), and a boiling stone in a 25-mL round-bottomed flask. Fit a water-cooled condenser in the flask and place a thermometer adapter in the top of the condenser with the gas trap set in the rubber thermometer holder [see Technique 3.3]. Begin heating the mixture gently with a heating mantle. After the solids dissolve, continue heating the mixture at a gentle reflux for 30 min. Remove the heating mantle and cool the reaction mixture for about 5 min before proceeding immediately with the hydrolysis of the anhydride product.

Hydrolysis of the Diels-Alder Product to the Dicarboxylic Acid

Remove the thermometer adapter and the NaOH-filled drying tube. Pour 4 mL of water down the condenser, add another boiling stone, and heat the mixture under reflux for 30 min. Cool the solution to room temperature. If crystallization of the product does not occur, add 3 or 4 drops of concentrated sulfuric acid, stir the contents of the flask, and cool the resulting mixture in an ice-water bath for 5 min. Collect the product by vacuum filtration. Wash the crystals twice with 1-mL portions of ice-cold water. The dicarboxylic acid is usually quite pure without recrystallization. If time permits, the product can be recrystallized from water [see Technique 5.3]. Allow the product to dry overnight before determining the melting point, percent yield, and IR spectrum.

Cleanup: Wearing gloves, remove the cotton from the top of the drying tube and pour the NaOH pellets into a 600-mL beaker containing 300 mL of water. Add the aqueous filtrate from the reaction mixture. Stir the mixture until the NaOH dissolves, add some crushed ice, then add 6 M HCl solution until the pH is between 6 and 8, as indicated by pH test paper. Wash the neutralized solution down the sink or pour it into the container for aqueous inorganic waste. Rinse the cotton with water before discarding it in the container for nonhazardous waste.

microscale Procedure

Techniques Drying tube used as a gas trap: Technique 3.3
IR Spectrometry: Spectrometric Method 1

SAFETY INFORMATION

Conduct this experiment in a hood, if possible.

Butadiene sulfone (3-sulfolene) is an irritant. Wear gloves and avoid contact with skin, eyes, and clothing. This compound emits toxic corrosive sulfur dioxide when it is heated. Be sure that the gas trap is positioned before you begin heating the reaction mixture.

Maleic anhydride is toxic and corrosive. Avoid breathing the dust and avoid contact with skin, eyes, and clothing.

Xylene is flammable.

Solid sodium hydroxide absorbs moisture from the atmosphere; spilled pellets can form droplets of concentrated NaOH solution quite rapidly.

Sodium hydroxide is corrosive and causes burns. Wear gloves and avoid contact with skin, eyes, and clothing. Clean up any spilled pellets immediately. Keep the jar tightly closed.

Synthesis of the Diels-Alder Adduct

Prepare a gas trap, using a drying tube. Place a small piece of cotton in the tube, fill it approximately three-fourths full with 20–40 mesh sodium hydroxide pellets, and put a piece of cotton on top of the NaOH pellets.

Combine 250 mg of butadiene sulfone, 150 mg of finely ground maleic anhydride, and 0.15 mL of xylene (a mixture of isomers is all right) in a 5-mL conical vial containing a spin vane. Fit a water-cooled condenser in the vial, place the gas trap in the top of the condenser, and clamp the apparatus in an aluminum heating block [see Technique 3.3]. Begin heating the mixture gently on an aluminum block; allow the temperature of the heating block to rise slowly to 150°C. After the solids dissolve, continue heating the mixture at a gentle reflux for 30 min. Remove the vial from the heating block and cool the reaction mixture for about 5 min before proceeding immediately with the hydrolysis of the anhydride product.

Hydrolysis of the Diels-Alder Adduct to the Dicarboxylic Acid

Remove the NaOH-filled drying tube. Pour 0.5 mL of water down the condenser and heat the mixture under reflux for 30 min. (The aluminum block should be heated to 125–140°C). Remove the spin vane with a tweezers and cool the solution to room temperature. If crystallization of the product does not occur, add 1 drop of concentrated sulfuric acid, stir the contents of the vial, and cool the resulting mixture in an ice-water bath for 5 min. Collect the product by vacuum filtration on a Hirsch funnel. Wash the crystals twice with a few drops of ice-cold water. The dicarboxylic acid is usually quite pure without recrystallization. If time permits, the product can be recrystallized from water [see Technique 5.5a]. Allow the product to dry overnight before determining the melting point, percent yield, and IR spectrum.

Cleanup: Wearing gloves, remove the cotton from the top of the drying tube and pour the NaOH pellets into a 250-mL beaker containing 100 mL of water. Add the aqueous filtrate from the reaction mixture. Stir the mixture until the NaOH dissolves, add some crushed ice, then add 6 M HCl solution until the pH is between 6 and 8, as indicated by pH test paper. Wash the neutralized solution down the sink or pour it into the container for aqueous inor-

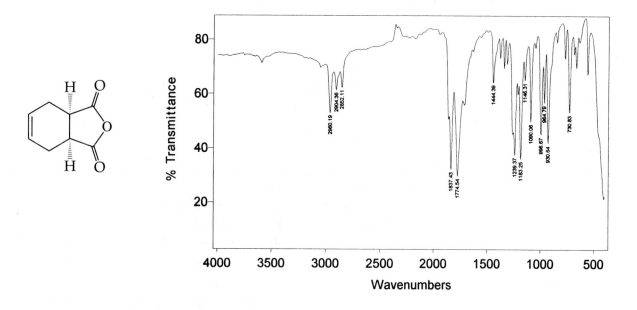

FIGURE 16.2 IR spectrum of 4-cyclohexene-cis-1,2-dicarboxylic anhydride.

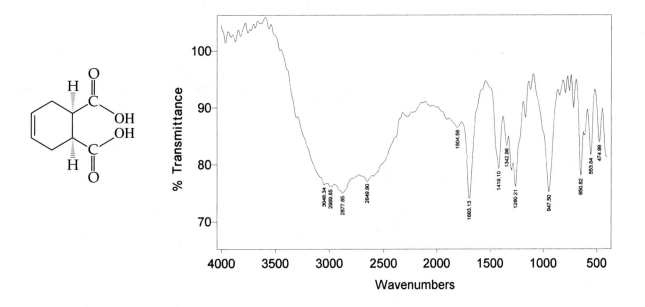

FIGURE 16.3 IR spectrum of 4-cyclohexene-*cis*-1,2-dicarboxylic acid.

ganic waste. Rinse the cotton with water before discarding it in the container for nonhazardous waste.

Reference

1. Sample, T. E.; Hatch, L. F. *J. Chem. Educ.* **1968,** *45,* 55–56.

Questions

1. Write an equation for the reaction that occurs in the gas trap between SO_2 and NaOH.
2. At what temperature was the Diels-Alder reaction done in Experiment 16.2? How were you able to predict this temperature?
3. What product is expected from the reaction of fumaric acid with 1,3-butadiene?

Fumaric acid

4. The reaction of 1,3-cyclopentadiene and maleic anhydride gives two products, called *exo* and *endo* (and usually the *endo* product is the favored product). Write the structures of these two products and identify them by making use of the fact that the reagents align as shown:

$\longrightarrow$ *endo* product

$\longrightarrow$ *exo* product

5. Unlike maleic acid, the reaction of fumaric acid with 1,3-cyclopentadiene gives only one adduct. Why?

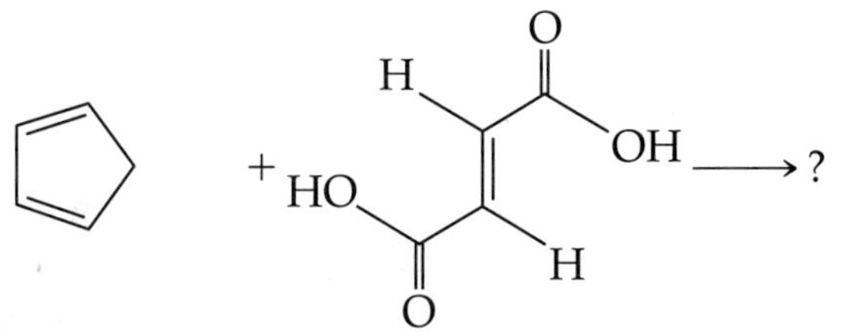

1,3-Cyclopentadiene Fumaric acid

Experiment 17 ELECTROPHILIC AROMATIC SUBSTITUTION

Investigate structure-reactivity relationships in the halogenation of aromatic compounds, focusing on substitution patterns and relative reaction rates.

Aromatic compounds such as benzene do not undergo the simple electrophilic addition reactions that alkenes do:

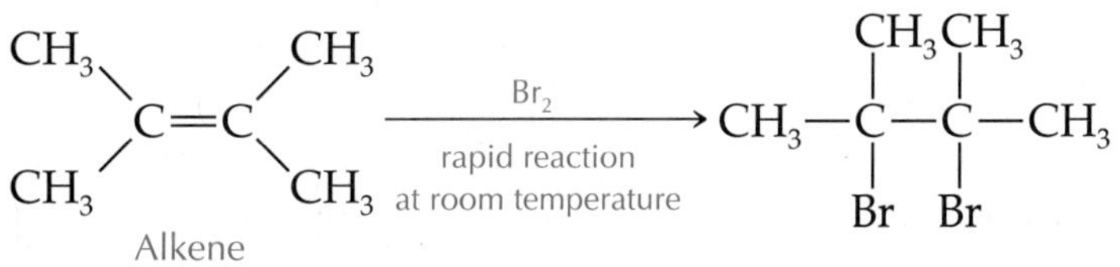

Alkene

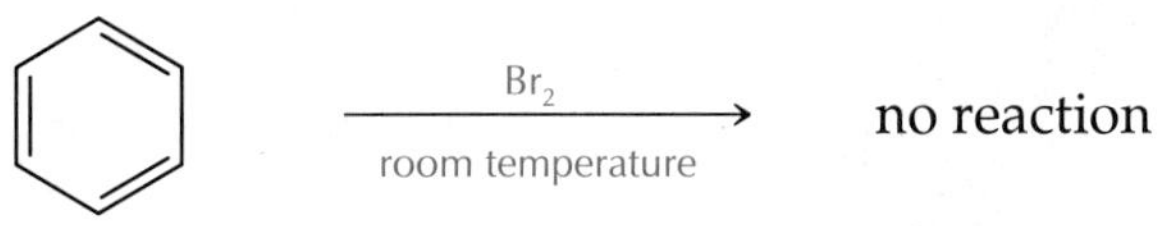

Benzene

Aromatic compounds fail to undergo electrophilic addition because their delocalized π-system is much less vulnerable to attack by an electrophile (E^+) than the localized π-system of an alkene double bond is:

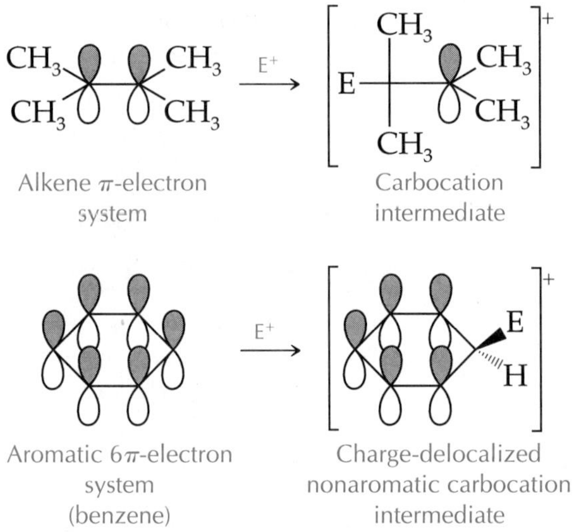

Alkene π-electron system → Carbocation intermediate

Aromatic 6π-electron system (benzene) → Charge-delocalized nonaromatic carbocation intermediate

However, as the preceding equation shows, aromatic compounds do undergo electrophilic substitution reactions:

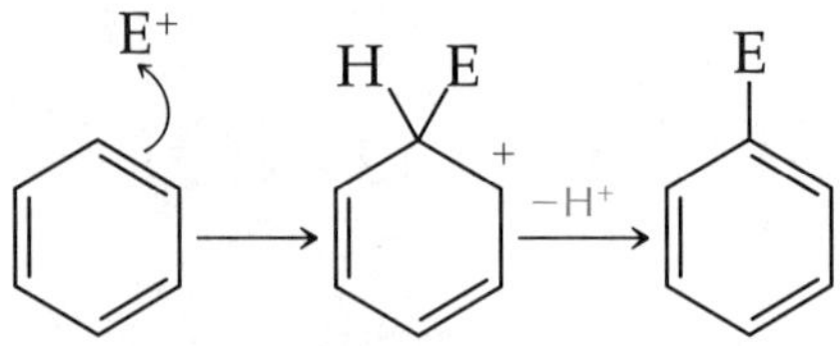

Because the conjugated 6π-electron system of the aromatic ring is so stable, the cationic intermediate loses a proton and regenerates the aromatic ring rather than reacting with a nucleophile. The positive charge of the cationic intermediate is centered mainly on the positions *ortho* and *para* to the carbon atom bonded to the electrophile (E):

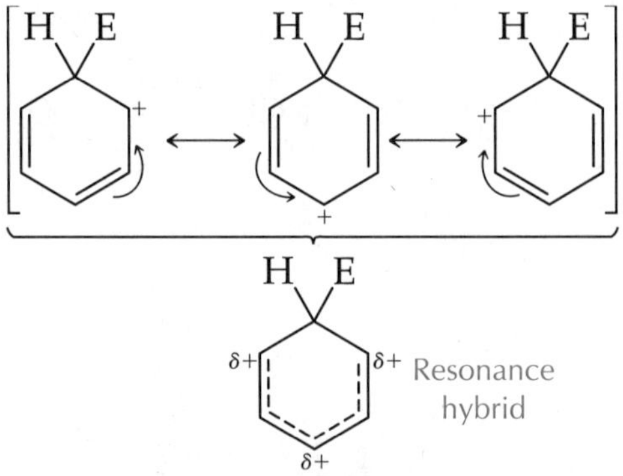

Resonance hybrid

The mechanism of electrophilic aromatic substitution has been thoroughly studied and includes recognition of the effect of aromatic ring substituents on the substitution reaction. If, for example, the first substituent is an electron-withdrawing group, formation of the carbocation intermediate is more difficult because the electron-withdrawing group removes electron density from a species that is already positively charged and thus quite electron deficient. The more strongly a group is electron withdrawing, the more it destabilizes the substitution intermediate and the more it slows the substitution process:

Carbocation intermediate destabilized by electron-withdrawing group (ewg)

As you might expect, electron-donating groups on the benzene ring speed up the substitution process by stabilizing the cationic intermediate.

In these experiments you will look at a range of electrophilic aromatic substitution reactions. In Experiment 17.1, the iodination of 1,2,4,5-tetramethylbenzene, you will examine the halogenation of a highly substituted hydrocarbon carried out under novel conditions. In Experiment 17.2, you will brominate acetanilide; this reaction reveals how modification of a highly activating substituent can lead to reduced reactivity of the ring system. And in Experiment 17.3, you will study the reaction rates for three substituted benzene compounds, diphenyl ether, acetylsalicylic acid (aspirin), and acetanilide and see how various substituents change the reactivity of the benzene ring toward electrophilic aromatic substitution.

17.1

Iodination of 1,2,4,5-Tetramethylbenzene Using Iodine and Periodic Acid

Synthesize an iodobenzene derivative, using novel redox chemistry.

$$7 \quad \text{(durene)} \; + 3I_2 + HIO_4 \cdot 2H_2O \xrightarrow[CH_3COOH]{H_2SO_4} \quad 7 \quad \text{(iododurene)} \; + 6H_2O$$

1,2,4,5-Tetramethylbenzene
(durene)
mp 79.2°C
MW 134.2

2,3,5,6-Tetramethyliodobenzene
(iododurene)
mp 78–80°C
MW 260.1

Usually when an aromatic compound is treated with molecular iodine, I_2, no reaction occurs. However, a variety of inorganic oxidizing agents can be used to produce I^+ from iodine; and with I^+ present, aromatic iodination can be carried out.

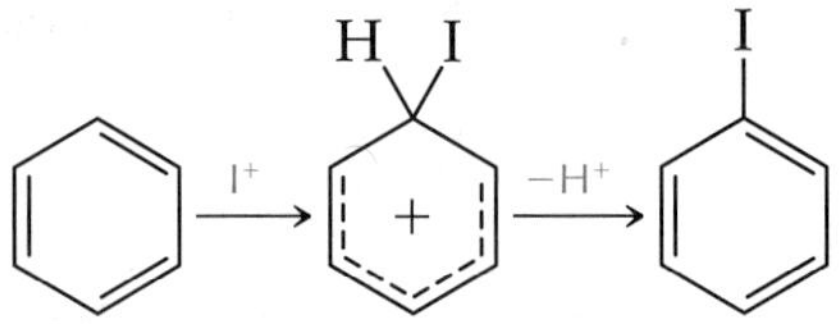

In this experiment, periodic acid is used to oxidize iodine to I^+. The novel aspect of this reaction is the fact that in the process of oxidizing iodine, the periodate is reduced to I^+ and also becomes a source of the electrophile necessary for substitution.

$$I_2 \xrightarrow{-2e^-} 2I^+$$

$$H^+ + IO_4^- \xrightarrow{+6e^-} I^+ + H_2O \quad \text{(unbalanced)}$$

Therefore, the only products are the substituted benzene derivative and water.

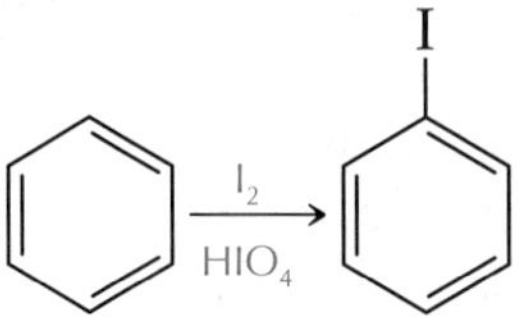

◊◊◊◊*Macroscale*
Procedure

Techniques   Recrystallization: Technique 5.3

NMR Spectrometry: Spectrometric Method 2

SAFETY INFORMATION

1,2,4,5-Tetramethylbenzene (durene) is a flammable solid.

Periodic acid is corrosive and a strong oxidizer. Avoid contact with skin, eyes, and clothing.

(continued on next page)

SAFETY INFORMATION *(continued)*

Iodine is a toxic, corrosive solid that sublimes readily. Wear gloves while weighing it, minimize the time that the container is open, and close the lid tightly. Weigh iodine in a hood, if possible.

The mixture of sulfuric acid, water, and glacial acetic acid is corrosive. Avoid contact with skin, eyes, and clothing.

2-Propanol (isopropyl alcohol) is flammable.

Place 1.35 g of 1,2,4,5-tetramethylbenzene, 0.46 g of periodic acid dihydrate, and 1.00 g of iodine in a 50-mL round-bottomed flask. Add 13 mL of an acid mixture containing 3:20:100 (v/v/v) concentrated sulfuric acid/water/glacial acetic acid. Put a magnetic stirring bar in the flask and fit a water-cooled reflux condenser to the flask. Clamp the flask in a water bath heated to 65–70°C and begin stirring the reaction mixture. Continue stirring until the brown iodine color disappears, usually about 75 min.

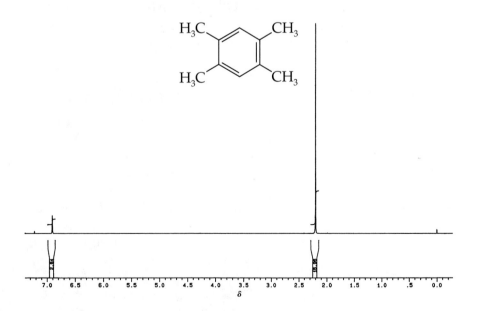

FIGURE 17.1 ¹H NMR spectrum (300 MHz) of 1,2,4,5-tetramethylbenzene (in CDCl₃).

When the reaction is complete, substitute a beaker of cold tap water for the hot-water bath and allow the reaction mixture to cool for 5 min. Continue stirring the reaction mixture during the cooling period. Pour 25 mL of water down the inside of the condenser and stir a few minutes longer until any lumps of product are reduced to fine particles.

Collect the crude product by vacuum filtration on a Buchner funnel. Wash the crystals four times with 10-mL portions of water. Recrystallize the product from a minimum volume of boiling 2-propanol [see Technique 5.3]. Allow the crystals to dry overnight before determining the melting point and percent yield. Determine the NMR spectrum of your dried product as directed by your instructor. Compare it to the NMR spectrum of 1,2,4,5-tetramethylbenzene shown in Figure 17.1.

We suggest you do a mixture melting-point determination of your product and the starting material [see Technique 6.4].

Cleanup: Neutralize the aqueous filtrate from the reaction mixture with solid sodium carbonate before pouring it down the sink or placing it in the container for aqueous inorganic waste. Pour the filtrate from the recrystallization into the container for halogenated waste.

m*icroscale* Procedure

Techniques   Microscale Reflux: Technique 3.3
Recrystallization: Technique 5.5a
NMR Spectrometry: Spectrometric Method 2

— SAFETY INFORMATION —

1,2,4,5-Tetramethylbenzene (durene) is a flammable solid.

Periodic acid is corrosive and a strong oxidizer. Avoid contact with skin, eyes, and clothing.

Iodine is a toxic, corrosive solid that sublimes readily. Wear gloves while weighing it, minimize the time that the container is open, and close the lid tightly. Weigh iodine in a hood, if possible.

The mixture of sulfuric acid, water, and glacial acetic acid is corrosive. Avoid contact with skin, eyes, and clothing.

2-Propanol (isopropyl alcohol) is flammable.

Place 200 mg of 1,2,4,5-tetramethylbenzene, 70 mg of periodic acid dihydrate, and 150 mg of iodine in a 10-mL round-bottomed flask. Add 2.5 mL of an acid mixture containing 3:20:100 (v/v/v) concentrated sulfuric acid/water/glacial acetic acid, measured with a graduated pipet. Put a magnetic stirring bar in the flask and fit a water-cooled reflux condenser to the flask [see Technique 3.3]. Clamp the flask in a water bath heated to 65–70°C and begin stirring the reaction mixture. Continue stirring until the color of iodine disappears; this usually requires about 75 min. When the reaction is complete, substitute a beaker of cold tap water for the hot-water bath and allow the reaction mixture to cool for 5 min. Continue stirring the reaction mixture during the cooling period. Remove the condenser and, with the stirrer running, add 5 mL of water to the reaction mixture. Stir a few minutes longer until any lumps of product are reduced to fine particles.

We suggest that you do a mixture melting-point determination of your product with the starting material [see Technique 6.4].

Collect the crude product by vacuum filtration on a Hirsch funnel. Wash the crystals four times with 2-mL portions of water. Transfer the solid to a 10-mL Erlenmeyer flask and recrystallize it from a minimum volume of boiling 2-propanol [see Technique 5.5a]. Collect the recrystallized product on a Hirsch funnel. Allow the crystals to dry overnight before determining the melting point and percent yield. Determine the NMR spectrum of your dried product as directed by your instructor. Compare it to the NMR spectrum of 1,2,4,5-tetramethylbenzene shown in Figure 17.1.

Cleanup: Neutralize the aqueous filtrate from the reaction mixture with solid sodium carbonate before pouring it down the sink or placing it in the container for aqueous inorganic waste. Pour the filtrate from the crystallization into the container for halogenated waste.

References

1. Suzuki, H.; Nakamura, K; Goto, R. *Bull. Chem. Soc. Jpn.* **1966,** *39,* 128–131.
2. Suzuki, H. *Organic Syntheses;* Wiley: New York, 1971, Coll. Vol. VI, pp. 700–701.

Questions

1. The active iodinating species (I^+) forms from both the oxidation of iodine by periodic acid and the reduction of periodic acid by iodine. Write the half-reactions and balance the redox equation for the following reaction:

$$I_2 + HIO_4 \cdot 2H_2O + H^+ \longrightarrow I^+ + H_2O \quad \text{(unbalanced)}$$

2. Why was it suggested that you do a mixture melting point of your product and the starting material? What would you expect to observe for the melting range of this mixture?
3. Explain why the NMR spectrum of your product differs from that of the starting material (Figure 17.1).
4. Salts such as silver sulfate and mercuric oxide have been used to oxidize molecular iodine to iodonium ion, I^+. Write equations for these oxidation-reduction processes.

(**17.2**)

Bromination of Acetanilide

Determine the directing effect of an acetamino group in an electrophilic aromatic substitution reaction.

HNCOCH$_3$

+   Br$_2$ $\xrightarrow{\text{CH}_3\text{COOH}}$ HNCOCH$_3$ + HBr

Br

Acetanilide	Bromine	MW 214.1
mp 114.3°C	bp 58.8°C	
MW 177.2	MW 159.8	
	density 3.12 g · mL^{-1}	

Which product is formed?
2-Bromoacetanilide   mp 99°C
OR
3-Bromoacetanilide   mp 87.5°C
OR
4-Bromoacetanilide   mp 168°C

Acetanilide, the starting reagent in this synthesis, is a mild analgesic (reduces pain) and a mild antipyretic (reduces fever). The antipyretic properties of acetanilide were discovered by accident when a sample, improperly labeled and thought to be naphthalene, was inadvertently administered to a patient in 1886.

Bromination of aniline (aminobenzene) suffers from lack of control. The electron-donating amino group activates the ring to such a great extent that usually only tribromoaniline can be isolated:

Aniline   2,4,6-Tribromoaniline

However, if aniline is converted to acetanilide, for example, by treatment with acetyl chloride or acetic anhydride, monosubstitution is easily achieved because the acetamido group [$HNC(=O)CH_3$] cannot activate the benzene ring toward electrophilic attack as well as the simple amino group does. The acetamido group is less effective in donating electron density to the benzene ring, because the electron pair on the nitrogen atom is delocalized by both the carbonyl group and the phenyl ring:

Acetanilide   Bromoacetanilide

It should be kept in mind, however, that the acetamino group is still an activating group.

In this experiment you will be able to deduce the structure of the bromoacetanilide produced by using melting points and NMR spectrometry to differentiate among the *ortho, meta,* and *para* possibilities of the disubstituted product. You can also predict the likely product(s) by using the mechanistic principles of electrophilic aromatic substitution. These can be found in nearly all organic chemistry textbooks.

If ^{1}H NMR is used to deduce the structure of the product, the symmetry of the aromatic protons is most useful. The symmetry in *para* compounds is such that the aromatic proton region is usually split into two signals, each integrating for two protons. Moreover, the two signals usually have a mirror plane of symmetry between them; thus each appears to be a doublet or a distorted doublet. If the *ortho* or *meta* isomer is obtained, the aromatic proton region of the ^{1}H NMR spectrum is less symmetrical.

The bromoacetanilide that you prepare can be hydrolyzed to the corresponding bromoaniline with hot aqueous acid followed by treatment with aqueous base:

Bromoacetanilide   Amine hydrochloride   Bromoaniline

The procedure for carrying out the hydrolysis step is provided as the optional experiment.

⁣⁣Macroscale Procedure

Technique   Mixed Solvent Recrystallization: Technique 5.2a

SAFETY INFORMATION

Acetanilide is toxic and an irritant. Avoid contact with skin, eyes, and clothing.

Glacial acetic acid is corrosive and causes burns. The vapor is extremely irritating to mucous membranes and the upper respiratory tract. Measure it in a hood and avoid contact with skin, eyes, and clothing.

Bromine is very corrosive and causes serious burns. The vapor is toxic and irritating to the eyes, mucous membranes, and respiratory tract. A solution of bromine also emits bromine vapor and should be used only in a well-ventilated hood. Wear gloves while measuring and transferring the bromine solution.

Ethanol is flammable.

Glacial acetic acid gets its name from the fact that it freezes at 16°C and looks like a glacier.

Place 2.0 g of acetanilide in a 50-mL Erlenmeyer flask. Add 7.5 mL of glacial acetic acid and swirl the flask until the solid dissolves. Put a magnetic stirring bar in the flask and clamp the flask on a magnetic stirrer.

Obtain 4.4 mL of a 1:4 (v/v) bromine/glacial acetic acid solution (4.1 M in Br_2). **(Caution: Wear gloves.)** Pour the bromine solution into a separatory funnel. Position the separatory funnel in a ring clamp, with the outlet tip of the funnel slightly inside the neck of the Erlenmeyer flask. Add the bromine solution slowly over a 5-min period. Stir the reaction mixture for another 15 min to complete the reaction. Pour the reaction mixture into a 150-mL beaker containing 60 mL of water. Rinse the flask with 15 mL of water and add this rinse to the beaker containing your crude product.

If the reaction mixture does not have a slight orange color after the addition of the bromine is completed, add an additional 0.5 mL of bromine solution.

— *SAFETY PROCEDURE* —

Carry out the bromination procedure in a hood. The rest of the synthesis can be performed outside of a hood.

Stir the precipitated solid well with a stirring rod to break up any chunks. If the solution still has an orange color from excess bromine, add solid sodium bisulfite in small portions with a spatula, stirring after each addition until the color is discharged. Collect the product by vacuum filtration and wash the solid four times with approximately 10-mL portions of cold water.

Record the total volume of ethanol needed to dissolve your product.

Transfer the solid to a 50-mL Erlenmeyer flask. Recrystallize your product by dissolving it in a minimum amount of hot 95% ethanol, which should be added in 1-mL increments [see Technique 5.2a]. When the solid is dissolved, continue heating the solution while adding a volume of water that is 80% of the ethanol volume used to dissolve the solid product. Swirl the flask to mix the solvents; heat the solution briefly until any crystals that formed have dissolved. Cool the flask to room temperature to initiate crystallization, then cool it in an ice-water bath for a few minutes. Collect the crystals by vacuum filtration. Allow the crystals to dry before determining the melting point and percent yield. Identify the product by comparing your experimental melting point to those listed on p. 208 for the possible products. Prepare a sample for NMR analysis as directed by your instructor. Analyze the aromatic region of the spectrum. Does the observed splitting pattern support your identification based on melting point?

Cleanup: Pour the aqueous filtrate from the crude product into a large beaker and neutralize the excess acid with solid sodium carbonate before washing the solution down the sink or placing it in the container for aqueous inorganic waste. Pour the filtrate from the recrystallization into the container for halogenated waste.

Optional Experiment

Removal of the Acetyl Protecting Group: Synthesis of a Bromoaniline

The acetyl group of the bromoacetanilide produced in Experiment 17.2 can be removed by a hydrochloric acid-catalyzed hydrolysis reaction to form the corresponding bromoaniline.

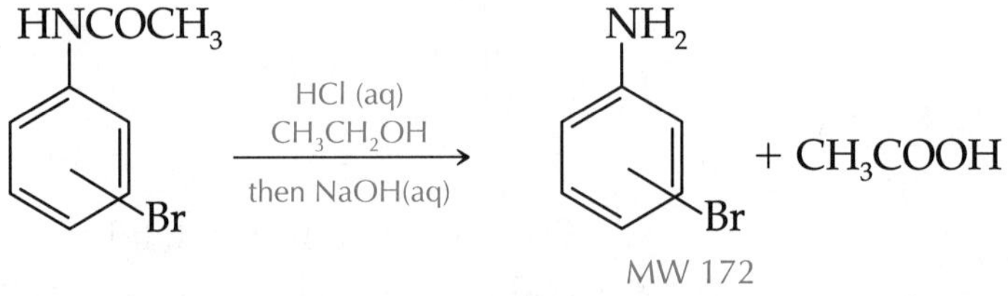

2-Bromoaniline mp 32°C
OR
3-Bromoaniline mp 18.5°C
OR
4-Bromoaniline mp 66°C

✎Macroscale Procedure

___ SAFETY INFORMATION ___

Concentrated **hydrochloric acid** is corrosive and causes burns. The vapor is extremely irritating to mucous membranes and the upper respiratory tract. Measure it in a hood and avoid contact with skin, eyes, and clothing.

Aqueous **sodium hydroxide solutions** are corrosive and cause burns.

Methanol is toxic and flammable. Pour it only in a hood.

Ethanol is flammable.

This experiment requires another laboratory period.

Use all of your remaining bromoacetanilide for this synthesis. The following amounts of reagents are for 1.0 g of bromoacetanilide. Calculate how much of each reagent you will need for your quantity of bromoacetanilide:

1.0 g of bromoacetanilide requires
    2.0 mL   95% ethanol
    1.2 mL   concentrated hydrochloric acid

Place your bromoacetanilide in a 50-mL round-bottomed flask. Add the calculated amount of ethanol and a boiling stone. Fit the flask with a water-cooled reflux condenser. Heat the flask gently with a heating mantle until the solid dissolves, then slowly add the calculated amount of concentrated hydrochloric acid down the condenser in small portions. Heat the reaction mixture under reflux for 35 min.

Remove the heating mantle and cool the flask for several minutes. Add 14 mL of water for each 1.0 g of bromoacetanilide and another boiling stone. Rearrange the apparatus for simple distillation [see Technique 7.3]. Use a 25-mL round-bottomed flask as the receiver. Collect distillate (a mixture of water and ethanol) until the thermometer reads 98–99°C.

Use at least 14 mL of water, even if you began with less than 1 g of bromoacetanilide.

Cool the flask briefly and pour the residual solution containing your product into a 150-mL beaker containing 8 mL of ice-cold water for each 1.0 g of bromoacetanilide. Neutralize the solution to pH 8–9 by dropwise addition of a 5% sodium hydroxide solution. Vigorously stir the solution with a stirring rod between additions of NaOH and use indicator paper such as pHydrion to monitor the pH. When the mixture becomes alkaline, the product will separate as an oil. Stir and cool the mixture in an ice-water bath until the oil crystallizes. Collect the solid product by vacuum filtration on a Buchner funnel.

Recrystallize the product from a minimum volume of a 50/50 (v/v) methanol/water solution [see Technique 4.3]. If the product begins to come out of solution as an oil while the crystallization solution is cooling, stir the solution and scratch the wall of the flask with a stirring rod to induce crystal formation. Cool the crystallized mixture briefly in an ice-water bath before collecting the crystals by vacuum filtration. When the product has dried, determine the percent yield and the melting point.

Cleanup: Wash the aqueous filtrate from the crude product down the sink or pour it into the container for aqueous inorganic waste. Pour the filtrate from the recrystallization into the container for halogenated waste.

Reference

1. Furniss, B.; Hannaford, A. J.; Smith, P. W. G.; Tatchell, A. R. *Vogel's Textbook of Practical Organic Chemistry;* 5th ed.; Longman Scientific: London, 1989, pp. 918–919.

Questions

1. Sodium bisulfite is a reducing agent. Write a balanced redox equation for the reaction that occurs between excess bromine and $NaHSO_3$ after the synthesis of bromoacetanilide.
2. Bromination of phenol with a solution of bromine in water gives a tribromo product. What isomer would you expect to be formed?
3. Write out the mechanism for the bromination of acetanilide in this experiment.
4. Write out the mechanism for acid-promoted hydrolysis of *p*-bromoacetanilide.

17.3

Rates of Electrophilic Aromatic Substitution

 Investigate how substituents on a benzene ring affect the rate of aromatic bromination.

The rate of electrophilic substitution in aromatic compounds depends on the substituents attached to the aromatic nucleus. Electron-donating groups attached to an aromatic ring in general facilitate all electrophilic substitution reactions, whereas electron-withdrawing groups retard the process. Although absolute rate constants of electrophilic aromatic substitution reactions are difficult to obtain, relative rate constants can easily be measured. Relative rate constants are sufficient to allow the estimation of the relative reactivities of a series of compounds. The relative rates are compared by measuring reactant concentration as a function of time for two systems under identical conditions.

In this experiment, the rate of bromination of a series of aromatic compounds will be determined by using a spectrophotometer to follow the bromine disappearance per unit time.

Three aromatic compounds will be used: acetylsalicylic acid (aspirin), diphenyl ether, and acetanilide.

| Acetylsalicylic acid | Diphenyl ether | Acetanilide |

Although the reaction kinetics are rather complex, there are distinct differences in the rates of bromination among these three aromatic compounds because of the different substituents attached to the aromatic nucleus.

ᴍmicroscale
Procedure

SAFETY INFORMATION

Acetanilide is toxic and an irritant. Avoid contact with skin, eyes, and clothing.

Glacial acetic acid is corrosive and causes burns. The vapor is extremely irritating to mucous membranes and the upper respiratory tract. Measure it in a hood and avoid contact with skin, eyes, and clothing.

Bromine is very corrosive and causes serious burns. The vapor is toxic and irritating to the eyes, mucous membranes, and respiratory tract. A solution of bromine also emits bromine vapor and should be used only in a well-ventilated hood. Wear gloves while measuring and transferring the bromine solution.

Do not use a metal spatula to transfer any of the reagents, because metal salts can catalyze these bromination reactions.

The following stock solutions should be made up before you begin the rate studies. Prepare 100 mL of 9:1 (v/v) acetic acid/water solution. Prepare 10 mL of a 0.2 M solution of each aromatic compound in the acetic acid/water solution. You will also need 10 mL of 0.05 M bromine solution in the same solvent. Prepare a 35°C water bath in a 250-mL beaker.

A Bausch and Lomb Spectronic 21, or a similar spectrometer, works well for this experiment.

Set the spectrophotometer at a wavelength of 520 nm, where the absorption of bromine is strong. Allow 2.0 mL of the 0.2 M stock solution of acetylsalicylic acid to equilibrate in a water bath at 35°C in a corked cuvet. This is your bromination reaction cuvet. Also place a corked test tube containing 8 mL of the bromine solution in the 35°C water bath.

Fill a second cuvet with the 0.2 M stock solution of acetylsalicylic acid. This cuvet serves as the reference solution used to adjust the instrument to zero absorbance (100% transmittance) before you begin the bromination reaction.

Adjusting the instrument to zero absorbance with the substrate solution ensures that any absorbance observed will be due to the change in bromine concentration, not to the concentration of substrate.

Rapidly carry out the following operations: Open the cuvet of acetylsalicylic acid that was equilibrated at 35°C, pipet 2.0 mL of the bromine solution that has also equilibrated at 35°C into the cuvet of acetylsalicylic acid solution, start the stopwatch, recork the cuvet, shake the cuvet to mix the solution, wipe the cuvet, remove the cork, and place the cuvet in the spectrophotometer. Record absorbances at 30, 60, 90, 120, 150, 180, 300, and 600 s. Continue to take readings of the absorbance of the sample at 15-min intervals until at least 4000 s. The corked cuvet should be maintained at 35°C in the water bath when readings are not being taken. If you find that your bromine solution is too concentrated and the bromination reaction is over too quickly, you can dilute the bromine solution with a 9:1 (v/v) acetic acid/water mixture and rerun the reaction to obtain good kinetic data.

With a little practice, sample preparation operations can be performed in less than 10 s.

Repeat the preceding experiment with diphenyl ether and then with acetanilide. For these experiments, the reference solution is the 0.2 M stock solution of the substrate being brominated: diphenyl ether or acetanilide. Take absorbance readings at 10-s intervals for the first minute and then at 30-s intervals until the change in absorbance between readings is small (about 5 min).

Treatment of Data   Plot the absorbance readings versus time for each bromination reaction. Extrapolate each graph to zero time to obtain A_0, the absorbance at time 0. Then plot relative absorbance (A/A_0) as a function of time for the three aromatic compounds. The relative rates of electrophilic substitution are the same as the relative slopes of the lines in your graphs if you use the same graph scale for each reaction.

Cleanup: Place all solutions containing bromine in the container for halogenated waste. Place the solutions from the reference cuvet in the container for flammable (organic) waste.

Optional Experiments

1. Devise and carry out a procedure to isolate and purify the organic product(s) from one of these bromination reactions. You will need to begin the synthesis with at least 1 g of the aromatic compound for a macroscale synthesis or 200 mg of the aromatic compound for a microscale synthesis.
2. If you had used brominated phenol in this experiment, what sort of rate would you expect? Test your prediction by running the experiment with a phenol solution.

Reference

1. Casanova, J. *J. Chem. Educ.* **1964,** *41,* 341–342.

Questions

1. How does the measurement of the absorbance of Br_2 give a measurement of the relative rates of bromination in this experiment?
2. Account for the relative rates that you experimentally determined by considering the structures of the aromatic compounds.
3. What should the products of bromination be in each reaction?
4. Predict the relative reactivities of the following three compounds when subjected to bromination conditions: methoxybenzene (anisole), benzene, nitrobenzene.

Experiment 18

ACYLATION AND ALKYLATION OF AROMATIC COMPOUNDS

Investigate a variety of Friedel-Crafts reactions and purify the products by column chromatography or recrystallization.

The Friedel-Crafts acylation and alkylation of aromatic compounds are specific examples of electrophilic aromatic substitution, which was discussed in Experiment 17. Friedel-Crafts reactions, named after the French and American chemists who discovered their synthetic importance over 100 years ago, lead to carbon-carbon bond formation. Acyl and alkyl groups can be substituted on aromatic rings by using acid catalysts, such as H_2SO_4, H_3PO_4, and HF, or Lewis acids, such as $AlCl_3$ and BF_3.

Friedel-Crafts chemistry is big business. For example, about nine billion pounds of ethylbenzene are produced in the United States each year by the reaction of benzene and ethene in the presence of either a proton or a Lewis acid catalyst. Most of it is dehydrogenated to form styrene, from which polystyrene (Experiment 29.1) is made:

| Benzene | Ethene | Ethylbenzene | Styrene |

In a Friedel-Crafts alkylation the alkyl electrophile can be prepared by many methods; the traditional one in undergraduate laboratories has been treatment of an alkyl halide with a Lewis acid, commonly aluminum trichloride. We will first review electrophilic alkylation, with a focus on the processes occurring in Experiments 18.2 and 18.3. Later we will discuss the acylation of ferrocene to produce acetylferrocene (Experiment 18.1).

In both 18.2 and 18.3, the electrophile is the *tert*-butyl cation. This cation is especially easy to produce because it is tertiary and thus more stable than either a secondary or a primary cation. Using benzene as the substrate, a simple rendition of the substitution (alkylation) mechanism is:

Delocalized cationic intermediate

tert-Butylbenzene (1,1-dimethylethyl) benzene

The delocalized cationic intermediate corresponds to three localized resonance forms:

Delocalized intermediate   Resonance forms

The process of delocalization, or distribution, of the positive charge over a large portion of the ring system stabilizes the cationic intermediate. When a proton is lost, the highly stable aromatic ring is regenerated.

Any additional groups on the benzene ring that stabilize the positive charge increase the rate of substitution. Moreover, the ability of such substituents to stabilize or destabilize positive charge can be used to predict the ability of a group to direct the substitution to either an *ortho, meta,* or *para* position. In Experiment 18.3, an alkoxy group (OR) provides an electronegative atom that has a nonbonding pair of electrons and is attached directly to the ring. This atom donates electrons to the ring, a process that stabilizes a positive charge.

Para substitution product

Thus, the "extra" resonance provided by the alkoxy group facilitates additional charge distribution, stabilizing the positive charge of the intermediate. Moreover, the direct interaction of the alkoxy group with positive charge causes electrophilic attack to occur in the *para* (as above) or *ortho* position.

Another type of Friedel-Crafts reaction is the acylation of aromatic compounds. In this electrophilic aromatic substitution reaction, an acyl derivative, such as an acyl chloride or acyl anhydride, reacts with the aromatic compound in the presence of acids, such as $AlCl_3$ or H_3PO_4. The product of the acylation reaction is a ketone:

Experiment 18.1 uses the novel aromatic compound ferrocene, an organometallic compound is composed of two planar five-membered rings that "sandwich" an iron atom:

The ferrocene sandwich compound can be named $Fe(\eta^5\text{-}C_5H_5)_2$, where the Greek letter η (eta) means that each ligand bonds to the metal atom through all five carbon atoms of the ring.

Ferrocene was originally made in 1951 by treating sodium cyclopentadienide with iron(II) chloride:

$$FeCl_2 \; + \; 2 \; \underset{\substack{\text{Sodium} \\ \text{cyclopentadienide}}}{\boxed{\;\ominus\;}} Na^+ \; \longrightarrow \; (C_5H_5)_2Fe + 2NaCl$$

Ferrous chloride   Sodium cyclopentadienide   Ferrocene

Ferrocene can be thought of as a compound formed by the bonding of an Fe^{2+} atom to two cyclopentadienide ligands, each bearing a negative charge. Each cyclopentadienide anion is aromatic because it has six π-electrons:

6 π-electrons

$C_5H_5^-$

Because the rings in ferrocene are aromatic, they undergo electrophilic aromatic substitution reactions, such as the acylation reaction in Experiment 18.1.

18.1
Acylation of Ferrocene

Acetylate a colored organometallic compound and purify it by column chromatography.

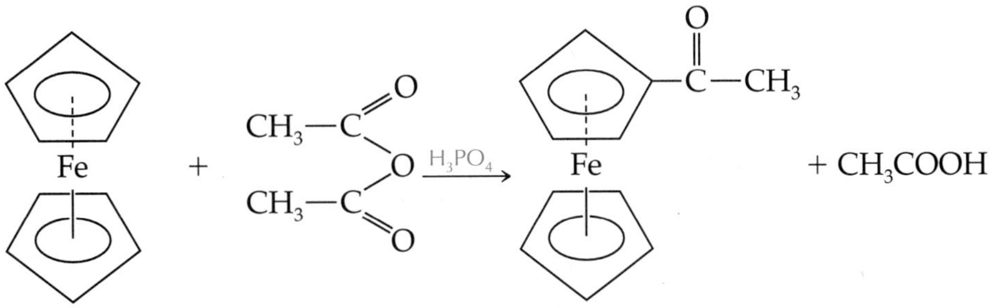

Ferrocene
mp 173°C
MW 186.0
yellow-orange color

Acetic anhydride
bp 139.5°C
MW 102.09
density 1.08 g · mL^{-1}

Acetylferrocene
mp 85–86°C
MW 228.1
orange-red color

This Friedel-Crafts acylation of ferrocene produces acetylferrocene by using acetic anhydride in the presence of a catalytic amount of phosphoric acid. A frequently used catalyst for such acylations is aluminum trichloride, but in this particular acylation that catalyst complicates the process by producing a disubstituted product: 1,1′-diacetylferrocene. The milder catalyst, phosphoric acid, works better. It generates the acylium ion electrophile by protonation and loss of acetic acid:

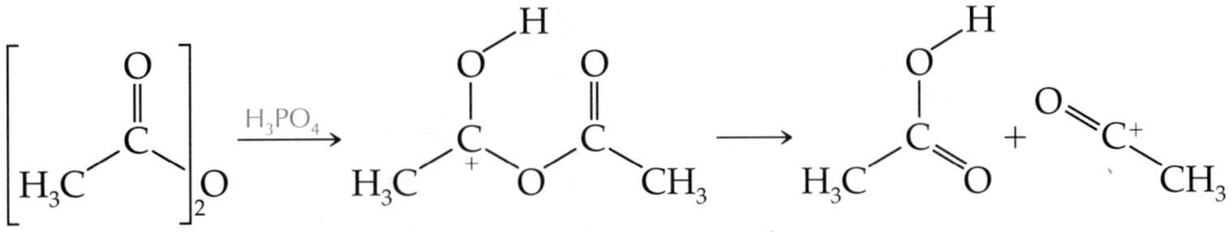

Acetic anhydride

Acylium ion
electrophile

The electrophile then attacks the ring, a reaction resulting in substitution of the acetyl group for a ring proton:

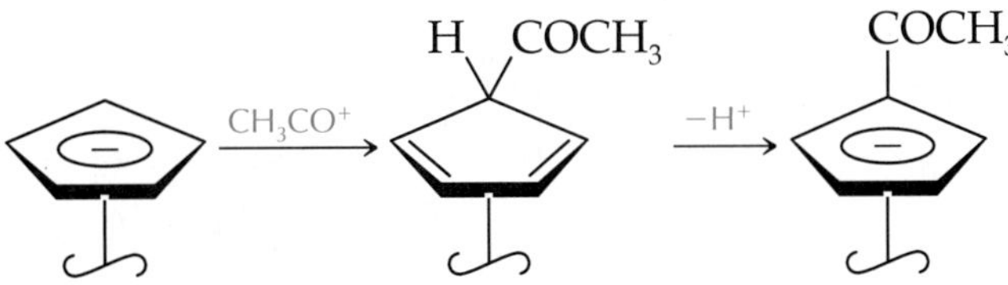

Acetylferrocene can be characterized by examining its IR (Figure 18.1) and ^{1}H NMR (Figure 18.2) spectra. Note the intense carbonyl stretching vibration in the IR spectrum at about 1670 cm^{-1}. The ^{1}H NMR spectrum of ferrocene shows 10 equivalent aromatic protons as a singlet at about δ 4.15. The ^{1}H NMR spectrum of acetylferrocene (Figure 18.2) shows the acetyl methyl group as a 3H singlet at δ 2.42. The unsubstituted ring yields a 5H singlet at

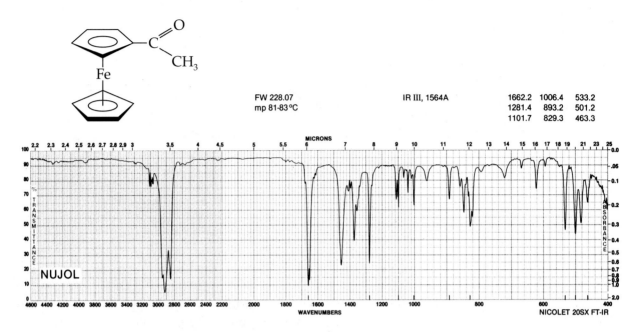

FW 228.07
mp 81-83°C

IR III, 1564A

1662.2	1006.4	533.2
1281.4	893.2	501.2
1101.7	829.3	463.3

NUJOL

NICOLET 20SX FT-IR

FIGURE 18.1 IR spectrum of acetylferrocene.

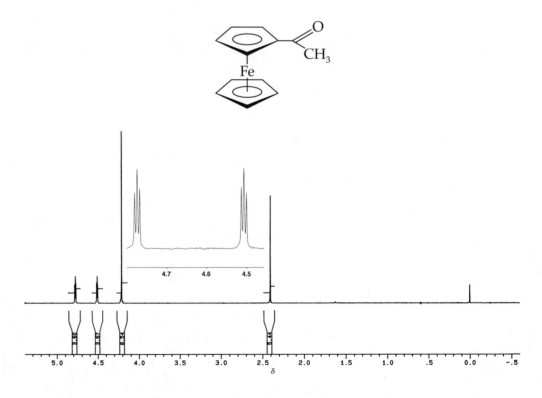

FIGURE 18.2 ^{1}H NMR spectrum (300 MHz) of acetylferrocene (in CDCl$_3$).

δ 4.22, and the substituted ring reveals a pair of 2H signals as apparent triplets, one at δ 4.5 and the other near δ 4.8.

**_microscale_
Procedure*** | **Techniques** Thin-Layer Chromatography: Technique 10
Column Chromatography: Technique 12
IR Spectrometry: Spectrometric Method 1
NMR Spectrometry: Spectrometric Method 2

— *SAFETY INFORMATION* —

Ferrocene is relatively nontoxic, but avoid contact with the skin.

Acetic anhydride is corrosive and a lachrymator (causes tears). Wear gloves and avoid contact with skin, eyes, and clothing. Dispense it in a hood.

Concentrated (85%) phosphoric acid is irritating to the skin and mucous membranes. Wear gloves. If you spill any phosphoric acid on your skin, wash it off immediately with copious amounts of water.

Aqueous sodium hydroxide solutions are corrosive and cause burns. Solutions as dilute as 9% (2.5 M) can cause severe eye injury. Avoid contact with skin, eyes, and clothing.

Hexane and diethyl ether are extremely volatile and flammable.

Alumina (Al_2O_3) is a lung irritant. Avoid breathing the dust.

***Preparation of
Acetylferrocene*** Fit a dry 10-mL round-bottomed flask with a screw cap and drying tube containing anhydrous calcium chloride [see Technique 3.4, Figure 3.6b (omit the condenser)]. Keep the drying tube on the flask except while you are adding reagents. Place 200 mg of ferrocene and 2.0 mL of acetic anhydride in the flask. Swirl the flask to mix these reagents. Slowly add 0.4 mL of 85% phosphoric acid (about 10 drops with a Pasteur pipet; the exact amount is not critical). Screw the drying tube on the flask and swirl the reaction mix-

*This procedure was developed by David Alberg, Department of Chemistry, Carleton College, Northfield, MN.

ture to thoroughly mix the reagents. Heat the flask on a steam bath or in a beaker of boiling water for 10 min with occasional swirling.

Acetylferrocene will appear as an orange-red spot ($R_f \approx 0.3$), and any remaining ferrocene appears as a yellowish spot at $R_f \approx 0.9$.

Remove the flask from the steam bath and check the progress of the reaction by thin-layer chromatography on silica gel plates. Also spot the plate with a 2% solution of ferrocene in ether. Use 25:75 (v/v) anhydrous diethyl ether/hexane as the TLC elution solvent. A UV lamp allows you to visualize traces of ferrocene. A trace of ferrocene is likely; but if you can see a substantial yellow spot of ferrocene without the aid of the UV lamp, heat your reaction mixture for an additional 2–5 min. If the amount of ferrocene is minimal, cool the reaction flask for a total of 10 min.

Pour the reaction mixture over about 10 g of ice in a 50-mL beaker. Use an additional 1 or 2 mL of water to complete the transfer of your mixture to the ice. Partially neutralize the mixture by adding 5 mL of 6 M sodium hydroxide in at least three portions. Determine the pH with pHydrion paper or other pH test paper. Continue adding 6 M NaOH dropwise until the pH is 7–8. Swirl the beaker after each addition to mix the contents. Cool the mixture to room temperature and collect the product by vacuum filtration on a Hirsch funnel. Use a few milliliters of water to complete the transfer of the tarry solid. With the vacuum on, pull air over the crude product on the Hirsch funnel for 15 min to dry the product while you prepare for the column chromatography.

Purification by Column Chromatography

Assemble all the equipment and reagents that you will need for the entire chromatography procedure before you begin to prepare the column.

Large-volume Pasteur pipets, available from Fisher Scientific, catalog no. 13678-8, have a capacity of 4 mL.

Read this procedure completely and review Technique 12 before you undertake this part of the experiment.

Obtain about 25 mL of hexane in a 50-mL Erlenmeyer flask fitted with a cork. Transfer your air-dried crude product to a 13 × 100 mm test tube and add about 1 mL of hexane. Much of the material will not dissolve in the hexane. Spot a thin-layer plate with this hexane mixture, then set the test tube and the thin-layer plate aside while you prepare the column.

Obtain a large-volume Pasteur pipet to use as the chromatography column and pack a small plug of glass wool down into the stem, using a wood applicator stick or a thin stirring rod. Clamp this pipet in a vertical position and place a 25-mL Erlenmeyer flask underneath it to collect the hexane that you will be adding to

Be sure that the alumina is covered with solvent at all times during the chromatographic procedure.

the column. Weigh approximately 3 g of Activity III alumina in a tared 50-mL beaker; add enough hexane to make a thin slurry. Transfer the alumina slurry to the column, using a regular Pasteur pipet. Continue adding slurry until the column is two-thirds full of alumina. Fill the column 4–5 times with hexane to pack it well. The eluted hexane can be reused for this purpose. After the alumina is packed, add a 2–3 mm layer of sand above the alumina by letting it settle through the hexane.

Allow the hexane level to almost reach the top of the alumina and place a flask labeled fraction 1 under the column. Begin transferring your crude product mixture to the column, using a Pasteur pipet and as many small portions of hexane (do not use the eluted hexane) as necessary to transfer all of your material to the top of the column. When all of the crude product is on the column and the hexane level is just above the top of the alumina, elute the column with 15 mL of hexane. You may see a faint yellow color in the hexane solution collecting in fraction 1; that color is due to ferrocene that did not react.

Next elute with 10 mL of 50:50 (v/v) hexane/anhydrous diethyl ether solution. You will see the orange-red acetylferrocene move rapidly down the column. Collect the eluent in fraction 1 until you see the orange-red solution in the column tip, then quickly change the collection flask to a clean, tared 50-mL Erlenmeyer flask labeled fraction 2. Continue adding 50:50 hexane/ether until the orange-red product has eluted from the column. (This elution requires about 10–15 mL of 50:50 hexane/ether.) Spot fraction 1, fraction 2, and pure ferrocene on the same thin-layer plate that you have already spotted with your crude acetylferrocene. Develop the thin-layer plate as you did previously. Record the results in your notebook.

Recover your purified acetylferrocene by evaporating the solvent from fraction 2 on a steam bath or with a stream of nitrogen in a hood. Alternatively, if a rotary evaporator is available, transfer fraction 2 to a tared 25- or 50-mL round-bottomed flask and remove the solvent under reduced pressure. Weigh your purified product, calculate the percent yield, and determine the melting point of your acetylferrocene. Prepare a sample for NMR or IR analysis as directed by your instructor. Assign all the major peaks but do not try to analyze the complex splitting patterns.

Cleanup: The aqueous filtrate from the crude product may be washed down the sink or placed in the container for aqueous inorganic waste. Pour any remaining TLC solvent and fraction 1 into the container for flammable (organic) waste. Place the thin-layer plates and the alumina from the column in the container for inorganic solid waste.

Questions

1. Any diacetylferrocene produced in this reaction remains near the top of the column under the chromatographic conditions used in this experiment. Explain the order of elution of ferrocene and acetylferrocene, and why the diacetylferrocene is retained by the column.

2. In the NMR spectrum of most aromatic compounds, the aromatic protons exhibit a chemical shift of $\delta = 7-8$ ppm. However, in ferrocene, the chemical shift of the aromatic protons is $\delta = 4.15$ ppm. Explain what factors cause the upfield shift.

3. Explain how the NMR spectrum of ferrocene supports the assigned sandwich structure rather than a structure in which the iron atom is bound to only one carbon atom of each ring.

4. Why is the acetylation of acetylferrocene faster on the unsubstituted cyclopentadienide ring?

5. There is only one isomer known for diacetylferrocene when each cyclopentadienide ring is monosubstituted. Explain why other isomers are not found.

18.2

Synthesis of 4,4′-Di-*tert*-Butylbiphenyl

Investigate a classic Friedel-Crafts reaction using AlCl$_3$ and an alkyl halide.

Biphenyl	2-Chloro-2-methylpropane	4,4′–Di-*tert*-butylbiphenyl
mp 69°C	(*tert*-butyl chloride)	mp 128–129°C
MW 154	bp 51°C	MW 266
	MW 92.6	
	density 0.85 g · mL^{-1}	

In this experiment an electrophile is produced by treating *tert*-butyl chloride with aluminum trichloride. Because aluminum trichloride has only six electrons in its valence shell, it is electron deficient and has Lewis acid properties. Therefore, aluminum trichloride will coordinate with *tert*-butyl chloride, leading to abstraction of chloride anion from the alkyl halide to give the *tert*-butyl cation

$$(CH_3)_3C\ddot{C}l: + AlCl_3 \longrightarrow (CH_3)_3\overset{\delta+}{C}\text{---}\overset{\delta+}{Cl}\text{---}\overset{\delta-}{AlCl_3} \xrightarrow{-AlCl_4^-} (CH_3)_3C^+$$

The electrophilic *tert*-butyl cation then attacks biphenyl, and a combination of electronic and steric effects causes *para* substitution in both rings:

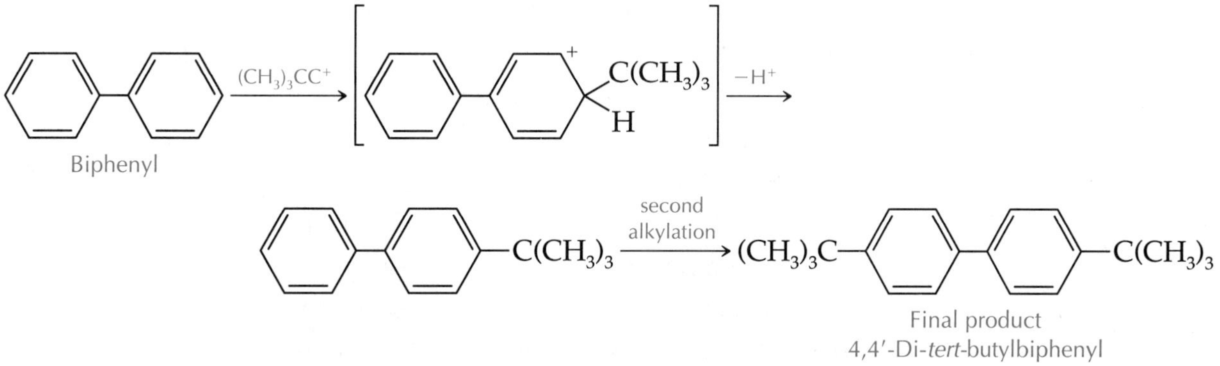

Biphenyl

Final product
4,4'-Di-*tert*-butylbiphenyl

In this reaction, the side product is HCl. Because this compound is a noxious and toxic gas, a good trap is essential.

⁄⁄⁄⁄⁄⁄Macroscale Procedure

Techniques Recrystallization: Technique 5.3

NMR Spectrometry: Spectrometric Method 2

Measure the required amount of AlCl₃ quickly and store the tightly closed reagent container in a desiccator.

— *SAFETY INFORMATION* —

Wear gloves while measuring the reagents and carrying out this experiment.

Aluminum chloride is corrosive, moisture sensitive, and forms HCl on contact with atmospheric moisture.

2-Chloro-2-methylpropane (*tert*-butyl chloride) is volatile and flammable.

Biphenyl is a skin irritant.

Nitromethane is toxic, absorbed through the skin, and flammable. Avoid contact with skin, eyes, and clothing.

(continued on next page)

If possible, work in a hood until after the addition of ice/water to the reaction mixture is completed.

Weigh 0.45 g of anhydrous aluminum chloride powder and quickly transfer it to a small, dry test tube. Cork or stopper the test tube *immediately* so that the $AlCl_3$ will not react with atmospheric moisture and become hydrated, thereby losing its catalytic effectiveness.

Place 1.5 g of biphenyl in a 125-mL filter flask, add 2.5 mL of 2-chloro-2-methylpropane (*tert*-butyl chloride) and 8 mL of nitromethane. Put a rubber stopper in the neck of the reaction flask. Prepare a gas trap by one of the following methods.

Procedure for laboratories equipped with water aspirators. If your laboratory is equipped with water aspirators, connect the side arm of the reaction flask to a guard flask that is connected to the vacuum takeoff of a water aspirator, as shown in Figure 18.3.

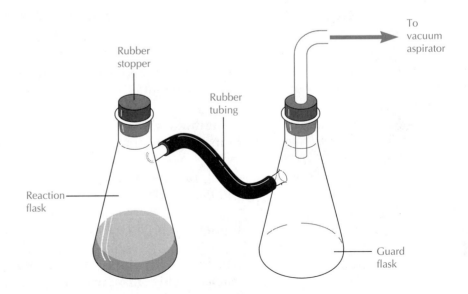

FIGURE 18.3 Reaction apparatus for a vacuum (water) aspirator.

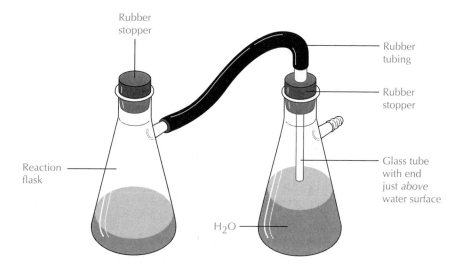

Rubber stopper

Rubber tubing

Rubber stopper

Reaction flask

Glass tube with end just *above* water surface

H₂O

FIGURE 18.4 Reaction apparatus with a gas trap.

Use a modest vacuum (by adjusting the water flow rate) to carry off the hydrogen chloride that will be formed in the reaction. If you apply too much vacuum, some of the 2-chloro-2-methyl-propane may be lost by evaporation and your product yield may be lowered.

Procedure for laboratories equipped with vacuum lines. If your laboratory has a central vacuum line system, a gas trap will be used rather than a vacuum trap. Assemble a gas trap as shown in Figure 18.4. Fill a 250-mL filter flask approximately one-third full of water. Place a one-hole rubber stopper containing a piece of glass tubing in the neck of the flask so that the top of the glass tubing is close to the surface of the water but *does not extend below* the water surface. Connect the upper end of the glass tubing to the side arm of the reaction flask with a piece of rubber tubing. Leave the side arm of the water-filled flask open to the atmosphere.

The Friedel-Crafts Reaction

Once your gas trap is positioned, begin addition of the aluminum chloride with a spatula. Remove the stopper briefly from the reaction flask and add approximately one-fourth of the $AlCl_3$. Return the cork to the test tube containing the aluminum chloride as quickly as possible.

─── *SAFETY PRECAUTION* ───────

Make sure that none of the AlCl$_3$ gets on the upper lip of the reaction flask, because it could hydrolyze in moist air to form toxic hydrogen chloride gas.

Swirl the reaction flask. Bubbles of HCl will be evolved for a few minutes and the surface of the AlCl$_3$ catalyst will turn yellow or orange. When the bubbling subsides, add another small portion of aluminum chloride to the reaction mixture. Continue this process until the addition of AlCl$_3$ is complete. The total time should be about 5–8 min. The rate of addition should be such that the reaction flask remains at room temperature during the entire addition of AlCl$_3$. If the reaction becomes warm at any time, a substantial amount of the low-boiling-point 2-chloro-2-methylpropane will be lost by evaporation. After the last addition of AlCl$_3$, allow the flask to stand for 15 min at room temperature to complete the reaction.

Prepare 15 mL of half ice and half water. Unstopper the reaction flask. Turn off the water aspirator or disconnect the rubber tubing connecting the gas trap to the side arm of the reaction flask. Pour the ice/water mixture into the reaction flask to hydrolyze the aluminum chloride and stir the mixture for several minutes. Add 7 mL of methanol and stir the mixture while cooling it in a salt/ice-water bath. The product should crystallize.

The product is relatively insoluble in cold methanol, and these washes should remove the nitromethane and aluminum salts.

Collect the crude product by vacuum filtration. Chill 10 mL of methanol in a test tube in the ice-water bath. Wash the crystals twice with 10 mL of ice-cold water, followed by two washes of 5-mL portions of ice-cold methanol. Draw air over the product for several minutes to evaporate the residual methanol. Pour the filtrate into a separatory funnel and set it aside for cleanup later.

Transfer the crude product to a 50-mL Erlenmeyer flask and recrystallize it from a minimum amount of a solvent solution consisting of 20% toluene and 80% methanol (v/v) [see Technique 5.3]. Use a steam bath as the heat source. Cool the product solution to room temperature, then in an ice-water bath before collecting the recrystallized product by vacuum filtration. Weigh your dried product, determine the melting point, and calculate the percent yield. Obtain an NMR spectrum of your product as directed by your instructor.

Cleanup: The filtrate from the reaction mixture, which you poured into a separatory funnel, consists of an upper aqueous phase and a milky lower organic layer of nitromethane. Remove the lower layer of nitromethane and place this solution in the container for flammable (organic) waste. The upper aqueous layer remaining in the separatory funnel contains aluminum salts; put this solution in the container for inorganic waste. If you used a water-filled flask as a gas trap, neutralize the HCl solution con-

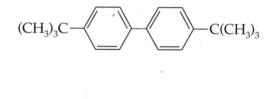

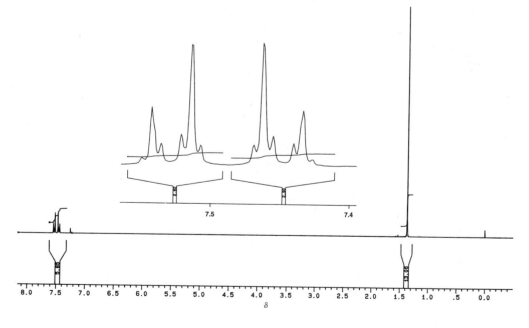

FIGURE 18.5 ^{1}H NMR spectrum (300 MHz) of 4,4'-di-*tert*-butyl-biphenyl (in CDCl$_3$).

tained in the flask with solid sodium carbonate before washing the solution down the sink or placing it in the container for aqueous inorganic waste. Pour the filtrate (toluene/methanol mixture) from the recrystallization into the container for flammable (organic) waste.

Questions

1. Write resonance forms to rationalize the formation of biphenyl with *para* substitution on both rings. Account for the lack of *ortho* substitution product.
2. To analyze the NMR spectrum of 4,4'-di-*tert*-butylbiphenyl, you need to consider its symmetry. How many kinds of protons does it have? Carbons?
3. When isobutylene (2-methylpropene) is treated with a Brønsted acid (for example, sulfuric acid) in the presence of benzene, *tert*-butylbenzene is formed. Write a mechanism that accounts for this observation.
4. Polyalkylation of benzene under Friedel-Crafts conditions is always a concern. That is, even when a simple monoalkylation product is desired, a dialkylation product may be formed. Use electronic theory to explain this.
5. Unless special conditions are used, it is generally not a good idea to try to convert benzene to 1-phenylbutane by using a mixture of 1-chlorobutane and aluminum trichloride. Explain, using mechanisms.

(18.3)
Synthesis of 1,4-Di-*tert*-Butyl-2,5-Dimethoxybenzene

Synthesize a tetrasubstituted benzene derivative, using an alcohol and H_2SO_4 for the electrophilic aromatic substitution reaction.

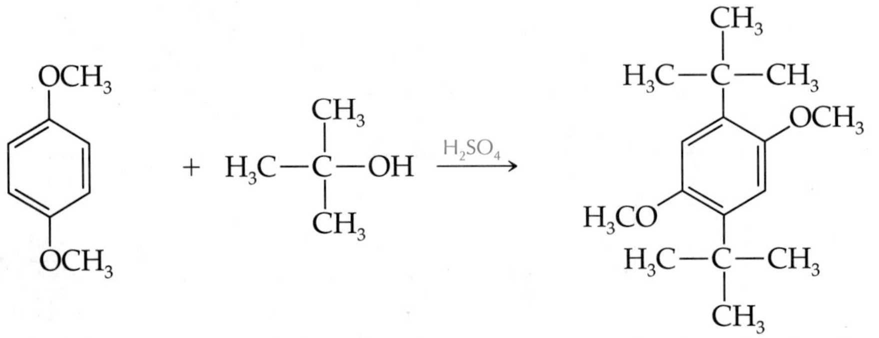

1,4-Dimethoxybenzene
mp 60°C
MW 138

2-Methyl-2-propanol
(*tert*-butyl alcohol)
bp 83°C

1,4-Di-*tert*-butyl-2,5-dimethoxybenzene
mp 104–105°C
MW 250

In this experiment a *tert*-butyl cation is generated by treatment of *tert*-butyl alcohol with sulfuric acid. This electrophilic alkyl cation then attacks 1,4-dimethoxybenzene to give a di-*tert*-butyl substituted final product. Because the methoxy groups on the disubstituted starting material are both activating, two electrophilic substitutions occur, producing a tetrasubstituted final product.

$$(CH_3)_3C-OH + H_2SO_4 \longrightarrow (CH_3)_3C-\overset{+}{O}H_2 \xrightarrow{-H_2O} (CH_3)_3C^+$$

tert-butyl cation

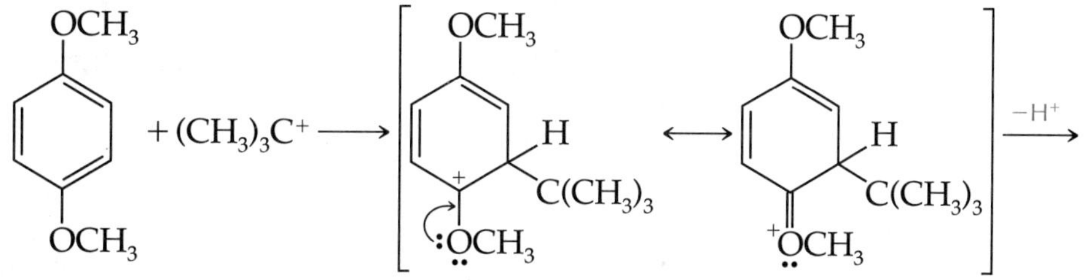

Two of the resonance forms

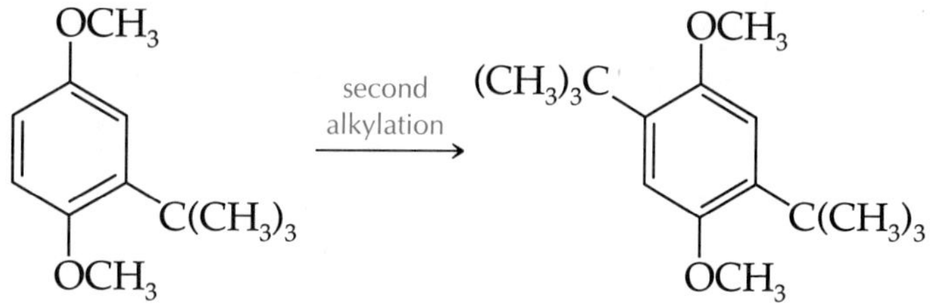

Tetrasubstituted final product

This double substitution occurs despite the fact that much of the reaction is conducted at reduced temperatures.

An advantage of this procedure over that of Experiment 18.2 is the fact that a toxic side product (HCl) is not produced; the side product here is water. Thus, a special trap is not necessary.

ᴍᴍmicroscale
Procedure

Techniques Microscale Reflux: Technique 3.3
Recrystallization: Technique 5.5a

─ *SAFETY INFORMATION* ─

Sulfuric acid is corrosive and causes severe burns. Measure the required amount carefully. Notify the instructor if any acid is spilled.

(continued on next page)

―― *SAFETY INFORMATION (continued)* ――

2-Methyl-2-propanol (*tert*-butyl alcohol) causes skin and eye irritation. Avoid contact with skin, eyes, or clothing. Wash thoroughly after handling it.

1,4-Dimethoxybenzene is a skin irritant. Wash thoroughly after handling it.

Acetic acid is a dehydrating agent and an irritant, and causes burns. Dispense it in a hood and avoid contact with skin, eyes, and clothing.

Methanol is toxic and flammable. Pour it only in a hood.

Place 150 mg of 1,4-dimethoxybenzene in a 10-mL Erlenmeyer flask, add 0.25 mL of 2-methyl-2-propanol (*tert*-butyl alcohol) and 0.5 mL of acetic acid. Warm the mixture briefly on a hot plate until the solid dissolves. Clamp the flask in a small ice-water bath and stir the contents with a short stirring rod. Obtain 1.0 mL of concentrated sulfuric acid in a 13-by-100-mm test tube and chill the acid in the ice-water bath. While stirring the mixture in the flask, add the sulfuric acid dropwise with a Pasteur pipet over a period of 1 min. Leave the reaction mixture in the ice-water bath for an additional 5 min, then allow the flask to stand at room temperature for 10 min.

Again cool the flask in the ice-water bath and add 5 mL of ice-cold water while stirring the contents. Chill 1 mL of methanol in a small test tube in the ice-water bath. Collect the crude product by vacuum filtration on a Hirsch funnel. Wash the crude product three times with 1-mL portions of water. Then wash the product twice with 0.5-mL portions of ice-cold methanol; to do this, disconnect the vacuum, cover the crystals with the cold methanol, then reconnect the vacuum to remove the methanol.

Transfer the crude product to a 10-mL Erlenmeyer flask. Recrystallize the product from a minimum amount of methanol, using a boiling stick to prevent bumping and a steam bath as the heat source [see Technique 5.5a]. Collect the product on a Hirsch funnel. After the crystals are dry, determine the melting point and percent yield. Prepare a sample of your dried product for NMR analysis as directed by your instructor.

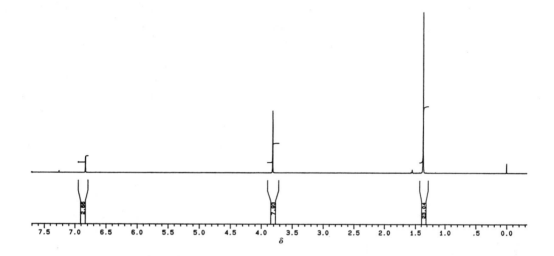

FIGURE 18.6 ^{1}H NMR spectrum (300 MHz) of 1,4-di-*tert*-butyl-2,5-dimethoxybenzene (in CDCl$_3$).

Cleanup: Pour the filtrate from the reaction mixture into a 250-mL beaker containing approximately 100 mL of water and neutralize the solution with solid sodium carbonate (**Caution: Foaming.**) before washing it down the sink or pouring it into the container for aqueous inorganic waste. Dispose of the methanol filtrate from the recrystallization in the container for flammable waste.

References

1. Hammond, C. N.; Tremelling, M. J. *J. Chem. Educ.* **1987,** *64,* 440–441.
2. Eaton, D. *Laboratory Investigations in Organic Chemistry;* McGraw-Hill: New York, 1989, pp. 454–456.

Questions

1. Why is it unlikely that significant amounts of 1,4-di-*tert*-butyl-2,3-dimethoxybenzene will form in this reaction? Does your reasoning also explain why little or no 1,4-dimethoxy-2,6-di-*tert*-butylbenzene is formed? Explain.

2. Draw all the resonance forms that account for the ring substitution of the second *tert*-butyl group on the 1,4-dimethoxybenzene ring system.

3. When 2-methylpropene (isobutylene) is treated with sulfuric acid, the resulting reaction mixture can be used to convert benzene to *tert*-butylbenzene. Write a mechanism that accounts for this conversion.

4. In Experiment 18.2 each aromatic ring gains one *tert*-butyl group, whereas in this experiment the ring gains two *tert*-butyl groups. Why?

5. What product(s) would be formed if 1-butanol were used instead of *tert*-butyl alcohol in this experiment? Would this be the best way to substitute an *n*-butyl group in a benzene ring? Why?

Experiment 19

REDUCTION REACTIONS OF 3-NITROACETOPHENONE

Investigate selective reduction reactions by determining, on the basis of melting point and IR spectra, which functional group of 3-nitroacetophenone is reduced.

The basic concepts of oxidation and reduction (redox reactions) were introduced in Experiment 15, Oxidation Reactions. Here we will consider reduction reactions.

The substrate in this experiment, 3-nitroacetophenone, has two potential sites for reduction, a nitro (NO_2) group and a keto carbonyl ($C=O$) group. Some reaction conditions favor reduction of the nitro group, whereas others lead to reduction of the ketone. You will use a different inorganic reducing agent in each experiment to determine which reagent reduces the nitro group and which one reduces the carbonyl group. Characterization of the products by melting point and infrared (IR) spectrometry will allow you to identify the product formed in each reaction. In Experiment 19.1 you will use tin (Sn) metal as the inorganic reducing agent under acidic conditions:

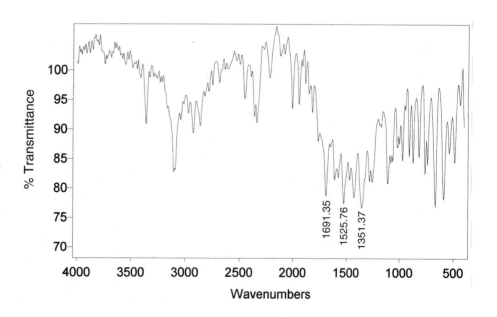

In Experiment 19.2 the inorganic reducing agent is sodium borohydride ($NaBH_4$):

To determine the outcome of these reduction reactions by infrared spectrometry, it is important to look for the IR bands corresponding to the functional groups of both the starting material (Figure 19.1) and the products. Here we are fortunate to find that all four functional groups of interest (C=O, OH, NO_2, NH_2) have polar bonds that, when stretched, give rise to significant molecular dipole changes and thus to reasonably intense IR bands [see Spectrometric Method 1].

Spectrum run on a sample dispersed on KBr using the diffuse reflectance technique on ATI Mattson Genesis Mattson Series Spectrometer.

FIGURE 19.1 IR spectrum of 3-nitroacetophenone.

First, let's assume that the reaction of interest is the reduction of the carbonyl group to the corresponding alcohol. Ketones with an aromatic or alkene π-system conjugated with the carbonyl group display strong C=O stretching bands at 1685 cm^{-1} or lower frequency. If the reducing agent converts the ketone to the corresponding alcohol, the C=O stretching band should disappear and bands due to the alcohol functional group should appear. Stretching of the O—H bond can give rise to two bands, a sharp band in the 3650–3580 cm^{-1} region and/or a strong, broad band in the 3550–3200 cm^{-1} region. The latter is due to the stretching of an O—H bond that is hydrogen bonded to other —OH groups. The former is due to the stretching of —OH groups free of such hydrogen bonding. IR sampling techniques such as mulls normally give rise to the broad "associated" band in the 3550–3200 cm^{-1} region. The sharp peak of the free —OH groups is rare and usually is observed only when the spectrum is obtained from a dilute solution. In theory, an alcohol group should also produce an observable C—O band. In practice, because C—O stretching bands occur in the fingerprint region of the IR, the presence of many other bands may make it difficult to identify.

Conversion of a nitro group to a primary amino group can also be conveniently monitored. The N—O bonds of nitro groups are polar, and two strong IR bands are observable, one in the 1550–1500 cm^{-1} region and another in the 1360–1290 cm^{-1} region. The two bands are due to the coupled stretches of the two N—O bonds in each molecule. One coupling interaction is asymmetric, and the other is symmetric:

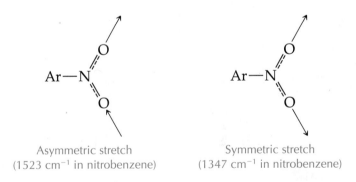

Asymmetric stretch
(1523 cm^{-1} in nitrobenzene)

Symmetric stretch
(1347 cm^{-1} in nitrobenzene)

Reduction of the nitro group to an amino group causes the disappearance of the two nitro IR bands. A pair of bands of moderate intensity due to the amino group should appear in the 3500–3070 cm^{-1} region. The band in the upper portion of that region is due to asymmetrically coupled stretching and that in the lower end is due to symmetrical coupling:

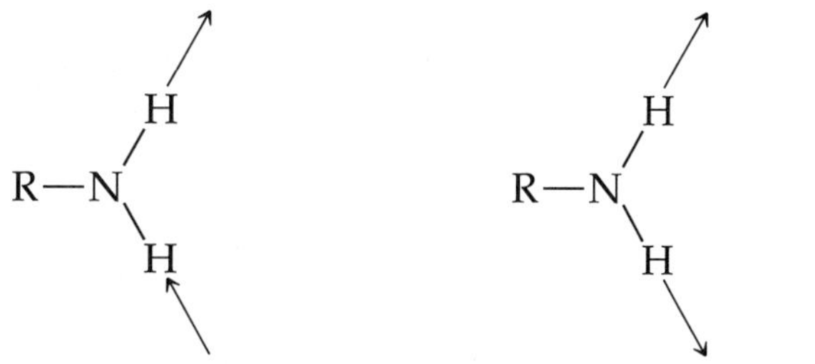

Asymmetric stretch
(3372 cm^{-1} in octylamine)

Symmetric stretch
(3290 cm^{-1} in octylamine)

In practice, inadequate instrument resolution, poor sampling technique, or complex hydrogen bonding may cause a primary amine to show from one to three IR bands in the N—H region. Thus, other physical properties, such as melting point, must also be used to ensure the correct structural assignment.

19.1

Reduction of 3-Nitroacetophenone Using Tin and Hydrochloric Acid

Investigate selective reduction, using Sn/HCl, and determine the structure of the reduction product by using melting point and IR data.

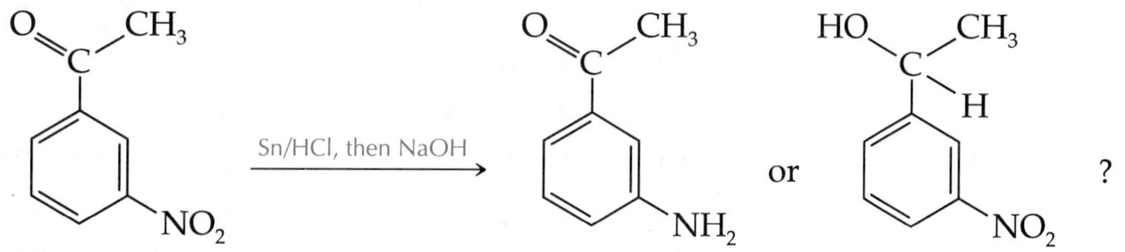

3-Nitroacetophenone
mp 81°C
MW 165

Sn/HCl, then NaOH

3-Aminoacetophenone
(3-acetylaniline)
mp 98–99°C
MW 135

or

1-(3-Nitrophenyl)ethanol
mp 60–61°C
MW 166

?

In this experiment you will selectively reduce 3-nitroacetophenone, using tin metal and HCl. These reagents represent one example of a general class of reactions known as dissolving metal reductions. Although the metal acts as a source of electrons, the reduction mechanism is quite complex. It is thought to proceed by a series of electron transfers from tin to the organic substrate. Tin is oxidized to Sn(II) during the reaction and is removed from the product mixture as the insoluble oxide, SnO, after the reaction solution is made basic.

Techniques   Microscale Filtration: Technique 5.5a
                IR Spectrometry: Spectrometric Method 1

SAFETY INFORMATION

3-Nitroacetophenone is an irritant. Wear gloves and avoid contact with skin, eyes, and clothing.

Hydrochloric acid is corrosive and irritates the skin, eyes, and mucous membranes.

30% Sodium hydroxide solution is corrosive and causes burns. Solutions as dilute as 9% (2.5 M) can cause severe eye injury.

While this reaction is in progress, begin Experiment 21.2.

Place 200 mg of 3-nitroacetophenone and 400 mg of granular tin (20 mesh) in a 25-mL Erlenmeyer flask and add 4.0 mL of 6 M hydrochloric acid. Place the flask on a steam bath for 25–30 min or until most of the tin has dissolved and no brown oil remains. Stopper the flask loosely with a cork to minimize evaporation during the heating period. The reaction time may vary, depending on the tin particle size.

At the end of the reaction period, cool the flask in an ice-water bath. Add 30% (10 M) sodium hydroxide dropwise until the pH reaches 10 when checked with pHydrion or other pH test paper. A thick yellow paste of tin salts and product will form; stir this mixture thoroughly while you are adding the NaOH solution. Heat the reaction flask for 10 min on a steam bath.

The 10-min heating period coagulates the tin oxide and dissolves the product in the hot solution. In the next step, the insoluble SnO is removed by vacuum filtration, leaving the product in the filtrate.

Thoroughly heat an inverted Buchner funnel on the steam bath. Heat approximately 5 mL of distilled water to boiling while the product mixture is being heated; do not let the water boil for an extended time or an appreciable volume will evaporate. Set up a vacuum filtration apparatus, using the hot Buchner funnel, a prewetted filter paper and a 25- or 50-mL filter flask. Immediately filter the hot product mixture and wash the precipitate (SnO) on the Buchner funnel with the boiling water. Cool the yellow filtrate to room temperature, then in an ice-water bath, before collecting the crystalline product by vacuum filtration on a Hirsch funnel. Wash the crystallized product four times with 0.5-mL portions of ice-cold water. Allow the product to dry completely before determining the melting point and percent yield. Use your experimentally determined melting point and IR spectrum to determine whether your product is 3-aminoacetophenone or 1-(3-nitrophenyl)ethanol. At the discretion of your instructor, the NMR spectrum may also be obtained.

Determine the IR spectrum in a mineral oil mull or in a CHCl₃ solution, if a solution cell is available [see Spectrometric Method 1.3].

Cleanup: Place the filter paper containing the SnO in the container for hazardous metals (or inorganic) waste. Neutralize the filtrate from the crystallized product with dropwise addition of 6 M hydrochloric acid until the pH reaches 6–7 before washing the solution down the sink or pouring it into the container for aqueous inorganic waste.

19.2

Reduction of 3-Nitroacetophenone Using Sodium Borohydride

Investigate selective reduction with NaBH₄ and determine the structure of the reduction product on the basis of its melting point and IR spectrum.

3-Nitroacetophenone
mp 81°C
MW 165

3-Aminoacetophenone
(3-acetylaniline)
mp 98–99°C
MW 135

1-(3-Nitrophenyl)ethanol
mp 60–61°C
MW 166

In this experiment you will reduce 3-nitroacetophenone with sodium borohydride (NaBH₄). Sodium borohydride is a useful and relatively safe reducing reagent. Unlike the more hazardous lithium aluminum hydride (LiAlH₄), which reacts explosively with water, NaBH₄ reacts only slowly with water. In fact, water and other hydroxylic solvents such as alcohols can be used as solvents for sodium borohydride reductions. When NaBH₄ is used as a reducing agent, the borohydride anion acts as a source of nucleophilic hydride ion.

<table>
<tr><td>

microscale
Procedure

</td><td>

Techniques   Microscale Extraction: Technique 4.5c
Microscale Separation of Drying Agent:
Technique 4.7b

</td></tr>
</table>

— SAFETY INFORMATION —

3-Nitroacetophenone is an irritant. Wear gloves and avoid contact with skin, eyes, and clothing.

Sodium borohydride is harmful if swallowed, inhaled, or absorbed through the skin. Avoid breathing the dust or contact with skin, eyes, and clothing. It decomposes to flammable, explosive hydrogen gas.

Ethanol is flammable.

Dichloromethane is toxic, an irritant, absorbed through the skin, and harmful if inhaled. Use it only in a hood, and wear gloves while doing the extractions and evaporation.

Sodium borohydride is moisture sensitive; keep the bottle tightly closed and store it in a desiccator.

The color of the solution may change from yellow to light brown during the addition of NaBH$_4$.

The end of hydrogen evolution (bubbling) indicates that the decomposition of excess sodium borohydride is complete.

Place 150 mg of 3-nitroacetophenone, 1.9 mL of absolute ethanol, and a magnetic spin vane in a 5-mL conical vial. Dissolve the solid by stirring and warming the vial in an aluminum block heated to 90°C on a hot plate-stirrer unit. Cool the solution at room temperature for 2–3 min; a fine suspension of solid may form. Stir the suspension at room temperature while adding in 55 mg of sodium borohydride powder in several portions over a 2- to 3-min period. Add 1.5 mL of distilled water and heat the contents of the vial to boiling on a 90°C aluminum block for 5 min. Cool the reaction solution for 10 min in an ice-water bath to produce a suspension of the product. If bubbling continues when the solution has been cooled, add 3 M hydrochloric acid dropwise to complete the decomposition process.

Extract the product mixture with three 1.5-mL portions of dichloromethane [see Technique 4.5c]. The upper aqueous layer remains in the conical vial during the extractions. Transfer the lower organic phase to a second conical vial (or small test tube) after each extraction. Dry the combined organic extracts with anhydrous magnesium sulfate for about 10 min.

Filter half of the solution into a tared 13 × 150 mm test tube, using a Pasteur pipet with cotton packed at the top of the tapered portion as a funnel and transferring the solution with a Pasteur filter pipet [see Technique 4.7b]. Evaporate the dichloromethane with a stream of nitrogen in a hood. When most of the solvent is gone, filter the other half of the product solution into the test tube and complete the evaporation of all solvent. The product usually forms a white solid after the removal of the solvent, although it may remain as a yellow oil.

Determine the IR spectrum in a mineral oil mull or in a CHCl₃ solution, if a solution cell is available [see Spectrometric Method 1.3].

Determine the mass of product and percent yield. Use the melting point and the product's IR spectrum to determine whether 3-aminoacetophenone or 1-(3-nitrophenyl)ethanol was formed in this reduction. Also determine the NMR spectrum if directed to do so by your instructor.

Cleanup: The aqueous solution remaining from the extraction may be washed down the sink or placed in the container for aqueous inorganic waste. Place the magnesium sulfate drying agent in the container for nonhazardous waste or inorganic waste.

Questions

1. What is the role of the sodium hydroxide added after the Sn/HCl reduction is complete? Write a balanced equation that supports your answer.
2. When butanoic acid is treated with sodium borohydride, 1-butanol is not obtained. There are, however, definite signs of reaction (bubbling, heat). What might the reaction be?
3. After determining which functional group of 3-nitroacetophenone is reduced by $NaBH_4$, describe the mechanism of the reduction.
4. Suggest an appropriate reducing reagent for the following substrates and predict the product of the reaction: (a) hexanal; (b) 2-hexanone; (c) ethyl 4-nitrobenzoate (reduction of the nitro group).

Experiment 20

PHOTOCHEMISTRY: PHOTOREDUCTION OF BENZOPHENONE AND REARRANGEMENT TO BENZPINACOLONE

Use ultraviolet light to carry out the photoreduction of benzophenone and use carbocation reaction conditions to effect the pinacol rearrangement.

$$2C_6H_5\!-\!\overset{O}{\underset{}{C}}\!-\!C_6H_5 + CH_3\overset{OH}{\underset{H}{CCH_3}} \xrightarrow{h\nu} CH_3\overset{O}{CCH_3} + C_6H_5\!-\!\overset{OH}{\underset{C_6H_5}{C}}\!-\!\overset{OH}{\underset{C_6H_5}{C}}\!-\!C_6H_5$$

Benzophenone
mp 49°C
MW 182

Benzpinacol
mp 188°C
MW 366

$$C_6H_5\!-\!\overset{OH}{\underset{C_6H_5}{C}}\!-\!\overset{OH}{\underset{C_6H_5}{C}}\!-\!C_6H_5 \xrightarrow[CH_3COOH]{I_2} C_6H_5\!-\!\overset{O}{C}\!-\!\overset{C_6H_5}{\underset{C_6H_5}{C}}\!-\!C_6H_5 + H_2O$$

Benzpinacolone
mp 179°C
MW 348

The creation of electronically excited states of organic compounds occurs when compounds absorb ultraviolet and visible radiation. Because these electronically excited molecules are of decidedly higher energy than their ground-state counterparts, they undergo many reactions that do not occur with molecules in their ground states. In effect, these reactions use light as a reagent to form highly reactive chemical intermediates.

In this experiment, we will consider the photochemical reduction of the ketone, benzophenone. Benzophenone absorbs ultraviolet radiation and uses the energy obtained as the driving force for its reduction by 2-propanol (isopropyl alcohol). The product, benzpinacol subsequently reacts with iodine in acetic acid to give the rearrangement product, benzpinacolone.

The electromagnetic spectrum is roughly divided as shown in Spectrometric Method 1, Figure 1.1. The shorter the wavelength of the radiation, the higher its energy. This relationship is summed up in the Planck equation:

$$E = \frac{hc}{\lambda}$$

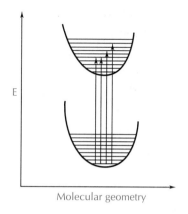

FIGURE 20.1
Electronic excitation.

where h is the Planck constant, λ is the wavelength, and c is the speed of light.

When a compound absorbs ultraviolet radiation, a quantized transition occurs and an electron is promoted to a higher energy orbital. An electron from the ground state S_0 is promoted to one of the many rotational and vibrational levels of the S_1 excited state (Figure 20.1). The energy changes associated with these transitions give rise to the absorption spectrum of the molecule, in which each wavelength is associated with a separate and distinct transition energy.

These electronic transitions occur in less than 10^{-12} s and take place with no change in the spin state of the electron. Because the electrons of the system are paired in the ground state and the excited state, both are called singlet states.

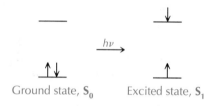

Once in the excited S_1 state, the molecule is no longer in equilibrium with its surroundings and is driven to return to a lower energy equilibrium condition. There are at least five processes that can return the S_1 state to a more stable arrangement. Four of these changes do not result in any new product but simply return the molecule to the ground state. One change converts S_1 to a triplet state, T_1, about which we will speak later.

The four changes of S_1 that produce S_0 are

1. fluorescence: a light emission process
2. energy transfer: transfer of the excitation energy to another molecule
3. radiationless decay: degradation of $S_1 \rightarrow S_0$ by converting the excited-state energy into thermal energy by collision
4. chemical reaction to produce a more stable intermediate

All of these processes occur rapidly, within 10^{-8} s. Although S_1 is easily formed, it is also rapidly destroyed. Perhaps the most important transition that S_1 undergoes is intersystem crossing, an electron-spin transition that produces the triplet, or T_1, state. In this excited state, the electron in the high-energy orbital has the same spin as the lower energy electron. This parallel spin condition gives T_1 a much longer lifetime, because return to S_0 is slow relative to most electronic processes:

↓   ↑

intersystem
⟶
crossing

↑   ↑

S_1 T_1

T_1 can also emit light, however, through a process called phosphorescence. In addition, T_1 can undergo chemical reaction or transfer energy to other molecules. All three of these processes have been extensively studied.

Photochemistry of Benzophenone: Theory and Practice

Benzophenone is one of the best-behaved compounds in photochemistry. The excited-state energy profile diagram for benzophenone is similar to Figure 20.2.

Benzophenone does not fluoresce; it intersystem crosses with unit efficiency to T_1 in less than 10^{-8} s. The T_1 state of benzophenone is moderately long lived, having a half-life of 5×10^{-5} s. In benzophenone, T_1 lies about 287 kJ·mol^{-1} above the ground state.

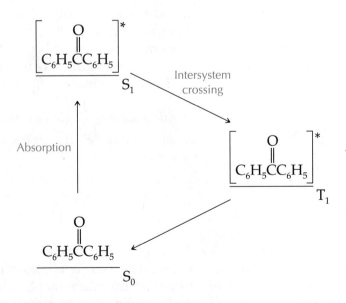

FIGURE 20.2 Excited-state energy profile diagram for benzophenone.

Benzophenone phosphoresces blue light when irradiated at low temperature.

Excited states of benzophenone are produced by promotion of a specific electron, one of the nonbonded electrons on the oxygen atom, into an antibonding orbital associated with the carbon-oxygen π-bond. Such a transition is called an $n \rightarrow \pi^*$ transition, indicating both the ground-state orbital and the excited-state orbital in the electronic transition. Under experimental conditions, triplet-state benzophenone reacts like a diradical at the odd-electron center on the oxygen atom.

In this experiment, the benzophenone T_1 state abstracts a hydrogen atom from 2-propanol:

Excitation

$$
\underset{}{C_6H_5\overset{\overset{\displaystyle O}{\|}}{C}C_6H_5} \xrightarrow{h\nu} \underset{\text{Singlet}}{C_6H_5\overset{\overset{\displaystyle O^*}{\|}}{C}C_6H_5} \xrightarrow[\text{crossing}]{\text{intersystem}} \underset{\text{Triplet}}{C_6H_5\overset{\overset{\displaystyle O^*}{\|}}{C}C_6H_5}
$$

Hydrogen abstraction

$$
\underset{\text{Triplet}}{C_6H_5\overset{\overset{\displaystyle O^*}{\|}}{C}C_6H_5} + CH_3\underset{\overset{\displaystyle |}{OH}}{\overset{\overset{\displaystyle H}{|}}{C}}CH_3 \longrightarrow
$$

$$
\underset{\underset{\text{radical}}{\text{Diphenylhydroxymethyl}}}{C_6H_5\overset{\displaystyle \cdot}{C}C_6H_5} + CH_3\underset{\overset{\displaystyle |}{OH}}{\overset{\overset{\displaystyle OH}{|}}{\overset{\displaystyle \cdot}{C}}}CH_3
$$

Hydrogen atom extraction from the alcohol occurs from the carbon atom alpha to the oxygen atom of 2-propanol, because this is the weakest bond toward homolytic scission. The resulting radical, called an α-hydroxyalkyl radical, is a good reducing agent and reacts with another ground-state benzophenone molecule to produce acetone and a diphenylhydroxymethyl radical:

$$
CH_3\underset{\overset{\displaystyle |}{OH}}{\overset{\displaystyle \cdot}{C}}CH_3 + C_6H_5\overset{\overset{\displaystyle O}{\|}}{C}C_6H_5 \longrightarrow CH_3\overset{\overset{\displaystyle O}{\|}}{C}CH_3 + \underset{\underset{\text{radical}}{\text{Diphenylhydroxymethyl}}}{C_6H_5\overset{\overset{\displaystyle OH}{|}}{\underset{\displaystyle \cdot}{C}}C_6H_5}
$$

Finally, two diphenylhydroxymethyl radicals combine to produce the observed product, benzpinacol:

$$2C_6H_5\overset{\overset{\displaystyle OH}{|}}{\underset{\displaystyle \cdot}{C}}C_6H_5 \longrightarrow C_6H_5 - \overset{\overset{\displaystyle C_6H_5}{|}}{\underset{\underset{\displaystyle OH}{|}}{C}} - \overset{\overset{\displaystyle C_6H_5}{|}}{\underset{\underset{\displaystyle OH}{|}}{C}} - C_6H_5$$

Benzpinacol

Photochemical reactions require both light and molecules that absorb the light. In the case of benzophenone, sunlight suffices. In other cases, artificial light sources are required for photoprocesses. Most of these artificial light sources are mercury-arc lamps, which emit radiation in four principal spectral regions: 182 nm, 254 nm, 313 nm, and 366 nm. Of these lines, only the latter three are important, because air completely absorbs the 182-nm spectral emission.

Pyrex absorbs radiation below 290 nm. Therefore, only the 313-nm and the 366-nm lines penetrate Pyrex glass, and only compounds that absorb light above 290 nm can be photolyzed in Pyrex glassware.

The absorption spectrum of benzophenone is shown in Figure 20.3. The principle $n \rightarrow \pi^*$ absorption occurs at 366 nm, so the photoreduction of benzophenone can be carried out in a Pyrex apparatus. The absorptions at shorter wavelengths are $\pi \rightarrow \pi^*$ bands that result from electronic transitions of the aromatic parts of the molecule.

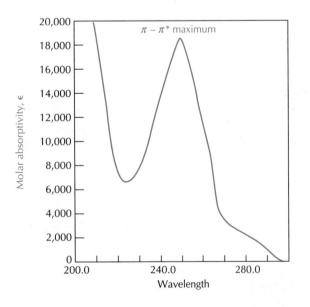
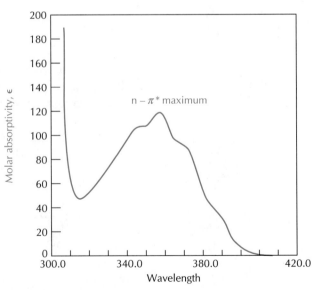

FIGURE 20.3 UV-VIS absorption spectrum of benzophenone.

Pinacol Rearrangement

Benzpinacol rearranges to benzpinacolone in an acid-catalyzed rearrangement. This kind of rearrangement was first reported in 1859, the year after Kekulé set forth his structural theory of organic chemistry, when pinacol was observed to rearrange to pinacolone.

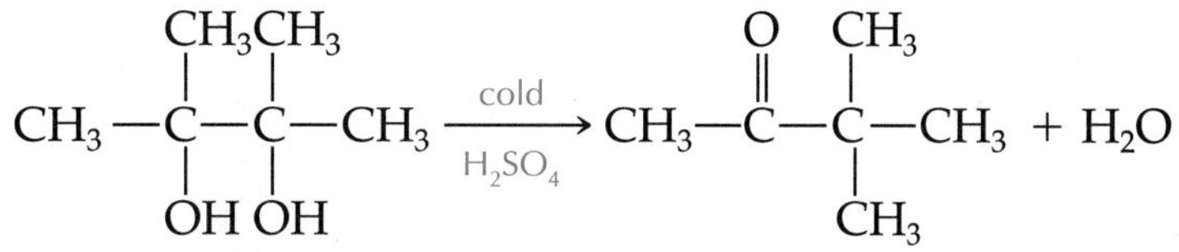

$$CH_3-\underset{\underset{OH}{|}}{\overset{\overset{CH_3}{|}}{C}}-\underset{\underset{OH}{|}}{\overset{\overset{CH_3}{|}}{C}}-CH_3 \xrightarrow[\text{H}_2\text{SO}_4]{\text{cold}} CH_3-\overset{\overset{O}{\|}}{C}-\underset{\underset{CH_3}{|}}{\overset{\overset{CH_3}{|}}{C}}-CH_3 + H_2O$$

Pinacol
(2,3-dimethyl-2,3-butanediol)

Pinacolone
(3,3-dimethyl-2-butanone)

Many other examples of the acid-catalyzed rearrangement of 1,2-diols to ketones or aldehydes were subsequently discovered and the name pinacol rearrangement became accepted for this whole class of transformations.

A great deal of experimental evidence points to the presence of carbocation intermediates in the pinacol rearrangement:

$$C_6H_5-\underset{\underset{C_6H_5}{|}}{\overset{\overset{OH}{|}}{C}}-\underset{\underset{C_6H_5}{|}}{\overset{\overset{OH}{|}}{C}}-C_6H_5 \xrightarrow{H^+} C_6H_5-\underset{\underset{C_6H_5}{|}}{\overset{\overset{\overset{+}{O}H_2}{|}}{C}}-\underset{\underset{C_6H_5}{|}}{\overset{\overset{OH}{|}}{C}}-C_6H_5 \xrightarrow{-H_2O}$$

$$\underset{C_6H_5}{\overset{C_6H_5}{\diagdown}}\overset{+}{\underset{}{C}}-\underset{\underset{C_6H_5}{|}}{\overset{\overset{OH}{|}}{C}}-C_6H_5 \xrightarrow{\text{1,2-aryl migration}} C_6H_5-\underset{\underset{C_6H_5}{|}}{\overset{\overset{C_6H_5}{|}}{C}}-\overset{\overset{+}{O}H}{C}-C_6H_5 \xrightarrow{-H^+} (C_6H_5)_3C-\overset{\overset{O}{\|}}{C}-C_6H_5$$

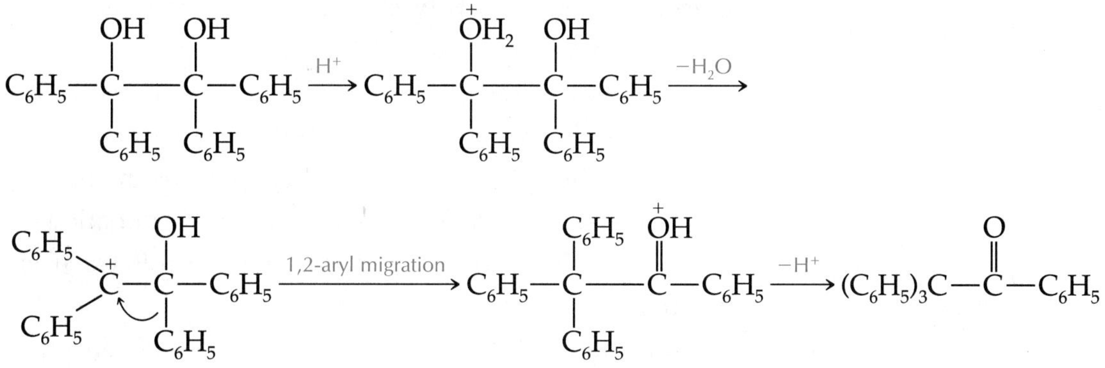

Even though the tertiary benzylic carbocation is a very stable one, the ion formed by rearrangement is even more stable, because the presence of the oxygen atom provides strong resonance stabilization.

Only a very weak acid is required to assist in the formation of the tertiary benzylic carbocation. In this experiment you will use iodine, a weak Lewis acid, to coordinate with the oxygen atom to make it a better leaving group. Once the tertiary benzylic carbocation forms, the reaction proceeds as shown above. Acetic acid is used as the solvent.

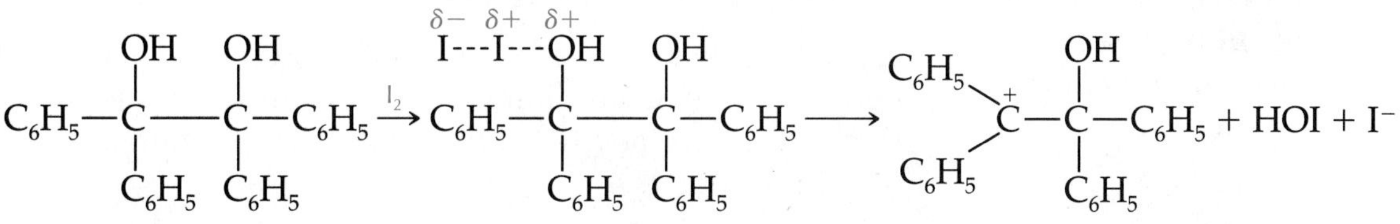

$$C_6H_5-\underset{\underset{C_6H_5}{|}}{\overset{\overset{OH}{|}}{C}}-\underset{\underset{C_6H_5}{|}}{\overset{\overset{OH}{|}}{C}}-C_6H_5 \xrightarrow{I_2} C_6H_5-\underset{\underset{C_6H_5}{|}}{\overset{\overset{\overset{\delta- \ \delta+ \ \delta+}{I---I---OH}}{|}}{C}}-\underset{\underset{C_6H_5}{|}}{\overset{\overset{OH}{|}}{C}}-C_6H_5 \longrightarrow \underset{C_6H_5}{\overset{C_6H_5}{\diagdown}}\overset{+}{\underset{}{C}}-\underset{\underset{C_6H_5}{|}}{\overset{\overset{OH}{|}}{C}}-C_6H_5 + HOI + I^-$$

Macroscale
Procedure

Photoreduction of Benzophenone

*This experiment requires
an irradiation period of
at least 24 h.*

In a 125-mL Erlenmeyer flask, dissolve 6.0 g of benzophenone in
30 mL of 2-propanol (isopropyl alcohol). Warming on a steam
bath may be necessary. Add a drop of glacial acetic acid to ensure
that no alkali is present. Label the flask, stopper it tightly with a
cork, and give it to your instructor to be irradiated for a day or
two. The irradiation will be carried out by focusing a strong sunlamp on the flask. Placing white paper under the flask speeds up
the reaction somewhat by reflecting the radiation. The progress of
the irradiation can easily be followed because the product, benzpinacol, is relatively insoluble in 2-propanol.

A second method of carrying out the photoreduction is to use
the ultraviolet rays of the sun. Solar irradiation for one or two
weeks produces high yields of benzpinacol in this experiment,
unless the skies are very cloudy during the irradiation or the air is
so cold in winter that benzophenone precipitates from solution
before it can react.

*You can tell if benzophenone
precipitates by its crystalline
shape—large prisms. The photoreduction product precipitates as small crystals.*

After irradiation is complete, cool the reaction mixture in ice
water for 10 min and collect the crystals on a Buchner funnel by
vacuum filtration. Draw air over the product for 15 min to dry it.
Determine its melting point.

Cleanup: Pour the filtrate into the flammable (organic) waste
container.

Dehydration of Benzpinacol to Benzpinacolone

In a 50-ml round-bottomed flask provided with a reflux condenser, place 0.05 g of iodine (two or three small crystals) and 15 ml of glacial acetic acid. Add 3.0 g of benzpinacol and heat the solution slowly until it boils gently. Reflux the red solution for 5–7 min. Benzpinacolone may precipitate during the reflux period. On cooling the reaction flask in an ice-water bath, benzpinacolone will precipitate as fine needles. Thin the pasty product with a few milliliters of *cold* ethanol before collecting the crystals by vacuum filtration on a Buchner funnel. Wash the product with ice-cold ethanol until it is free of the iodine color. Dry the benzpinacolone, calculate the percent yield of your dried product, and determine its melting point. Obtain the NMR spectrum [see Spectrometric Method 2] and/or the infrared spectrum in a mineral oil mull or KBr pellet [see Spectrometric Method 1.3], as directed by your instructor.

Cleanup: Pour the filtrate into a 600-mL beaker containing 100 mL of water. Add solid sodium carbonate until the acetic acid is neutralized. **(Caution: Foaming.)** Wash the neutralized solution down the sink or place it in the container for aqueous inorganic waste.

Questions

1. Why does the photochemical reaction have a faster rate when a strong sunlamp is used rather than direct solar energy?
2. The T_1 excited state of benzophenone is 287 kJ·mol^{-1} above the ground state. Calculate the wavelength of its phosphorescence.
3. Propose a mechanism for the photoreduction of acetophenone. How many products can be formed? Draw their structures.
4. Benzophenone shows the following ^{1}H NMR characteristics (300 MHz, CDCl$_3$ solvent). Use this information to assign the ^{1}H NMR signals for your products.

O
‖
C—C$_6$H$_5$

δ 7.81, 4H, d
δ 7.48, 4H, t
δ 7.58, 2H, distorted t

s = singlet
d = doublet
t = triplet

ALDOL CONDENSATION: REACTION OF BENZALDEHYDE AND ACETONE

Identify the product that forms in an aldol condensation by using the melting point of its derivative formed with 2,4-dinitrophenylhydrazine.

$$CH_3-\overset{\overset{\displaystyle O}{\|}}{C}-CH_3 \ + \ \text{(benzaldehyde)} \xrightarrow{\text{NaOH}} \text{(benzalacetone)} \quad \text{or}$$

Acetone
MW 58
density 0.791 g · mL^{-1}

Benzaldehyde
MW 106
density 1.044 g · mL^{-1}

1-phenyl-1-buten-3-one
(benzalacetone)
mp 42°C
MW 146

1,5-Diphenyl-3-pentadienone
(dibenzalacetone)
mp 112°C
MW 234

Benzaldehyde has a distinctive aroma. It occurs in the kernels of bitter almonds and has been used as the artificial essence of "oil of almonds." This use is interesting in view of the fact that benzaldehyde is considered to be toxic in its pure state.

The aldol condensation is a cornerstone of the methods for the synthesis of carbon-carbon bonds. The synthesis is based on the formation of an enolate anion and its subsequent reaction with another aldehyde or ketone molecule to form a carbon-carbon bond.

In the first step of the aldol condensation in this experiment, a proton is abstracted from acetone to form its conjugate base, the resonance-stabilized enolate anion. This enolate then attacks the carbonyl carbon of benzaldehyde in the carbon-carbon bond-forming step that leads eventually to the aldol product. An outline of the condensation steps follows:

This process is a "crossed" or "mixed" aldol condensation. The enolate species attacks a carbonyl compound that is not the precursor of the enolate. It succeeds here because benzaldehyde has no α-hydrogen atoms and therefore cannot form an enolate anion. In addition, benzaldehyde undergoes nucleophilic attack by the enolate anion faster than a second molecule of acetone does.

The aldol, which contains the functional elements of both an alcohol and a ketone, is frequently not the final product. Often it spontaneously dehydrates to form a conjugated enone. In this experiment, the dehydration is readily accomplished because the resulting product is stabilized by additional conjugation between the benzene ring and the alkene double bond:

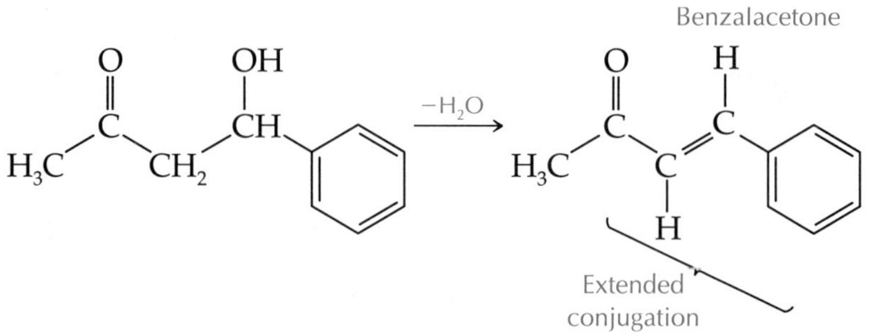

Benzalacetone

Extended
conjugation

The remaining methyl group of benzalacetone is also acidic and in principle can undergo a second aldol condensation to form dibenzalacetone. A third condensation cannot occur because there are no remaining acidic protons. It will be your responsibility to discover whether, in fact, one or two aldol condensations/dehydrations take place through analysis of the reaction conditions and characterization of your products.

A convenient way to identify which condensation product forms employs derivative formation using 2,4-dinitrophenylhydrazine. This approach is particularly effective here because the 2,4-dinitrophenylhydrazones of benzalacetone and dibenzalacetone have melting points that differ by 44°C. This large difference in melting points allows ready identification of the product.

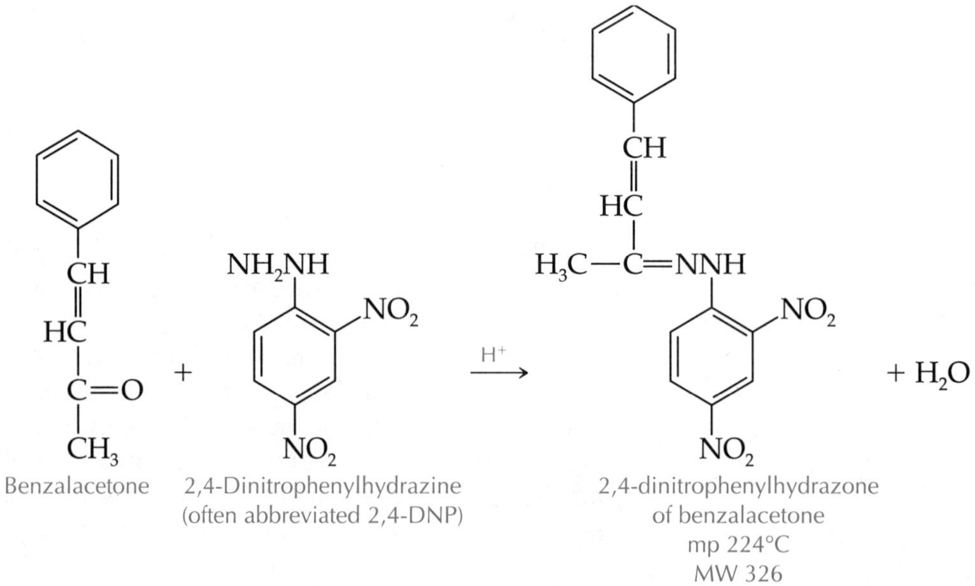

Benzalacetone   2,4-Dinitrophenylhydrazine
(often abbreviated 2,4-DNP)

2,4-dinitrophenylhydrazone
of benzalacetone
mp 224°C
MW 326

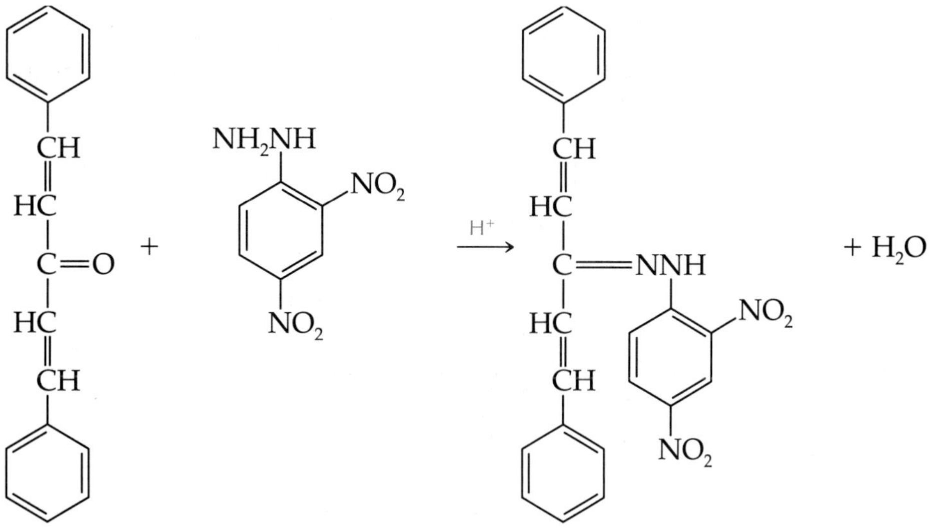

Dibenzalacetone   2,4-Dinitrophenylhydrazine   2,4-dinitrophenylhydrazone
of dibenzalacetone
mp 180°C
MW 414

The reagent, 2,4-dinitrophenylhydrazine, is especially useful for derivative preparation because hydrazones form readily from most aldehydes and ketones, and the products are often high-melting-point solids. Moreover, the dinitro substitution on the phenyl group causes the hydrazones to be highly colored, ranging from a deep orange to red. The color of the product is a rough indication of the extent of the conjugation in the derivative.

This arylhydrazone derivatization process is an example of the condensations that aldehydes and ketones undergo with substituted ammonia compounds. It begins with nucleophilic attack of the terminal amino group on the carbonyl carbon atom, resulting eventually in the formation of a substituted α-hydrazinoalcohol. This hydrazinoalcohol is readily dehydrated to form the imine ($C = N$) bond of the hydrazone:

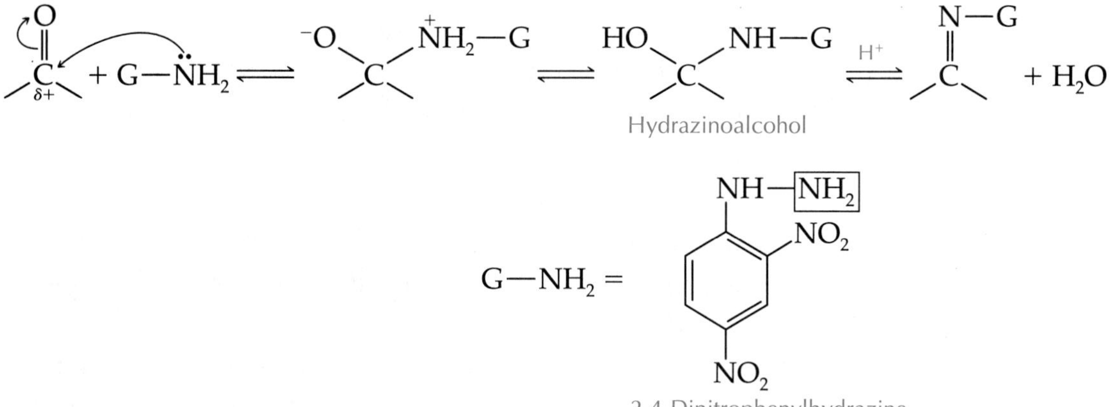

Hydrazinoalcohol

$$G-NH_2 = $$

2,4-Dinitrophenylhydrazine

Techniques   Microscale Recrystallization: Technique 5.5a
Microscale Extraction: Technique 4.5d

SAFETY INFORMATION —————

Wear gloves throughout the experimental procedure.

Benzaldehyde is toxic. The reagent solution used in this experiment also contains 0.25 M **sodium hydroxide**. Avoid contact with skin, eyes, and clothing.

Acetone is flammable, volatile, and an irritant. **Diethyl ether** (ether) is also extremely flammable and volatile. **Ethanol** is flammable. Do not use any of these reagents near open flames or hot electrical devices.

2,4-Dinitrophenylhydrazine is an irritant and stains the skin. The reagent solution also contains sulfuric acid. Wear gloves, and avoid contact with skin, eyes, and clothing.

Reaction 1.   Place 5.0 mL of benzaldehyde/sodium hydroxide solution (0.5 M benzaldehyde and 0.25 M NaOH in 1:1 (v/v) ethanol/water) in a 10-mL Erlenmeyer flask. Add 0.10 mL of acetone and stir the reaction mixture well. Allow the reaction mixture to stand for 10 min at room temperature.

Collect the precipitated product by vacuum filtration on a Hirsch funnel [see Technique 5.5a]. Wash the crude product twice by covering the crystals with cold water and removing the water from the crystals by applying suction. Transfer the crystals to a test tube. Recrystallize the crude product in a minimum volume of hot ethanol on a steam bath or in a boiling water bath on a hot plate. Add water dropwise to the hot solution until turbidity occurs, then warm the mixture briefly until a clear solution results (the addition of 1 or 2 drops of ethanol may be necessary). Set the test tube aside for crystallization to occur slowly while you begin Reaction 2.

Use a boiling stick to prevent bumping during the recrystallization.

After crystallization is well under way, cool the test tube for several minutes in an ice-water bath before collecting the recrystallized product on a Hirsch funnel. Allow the product to dry overnight before determining its mass and melting point. Use your experimental melting point to identify your product as benzalacetone or dibenzalacetone. At your instructor's discretion, you may also determine the NMR spectrum of your product [see Spectrometric Method 2]. Once you have identified your product, calculate the theoretical yield and your percent yield.

Reaction 2. Place 5.0 mL of the benzaldehyde/NaOH solution in a 15-mL centrifuge tube with a tight-fitting cap and add 1.0 mL of acetone. Cap the tube and shake the mixture briefly. Allow the reaction mixture to stand for 10–15 min while you filter the recrystallized product from Reaction 1.

At the end of the reaction period, add 4 mL of water, a small piece of ice, and 3 mL of diethyl ether to the centrifuge tube. Cap the tube and shake it to mix the two phases thoroughly. Allow the layers to separate and remove the lower aqueous phase with a Pasteur filter pipet [see Technique 2.4]. Wash the organic layer remaining in the centrifuge tube with 4 mL of water [see Technique 4.5d]. Remove the aqueous phase with the Pasteur filter pipet. Transfer the ether solution to a 10-mL Erlenmeyer flask with a clean Pasteur pipet. Working in a hood, evaporate the ether on a steam bath or with a stream of nitrogen. A small amount of yellow oil (your product) should remain in the flask after the solvent has evaporated.

Directions for the preparation of 2,4-dinitrophenylhydrazine reagent solution are on p. 530.

Add 1.0 mL of ethanol to the oil remaining in flask, swirl thoroughly, then add 1.0 mL of 2,4-dinitrophenylhydrazine reagent solution. Allow the mixture to stand for 10 min until precipitation is complete. Collect the product by vacuum filtration on a Hirsch funnel. Thoroughly wash the crystals four times by covering them with approximately 1 mL of cold ethanol each time and drawing the solvent through the funnel with the vacuum line. If you are going to determine the melting point immediately, draw air over the crystals for 10 min with the vacuum source turned on, otherwise place the crystals on a watch glass in your laboratory drawer to dry. Compare your experimental melting point to those listed in order to identify your product as the 2,4-dinitrophenylhydrazone of either benzalacetone or dibenzalacetone.

A thorough washing of the product with cold ethanol is necessary to remove all of the sulfuric acid that was present in the reagent solution. Any remaining acid catalyzes decomposition of the dinitrophenylhydrazone while the melting point is being determined. The decomposition products lower and broaden the melting range.

Cleanup: In a 100-mL beaker, combine the aqueous filtrate from the crude product of Reaction 1 with the aqueous phase separated from the ether solution in Reaction 2. Neutralize the mixture with dropwise addition of 3 M hydrochloric acid, using pH test paper to determine the pH. Wash the neutralized solution down the sink or put it in the container for aqueous inorganic waste, as directed by your instructor. Pour the filtrate from the recrystallization of Reaction 1 and the filtrate from the preparation and washing of the 2,4-dinitrophenylhydrazone in Reaction 2 into the container for flammable (organic) waste.

Questions

1. Calculate the molar ratios of benzaldehyde to acetone used in each of Reactions 1 and 2. How does a comparison of these ratios allow you to conclude whether dicondensation or monocondensation should be favored in each reaction?

2. What is the role of the sodium hydroxide used in these condensation reactions?

3. The order of addition of reagents is a factor that can be used to favor crossed-aldol condensations. For example, it is not uncommon to find that slow addition of the organic compound bearing α-hydrogens to an alkaline solution of the compound bearing no α-hydrogens is a good approach. Explain, using the basic principles of reaction rates, why crossed- rather than self-condensation would be favored by such an approach.

4. A student wishes to identify two ketones, 1-phenylpropanone (benzyl methyl ketone) and methyl *o*-tolyl ketone by derivative formation in a manner similar to that used in this experiment. The 2,4-dinitrophenylhydrazones show melting points of 156°C for 1-phenylpropanone and 159°C for methyl *o*-tolyl ketone; the semicarbazones show melting points of 198°C for 1-phenyl-propanone and 206°C for methyl *o*-tolyl ketone. Would you use the dinitrophenylhydrazone or semicarbazone derivative to differentiate between the two ketones? Explain your choice.

5. An acid should assist the dehydration step leading to the 2,4-dinitrophenylhydrazone from the hydrazinoalcohol in much the same fashion as it does in the dehydration of a tertiary alcohol. However, rather than using an acid solution of extremely low pH, investigators have found that a solution of moderate acidity is usually more effective. Considering all of the reagents used in this derivatization, why might this be the case?

SYNTHESIS OF 2,3,4,5-TETRAPHENYLCY-CLOPENTADIENONE

Use aldol condensations to synthesize a colored cyclopentadienone; characterize the product by UV-VIS spectrometry.

Benzil	1,3-Diphenyl-2-propanone
mp 95–96°C	(1,3-diphenylacetone)
MW 210.2	mp 35°C
	MW 210.3

$C_6H_5CH_2N^+(CH_3)_3OH^-$ / triethylene glycol

2,3,4,5-Tetraphenylcyclopentadienone
mp 219°C
MW 384.45

In this experiment you will carry out a cyclization by condensing benzil (a dione) with 1,3-diphenylacetone. The latter compound has two acidic methylene (CH_2) groups. Base is used to abstract a proton from one of the acidic sites of diphenylacetone to form an enolate anion, which in turn attacks one of the benzil carbonyl car-

bons. This reaction leads to an aldol product, which readily dehydrates to form a more highly conjugated system:

Proton abstraction to form the enolate anion:

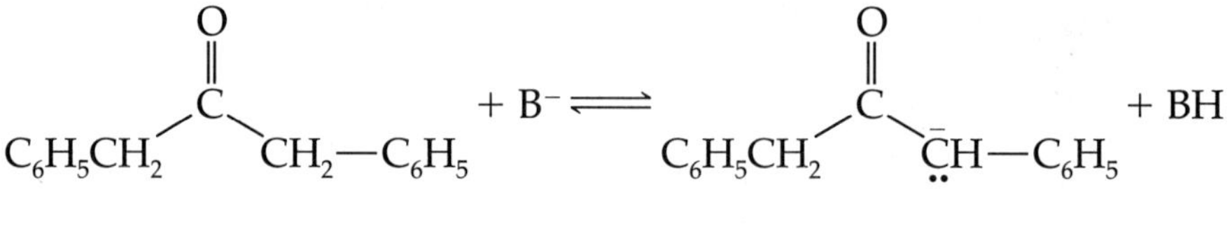

Nucleophilic attack on carbonyl carbon:

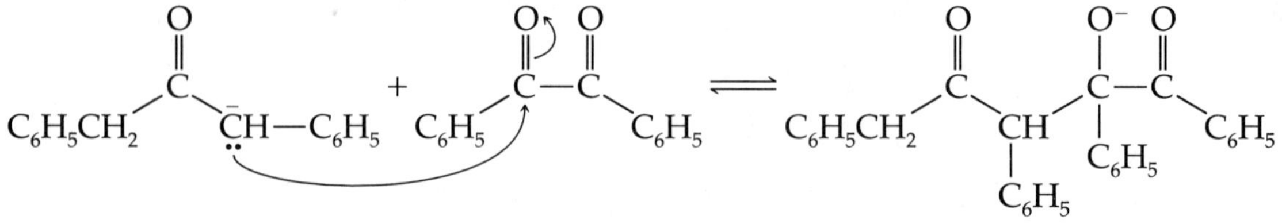

Protonation to form the aldol product:

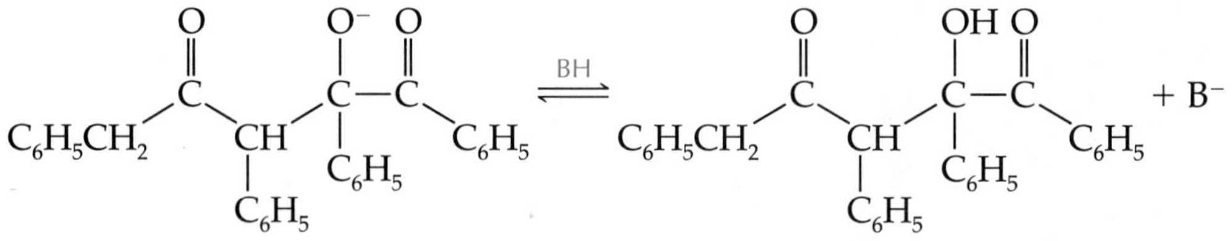

Dehydration of aldol product to form an enone:

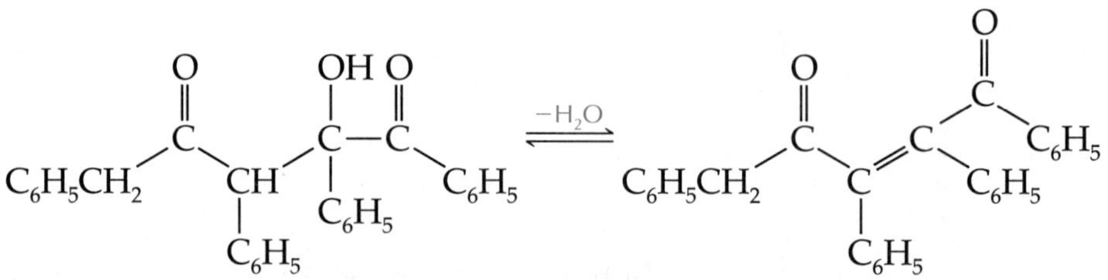

This process is repeated to also form the other side of the central five-membered ring of the final product. The reaction is promoted by a strong base, and the appearance of the final product is

readily shown by the deep purple color of the highly conjugated substituted cyclopentadienone.

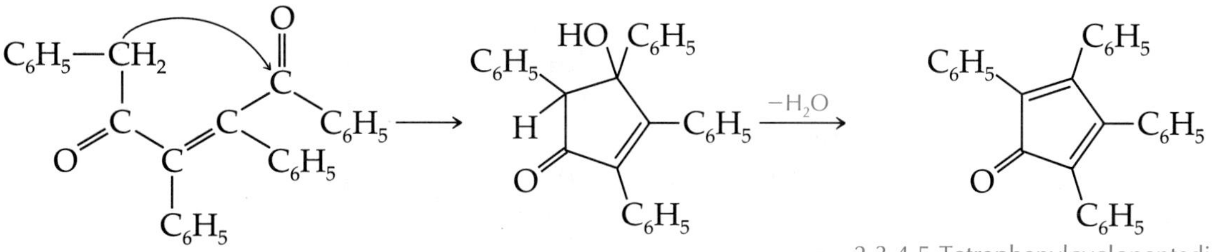

2,3,4,5-Tetraphenylcyclopentadienone
(the final product)

In this reaction you will use the strong base benzyltrimethyl-ammonium hydroxide (Triton B). The use of a strong base drives the following equilibrium farther to the right, thereby providing a greater concentration of the enolate anion necessary for the condensation:

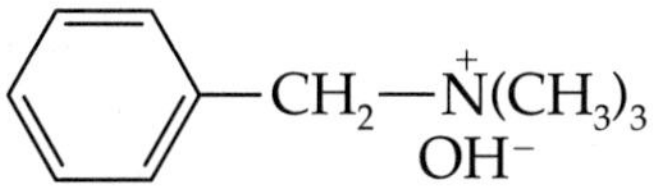

Triton B

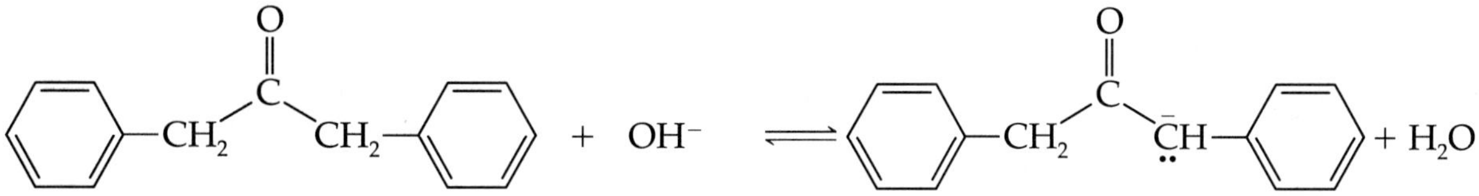

(from Triton B)

Another important feature of this reaction is the solvent, triethylene glycol ($HOCH_2CH_2OCH_2CH_2OCH_2CH_2OH$). Because this compound has a fairly high molecular weight and can form an extended hydrogen-bonded network, it has a high boiling point (285°C). Moreover, because it is a moderately polar organic compound, it dissolves both the organic starting materials and the highly polar base.

Very often, we use the reflux point of a solvent as a method of controlling the temperature of the reaction. That is *not* the case here, however. You will be conducting the experiment at a temperature no higher than 105–110°C, well below the boiling point of triethylene glycol.

As already mentioned, the final product is deeply colored because it is a highly conjugated dienone. Color arises as the result of electronic transitions similar to the transitions causing the absorption of ultraviolet radiation [UV-VIS Spectrometry, Spectrometric Method 4]. Ultraviolet and visible radiation are absorbed by compounds, resulting in the promotion (transition) of an electron from a bonding or nonbonding orbital to a higher energy antibonding orbital.

The wavelength of this absorption is affected by conjugation, defined as an extended π-bonding system. For example, C=C—C=C units are conjugated, whereas C=C—C—C=C units are not because the double bonds do not interact with each other. Extending the conjugation decreases the magnitude of the energy gap for the electronic transition, and this effect increases the wavelength of maximum absorption (λ_{max}). Thus, ethene (CH_2=CH_2) has a λ_{max} of 171 nm (nanometers) for its UV band, and 1,3-butadiene (CH_2=CH—CH=CH_2) has a λ_{max} of 217 nm for the corresponding absorption. This increase in wavelength continues with extended conjugation until it is sufficient to cause absorption in the visible region, that is, at wavelengths between 400 and 750 nm.

One well-known compound that is sufficiently conjugated to cause the absorption in the visible region is the highly colored substance β-carotene:

β-Carotene

β-Carotene, which can be isolated from carrots, has a distinct yellow-orange color in solution corresponding to absorption maxima at 483 and 453 nm. The visible light region that human eyes can sense is 400–700 nm, and it can be broken down into the wavelength regions and the corresponding color of the light listed in Table 22.1. It is important to realize that when light of a particular absorption region (and thus of a certain color) is absorbed, the solution will transmit (and thus appear to be) the complementary color. For example, solutions that absorb violet light will have a yellow appearance. A solution that absorbs yellow light (λ_{max} = 570–580 nm) will appear violet.

Although compounds that have maxima in the visible region (as does β-carotene) will certainly be colored, a maximum in the

Table 22.1. **Visible light and solution colors**

λ (nm)	Color of light[a]	Color of solution[a]
400	Violet	Yellow
430	Blue	Orange
500	Green	Red
580	Yellow	Violet
600	Orange	Blue-violet
650	Red-orange	Blue-green
700	Red	Green

a. For example, a solution that absorbs yellow light (λ_{max} = 570–580 nm) will transmit, or appear to have, a violet color.

400–700 region is not required for color. Compounds that have maxima in the UV region (200–400 nm) but whose absorptions tail into the visible region (over 400 nm) will have a yellowish hue.

Macroscale Procedure

Technique   UV-VIS Spectrometry: Spectrometric Method 4

— SAFETY INFORMATION —

Benzyltrimethylammonium hydroxide (Triton B) is highly toxic and corrosive; wear gloves while handling it.

Methanol is volatile, toxic, and flammable. Pour it in a hood, if possible, and do not pour it near a hot electrical device.

Begin heating an aluminum block on a hot plate to 160°C. Weigh 1.00 g of benzil and place it in a 20 × 150 mm test tube. Measure

5.0 mL of 0.95 M 1,3-diphenylacetone in triethylene glycol and add this solution to the test tube containing the benzil, using the solution to rinse any benzil adhering to the wall of the test tube.

Clamp the test tube in the aluminum block and place a thermometer in the test tube. Stirring gently with the thermometer, heat the mixture until the benzil dissolves and the temperature reaches 105–110°C.

Remove the reaction test tube from the aluminum block. When the solution cools to 100°C, add 2.0 mL of a 40% solution of benzyltrimethylammonium hydroxide (Triton B) measured with a graduated 1-mL pipet. Stir to mix. Crystals begin to form as the mixture cools. When the temperature falls below 80°C, cool the test tube in a 250-mL beaker of water until it reaches 40°C, then add 5 mL methanol. Stir the mixture briefly, cool it in a cold-water bath, and collect the crystals by vacuum filtration. Wash the crystals with methanol until the filtrate is purple-pink, not brown. Three washes are usually sufficient. Draw air over the crystals with the vacuum source turned on for 5 min to dry them. Determine the mass and melting point of your tetraphenylcyclopentadienone.

The product should be of sufficient purity to use without further treatment in the synthesis of dimethyl tetraphenylphthalate (Experiment 16.1).

UV-VIS Spectrum

Determine the UV-visible spectrum of your product between 200 and 800 nm, according to your instructor's directions, or use the following method: In a dry 25-mL Erlenmeyer flask, labeled "dilution 1," dissolve 10 mg of your tetraphenylcyclopentadienone in 10.0 mL of dichloromethane, measured carefully with a dry graduated cylinder. Transfer 0.50 mL of this solution to another dry 25-mL Erlenmeyer flask, labeled "dilution 2," and add 9.5 mL of dichloromethane. Cork both flasks. Obtain the UV-visible spectrum of your product as directed by your instructor, using the solution in the "dilution 2" flask.

Cleanup: The filtrate and all washings contain Triton B; pour them into the waste container labeled Triton B. Pour the dichloromethane solutions remaining from the UV-VIS spectrum determination into the container for halogenated waste.

microscale
Procedure

Technique UV-VIS Spectrometry: Spectrometric Method 4

> — *SAFETY INFORMATION* —
>
> **Benzyltrimethylammonium hydroxide (Triton B)** is highly toxic and corrosive; wear gloves while handling it.
>
> **Methanol** is volatile, toxic, and flammable. Pour it in a hood, if possible, and do not pour it near a hot electrical device.

Begin heating an aluminum block on a hot plate to 160°C. Weigh 0.200 g of benzil and place it in a 5-mL conical vial containing a magnetic spin vane. Add 1.0 mL of 0.95 M 1,3-diphenylacetone in triethylene glycol. Fit an air condenser to the vial and set the apparatus in the heated aluminum block. Fasten the air condenser with a microclamp. While slowly stirring the mixture, heat it until the benzil dissolves. Dissolution requires about 5–10 min.

Remove the reaction vial from the aluminum block and add, through the top of the condenser, 0.40 mL of a 40% solution of benzyltrimethylammonium hydroxide, measured with a graduated 1-mL pipet or automatic pipettor. Shake the mixture gently to combine the solutions. Heat the vial for 2 min, then cool it. The cooling can be hastened by using a beaker of tap water. Crystals begin to form as the mixture cools. Remove the condenser and add 1.0 mL of methanol. Stir the mixture briefly and cool it in an ice-water bath for 5–10 min. Collect the crystals by vacuum filtration on a Hirsch funnel [see Technique 5.5a]. Use a few drops of cold methanol to rinse the remaining crystals from the reaction vial. Wash the crystals with dropwise additions of cold methanol until the filtrate is purple-pink, not brown. Draw air over the crystals for 5 min to dry them. Determine the mass and melting point of the product.

The product should be of sufficient purity to use without further treatment in the synthesis of dimethyl tetraphenylphthalate (Experiment 16.1).

Determine the UV-visible spectrum of your product between 200 and 800 nm, according to your instructor's directions, or follow the procedure given in the preceding macroscale synthesis.

Cleanup: The filtrate and all washings contain Triton B; pour them in the waste container labeled Triton B. Pour the dichloromethane solutions remaining from the UV-VIS spectrum determination into the container for halogenated waste.

Questions

1. Is your measured UV-VIS absorption maximum consistent with the color of tetraphenylcyclopentadienone? Explain.

2. Explain why pure benzene, naphthalene, and anthracene are colorless, whereas tetracene (or naphthacene) has an orange color.

Benzene   Naphthalene   Anthracene   Tetracene
(naphthacene)

3. Outline a mechanism for the following transformation. Use base to form an enolate anion from the starting material to initiate this series of steps.

4. A base that has gained popularity as a reagent for aldol condensation reactions is LDA (lithium diisopropylamide). Because LDA is the conjugate base of the very weak acid diisopropyl amine $((CH_3)_2CH)_2NH$, it is a very strong base and thus would be expected to greatly enhance base-promoted condensation reactions. Briefly explain this.

LDA

5. Two solvents that have been used for a number of reactions are "glyme" (ethylene glycol dimethyl ether, $CH_3OCH_2CH_2OCH_3$) and diglyme (diethylene glycol dimethyl ether, $CH_3OCH_2CH_2OCH_2CH_2OCH_3$). The latter solvent boils at 161°C; thus, it has been especially useful for carrying out reactions at high temperatures. Briefly discuss what you might expect for solvent properties for these two compounds.

6. Triton B has been used for the following condensation reaction in 97% yield. This reaction takes advantage of the fact that the cyano (—C≡N) group in the nitrile starting material can stabilize an adjacent negative charge in much the same way as the carbonyl group does. Propose the structure of the condensation product and outline a mechanism that rationalizes its formation.

$$(C_6H_5)_2CH—CN + HCHO \xrightarrow{\text{Triton B}} ?$$

Experiment 23

HORNER-EMMONS-WITTIG SYNTHESIS OF METHYL *E*-4-METHOXYCINNAMATE

Synthesize a conjugated alkene, using a Wittig reagent.

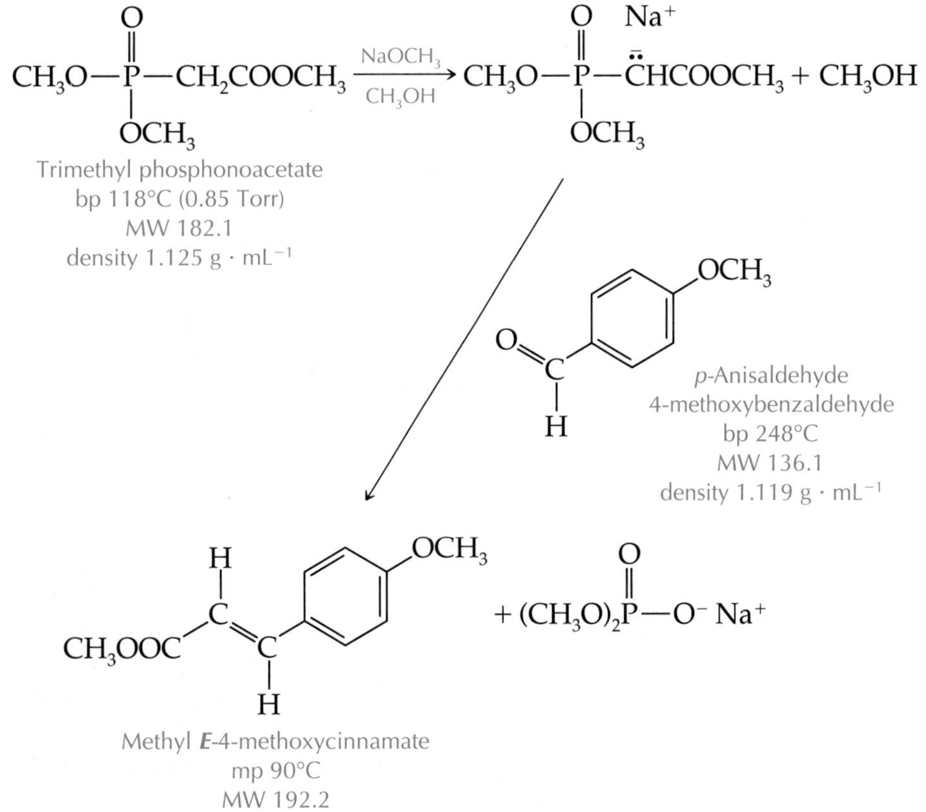
O
CH3O—P—CH2COOCH3
OCH3
Trimethyl phosphonoacetate
bp 118°C (0.85 Torr)
MW 182.1
density 1.125 g · mL−1
NaOCH3
CH3OH
O Na+
CH3O—P—CHCOOCH3 + CH3OH
OCH3
OCH3
O
C
H
p-Anisaldehyde
4-methoxybenzaldehyde
bp 248°C
MW 136.1
density 1.119 g · mL−1
H
C
CH3OOC
C
H
OCH3
Methyl E-4-methoxycinnamate
mp 90°C
MW 192.2
+ (CH3O)2P—O− Na+

As the name suggests, *p*-anisaldehyde has an odor somewhat reminiscent of anise. The *p*-methoxycinnamate product that you obtain is expected to be largely the *E*-compound (*trans*), which is related to the compound cinnamaldehyde in Chinese cinnamon.

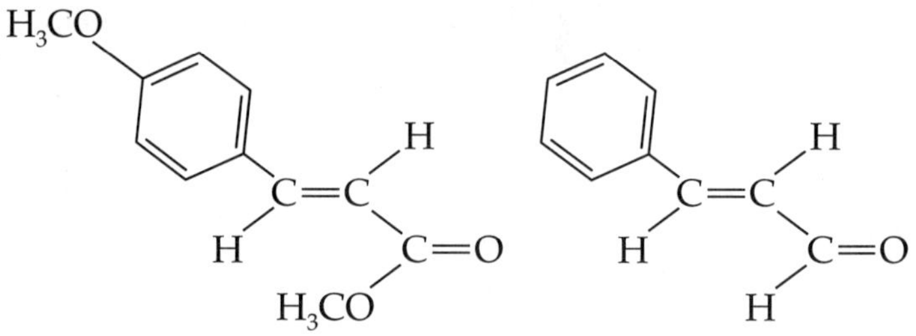

Methyl *E*-4-methoxycinnamate   *E*-Cinnamaldehyde

The Wittig reaction has proved to be a very useful synthetic method for preparing alkenes (olefins). As a result of his studies revolving around this reaction, Georg Wittig was one of the winners of the 1979 Nobel Prize in Chemistry. In the reaction that bears his name, a carbonyl compound (an aldehyde or ketone) is treated with a Wittig reagent to produce an alkene:

Cyclopentanone   Methylenetriphenylphosphorane   Methylenecyclopentane   Triphenylphosphine oxide
                 (a Wittig reagent)

The value of the Wittig method becomes clear when we consider alternative methods for preparing alkenes. For example, treatment of alkyl halides with base gives rise to alkenes, but even this simple example (2-bromobutane) shows that complex product mixtures arise:

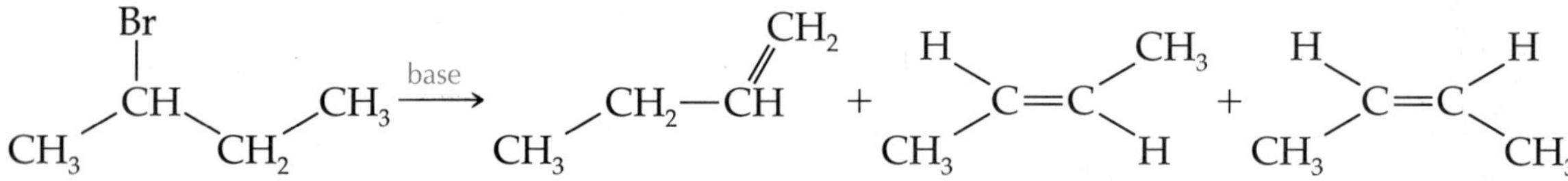

2-Bromobutane   1-Butene   *E*-2-Butene   *Z*-2-Butene
                                                    (*trans*)   (*cis*)

If we wish to prepare just one of the three products—say, 1-butene—we can do so in an efficient fashion (excluding other alkene products) by using the proper choice of Wittig reagent and carbonyl compound:

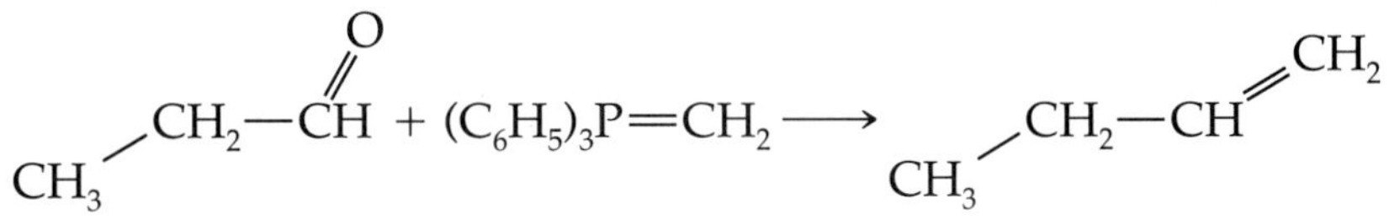

$$CH_3CH_2-\overset{\overset{\displaystyle O}{\|}}{CH} + (C_6H_5)_3P{=}CH_2 \longrightarrow CH_3CH_2-CH{=}CH_2$$

1-Butene

To understand the chemistry of the Wittig reagent, let us examine how it is formed. When alkyl halides are treated with phosphines, they undergo direct displacement (S_N2) reactions to form phosphonium salts:

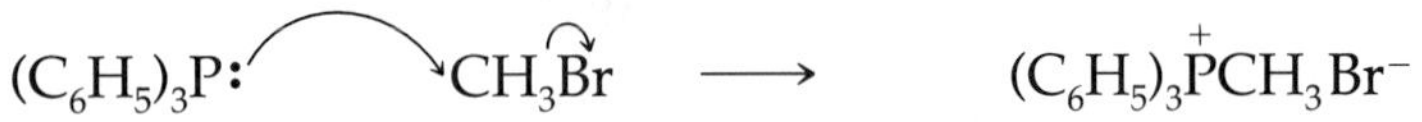

$$(C_6H_5)_3P\overset{\frown}{:} \quad CH_3Br \longrightarrow (C_6H_5)_3\overset{+}{P}CH_3\,Br^-$$

Triphenylphosphine   Bromomethane   Methyltriphenylphosphonium bromide

When phosphonium salts are treated with strong bases (butyl-lithium is commonly used), the Wittig reagent is formed:

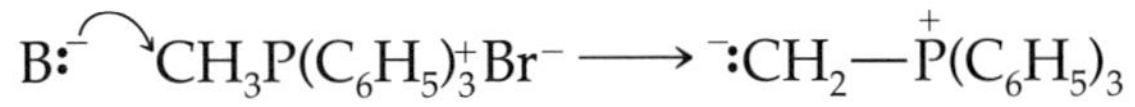

$$B\overset{\frown}{:} \quad CH_3P(C_6H_5)_3^+Br^- \longrightarrow {^-}{:}CH_2{-}\overset{+}{P}(C_6H_5)_3$$

Base   Wittig reagent

There are two resonance forms of Wittig reagents that should be examined. The first is merely the dipolar form we would expect as a result of simple proton extraction. The second has an expanded octet of 10 valence electrons around phosphorus and no formal charge separation. This behavior is not uncommon for second-row atoms bearing *d*-orbitals that can participate in bonding:

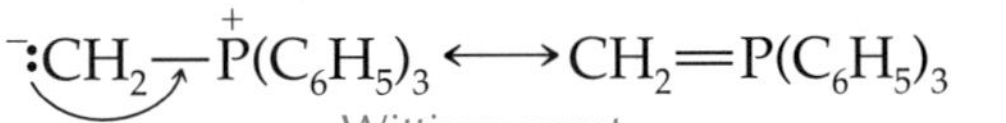

$$^-{:}CH_2{-}\overset{+}{P}(C_6H_5)_3 \longleftrightarrow CH_2{=}P(C_6H_5)_3$$

Wittig reagent

The mechanism of the Wittig reaction is complex, and we will include only some major details here. The first step is attack of the Wittig reagent on the carbonyl-containing substrate to form a charged intermediate called a betaine, which can close to form a four-membered ring. This ring then opens to form the alkene product and a phosphine oxide side product:

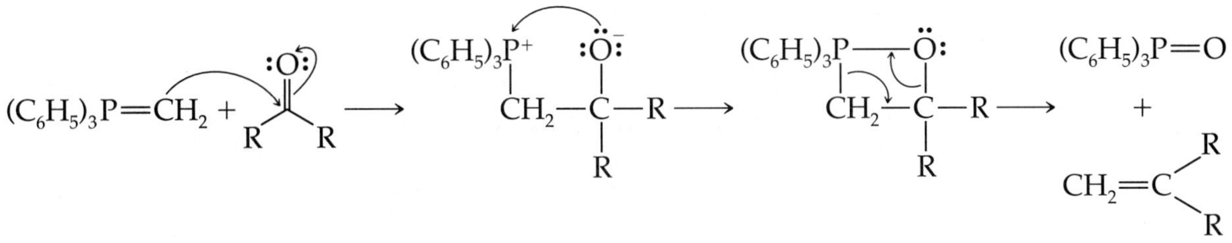

In this experiment you will use a modified phosphorus-containing reagent called the Horner-Emmons-Wittig reagent. This reagent is prepared from trimethyl phosphonoacetate, a commercially available compound that is much more acidic than simple phosphonium salts. This enhanced acidity means that the base sodium methoxide is strong enough to produce the Horner-Emmons-Wittig reagent. Perhaps even more important, the final phosphorus-containing side product is water-soluble, a great advantage during isolation and purification of the product. Presumably, the mechanism of the Horner-Emmons-Wittig reaction parallels that for the simple Wittig reaction:

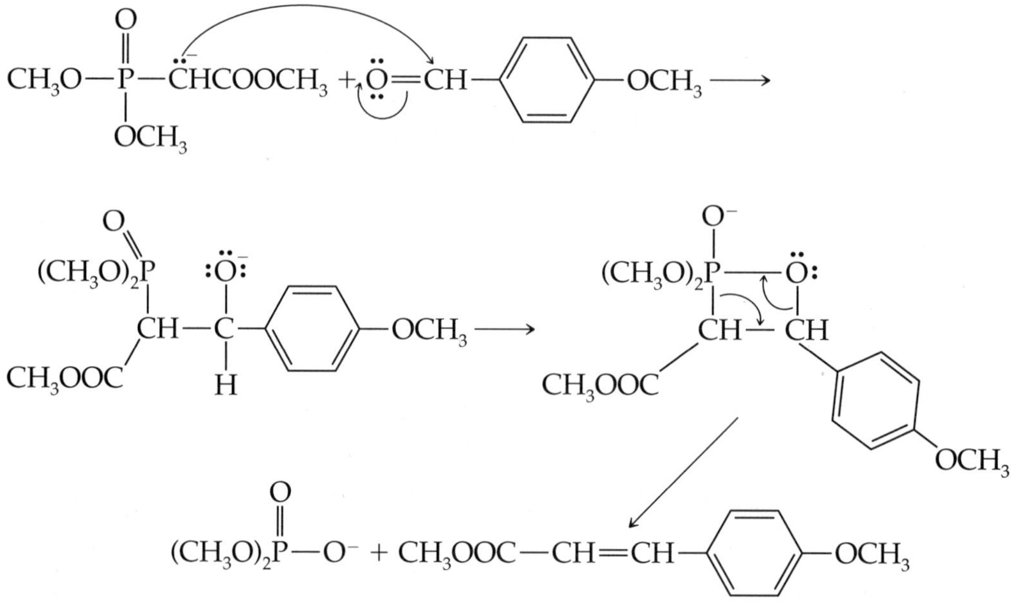

$$CH_3O-\overset{\overset{O}{\|}}{\underset{\underset{OCH_3}{|}}{P}}-\overset{..}{C}HCOOCH_3 + \overset{..}{O}=CH-\underset{}{\bigcirc}-OCH_3 \longrightarrow$$

$$\overset{\overset{O}{\|}}{(CH_3O)_2P}\overset{}{\underset{CH_3OOC}{\diagup}}\overset{..}{\underset{}{O}}\overset{:\overset{..}{O}:}{\underset{H}{|}}-OCH_3 \longrightarrow \overset{\overset{O^-}{|}}{(CH_3O)_2P}-\overset{..}{O}: \quad CH\overset{}{-}CH\underset{}{\bigcirc}-OCH_3$$

$$(CH_3O)_2\overset{\overset{O}{\|}}{P}-O^- + CH_3OOC-CH=CH-\underset{}{\bigcirc}-OCH_3$$

Macroscale
Procedure

Techniques   Mixed Solvent Recrystallization: Techniques 5.2a and 5.3
NMR Spectrometry: Spectrometric Method 2

SAFETY INFORMATION

Syringe needles are a puncture hazard.

Sodium methoxide in methanol solution is flammable, corrosive, and moisture sensitive. Wear gloves and avoid contact with skin, eyes, and clothing. Keep the container tightly closed and store it in a desiccator.

(continued on next page)

Place 4.0 mL of anhydrous methanol, 1.60 mL of 25 wt % sodium methoxide/methanol solution (4.4 mmol/mL), and 1.75 mL of trimethyl phosphonoacetate in a 50-mL Erlenmeyer flask containing a magnetic stirring bar. Quickly close the flask with a rubber septum and insert a syringe needle through the septum to act as a vent. Set the flask on a magnetic stirrer and clamp it to a ring stand.

In a small, dry test tube, prepare a solution of 0.80 mL of *p*-anisaldehyde and 2.0 mL of anhydrous methanol. Draw this solution into a syringe. Insert the syringe through the rubber septum and add the solution slowly to the reaction mixture over a period of 10 min. Remove the syringe. Continue to stir the reaction mixture at room temperature for 1 h. During this time, precipitation occurs and the solution may become light brown.

At the end of the reaction period, remove the venting needle and rubber septum. Add 8.0 mL of water and stir thoroughly. Collect the crude product by vacuum filtration. Rinse the flask with 1–2 mL of water and add this rinse to the Buchner funnel.

Recrystallize the product in 95% ethanol [see Technique 5.2]. When the crystals have dissolved completely, add water dropwise to the hot solution until cloudiness occurs, then add ethanol dropwise until the solution clears [see Technique 5.3]. Allow the solution to cool slowly to room temperature before cooling it in an ice-water bath. Collect the purified product by vacuum filtration. Allow the product to dry thoroughly by drawing air over the crystals for 30 min with the vacuum source turned on or leaving them in your desk overnight. Determine the melting point, the percent yield, and obtain the NMR spectrum of your dried product as directed by your instructor.

Cleanup: Dilute the aqueous filtrate from the crude product with water and neutralize it with 5% HCl solution before washing it down the sink, or pouring it into the container for aqueous inorganic waste. Pour the filtrate from the recrystallization into the container for flammable (organic) waste.

microscale Procedure

Techniques   Craig Tube Recrystallization: Technique 5.5b
NMR Spectrometry: Spectrometric Method 2

─ *SAFETY INFORMATION* ──────

Syringe needles are a puncture hazard.

Sodium methoxide in methanol solution is flammable, corrosive, and moisture sensitive. Wear gloves and avoid contact with skin, eyes, and clothing. Keep the container tightly closed and store it in a desiccator.

Trimethyl phosphonoacetate presents no particular hazards.

***p*-Anisaldehyde (4-methoxybenzaldehyde)** is an irritant. Wear gloves and avoid contact with skin, eyes, and clothing.

Methanol is flammable and toxic.

Ethanol is flammable.

Place 1.0 mL of anhydrous methanol, 0.40 mL of 25 wt % sodium methoxide/methanol solution (4.4 mmol/mL) and 0.43 mL of trimethyl phosphonoacetate in a 5-mL conical vial containing a magnetic spin vane. Quickly close the vial with a screw cap and septum, then insert a syringe needle through the septum to act as a vent. Clamp the vial so that it is positioned on a magnetic stirrer.

In a small, dry test tube, prepare a solution of 0.20 mL of *p*-anisaldehyde and 0.50 mL of anhydrous methanol. Draw this solution into a syringe. Insert the syringe through the septum and add the solution slowly to the reaction mixture over a period of 10 min. Remove the syringe. Continue to stir the reaction mixture at room temperature for 1 h. During this time, precipitation occurs and the solution may become light brown.

At the end of the reaction period, remove the venting needle and open the vial. Add 2.0 mL of water, cap the vial, and shake it to mix the contents thoroughly. Collect the crude product by vacuum filtration on a Hirsch funnel [see Technique 5.5a]. Rinse the vial with a few drops of water and add this rinse to the Hirsch funnel.

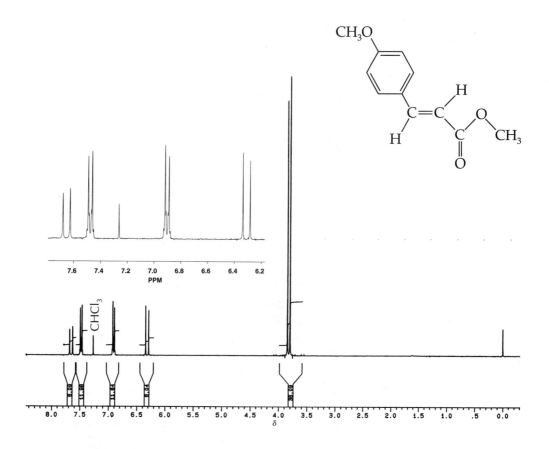

FIGURE 23.1 ¹H NMR spectrum (300 MHz) of methyl *E*-4-methoxycinnamate (in CDCl₃).

Recrystallize the product in 95% ethanol in a Craig tube [see Technique 5.5b]. When the crystals have dissolved completely, add water dropwise to the hot solution until cloudiness occurs, then add ethanol dropwise until the solution clears [see Technique 5.3]. Set the Craig tube in a 25-mL Erlenmeyer flask to cool slowly to room temperature before cooling it in an ice-water bath. Centrifuge the Craig apparatus to remove the solvent. Allow the product to dry thoroughly before determining the melting point, the percent yield, and the NMR spectrum.

Cleanup: Dilute the aqueous filtrate from the crude product with water and neutralize it with 5% HCl solution before washing it down the sink, or pouring it into the container for aqueous inorganic waste. Pour the liquid remaining in the centrifuge tube from the recrystallization into the container for flammable (organic) waste.

References

1. Crandall, J. K.; Mayer, C. F. *J. Org. Chem.* **1970,** *35,* 3049–3053.
2. Bottin-Strzalko, T. *Tetrahedron* **1973,** *29,* 4199–4204.

Questions

1. Suggest a reason why the phosphorus-containing side product from the Horner-Emmons-Wittig reaction is more water-soluble than the triphenylphosphine oxide side product obtained from the standard Wittig reaction.
2. Sodium methoxide is a weaker base than butyllithium. Suggest a reason why it would not be surprising to find that a weaker base is sufficient to prepare the Horner-Emmons-Wittig reagent.
3. The *E* configuration of methyl 4-methoxycinnamate prepared in this experiment might very well be established in the cyclic intermediate proposed in the mechanism as arising from the charged intermediate formed by attack of the Wittig reagent on *p*-anisaldehyde. Suggest a reason why this intermediate might have two important groups in a *trans* relationship that could give rise to the *E*-alkene.

Experiment 24

KINETIC VERSUS THERMODYNAMIC CONTROL IN CHEMICAL REACTIONS

Explore kinetic and equilibrium control of organic chemical reactions.

When two or more products are possible in a chemical reaction and one is favored simply because it forms more rapidly, the favored product is said to be the result of kinetic control. If a product is favored because it is more stable, it is said to be the result of thermodynamic or equilibrium control. The products of kinetic and thermodynamic control sometimes are and sometimes are not the same compound.

The energy diagram shown in Figure 24.1 may help you to visualize the concepts of thermodynamic and kinetic control. In this case the products of these two processes are different. The energy barrier to the product on the left is lower, and it is formed faster (kinetic control). The product on the right has lower energy than the product on the left; its greater stability reveals it to be the product of thermodynamic control.

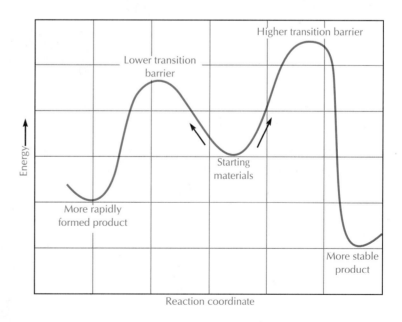

FIGURE 24.1 Energy diagram for a reaction illustrating both kinetic and thermodynamic control.

In a process behaving as shown in Figure 24.1, the lower energy product will not be formed first. It takes reaction conditions that allow an equilibrium to develop so that the product of kinetic control can return to the starting materials many times. Every so often the thermodynamic product forms, and once it does, it is much harder for the low-energy (more stable) product to get over the high-energy barrier of the reverse reaction and return to the starting materials. It is not surprising, therefore, that thermodynamic control is often called equilibrium control. Reactions run at lower temperatures often favor kinetically controlled products; but when there is sufficient energy to overcome all transition-state barriers, the outcome of the reaction is determined by the relative energy of the products, that is, by the equilibrium position.

In Experiment 24.1, you will explore which semicarbazone forms from a mixture of cyclohexanone and 2-furaldehyde when it is treated with semicarbazide at several temperatures and pH levels. In Experiment 24.2, you will investigate the effect of modifying the reaction time and the amount of the Lewis acid, AlCl$_3$, on the ratio of monosubstituted isomers produced in a Friedel-Crafts alkylation of chlorobenzene. Gas chromatographic analysis of the product mixture will allow you to draw conclusions about kinetic versus thermodynamic control of the alkylation.

24.1
Semicarbazones of Cyclohexanone and 2-Furaldehyde

Explore the effect of reaction conditions on kinetic and thermodynamic control in a teamwork situation.

Cyclohexanone
bp 155°C
MW 98.2
density 0.947 g · mL^{-1}

Semicarbazide hydrochloride
mp 173°C
MW 111.5

Cyclohexanone semicarbazone
mp 166°C

2-Furaldehyde
bp 161°C
MW 96.1
density 1.16 g · mL^{-1}

Semicarbazide hydrochloride
mp 173°C
MW 111.5

2-Furaldehyde semicarbazone
mp 202°C

We recommend that you work as part of a team for this experiment. You will prepare the semicarbazones of cyclohexanone and 2-furaldehyde by treating these carbonyl compounds with semicarbazide hydrochloride in the presence of a buffer. You will study the conditions under which each of these semicarbazones forms and undergoes reversible interconversion, in order to determine which semicarbazone is the product of kinetic control and which is the product of thermodynamic control.

The conversion of aldehydes and ketones to semicarbazones is carried out by treating semicarbazide hydrochloride with various buffers to free the semicarbazide nucleophile:

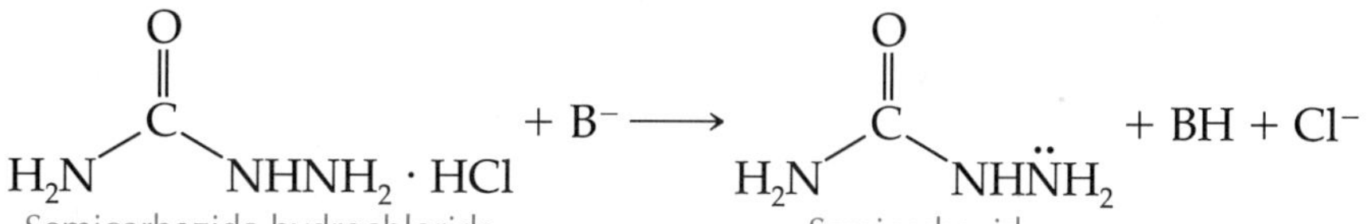

Semicarbazide hydrochloride — Semicarbazide

The free semicarbazide is sufficiently nucleophilic to attack the carbonyl carbon of aldehydes or ketones, inducing an addition-elimination reaction and resulting ultimately in a semicarbazone:

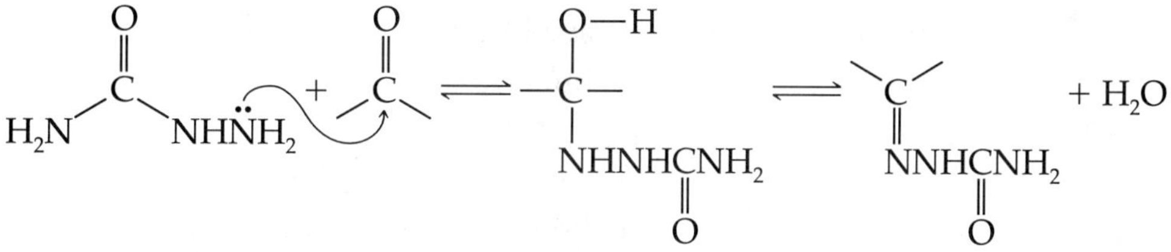

A semicarbazone

The effect of pH on the reaction is difficult to predict. Whereas we might expect a higher pH to enhance reactivity because higher concentrations of the reactive nucleophile would be formed, lower pH could also enhance reactivity because protonation of the carbonyl group should enhance its reactivity toward the semicarbazide nucleophile and catalyze the dehydration step as well. The outcome depends on which steps are rate determining under the conditions used.

Because all the steps in the semicarbazone formation process are reversible, conditions can be found where one semicarbazone converts to another:

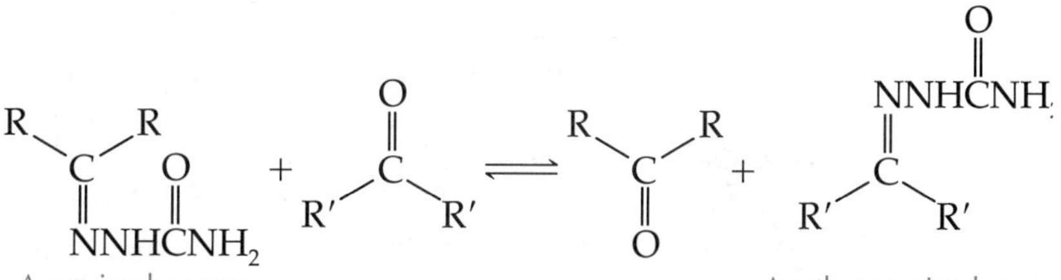

A semicarbazone — Another semicarbazon

Carry out the experiments and carefully consider the conditions under which a specific semicarbazone is formed. You can identify products by melting point, NMR spectrometry, and any other analytical technique recommended by your instructor. You need to find out what kinds of reaction conditions are associated with both kinetic and thermodynamic control. Then you can see whether your experimental conclusions are consistent with your understanding of the mechanisms for the formation of each of the semicarbazones.

mmm*icroscale* Procedure

SAFETY INFORMATION

Both cyclohexanone and 2-furaldehyde are toxic.

Semicarbazide hydrochloride is toxic. Wear gloves while carrying out the experimental procedures.

Work in teams of two to four students (depending on the time available) and divide the work in an equitable manner.

For the bicarbonate solution: Dissolve the sodium bicarbonate in distilled water, then add the solid semicarbazide hydrochloride slowly to prevent excess foaming.

You will need to prepare the following solutions before beginning the experiments.

Aqueous semicarbazide/buffer solutions. Prepare these solutions in 50-mL beakers. Stir the solutions until the solids dissolve, then determine the pH of each solution, using a pH meter or a universal indicator paper such as pHydrion.

Buffer	Solution A K_2HPO_4 (dibasic)	Solution B $NaHCO_3$
Amount of buffer	2.00 g	1.00 g
Semicarbazide · HCl	1.00 g	0.50 g
Water	25 mL	12.5 mL

The reagent volumes should be measured with graduated pipets for all experiments.

Ethanolic cyclohexanone/2-furaldehyde solution. Combine 1.5 mL of cyclohexanone, 1.2 mL of freshly distilled 2-furaldehyde, and 7.5 mL of 95% ethanol in a 18 × 150 mm test tube. Stir thoroughly to ensure a homogeneous solution.

24.1a Semicarbazones of 2-Furaldehyde and Cyclohexanone

Place 5.0 mL of solution A (K_2HPO_4/semicarbazide) in a 10-mL Erlenmeyer flask, add 0.90 mL of 95% ethanol and 0.17 mL of freshly distilled 2-furaldehyde. Stir or swirl the solution until crystals begin to form, then cool the flask in an ice-water bath for 5 min. Collect the solid product by vacuum filtration on a Hirsch funnel. Remove the filtrate from the filter flask. With the vacuum source turned on, wash the product four times by covering the crystals with approximately 1-mL portions of ice-cold water. Dry the crystals in a small beaker or Petri dish in a 50°C oven for 20 min. Determine the weight and melting point.

Prepare the semicarbazone of cyclohexanone according to the preceding procedure, substituting 0.21 mL of cyclohexanone for 2-furaldehyde.

Cleanup: Pour the filtrate from the reaction mixture into the container for flammable (organic) waste. The aqueous filtrate from washing the crystals can be washed down the sink or placed in the container for aqueous inorganic waste.

General Instructions for Experiments 24.1b through 24.1d

Collect the solid product by vacuum filtration on a Hirsch funnel. Remove the filtrate from the filter flask. With the vacuum source turned on, wash the product four times with approximately 1-mL portions of ice-cold water.

Dry the crystals in a small beaker or Petri dish in a 50°C oven for 20 min, if you will be determining the melting point during the laboratory period. Otherwise, allow the crystals to dry overnight before determining their melting point.

24.1b Reversibility of Semicarbazone Formation

Test 1. Combine the following in a 10-mL Erlenmeyer flask: 0.100 g of dry 2-furaldehyde semicarbazone (prepared in Experiment 24.1a), 3.5 mL of water, 0.70 mL of ethanol, and 0.10 mL of cyclohexanone. Warm the mixture in a hot-water bath at 80°C for 5 min, swirling the flask until the solid dissolves. Cool the flask to room temperature, then cool it in an ice-water bath. Collect and dry the crystals as directed in the general instructions given earlier before determining the melting point.

Test 2. Repeat the procedure in Test 1, using 0.100 g of dry cyclo-hexanone semicarbazone (prepared in Experiment 24.1a) and 0.10 mL of 2-furaldehyde.

24.1c Competitive Semicarbazone Formation in (Dibasic) Potassium Hydrogen Phosphate (K_2HPO_4) Buffer Solution

Measure two sets of the following solutions for Tests 3 and 5. Place 3.6 mL of solution A (K_2HPO_4/semicarbazide) in a 10-mL Erlenmeyer flask and 1.0 mL of the ethanolic cyclohexanone/2-furaldehyde solution in a small test tube.

Test 3—Reaction at 0°C. Prepare a vacuum filtration apparatus with a Hirsch funnel before beginning the reaction. Cool both solutions in an ice-water bath for several minutes before pouring the ethanolic ketone/aldehyde solution from the test tube into the Erlenmeyer flask containing solution A. Stir the mixture for 20 s, keeping the flask in the ice-water bath. Immediately collect the solid product on a Hirsch funnel by vacuum filtration. Rinse and dry the crystals as directed in the preceding general instructions before determining the melting point.

Test 4—Reaction at room temperature. Place 3.6 mL of solution A (K_2HPO_4/semicarbazide) and 1.0 mL of the ethanolic cyclo-hexanone/2-furaldehyde solution in a 10-mL Erlenmeyer flask. Swirl to mix and allow the reaction mixture to stand at room temperature for 5 min. Cool the flask in an ice-water bath for 5 min before collecting the product on a Hirsch funnel by vacuum filtration. Rinse and dry the crystals as directed in the preceding general instructions before determining the melting point.

Test 5—Reaction at 80°C. Heat both solutions in a water bath at 80°C for several minutes before pouring the ethanolic ketone/aldehyde solution from the test tube into the Erlenmeyer flask containing solution A. Stir the mixture and heat the flask at 80°C for 15 min. Cool the flask to room temperature, then in an ice-water bath for several minutes. Collect and dry the crystals as directed in the preceding general instructions before determining the melting point.

24.1d Competitive Semicarbazone Formation in Sodium Bicarbonate Buffer Solution

Test 6—Reaction at room temperature. Place 3.6 mL of solution B (NaHCO$_3$/semicarbazide) and 1.0 mL of ethanolic cyclohexanone/2-furaldehyde solution in a 10-mL Erlenmeyer flask. Swirl to mix and allow the reaction mixture to stand at room temperature for 5 min. Cool the flask in an ice-water bath for 5 min. Collect and dry the crystals as directed in the preceding general instructions before determining the melting point.

Test 7—Reaction at 80°C. Place 3.6 mL of solution B (NaHCO$_3$/semicarbazide) in a 10-mL Erlenmeyer flask and 1.0 mL of the ethanolic cyclohexanone/2-furaldehyde solution in a small test tube. Heat both solutions in a water bath at 80°C for several minutes before pouring the ethanolic ketone/aldehyde

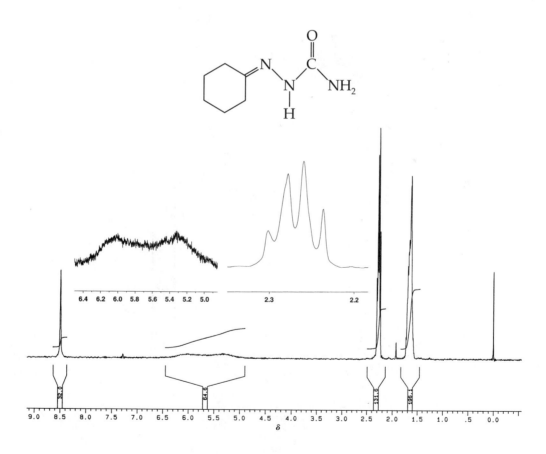

FIGURE 24.2 ^{1}H NMR spectrum (300 MHz) of cyclohexanone semi-carbazone (in acetone-d$_6$).

When you have completed the analysis of your products, submit them as directed by your instructor.

solution into the Erlenmeyer flask containing solution B. Stir the mixture and heat the flask at 80°C for 15 min. Cool the flask to room temperature, then in an ice-water bath for several minutes. Collect and dry the crystals as directed in the preceding general instructions before determining the melting point.

Cleanup for Experiments 24.1b through 24.1d: Pour the filtrate from each reaction mixture and any remaining ethanolic solution of cyclohexanone and 2-furaldehyde into the container for flammable (organic) waste. The aqueous filtrate from washing the crystals and any remaining buffer/semicarbazide solutions can be washed down the sink or placed in the container for aqueous inorganic waste.

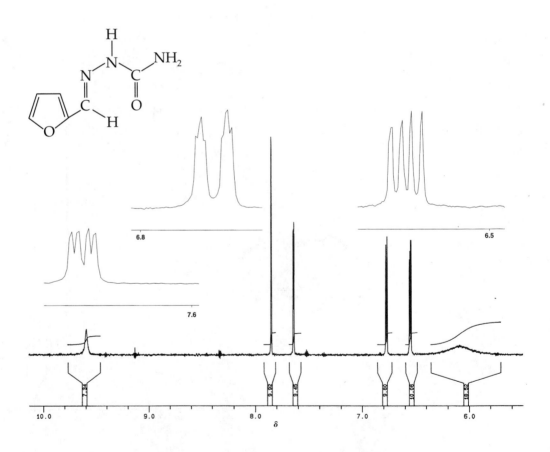

FIGURE 24.3 ^{1}H NMR spectrum (300 MHz) of 2-furaldehyde semicarbazone (in acetone-d_6).

Product Analysis and Interpretation of Results

Remember that only a small amount of impurity can significantly depress a melting point.

Identify the product of each experiment (24.1b through 24.1d) by its melting point. Summarize the results for your group in a tabular format. In some instances the observed melting point may indicate a mixture of the two semicarbazones formed in the reaction under consideration. Your instructor may ask you to determine the composition of a particular product by NMR analysis. Prepare the NMR sample(s) as directed by your instructor. Compare the times required for product formation in Experiments 24.1c and 24.1d. Discuss the results of your experiments. Which product formation is kinetically controlled and which is thermodynamically controlled?

Reference

1. Conant, J. B.; Bartlett, P. D. *J. Am. Chem. Soc.* **1932,** *54,* 2881–2899. Dr. Conant was president of Harvard University and a science advisor to Franklin D. Roosevelt.

Questions

1. Which semicarbazone is the product of kinetic control? Of thermodynamic control? What is the experimental basis for the foregoing conclusions?
2. Propose a mechanistic or structural rationale for your conclusions in Question 1.
3. Draw an energy diagram similar to that in Figure 24-1, positioning specific starting materials and products from this experiment on the curve.
4. There are three nitrogen atoms in semicarbazide and yet only one of these is the nitrogen that nucleophilically attacks the carbonyl compound? Which and why?
5. Propose a mechanism for the interconversion of one semicarbazone to another by taking advantage of the fact that the formation of semicarbazones in water is a reversible process. Keep in mind that as reactants at various stages in this process you may use each of the two semicarbazones, the two carbonyl compounds, water, semicarbazide, semicarbazide hydrochloride, and the components of the buffer (here represented by the BH^+/B pair).

24.2

Friedel-Crafts Alkylation of Chlorobenzene

 Investigate kinetic and thermodynamic control in an electrophilic aromatic substitution reaction, using GC analysis.

2-Chloropropane
(isopropyl chloride)
bp 36°C

Chlorobenzene
bp 132°C
MW 112.6

Kinetic product
(ratio of *o-/m-/p*-isomers to be determined)

+ HCl

Kinetic product

AlCl$_3$
7 days at 25°C

Thermodynamic product
(ratio of isomers to be determined)

The fundamental aspects of electrophilic aromatic substitution have already been covered in some detail in Experiments 17 and 18. In this experiment we will take advantage of the fact that the second substitution on a monosubstituted benzene ring can give rise to any or all of three isomers: *ortho, meta,* and *para:*

E$^+$ = electrophile

Commonly, this reaction is subject to electronic control because the key intermediate carries a positive charge:

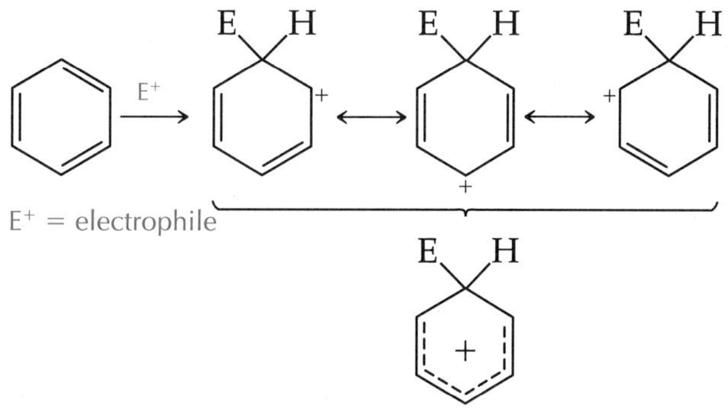

E⁺ = electrophile

The stability of this intermediate is sensitive to electronic effects, and the electronic characteristics of a substituent, G, are important. Most often when G is an electron-withdrawing group (ewg), the *meta* product is favored; whereas any time G is an electron-donating group (edg), *ortho* and *para* products are favored. If the group is electron withdrawing, destabilization of the intermediate will result unless the group is positioned *meta* to the point of attachment of the electrophile. If the group is electron donating, positioning the group either *ortho* or *para* will stabilize the intermediate, thus favoring *ortho* and *para* products:

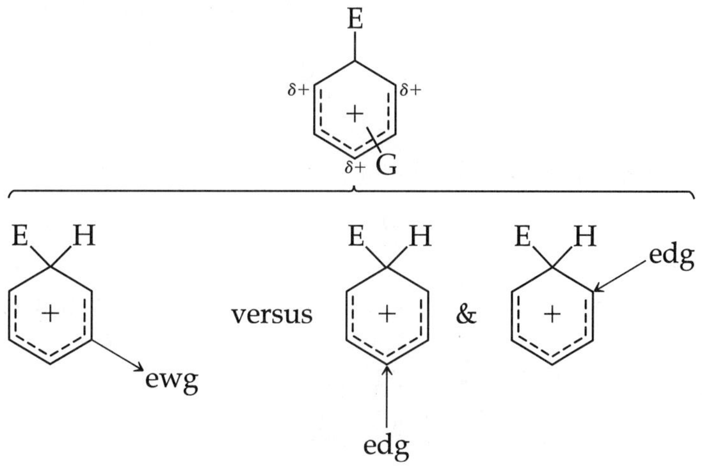

When enough energy is available to overcome all energy barriers (equilibrium control), the most stable thermodynamic product is formed. Otherwise, the substitution products are those that form fastest (kinetic control).

You will use a nitromethane/aluminum trichloride mixture in the first experiment to temper the reactivity of the electrophile, thus gaining selectivity in the reaction:

$$CH_3 \text{—} NO_2 + AlCl_3 \longrightarrow [CH_3 \text{—} NO_2 \cdots\cdots AlCl_3]$$

Lewis acid-base
complex

Under such conditions, kinetically controlled product(s) are expected to dominate. In the second reaction you will omit the nitromethane and use a much longer reaction time; under such conditions, thermodynamic control is expected.

mmmmicroscale Procedure*

Techniques   Microscale Extraction: Technique 4.5b
Gas Chromatography: Technique 11

SAFETY INFORMATION

Wear gloves while carrying out the experimental procedures.

Both **chlorobenzene** and **2-chloropropane** (isopropyl chloride) are flammable and are irritants. Avoid contact with skin, eyes, and clothing.

Aluminum chloride is corrosive and moisture sensitive. It reacts with atmospheric moisture to form HCl.

Nitromethane is flammable, moisture sensitive, and toxic.

Keep the bottle of AlCl₃ tightly capped, and store it in a desiccator.

Kinetic Product

Prepare two Pasteur filter pipets during the reaction period [see Technique 2.4].

Place 1.40 mL of a solution containing 0.35 mmol·mL of aluminum chloride and 0.70 mmol·mL^{-1} of nitromethane in chlorobenzene in a 5-mL conical vial. Put a spin vane in the vial and cap it with a septum and screw cap. Insert a syringe needle to act as a vent. Clamp the vial in an ice-water bath and stir the solution for several minutes. Prepare a solution containing 0.15 mL of 2-chloropropane and 0.80 mL of dry chlorobenzene in a small test tube or vial. Draw this solution into a 1-mL syringe.

When the solution in the vial is thoroughly chilled, insert the syringe through the septum and add the 2-chloropropane solution dropwise over a period of 5 min. Remove the syringe and the

---

*This experiment requires one laboratory period to prepare the kinetic product, a 7-day reaction period for conversion of the kinetic product to the thermodynamic product, and approximately 1 h of another laboratory period to work up the thermodynamic product and carry out gas chromatographic analysis of it.

venting needle. Continue stirring the reaction mixture for 40 min while the ice melts and the cooling bath gradually warms to room temperature.

Open the vial and add 2.0 mL of water. Stir the mixture for 1–2 min, then cap the vial and shake it. Allow the layers to separate. Transfer the lower organic phase to another 5-mL conical vial with a Pasteur filter pipet [see Technique 4.5b, Method B].

The densities of the two phases are quite similar, with the organic layer being slightly more dense; this similarity in densities may cause the cloudy organic layer to be partially suspended in the aqueous phase rather than completely separated. Should this happen, draw some of the mixture into the filter pipet; the two phases will separate and you can drain the lower organic portion slowly into the second conical vial, stopping when the interface between the phases reaches the cotton plug at the pipet tip. Return the aqueous phase to the reaction vial. Repeat this process until the organic phase is completely transferred to the second vial.

— **SAFETY PRECAUTION** —

There will be some CO_2 formed during the Na_2CO_3 wash; open the screw cap slowly to vent the vial.

Drying the product mixture twice with small amounts of drying agent is more effective than using a larger amount initially.

Wash the organic phase, now in the second vial, with 1.0 mL of 5% sodium carbonate solution. After this washing, transfer the organic phase to a dry 12 × 75 mm test tube. Add approximately 150–200 mg of anhydrous potassium carbonate to the test tube and allow the mixture to stand for 5 min. With a dry Pasteur filter pipet, transfer the product solution to another test tube; rinse the drying agent with 0.5 mL of chlorobenzene and transfer the rinse to the test tube containing the product. Dry the product a second time with potassium carbonate for 10 min.

Separate the product from the drying agent with a clean Pasteur filter pipet, transferring it to a small screw-capped vial (not a conical vial). Prepare a sample for GC analysis as directed on p. 288, then use the product solution remaining in the vial to prepare the thermodynamic product according to the following directions.

Cleanup: Combine the aqueous solutions from the extractions and neutralize the solution with solid sodium carbonate before washing it down the sink, or placing it in the container for aqueous inorganic waste. The spent potassium carbonate drying agent is coated with halogenated organic compounds; place it in the container for hazardous solid waste.

Thermodynamic Product

Weigh a sample of aluminum chloride between 40–50 mg as quickly as possible and add it to the kinetic product contained in a screw-capped vial. Cap the vial tightly and swirl it occasionally for a period of 15 min. Place the vial in your drawer for 7 days.

After 7 days, add 2.0 mL of water to the product mixture, swirl the vial to dissolve the aluminum chloride and transfer the entire contents of the screw-capped vial to a 5-mL conical vial. Carry out the separation, washing, and drying of the organic phase by the same procedure used for the kinetic product. Analyze the product mixture by gas chromatography. Submit any remaining product mixture to your instructor.

Cleanup: Follow the same cleanup procedure used for the kinetic product.

Gas Chromatographic Analysis of the Product Solutions

For a packed column gas chromatograph, use a 15% polyphenyl ether on Chromasorb W (acid washed, 60/80 mesh) column at 110°C or a nonpolar column such as OV-101 at 100°C (Ref. 1). The product solution can be analyzed without further dilution. Consult your instructor about the appropriate sample size for the packed column instruments in your laboratory. For a capillary column chromatograph, use a nonpolar column such as polydimethylsiloxane (SE-30 or OV-1) and a column temperature of 100°C. Prepare a solution for GC analysis by combining 2 drops of your product mixture and 0.5 mL of diethyl ether; inject 1 μL of this ether solution into the chromatograph.

In analyzing your chromatograms you will find a peak for chlorobenzene and peaks for the *ortho*, *meta*, and *para* isomers of chloro(isopropyl)benzene. Identify the chlorobenzene peak by the peak enhancement method [see Technique 11.6]. The chloro(iso-

Isomeric compounds usually have virtually identical molar response factors on GC.

propyl)benzenes elute in close succession in the order of *ortho*, *meta*, and *para* isomers. Calculate the relative percentages of the three isomers in both product mixtures, using the integration data from the recorder or the peak areas [see Technique 11.6].

Optional Experiments

After consulting with your instructor, carry out the experiments you design in answer to Questions 4 and 5, which follow.

Reference

1. Kolb, K. E.; Standard, J. M.; Field, K. W. *J. Chem. Educ.* **1988,** *65,* 367.

Questions

1. Why is the product mixture washed with 5% sodium carbonate? Write a balanced equation for the reaction that occurs.
2. Which isomer(s) is the kinetic product? Why does this isomer(s) form faster?
3. Which isomer(s) is the thermodynamic product?
4. Why must the aluminum chloride catalyst be deactivated with nitromethane to form the kinetic product? Outline an experiment to test your answer to this question.
5. Outline an experiment designed to test whether the isomerization of the kinetic product to the thermodynamic product occurs only with the uncomplexed aluminum chloride catalyst.

Experiment 25
SYNTHESIS OF ESTERS

Synthesize an ester from an alcohol and a carboxylic acid.

The conversion of a carboxylic acid and an alcohol into the corresponding ester is often catalyzed by a mineral acid, such as sulfuric acid:

$$RCOOH + R'OH \overset{H^+}{\rightleftharpoons} RCOOR' + H_2O$$

Carboxylic   Alcohol   Ester
acid

Because the reaction is readily reversible, we must consider the magnitude of the equilibrium constant. When it has a value close to 1—often the case for these reactions—the use of a 1:1 stoichiometric mixture of the two starting materials can give only a modest yield of product. It is, however, possible to make use of the Le Chatelier principle to manipulate the equilibrium and increase the yield of ester. For example, synthesis of an ethyl or methyl ester can be carried out in ethanol or methanol as the solvent. When the alcohol is in high concentration, the equilibrium is driven to the product side:

$$R'OH + RCOOH \rightleftharpoons RCOOR' + H_2O$$

If the yield is then measured in terms of moles of carboxylic acid converted to ester, an excellent yield can be realized. Moreover, ethanol and methanol are convenient solvents, because they are quite volatile and excess alcohol is easily removed after the reaction is done.

Azeotropes are discussed in Technique 7.5.

Another technique for manipulating the equilibrium involves continually removing one of the products. For example, a solvent such as toluene forms an azeotrope (constant boiling mixture) with water. If the reaction is conducted in toluene, the azeotrope can be distilled off and collected. This technique drives the equilibrium to the product side; and because water is not soluble in toluene, the volume of water that has been removed can be monitored in the collection device. When the volume of water collected roughly corresponds to that expected for a 100% yield of the ester, the reaction can be stopped.

The mechanism of acid-catalyzed ester formation is well understood. The first step is protonation of the carboxylic acid by the sulfuric acid catalyst to produce the conjugate acid of the carboxylic acid:

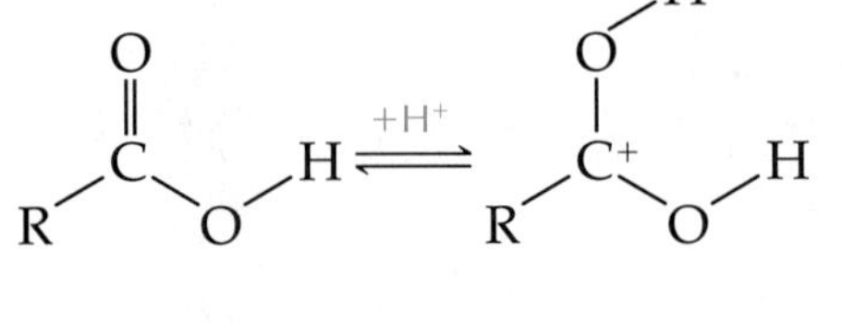

This protonated intermediate is vulnerable to attack by the alcohol molecule even though alcohols are only moderately nucleophilic:

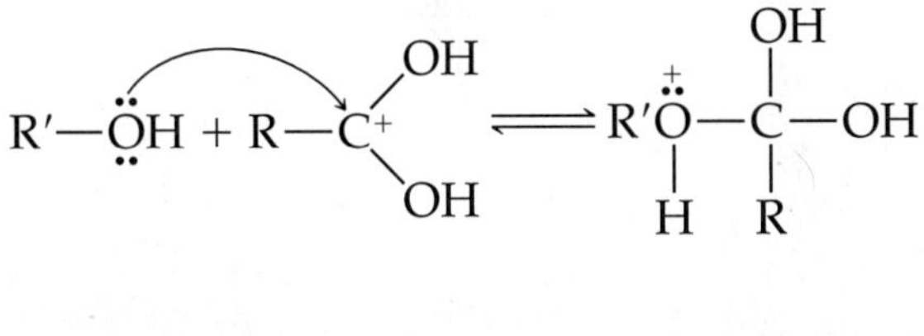

A proton shift produces a new intermediate from which water can be lost:

$$R'-\overset{+}{\underset{\underset{H}{|}}{\ddot{O}}}-\overset{\overset{OH}{|}}{\underset{\underset{R}{|}}{C}}-\ddot{O}H \rightleftharpoons^{-H^+} R'\ddot{O}-\overset{\overset{OH}{|}}{\underset{\underset{R}{|}}{C}}-\ddot{O}H \rightleftharpoons^{+H^+} R'\ddot{O}-\overset{\overset{OH}{|}}{\underset{\underset{R}{|}}{\overset{+}{C}}}\overset{+}{\ddot{O}}-H \rightleftharpoons^{-H_2O} R'\ddot{O}-\overset{OH}{\underset{R}{\overset{+}{C}}}$$

Finally, this new protonated species loses a proton to form the ester.

$$R'\ddot{O}-\overset{\overset{\ddot{O}-H}{|}}{\underset{R}{\overset{+}{C}}} \rightleftharpoons^{-H^+} R'\ddot{O}-\overset{\overset{O}{||}}{\underset{R}{C}}$$

Ester

In Experiment 25.1 you will use an esterification procedure to prepare isopentyl acetate, a compound that has the odor of banana oil. In Experiment 25.2 the same procedure will be used to prepare the acetate of an unknown alcohol. You will use the boiling point and the ^1H NMR spectrum of the alcohol to identify both it and the acetate ester.

25.1
Synthesis of Isopentyl Acetate (Banana Oil)

Make an ester with a fruity odor, using a variety of standard lab techniques.

$$\underset{\substack{\text{Acetic acid} \\ \text{bp 118°C} \\ \text{MW 60.0} \\ \text{density 1.05 g} \cdot \text{mL}^{-1}}}{\overset{\overset{O}{||}}{CH_3C}-OH} + \underset{\substack{\text{Isopentyl alcohol} \\ \text{(3-methyl-1-butanol)} \\ \text{bp 129°C} \\ \text{MW 88.1} \\ \text{density 0.81 g} \cdot \text{mL}^{-1}}}{HOCH_2CH_2\underset{\underset{CH_3}{|}}{\overset{\overset{CH_3}{|}}{CH}}} \overset{H^+}{\rightleftharpoons} \underset{\substack{\text{Isopentyl acetate} \\ \text{(3-methyl-1-butyl acetate)} \\ \text{bp 142°C} \\ \text{MW 130.2} \\ \text{density 0.867 g} \cdot \text{mL}^{-1}}}{CH_3\overset{\overset{O}{||}}{C}OCH_2CH_2\underset{\underset{CH_3}{|}}{\overset{\overset{CH_3}{|}}{CH}}} + H_2O$$

In this experiment you will prepare an ester from a mixture of isopentyl alcohol and acetic acid with a small amount of sulfuric acid as catalyst.

Acetic acid (CH_3COOH) is a product of the oxidation of ethanol. This carboxylic acid is the major organic component of vinegar. If the sugars in apple cider or grape juice are allowed to ferment, ethanol is formed; but prolonged fermentation can produce acetic acid—and a vinegary taste. Carboxylic acids frequently have obnoxious odors, but the corresponding esters commonly have pleasant fruity odors. The target compound for Experiment 25.1, isopentyl acetate, has the distinct odor of bananas; and it is, in fact, a major component of banana oil.

A few words about nomenclature may be in order here, because if you do one of the optional experiments, you will need to use a handbook to search out the ester that you will synthesize. The listing may well be an alternate name to the one that you are using. The alcohol in this experiment, $(CH_3)_2CHCH_2CH_2OH$, is now usually referred to by its systematic IUPAC name, 3-methyl-1-butanol. Earlier common names were isopentyl alcohol and isoamyl alcohol. The carboxylic acid used in this experiment is acetic acid; its systematic IUPAC name is ethanoic acid. Thus, the ester could be listed as 3-methyl-1-butyl ethanoate, 3-methyl-1-butyl acetate, isopentyl acetate, or isoamyl acetate. Alternatively, esters may be listed as derivatives of the parent acid. In this case you would look up acetic acid (or ethanoic acid) and then locate 3-methyl-1-butyl ester in the derivatives listed under the parent acid name.

Acetic acid dissolves most simple organic compounds, and it has a convenient boiling point (118°C). In this experiment acetic acid is both solvent and reactant. Using an excess of acetic acid also helps to drive the equilibrium toward the desired product.

The name amyl *comes from the Greek and Latin names for starch and was used for* $C_5H_{11}OH$ *because fermentation of potato starch gave rise to the first alcohol known to have five carbon atoms.*

⁀Macroscale Procedure

Techniques   Reflux: Technique 3.3
Extraction: Technique 4.2
Simple Distillation: Technique 7.3
Gas Chromatography: Technique 11

SAFETY INFORMATION ───

Isopentyl alcohol (3-methyl-1-butanol) and the alcohols used in the optional experiments are flammable and are irritants to the skin and eyes. Wear gloves and avoid contact with skin, eyes, and clothing.

Both glacial acetic acid and sulfuric acid are corrosive and cause severe burns, especially when they are hot. Acetic acid also emits irritating fumes, so it should be dispensed in a hood.

Glacial acetic acid gets its name from the fact that, when it is frozen (mp 16°C), it looks like an icy glacier.

A brownish color may develop during the reflux period because of acid-catalyzed polymerization reactions.

Be sure to vent the separatory funnel often because CO_2 evolution will cause a pressure buildup.

When a drop of the lower aqueous extract turns red litmus paper blue, you have neutralized all the acid in the product layer.

Stop the distillation before the distilling flask reaches dryness.

Pour 16 mL (0.147 mol) of isopentyl alcohol (3-methyl-1-butanol) and 22 mL (0.38 mol) of glacial acetic acid into a 100-mL round-bottomed flask. Carefully, with swirling, add 1.0 mL of concentrated sulfuric acid to the contents of the flask. Add one or two boiling stones.

Assemble a reflux apparatus [see Technique 3.3], and heat the reaction mixture at reflux for 1 h. Remove the heating mantle and let the mixture cool to room temperature. You can speed this process by using a water bath of room-temperature water after the flask has cooled slightly.

Pour the cooled mixture into a separatory funnel and carefully add 50 mL of cold distilled water. Rinse the reaction flask with 5–10 mL of cold water and also add this to the separatory funnel. Stopper the funnel and invert it several times before separating the lower aqueous layer from the upper organic layer [see Technique 4.2]. Set the aqueous phase aside for treatment later.

Pour 25 mL of a 0.5 M sodium bicarbonate solution into the separatory funnel, stopper it, then carefully turn it upside down and *immediately* vent the CO_2 gas that forms. Shake the funnel until no more gas is evolved when the funnel is vented. Remove the lower aqueous layer and repeat the extraction with another 25 mL of 0.5 M sodium bicarbonate solution. Continue doing $NaHCO_3$ extractions until the lower layer remains basic to litmus paper after the extraction. Wash the organic layer a last time with 20 mL of 4 M sodium chloride solution. Remove the aqueous layer again and pour the ester into a dry 50-mL Erlenmeyer flask. Add a small amount of anhydrous calcium chloride and allow the product to stand over the drying agent for 10–15 min.

Decant the product from the $CaCl_2$ or filter it through a small plug of glass wool into a 50-mL round-bottomed flask, and assemble a simple distillation apparatus [see Technique 7.3]. Collect the fraction boiling between 134°C and 141°C in a tared 50-mL round-bottomed flask. Weigh the receiving flask and product, determine your yield of isopentyl acetate, and calculate the percent yield.

Product Analysis by Gas Chromatography

Determine the purity of your product by gas chromatographic analysis, using a nonpolar methylsilicone column. For a packed column GC, a column temperature of 120–140°C works well.

Inject a 1- to 2-μL sample of your product. For a capillary column GC, the column temperature should be 100–110°C; inject 1 μL of a solution containing 1 drop of your product in 0.5 mL of diethyl ether. If your product is impure, the impurity is most probably unreacted isopentyl alcohol, which should have a shorter retention time on a nonpolar column than the ester. You may check this by the peak enhancement method, using a known sample of isopentyl alcohol [see Technique 11.6].

Nonpolar GC columns separate on the basis of boiling point.

Cleanup: Combine the aqueous phases left from the extractions. If the pH as determined by pH test paper is below 7, neutralize the solution with solid sodium carbonate before washing it down the sink, or pouring it into the container for aqueous inorganic waste. Place the spent drying agent in the container for nonhazardous solid waste. Pour the residue remaining in the distilling flask into the container for flammable (organic) waste.

ⅲⅲⅲ*microscale* Procedure

Techniques   Microscale Reflux: Technique 3.3
Microscale Extraction: Technique 4.5d
Microscale Distillation: Technique 7.3b
Gas Chromatography: Technique 11

SAFETY INFORMATION

Isopentyl alcohol (3-methyl-1-butanol) and the **alcohols** used in the optional experiments are flammable and are irritants to the skin and eyes. Avoid contact with skin, eyes, and clothing.

Both **glacial acetic acid** and **sulfuric acid** are corrosive and cause severe burns, especially when they are hot. Acetic acid fumes are also irritating, so the acid should be dispensed in a hood.

Begin heating an aluminum block to 150°C on a hot plate. Use graduated pipets to measure 1.6 mL (15 mmol) of isopentyl alcohol (3-methyl-1-butanol) and 2.2 mL (38 mmol) of glacial acetic acid into a 10-mL round-bottomed flask. Carefully, with swirling, add 1 drop of concentrated sulfuric acid to the flask. Also add a boiling stone.

A brownish color may develop during the reflux period.

Assemble a reflux apparatus [see Technique 3.3], using a water-cooled condenser, and heat the reaction mixture at reflux for 50–60 min. Remove the apparatus from the heating block, let it cool briefly, then put the flask in a beaker of water until the reaction mixture reaches room temperature.

Set the 10-mL round-bottomed flask and the conical vial in 30- and 50-mL beakers to keep them upright.

Add 5.0 mL of water to the reaction mixture and carefully stir the contents of the flask. Transfer approximately half of the reaction mixture to a 5-mL conical vial, using a Pasteur filter pipet [see Technique 2.4]. Allow the phases to separate and transfer the lower aqueous layer to a 100-mL beaker [see Technique 4.5d]. Transfer the rest of the reaction mixture to the conical vial and again remove the lower aqueous layer, adding it to the beaker.

SAFETY PRECAUTION

Be sure to vent the vial by frequently loosening the cap, because CO_2 evolution will cause a pressure buildup.

Wash the organic phase remaining in the conical vial with 2.5-mL portions of 0.5 M sodium bicarbonate solution until the lower aqueous layer remains basic to litmus paper after the extraction. **Vent the vial frequently by loosening the cap.** After each washing, remove the lower aqueous phase. Wash the organic layer a last time with 2.0 mL of 4 M sodium chloride solution. Remove the aqueous layer and add a small amount of anhydrous calcium chloride. Allow the product to dry for 10–15 min.

Read Technique 7.3b before beginning the distillation.

Transfer the dried product to a dry 5-mL conical vial containing a spin vane. Assemble the apparatus for microscale distillation, using an air condenser above the Hickman distilling head [see Technique 7.3b]. Place the auxiliary blocks around the vial as shown in Technique 3, Figure 3.2. Turn on the stirrer and begin heating the aluminum block to 180–185°C. When the well fills with distillate, remove the distillate by inserting a syringe through the septum in the port, or opening the port and removing the distillate with a Pasteur pipet. Place the distillate in a tared vial. Record the boiling point. Stop the distillation by lifting the apparatus out of the aluminum block before the vial reaches dryness. Weigh your product and determine the percent yield. Carry out the GC analysis described in the macroscale experimental procedure.

Optional Experiments

Many other esters can be synthesized by procedures very similar to that used for isopentyl acetate. Develop and carry out the synthesis of propyl acetate (the aroma of pears) from acetic acid and 1-propanol, *or* isobutyl propionate (the odor of rum) from propionic acid and isobutyl alcohol (2-methyl-1-propanol), using sulfuric acid as the catalyst. Discuss your procedure with your instructor before beginning any experimental work.

Propyl Ethanoate (Propyl Acetate)

Use 0.148 mol (macroscale) or 14.8 mmol (microscale) of 1-propanol and the amounts of other reagents suggested in the isopentyl acetate synthesis.

2-Methyl-1-Propyl Propanoate (Isobutyl Propionate)

Use 0.148 mol (macroscale) or 14.8 mmol (microscale) of isobutyl alcohol (2-methyl-1-propanol), 0.38 mol (macroscale) or 38 mmol (microscale) of propanoic acid (propionic acid), and the amounts of other reagents suggested in the isopentyl acetate synthesis.

For both syntheses. The reaction time should remain the same, but rather than adding cold water to the reaction mixture, use an equal volume of 2 M sodium chloride solution. Also, use 1.0 M rather than 0.5 M sodium bicarbonate in the later extractions. The distillation temperature will have to be modified to fit the boiling point of your product, which you can find in the *Dictionary of Organic Compounds*, 4th (or later) ed., Oxford University Press. The densities that you need can be found in the *Handbook of Chemistry and Physics*.

Questions

1. The ester is washed with 0.5 M sodium bicarbonate. Why? Explain with chemical reactions.
2. Ethyl acetate is conveniently prepared by using ethanol as solvent and reactant. Explain why this procedure would be reasonable.
3. Commercial suppliers can sell compounds under either systematic or common names. Write another name for each of the following: (a) ethyl acetate; (b) isopropyl alcohol; (c) 2-butanol.
4. The compounds isoamyl valerate, butyl butyrate, and isobutyl propionate have the pleasant odors associated with apples, pineapples, and rum, respectively, whereas the corresponding carboxylic acids have obnoxious odors. Write structures and systematic names for each ester.
5. Assume that you have a sample of isopentyl alcohol that has its hydroxyl group enriched with ^{18}O. Use the mechanism for the esterification reaction to find out where the ^{18}O enrichment will appear in the final products.
6. What physical properties of methanol suggest that it would be a good solvent to use for an esterification reaction.
7. Odor and volatility are related. Conversion of a long-chain carboxylic acid to an ethyl ester often enhances its volatility. For example, acetic acid has a bp of 118°C, whereas the bp of ethyl acetate is 77°C. Why?

25.2
Synthesis and Identification of an Ester from an Unknown Alcohol and Acetic Acid

Synthesize and identify the structures of an unknown alcohol and its esterification product.

$$CH_3-\overset{\overset{\displaystyle O}{\|}}{C}-OH \ + \ R-O-H \ \underset{}{\overset{H^+}{\rightleftharpoons}} \ CH_3-\overset{\overset{\displaystyle O}{\|}}{C}-O-R \ \ + \ H_2O$$

Acetic acid $R = C_5H_{11}$ or C_6H_{13} Acetate ester of unknown alcohol
bp 118°C   bp to be determined   bp 12–16°C above bp of unk. alcohol
MW 60.0   experimentally
density 1.05 g · mL^{-1}

In this experiment the class will be provided with samples of unknown alcohols, each with a molecular formula of either $C_5H_{11}OH$ or $C_6H_{13}OH$. You will prepare the acetate ester of your alcohol unknown by treating it with acetic acid in the presence of sulfuric acid. You will use the boiling point and the ^1H NMR spectrum of the alcohol and the boiling point of the ester that you synthesize to identify the structure of the alcohol and the ester product.

Because the esterification procedure is virtually the same as that used in Experiment 25.1, it is important that you carefully read the background material for the earlier experiment, and you should also scrutinize the mechanism for this acetylation reaction (pp. 291–292).

To help you with the interpretation of the NMR spectrum of your alcohol starting material, it may be useful to consider the NMR spectra of all four isomers of C_4H_9OH, the butanol family (Figures 25.1 through 25.4). We will find in all cases that only vicinal splitting (splitting due to CH—CH coupling between non-equivalent protons on neighboring carbon atoms) occurs.

The spectrum of 1-butanol (butyl alcohol) is a good starting point (Figure 25.1). It shows a 3H triplet at δ 0.9, establishing the presence of a methyl group attached to a methylene (CH_2) group. Other signals that readily lend themselves to simple first-order

$$CH_3CH_2CH_2CH_2OH$$

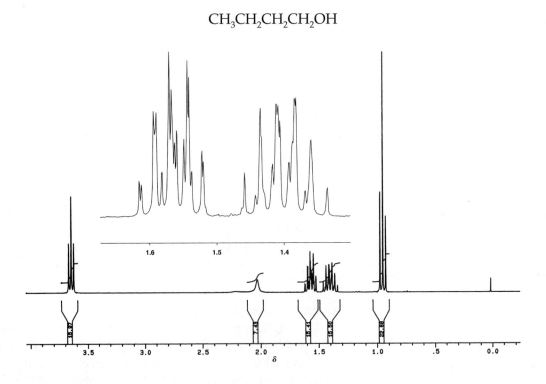

FIGURE 25.1 ¹H NMR spectrum (300 MHz) of 1-butanol (in $CDCl_3$).

analysis are the 2H triplet at δ 3.6 and the singlet at δ 2.05. The triplet is due to the protons on the methylene bearing the —OH group. The attached oxygen deshields these protons, thus causing this signal to be at a relatively low field (high δ). The triplet character is due to splitting by the two vicinal protons on the C-2 methylene group. The 1H singlet at δ 2.05 is due to the —OH proton. This assignment is supported by the observation that it shows no coupling, a fact consistent with the observation that this spectrum was run in $CDCl_3$ solvent. Usually $CDCl_3$ contains a trace of the acid DCl (or HCl), which promotes rapid —OH proton exchange, thereby preventing coupling to nearby protons. The δ 2.05 signal can be readily confirmed as an —OH signal by running the spectrum at a different concentration. This change causes the —OH signal to occur at a different chemical shift (δ value) because the importance of intermolecular H-bonding will depend on the concentration. The four-proton signal (2H + 2H) running from δ 1.3 to 1.7 is due to the C-2 and C-3 methylene protons, with

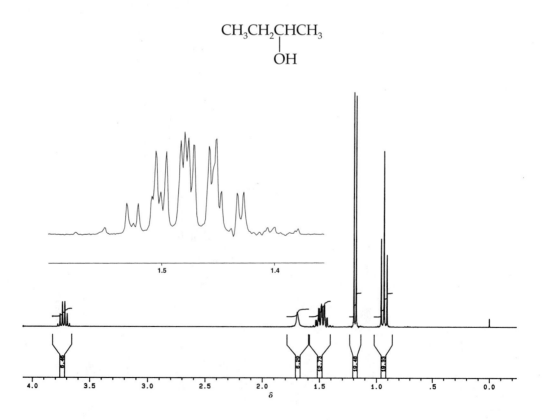

FIGURE 25.2 ¹H NMR spectrum (300 MHz) of 2-butanol (in $CDCl_3$).

the —CH$_2$ group downfield due to the group closer to the oxygen atom.

These same principles can be used to interpret the other three ^{1}H NMR spectra (Figures 25.2 to 25.4). The spectrum of 2-butanol (*sec*-butyl alcohol) reveals two methyl signals, a 3H triplet at δ 0.93 due to the C-4 methyl group and a 3H doublet at about δ 1.2 due to the C-1 methyl. The latter is more downfield because it is closer to the deshielding —OH group. Moreover, this is a doublet, because there is only one vicinal proton on C-2, whereas the two protons on C-3 cause the C-4 methyl to be a triplet. The 1H multiplet at δ 3.72 is due to the C-2 methine (CH) proton, and the 1H singlet at δ 3.98 is due to the —OH proton. The 2H signal at δ 1.48 is due to the two nonequivalent protons at C-3. These two protons are diastereotopic, and thus nonequivalent, because C-2 is a stereocenter (or chiral center). These two protons couple to each other as well as to their vicinal neighbors and give rise to a complex signal.

(CH$_3$)$_2$CHCH$_2$OH

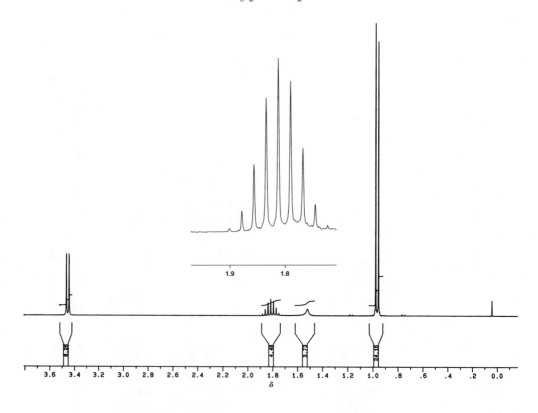

FIGURE 25.3 ^{1}H NMR spectrum (300 MHz) of 2-methyl-1-propanol (in CDCl$_3$).

The spectrum of 2-methyl-1-propanol (isobutyl alcohol) contains a 6H methyl doublet at δ 0.92 due to the two C-2 methyl groups; the 6 H's reveal the isopropyl "tail" of the isobutyl structure. The 2H doublet at δ 3.38 is due to the methylene protons at C-1, as suggested by the fact that this signal is more downfield than any other signal except the —OH proton (the 1H singlet at δ 3.92). Finally, the 1H signal at δ 1.76 is the C-2 methine proton, which couples with the vicinal methyl and methylene groups.

The spectrum of 2-methyl-2-propanol (*tert*-butyl alcohol) is exceedingly simple. There are no vicinal neighboring protons; thus, both signals are singlets. The 9H signal at δ 1.29 is due to the unsplit protons of three equivalent methyl groups. The 1H signal at δ 1.67 is due to the —OH proton.

This same analytical approach can readily be applied to the ^{1}H NMR spectra arising from the isomeric pentanols and hexanols used in this experiment.

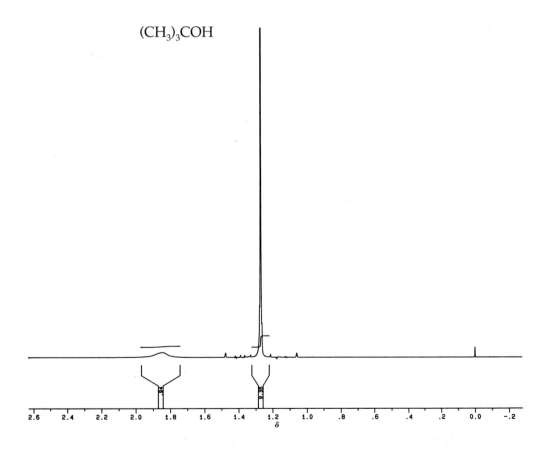

FIGURE 25.4 ^{1}H NMR spectrum (300 MHz) of 2-methyl-2-propanol (in CDCl$_3$). The small peaks about the δ1.29 signal are spinning side bands.

Prelaboratory Assignment

Draw the structures for all of the isomeric alcohols defined by the formulas $C_5H_{12}O$ and $C_6H_{14}O$ (8 and 17, respectively). List the isomers in groups of primary, secondary, and tertiary alcohols.

Macroscale Procedure

Technique   Determination of Boiling Point: Technique 7.1a

> ### SAFETY INFORMATION
>
> The alcohols used in this experiment are flammable and are irritants to the skin and eyes. Wear gloves, and avoid contact with skin, eyes, and clothing.
>
> Both glacial acetic acid and sulfuric acid are corrosive and cause severe burns, especially when they are hot. Acetic acid also emits irritating fumes, so it should be dispensed in a hood.

You will be given a vial containing a volume of an unknown alcohol (either a $C_5H_{12}O$ or $C_6H_{14}O$ alcohol) equivalent to 0.15 mol plus 1.0 mL. Transfer 1.0 mL of the alcohol to a small test tube and use the rest for the synthesis; cork the test tube to prevent the alcohol from evaporating. Follow the macroscale experimental procedure given in Experiment 25.1, substituting your unknown alcohol for isopentyl alcohol.

During the reflux period, use the remaining amount of your unknown alcohol to prepare a sample for NMR analysis and to determine the boiling point of the alcohol. Prepare the sample for NMR analysis as directed by your instructor. Ascertain the boiling point of the unknown alcohol by a microscale boiling-point determination [see Technique 7.1a].

The boiling point of the acetate ester will be 12–16°C above the boiling point that you determine for the alcohol. Use this information to decide whether a water-cooled or an air-cooled condenser should be used for the final distillation of the ester.

Identification of Your Unknown Alcohol

Use all your evidence, including the boiling points of both the alcohol and the ester plus the NMR interpretation, to decide the identity of your alcohol. You should consult a handbook such as the *CRC Handbook of Chemistry and Physics* for the reported boiling point of a possible alcohol and its corresponding acetate ester. For some unknowns, it may not be possible to narrow the choice to one alcohol. In that case, list all the possible structures that fit your experimental data and NMR spectrum.

You need to consider the chemical shift, relative integration value, and multiplicity of each signal in the NMR spectrum to determine a possible structure for your unknown alcohol. Use the CH or CH_2 multiplet furthest downfield as the base integral and the relative integration values for the other signals to determine whether your unknown is a 5-carbon or a 6-carbon alcohol. The integration of this multiplet relative to the other proton signals will also enable you to classify the alcohol as primary, secondary, or tertiary. At what chemical shift would you expect to find the signal for terminal methyl groups? The integrals for these signals will allow you to determine the number of terminal methyl groups in your unknown. For many of the possible alcohols, some protons may produce only one complex multiplet rather than distinct multiplets for each type of proton.

Once you have identified your unknown alcohol, calculate the theoretical yield for your synthesis and determine your percent yield.

If you have a C_5 alcohol, use a density of 0.81 $g \cdot mL^{-1}$. If you have a C_6 alcohol, use a density of 0.82 $g \cdot mL^{-1}$ in the yield calculations.

microscale Procedure

Technique   Determination of Boiling Point: Technique 7.1a

─ SAFETY INFORMATION ─

The **alcohols** used in this experiment are flammable and are irritants to the skin and eyes. Wear gloves, and avoid contact with skin, eyes, and clothing.

Both **glacial acetic acid** and **sulfuric acid** are corrosive and cause severe burns, especially when they are hot. Acetic acid also emits irritating fumes, so it should be dispensed in a hood.

You will be given a vial containing 3 mL of your unknown alcohol (either a $C_5H_{12}O$ or $C_6H_{14}O$ alcohol). Using a graduated pipet, measure 1.6 mL of the alcohol for your synthesis. Follow the microscale experimental procedure given in Experiment 25.1, substituting your unknown alcohol for isopentyl alcohol.

During the reflux period, use the remaining amount of your unknown alcohol to prepare a sample for NMR analysis and to determine the boiling point of the alcohol. Prepare the sample for NMR analysis as directed by your instructor. Ascertain the boiling point of the unknown alcohol by a microscale boiling-point determination [see Technique 7.1a].

The boiling point of the acetate ester will be 12–16°C above the boiling point that you determine for the alcohol. Use this information to decide whether a water-cooled or an air-cooled condenser should be used for the final distillation of the ester.

The temperature observed on the thermometer in the Hickman distilling head is sometimes less than the actual boiling point, because not enough vapor is surrounding the thermometer bulb. Therefore, verify the boiling point of your distilled product by a microscale boiling-point determination [see Technique 7.1a].

If you have a C_5 alcohol, use a density of 0.81 $g \cdot mL^{-1}$. If you have a C_6 alcohol, use a density of 0.82 $g \cdot mL^{-1}$ in the yield calculations.

Follow the procedure given in the macroscale directions for identifying your unknown alcohol. Once you have identified your alcohol, calculate the theoretical yield for your product.

Reference

1. Branz, S. E. *J. Chem. Educ.* **1985,** *62,* 899–900.

Questions

1. All acetate esters show an ^1H NMR singlet at roughly δ 2.0. What causes this signal?
2. Samples of isobutyl acetate and butyl acetate are subjected to ^1H NMR analysis. One shows an upfield (approximately δ 1.0) doublet and the other does not. Which is which? Explain.
3. A compound that has the odor of well-used gym socks shows ^1H NMR peaks running from δ 1.0–2.0 that are, from high field to low field, a triplet (3H), a multiplet (2H), a multiplet (2H), and finally a triplet (2H). Moreover a fifth signal, a somewhat broadened singlet (1H), occurs at different chemical shifts depending on the concentration in $CDCl_3$ solvent. What is the structure of this compound?

AMIDE CHEMISTRY

Explore the synthesis of amides from acid chlorides, using examples of biological and commercial relevance.

Carboxylic acids form a variety of derivatives, and each of these is a significant class of organic molecules in its own right.

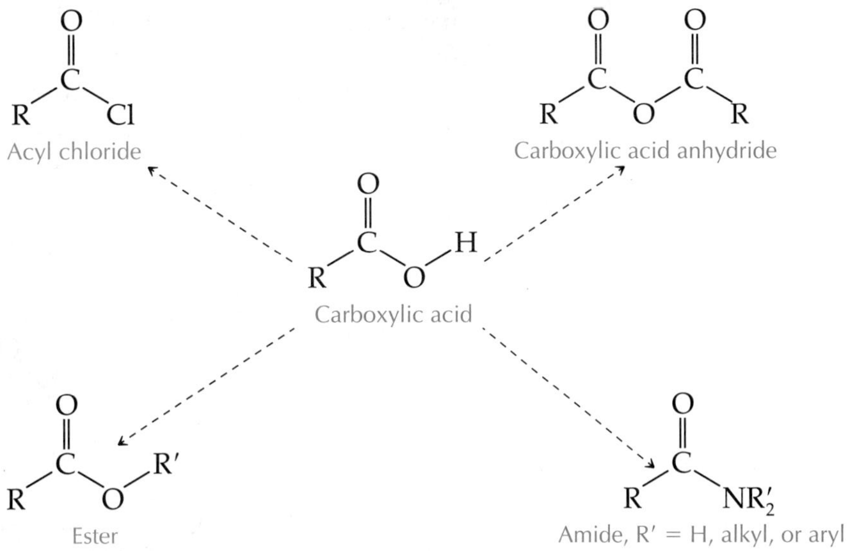

The derivatives of carboxylic acids have a huge range of reactivity toward nucleophilic attack at their carbonyl carbon atoms. It is instructive to consider their relative reactivity toward hydrolysis, the reaction with water, to produce the parent carboxylic acid:

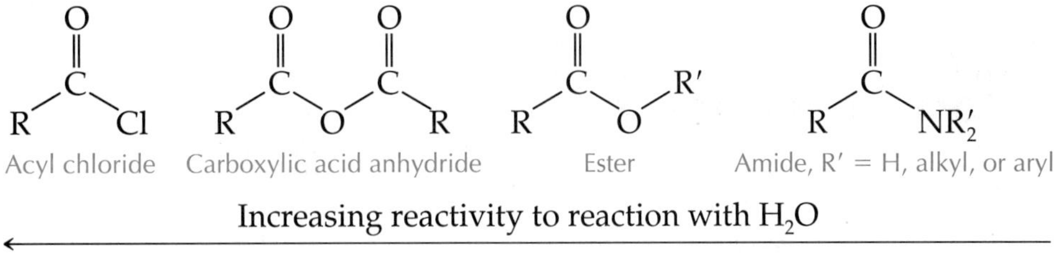

Increasing reactivity to reaction with H_2O

The great strength of the bond between the carbonyl carbon and the amide nitrogen is one of the main reasons why amides are among the most stable and least reactive compounds in the preceding series. The resonance forms that contribute to the amide bond are displayed below, and the stability of this linkage is at least partially rationalized by the existence of a significant degree of double bond character in the carbon-nitrogen bond:

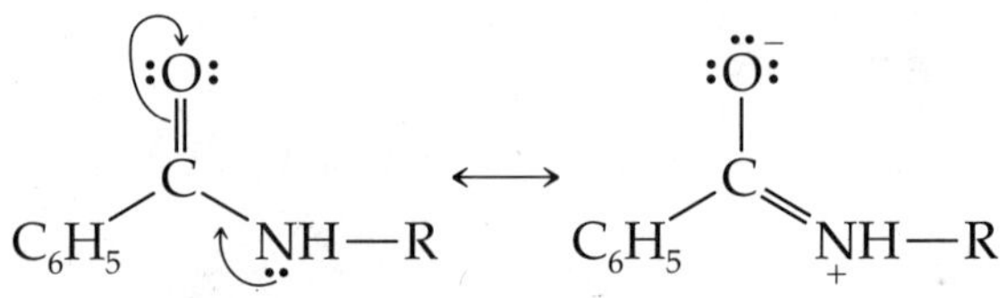

Amide bond resonance

We are concerned with the conversion of carboxylic acids to amides in this experiment. There are one-step methods of converting carboxylic acids directly to amides, but it is generally necessary to carry out a two-step sequence. The first step usually involves initial conversion of the carboxylic acid to an acid chloride, which is highly reactive toward nucleophilic attack at carbonyl carbon. Acid chlorides are readily converted to amides by treatment with ammonia or the appropriate amine:

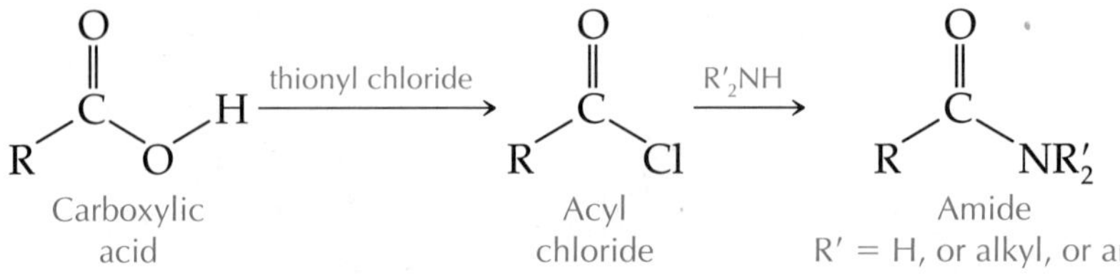

Carboxylic acid		Acyl chloride		Amide
				R′ = H, or alkyl, or aryl

Experiment 26.1 uses a commercially available acyl chloride, benzoyl chloride (the acid chloride of benzoic acid), to convert an amino acid, glycine, to hippuric acid. Experiment 26.2 uses the reaction with thionyl chloride or *bis*(trichloromethyl) carbonate to convert a carboxylic acid to an acid chloride, which, in turn, is converted to *N,N*-diethyl-*m*-toluamide ("DEET"), a commonly used insect repellent.

(26.1)
Synthesis of Hippuric Acid

Synthesize hippuric acid from glycine and benzoyl chloride.

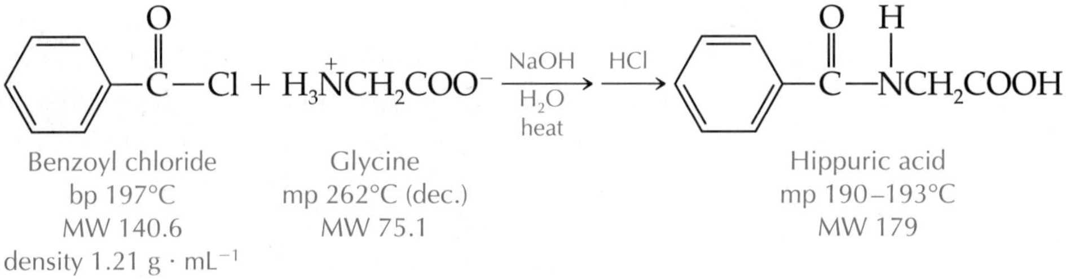

Benzoyl chloride
bp 197°C
MW 140.6
density 1.21 g · mL^{-1}

Glycine
mp 262°C (dec.)
MW 75.1

Hippuric acid
mp 190–193°C
MW 179

The in vitro synthesis of hippuric acid that you will be doing, is like the in vivo synthesis of hippuric acid from benzoic acid and glycine. Benzoic acid is present in many fruits and vegetables and is a common preservative in food products such as soft drinks and canned foods. Sodium benzoate is also sometimes used as an anti-fungal agent to preserve bread.

An adult human being excretes about 0.7 g of hippuric acid daily. The enzyme-catalyzed biosynthesis of hippuric acid in the liver proceeds by reaction of glycine with a thioester, benzoyl CoA. Hippuric acid is more water-soluble than benzoic acid; this property allows hippuric acid to be excreted more efficiently. The conversion of relatively water-insoluble substances to more soluble products is a general process for the elimination of foreign substances from our bodies. The metabolism of sodium benzoate is sometimes used in hospitals as a test of liver function.

Your synthesis of hippuric acid starts with glycine, an amino acid that is commonly found in proteins. The reaction of glycine with benzoyl chloride in the presence of an alkaline catalyst yields the amide, benzoylglycine (hippuric acid). The name comes from the Greek word for horse, *hippos,* because hippuric acid was first isolated from the urine of horses.

When the amino acid glycine is in the zwitterionic (dipolar) form, the nitrogen has no nonbonding electron pair and is not nucleophilic. Treatment of zwitterionic glycine with NaOH converts the amino acid to a nucleophilic form, and the amino group bearing the electron pair readily attacks benzoyl chloride to initiate amide formation:

Loss of the elements of HCl yields a carboxylate salt, which, when acidified, produces hippuric acid:

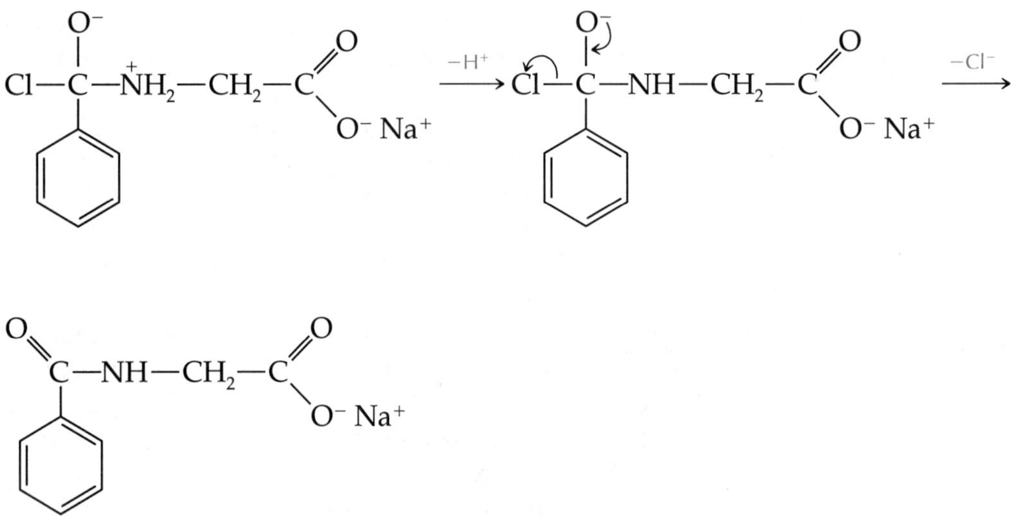

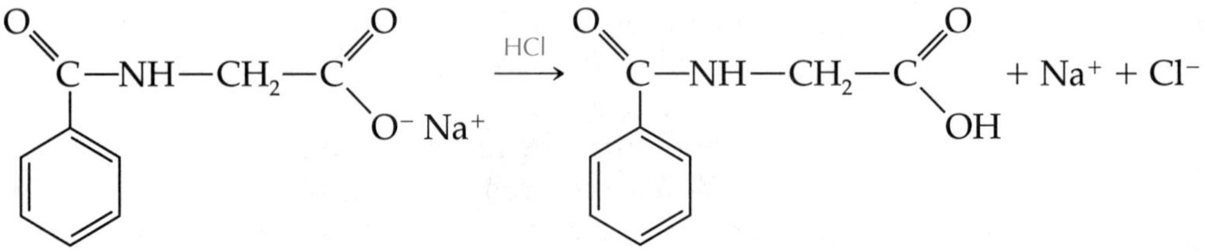

Hippuric acid

(continued on next page)

Macroscale Procedure

Technique   Recrystallization, Using Activated Charcoal: Technique 5.3a

SAFETY INFORMATION

If possible, conduct the entire synthesis in a hood.

Benzoyl chloride is a lachrymator (causes tears) and a skin irritant. Wear gloves and pour it only in a hood.

Aqueous **sodium hydroxide** solutions are corrosive and cause burns. Solutions as dilute as 9% (2.5 M) can cause severe eye injury.

Diethyl ether is extremely flammable. Be certain that no flames are used in the laboratory and that hot electrical devices are not in the vicinity where ether is being used.

Place 2.1 g (0.028 mol) of crystalline glycine in a 125-mL Erlenmeyer flask that contains a mixture of 22 mL of distilled water and 3.0 mL of 6 M sodium hydroxide solution. Swirl the mixture until a clear solution is obtained.

Wearing gloves and working in a hood, measure 3.7 mL of benzoyl chloride in a graduated cylinder and quickly add it to the glycine solution. While still working in the hood, rinse the graduated cylinder with 3.0 mL of 6 M NaOH solution and also add this rinse to the reaction flask. Stopper the flask tightly with a solid rubber stopper.

Monitor the pH by putting the tip of a stirring rod into the solution, and then touching it to a piece of pH test paper.

Shake the Erlenmeyer flask until the layers are completely mixed. The reaction mixture will become warm, so vent it cautiously in the hood by removing the stopper. If the reaction mixture becomes hot, cool it in a water bath. Monitor the pH, then replace the stopper immediately. When the mixture becomes acidic, add another 6.0 mL of 6 M NaOH solution over a 2- to 3-min period, shaking the flask after each addition. It may be necessary to add an additional 1–2 mL of 6 M NaOH toward the end of the reaction to maintain an alkaline pH.

The reaction is complete when it forms a clear yellow solution and when no oil droplets remain at the bottom of the flask. If any precipitate is present at this point, filter the mixture, using vacuum filtration. Save the filtrate (your product) and discard any solids on the filter paper.

Place approximately 15 g of ice in a 250-mL beaker and carefully pour 15 mL of concentrated hydrochloric acid over the ice. Pour your reaction solution into the acid/ice mixture with stirring. Collect the solid product by vacuum filtration. Remove the filtrate from the filter flask and set it aside in a beaker for treatment later.

To remove any benzoic acid formed by hydrolysis of excess benzoyl chloride, turn off the vacuum source and cover the crystals with 6 mL of ether. **(Caution: Flammable.)** Remove the ether by reapplying the suction. Repeat this ether washing two more times. Allow the hippuric acid to dry before weighing it and determining the melting point. Calculate your percent yield.

The product can be recrystallized if colored impurities are present (consult your instructor). Use a 1:3 (v/v) ethanol/water solution as the recrystallizing solvent and a steam bath or a hot plate as the heat source. Add decolorizing carbon (2–3 spatulas full) and swirl the hot mixture [see Technique 5.3a]. Remove the decolorizing carbon by filtering the hot solution into an Erlenmeyer flask through a fluted filter paper placed in a heated stemless funnel or plastic filling funnel. Heat the receiving flask on a steam bath during the filtration, as a way of minimizing crystallization in the funnel. Upon cooling the filtrate, the hippuric acid will crystallize and can be collected by vacuum filtration.

Determine the melting point, mass, and percent yield of your product. At the discretion of your instructor, you may analyze your hippuric acid by IR or NMR spectrometry.

Cleanup: Neutralize the aqueous filtrate from the crude product with solid sodium carbonate **(Caution: Foaming.)** before washing the solution down the sink or pouring it into the container for aqueous inorganic waste. Pour the ether filtrate into the container for flammable (organic) waste. The aqueous ethanol filtrate from the recrystallization can be washed down the sink or placed in the flammable (organic) waste container. Place the filter paper containing the decolorizing carbon in the container for nonhazardous solid waste.

microscale
Procedure

Techniques   Recrystallization, Using Activated Charcoal: Technique 5.3a

Microscale Recrystallization: Technique 5

— *SAFETY INFORMATION* —

Benzoyl chloride is a lachrymator (causes tears) and a skin irritant. Wear gloves and dispense it only in a hood.

Aqueous sodium hydroxide solutions are corrosive and cause burns. Solutions are dilute as 9% (2.5 M) can cause severe eye injury.

Diethyl ether is extremely flammable. Be certain that no flames are used in the laboratory and that hot electrical devices are not in the vicinity where ether is being used.

Concentrated hydrochloric acid is corrosive and causes burns. The vapor is extremely irritating to mucous membranes and the upper respiratory tract. Measure it in a hood and avoid contact with skin, eyes, and clothing.

Place 150 mg (2.0 mmol) of crystalline glycine in a 5-mL conical vial that contains a mixture of 1.6 mL of distilled water and 0.40 mL of 6 M sodium hydroxide solution. Cap the vial and shake the mixture until a clear solution is obtained.

Wearing gloves and working in a hood, measure 0.26 mL of benzoyl chloride with a graduated pipet and quickly add it to the glycine solution. Cap the vial and shake it until the layers are completely mixed. The reaction mixture may become warm, so vent it cautiously in the hood by removing the cap. Monitor the pH, then replace the cap immediately. When the mixture becomes acidic, add another 0.40 mL of 6 M NaOH solution. Continue shaking the vial. It may be necessary to add an additional 0.10 mL of 6 M NaOH toward the end of the reaction to maintain an alkaline pH.

Monitor the pH by putting the tip of a stirring rod into the solution and then touching it to a piece of pH test paper.

The reaction is complete when it forms a clear yellow solution and there are no oil droplets at the bottom of the flask. If any precipitate is present at this point, filter the mixture, using a Hirsch funnel and vacuum filtration. Save the filtrate (your product) and discard any solids on the filter paper.

Place 0.40 mL of water in a 30-mL beaker, cool the beaker in an ice-water bath, and add 0.40 mL of concentrated hydrochloric acid. Pour your reaction solution into the acid/water solution with stirring. Collect the solid product by vacuum filtration on a Hirsch funnel [see Technique 5.5a]. Remove the filtrate from the filter flask and set it aside in a beaker for treatment later.

To remove any benzoic acid formed by hydrolysis of excess benzoyl chloride, turn off the vacuum source and cover the crystals with approximately 1 mL of ether. **(Caution: Flammable.)** Remove the ether by reapplying the suction. Repeat this ether washing two more times. Allow the hippuric acid to dry before weighing it and determining the melting point. Calculate your percent yield.

If colored impurities are present, recrystallize the crude product, using 1:3 (v/v) ethanol/water solution as the solvent. Place the crystals and a boiling stick in a 10×125 mm test tube; use a steam bath or a beaker of boiling water as the heat source. Add 10–15 Norit pellets (decolorizing carbon) and heat the mixture for another 1–2 min [see Technique 5.3a]. Carefully transfer the hot solution to a Craig tube with a Pasteur filter pipet [see Technique 2.4] warmed in the steam bath or hot water bath. Upon cooling, the hippuric acid will crystallize and can be collected by centrifugation [see Technique 5.5b].

Determine the melting point, mass, and percent yield of your product. At the discretion of your instructor, you may analyze your hippuric acid by IR or NMR spectrometry.

Cleanup: Neutralize the aqueous filtrate from the crude product with solid sodium carbonate before washing the solution down the sink or pouring it into the container for aqueous inorganic waste. Pour the ether filtrate into the container for flammable (organic) waste. The aqueous ethanol filtrate from the recrystallization can be washed down the sink or placed in the container for flammable (organic) waste. Put the Norit into the container for nonhazardous solid waste.

Optional Experiment

Solubilities of Benzoic Acid and Hippuric Acid

The biological activity of a drug or other substance metabolized by the body depends to a considerable extent on its solubility in body tissues and how fast the metabolites are excreted in the urine. Body tissues are usually fatty, whereas urine is an aqueous solution. Our bodies convert benzoic acid to hippuric acid. The partitioning of benzoic acid and hippuric acid between water and ether provides a simple model system for ascertaining the relative solubilities of benzoic acid and hippuric acid in water (a substitute for urine) and ether (a substitute for fatty tissue). Determine the partitioning of benzoic acid and hippuric acid between water and ether by the method described in Reference 1.

SAFETY PRECAUTION

Ether is extremely flammable. Do the extraction in a hood.

Reference

1. Goldsmith, R. H. *J. Chem. Educ.* **1971,** *51,* 272–273.

Questions

1. Why is the hippuric acid product soluble in the water until hydrochloric acid is added?
2. Why must sodium hydroxide be present for the reaction between glycine and benzoyl chloride to take place at a reasonable rate?
3. A possible alternative procedure would be to add all the 6 M NaOH solution at the beginning of the reaction. What disadvantage could this have?

26.2
Synthesis of a Mosquito Repellent: *N,N*-Diethyl-*meta*-Toluamide ("DEET")

Synthesize DEET by a two-step process and purify it by column chromatography.

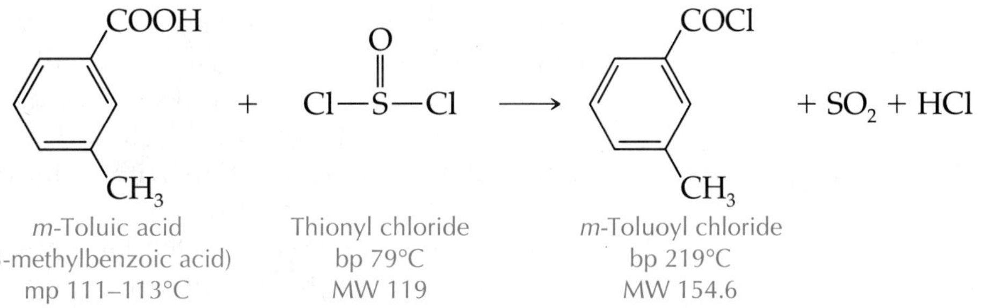

m-Toluic acid	Thionyl chloride	*m*-Toluoyl chloride
(3-methylbenzoic acid)	bp 79°C	bp 219°C
mp 111–113°C	MW 119	MW 154.6
MW 136.2	density 1.66 g · mL^{-1}	density 1.17 g · mL^{-1}

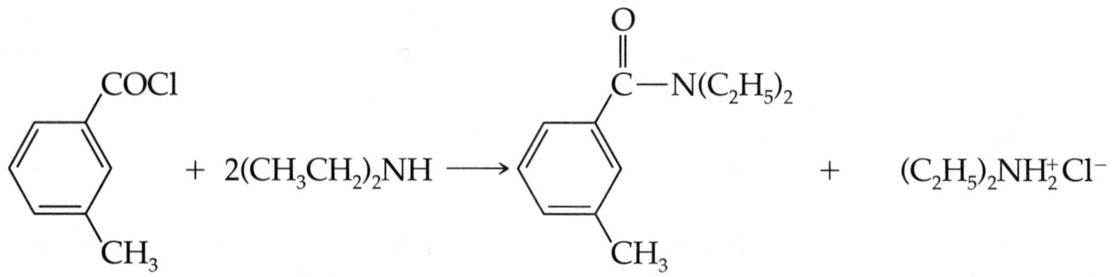

Diethylamine
bp 56°C
MW 73.1
density 0.71 g · mL⁻¹

N,N-Diethyl-*m*-toluamide
("DEET")
bp 160°C (at 19 mmHg)
MW 191.3
density 1.00 g · mL⁻¹

Diethylammonium chloride
(diethylamine hydrochloride)
mp 227–230°C
MW 109.6

Two compounds have been extensively used as insect repellents: *N,N*-diethyl-*meta*-toluamide (abbreviated DEET) and 2-ethyl-1,3-hexanediol. You will synthesize DEET in this experiment, using a two-step procedure. In the first step you will convert *m*-toluic acid to *meta*-toluoyl chloride, using thionyl chloride ($SOCl_2$); and in the second step you will treat the *m*-toluoyl chloride produced in the first step with two equivalents of diethylamine to produce *N,N*-diethyl-*meta*-toluamide.

It is thought that the formation of an acyl halide involves preliminary formation of a sulfur-containing chloride intermediate, one that is very reactive to attack by chloride ion:

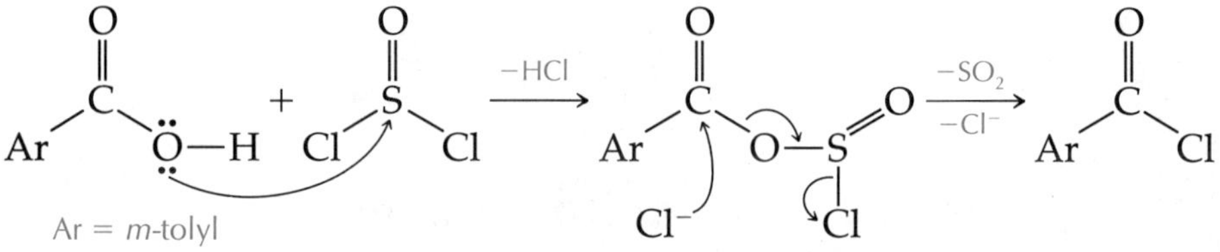

It is common to use the acid chloride produced in this way with no further purification. This practice deserves special comment because synthetic intermediates are normally purified before carrying out subsequent steps. Why does the simpler approach work here? First, the side products, sulfur dioxide and hydrogen chloride, are gases and should be readily lost from the reaction mixture. Second, because acid chlorides are highly reactive, the amine may undergo reaction with them more quickly than with any impurities in the mixture.

In the second step the nucleophilic amine attacks the acid chloride to form a tetrahedral intermediate that readily loses the elements of HCl to form the amide ("DEET") product:

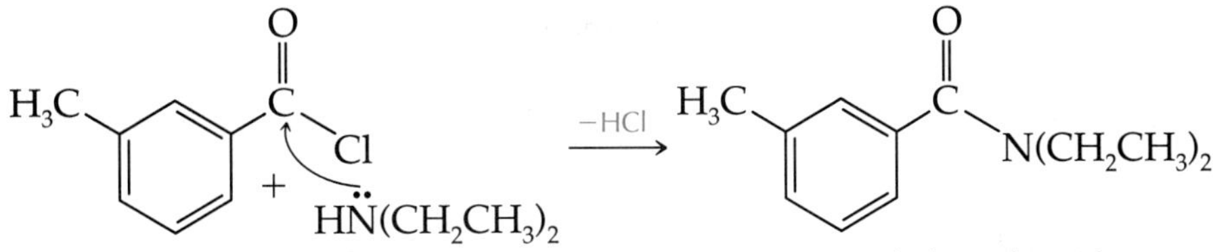

Ar(C=O)Cl   *N,N*-Diethyl-*m*-toluamide ("DEET")

The second mole of diethylamine serves as a base to neutralize the side product HCl as it forms:

$$(CH_3CH_2)_2NH + HCl \longrightarrow (CH_3CH_2)_2NH_2^+ \ Cl^-$$

Diethylammonium chloride

Because both diethylamine and hydrogen chloride are volatile, they often produce a significant portion of the ammonium chloride in the air above the reaction mixture, causing a white "smoky" appearance.

In the microscale synthesis of DEET, you will use bis(trichloromethyl) carbonate (triphosgene), rather than thionyl chloride, to produce *m*-toluoyl chloride from *m*-toluic acid.

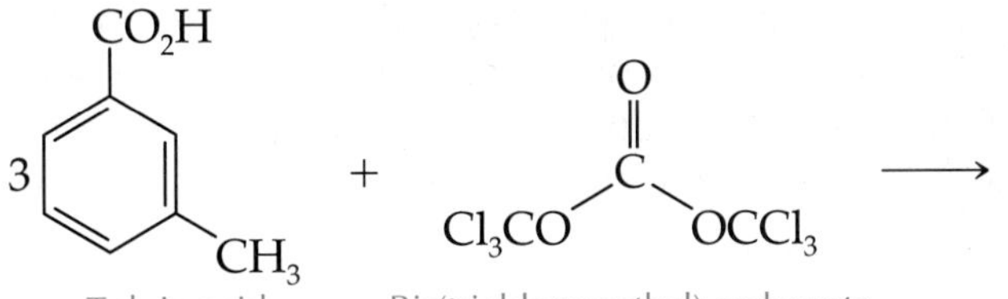

m-Toluic acid
(3-methylbenzoic acid)
mp 111–113°C
MW 136.2

Bis(trichloromethyl) carbonate
(triphosgene)
mp 79°C
MW 296.7

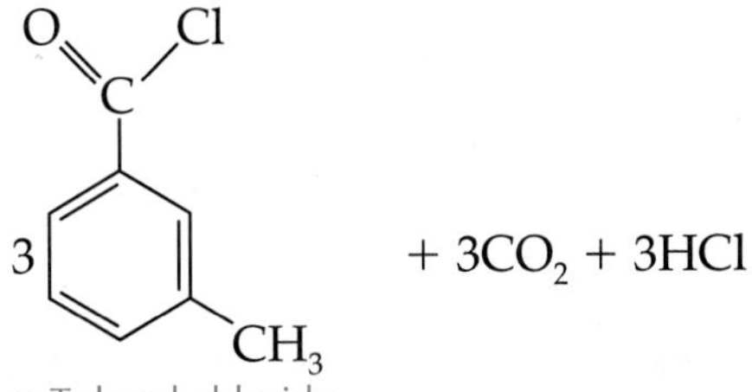

m-Toluoyl chloride
(3-methylbenzoyl chloride)
bp 219°C
MW 154.6
density 1.17 g · mL^{-1}

$+ 3CO_2 + 3HCl$

While the mechanism of this reaction is quite complex, it is known how *N,N*-dimethylformamide (DMF) acts as a catalyst to produce the actual chlorinating species from triphosgene:

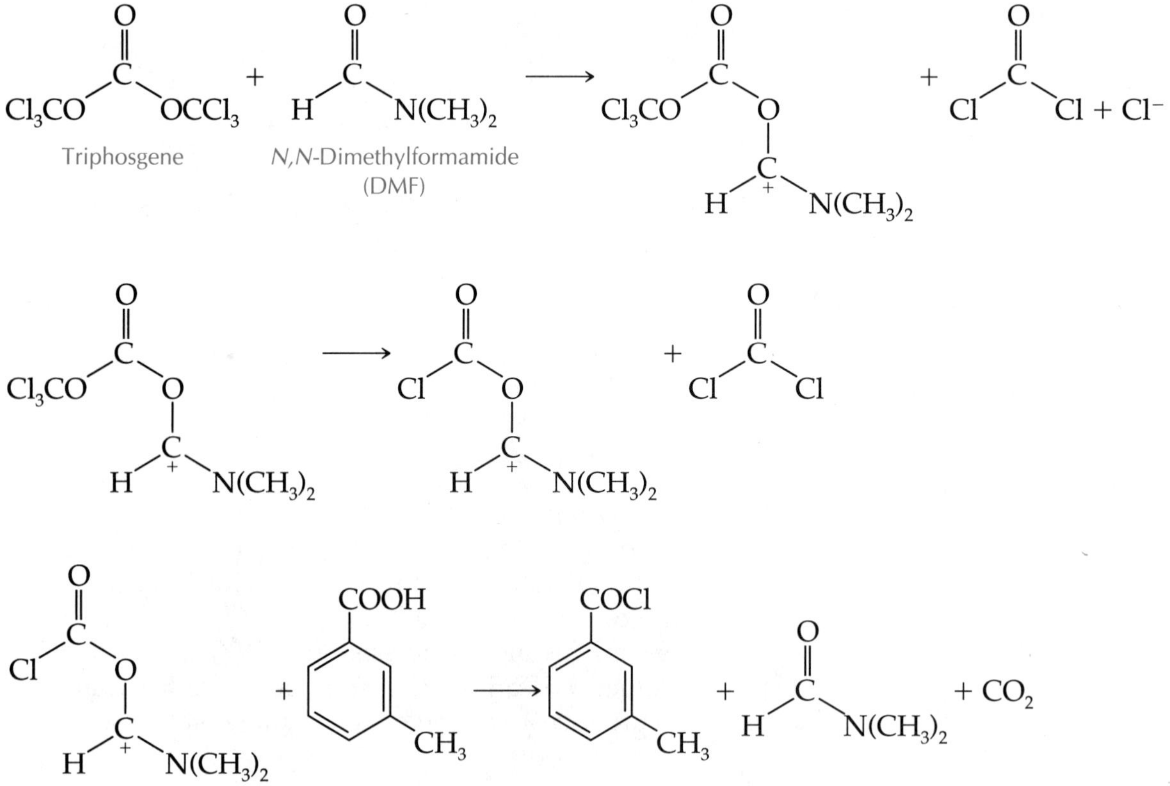

The two molecules of phosgene ($COCl_2$) that are made in the first two steps probably also react with *m*-toluic acid to produce two additional molecules of *m*-toluoyl chloride, in a mechanism analogous to the one given above for thionyl chloride.

Macroscale Procedure

Techniques   Trapping Noxious Gases (described in this experiment)

Extraction: Technique 4.3

Liquid Chromatography: Technique 12

NMR Spectrometry: Spectrometric Method 2

— SAFETY INFORMATION —

Wear gloves throughout the procedure and conduct the experiment in a hood.

(continued on next page)

Thionyl chloride is a lachrymator and causes serious burns. It is also volatile and undergoes reaction with atmospheric moisture, forming hydrogen chloride and sulfur dioxide. Wear gloves while handling it and dispense it only in a hood.

Diethylamine is flammable, toxic, and has a noxious odor. Wear gloves while handling it and dispense it only in a hood.

Diethyl ether is very flammable. Use it only in a hood and keep it away from flames or hot electrical heating devices.

Alumina is an eye and respiratory irritant. Avoid breathing any fine particles while handling it.

Heptane is extremely flammable.

The product, *N,N*-diethyl-*meta*-toluamide, is a skin irritant in high concentrations.

Hydrogen chloride and sulfur dioxide, produced as side products in this synthesis, are noxious, toxic gases that must be trapped by dissolution in water.

Step 1: Synthesis of m-Toluoyl Chloride

The reaction apparatus is shown in Figure 26.1. Both thionyl chloride (SOCl₂) and the acid chloride formed in this procedure react violently with water, so your apparatus and all reagents must be completely dry. Use a 250-mL, three-neck round-bottomed flask as the reaction vessel. Carefully insert an eyedropper or gas bleed tube through your rubber thermometer adapter; place this apparatus at the top of a water-cooled condenser; close the other two necks with glass stoppers for the first part of the reaction.

Use a piece of rubber tubing to connect the bleed tube to a short glass tube inserted in a rubber stopper that fits the neck of a 125-mL filter flask. Use another piece of rubber tubing to connect the side arm of the filter flask to a funnel that is submerged in approximately 75 mL of water in a 250-mL beaker. This system will trap the sulfur dioxide and hydrogen chloride vapors given off by the reaction. The filter flask guards against siphoning of water into the reaction flask.

Place 2.5 g (0.018 mol) of *m*-toluic acid in the 250-mL round-bottomed flask. Add one or two boiling stones and 2.0 mL of thionyl chloride to the reaction flask. Heat the mixture in a water

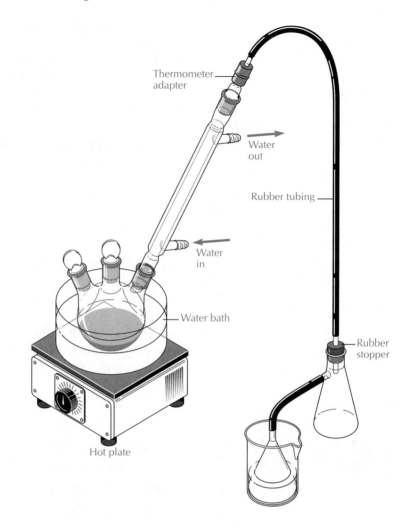

FIGURE 26.1 Apparatus for the synthesis of *m*-toluoyl chloride.

The reaction will turn from a creamy mixture to a light brown solution as it goes to completion.

bath at 70°C, submerging the flask just enough to keep the mixture bubbling gently. Continue to heat the reaction until the bubbling slows and the reaction mixture is a clear liquid. The entire heating period should take about 15 min.

Remove the funnel from the water trap and then remove the water bath from underneath the reaction flask. This procedure prevents water from siphoning back into the reaction as it cools. While the reaction mixture is cooling, remove the thermometer adapter from the condenser. To prevent irritating acid fumes from getting into the laboratory, open the filter flask while it is still in the hood, and rinse it with approximately 50 mL of cold water. Pour this rinse and the acidic solution in the 250-mL beaker into an 800-mL beaker; set this waste aside for later disposal.

Step 2: Synthesis of N,N-*Diethyl*-m-Toluamide

The diethylamine should be a colorless liquid, uncontaminated by colored oxidation products; otherwise your final product may be highly colored.

When your reaction mixture reaches room temperature, pour 30 mL of anhydrous ether into the reaction flask and place a separatory funnel in the center neck of the flask. Be sure that the stopcock of the separatory funnel is closed. The rest of the apparatus remains unchanged. Pour 10 mL of anhydrous diethyl ether into the separatory funnel and add 5.0 mL of diethylamine to the ether in the funnel.

Add the diethylamine/ether solution dropwise over a period of about 10 min. Large amounts of white smoke will be evolved, possibly accompanied by sizzling noises. It is quite dramatic but not dangerous. After the addition is complete and the reaction has ceased, swirl the flask vigorously to ensure thorough mixing. Add 15 mL of 2.5 M sodium hydroxide solution and swirl the flask to dissolve the solid.

Transfer the reaction mixture to a separatory funnel and remove the lower aqueous layer; set this aqueous layer aside in a separate flask. Wash the ether layer with 15 mL of *cold* 3 M hydrochloric acid and then with 15 mL of water [see Technique 4.3].

SAFETY PRECAUTION

The reaction of hydrochloric acid and residual sodium hydroxide in the separatory funnel produces a considerable amount of heat, which may vaporize the diethyl ether. Before you stopper the funnel, swirl the ether and hydrochloric acid layers together to complete most of the acid-base reaction. If the solution becomes warm, add a few pieces of crushed ice. After you stopper the funnel, immediately invert it and open the stopcock to relieve any pressure buildup.

After removing the final aqueous solution from the separatory funnel, transfer the upper ether layer to a clean Erlenmeyer flask and dry it over anhydrous magnesium sulfate for at least 10 min [see Technique 4.6]. Decant the ether solution through a fluted filter paper into a tared 100-mL round-bottomed flask. Rinse the magnesium sulfate remaining in the Erlenmeyer flask with a few milliliters of ether and pour the rinse through the filter paper. Remove the ether by simple distillation [see Technique 7.3], using a steam bath as the heat source, or with a rotary evaporator [see Technique 4.8]. You should have a viscous tan liquid left after the ether is removed.

Column Chromatography and NMR Spectrometry

Be sure that the solvent level never falls below the top of the adsorbent throughout the entire chromatography process.

Your product can be purified by column chromatography [see Technique 12]. Following the procedure in Technique 12.4, make an alumina column, using heptane as the solvent. You should use 20–25 g of alumina and a chromatographic column approximately 1.6 cm in diameter.

After the column has been made, allow the heptane to drip into an Erlenmeyer flask from the bottom of the column until its level is at the top of the sand above the adsorbent. Close the stopcock and transfer your product to the column with a Pasteur pipet. Rinse the round-bottomed flask with 2–3 mL of heptane and also transfer this solution into the column. Start the column flowing again and, when the solution of your product is just level with the sand, add about 60 mL of heptane to the column. This eluent can be the same heptane that you used to make your column.

Elute the product from the column with heptane and collect the faintly yellow amide fraction. You can add a few milliliters more heptane to the top of the column if you have not added enough to elute all the *N,N*-diethyl-*m*-toluamide. If you have doubts as to when all the amide has been eluted, use a watch glass to collect a few drops of the liquid draining from the column. Stop the flow of the column and evaporate the liquid on a steam bath. If the amide is still eluting, a viscous residue will remain on the watch glass after the heptane has evaporated.

Remove the heptane from your product by simple distillation [see Technique 7.3] or with a rotary evaporator [see Technique 4.8]. When the distillation slows down (as the removal of heptane nears completion), pour the heptane out of the receiving flask, replace the receiving flask, and apply full water-aspirator suction at the vacuum adapter nipple to complete the removal of heptane from your product [see Technique 7.7]. If your laboratory is not equipped with water aspirators, work in a hood and remove the residual heptane by blowing a stream of dry nitrogen over the product until the flask and product have reached a constant mass (consecutive weights within 0.05 g). The viscous, high-boiling-point *N,N*-diethyl-*m*-toluamide will remain as a colorless or light tan liquid. Weigh your product and calculate the percent yield.

Determine the NMR spectrum of your product as directed by your instructor. Verify the structure of your product from its NMR spectrum. Assign each peak in the spectrum, and account for its chemical shift and spin-spin splitting pattern. Are the peak areas consistent with your assignments of the peaks?

Your instructor may post the NMR spectrum of *m*-toluic acid. If so, compare your amide NMR spectrum with that of the acid. Account for the differences in chemical shifts and splitting of the peaks.

Cleanup: The aqueous phase initially separated from the reaction mixture contains diethylamine; pour this into the container for flammable (organic) waste. Combine the HCl and water washes remaining from the extractions with the acid solution from the gas trap in a 800-mL beaker. Neutralize the solution with solid sodium carbonate **(Caution: Foaming.)** before washing it down the sink or pouring it into the container for aqueous inorganic waste. Place the recovered ether and recovered heptane into the bottle for flammable (organic) waste. Place the spent drying agent (magnesium sulfate) into the container for inorganic or nonhazardous solid waste, as directed by your instructor.

microscale Procedure

Techniques Anhydrous Reaction Conditions: Technique 3.3a

Microscale Extraction: Technique 4.5b

Flash Chromatography: Technique 12.8

Thin-Layer Chromatography: Technique 10

NMR: Spectrometric Method 2

— *SAFETY INFORMATION* —

Wear gloves and conduct this synthesis in a hood.

3-Methylbenzoic acid (*meta*-toluic acid) presents no particular hazards.

Bis(trichloromethyl) carbonate (triphosgene) is a lachrymator (causes tears). Use it in a hood. The compound is moisture sensitive; keep it in a desiccator.

Diethyl amine is flammable and corrosive. Wear gloves while working with it.

Dichloromethane is toxic, an irritant, absorbed through the skin, and harmful if inhaled. Use it only in a hood and wear gloves while doing the extractions.

Aqueous **sodium hydroxide** solutions are corrosive and cause burns. Solutions as dilute as 9% (2.5 M) can cause severe eye injury.

Synthesis of* m-*Toluoyl Chloride

Fill a micro drying tube with calcium chloride [see Technique 3.3a]. Set a water-jacketed condenser, the drying tube filled with CaCl$_2$, a Claisen adapter, and a 3-mL conical vial containing a spin vane in a 250-mL beaker and dry this glassware in a 125°C oven for 20 min. Cool the glassware in a desiccator to room temperature.

Place 136 mg of *m*-toluic acid in the 3-mL conical vial and add 0.50 mL of dichloromethane. Stir the mixture until the acid dissolves, then add 100 mg of bis(trichloromethyl) carbonate (triphosgene) and 1 drop of dry dimethylformamide (DMF). Fit the Claisen adapter to the vial and close the opening directly above the vial with a septum and screw cap. Fit the water-cooled condenser to the other opening of the Claisen adapter and put the drying tube in the top of the condenser [see Technique 3.3a, Figure 3.8]. Clamp the apparatus on the condenser and set it in an aluminum block on a hot plate-stirrer unit. Prepare an HCl trap by putting a piece of Tygon or rubber tubing over the end of the drying tube and connecting it as shown in Figure 26.1 to the HCl trap.

Use 15 mL of 10% NaOH solution in a 100-mL beaker and position the funnel just below the surface of the NaOH solution. Heat the reaction mixture, with stirring, in an aluminum block heated to 90–95°C on a hot plate. Turn on the water to the condenser. You should observe bubbles of CO_2 and HCl in the gas trap as the reaction begins. At the end of the heating period, **remove the funnel from the NaOH solution** *before* **you turn off the heat** and lift the apparatus out of the aluminum block. The acid chloride is used immediately in the next step.

To prevent NaOH solution from being drawn into your reaction apparatus, remove the funnel from the NaOH solution before *you turn off the heat.*

Synthesis of N,N-Diethyl-m-Toluamide

During the heating period, mix 0.50 mL of dichloromethane and 0.22 mL of freshly distilled diethylamine in a dry 1-mL conical vial or a 1-dram vial. Cap the vial and cool it in an ice-water bath.

Cool the reaction vial (with the Claisen adapter, condenser, and drying tube still in place) for 5 min at room temperature, then for 10 min in an ice-water bath. Draw the dichloromethane/diethylamine solution into a dry 1-mL syringe and insert the syringe needle through the septum in the Claisen adapter. Add the diethylamine solution dropwise over 2–3 min to the cooled reaction mixture. A white precipitate of diethylamine hydrochloride forms immediately. After all of the diethylamine has been added, remove the septum and then remove a drop of the reaction mixture with a Pasteur pipet and test it for basicity with litmus paper or pH test paper. If the mixture is still acidic, add diethylamine dropwise until the solution is basic. Remove the vial from the ice-water bath and allow it to warm to room temperature.

Wash the reaction mixture with 1 mL of 10% NaOH solution, then with 1 mL of 3 M HCl solution, and finally with 1 mL of water [see Technique 4.5b]. Dry the dichloromethane/product solution with 150 mg of anhydrous sodium sulfate.

Transfer the dried solution to a tared 10-mL Erlenmeyer flask. Rinse the Na_2SO_4 with 0.5 mL of fresh dichloromethane and add this rinse to the flask. Evaporate the dichloromethane by leaving the flask open in a hood until the next laboratory period. Alternatively, evaporate the solvent with a stream of nitrogen in a hood. Weigh the flask and determine the percent yield of *N,N*-diethyl-*m*-toluamide.

Cleanup: Combine the aqueous washes in a beaker and determine the pH with pH test paper. Adjust the pH to approximately 7 with either solid sodium carbonate or a few drops of 3 M HCl. Wash the neutralized solution down the sink or pour it into the container for aqueous inorganic waste. Allow the solvent to evaporate before putting the solid Na_2SO_4 in the container for inorganic waste.

A good stopping point.

Purification of N,N-*Diethyl*-m-*Toluamide by Flash Chromatography*

— **SAFETY INFORMATION** —————————

Ethyl acetate and **petroleum ether** are very volatile and flammable.

Obtain 250 mL of 2:3 (v/v) ethyl acetate/petroleum ether (bp 30–60°C). Dissolve your crude *N,N*-diethyl-*meta*-toluamide by adding 1 mL of the ethyl acetate/petroleum ether solvent.

Use a 30 cm × 7 mm i.d. chromatography column. With the stopcock open, prepare the column with approximately 15 g of silica gel 60 (230–400 mesh) according to the procedure described in Technique 12.8. Fill the column to the top with solvent and close the stopcock. Place a number-two rubber stopper, fitted with a glass Y joint in the hole of the stopper, at the top of the column as a simple air-flow controller. Connect one branch of the Y to an air or nitrogen line with a piece of Tygon or rubber tubing. Open the stopcock again, turn on the air valve slightly, and use your index finger on the other branch of the Y to manipulate the pressure so that the silica gel packs tightly, forcing all entrapped air out the bottom. Control the pressure so that solvent flows through the column at a rate of 2 in. (~5 cm) per minute. Never let the solvent level fall below the top of the column; add more solvent, if necessary. When the column is packed tightly, allow the solvent level to drop almost to the top of the sand, close the stopcock, release the pressure, and add the solution of your product to the column, using a Pasteur pipet for the transfer. Allow the solution to sink into the sand at the top of the column.

Fill the column with solvent and adjust the air pressure so that the solvent flows through the column at a rate of 2 in. (~5 cm) per minute. Collect 20 fractions of 10 mL each in labeled test tubes. You will need to stop and refill the column several times with solvent while you are collecting the fractions.

Analyze the fractions, using TLC, to determine which fractions contain *N,N*-diethyl-*m*-toluamide [see Technique 10]. Develop the thin-layer plates in the same solvent as was used for the column chromatography; visualize the chromatograms with iodine [see Technique 10.5]. Combine the fractions containing *N,N*-diethyl-*meta*-toluamide and evaporate the solvent, using a rotary evaporator [see Technique 4.8], or by a simple distillation, using a steam bath as the heat source [see Technique 7.3] to yield pure *N,N*-diethyl-*m*-toluamide as a colorless oil.

Confirm the identity of your purified product with a ^{1}H NMR spectrum and/or an IR spectrum.

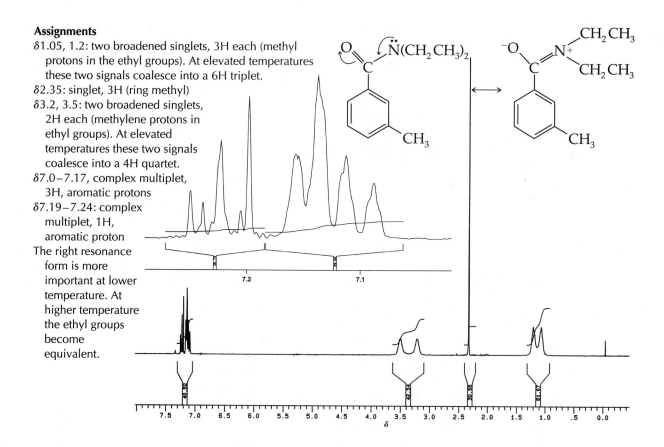

Assignments

δ1.05, 1.2: two broadened singlets, 3H each (methyl protons in the ethyl groups). At elevated temperatures these two signals coalesce into a 6H triplet.

δ2.35: singlet, 3H (ring methyl)

δ3.2, 3.5: two broadened singlets, 2H each (methylene protons in ethyl groups). At elevated temperatures these two signals coalesce into a 4H quartet.

δ7.0–7.17, complex multiplet, 3H, aromatic protons

δ7.19–7.24: complex multiplet, 1H, aromatic proton

The right resonance form is more important at lower temperature. At higher temperature the ethyl groups become equivalent.

FIGURE 26.2 ^{1}H NMR spectrum (300 MHz) of *N,N*-diethyl-*m*-toluamide (in CDCl$_3$).

Cleanup: Pour the recovered ethyl acetate/petroleum ether solution into the container for flammable (organic) waste. Place the silica gel from the column in the container for solid inorganic waste.

References

1. Wang, B. J.-S. *J. Chem. Educ.* **1974,** *51,* 631.
2. LeFevre, J. W. *J. Chem. Educ.* **1990,** *67,* A278–A279.
3. Echkert, H.; Forster, B. *Ang. Chem. Int. Ed. Engl.* **1987,** *26*(9), 894–895.

Questions

1. Write the balanced chemical equation that describes the reaction of *m*-toluoyl chloride with water.
2. Diethyl ether is used as a solvent for the reaction of diethylamine with *m*-toluoyl chloride in the macroscale experiment. Why is a solvent necessary for this reaction?
3. If the hot-water bath is abruptly removed during the synthesis of the acid chloride, the water from the trap might siphon into the reaction flask. Why? What chemical reaction will occur if water enters the reaction mixture?
4. What is the white solid that precipitates from the reaction mixture as diethylamine is added to *m*-toluoyl chloride in the macroscale experiment? Why would this compound not be soluble in ether? Write an equation for this reaction.
5. When there is excess thionyl chloride present in the macroscale reaction mixture, it also reacts with diethylamine. Write an equation for this reaction.
6. Write a mechanism for the reaction of *m*-toluic acid and phosgene to form *m*-toluoyl chloride.
7. Why is it crucial that a chromatography column never go dry after preparation or during use?
8. Arrange the following compounds in order of their decreasing ease of elution on silica gel chromatography: (a) 2-octanol; (b) *m*-dichlorobenzene; (c) *tert*-butylcyclohexane; (d) benzoic acid.
9. When ^{1}H NMR analysis is conducted on the DEET product, it is found that a clean quartet occurs for the methylene groups at 95°C, whereas at 30°C the methylene signal has lost its definition and occurs as two broad, ill-resolved signals (2H each). Explain these observations.

Experiment 27

POLYPHOSPHORIC ACID-CATALYZED BECKMANN REARRANGEMENT OF ACETOPHENONE OXIME

Determine which isomeric product forms in a rearrangement re-action and explore the mechanism of the Beckmann rearrange-ment, using NMR.

Acetophenone
bp 202°C
MW 120.2
density 1.026 g · mL⁻¹

Hydroxylamine hydrochloride
MW 69.5

Acetophenone oxime intermediate

Acetanilide
mp 113–114°C
MW 135.2

N-Methylbenzamide
mp 82°C
MW 135.2

Although acetophenone is converted to an amide without inter-ruption in this experiment, the process occurs in two major steps. In the first, acetophenone is converted to an oxime; the second step involves a rearrangement of the oxime to an amide. Polyphosphoric acid (PPA) serves as the solvent and promotes the conversion of the intermediate oxime to the amide. Rearrangement of acetophenone oxime could lead to two possible amides. You can determine which amide is formed by the melting point and NMR spectrum of your product.

Oxime formation in the first step is one example of the many reactions that occur when nitrogen-containing nucleophiles undergo reactions with aldehydes and ketones. Hydroxylamine converts acetophenone to its oxime by an addition-elimination process:

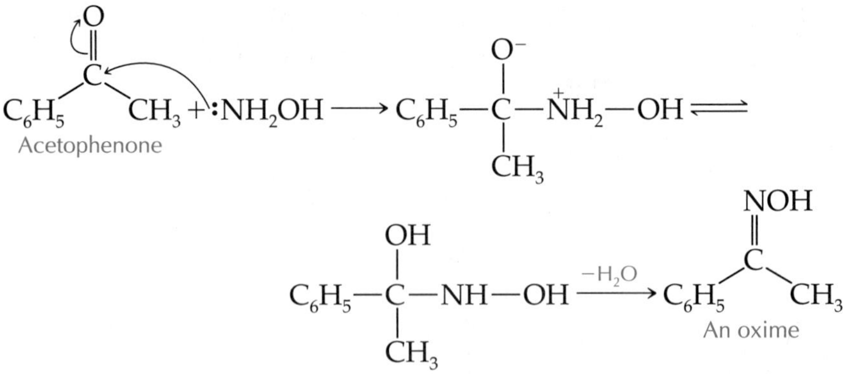

The geometry of the C=N unit of the oxime deserves scrutiny, because it is an important factor in the subsequent rearrangement reaction. The C=N is not unlike the double bond of an alkene in that geometrical isomers are possible. In the oxime, the lone pair of electrons on the nitrogen atom, which occupies an sp^2 orbital lobe, is the fourth "group" on the oxime double bond. Rotation about this double bond is restricted, and it is possible to have either the phenyl or the methyl group on the same side of the C=N as the —OH group (called *syn* to the —OH). In like fashion, a group can be on the opposite side of the double bond (*anti* to the —OH). Therefore, two stereoisomers of the oxime are possible:

Syn and anti *isomers are possible.*

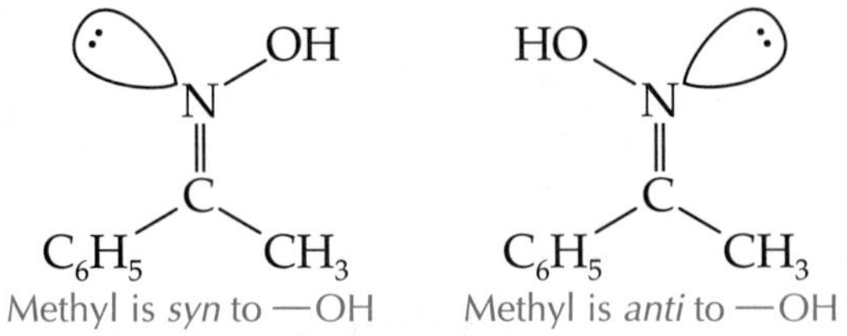

Methyl is *syn* to —OH   Methyl is *anti* to —OH

In the second step, a Beckmann rearrangement, named after the chemist who discovered it, is induced by the polyphosphoric acid solvent. Polyphosphoric acid has a polymeric structure and has proved to be a useful catalyst for many organic reactions.

$$HO-\overset{\overset{\displaystyle O}{\|}}{\underset{\underset{\displaystyle OH}{|}}{P}}\left[O-\overset{\overset{\displaystyle O}{\|}}{\underset{\underset{\displaystyle OH}{|}}{P}}\right]_x O-\overset{\overset{\displaystyle O}{\|}}{\underset{\underset{\displaystyle OH}{|}}{P}}-OH$$

Polyphosphoric acid
(PPA)

The acidic properties of polyphosphoric acid make it a reasonable alternative to sulfuric acid, but PPA is not as strong an oxidizing agent as sulfuric acid. In addition, PPA does not undergo reaction with aromatic rings, a reaction that can occur with sulfuric acid. PPA has moderately good solvent properties, especially when heated to at least 90°C. One disadvantage of PPA is the fact that hydrolysis of the polymeric bonds occurs somewhat slowly, and thorough mixing is required in the water-treatment step following the reaction.

The Beckmann rearrangement process leading to amide formation is influenced by both the geometry of the oxime and by the ease with which the methyl and phenyl groups migrate. The controlling feature seems to be the configuration about the oxime double bond. After protonation of the —OH, the group *anti* to —OH$_2^+$ migrates, directly displacing it. This stereochemistry determines which amide is formed:

Rearrangement step

$$R-N=\overset{\overset{\displaystyle OH}{|}}{C}-R' \xrightarrow{\text{tautomerization}} R-NH-\overset{\overset{\displaystyle O}{\|}}{C}-R'$$

Amide

Thus we would expect that the particular amide formed is controlled by the stereochemistry of the oxime precursor: Acetanilide would be formed by phenyl migration *anti* to the —OH$_2^+$ group and *N*-methylbenzamide would be formed by methyl migration *anti* to the —OH$_2^+$ group. This, in turn, suggests the stereochemistry of the precursor oxime. You will use the melting point of your amide product and its ^{1}H NMR spectrum to determine its identity.

Although the ^{1}H NMR spectra of acetanilide and *N*-methylbenzamide are similar, there are enough differences to allow identification of these two amides by NMR. The ^{1}H NMR spectrum of acetanilide shows a singlet at δ 2.2 (3H) and a broadened singlet at δ 9.0 (1H). The former is due to the methyl group, and the latter to the N—H proton. Acetanilide's aromatic protons occur at δ 7.5, 2H, doublet; δ 7.3, 2H, triplet; and δ 7.1, 1H, triplet. The ^{1}H NMR spectrum of *N*-methylbenzamide reveals two aromatic protons at considerably lower field (approximately δ 7.8); these are the protons *ortho* to and thus deshielded by the nearby carbonyl group. The existence of a doublet near δ 7.8 in the aromatic proton region supports a structure with a carbonyl attached to the benzene ring and thus the *N*-methylbenzamide structure. The absence of such a signal and aromatic protons only at higher field (lower δ) supports the acetanilide structure.

Macroscale Procedure

Techniques   Recrystallization: Technique 5.3

Removing Colored Impurities: Technique 5.3a

SAFETY INFORMATION

Wear gloves and, if possible, conduct the reaction in a hood until after the water is added to the cooled reaction mixture.

Acetophenone is flammable and an irritant to skin and eyes.

Polyphosphoric acid is corrosive and causes burns. Neutralize any spills immediately with solid sodium carbonate.

Hydroxylamine hydrochloride is corrosive and toxic. Avoid contact with your skin, eyes, or clothing.

The reaction emits **hydrogen chloride,** a toxic corrosive gas; use a gas trap.

Polyphosphoric acid is very viscous and pours more easily if the bottle is warmed before measuring the acid.

Carefully pour 20 g ($\pm$ 0.5 g) of polyphosphoric acid directly into a tared 125-mL filter flask. Add 2.1 g of hydroxylamine hydrochloride and 1.2 mL (1.2 g) of acetophenone to the flask. Stir the mixture with a stirring rod and gently break up any lumps of hydroxylamine hydrochloride. Assemble a gas trap as shown in Experiment 18, Figure 18.3 if your laboratory is equipped with water aspirators or as shown in Figure 18.4 if your laboratory does not have water aspirators. Close the top of the flask containing the reaction mixture with a rubber stopper. Firmly clamp the flask in a 400-mL beaker of hot water. Add a boiling stone or boiling stick to the beaker to prevent bumping, and bring the water to boiling. Heat the reaction mixture for 45 min, removing the stopper frequently to stir the mixture, which will turn green and foam as the reaction proceeds.

Cool the reaction flask to room temperature and add 50 mL of water. Stir it thoroughly until the dark viscous reaction mixture forms a milky suspension in the water. Cool the mixture in an ice-water bath for 5–10 min before collecting the crude solid product by vacuum filtration. Wash the product three times with 10-mL portions of ice-cold water.

Transfer the crude product to a 125-mL Erlenmeyer flask and recrystallize it from boiling water [see Technique 5.3]. If there is a brown film on top of the recrystallization solution or if the solution is colored, add 20–30 mg of activated charcoal to the hot (not boiling) solution [see Technique 5.3a]. Reheat the mixture to boiling and gravity filter the hot solution through a fluted filter paper set in a stemless funnel or a plastic filling funnel. Collect the filtrate in an Erlenmeyer flask kept warm on a steam bath (or hot plate heated at the lowest setting) during the filtration. Cool the filtrate to room temperature, then cool it further in an ice-water bath before collecting the product by vacuum filtration. Allow the product to dry overnight before determining the melting point and percent yield.

Identify the rearrangement product by comparing your experimental melting point to those listed at the beginning of the experiment for the two possible amides. Your instructor may also have you confirm your identification by obtaining and analyzing the NMR spectrum of your product.

Cleanup: Pour the filtrate from the crude product into a large beaker, add the acidic solution from the gas trap flask (if you used the setup in Figure 18.4) and neutralize the combined solution with sodium carbonate **(Caution: Foaming.)** before washing it down the sink or pouring it into the container for aqueous inorganic waste. The filtrate from the recrystallization can be washed down the sink or placed in the aqueous waste container, as directed by your instructor.

References

1. Uhlig, F.; Snyder, H. R. *Advances in Organic Chemistry*; Interscience: New York, 1960; Vol. 1, pp. 35–81.
2. Fieser, L.; Fieser, M. *Reagents for Organic Synthesis*; Wiley: New York, 1967; Vol. 1, pp. 894–904.

Questions

1. What gas is most likely to produce the foaming observed in the reaction flask?
2. On the basis of your identification of the Beckmann rearrangement product, draw the structure of your acetophenone oxime intermediate. Did the methyl or the phenyl group migrate in the rearrangement?
3. As stated earlier, acetanilide's aromatic protons occur at δ 7.5, 2H, doublet; δ 7.3, 2H, triplet; and δ 7.1, 1H, triplet. Assign these three signals to the specific ring protons. Assume that only *ortho* coupling is significant.
4. Phenylhydrazones of aldehydes and ketones are prepared in a manner that is very similar to the preparation of oximes. Would you expect *syn* and *anti* isomeric possibilities for the phenylhydrazone of acetophenone?

$$C_6H_5-C\overset{\displaystyle N-NH-C_6H_5}{\underset{\displaystyle CH_3}{\Big\langle}}$$

Phenylhydrazone of
acetophenone

5. It is common to derivatize aldehydes and ketones as 2,4-dinitrophenylhydrazones. They are formed in addition-elimination processes, as are oximes. For example, treatment of acetophenone with 2,4-dinitrophenylhydrazine in an aqueous ethanol solution of sulfuric acid results in the hydrazone:

Acetophenone   2,4-Dinitrophenylhydrazine   2,4-Dinitrophenylhydrazone of acetophenone

(a) Write a complete mechanism for hydrazone formation.
(b) Can *syn* and *anti* isomeric hydrazone structures be obtained from acetophenone? From benzophenone?

Experiment 28

SYNTHESIS OF *trans*-1,2-DIBENZOYLCYCLOPROPANE BY RING CLOSURE

Synthesize a cyclopropane-ring compound, using a base-promoted S_N2 reaction.

1,3-Dibenzoylpropane
(1,5-diphenyl-1,5-pentanedione)
mp 67.5°C
MW 252

trans-1,2-Dibenzoylcyclopropane
mp 103–104°C
MW 250

In this experiment the cyclization of a diketone to form a cyclo-propane ring takes advantage of the significant acidity of the α-hydrogen atoms of ketones and the nucleophilic character of eno-late anions. One of the protons of the starting diketone is removed by NaOH, forming an enolate anion, which then undergoes an S_N2-like reaction on molecular iodine to form an iodoketone:

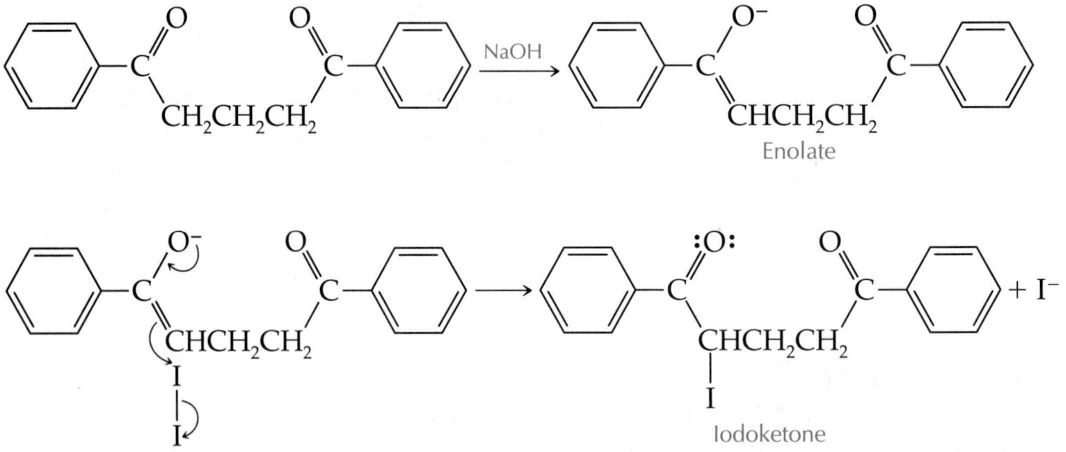

Abstraction of a proton by NaOH transforms the intermediate iodoketone to another enolate anion, which then undergoes ring closure by an S_N2 reaction to form *trans*-1,2-dibenzoylcyclo-propane:

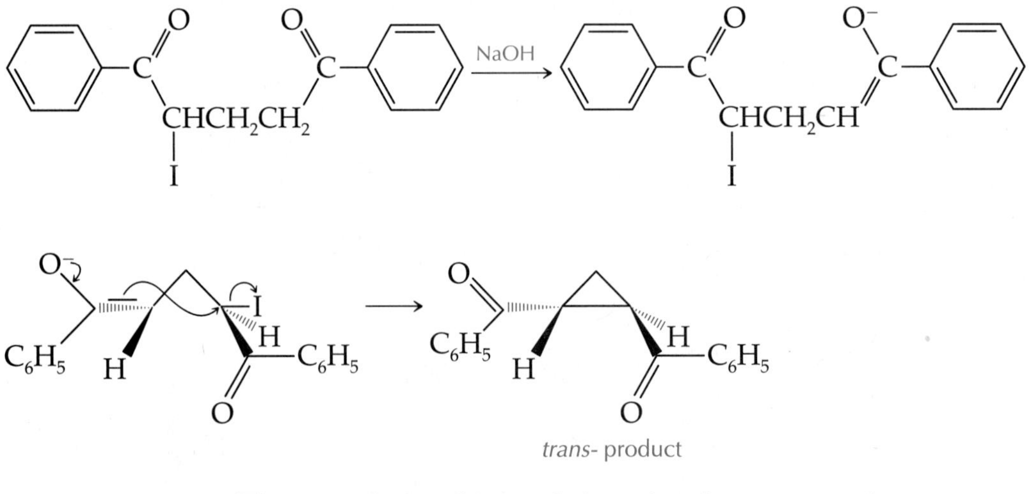

trans- product

The organic synthesis of rings involves two major concepts: ring strain, and ease of formation of rings based on entropy (or "ordering") effects. Rings smaller than cyclohexane are destabi-lized because they have bond angles smaller than that stable for the sp^3 hybridization of their carbons. Ring strain is at a maximum for cyclopropane, and it would seem that such a ring should be a challenge to make.

However, ring strain factors are to some degree counterbalanced by entropy effects. Specifically, the formation of a cyclopropane ring, the smallest ring, has an entropic advantage because the increase in order needed to form the ring is at a minimum. Keep in mind that the greater the increase in order during the course of a reaction, the smaller the driving force for that reaction.

The NMR spectrum of *trans*-1,2-dibenzoylcyclopropane has some interesting features. The carbonyl (C=O) group has a powerful deshielding effect, especially when attached to a phenyl group, which deshields the *ortho* protons of the phenyl group most strongly. The cyclopropane protons are upfield (shielded) relative to the usual chemical shifts of alkyl protons. This behavior is due to the fact that there is a quasi π-system for cyclopropane, which yields a moderately strong ring current. Because the cyclopropane ring protons are above and below the plane of the ring, these protons are shielded. This shielding contrasts with that of protons on benzene rings; these protons are deshielded because they are in the plane of the ring system:

Chemical shifts of methylene protons:

△ δ 0.22

⬡ δ 1.44

⋀ δ 1.3

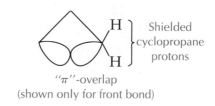

"π"-overlap
(shown only for front bond)

H } Shielded cyclopropane protons

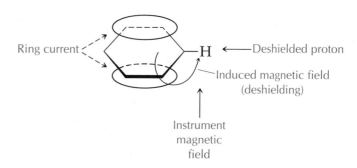

Ring current

—H ← Deshielded proton

Induced magnetic field (deshielding)

Instrument magnetic field

Macroscale Procedure

Techniques Recrystallization: Technique 5.3
NMR Spectrometry: Spectrometric Method 2

SAFETY INFORMATION

Methanol is volatile, toxic, and flammable. **Iodine** is toxic and corrosive. Wear gloves while working with the **iodine/methanol** solution and pour the solution in a hood.

(continued on next page)

— *SAFETY INFORMATION (continued)* —

Avoid contact with the **sodium hydroxide/methanol** solution, which is corrosive.

1,3-Dibenzoylpropane and the product require no special precautions.

Place 1.0 g (4.0 mmol) of 1,3-dibenzoylpropane and 12 mL of 0.67 M sodium hydroxide in methanol solution in a 50-mL round-bottomed flask containing a magnetic stirring bar. Assemble a Claisen adapter, separatory funnel, and water-cooled condenser as shown in Technique 3, Figure 3.7a, omitting the drying tubes at the top of both the separatory funnel and the condenser. Pour 6.0 mL of 0.67 M iodine in methanol solution into the separatory funnel. Put a glass stopper in the separatory funnel with a small strip of paper inserted between the stopper and the neck of the funnel to prevent an airlock.

Control the rate of iodine addition so that the brown color of the iodine disappears before the next drop is added.

Place a 40°C water bath under the round-bottomed flask and begin stirring the mixture. When the solid dissolves, begin dropwise addition of the iodine solution. After the addition is complete, continue to heat the reaction mixture for 45–60 min, until precipitation of white solid ceases.

Dismantle the reaction apparatus, cool the flask to room temperature, and then cool it further in an ice-water bath. Collect the crude product by vacuum filtration. Pour the methanol filtrate out of the filter flask into a 50-mL beaker *before* washing the product, which remains on the Buchner funnel. Wash the crude product on the funnel three times with 5-mL portions of water. You may collect a second crop of crude product by evaporating the methanol filtrate on a steam bath in the hood. Stir the resulting reddish residue with 5 mL of 10% sodium bisulfite solution to reduce any iodine present. Collect the solid again by vacuum filtration and wash it as directed above. Recrystallize the combined crude product from methanol, using a steam bath as the heat source [see Technique 5.3]. Allow the recrystallized product to dry before determining the melting point and percent yield. Obtain and analyze the NMR spectrum of your dried product as directed by your instructor.

If time permits, you may wish to recrystallize the second crop of crude product separately from the first, because the second crop usually contains more impurities than the first.

Cleanup: If you did not collect a second crop from the methanol filtrate, pour this solution into the container for halogenated waste. Pour the filtrate from the recrystallization into the container for flammable waste. The aqueous filtrate from washing the crude product can be washed down the sink or placed in the aqueous waste container, as directed by your instructor.

mmmicroscale Procedure

Techniques Microscale Recrystallization: Technique 5.5b
NMR Spectrometry: Spectrometric Method 2

SAFETY INFORMATION

Methanol is volatile, toxic, and flammable. **Iodine** is toxic and corrosive. Wear gloves while working with the **iodine/methanol** solution and pour the solution in a hood.

Avoid contact with the **sodium hydroxide/methanol** solution, which is corrosive.

1,3-Dibenzoylpropane and the product require no special precautions.

Place 200 mg (0.79 mmol) of 1,3-dibenzoylpropane and 2.4 mL of 0.67 M sodium hydroxide in methanol solution in a 10-mL round-bottomed flask containing a magnetic stirring bar. Assemble a Claisen adapter, septum cap, and water-cooled condenser as shown in Technique 3, Figure 3.8, omitting the drying tube at the top of the condenser. Measure 1.2 mL of 0.67 M iodine in methanol solution into a small screw-capped vial. Draw this iodine solution into a syringe and insert the syringe needle through the septum cap.

Place a 40°C water bath under the round-bottomed flask and begin stirring the mixture. When the solid dissolves, begin drop-wise addition of the iodine solution. After the addition is complete, continue to heat the reaction mixture for 35–40 min. A white precipitate may form.

Control the rate of iodine addition so that the brown color of the iodine disappears before the next drop is added.

Dismantle the reaction apparatus, cool the flask to room temperature, and then cool it further in an ice-water bath. Collect the crude product by vacuum filtration on a Hirsch funnel. Wash the product three times with 0.75-mL portions of water. A second crop of crude product will precipitate in the filtrate. This second crop should also be collected and washed as above.

Transfer the combined crude product to a Craig tube, insert a boiling stick, and recrystallize the product from a minimum amount of methanol, using a steam bath as the heat source [see Technique 5.5b]. After the product has been dissolved in the methanol, add water dropwise (only a few drops are needed) until the hot solution is cloudy, then warm the solution briefly to boiling and add methanol dropwise until the cloudiness clears. Place the plug in the tube and set the Craig apparatus in a 25-mL Erlenmeyer flask to cool slowly. When crystallization is well under way and the Craig tube has cooled to room temperature, place it in an ice-water bath for a few minutes to complete crystallization.

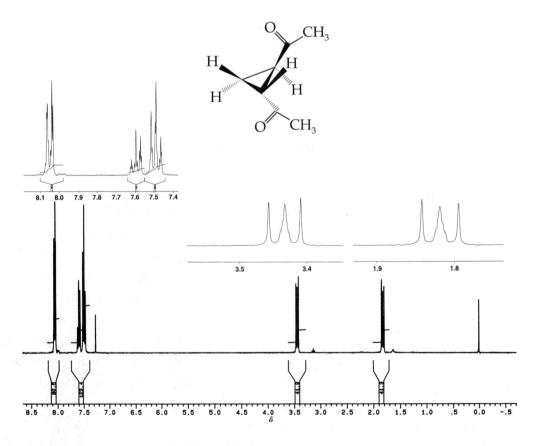

FIGURE 28.1 ^{1}H NMR spectrum (300 MHz) of 1,2-dibenzoylcyclopropane (in CDCl$_3$).

Remove the mother liquor from the recrystallized product by centrifugation [see Technique 5.5b, Figure 5.9]. Allow the product to dry before determining the melting point and percent yield. Obtain and analyze the NMR spectrum (Figure 28.1) of your dried product as directed by your instructor.

Cleanup: Pour the filtrates from the crude product and the recrystallization into the container for flammable waste.

Reference

1. Colon, I.; Griffin, G. W.; O'Connell, Jr., E. J. *Organic Syntheses;* Wiley: New York, 1988; Collect. Vol. VI, pp. 401–402.

Questions

1. Sodium bisulfite was used to reduce any residual iodine remaining in the second crop of crude product. Write a balanced equation for the reaction of sodium bisulfite with I_2.
2. How would the NMR spectrum of 1,3-dibenzoylpropane differ from that of the product, *trans*-1,2-dibenzoylcyclopropane? Would the chemical shifts of the protons α and β to the carbonyl groups be appreciably different? Explain.
3. In this experiment you are asked to heat the reaction mixture until the formation of a white solid ceases. What do you think the white solid is?
4. Rationalize the formation of *trans*-diketone here (rather than *cis*) by examining the S_N2 transition state for the ring closure. What difficulties arise when rationalizing closure to the *cis*-compound?

Experiment 29
POLYMER CHEMISTRY

Explore the synthesis and properties of polymers.

Polymers (from Greek *poly* = many, *meros* = parts) are compounds containing long chains of atoms. A single polymer molecule can have thousands or even millions of atoms, all covalently bonded together. Polymers are conveniently divided into two major groups, biopolymers and synthetic polymers.

Biopolymers occur naturally in living beings. They include the nucleic acids (DNA and RNA), which are crucial to the processing and transmission of genetic information. Another class of biopolymers are the proteins, which serve many biological functions— from the catalysis of metabolic reactions (enzymes) to making up muscles and tendons. Polysaccharides, a third major group of biopolymers, are structural components of plants and are intimately involved in energy metabolism; cellulose and starch are examples of polysaccharides. Starch is the focus of Experiment 29.2.

Synthetic polymers are an important part of materials science, an area of technology that is becoming more significant every day and one in which chemistry plays an important role. Synthetic polymers are made from relatively small organic compounds, called monomers, which are chemically linked to produce large polymeric molecules. For example, under the right conditions, a large number of ethylene molecules can combine to form polyethylene:

$$n\,CH_2\!=\!CH_2 \xrightarrow{\text{catalyst}} \text{---}(CH_2\!-\!CH_2)\text{---}_n$$

The subscript n refers to the number of identical repeating units in the polymer, that is, the degree of polymerization. It is not uncommon to find polymeric compounds that have molecular weights in the tens to hundreds of thousands, and even higher.

Polymers can be broken down into classes that are defined by the manner in which they are formed. One class is called addition polymers, and polyethylene is an example. Addition polymers are formed by the linkage of intact monomer units. Another class is condensation polymers, which are formed by the chemical linking of monomers that is accompanied by the loss of a small molecule such as water. A common example of a condensation polymer is a polypeptide that is formed by the condensation of amino acid units, accompanied by the elimination of water:

An amino acid   Another amino acid   A dipeptide

$$H_2N \overset{\displaystyle \overset{R}{|}}{\underset{\displaystyle \underset{\displaystyle O}{\parallel}}{CH}} \cdots NH \overset{\displaystyle \overset{R'}{|}}{CH} \cdots COOH \xrightarrow{\text{more amino acids}} \longrightarrow \longrightarrow \text{a polypeptide}$$

Polymers are significant commercial materials. Polymers of commercial interest include polystyrene (Experiment 29.1), which is used in the commonly encountered white packing "bubbles" and insulated coffee cups. Teflon is an exceedingly useful polymer because it has a very low coefficient of friction, a high melting point (327°C), and very good resistance to corrosion. It has found application as a surface coating for the inside of pots and pans and in valves for heart repairs.

$$n\,CF_2{=}CF_2 \longrightarrow {+}CF_2{-}CF_2{\,)_n}$$

Tetrafluoroethylene   Polytetrafluoroethylene
(Teflon)

Polyvinyl chloride is formed from the head-to-tail polymerization of vinyl chloride (or chloroethene). It is used in shower curtains, drainpipes, and many other plastic products.

$$n\,CH_2{=}CHCl \longrightarrow {+}CH_2{-}\underset{\displaystyle \underset{Cl}{|}}{CH}{\,)_n}$$

Vinyl chloride   Polyvinyl chloride (PVC)

The properties of a polymer can be modified by forming a copolymer, a polymer containing two different monomers. Saran wrap is a copolymer of vinyl chloride and 1,1-dichloroethylene:

$$m\,CH_2{=}CCl_2 + n\,CH_2{=}CHCl \longrightarrow {+}CH_2{-}\overset{\displaystyle \overset{Cl}{|}}{\underset{\displaystyle \underset{Cl}{|}}{C}}{-}CH_2{-}\underset{\displaystyle \underset{Cl}{|}}{CH}{\,)_{m+2n}}$$

1,1-Dichloroethylene   Vinyl chloride   Saran wrap

It should be noted that the monomeric units of Saran wrap are not necessarily a regular alternation of 1,1-dichloroethylene and vinyl chloride units.

The properties of polymers can be affected by cross-linking. For example, the reaction of glycerol (1,2,3-trihydroxypropane) in the presence of phthalic anhydride allows the formation of glyptal (a polyester), which is highly cross-linked. Cross-linking here is made possible by the presence of the third —OH group in glycerol:

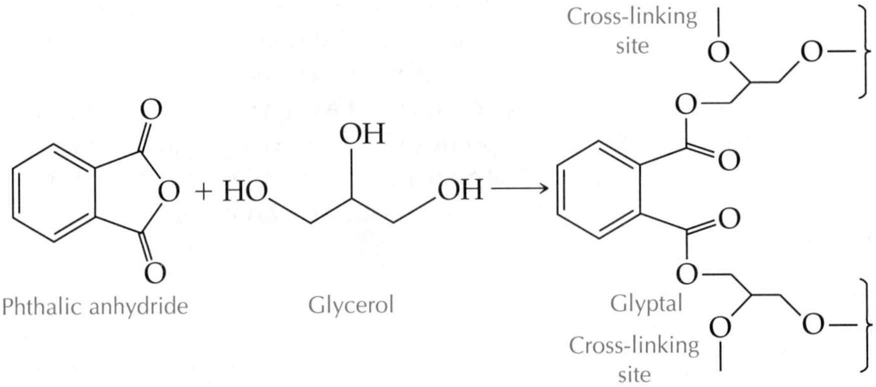

Phthalic anhydride   Glycerol   Glyptal

The properties of glyptal include high rigidity relative to the flexibility associated with the long, linear chains of polyethylene.

29.1 Synthesis of Polystyrene

Synthesize polystyrene and use IR to analyze the polymers made by different methods.

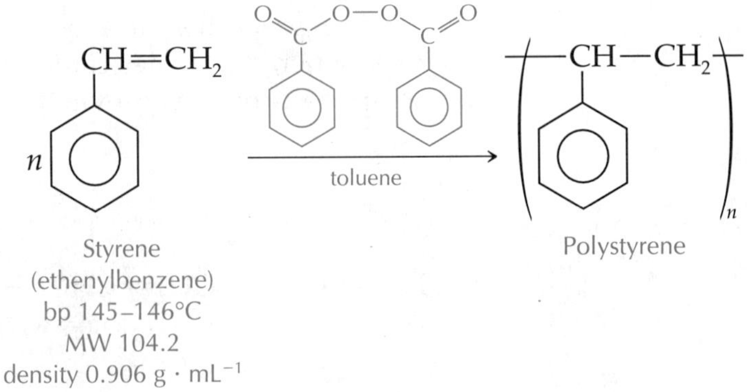

Styrene
(ethenylbenzene)
bp 145–146°C
MW 104.2
density 0.906 g · mL^{-1}

Polystyrene

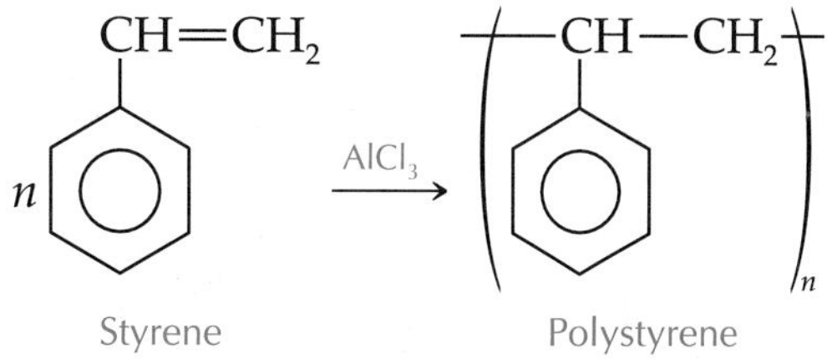

Treatment of styrene with either a free radical or ionic catalyst results in the formation of polystyrene. The polystyrene that is formed varies significantly (molecular weight, degree of polymerization, etc.) with the nature of the catalyst and the other conditions used to form it.

Treatment of styrene with benzoyl peroxide produces free radical polymerization, because the weak oxygen-oxygen bond of peroxides readily undergoes homolysis to form two equivalent benzoyloxy radicals:

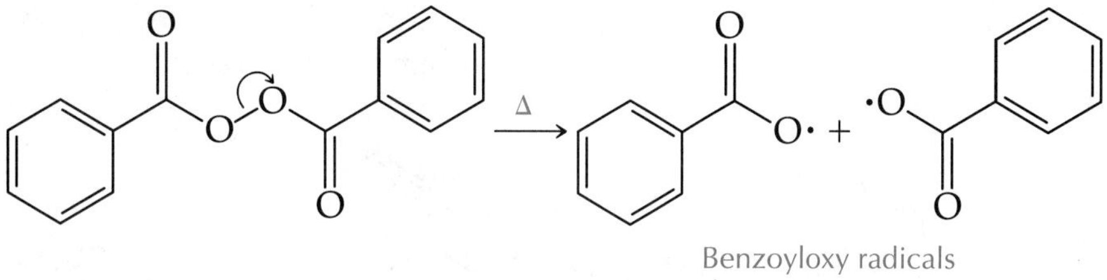

Benzoyloxy radicals

The benzoyloxy radicals initiate further radical formation by adding to a styrene molecule:

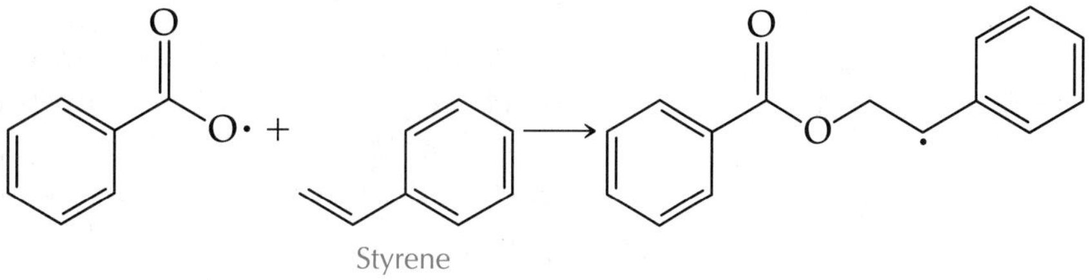

Styrene

The carbon radicals can then bond successively with additional styrene molecules to form long chains:

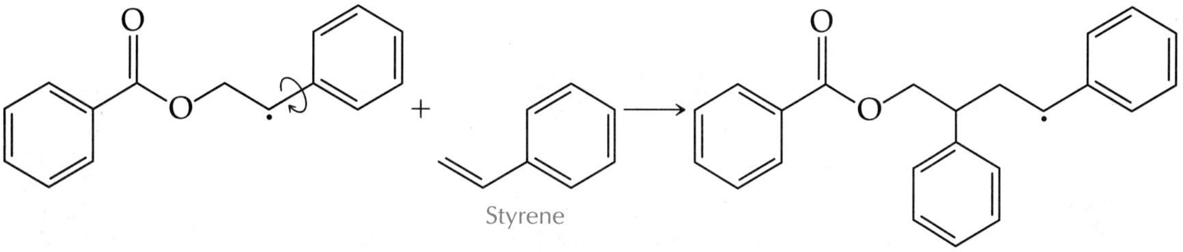

Styrene

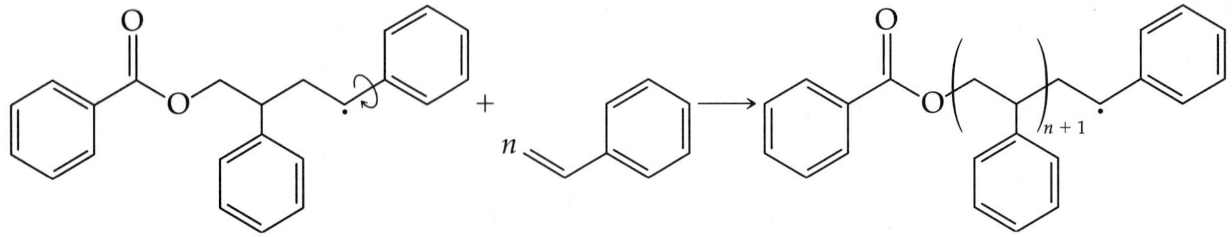

A major driving force for this reaction is the formation of free radical sites adjacent to phenyl groups that can be stabilized by resonance with the aromatic π-system:

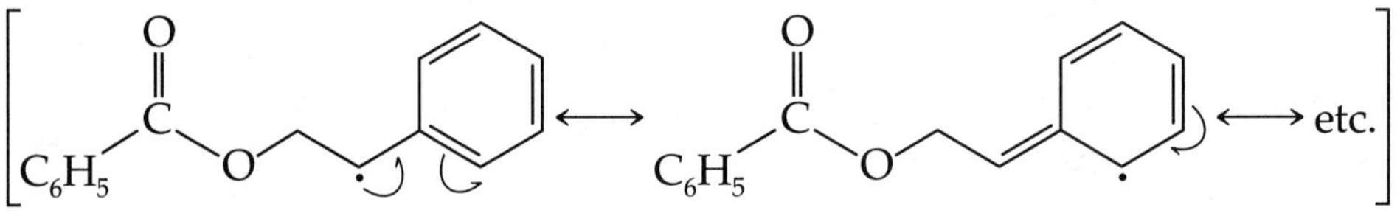

The reaction is terminated by the combination of the chain with any other radical present:

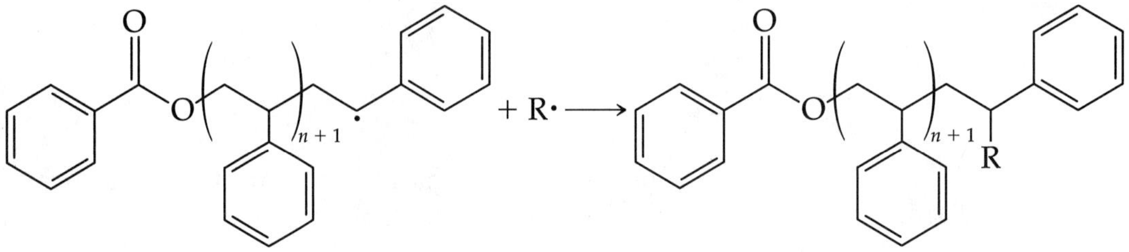

Polystyrene can also be formed by ionic methods. When styrene is treated with aluminum trichloride, a vigorous reaction occurs. The mechanism of the reaction is likely initiated by autoionization of the aluminum trichloride:

$$2AlCl_3 \rightleftharpoons AlCl_2^+ + AlCl_4^-$$

The $AlCl_2^+$ electrophile is highly reactive and will attack styrene to produce a secondary carbocation, again stabilized by resonance with the benzene ring:

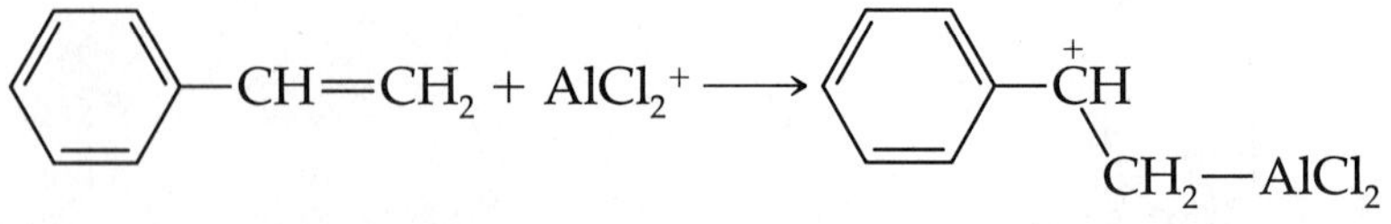

The newly formed benzylic carbocation can react with a series of polystyrene molecules to form a long chain structure:

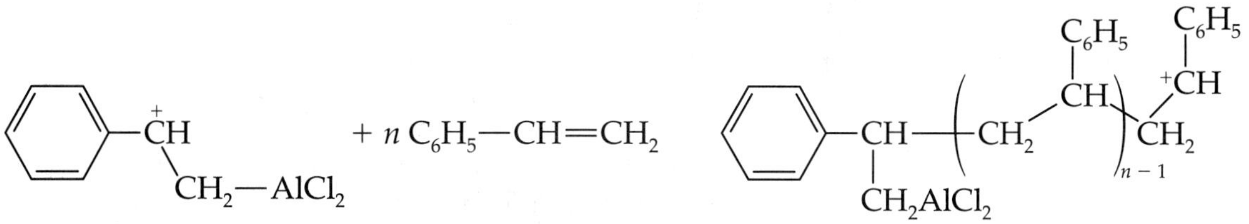

This cationic polymeric chain can be terminated by the attack of any reasonable acid, for example, HCl:

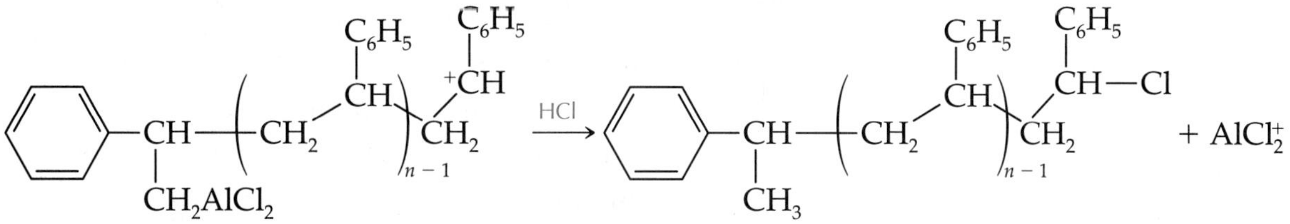

Macroscale Procedure

Techniques Liquid Chromatographic Column: Technique 12.6

IR Spectrometry: Spectrometric Method 1

SAFETY INFORMATION

Wear gloves and work in a hood, if possible, while doing these procedures.

Styrene is a flammable liquid and an irritant to eyes and mucous membranes. Use it only in a hood and avoid contact with skin, eyes, and clothing.

Benzoyl peroxide is a strong oxidizer and may explode if it is heated above its melting point or if the dry compound is subjected to friction or shock. Avoid contact with skin, eyes, and clothing. Do not transfer it with a metal spatula.

Aluminum chloride is corrosive, moisture sensitive, and forms HCl on contact with atmospheric moisture. Measure the required amount quickly and store the reagent container tightly closed in a desiccator.

Methanol is volatile, toxic, and flammable. Use it only in a hood.

2-Butanone (methyl ethyl ketone) is flammable and an irritant.

Polymerization of Styrene with Benzoyl Peroxide

Reactive alkenes are doped with inhibitors to prevent their polymerization.

The styrene available in your laboratory will likely contain a free radical inhibitor such as an alkylated phenol. The inhibitor must be removed before the styrene is used in this procedure. To do this, prepare a microscale chromatography column in a Pasteur pipet [see Technique 12.6]. Pack a small plug of cotton in the upper stem of the pipet and add about 1.8 g of dry alumina on top of the cotton. Tap the pipet gently on the bench top to settle the alumina. Clamp the pipet in an upright position and place a tared 18 × 150 mm test tube under the tip. Transfer about 2 mL of styrene to the top of the column with a Pasteur pipet and allow the styrene to flow through the column into the test tube. You may use a rubber bulb on the top of the column to exert a moderate pressure. Weigh the test tube and the eluted styrene that it contains; you should have 1.0–1.2 g of styrene. Record the weight of styrene for your yield calculation.

Add 6 mL of toluene and 50 ± 5 mg of benzoyl peroxide to the test tube containing the styrene. Clamp the test tube in a hot-water bath (90–95°C) and heat it for 1 h (carry out the polymerization of styrene with aluminum chloride during this time). Cool the test tube for 5 min, note the viscosity of the solution, and pour the solution into a 100-mL beaker containing 40 mL of methanol. Collect the precipitated white polymer by vacuum filtration and wash the precipitate on the Buchner funnel with 10 mL of methanol. Spread the polymer on a piece of filter paper to dry.

Cleanup: Pour the methanol filtrate into the container for flammable (organic) waste. Place the alumina from the chromatography column in the container for inorganic waste.

Polymerization of Styrene with Aluminum Chloride

Pour 2.0 mL of styrene (the inhibitor does not need to be removed) into a 50-mL beaker. Obtain 80 ± 5 mg of anhydrous aluminum chloride in a small tared test tube. Cork the test tube immediately after adding the aluminum chloride. Set the beaker toward the back of the hood and add the aluminum chloride in four or five portions, stirring the reaction mixture after each addition. The reaction is extremely vigorous, and the mixture becomes dark brown. Allow the mixture to stand for 15 min, then add 15 mL of methanol and stir to mix. Heat the mixture to boiling on a steam bath while continuing to stir, and then set the beaker aside until it cools to room temperature. Decant the methanol and allow the polymer to dry a few minutes in the hood. Describe the properties of the product and compare them to the properties of polystyrene prepared by benzoyl peroxide catalysis.

IR Analysis of Polystyrene

Weigh a sample of polymer from each procedure that is approximately 50 mg. Place each sample in a separate 13 × 100 mm test tube and add 1 mL of 2-butanone (methyl ethyl ketone). Stir until the polymer dissolves (this may require several minutes).

A thin film (~0.01–0.03 mm) of polystyrene can be cast directly onto a sodium chloride IR plate. Working in a hood, place 1 drop of the polymer solution in the center of the salt plate and allow the solvent to evaporate. Repeat this process five or six times until a uniform film is built up on the plate. Mount the salt plate (do not use a second plate on top of the film) in the spectrometer holder and obtain the infrared spectrum. Compare the spectra of the two polymerization products. Compare these spectra to the spectrum of polystyrene film used to calibrate the IR spectrometer shown in Figure 29.1.

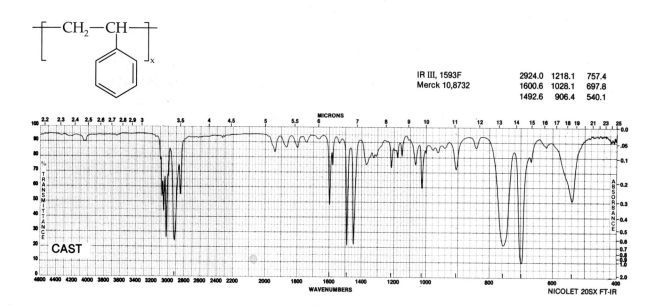

FIGURE 29.1 IR spectrum of polystyrene film used to calibrate IR spectrometers. (Provided by Aldrich Chemical Co., Inc., Milwaukee, WI.)

Cleanup: The polystyrene film can be removed by rinsing the salt plate with dichloromethane. Pour the residual dichloromethane into the container for halogenated waste. Pour any remaining polystyrene solutions into the container for flammable (organic) waste.

Optional Experiment

Bring to the laboratory commercial products made of polystyrene, such as transparent food containers, hot drink cups, packing "bubbles," or any other product that is labeled polystyrene. For each product, prepare a thin film casting as previously described and obtain the IR spectrum. Compare the spectra and describe any similarities or differences.

Use 2-butanone to prepare the casting.

The "window" in mailing envelopes is frequently made from polystyrene. Mount a piece of the window over a hole in a piece of cardboard that is sized to fit in the spectrometer sample holder. Obtain the IR spectrum.

Reference

1. Selinger, B. *Chemistry in the Market Place*, 4th ed.; Harcourt Brace: Sydney, 1994.

Questions

1. What would happen if the benzoyl peroxide is omitted in the first experiment?
2. What could happen if the inhibitor is not removed from the styrene in the first experiment?
3. Why must the free radical polymerization be heated but not the ionic polymerization?
4. If *p*-divinylbenzene is used rather than styrene in the first experiment, a cross-linked polymer would occur. Show why this would be the case with structures and a brief mechanism.

p-Divinylbenzene

29.2
Biodegradable Starch Films

Change the properties of starch by chemical modification of a biopolymer.

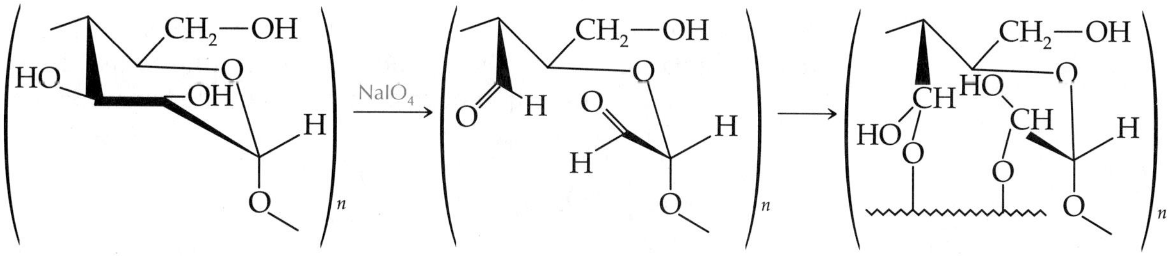

Starch   Starch dialdehyde   Cross-linked starch

Starch is a polymer formed metabolically in plants by the condensation polymerization of glucose, accompanied by the elimination of water. In the starch polymer the glucose units commonly are held together by covalent carbon-oxygen bonds to C-4 and to C-1, the anomeric carbon of the glucose unit. Human beings can metabolize starch because we have the enzymes, called amylases, that can efficiently catalyze the hydrolysis of α-linkages at C-1, which bond the glucose units together.

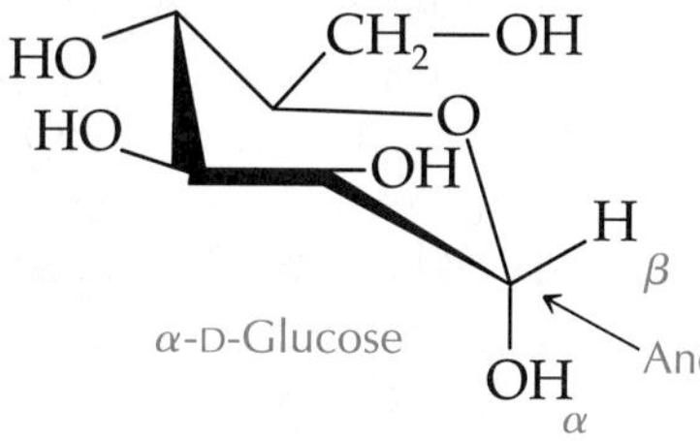

α-D-Glucose

Anomeric carbon (C-1)

A vicinal diol:

OH OH
| |
—C—C—
| |

(1,2-diol)

Starch is an inexpensive substance and biodegradable as well. Beyond its food value, starch is used in adhesives and in the manufacture of textiles and paper. For certain applications, such as starch films, however, greater strength in the starch polymer is desirable, and this extra strength can be achieved by cross-linking glucose units in the polymer. Cross-linking is accomplished by oxidation of the starch by periodate (IO_4^-), breaking apart C—C bonds where each carbon is bonded to an —OH group. Periodate oxidizes these vicinal or 1,2-diols to dialdehydes, which readily undergo addition reactions with —OH groups on adjacent strands of starch to form hemiacetals. When this occurs, a stronger cross-linked polymer is obtained:

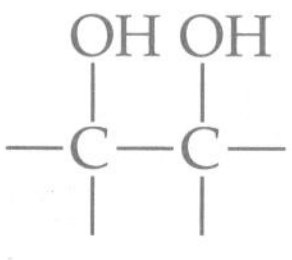

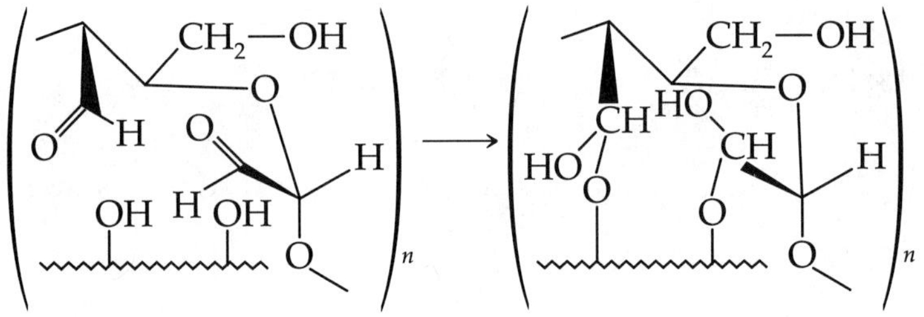

The oxidation of vicinal —OH groups occurs easily because of the ready formation of a cyclic iodine-containing ring when the two —OH groups are on adjacent carbon atoms of the substrate molecule:

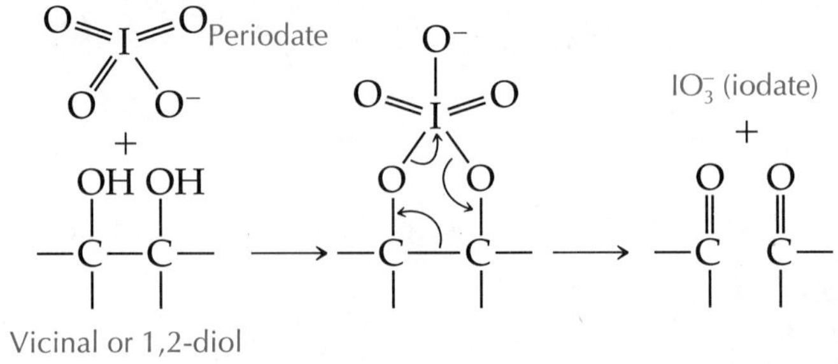

\\\\\\Macroscale Procedure

SAFETY INFORMATION

Sodium periodate is a strong oxidant, an irritant, and combustible. Wear gloves and avoid contact with skin, eyes, and clothing. Rinse the weighing paper with water before placing the paper in the wastebasket.

The **dialdehyde starch** prepared by the oxidation of potato or corn starch may be harmful; avoid contact with skin, eyes, and clothing.

The other compounds used in this procedure present no particular hazards.

Preparation of a Starch Film

Begin heating a sand bath or an oil bath to 105–110°C on a hot-plate-stirrer unit. While the bath is warming, place 1.25 g of corn starch or potato starch in a 50-mL Erlenmeyer flask and add 1.0 mL of 50/50 (v/v) glycerol/water solution, 1.5 mL of 0.1 M hydrochloric acid solution and 13 mL of distilled water. If you would like to make a colored film, add 0.5 mL of an aqueous food color solution. Put a magnetic stirring bar in the flask. Heat and stir the contents of the flask for 15 min at 105–110°C. The mixture will become relatively homogeneous (but not clear) during this time.

Remove the flask from the heating bath and add 1.5 mL of 0.1 M sodium hydroxide to the hot mixture. Stir the mixture briefly and pour the clear solution on a 15 × 15 cm acrylic plastic plate. Place the plastic plate in an oven at 105–110°C for about 90 min or until the film is no longer tacky.

Oxidation of Starch to Dialdehyde Starch

Combine 2.0 g of corn starch with 10 mL of distilled water (or 2.0 g of potato starch with 12.5 mL of water) in a 125-mL Erlenmeyer flask. Put a magnetic stirring bar in the flask. Prepare a solution of 2.5 g of sodium periodate and 35 mL of water in a small beaker. Pour the solution into a separatory funnel that is suspended above the flask containing the starch. Add the sodium periodate solution dropwise to the stirred starch solution over a period of 30 min. Continue stirring the reaction mixture at room temperature for 24 h.

This synthesis requires a 24-h stirring period before the dialdehyde starch film can be prepared.

Collect the dialdehyde starch on a Buchner funnel by vacuum filtration. Rinse the product three times with 10-mL portions of distilled water, drawing off each rinse with suction before adding the next. The wet dialdehyde starch, containing about 75% water, can be used immediately to prepare a starch film.

Preparation of Dialdehyde Starch Film

Follow the preceding procedure for the preparation of a starch film, substituting 4.0 g of wet dialdehyde starch for the 125 g of dry corn starch or potato starch. The quantities of the other reagents remain the same.

Properties of the Starch Films

Compare the properties of your dialdehyde starch film to your corn starch or potato starch film. Has the cross-linking changed the properties? Place a piece of each film in separate beakers of water. What happens to each film?

Cleanup: The filtrate from the oxidation procedure contains sodium iodate and possibly some excess sodium periodate. Add a 10% solution of sodium bisulfite in approximately 1-mL increments with stirring until the black precipitate that initially formed dissolves and a colorless to pale yellow solution forms. Wash this solution down the sink with additional water or pour it into the container for aqueous inorganic waste.

Reference

1. Sommerfeld, H.; Blume, R. *J. Chem. Educ.* **1992,** *69,* A151–A152.

Questions

1. Compare the properties of starch film to those of the cross-linked aldehydic starch film. Do they have the same degree of brittleness? Rigidity? How do other properties compare? What kind(s) of functional groups are introduced as the result of cross-links?

2. Write a balanced equation for the reaction by which the dialdehydic starch is produced upon periodate treatment of starch (the reduction product is iodate, IO_3^-).

3. Cellulose has a structure that is very similar to that of starch, yet cellulose is not digestible by human beings whereas starch is. Conversely, cows chew their cuds of grass (cellulose) and obtain nutrition. It is known that all glucose linkages in starch are α-links, whereas those in cellulose are β-links. The α-links in starch are vulnerable to enzymes found in humans. Cows have bacteria in their system that break down the cellulose β-links to produce glucose. Draw a simple structure for cellulose and compare it to the corresponding structure for starch.

4. Periodate treatment of *cis*-1,2-cyclopentanediol is expected to proceed more rapidly than the same reaction for the corresponding *trans*-isomer. Explain, using a simple mechanism.

5. Using the planar representations of cyclohexane-1,2-diols, it might be expected that the *cis*-form would be cleaved by periodate to a dialdehyde much more rapidly than would the *trans*-form. If, however, chair cyclohexanes are used, it might be expected that there would be little difference in reactivity between the two isomers. Explain, using mechanisms containing stereochemical details.

Projects

Part 2 contains 13 experimental projects designed to cover levels from the introductory to the advanced. We have found that multiweek projects provide an experimental context that is often missing in single-week experiments. Projects allow time to design experimental approaches or probe an area in ways that increase motivation and lead to active learning.

Many of the projects are multistep syntheses in which the product of one step becomes the starting reagent for the next step. Other projects are discovery based, that is, you are asked to answer a question about the chemical system under study, using your experimental data. Four projects (8, 10, 11, and 13) are open-ended, offering opportunities to design your own experiments within the framework of investigating a particular type of reaction. We have made an effort to provide projects that are interesting and enjoyable, and reflect how the science of organic chemistry is actually done.

HYDROLYSIS OF AN UNKNOWN ESTER

Project 1 concerns the hydrolysis of an unknown ester, a reaction that yields an alcohol and a carboxylic acid as the products. In carrying out this reaction, you will learn many of the techniques that chemists use in the synthesis, purification, and characterization of organic compounds. These techniques include reflux for the reaction itself, then distillation, extraction, and recrystallization to separate and purify the products. Finally, you will identify both products—the alcohol by its boiling point and refractive index, and the carboxylic acid by its melting point.

$$R-C{\overset{\displaystyle O}{\underset{OR'}{\big\langle}}} + H_2O \rightleftharpoons R-C{\overset{\displaystyle O}{\underset{OH}{\big\langle}}} + R'-OH$$

Ester   Carboxylic acid   Alcohol

Esters undergo a fundamental organic reaction that involves the splitting apart of the molecule by the action of water, a process called *hydrolysis.* Hydrolysis of an ester yields a molecule of an alcohol and a molecule of a carboxylic acid for each molecule of ester that is hydrolyzed, as shown in the above equation.

The sodium hydroxide is not a true catalyst here because it is irreversibly consumed in this reaction.

Hydrolysis can be promoted by either acid or base. If the hydrolysis is base promoted, the reaction is called *saponification* (literally, "soap making"). Instead of a carboxylic acid, saponification yields a carboxylate anion, together with an alcohol. The formation of the carboxylate anion causes the reaction to be irreversible:

$$R-C{\overset{\displaystyle O}{\underset{OR'}{\big\langle}}} \xrightarrow[H_2O]{OH^-} R-C{\overset{\displaystyle O}{\underset{O^-}{\big\langle}}} + R'-OH$$

Carboxylate anion

Although hydrolysis of an ester involves a functional group that you will study much later in organic chemistry, you can think of the reaction in a simple way and not be too far afield. Reactions of two chemical species are often dominated by charge complementarity, that is, plus attracts minus and vice versa. The ester and the hydroxide ion are the two reagents in the key step of ester

Reaction mechanisms for both acid- and base-promoted hydrolysis have been well studied and are discussed in detail in organic chemistry texts.

hydrolysis, and base-promoted hydrolysis requires a molecule of base for each ester functional group. The negative hydroxide ion reacts at the carbonyl carbon atom, the positive site of the ester. The result is a displacement of the alkoxide anion, RO^-. The alkoxide anion, a strong base, then forms the alcohol product by removal of the acidic proton from the carboxylic acid to yield the corresponding carboxylate anion:

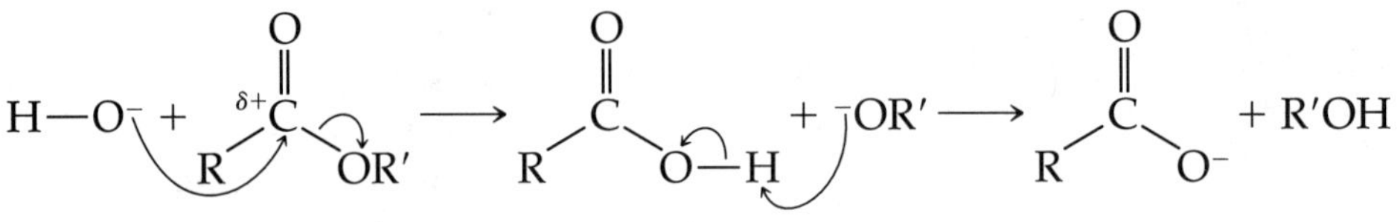

Subsequent addition of acid (hydrochloric acid in this experiment) converts the carboxylate anion to the carboxylic acid:

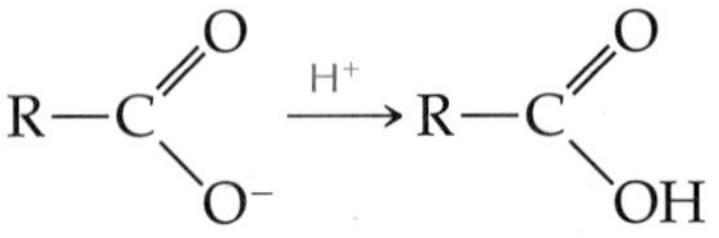

Soap is an example of a carboxylate salt formed by a saponification reaction. Soap consists of the sodium or potassium salts of long-chain fatty acids (12–18 carbons), produced by the saponification of esters called triglycerides that occur in fats and oils:

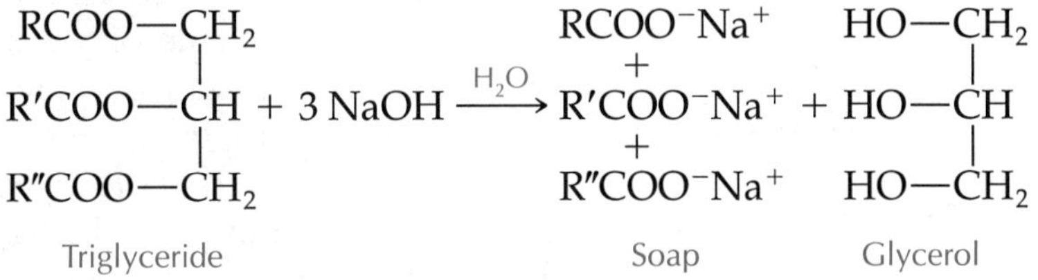

| Triglyceride | Soap | Glycerol |

Soap molecules have both polar (carboxylate ion) and nonpolar (long-chain R group) portions, and these properties relate directly to their use as cleaning agents. Sodium palmitate, $CH_3(CH_2)_{14}COO^-Na^+$, is an example of a carboxylate salt found in soap:

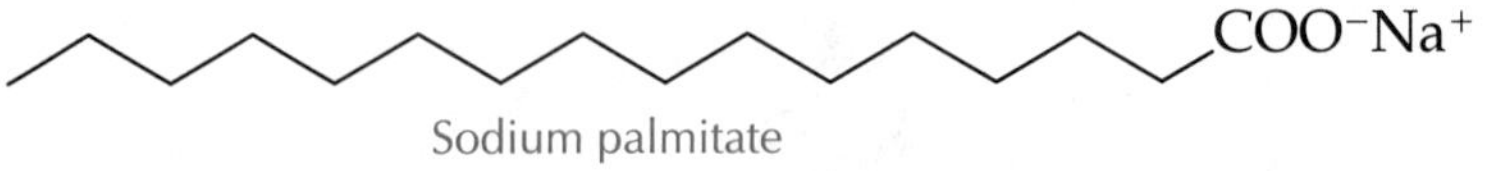

Sodium palmitate

Soap action occurs as a result of micelle formation by the aggregation of the hydrophobic (water-fearing; or oleophilic = fat-loving) long-chain R groups of the carboxylate salt (soap). The aggregation of the hydrophobic R groups around the oil drop produces an outer surface of the micelle that is hydrophilic (water-loving; or

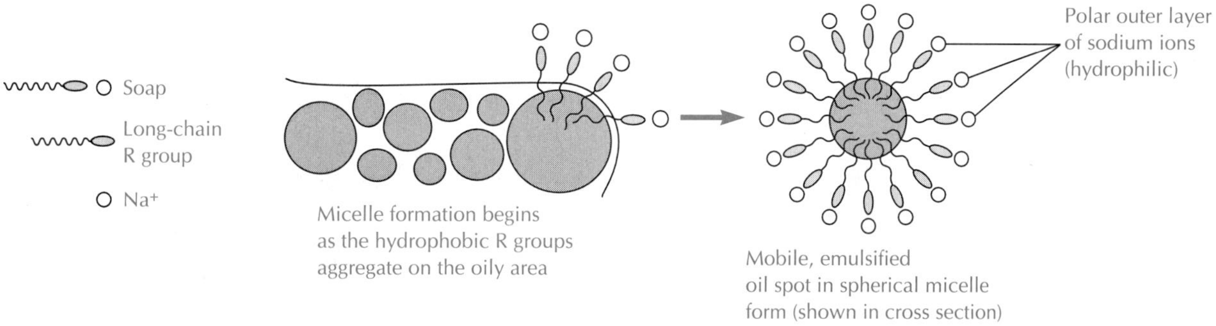

Soap

Long-chain
R group

○ Na⁺

Polar outer layer
of sodium ions
(hydrophilic)

Micelle formation begins
as the hydrophobic R groups
aggregate on the oily area

Mobile, emulsified
oil spot in spherical micelle
form (shown in cross section)

FIGURE 1.1 Soap action by micelle formation.

oleophobic) and allows the micelle to be dispersed in water as an emulsion (Figure 1.1). Micelle formation effectively provides an avenue of mobility for the oily dirt that allows it to be carried off the soiled surface and dispersed in water by mechanical action, such as the agitation in a washing machine.

In this project you will use a hydrolysis or saponification reaction to convert an unknown ester to an alcohol and the conjugate base of a carboxylic acid. The organic solvent, ethylene glycol (antifreeze), is useful because in a mixture with water it will dissolve both the organic compound (the ester) and sodium hydroxide, a polar inorganic compound. The alcohol product is much more volatile than the ionic carboxylate product; therefore, it can be removed from the reaction mixture by distillation; here azeotropic distillation [see Technique 7.5] is used.

This experiment introduces a number of synthetic organic techniques, including a reflux procedure, simple distillation, extraction (with diethyl ether), azeotropic distillation, and recrystallization. You will use refractometry and boiling points to identify the alcohol and melting points to identify the carboxylic acid formed.

1.1
Hydrolysis and Azeotropic Distillation

First laboratory period

As a high-boiling-point solvent, ethylene glycol increases the boiling point of the reaction mixture and consequently reduces the reaction time.

In the first laboratory period you will hydrolyze the unknown ester to form an alcohol and a carboxylate salt by heating the ester in a solution of sodium hydroxide, water, and ethylene glycol. After the heating period, the reaction flask contains the sodium salt of the unknown acid, the unknown alcohol, excess sodium hydroxide, water, ethylene glycol, and a small amount of unhydrolyzed ester. Then you will use an azeotropic distillation to separate the unknown alcohol from the reaction mixture. Figure 1.2 shows the steps for the first week of the project.

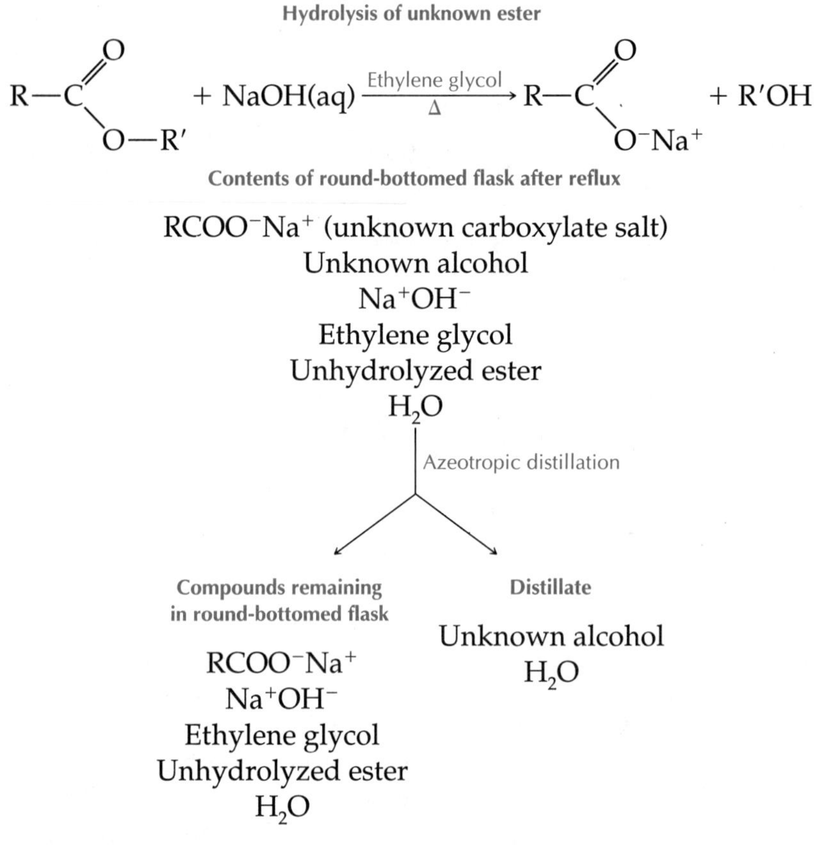

Hydrolysis of unknown ester

$$R-C{\overset{\displaystyle O}{\underset{\displaystyle O-R'}{\big|}}} + NaOH(aq) \xrightarrow[\Delta]{\text{Ethylene glycol}} R-C{\overset{\displaystyle O}{\underset{\displaystyle O^-Na^+}{\big|}}} + R'OH$$

Contents of round-bottomed flask after reflux

RCOO⁻Na⁺ (unknown carboxylate salt)
Unknown alcohol
Na⁺OH⁻
Ethylene glycol
Unhydrolyzed ester
H_2O

Azeotropic distillation

Compounds remaining
in round-bottomed flask

RCOO⁻Na⁺
Na⁺OH⁻
Ethylene glycol
Unhydrolyzed ester
H_2O

Distillate

Unknown alcohol
H_2O

FIGURE 1.2 Flow chart for hydrolysis and azeotropic distillation.

〰〰Macroscale Procedure

Techniques   Boiling Stones: Technique 3.1
Reflux: Technique 3.3
Azeotropic Distillation: Technique 7.5
Simple Distillation Apparatus: Technique 7.3

Solid sodium hydroxide is hygroscopic and rapidly absorbs water from the atmosphere; keep the reagent bottle tightly closed.

─── *SAFETY INFORMATION* ───

Sodium hydroxide solutions are corrosive and cause severe burns. The solution prepared in this experiment is very concentrated. Wear gloves while weighing and transferring NaOH pellets. Avoid contact with skin, eyes, and clothing. Notify the instructor if any solid is spilled.

Ethylene glycol and the **unknown ester** may cause skin irritation. Avoid contact with skin, eyes, and clothing. Wash hands thoroughly after handling.

Hydrolysis

Clamp a 100-mL round-bottomed flask in an upright position. Place 7.0 g of sodium hydroxide pellets in the flask and add, *in order*, 2 mL of water, 20 mL of ethylene glycol, 15 mL of the unknown ester, and two boiling stones. Fit a water-cooled condenser above the flask [see Technique 3.3] and position a heating mantle under the flask. Connect the heating mantle to the variable transformer and plug the transformer into the electrical outlet [see Figure 3.1, Technique 3.2]. Have the instructor check your apparatus before you begin heating. After the reaction mixture begins to boil, continue heating at reflux for 30 min. Adjust the transformer, if necessary, to maintain gentle boiling.

A cake of white solid forms (see Question 2) during the reflux period and the upper layer of liquid ester should disappear.

Azeotropic Distillation

At the end of the reflux period, lower the heating mantle from the flask to stop the boiling. Allow the flask to cool for 10 min and add 35 mL of water through the top of the condenser. Add two more boiling stones. Set up a simple distillation apparatus [see Technique 7.3], using a 50-mL Erlenmeyer flask as the receiving flask. Keep a slow stream of cold water running through the condenser. Fit the heating mantle under the flask. Have the instructor check your apparatus before you begin heating. The unknown alcohol forms an azeotrope with water that boils below 100°C [see Technique 7.5]; when the azeotrope has completely distilled, the boiling point will rise to that of water. Adjust the rate of heating so that the distillate drops from the condenser at a rapid rate of 20–30 drops per minute. Record the temperature of the vapor when distillate begins to collect in the receiving flask. Continue the distillation until the temperature reaches 100°C or until about 20 mL of distillate are collected. Stop the heating process by turning off the transformer and lowering the heating mantle from the flask.

Water saturated with sodium chloride has a lower solubility in the unknown alcohol than water without dissolved salt.

Add 4.0 g of sodium chloride to the flask containing the distillate that you collected. Cork the flask and swirl it to dissolve most of the NaCl. Label the flask and store it in your drawer until the next laboratory period. Allow the mixture remaining in the round-bottomed flask to cool. Then label the flask, cork it tightly, and place it in a beaker so that it will not tip over. Store the round-bottomed flask in your drawer until the next laboratory period.

1.2

Extraction and Recrystallization

Second laboratory period

The mixture remaining in the round-bottomed flask contains an aqueous solution of the sodium salt of the carboxylic acid, excess sodium hydroxide, ethylene glycol, and a small amount of unhydrolyzed ester. In this procedure you first remove the unhydrolyzed ester by an extraction with diethyl ether. All the other compounds remain dissolved in the aqueous phase. The addition of hydrochloric acid to this aqueous solution converts the carboxylate salt to the carboxylic acid, which is insoluble and precipitates. The hydrochloric acid also neutralizes the excess sodium hydroxide. The ethylene glycol remains dissolved in the aqueous solution. The precipitated unknown acid is recovered by vacuum filtration. Figure 1.3 illustrates the steps in the recovery of the unknown acid.

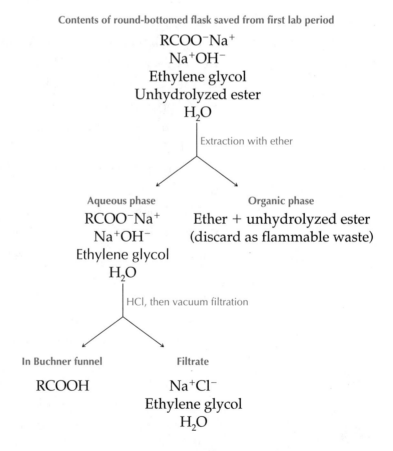

Contents of round-bottomed flask saved from first lab period

$RCOO^-Na^+$
Na^+OH^-
Ethylene glycol
Unhydrolyzed ester
H_2O

Extraction with ether

Aqueous phase
$RCOO^-Na^+$
Na^+OH^-
Ethylene glycol
H_2O

Organic phase
Ether + unhydrolyzed ester
(discard as flammable waste)

HCl, then vacuum filtration

In Buchner funnel
$RCOOH$

Filtrate
Na^+Cl^-
Ethylene glycol
H_2O

FIGURE 1.3 Flow chart for recovery of the unknown acid, RCOOH.

Macroscale Procedure

Techniques   Extraction: Technique 4.2
Drying Agents: Technique 4.6 and 4.7
Recrystallization: Technique 5.3

SAFETY INFORMATION

Diethyl ether is extremely volatile and flammable. Be certain that there are no flames in the laboratory. Also be sure that there are no hot electrical devices in the vicinity while you are pouring ether or performing the extractions. Diethyl ether should be used in a hood, if possible.

Recovery of the Carboxylic Acid

If the mixture in the round-bottomed flask contains any solid, add 10–20 mL of water and stir to dissolve the solid before transferring the mixture to the separatory funnel. Working in a hood, if possible, pour the aqueous mixture from the round-bottomed flask into a 125-mL separatory funnel; use a funnel in the neck of the separatory funnel to prevent spills. Rinse the flask with 5–10 mL of water and also add this to the separatory funnel. Then add 10 mL of ether to the separatory funnel and place the stopper in it. Alternately shake gently to mix the layers and vent the funnel [see Technique 4.2]. Allow the two phases to separate and drain the aqueous phase into a clean 125-mL Erlenmeyer flask.

What is the density of ether? Will the ether layer be the upper or the lower one in the extraction?

Add 30 mL of 6 M hydrochloric acid to the aqueous solution in the Erlenmeyer flask and stir. This amount should be enough acid to neutralize both the excess sodium hydroxide and the sodium salt of the acid, giving a copius white precipitate. Check the pH, using pH test paper such as pHydrion. Add an additional 1 or 2 mL of acid if the pH is greater than 1. Cool the mixture in an ice-water bath before collecting the crystals by vacuum filtration. [See the filtration apparatus in Figure 5.4, Technique 5.3.]

To test the pH, dip the tip of a stirring rod in the mixture, then touch the tip to the pH test paper.

Cleanup: Discard the ether solution remaining in the separatory funnel in the container for flammable (organic) waste. The aqueous solution in the filter flask should be neutralized with solid sodium carbonate **(Caution: Foaming.)** before washing it down the sink or pouring it into the container for aqueous inorganic waste.

Recrystallization

Using a hot plate, heat about 200 mL of water to boiling in an Erlenmeyer flask [see Technique 5.3]. Place the solid in a 250-mL Erlenmeyer flask, add a boiling stick or boiling stone, and carefully pour about 75 mL of the hot water over the solid. Bring the mixture to a boil. Continue adding hot water in 10- to 15-mL increments (estimate this volume; it does not need to be known exactly), allowing the mixture to boil after each addition, until the solid and the oily layer in the bottom of the flask just dissolve. When all of the solid and oil has dissolved, add an excess of about 20–25 mL of hot water. Estimate the total volume of water used to dissolve the solid.

Dissolution is not an instantaneous process, so let the solution boil a minute or two before concluding that more water is needed.

Set the flask on the desk top to cool slowly; cooling its contents to room temperature requires at least 20 min. Proceed with the extraction of the alcohol while you are waiting for the acid solution to cool.

When the flask has cooled to room temperature and crystals have formed throughout the solution, cool the flask in an ice-water bath for at least 10 min. Collect the crystals by vacuum filtration and place them on a watch glass to dry in your drawer until the next laboratory period.

Cleanup: The filtrate remaining in the filter flask can be washed down the sink or poured into the container for aqueous inorganic waste.

Extraction and Drying of the Alcohol

Sodium chloride, added to the alcohol/water mixture in the previous laboratory period, dissolves in the water and reduces its solubility in the unknown alcohol.

The distillate saved from the last laboratory period contains two liquid phases, the unknown alcohol and water. Each compound has a slight solubility in the other. In the following procedure, the addition of diethyl ether extracts the alcohol from the water layer, thereby separating one from the other. The solution of ether and unknown alcohol that results from the extractions is saturated with water, because water has a slight solubility in ether. This solution is dried by allowing it to stand over an anhydrous salt, sodium sulfate in this instance [see Technique 4.6]. Figure 1.4 shows the steps involved in the separation.

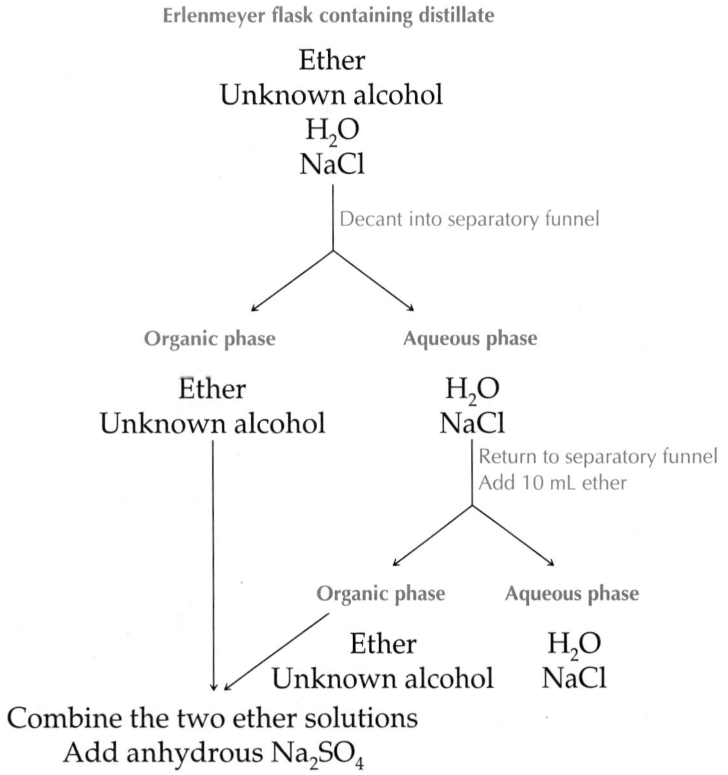

Erlenmeyer flask containing distillate

Ether
Unknown alcohol
H_2O
NaCl

Decant into separatory funnel

Organic phase   Aqueous phase

Ether H_2O
Unknown alcohol   NaCl

Return to separatory funnel
Add 10 mL ether

Organic phase   Aqueous phase

Ether H_2O
Unknown alcohol   NaCl

Combine the two ether solutions
Add anhydrous Na_2SO_4

FIGURE 1.4 Flow chart for extraction and drying of the unknown alcohol, ROH.

Macroscale Procedure

SAFETY INFORMATION

Ether vapors pose a fire hazard. The hot plate used for the recrystallization may still be very hot; remove it from your work area before beginning the extraction.

Working in a hood, if possible, add 15 mL of diethyl ether to the distillate of unknown alcohol and water saved from the previous laboratory period. Cork the flask again and swirl it. Carefully decant the ether/water mixture into a separatory funnel so that any undissolved NaCl remains in the flask. Stopper the separatory funnel. Gently shake it to mix the layers and allow the phases to

When performing extractions, it is prudent to save all solutions until completing the procedure. Most organic and aqueous solutions are colorless and it is easy to become confused about which flask contains the organic phase and which the aqueous. Labeling all flasks is essential.

The amount of drying agent required depends on how much water is in your particular ether solution. You need enough drying agent to remove the water, but you want to avoid an excess of anhydrous salt because some of your product may be adsorbed on the salt crystals and cause a lowered yield.

separate [see Technique 4.2]. Drain the aqueous layer into a clean, labeled flask and pour the ether layer out of the top of the funnel into another clean, labeled flask. Return the aqueous phase to the separatory funnel and extract it with another 10-mL portion of ether. Separate the layers, drain the aqueous layer into the same flask that you previously used for it, then pour the ether phase into the flask that contains the first portion of ether. Set the flask containing the aqueous solution aside.

Dry the ether solution with anhydrous sodium sulfate. Add sodium sulfate in small scoops *one* at a time, swirling the flask between additions. When some of the sodium sulfate moves freely, that is, when it is not clumped together or stuck to the glass, you have added enough [see Technique 4.6]. Cork the flask tightly and store the solution until the next laboratory period. Consult your instructor about whether to store the flask in your drawer or in some other designated place in the laboratory, because ether fumes may be emitted.

Cleanup: The aqueous solution remaining from the extraction may be poured down the sink or poured into the container for aqueous inorganic waste.

1.3
Distillation, Melting Points, and Identification

Third laboratory period   **Techniques**   Simple Distillation: Technique 7.3
Melting Points: Technique 6
Refractive Index: Technique 9

Simple Distillation of the Alcohol

— SAFETY INFORMATION —
Ether vapors pose a fire hazard. Perform the evaporation on a steam bath in a well-ventilated hood.

Place a small ball of cotton in the bottom of a small funnel and set the funnel into a 50-mL round-bottomed flask that is clamped firmly. Decant the ether/alcohol solution into the funnel, rinse the sodium sulfate remaining in the Erlenmeyer flask with 5 mL of

Alternatively, the ether may be removed with a rotary evaporator [see Technique 4.8]

ether, and add this to the funnel. Put a boiling stone in the round-bottomed flask and evaporate the solution on a steam bath *in the hood* until the volume is reduced by approximately two-thirds. Then set up a simple distillation apparatus [see Technique 7.3, Figure 7.7], adding another boiling stone and using a 25-mL round-bottomed flask cooled in an ice-water bath for the receiver. Collect a first fraction until the boiling point reaches 100°–110°C; this fraction is mainly residual ether. Change the receiver to a tared (weighed) 50-mL Erlenmeyer flask or wide-mouth vial. (This receiver does not need to be cooled in an ice-water bath.) Collect the unknown alcohol fraction. The temperature will stabilize while most of the alcohol distills—this temperature is the boiling point of the unknown alcohol.

Stop the distillation just before the boiling flask reaches dryness. Determine the weight of the alcohol and its refractive index [see Technique 9.3].

— **SAFETY PRECAUTION** ———————

Be certain to stop the distillation before the boiling flask reaches dryness. Diethyl ether may form peroxides. If the flask goes to dryness, the temperature inside it rises rapidly, and the peroxides could explode.

Cleanup: Pour the first fraction (boiling under 110°C) and residue remaining in the boiling flask into the container for flammable (organic) waste. Place the sodium sulfate in the container for nonhazardous waste or inorganic waste.

Identification of the Products

If you are doing this experiment in a laboratory at an altitude much above sea level, the boiling point of the unknown alcohol could be a few degrees lower than the boiling points listed in Table 1.1.

Compare your experimental data for the unknown alcohol with the list of boiling points and refractive indices in Table 1.1. Select the compound whose properties most closely fit your data. The refractive index of diethyl ether is 1.3555 at 15°C and that of water is 1.3330 at 20°C; if either remains in your product, its presence will lower the observed refractive index of the alcohol. Refractive index is a temperature-dependent property. The values given in Table 1.1 are for 20°C. Refer to Technique 9.4 for calculating the

Table 1.1 Possible compounds for the unknown alcohol

Compound	Boiling point, °C	n_D^{20}	MW
1-Propanol	97.4	1.3850	60
3-Methyl-2-butanol	112.9	1.4089	88
1-Butanol	117.6	1.3993	74
3-Methyl-3-pentanol	122.4	1.4186	102

temperature correction if you measured the refractive index at a temperature other than 20°C.

Weigh the recrystallized unknown acid. Determine the melting-point range; ideally this range will be 2° or less [see Technique 6.3]. Allow the melting-point apparatus to cool at least 20°, prepare a new sample, and do a second determination of the melting point to verify your first finding. The two determinations should be within 1° of each other; if not, continue making determinations until you do have two melting-point ranges that agree. Consult Table 1.2 and select the carboxylic acid with the melting point that most closely matches your experimental melting point.

When you have completed the identification of your products, submit them to your instructor as directed.

Table 1.2 Possible compounds for the unknown carboxylic acid

Compound	Melting point, °C	MW
2-Methylbenzoic acid (2-toluic acid)	104–105	136
3-Methylbenzoic acid (3-toluic acid)	111	136
α-Hydroxyphenylacetic acid	118	152
Benzoic acid	121–122	122
2-Benzoylbenzoic acid	127–129	226

Questions

1. Explain why a reaction mixture can be heated under reflux for an extended period of time without losing the solvent.
2. What is the white precipitate that forms while the reaction mixture is heated?
3. Why did the observed boiling point rise to 100°C during the azeotropic distillation?
4. If you were doing this experiment in, say, Denver, Colorado, the observed boiling point would never reach 100°C during the azeotropic distillation. Why not?
5. Explain why the unknown alcohol cannot be separated from water by simple distillation.
6. Discuss the principles of recrystallization and how the process purifies a solid product.
7. Show, by calculations, why 30 mL of 6 M HCl is enough to neutralize both the excess sodium hydroxide and the sodium salt of the carboxylic acid.
8. Suppose the solubility of your unknown acid is 6.8 g/100 mL water at 95°C and 0.17 g/100 mL water at 0°C. Calculate the amount of acid that would remain in your recrystallization solution at 0°C on the basis of the volume of water that you used for the recrystallization.
9. The solubility of an unknown acid is 67 g/100 mL in boiling ethanol and 43 g/100 mL in cold ethanol. Would this be a good recrystallization solvent? Explain.
10. Draw the structure of your unknown ester on the basis of your identification of the carboxylic acid and the alcohol. Give its molecular weight. Write the balanced equation for the hydrolysis of this ester, using structures for the ester, the alcohol, and the carboxylic acid.
11. Look up the density of your now-identified ester in a handbook. Calculate the theoretical yields for both the acid and the alcohol that you identified as the hydrolysis products. (Refer to section I.7 for a discussion of theoretical yield.)
12. Calculate the percent yields for the acid and for the alcohol that you recovered.
13. Acid-catalyzed hydrolysis is an equilibrium situation and is subject to the laws of mass action. Calculate the concentrations of alcohol and carboxylic acid arising from hydrolysis of a 0.100 M solution of an ester whose hydrolysis equilibrium constant is 1.00.

SYNTHESIS OF 1-BROMOBUTANE AND A GRIGNARD SYNTHESIS OF AN ALCOHOL

This project is a sequential synthesis in which you will prepare and purify an alkyl halide, 1-bromobutane, convert it to a Grignard reagent by reaction with magnesium in ether solution, and then add the Grignard reagent to an aldehyde or ketone of your choice. The final product will be a secondary alcohol, if you use an aldehyde, or a tertiary alcohol, if you use a ketone. You can characterize your final product by boiling point, gas chromatographic analysis, and IR and NMR spectrometry.

Step 1. Synthesis of 1-bromobutane

$$CH_3CH_2CH_2CH_2OH \xrightarrow[\text{H}_2\text{SO}_4]{\text{NaBr}} CH_3CH_2CH_2CH_2Br$$

1-Butanol   1-Bromobutane

Step 2. Synthesis of Grignard reagent, 1-butylmagnesium bromide

$$CH_3CH_2CH_2CH_2Br + Mg \xrightarrow{\text{ether}} CH_3CH_2CH_2CH_2MgBr$$

1-Butylmagnesium bromide

Step 3. Addition of carbonyl compound to Grignard reagent

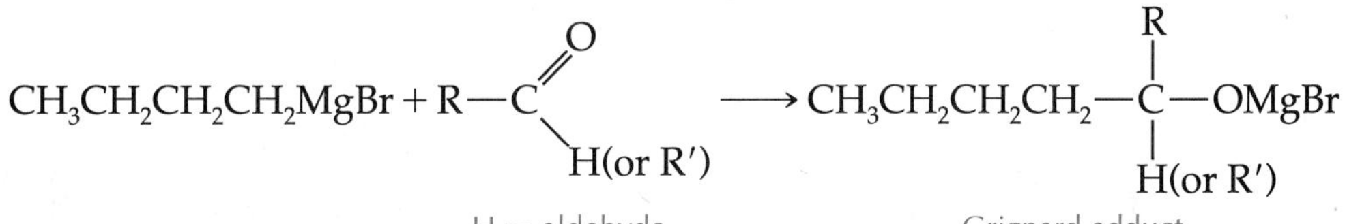

$$CH_3CH_2CH_2CH_2MgBr + R-\overset{\displaystyle O}{\underset{\displaystyle H(\text{or } R')}{C}} \longrightarrow CH_3CH_2CH_2CH_2-\overset{\displaystyle R}{\underset{\displaystyle H(\text{or } R')}{C}}-OMgBr$$

H = aldehyde
R′ = ketone

Grignard adduct

Step 4. Hydrolysis of Grignard adduct to form alcohol product

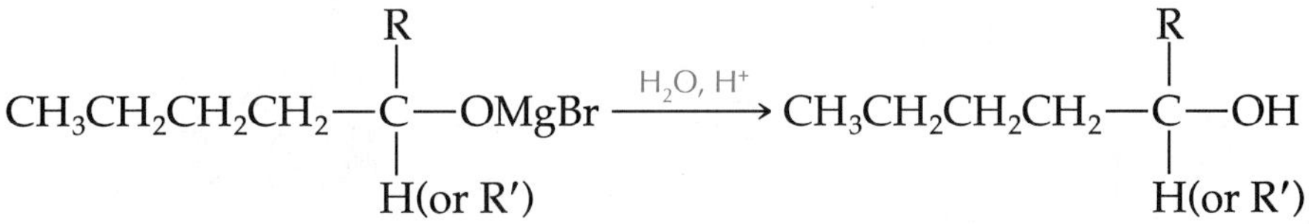

$$CH_3CH_2CH_2CH_2-\overset{\displaystyle R}{\underset{\displaystyle H(\text{or } R')}{C}}-OMgBr \xrightarrow{\text{H}_2\text{O, H}^+} CH_3CH_2CH_2CH_2-\overset{\displaystyle R}{\underset{\displaystyle H(\text{or } R')}{C}}-OH$$

H = 2° alcohol
R′ = 3° alcohol

These reactions are separate sequential procedures. You will work up and characterize the product of the first step, 1-bromobutane, before you use it to prepare the intermediate Grignard reagent, 1-butylmagnesium bromide. This intermediate is not isolated but is immediately combined with the aldehyde or ketone that you have selected to produce, after hydrolysis of the Grignard adduct, a secondary or tertiary alcohol, respectively.

2.1 Synthesis of 1-Bromobutane from 1-Butanol

$$CH_3CH_2CH_2CH_2OH + NaBr + H_2SO_4 \longrightarrow CH_3CH_2CH_2CH_2Br + H_2O + NaHSO_4$$

1-Butanol	1-Bromobutane
bp 117.2°C	bp 101.3°C
MW 74	MW 137
density 0.810 g·mL^{-1}	density 1.27 g·mL^{-1}

The conversion of alcohols to organic halides is an important first step in a variety of syntheses. In this synthesis you will use a mixture of a strong acid, sulfuric acid, and sodium bromide to convert 1-butanol to 1-bromobutane by an S_N2 reaction. Because the hydroxide anion is such a poor leaving group, it is necessary to protonate the alcohol to produce a better leaving group, H_2O. Bromide ion is sufficiently nucleophilic to displace the protonated hydroxyl group:

$$R-CH_2-\ddot{O}H + H^+ \longrightarrow R-CH_2-\overset{+}{\underset{\cdot\cdot}{O}}H_2 \xrightarrow{Br^-} [\overset{\delta-}{Br}---\overset{\overset{H\ \ H}{|}}{\underset{|}{C}}---\overset{\delta+}{\underset{\cdot\cdot}{O}}H_2]^{\ddagger} \longrightarrow R-CH_2-Br + H_2\ddot{\ddot{O}}$$

$$\underset{R}{}$$

S_N2 transition state

A variety of fates are possible for the protonated alcohol. In fact, it may seem surprising at first glance that you can obtain a reasonable yield of 1-bromobutane in this experiment. For example, a secondary carbocation could form by a 1,2-hydride shift. This carbocation could react with a bromide anion or elimination could occur to form a mixture of alkenes:

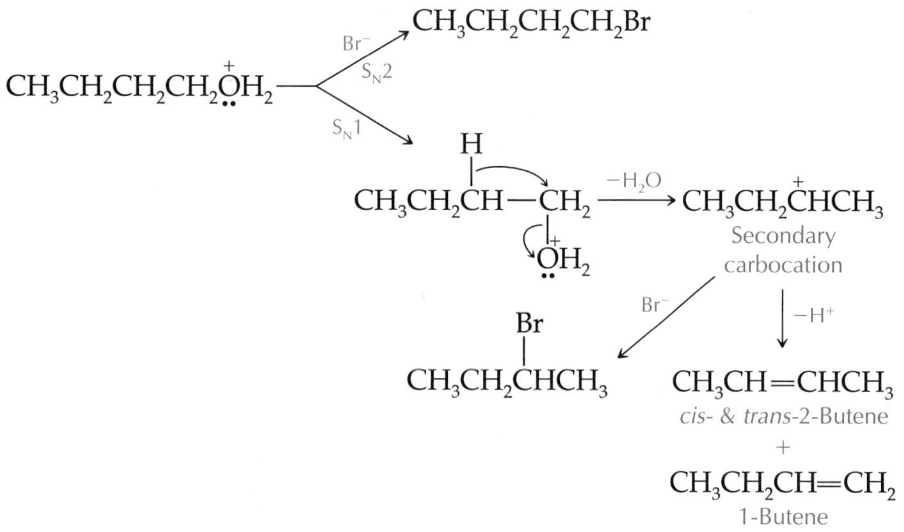

$$CH_3CH_2CH_2CH_2\overset{+}{\underset{\cdot\cdot}{O}}H_2$$

$$\xrightarrow[S_N2]{Br^-} CH_3CH_2CH_2CH_2Br$$

$$\xrightarrow{S_N1}$$

$$CH_3CH_2\overset{H}{\underset{\underset{\overset{|}{\underset{\cdot\cdot}{O}}H_2}{|}}{CH}}\!\!-\!\!CH_2 \xrightarrow{-H_2O} CH_3CH_2\overset{+}{C}HCH_3$$

Secondary carbocation

$$\overset{Br}{\underset{|}{CH_3CH_2CHCH_3}}$$

$$\xleftarrow{Br^-}$$

$$\downarrow -H^+$$

$$CH_3CH=CHCH_3$$
cis- & *trans-*2-Butene
+
$$CH_3CH_2CH=CH_2$$
1-Butene

There are three reasons why a good yield of 1-bromobutane can be obtained. The first is that the S_N2 reaction occurs at a primary carbon atom where steric hindrance is minimal. The second is that the bromide ion is a potent nucleophile. The third is that the secondary carbocation that would form by a hydride shift is not particularly stable and doesn't form very fast. In situations where steric hindrance is greater or rearrangement is more likely, other methods are used to synthesize the desired bromoalkane, usually using phosphorus tribromide.

〰️Macroscale
Procedure

Techniques Reflux: Technique 3.3
Steam Distillation: Technique 7.6
Extraction: Technique 4.2
Drying Agents: Techniques 4.6 and 4.7
Simple Distillation: Technique 7.3

― *SAFETY INFORMATION* ―

The **1-butanol** is flammable and an irritant to skin and eyes. Avoid contact with skin, eyes, and clothing.

The **1-bromobutane** synthesized is harmful if inhaled, ingested, or absorbed through the skin. Wear gloves when handling it and work in a hood, if possible.

(continued on next page)

Wear gloves while adding the sulfuric acid.

Pour 70 mL of 8.7 M (0.61 mol) sulfuric acid into a 125-mL Erlenmeyer flask and cool the flask in an ice-water bath. Using a conical funnel, pour 18.2 mL (0.20 mol) of 1-butanol into a 250-mL round-bottomed flask that is firmly clamped at the neck to a rack or ring stand. Using a powder funnel, add 40.8 g (0.40 mol) of sodium bromide to the flask. Add one or two boiling stones or use magnetic stirring. Attach a water-cooled condenser to the flask, as shown for a reflux apparatus [see Technique 3.3]. Place a conical funnel in the top of the condenser and pour the chilled sulfuric acid solution down the condenser in three portions. Gently swirl the flask between each addition.

Place a heating mantle under the flask and bring the mixture to a boil. Heat the reaction mixture under reflux for 45 min; the flask contents should be boiling vigorously so that the two phases mix well. Remove the heating mantle, let the flask cool for 5 min.

The addition of water allows a steam distillation to be carried out [see Technique 7.6].

Pour 60 mL of water down the condenser. Add another boiling stone and set up the apparatus for a simple distillation [see Technique 7.3]. Use a 125-mL Erlenmeyer as the receiver. Collect 40–45 mL of distillate, or collect distillate until it no longer contains water-insoluble droplets. When the distillation is complete, remove the heating mantle from the round-bottomed flask. During this steam distillation, place 32 mL of concentrated hydrochloric acid in a corked flask and chill the acid in an ice-water bath until you need it during the extraction procedure.

Test for completeness of the distillation by removing the receiving flask and collecting several drops of distillate in a test tube containing about 1 mL of water.

After you finish the distillation, wash the distillation head, thermometer adapter, condenser, and vacuum adapter. Place them in a drying oven for 10–15 min, and allow them to cool to room temperature before assembling the final distillation apparatus. Alternatively, the glassware can be dried by rinsing it with a small amount of acetone and placing the glassware in a hood until the acetone evaporates. **Caution: If you use acetone, do not put the glassware in an oven.**

SAFETY PRECAUTIONS

Wear gloves while performing the extractions and work in a hood to keep hydrochloric acid vapors out of the laboratory.

Foaming from CO_2 will occur in the extraction with sodium bicarbonate. Vent the funnel frequently to relieve the gas pressure.

Review the extraction procedure in Technique 4.2 before you begin this part of the procedure.

In these extractions, the organic phase will be the lower layer.

Transfer the distillate, which contains water, 1-bromobutane, and any 1-butanol that did not react, to a separatory funnel, using a conical funnel. Allow the phases to separate and drain the lower organic phase into an Erlenmeyer flask labeled organic phase. Pour the upper aqueous phase out of the top of the separatory funnel into a flask labeled aqueous phase. Return the organic phase to the separatory funnel. With each extraction, it is necessary to repeat this process of removing both layers from the funnel and returning the organic phase for the next extraction. The same flask can be used to hold the organic phase after each separation. Wash the organic phase with two 16-mL portions of chilled concentrated hydrochloric acid, then with 15 mL of cold water. Finally, wash the organic phase with 12 mL of saturated sodium bicarbonate solution. After the phases separate, drain the lower organic phase into a clean, dry 50-mL Erlenmeyer flask. Add anhydrous potassium carbonate and dry the product for at least 10 min [see Technique 4.6].

Any unreacted 1-butanol would be difficult to remove from your product by distillation. Using hydrochloric acid helps to extract it into the aqueous phase.

$$ROH + H_3O^+ \longrightarrow$$
$$R\overset{+}{O}H_2 + H_2O$$

Filter the dried 1-bromobutane through a dry conical funnel containing a small cotton plug into a clean, dry 50-mL round-bottomed flask [see Figure 4.14, Technique 4.7]. Add two boiling stones and assemble the apparatus for simple distillation. Collect the product fraction from 97°–102°C in a tared (weighed) 20-mL screw-capped vial. **Be sure to remove the heating mantle before the distillation flask reaches dryness.** Weigh your product and determine the percent yield. Obtain an NMR spectrum as directed by your instructor [see Spectrometric Method 2]. Alternatively, you can analyze your product by IR spectrometry or gas chromatography.

This extra treatment with drying agent ensures that no water will be present in your 1-bromobutane when you use it in the Grignard reaction.

After determining the mass of the product and its NMR spectrum, add just enough anhydrous potassium carbonate to cover the bottom of the vial. Cap the vial tightly and store it upright in a small beaker in your desk.

Cleanup: The residue remaining in the distillation flask following the first distillation contains sulfuric and hydrobromic acids. Carefully pour it into a large beaker containing 300 mL of water. Add the hydrochloric acid and other aqueous washes from the extractions to the beaker. Add sodium carbonate in small portions **(Caution: Foaming.)** until the acid is neutralized. The solution can then be washed down the sink or poured into the container for aqueous inorganic waste. The residue remaining in the distillation flask after the final distillation should be poured into the container for halogenated waste. The potassium carbonate used as the drying agent is saturated with 1-bromobutane; discard it in the container for hazardous solid waste.

Optional Syntheses of Alkyl Halides

Substitute one of the following alcohols for 1-butanol: 21.6 mL (0.20 mol) of 1-pentanol, 21.6 mL (0.20 mol) of 3-methyl-1-butanol, or 15.0 mL (0.20 mol) of 1-propanol. The rest of the procedure remains the same. You should locate the boiling point of your product in *The Merck Index, The Aldrich Catalog,* or the *CRC Handbook of Chemistry and Physics* as part of your prelaboratory preparation.

2.2
Grignard Synthesis of Secondary and Tertiary Alcohols

$$CH_3CH_2CH_2CH_2Br + Mg \xrightarrow{\text{ether}} CH_3CH_2CH_2CH_2MgBr$$

1-Bromobutane   1-Butylmagnesium bromide

$$R-C{\overset{O}{\underset{H(\text{or } R')}{\big\|}}}$$

H = aldehyde
R′ = ketone

$$CH_3CH_2CH_2CH_2-\overset{\displaystyle R}{\underset{\displaystyle H(\text{or } R')}{C}}-OMgBr \xrightarrow{\text{H}_2\text{O, H}^+} CH_3CH_2CH_2CH_2-\overset{\displaystyle R}{\underset{\displaystyle H(\text{or } R')}{C}}-OH$$

H = 2° alcohol
R′ = 3° alcohol

Organometallic compounds are versatile synthetic intermediates for producing carbon-carbon bonds. Among the most important organometallic reagents are the alkyl- and arylmagnesium halides, which are almost universally called Grignard reagents after the French chemist Victor Grignard, who first realized their tremendous potential in organic synthesis. The formation of Grignard reagents and their use in syntheses are discussed in detail in the introduction to Experiment 13; it will be worthwhile for you to read this material before undertaking this part of Project 2.

A synthesis involving a Grignard reagent has three procedural steps: the formation of the Grignard reagent, RMgX, from an alkyl or aryl halide, the addition of the Grignard reagent to a carbonyl compound (an aldehyde or ketone in this synthesis), and the hydrolysis of the Grignard adduct by an aqueous acid solution to yield the alcohol product. In this part of Project 2 you have a choice of the aldehyde or ketone that undergoes reaction with the Grignard reagent prepared from the 1-bromobutane synthesized in Project 2.1. If you choose an aldehyde, your synthesis will yield a secondary alcohol, whereas using a ketone will lead to a tertiary alcohol.

One of the side reactions that may interfere with a Grignard synthesis is the formation of an alkane by a coupling reaction. The coupling reaction takes place at the active surface of the solid magnesium when two carbon-free radicals react together to form a stable hydrocarbon dimer:

$$2 \text{ RX} + \text{Mg} \longrightarrow \text{R—R} + \text{MgX}_2$$

What is the alkane that forms as the side product from 1-bromobutane in this experiment?

A high concentration of alkyl halide favors this side reaction because the possibility of two free radicals forming near each other at the metal surface is more likely. However, even with a fairly large volume of ether present in the reaction flask, the formation of alkanes as a side product from the coupling of two alkyl free radicals usually occurs to some extent.

It is critical to maintain anhydrous conditions throughout the formation and reaction of the Grignard reagent.

The presence of water or other acids inhibits the initiation of the reaction and destroys the organometallic reagent once it forms. Take particular note of the anhydrous reaction conditions for a Grignard synthesis that are discussed in Experiment 13, p. 137.

Select an aldehyde or ketone from the following lists and write an equation for the Grignard synthesis of the secondary or tertiary alcohol that you will synthesize by its reaction with 1-bromobutane (or other alkyl halide). *Your product needs to have eight carbons or less* to have a boiling point under 200°C, thus allowing it to be distilled at atmospheric pressure without decomposition.

Aldehydes	**Ketones**
Ethanal (acetaldehyde)	2-Propanone (acetone)
Propanal (propionaldehyde)	2-Butanone (methyl ethyl ketone)
	3-Pentanone (diethyl ketone)

You will determine the purity of your alcohol product by gas chromatography. The competing reaction that produces the coupled alkane from your bromoalkane may occur to a significant degree in this Grignard reaction. The boiling point of the coupled alkane lies within 30° of the alcohol product's boiling point for many of the possible combinations of bromoalkane and carbonyl compounds. This small difference in boiling points means that you will not be able to separate the alkane from the alcohol by simple distillation. As part of your prelaboratory preparation, look up the boiling points of your alcohol product and the coupled alkane that arises from the Grignard reaction with your bromoalkane in a *CRC Handbook of Chemistry and Physics* or *The Dictionary of Organic Compounds*.

〰〰〰Macroscale〰〰 Procedure

Techniques   Reflux under Anhydrous Conditions: Technique 3.3a

Extraction: Technique 4.2

Drying Agents: Techniques 4.6 and 4.7

Simple Distillation: Technique 7.3

Diethyl ether is extremely flammable. Make sure that there are no flames in the laboratory, and use ether only in a hood.

The alkyl halides used in this experiment are harmful if inhaled, ingested, or absorbed through the skin. Wear gloves when handling them and use them in a hood, if possible. They are also very flammable.

The aldehydes are flammable and corrosive. Wear gloves when handling them and dispense them in a hood. Acetaldehyde is also a lachrymator and a suspected carcinogen, and may be an immune-response sensitizer.

The ketones are flammable.

All glassware and reagents need to be thoroughly dry before beginning the experiment.

This synthesis is based on 0.10 mol of bromoalkane. If you have more or less than 0.10 mol of your 1-bromobutane (or other alkyl halide), you must adjust proportionally the amounts of *all* other reagents. Consult your instructor before you begin the procedure.

Synthesis of the Grignard Reagent

Keep the ether can tightly closed as much as possible and do not let your ether stand in an open container, because water from the air will dissolve into it.

Fit a 250-mL three-necked round-bottomed flask with a dropping funnel (a 125-mL separatory funnel) and a condenser; close the third opening with a glass stopper. If you are using a one-necked flask, fit it with the Claisen adapter and insert the dropping funnel into one neck and the condenser into the other. Place a drying tube filled with calcium chloride in the top of the condenser and another in the top of the dropping funnel [see Technique 3.3a].

Weigh 2.4 g of magnesium turnings; crush several pieces with a glass stirring rod to expose a fresh surface. Place the magnesium turnings in the round-bottomed flask. Obtain 75 mL of anhydrous ether in a corked graduated cylinder. Cover the magnesium turnings with 35 mL of anhydrous ether. Put a magnetic stirring bar into the flask.

Measure 0.10 mol of your bromoalkane (check the quantity with your instructor) and pour it into the dropping funnel (be sure the stopcock is closed). Pour the remaining 40 mL of ether into the dropping funnel, stopper the funnel, and shake it thoroughly to produce a homogeneous solution; put the drying tube filled with calcium chloride back in the top of the separatory funnel. Add about 3 mL of the ether-bromoalkane solution to the reaction flask. Do not stir the contents of the flask; initiation of the reaction occurs more rapidly if there is a high concentration of halide in the vicinity of the magnesium. Warm the flask for a few minutes in a water bath at 45°–50°C. The ether will begin to boil. Within 5–10 min, you should see tiny bubbles forming at the magnesium surface, and a small amount of cloudy precipitate may appear.

The precipitate is probably some magnesium hydroxide, that is generated from any moisture in the flask or the reagents.

Once the reaction is well under way, remove the warm-water bath and begin stirring the reaction mixture slowly. Add the bromoalkane solution to the flask at a rate that allows for moderate reflux to take place. The condensing ether vapor should be in the lower half of the condenser. Have a beaker of ice water available to cool the flask, if the reaction becomes too vigorous, by briefly immersing it; too much cooling will stop the reaction.

When all of the bromoalkane has been added, continue stirring the reaction mixture until it ceases to reflux. Then heat the mixture at reflux for 15 min, using a water bath at 45°–50°C.

Addition of the Carbonyl Compound

Pour 0.098 mol of the aldehyde (or ketone) that you selected (check the quantity with your instructor) into the dropping funnel, rinse the graduated cylinder used for the aldehyde (or ketone) with 10 mL of anhydrous ether, and also add this ether rinse to the dropping funnel. Place an ice-water bath under the reaction flask and spin the stirrer at a moderate speed. Add the aldehyde (or ketone) solution slowly over a period of 15–20 min. This reaction may be vigorous. Use a rate of addition that maintains the condensing ether vapors in the lower half of the condenser. After the addition is complete and the reaction is no longer refluxing on its own accord, reflux it by heating it in a water bath at 50°C for 15 min. Cool the reaction mixture to room temperature before proceeding with the hydrolysis.

Instead of heating, the reaction mixture may be allowed to stand until the next laboratory period; stopper the flask tightly and store it as directed by your instructor.

Hydrolysis of the Grignard Adduct

SAFETY PRECAUTION

Ether vapors will be present while conducting the hydrolysis and extractions. Be sure that there are no flames in the laboratory or hot electrical heating devices in your work area. Work in a hood, if possible.

Pour the reaction mixture slowly into a 400-mL beaker containing 50 g of crushed ice and 50 mL of 10% sulfuric acid in the hood. Stir or swirl the mixture until the precipitate dissolves. If the ether begins to boil, add more ice.

When the residual magnesium has all dissolved, transfer the mixture to a separatory funnel and separate the layers [Technique 4.2]. Keep both layers. Return the aqueous layer to the funnel and extract it twice with 25-mL portions of ether. Combine the ether extracts with the ether layer from the first separation and dry the combined ether solution with anhydrous potassium carbonate. Set the aqueous layer aside for later neutralization.

Anhydrous ether is not necessary for the extractions because you are adding it to an aqueous solution.

Filter some of the dried ether solution into a 100-mL round-bottomed flask, filling the flask no more than two-thirds full [Technique 4.7]. Remove the ether by simple distillation over a steam bath [Technique 7.3]; place the receiving flask in an ice-water bath. Stop the distillation and filter the remaining ether solution into the round-bottomed flask; add another boiling stone. Continue the distillation until the ether stops distilling. Pour the distilled ether into the recovered ether bottle in the hood. Transfer the liquid remaining in the flask to a 50-mL round-bottomed flask, rinse the larger flask with the 1 or 2 mL of ether and add the rinse to the 50-mL flask. Set up the apparatus for simple distillation, using a heating mantle as the heat source.

Alternatively, the ether may be removed with a rotary evaporator [see Technique 4.8].

Collect as a first fraction the residual ether. Use the boiling points of your alcohol and the coupled alkane to determine the temperature range for collecting the product fraction. Collect your product fraction in a tared (weighed) Erlenmeyer flask. Stop the distillation when only a small amount of liquid remains in the boiling flask but be very careful not to let the flask boil dry. Record the boiling-point range for the product fraction.

— SAFETY PRECAUTION —

Ether may form peroxides that can explode if the flask goes to dryness at the end of the distillation.

Determine the purity of your product by gas chromatographic analysis. For capillary GC, dissolve 1 drop of your product in

0.5 mL of ether and inject 1 μL of the solution into the chromato-graph. Consult your instructor about sample preparation for a packed column GC. The alkane and alcohol elute in order of increasing boiling point from a nonpolar column. Obtain an IR spectrum of your product [see Spectrometric Method 1].

Cleanup: Pour the residue remaining in the distillation flask into the container for flammable (organic) waste. Place the recovered ether in the "recovered ether" container or the container for flammable (organic) waste, as directed by your instructor. Neutralize the aqueous solution remaining from the extractions with sodium carbonate **(Caution: Foaming.)** before washing it down the sink or pouring it into the container for aqueous inorganic waste.

Optional Synthesis

Secondary alcohol products from this synthesis may be oxidized to ketones, using the procedure in Experiment 15.2.

Questions

1. Another procedure for the preparation of 1-bromobutane in-volves direct chemical reaction of hydrobromic acid (HBr) with 1-butanol. (a) Write an equation that shows how HBr could be formed from the inorganic reagents used in this experiment. (b) Write a mechanism for the formation of 1-bromobutane from 1-butanol by direct reaction with HBr.
2. Treatment of 1-butanol with phosphoric acid (H_3PO_4) mixed with sodium chloride should result in formation of 1-chlorobutane when done at elevated temperatures. Write a mechanism for this reaction.
3. Comment on the viability of each of the following reactions for its ability to lead to 1-bromobutane:
   (a) 1-butanol + NaBr / H_2O
   (b) 1-butanol + NaBr / H_3PO_4
   (c) 1-butanol + KBr / H_2SO_4
   (d) 1-butanol + NaBr / HCl
   (e) 1-butanol + NaBr / acetic acid
4. Treatment of 2-methyl-2-propanol (*tert*-butyl alcohol) with HCl leads to a reasonable yield of 2-chloro-2-methylpropane when done at reduced temperatures. Write a mechanism for this reaction.

$CH_3CH_2CH_2CH_2Br$

FW 137.03	d 1.276	IR III, 42H	2961.8 1261.8 740.6
mp -112°C	Fp 75°F	NMR II, 1,60C	1465.4 915.0 644.0
bp 100-104°C	nβ 1.4394	Merck 10,1526	1380.5 866.8 563.4

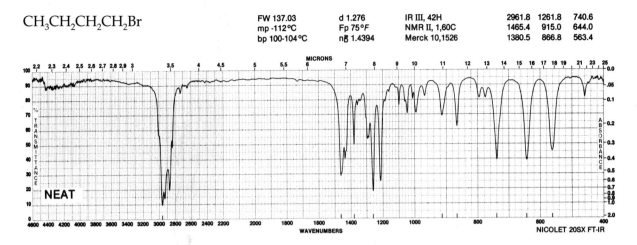

FIGURE 2.1 IR spectrum of 1-bromobutane. (Provided by Aldrich Chemical Company, Inc., Milwaukee, WI.)

5. Why was the product washed with aqueous sodium bicarbonate in the 1-bromobutane synthesis?

6. The IR spectrum of 1-bromobutane is shown in Figure 2.1. (a) What causes the peaks in the 3000–2800 cm^{-1} region? (b) Where do you expect to find C—Br stretching? (c) What would be the most obvious indication of unreacted 1-butanol in the IR spectrum?

7. Assign and explain all the ^{1}H peaks for 1-bromobutane shown in the NMR spectrum (Figure 2.2).

8. What is the copious white precipitate that forms when the carbonyl compound is added to an ether solution of 1-butylmagnesium bromide?

9. If water is present in the Grignard reaction mixture, what organic product will form?

10. Which of the following will undergo reaction with a Grignard reagent? What will the products be if a reaction occurs?

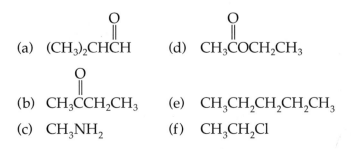

(a) $(CH_3)_2CHCH$ (d) $CH_3COCH_2CH_3$

(b) $CH_3CCH_2CH_3$ (e) $CH_3CH_2CH_2CH_2CH_3$

(c) CH_3NH_2 (f) CH_3CH_2Cl

$$CH_3CH_2CH_2CH_2Br$$

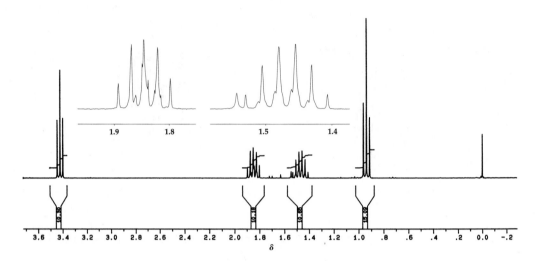

FIGURE 2.2 ^{1}H NMR (300 MHz) spectrum of 1-bromobutane.

11. Suggest organometallic syntheses for the following compounds.

(a) $\underset{\displaystyle CHCH_3}{\overset{\displaystyle OH}{|}}$ (phenyl)

(b) $CH_3\underset{|}{\overset{OH}{C}}HCH_3$

(c) $CH_3CH_2CH_2CH_2COOH$

(d) $CH_3\underset{|}{\overset{CH_3}{C}}HCH_2CH_3$

(e) $CH_3\underset{|}{\overset{CH_3}{C}}HCH_2CH_2CH_2OH$

SYNTHESIS AND DEHYDRATION OF 3,3-DIMETHYL-2-BUTANOL

Project 3 involves a two-step reaction sequence, the first step of which is the reduction of the ketone, 3,3-dimethyl-2-butanone (also called pinacolone), to produce a secondary alcohol, 3,3-dimethyl-2-butanol. This reduction reaction is followed by acid-catalyzed dehydration of the secondary alcohol to yield a mixture of three alkenes, two of which form by carbon-skeleton rearrangement. The composition of the alkene mixture will be determined by gas chromatography.

3,3-Dimethyl-2-butanone   3,3-Dimethyl-2-butanol

2,3-Dimethyl-2-butene   2,3-Dimethyl-1-butene   3,3-Dimethyl-1-butene

These reactions are two separate, sequential procedures. You work up and characterize the intermediate, 3,3-dimethyl-2-butanol, before carrying out the dehydration reaction.

Reduction reactions using sodium borohydride will be discussed briefly in the next section. You can also find information about sodium borohydride reductions in Experiment 19.2 and Project 4. Dehydration of alcohols with acid catalysts is described in Experiment 11. In this project we will focus on the carbocation aspects of the dehydration reaction as they relate to rearrangement.

3.1

Synthesis of 3,3-Dimethyl-2-Butanol by Reduction with Sodium Borohydride

$$4H_3C-\underset{\underset{CH_3}{|}}{\overset{\overset{CH_3}{|}}{C}}-\overset{O}{\overset{\|}{C}}-CH_3 \quad + \quad NaBH_4 \quad \xrightarrow[\text{methanol}]{NaOCH_3} \quad 4H_3C-\underset{\underset{CH_3}{|}}{\overset{\overset{CH_3}{|}}{C}}-\underset{\underset{H}{|}}{\overset{\overset{OH}{|}}{C}}-CH_3 + NaBO_2$$

3,3-Dimethyl-2-butanone   Sodium borohydride   3,3-Dimethyl-2-butanol
(pinacolone)   MW 37.83   (pinacolyl alcohol)
bp 106°C   bp 120.1°C
MW 100.2   MW 102.2
density 0.801 g · mL^{-1} density 0.812 g · mL^{-1}

Sodium borohydride ($NaBH_4$) has become a popular reagent for reducing the carbonyl groups of aldehydes and ketones. One of the reasons for this popularity is the fact that borohydride reductions occur readily within a reasonable length of time, yet $NaBH_4$ is far less dangerous to use than lithium aluminum hydride ($LiAlH_4$). Sodium borohydride reduces carbonyl groups significantly faster than it reacts with hydroxylic solvents, such as alcohols; whereas $LiAlH_4$ can react explosively with alcohols. BH_4^- is a good source of hydride ($:H^-$), which reduces the carbonyl group to an oxyanion intermediate by a nucleophilic addition pathway:

$$R-\underset{\underset{R'}{|}}{\overset{\overset{O}{\|}}{C}} + NaBH_4 \longrightarrow R-\underset{\underset{R'}{|}}{\overset{\overset{H}{}}{C}}\overset{\overline{O}BH_3\ Na^+}{} \xrightarrow[-OCH_3]{CH_3OH} R-\underset{\underset{R'}{|}}{\overset{\overset{H}{}}{C}}\overset{OH}{}$$

The borohydride species that forms can further reduce another molecule of the ketone. Overall, one mole of sodium borohydride can reduce four moles of the ketone.

You can characterize your intermediate reduction product, 3,3-dimethyl-2-butanol, by its boiling point and its IR spectrum, and assess its purity by gas chromatographic analysis.

Macroscale
Procedure

Techniques   Extraction: Technique 4.2

Simple Distillation: Technique 7.3

Gas Chromatography: Technique 11

IR Spectrometry: Spectrometric Method 1

— SAFETY INFORMATION —

3,3-Dimethyl-2-butanone (pinacolone) is flammable.

Sodium methoxide in methanol solution is flammable, corrosive, and moisture sensitive. Avoid contact with skin, eyes, and clothing.

Sodium borohydride is harmful if swallowed, inhaled, or absorbed through the skin. Avoid breathing the dust; and avoid contact with skin, eyes, and clothing. It decomposes to flammable, explosive hydrogen gas.

Methanol is volatile, flammable, and toxic.

Diethyl ether is very flammable. Use it only in a hood and keep it away from flames or electrical heating devices.

Hydrochloric acid solution (6 M) is a skin irritant.

Keep the container of sodium methoxide solution tightly closed and store the reagent in a desiccator.

NaBH$_4$ is moisture sensitive; keep the bottle tightly closed. Store the reagent in a desiccator.

Pour 12 mL of 3,3-dimethyl-2-butanone (pinacolone) and 20 mL of methanol into a 125-mL Erlenmeyer flask. Place the flask in an ice-water bath.

Combine 27 mL of methanol and 1.6 mL of 25 wt % sodium methoxide in methanol in another Erlenmeyer flask. Add 3.0 g of sodium borohydride and stir until the large particles are completely dispersed (the mixture will still be cloudy). Transfer the borohydride mixture to a separatory funnel that is suspended above the flask containing the 3,3-dimethyl-2-butanone solution (Figure 3.1). Add the borohydride solution in small portions over a period of 10 min, swirling the reaction flask after each addition. Allow the reaction mixture to stand in the ice-water bath for an additional 20 min, with occasional swirling of the flask.

The separatory funnel serves as a dropping funnel during the addition of the NaBH$_4$ solution.

With the flask still in the ice-water bath, add 40 mL of water and allow the mixture to stand for another 5 min. Add 8.5 mL of 6 M hydrochloric acid *dropwise* with swirling after each addition. If hydrogen is still being formed with the last portion of HCl, add an additional 0.5–1.0 mL of HCl solution.

Gaseous hydrogen is evolved during the addition of HCl, and the mixture will foam out of the flask if the acid is added too quickly.

Separate the precipitated borate salt from the product mixture by vacuum filtration. Transfer the filtrate to a separatory funnel

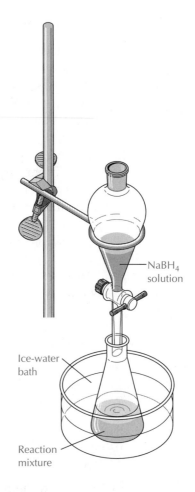

NaBH$_4$ solution

Ice-water bath

Reaction mixture

FIGURE 3.1 Setup for running the synthesis.

and rinse the filter flask with 20 mL of diethyl ether. Add the ether rinse to the separatory funnel. Shake the separatory funnel to extract the product into the ether layer [see Technique 4.2]. Drain the lower aqueous phase into a labeled Erlenmeyer flask and pour the upper organic phase out of the top of the separatory funnel into a labeled 125-mL Erlenmeyer flask. Return the aqueous phase to the separatory funnel and extract it with a second 20-mL portion of ether. Separate the aqueous phase. Combine the two ether solutions in the separatory funnel. Wash the combined ether phase with 20 mL of water, then with 20 mL of saturated sodium chloride solution, removing the aqueous phase each time. Pour the ether solution into a clean Erlenmeyer flask and dry it with anhydrous magnesium sulfate for at least 10 min [see Technique 4.6].

Filter the ether solution through a fluted filter paper into a 100-mL round-bottomed flask. Assemble the apparatus for a simple

Extraction with saturated NaCl solution partially dries the ether solution because the NaCl solution has a lower solubility in ether than water does.

Alternatively, the ether and methanol can be removed on a rotary evaporator [see Technique 4.8].

distillation, using a steam bath as the heat source and a 50-mL round-bottomed flask cooled in an ice-water bath as the receiver [see Technique 7.3]. Collect distillate until the temperature reaches 90°C or until distillation stops. After the ether and any remaining methanol are removed, change the heat source to an electric heating mantle. Collect the product fraction from 100°–122°C in a tared flask.

> ─ **SAFETY PRECAUTION** ─
>
> Be sure to stop the distillation before the boiling flask reaches dryness. Diethyl ether may form peroxides. If the flask goes to dryness, the temperature inside it rises rapidly and the peroxides could explode.

Store your product in a tightly corked flask for use in Project 3.2.

Weigh your alcohol product and calculate the percent yield. Use gas chromatographic analysis to assess the percent purity of your product. Any starting material present in the product can be identified by the peak enhancement method [see Technique 11.6]. Obtain an IR spectrum of your product, if directed to do so by your instructor [see Spectrometric Method 1].

Cleanup: Pour the aqueous phase from the extractions into the container for aqueous inorganic waste. Place the solid borate residue and the spent drying agent in the container for inorganic waste. Pour the recovered ether and the residue left in the boiling flask into the container for flammable (organic) waste.

3.2

Acid-Catalyzed Dehydration of 3,3-Dimethyl-2-Butanol with Rearrangement

3,3-Dimethyl-2-butanol
(pinacolyl alcohol)
bp 120.1°C
MW 102.2
density 0.812 g · mL⁻¹

2,3-Dimethyl-2-butene
bp 73.2°C
MW 84.2
density 0.708 g · mL⁻¹

2,3-Dimethyl-1-butene
bp 55.7°C
MW 84.2
density 0.68 g · mL⁻¹

3,3-Dimethyl-1-butene
bp 41.2°C
MW 84.2
density 0.65 g · mL⁻¹

As described in Experiment 11, dehydration promoted by sulfuric acid often involves carbocation intermediates. These intermediates are prone to rearrangement whenever a pathway is available. For example, a secondary carbocation can undergo a hydride or methyl migration and thereby rearrange to a more stable tertiary carbocation, as occurs in this experiment.

3,3-Dimethyl-2-butanol

Secondary carbocation   Tertiary carbocation

The tertiary carbocation then readily deprotonates to yield an alkene:

Tertiary carbocation   2,3-Dimethyl-2-butene

The same tertiary carbocation can deprotonate in a different fashion to form 2,3-dimethyl-1-butene:

Tertiary carbocation   2,3-Dimethyl-1-butene

We might expect more 2,3-dimethyl-2-butene to form because it is more stable, having more alkyl groups directly attached to the sp^2 carbon atoms.

The secondary carbocation formed earlier in the mechanistic sequence can also deprotonate to form a third alkene, 3,3-dimethyl-1-butene:

$$CH_3-\underset{\underset{CH_3}{|}}{\overset{\overset{CH_3}{|}}{C}}-\overset{+}{CH}\overset{\curvearrowleft}{\underset{CH_2-H}{}} \longrightarrow CH_3-\underset{\underset{CH_3}{|}}{\overset{\overset{CH_3}{|}}{C}}-CH=CH_2$$

Secondary carbocation   3,3-Dimethyl-1-butene

The reaction actually produces a mixture of all three butenes. You can determine the relative amounts of each alkene by analyzing your product with gas chromatography. This information will allow you to investigate the relative likelihood of the competing reaction pathways in this acid-catalyzed dehydration reaction.

Macroscale Procedure

Techniques   Short-Path Distillation: Technique 7.3a
Pasteur Filter Pipets: Technique 2.4
Gas Chromatography: Technique 11

The alkenes produced in this reaction have a strong, musty odor. Conduct the experiment in a hood, if possible.

SAFETY INFORMATION

3,3-Dimethyl-2-butanol and heptane are flammable.

Concentrated sulfuric acid is extremely corrosive and causes severe burns. Avoid contact with skin, eyes, and clothing.

Glass beads serve the same purpose as boiling stones, which tend to disintegrate in hot concentrated sulfuric acid.

Place 3.0 mL of 3,3-dimethyl-2-butanol that you prepared in Project 3.1 in a 50-mL round-bottomed flask and add 8 drops of concentrated sulfuric acid. Put four or five glass beads in the flask. Assemble the apparatus for short-path distillation, using a 25-mL round-bottomed flask for the receiver [see Technique 7.3a]. Place an ice-water bath around the receiving flask so that the alkene products do not evaporate. Heat the reaction flask gently and distill the product mixture at a moderate rate until the temperature reaches 100°C or until distillation ceases. A brown residue will remain in the distilling flask.

The alkene products evaporate quite easily at room temperature. Keep the flask they are in tightly corked.

Dry the product mixture with anhydrous potassium carbonate. Cork the flask so that the volatile alkenes do not evaporate during the drying process. Transfer the dried product to a tared (weighed) vial, using a Pasteur filter pipet [see Technique 2.4]. Weigh the product mixture and determine the percent yield.

Analyze the product mixture by gas chromatography [see Technique 11]. For a packed column instrument, use a 20% Carbowax 20M on Chromosorb P column at 75°C and a gas flow rate of 50 mL · min.$^{-1}$ (Ref. 1). Consult your instructor about sample preparation for a packed column gas chromatograph. The alkenes will elute in order of increasing boiling point.

For a capillary column gas chromatograph with a nonpolar column, such as polydimethylsiloxane, use a column temperature no higher than 50°C. Prepare the sample by dissolving 2 drops of the product mixture in 0.5 mL of heptane; inject 1 μL of this solution into the GC. The alkenes elute from a nonpolar column in order of increasing boiling point, with the solvent peak (heptane) occurring *after* the product peaks rather than before, as is usually the case.

Identification of the alkenes can be done by the peak enhancement method, because all three products are commercially available [see Technique 11.6]. Submit the remaining product to your instructor.

Determine the relative amounts of each alkene from the peak areas or the peak heights on your chromatogram. Explain the formation of all observed products and discuss the observed product distribution.

Cleanup: Carefully dilute the residue in the distilling flask with 20 mL of water. (**Caution: The residue contains sulfuric acid.**) Add solid sodium carbonate in small portions until the acid is neutralized. (**Caution: Foaming.**) Pour the neutralized solution on a Buchner funnel to recover the glass beads. The filtrate may be washed down the sink or poured into the container for aqueous inorganic waste. Put the glass beads in the appropriate container. Pour the heptane solution remaining from the GC into the container for flammable waste.

Reference

1. Sayed, Y.; Ahlmark, C. A.; Martin, N. H. *J. Chem. Educ.* **1989,** *66,* 174–175.

Questions

1. Why does the solvent peak (heptane) show a longer GC retention time than the alkene products? Why is heptane a better choice for the solvent than diethyl ether in this GC analysis?

2. Acid-catalyzed dehydration of an alcohol occurs under equilibrating conditions that favor the thermodynamically more stable product(s). Identify the thermodynamic product(s) and the kinetic product(s) of your dehydration reaction.

3. Why did the observed boiling point in the short-path distillation rise above the boiling points of the alkene products?

4. Because acid-catalyzed dehydration is an equilibrium situation, how was the reaction forced to completion in Project 3.2?

5. Predict the products of the reaction:

$$H_3C - \overset{\overset{\displaystyle CH_3}{|}}{\underset{\underset{\displaystyle CH_3}{|}}{C}} - \overset{\overset{\displaystyle OH}{|}}{CH} - CH_3 \xrightarrow{\text{HCl}}$$

Use a mechanism to explain your prediction.

6. Why might a poor yield of alkenes be realized if the dehydration were carried out with hydrochloric acid rather than H_2SO_4?

7. With the knowledge that $NaBH_4$ is a base, suggest a reaction that would make clear why the presence of any Brønsted acid of significant strength would be deleterious to a sodium borohydride reduction.

8. Methanol is known to undergo autoprotolysis to form $CH_3OH_2^+$ and CH_3O^-. Suggest how the Le Chatelier principle can be used to explain why added methoxide ion would adjust the autoprotolysis equilibrium reaction for methanol in a way favorable to carbonyl reduction.

9. Explain why the necessary reaction conditions would be successively milder (lower temperatures, lower acid concentrations) for the dehydration of the following alcohols: 1-heptanol, 2-heptanol, 1-methylcyclohexanol.

10. Write a mechanism for the following dehydration and rearrangement reaction:

11. Write a mechanism for the following reaction:

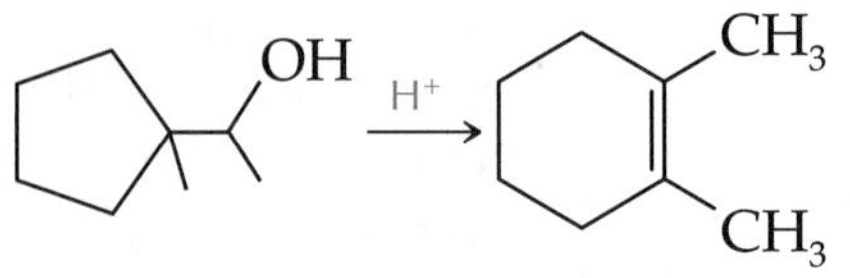

12. Potassium carbonate is used as the drying agent for the alkene product to prevent acid-catalyzed rearrangements. This is important because you need to measure the alkene product ratio arising from the dehydration process, rather than from an isomerization that occurs after dehydration. Sketch mechanisms that rationalize the acid-catalyzed interconversion of 2,3-dimethyl-2-butene, 2,3-dimethyl-1-butene, and 3,3-dimethyl-1-butene.

13. What band would be present in the IR spectrum of 3,3-dimethyl-2-butanol if unreacted 3,3-dimethyl-2-butanone were present?

Project 4

INTERCONVERSION OF 4-*tert*-BUTYLCYCLOHEXANOL AND 4-*tert*-BUTYLCYCLOHEXANONE

In this project you will investigate two oxidation-reduction reactions involving the interconversion of a mixture of *cis*- and *trans*-4-*tert*-butylcyclohexanol and 4-*tert*-butylcyclohexanone. You will use sodium hypochlorite to oxidize the 4-*tert*-butylcyclohexanols to 4-*tert*-butylcyclohexanone in the first reaction; then in the second reaction you will use sodium borohydride to reduce 4-*tert*-butylcyclohexanone to 4-*tert*-butylcyclohexanols. You will follow the progress of both reactions by thin-layer chromatography and determine the ratio of *cis* and *trans* isomers formed in the reduction reaction by gas chromatography or NMR spectrometry.

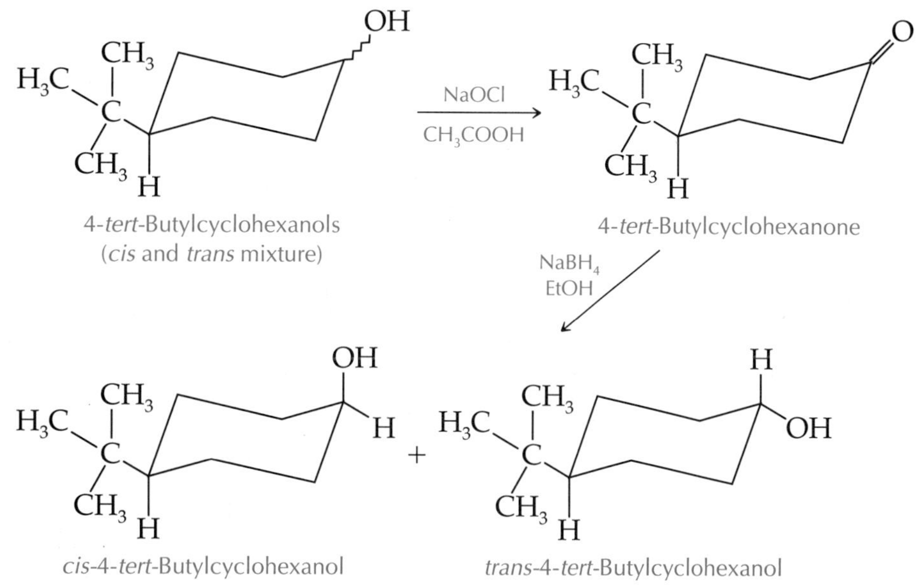

4-*tert*-Butylcyclohexanols
(*cis* and *trans* mixture)

4-*tert*-Butylcyclohexanone

cis-4-*tert*-Butylcyclohexanol

trans-4-*tert*-Butylcyclohexanol

The first reaction is oxidation of a commercially available mixture of *cis*- and *trans*-4-*tert*-butylcyclohexanols, using sodium hypochlorite as the oxidizing agent, to yield 4-*tert*-butylcyclohexanone. In the second step, 4-*tert*-butylcyclohexanone is reduced by sodium borohydride to give a mixture of *cis* and *trans* isomers of 4-*tert*-butylcyclohexanol, with gas chromatography or NMR spectrometry employed to ascertain the ratio of 4-*tert*-butylcyclohexanol isomers formed in the reaction. You will find a general discussion of oxidation-reduction reactions as they pertain to organic compounds in the introduction to Experiment 15, p. 171.

A question that you can consider after you have analyzed your experimental data from this project is whether the mixture of *cis*- and *trans*-4-*tert*-butylcyclohexanols that you used in the oxidation reaction is reproduced by your sodium borohydride reduction of 4-*tert*-butylcyclohexanone. Or is it more likely that it is synthesized by another route, such as the reduction of the benzene ring of 4-*tert*-butylphenol?

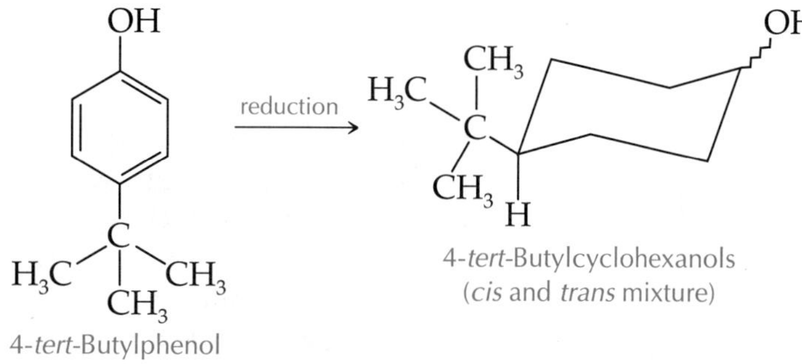

4-*tert*-Butylphenol

reduction

4-*tert*-Butylcyclohexanols
(*cis* and *trans* mixture)

4.1 Sodium Hypochlorite Oxidation of 4-*tert*-Butylcyclohexanol to 4-*tert*-Butylcyclohexanone

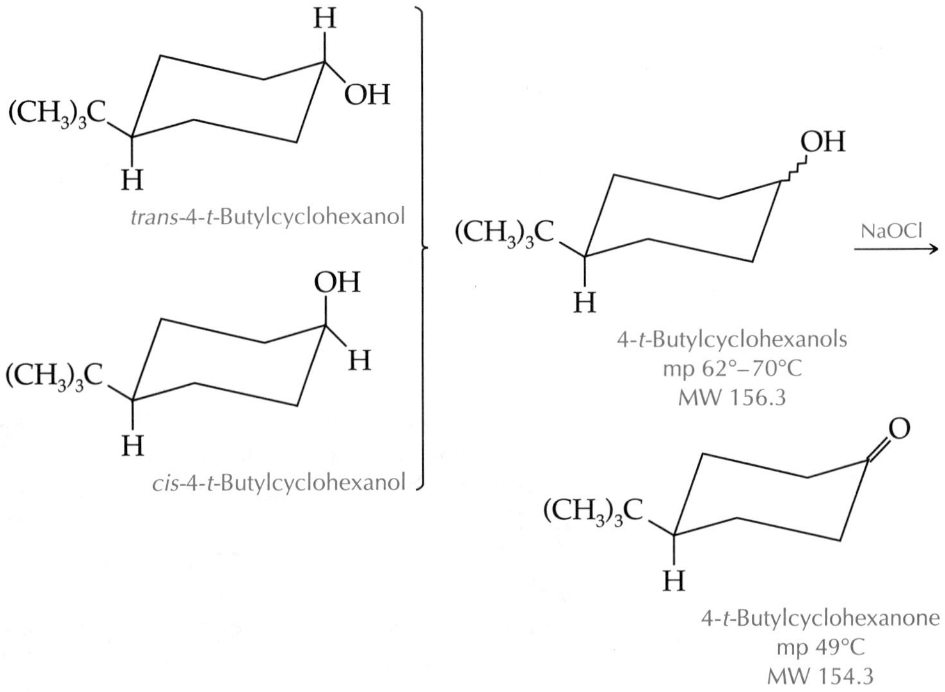

trans-4-*t*-Butylcyclohexanol

cis-4-*t*-Butylcyclohexanol

4-*t*-Butylcyclohexanols
mp 62°–70°C
MW 156.3

4-*t*-Butylcyclohexanone
mp 49°C
MW 154.3

NaOCl

Alcohols with at least one α-hydrogen can easily be oxidized to the corresponding carbonyl compounds.

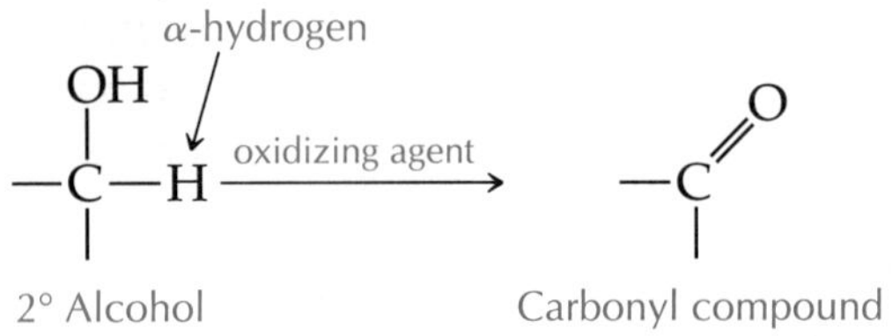

α-hydrogen

OH

oxidizing agent

2° Alcohol

Carbonyl compound

Although the aldehydes that are produced by the oxidation of primary alcohols are also easily oxidized further to carboxylic acids, the oxidation of secondary alcohols to ketones poses no such problem. A large variety of inorganic oxidizing agents can be used. Chromium (VI) is probably the most common oxidizing agent; however, chromium compounds are toxic and pose a hazard to the environment. An excellent alternative is sodium hypochlorite (NaOCl), which is the active ingredient of household bleach. This reagent has low toxicity and also gives products that are nontoxic.

Most chlorine bleach is prepared by adding chlorine gas to aqueous sodium hydroxide solution:

$$Cl_2 + 2\,NaOH \rightleftharpoons NaOCl + NaCl + H_2O$$

When sodium hypochlorite is added to acetic acid, the following acid-base reaction occurs:

$$NaOCl + CH_3COOH \rightleftharpoons HOCl + CH_3COO^-Na^+$$

NaOCl, Cl_2, and hypochlorous acid (HOCl) are all sources of the oxidizing agent positive chlorine (Cl^+), which is very different from the chloride anion (Cl^-). As you might expect from the deficiency of electrons in the valence shell of Cl^+, positive chlorine is a strong electrophile. Although evidence for the discrete existence of Cl^+ in aqueous solution has never been found, a key step in reactions of positive chlorine oxidizing agents is the transfer of Cl^+ from Cl_2 or HOCl to the substrate. The mechanism of the hypochlorite oxidation of an alcohol is illustrated here by the oxidation of cyclohexanol. It is reasonable to expect that it proceeds first by exchange of positive chlorine with the hydroxyl proton of the alcohol, followed by elimination of HCl from the resulting alkyl hypochlorite to form cyclohexanone:

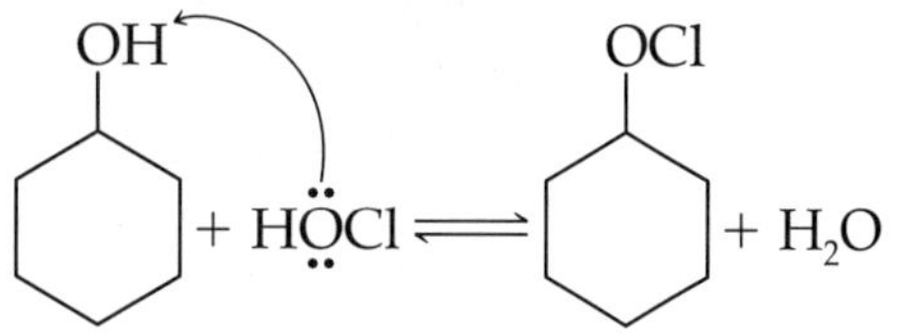

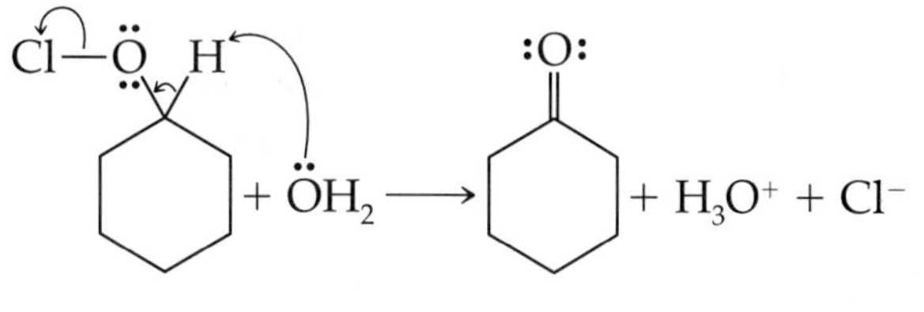

In the first reaction Cl^+ is transferred to the substrate, and in the second, Cl^- is lost. This is a reduction by two electrons. Cyclohexanol provides the two electrons and is thereby oxidized to cyclohexanone, or as is the case in this experiment, 4-*tert*-butylcyclohexanol is oxidized to 4-*tert*-butylcyclohexanone.

The chair conformations of the cyclohexane ring are a reasonable way to visualize the *cis* and *trans* isomers of 4-*tert*-butylcyclohexanol. The bulky *tert*-butyl group is by far the largest substituent, and this bulk holds the ring in the chair conformation, with the *tert*-butyl group in an equatorial position:

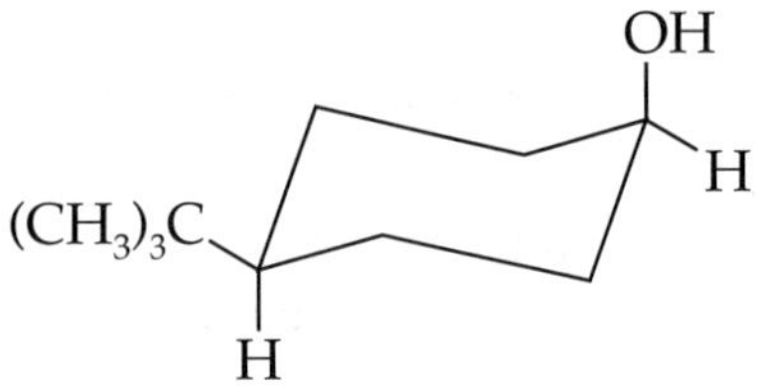

cis-4-*tert*-Butylcyclohexanol

Equatorial *tert*-butyl group
Axial hydroxyl group

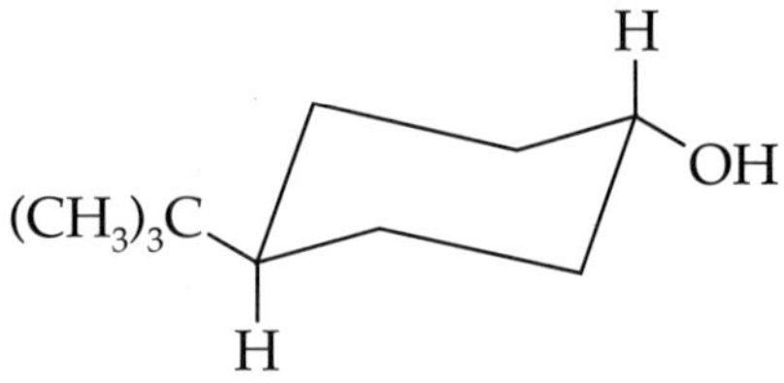

trans-4-*tert*-Butylcyclohexanol

Equatorial *tert*-butyl group
Equatorial hydroxyl group

The mixture of *cis* and *trans* starting materials forms only one product, because the alcohol functional group (source of the geometrical isomerism in the starting materials) is lost on oxidation:

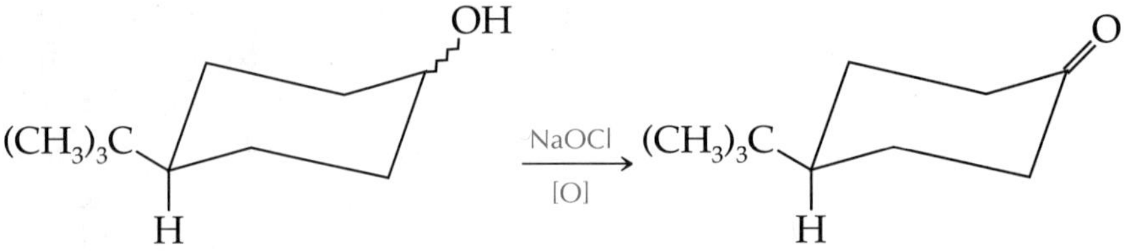

You can use ^{1}H NMR spectrometry to analyze the composition of the starting material because the methine hydrogen, the proton on the carbon bearing the hydroxyl group, is usually shifted downfield from the other protons. In the *trans* isomer the proton is axial and can be observed at δ 3.5. The corresponding proton in the *cis* isomer is at δ 4.03. Because each isomer has only one of these protons, the ratio of the integrated areas of these methine hydrogens can be used to determine the isomer distribution. You can use either the heights of the integration curves corresponding to each of the two signals or the numerical data proportional to the integrated areas for each. Thus,

$$\% \; cis = \frac{\text{area of the } \delta \, 4.03 \text{ signal}}{\text{sum of areas of the } \delta \, 4.03 \text{ and } \delta \, 3.5 \text{ signals}} \times 100$$

**Macroscale
Procedure**

Techniques   Thin-Layer Chromatography: Technique 10
Extraction: Technique 4.2
Drying Organic Liquids: Technique 4.6
Simple Distillation: Technique 7.3
IR Spectrometry: Spectrometric Method 1

———— *SAFETY INFORMATION* ————

Conduct this experiment in a hood, if possible.

4-*tert*-Butylcyclohexanol may be an irritant. Wash thoroughly after handling it.

Acetone is extremely flammable, volatile, and a severe eye irritant; and it may cause skin irritation. Avoid contact with skin, eyes, and clothing.

Sodium hypochlorite solution emits chlorine gas, which is a respiratory and eye irritant. Use it in a hood, if possible.

Glacial acetic acid is a dehydrating agent and irritant, and causes burns. Dispense it only in a hood. Wear gloves. Avoid contact with skin, eyes, and clothing.

Diethyl ether is extremely volatile and flammable. Be certain that there are no flames in the laboratory or hot electrical devices in the vicinity while you are pouring ether or performing the extractions. Use ether in a hood, if possible.

Monitoring the Reaction by Thin-Layer Chromatography

Before using TLC [Technique 10] to follow the course of this reaction, it is necessary for you to know the R_f values for 4-*tert*-butylcyclohexanol and 4-*tert*-butylcyclohexanone. Standard solutions of 2% 4-*tert*-butylcyclohexanol and 2% 4-*tert*-butylcyclohexanone in dichloromethane are available in the laboratory. Spot a glass-backed or aluminum-backed silica gel thin-layer plate with each compound and develop the plate in diethyl ether. Mark the solvent front with a pencil as soon as the plate is removed from the developing chamber and allow the ether to evaporate in a hood.

Briefly dip the thin-layer plate in a jar containing *p*-anisaldehyde visualizing reagent [see Technique 10.5]. Allow the excess to drip back into the jar and wipe the back of the thin-layer plate with a tissue. In a hood, heat the thin-layer plate on a hot plate heated to a medium setting; the hot plate setting should produce colored spots within about 1 min without burning or excessively darkening the entire thin-layer plate. Calculate the R_f values for both isomers of 4-*tert*-butylcyclohexanol (the commercially avail-

Note to instructor: Prepare p-anisaldehyde reagent solution by mixing together 5.1 mL of p-anisaldehyde, 2.1 mL of acetic acid, 6.9 mL of concentrated sulfuric acid, and 186 mL of 95% ethanol.

able alcohol is a mixture of *cis* and *trans* isomers) and for 4-*tert*-butylcyclohexanone. Use the same developing and visualizing procedure to monitor the reaction.

Oxidation Reaction

Use recently purchased household bleach, such as Clorox or a supermarket brand. Bleach solutions that have been stored for many months may have decreased concentrations of sodium hypochlorite.

Place 1.5 g of 4-*tert*-butylcyclohexanol, 1.20 mL of glacial acetic acid, and 20 mL of acetone in a three-necked 100-mL or 250-mL round-bottomed flask that is firmly clamped to a ring stand or metal support. Put a magnetic stirring bar in the flask and insert a water-cooled condenser in one neck, a separatory funnel in the second neck, and a glass stopper in the third neck. Pour 13 mL of 5.25% (0.75 M) sodium hypochlorite solution into the separatory funnel. Place the flask in a room-temperature water bath and begin stirring the solution. Add sodium hypochlorite solution dropwise over a period of approximately 15 min. A moderate to fast stirring rate is necessary to ensure thorough mixing of the two-phase system, because the acetone solution of 4-*tert*-butyl-cyclohexanol is not totally miscible with aqueous hypochlorite solution.

Monitor the reaction by TLC at the midpoint of the hypochlorite addition and after the addition is complete. To do this, turn off the stirrer and allow the two phases to separate. Touch the top of the upper (acetone) layer with a capillary pipet suitable for spotting thin-layer plates. Spot a thin-layer plate with the reaction solution in one channel and 2% 4-*tert*-butylcyclohexanol solution in a second channel. Develop the thin-layer plate, using the same procedure that you previously used on the known solutions.

When the addition of hypochlorite solution is complete, test the lower layer for the presence of excess hypochlorite ion. Turn off the stirrer and let the two phases separate. Expel the air from a rubber bulb on a Pasteur pipet *before* inserting the pipet to the bottom of the reaction flask. Release the pressure only enough to draw a few drops of lower aqueous solution into the pipet and remove the pipet from the flask. Put a drop of the solution on a strip of wet starch-iodide test paper. (If the bleach is too concentrated, colorless iodate may form.) A bluish black color indicates excess hypochlorite ion. If no color appears, put an additional 2 mL of bleach into the separatory funnel and add the solution dropwise to the stirred reaction mixture.

Continue monitoring the reaction by TLC until complete conversion of 4-*tert*-butylcyclohexanol to 4-*tert*-butylcyclohexanone has occurred. When TLC analysis indicates that the reaction is complete, cool the reaction mixture to room temperature and add 2 mL of saturated sodium bisulfite to reduce any excess hypochlorite, stir the mixture briefly, and test as directed above for excess hypochlorite ion. If a blue color appears on the test paper, add an additional 2 mL of NaHSO$_3$ solution and repeat the hypochlorite test. If no color appears on the test paper, proceed to the next step.

In each extraction, remove the lower aqueous phase before adding the next aqueous solution; the ether solution remains in the funnel during all the washings.

Transfer the contents of the flask to a separatory funnel. Rinse the flask with 20 mL of ether and add the ether to the separatory funnel. Shake the funnel thoroughly to mix the layers [see Technique 4.2]. Remove the lower aqueous layer and wash (extract) the organic phase consecutively with 10 mL water, 10 mL of saturated sodium bicarbonate solution **(Caution: Foaming may occur.)**, and 10 mL of distilled water. Pour the ether solution into a dry 50-mL Erlenmeyer flask and add anhydrous magnesium sulfate to dry the solution [see Technique 4.6]. Place a cork in the Erlenmeyer flask and allow the solution to stand with the drying agent for at least 10 min.

When performing extractions, it is prudent to save all *solutions until the procedure is complete.*

Place a fluted filter paper in a conical funnel and filter the dried solution through the filter paper into a dry 100-mL round-bottomed flask. Rinse the drying agent with approximately 2 mL of ether and add this ether rinse to the funnel. Remove most of the ether by simple distillation on a steam bath [see Technique 7.3] or with a rotary evaporator [see Technique 4.8]. Transfer the residue to a tared 50-mL Erlenmeyer flask. Rinse the round-bottomed flask with approximately 2 mL ether and add this rinse to the Erlenmeyer flask. Working in a hood, finish the evaporation of ether, using a stream of nitrogen until the product forms a dry white solid. Weigh the flask and calculate the percent yield. Determine the IR spectrum of your product [see Spectrometric Method 1].

Tightly cork the flask containing your 4-*tert*-butylcyclohexanone and store it in your laboratory drawer until you are ready to perform the reduction procedure given in Project 4.2.

Cleanup: The aqueous solutions remaining from the extractions should be combined and neutralized with sodium carbonate **(Caution: Foaming.)** before being washed down the sink or poured into the container for aqueous inorganic waste. Allow the solvent to evaporate from the solid magnesium sulfate in a hood before placing it in the container for nonhazardous waste or inorganic waste. Pour any ether remaining in the TLC developing jar into the container for flammable (organic) waste. Place glass-backed thin-layer plates in the appropriate container for glass waste, as directed by your instructor; aluminum-backed thin-layer plates should be placed in a container for metal waste.

4.2 Sodium Borohydride Reduction of 4-*tert*-Butylcyclohexanone

4-*tert*-Butylcyclohexanone
mp 49° C
MW 154.2

cis-4-*tert*-Butylcyclohexanol
mp 82°–83° C

trans-4-*tert*-Butylcyclohexanol
mp 81°–82° C

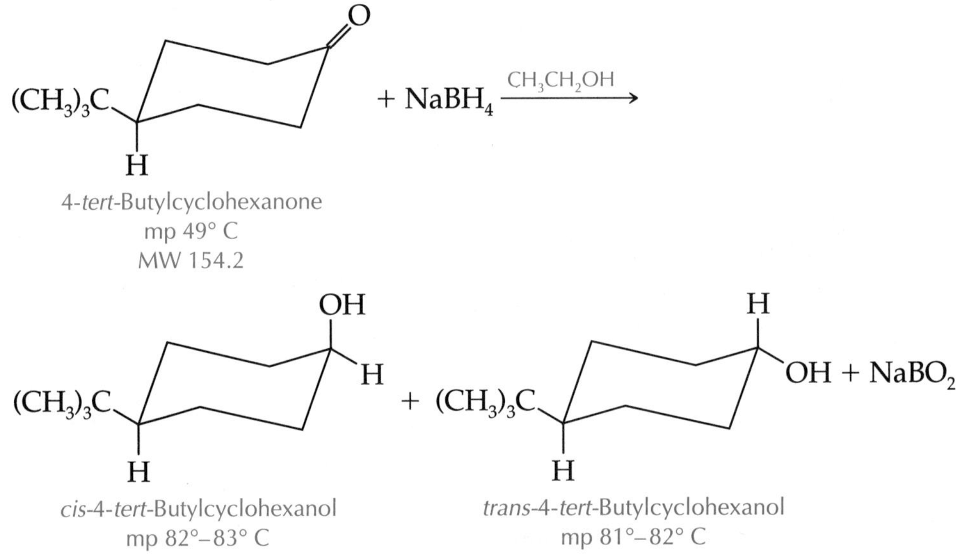

The complex metal hydride, sodium borohydride ($NaBH_4$), is a useful reducing agent for organic compounds. Aldehydes, ketones, acyl chlorides, and some lactones are readily reduced at room temperature, whereas other functional groups are not reduced except under more vigorous conditions. Sodium borohydride can be used in a range of solvents. It reacts only slowly with water to form hydrogen. The addition of alkali greatly decreases the rate of hydrogen evolution; acid, on the other hand, catalyzes it. Sodium borohydride is stable in water solution above pH 10. It is soluble in methanol and ethanol and is often used in alcohol solvents.

The probable reaction mechanism involves a hydride-ion transfer from boron to the electropositive carbonyl carbon of the ketone:

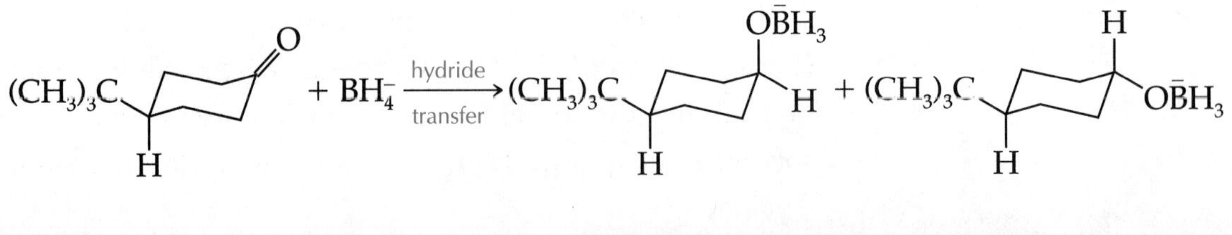

Substitution of an R_2CHO group for a hydrogen on boron does not seriously alter the reducing ability of the borohydride, and the overall stoichiometry is

$$4R-\overset{\overset{\displaystyle O}{\|}}{C}-R + Na^+BH_4^- \longrightarrow (R_2CHO)_4B^-Na^+$$

Hydrolysis of the alkoxyborane follows:

$$(R_2CHO)_4B^-Na^+ + 2H_2O \longrightarrow 4R_2CHOH + NaBO_2$$

One of the more intriguing questions in stereochemistry has been the three-dimensional structure of six-membered saturated rings. There are many natural products which contain cyclohexane rings, such as the steroid cholesterol and its analogs.

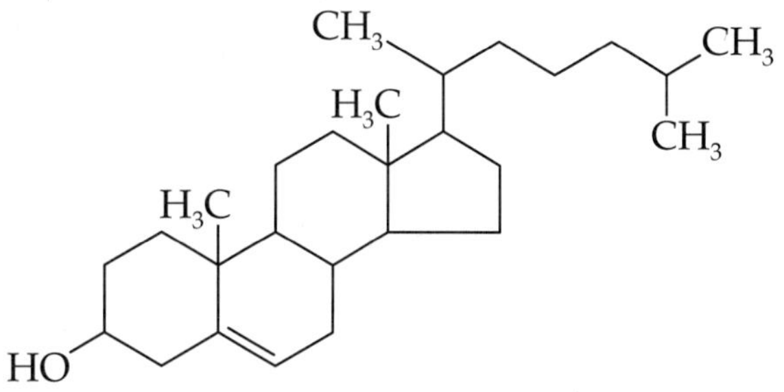

Cholesterol

The most stable conformational isomer of cyclohexane is a structure in the form of a chair. In the chair form, all the carbon-carbon bond angles are 109°, the stable tetrahedral angle.

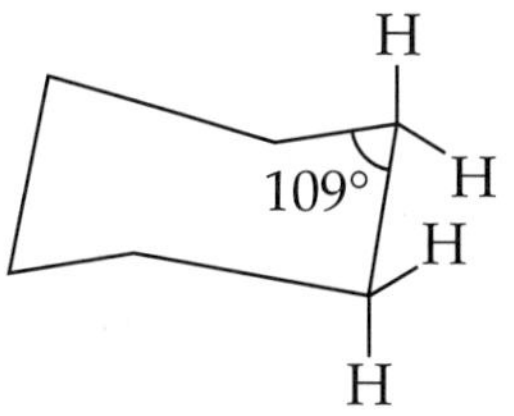

Chair form

Under ordinary conditions, a single chair form has a finite, but short existence, before it undergoes a ring-flipping process to another conformational isomer of equal stability.

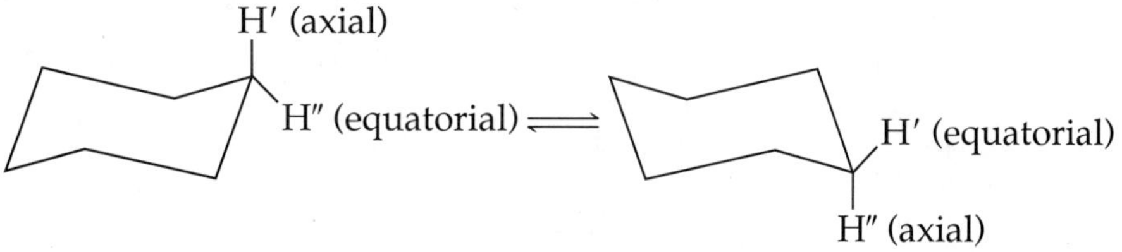

This ring inversion in cyclohexane converts axial hydrogens to equatorial ones and vice versa.

There are ways to freeze a cyclohexane ring in a particular conformation. One way is to put a large bulky group on the cyclohexane ring so that steric interference between the bulky group and other atoms in the molecule becomes so large that one conformer is completely favored over the other. Such a group is the *tert*-butyl group.

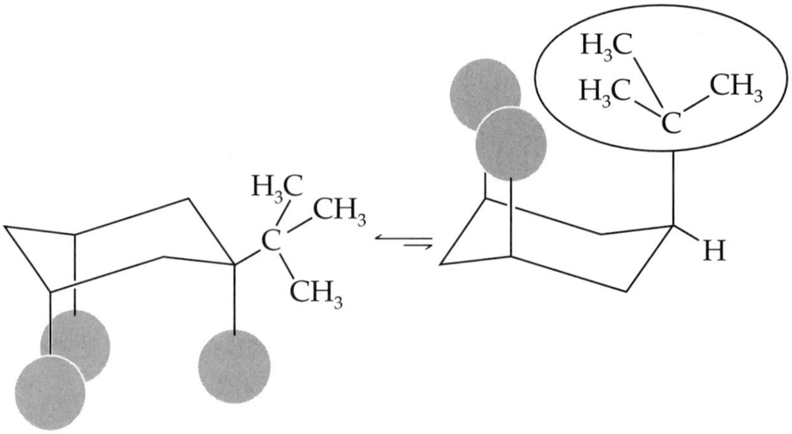

Equatorial-*tert*-butyl
no 1,3-diaxial interactions
>99.9%

Axial-*tert*-butyl
severe 1,3-diaxial interactions
<0.1%

The large *tert*-butyl group freezes the conformation of the cyclohexane ring, because it strongly prefers the equatorial position. The hydroxyl group in *cis*-4-*tert*-butylcyclohexanol is therefore held in the axial position, while both large groups are in equatorial positions in the more stable *trans* compound.

You will follow the course of the reduction reaction by using TLC and determine the ratio of *cis* to *trans* isomers formed in the reaction by using either gas chromatographic or NMR analysis.

Macroscale Procedure

Techniques   Extraction: Technique 4.2
Drying Organic Liquids: Technique 4.6
Thin-Layer Chromatography: Technique 10
Gas Chromatography: Technique 11
NMR: Spectrometric Method 2

___ *SAFETY INFORMATION* ___

4-*tert*-**Butylcyclohexanone** may be an irritant. Avoid breathing the dust. Avoid contact with skin, eyes, and clothing.

(continued on next page)

━━ *SAFETY INFORMATION (continued)* ━━

NaBH₄ is moisture sensitive; keep the bottle tightly closed. Store the reagent in a desiccator.

Sodium borohydride is harmful if swallowed, inhaled, or absorbed through the skin. Avoid breathing the dust. Avoid contact with skin, eyes, and clothing. It decomposes to flammable, explosive hydrogen gas.

Diethyl ether is extremely volatile and flammable. Use it in a hood, if possible. Be sure that there are no flames in the laboratory or that there are no hot electrical devices in the vicinity while you are pouring ether or performing the extractions.

If you are using the 4-*tert*-butylcyclohexanone that you synthesized in Project 4.1, proportionally adjust the quantities of all reagents for your amount of ketone. Carry out the reaction in the Erlenmeyer flask containing your 4-*tert*-butylcyclohexanone.

Reduction Reaction   Dissolve 0.80 g of 4-*tert*-butylcyclohexanone in 6.0 mL of 95% ethanol in a 50-mL Erlenmeyer flask. Slowly add 0.60 g of sodium borohydride. The mixture may become warm, so have a cold-water bath handy. Keep the reaction temperature between 25°–35°C. You may see a slow evolution of hydrogen gas during the reaction. After 15 min, check for completeness of reaction (indicated by the disappearance of the 4-*tert*-butylcyclohexanone spot) by the TLC method used in Project 4.1 [also see Technique 10]. When the reaction is complete, add 10 mL of water and then carefully add 1.0 mL of 3 M hydrochloric acid solution. After the evolution of hydrogen gas becomes very slow or ceases, heat the mixture to the boiling point and then cool the mixture to room temperature or lower. Add 10 mL of diethyl ether and stir the mixture. If a white inorganic solid forms, decant the liquid away from the solid into a separatory funnel. If no solid is present, simply pour the mixture into a separatory funnel. Rinse the Erlenmeyer flask with 5 mL of ether and add this rinse to the separatory funnel.

Working in a hood, shake the separatory funnel to mix the phases and drain the lower aqueous layer into a labeled flask [see Technique 4.2]. Pour the ether layer into another labeled flask.

Return the aqueous phase to the separatory funnel and extract it a second time with 10 mL of ether. Remove the lower aqueous phase and combine the two ether solutions in the separatory funnel. Wash the combined ether layers with two 8-mL portions of water to remove most of the ethanol that remains dissolved in the ether layer. Transfer the ether solution to a dry 50-mL Erlenmeyer flask after the last water wash. Dry the ether/product solution with anhydrous magnesium sulfate [see Technique 4.6].

Be sure to separate the first portion of water before adding the second.

Filter the product solution into a tared (weighed) 50-mL Erlenmeyer flask, using a fluted filter paper [see Technique 4.7], and evaporate the ether on a steam bath in a hood, using a boiling stick to prevent bumping. Alternatively, the ether may be evaporated in a hood, using a stream of nitrogen, or evaporated, using a rotary evaporator [see Technique 4.8]. Weigh your product after all of the ether has been removed and calculate your percent yield.

GC or NMR Analysis of the Product

Analyze the product composition by quantitative GC analysis [see Technique 11.6]. Use a nonpolar column at 120–125°C. Consult your instructor about sample preparation and sample size for a packed column chromatograph. For a capillary column instrument, prepare a solution containing 5 mg of your product in 1.0 mL of ether; inject a 1-μL sample of this solution into the chromatograph. On both a packed Carbowax column and a nonpolar capillary column, *cis*-4-*tert*-butylcyclohexanol has a shorter retention time than does *trans*-4-*tert*-butylcyclohexanol. On a packed column, 4-*tert*-butylcyclohexanone and the *cis* alcohol may have very similar retention times. On a nonpolar capillary column, the compounds elute in the following order: *cis*-4-*tert*-butylcyclohexanol, *trans*-4-*tert*-butylcyclohexanol, 4-*tert*-butylcyclohexanone. If you have two peaks on your gas chromatogram, demonstrate that one of them is or is not 4-*tert*-butylcyclohexanone by using the peak enhancement method [see Technique 11.6].

The product may also be analyzed by NMR spectrometry [see Spectrometric Method 2]. Prepare an NMR sample as directed by your instructor. Refer to p. 396 for a discussion of NMR analysis that uses the alcohol peaks of the *cis* and *trans* isomers to quantitatively determine their ratio in the product mixture. Alternatively, the peaks from the *tert*-butyl group of each isomer can be expanded and integrated to give the ratio of the two isomers.

(What chemical shift would you expect for the protons on a *tert*-butyl group?)

Optional GC or NMR Experiment

The 4-*tert*-butylcyclohexanol available commercially is an unspecified mixture of *cis* and *trans* isomers. Analyze the composition by quantitative GC analysis, using the chromatographic conditions specified in Project 4.2.

Alternatively, the isomeric mixture in commercial 4-*tert*-butylcyclohexanol may be analyzed by NMR spectrometry. Prepare an NMR sample as directed by your instructor. The methine proton on the carbon containing the hydroxyl group occurs at 3.5 ppm for *trans*-4-*tert*-butylcyclohexanol and at 4.0 ppm for *cis*-4-*tert*-butylcyclohexanol. Integration of these peaks will give the ratio of isomers [see Spectrometric Method 2].

Compare the ratio of *cis* and *trans* isomers found in commercially available 4-*tert*-butylcyclohexanol with the isomeric ratio that you found in your product from the sodium borohydride reduction. What does this suggest about the method used to produce commercial 4-*tert*-butylcyclohexanol?

Cleanup: The aqueous solutions remaining from the extractions may be washed down the sink or poured into the container for aqueous inorganic waste. Put any white solid formed in the reaction mixture into the container for inorganic waste. Place the magnesium sulfate drying agent in the container for nonhazardous waste or inorganic waste. Place the glass-backed thin-layer plates in the appropriate glass waste container, as directed by your instructor. Place aluminum-backed thin-layer plates in the metal waste container. If you prepared ether solutions for the GC analysis, pour any remaining solutions into the container for flammable (organic) waste.

〜〜〜Macroscale
       Procedure

Techniques   Microscale Extraction: Technique 4.5c
Drying Organic Solutions: Technique 4.6
Thin-Layer Chromatography: Technique 10
Gas Chromatography: Technique 11
NMR: Spectrometric Method 2

$NaBH_4$ is moisture sensitive; keep the bottle tightly closed. Store the reagent in a desiccator.

Note to instructor: Prepare the sodium borohydride solution just prior to use in a quantity sufficient for the entire class. The solution will be cloudy and starts to decompose slowly; it should be used as soon as possible after preparation.

Place 200 mg ± 5 mg of 4-*tert*-butylcyclohexanone in a 5-mL conical vial. Add 1.0 mL of 95% ethanol and dissolve the solid. Set the vial in a 50-mL beaker containing enough water to reach the level of liquid in the vial. Pipet 0.80 mL of a freshly prepared sodium borohydride solution that contains 150 mg of $NaBH_4$ per milliliter of 95% ethanol into a 3-mL conical vial and cap the vial. Using a Pasteur pipet, make slow dropwise additions of the sodium borohydride solution to the reaction vial containing the 4-*tert*-butylcyclohexanone solution. The initial reaction may be vigorous. After the addition is complete, remove the reaction vial from the water bath. Monitor the reaction by the TLC method used in Project 4.1 [also see Technique 10]. Continue stirring for 15 min and check for completeness of reaction (indicated by the disappearance of the 4-*tert*-butylcyclohexanone spot) by TLC.

When the reaction is complete, add 1.0 mL of water to the reaction mixture, then slowly add 0.25 mL of 3 M hydrochloric acid. **(Caution: If you add the HCl too quickly, it may cause the reaction mixture to bubble out of the vial.)** When the evolution of hydrogen gas ceases, add 1.0 mL of ether to the vial. Cap the vial and shake it to mix the layers thoroughly. Open the cap cautiously to vent the vial. Remove the lower aqueous layer, using a Pasteur filter pipet [see Technique 4.5c] and transfer it to a clean 5-mL conical vial. Repeat the extraction of the aqueous layer, using another 1.0-mL portion of ether. Remove the lower aqueous phase and transfer it to a small test tube. Transfer the first ether extract to the vial containing the second ether extract. Then wash the combined ether layers successively with two 1.0 mL portions of water. Dry the ether solution with 100 mg of anhydrous magnesium sulfate [see Technique 4.6].

The water washes remove most of the ethanol that is dissolved in the ether solution.

Transfer the dried ether solution to a tared 10 × 75 mm test tube or 3-mL conical vial, using a clean Pasteur filter pipet. Under a hood, evaporate all the ether from the product mixture, using a gentle stream of nitrogen. Weigh the vial (or test tube) and product to determine the percent yield.

Analyze the product composition by quantitative GC or NMR spectrometry as directed in the macroscale procedure.

Cleanup: The aqueous solutions remaining from the extractions may be flushed down the sink or poured into the container for aqueous inorganic waste. Place the magnesium sulfate drying agent in the container for nonhazardous or inorganic waste. Place the glass-backed thin-layer plates in the appropriate container for glass waste, as directed by your instructor; place aluminum-backed thin-layer plates in the container for metal waste. If you prepared ether solutions for the GC analysis, pour any remaining solutions into the container for flammable (organic) waste.

Reference

1. Mohrig, J. R.; Nienhuis, D. M.; Linck, C. F.; Van Zoeren, C.; Fox, B. G.; Mahaffy, P. G. *J. Chem. Educ.* **1985**, *62*, 519–521.

Questions

1. What IR band is most likely to be observed for unreacted starting material in the oxidation procedure?
2. Reference 1 (Mohrig *et al.,*) suggests that 2-propanol, rather than sodium bisulfite ($NaHSO_3$), can also be used to destroy the excess sodium hypochlorite. What organic compound would be formed and why would it not be a contaminant here?

3. Balance the redox reaction that occurs between bisulfite and hypochlorite ions. The products are sulfate and chloride ions.
4. How might you minimize the hydrogen evolution when using $NaBH_4$ in water?
5. How does the temperature affect $NaBH_4$ reactivity?
6. Which isomer, *cis* or *trans*, of 4-*tert*-butylcyclohexanol was formed in greater yield in your experiment? Explain why.
7. Reductions using $LiAlH[OC(CH_3)_3]_3$ are much milder reactions than reductions using $LiAlH_4$. Explain.

Project 5

ALDOL CONDENSATION USING AN UNKNOWN ALDEHYDE AND AN UNKNOWN KETONE

Derivative Formation and Identification of Unknowns

In this project you will carry out an aldol condensation between an unknown aldehyde and an unknown ketone. Your task is to identify the aldehyde and ketone by determining the melting point and analyzing the NMR spectrum of the condensation product. You also prepare derivatives of the aldehyde and ketone and use their melting points as additional data in determining their identity.

The formation of carbon-carbon bonds is of fundamental importance in synthetic organic chemistry, and the aldol condensation, including its many modifications, has a long and successful history as a method of carbon-carbon bond formation. The base-promoted condensation of two molecules of acetaldehyde (ethanal) represents a typical aldol condensation. The aldol product can subsequently dehydrate to form the conjugated aldehyde product:

Acetaldehyde   "Aldol"   E-2-butenal
(ethanal)   (3-hydroxybutanal)

The mechanism of this reaction involves base-catalyzed abstraction of an α-hydrogen of acetaldehyde, followed by condensation of this resonance-stabilized enolate with another molecule of acetaldehyde:

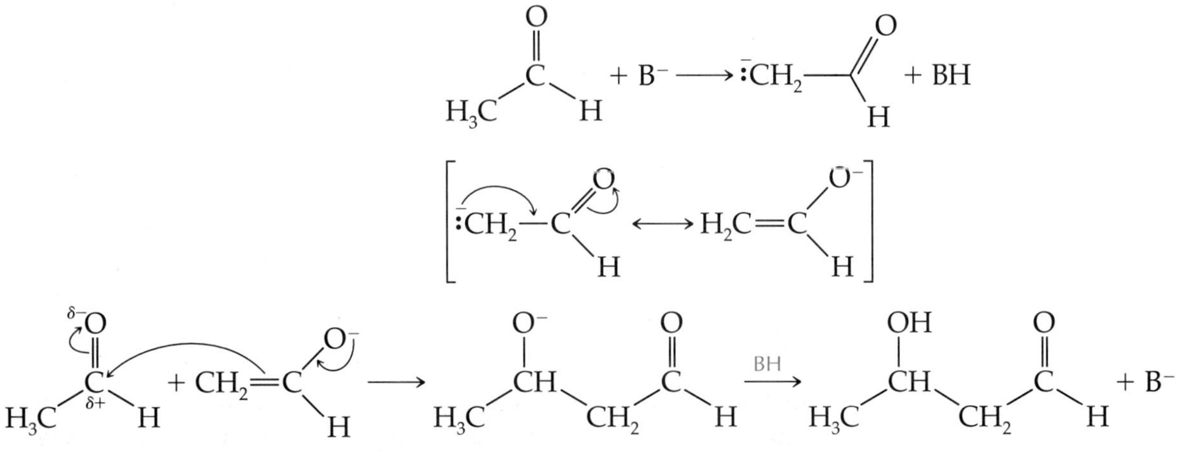

In the presence of base, dehydration occurs in two steps. The first step is the formation of an enolate anion by base-catalyzed abstraction of an α-proton from the aldol condensation product. The second is the expulsion of the hydroxide anion to form the conjugated dehydration product:

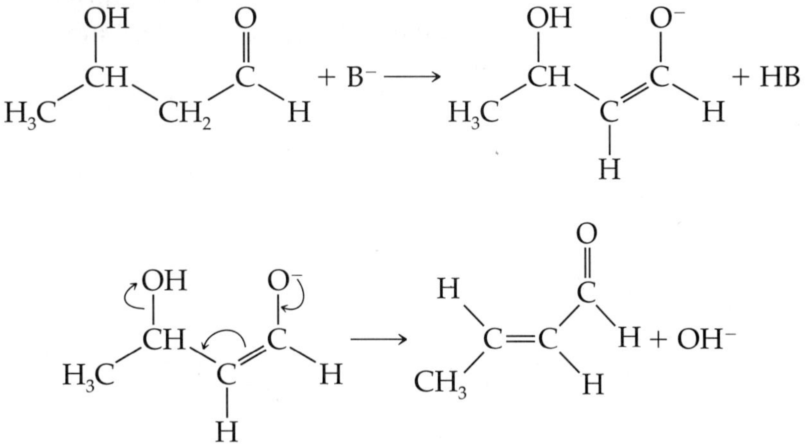

It can be difficult to predict whether the initially formed aldol condensation product will spontaneously dehydrate or not, but the equilibrium position for dehydration is made more favorable by formation of a product with extended conjugation, as well as by precipitation of the dehydration product.

For all the compounds used in this experimental study, not only does dehydration occur under the reaction conditions, but a "double condensation" also occurs, because two molecules of aldehyde condense with one molecule of ketone to form a doubly conjugated ketonic product:

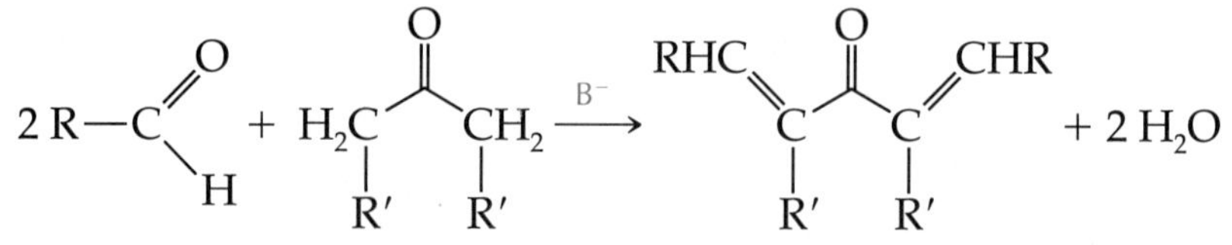

In this experiment, R′ = H or the alkyl portion of a cyclic ketone. For example, cyclopentanone gives the condensation product at the right.

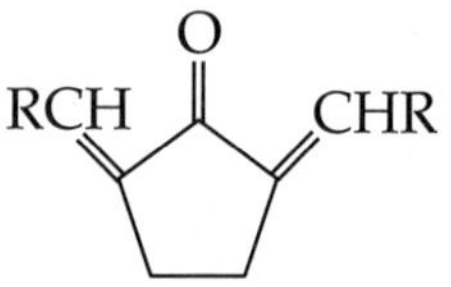

In your study of aldol condensations, you will be provided with a pair of carbonyl compounds (one aldehyde and one ketone) and asked to carry out a condensation between these two compounds. You will use an aldehyde that bears no reactive α-hydrogens and has a good potential for carrying out a crossed aldol condensation with the ketone.

To characterize your compounds, you need to carry out certain chemical reactions, and you should also obtain an NMR spectrum of your aldol product. First, you will make use of chromic acid to differentiate between aldehyde and ketone structures of your starting materials. When chromium trioxide (CrO_3) is dissolved in aqueous sulfuric acid, chromic acid is formed:

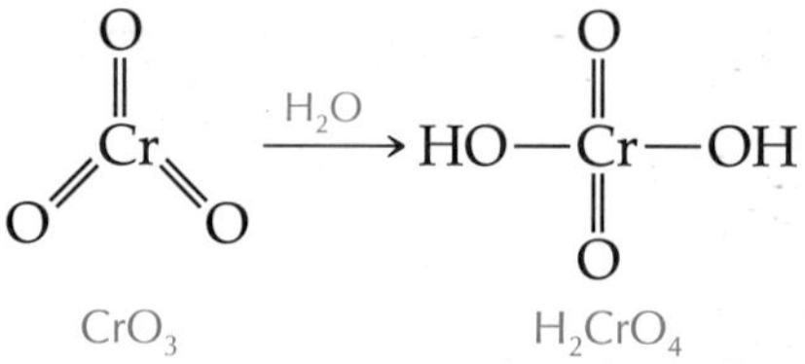

CrO_3 H_2CrO_4 (chromic acid)

Like a Cr(VI) compound, chromic acid will oxidize compounds containing a hydroxyl group on a carbon also bearing a hydrogen atom:

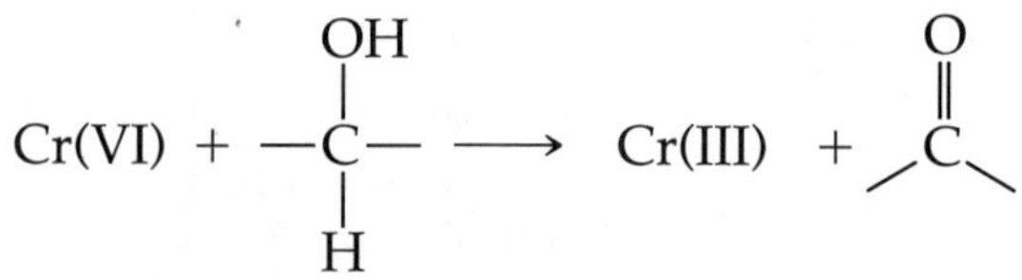

(red-orange) (blue-green)

Aldehydes in aqueous acid form *gem*-diols (compounds with two —OH groups on the same carbon), and these compounds are oxidized to carboxylic acids, reducing Cr(VI) to an opaque blue-green solution of Cr(III):

gem-Diol

A color change from orange-red to blue-green is taken as a positive test for aldehydes, but not ketones. In fact, acetone is used as the solvent for this procedure. Ketones undergo oxidation with Cr(VI) much more slowly than aldehydes do.

Both aldehydes and ketones usually undergo rapid reactions with arylhydrazines:

Phenylhydrazine   A phenylhydrazone

2,4-Dinitrophenylhydrazine   A dintrophenylhydrazone

These hydrazones normally occur as high-melting-point solids (especially in the case of the deeply colored 2,4-dinitrophenylhydrazones), and their melting points can serve as aids in identifying your starting carbonyl compounds. Normally, an organic chemist has a large degree of flexibility in choosing which derivative to prepare. In this experiment you will find it useful to prepare the phenylhydrazone as the derivative of your aldehyde and the 2,4-dinitrophenylhydrazone as the derivative of your ketone.

You will use ^{1}H NMR to characterize your aldol product. Analysis of its spectrum will allow you to draw conclusions about the structure of your aldol condensation-dehydration product and, by deduction, the structures of the aldehyde and ketone you used as starting materials for the reaction. The ^{1}H NMR spectra for all the aldehyde and ketone starting materials are available in the *Aldrich Library of* ^{13}C *and* ^1H *FT-NMR Spectra,* and you should use them to deduce the origin and the identity of your aldol product. Normally, we can expect that the NMR characteristics of the starting material will be reflected to a reasonable degree in the products. For example, the pattern of the protons on an aromatic ring will be similar in the NMR spectra of a starting aldehyde and its aldol product. More ^{1}H NMR theory and practical advice on reading spectra are discussed in Spectrometric Method 2.

You will be given an aldehyde and a ketone from the following list as the starting reagents for the aldol condensation.

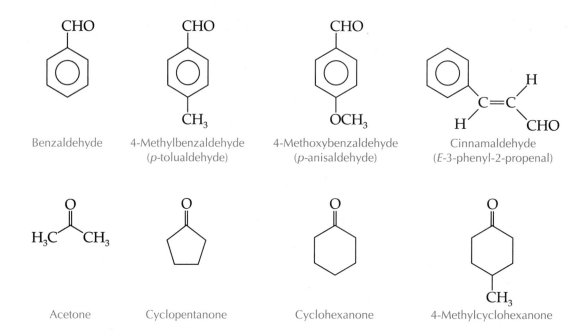

| Benzaldehyde | 4-Methylbenzaldehyde (*p*-tolualdehyde) | 4-Methoxybenzaldehyde (*p*-anisaldehyde) | Cinnamaldehyde (*E*-3-phenyl-2-propenal) |

| Acetone | Cyclopentanone | Cyclohexanone | 4-Methylcyclohexanone |

Table 5.1 lists the melting points of the aldol products formed from these aldehydes and ketones. The identification of the condensation-dehydration product deduced from this melting point table and from the ^{1}H NMR data should, of course, be the same.

In summary, the following strategy will be useful in your study:

1. Carry out the aldol condensation between an unknown aldehyde and an unknown ketone.

2. Determine the melting point and ^{1}H NMR spectrum for this product.

3. Differentiate the aldehyde and ketone starting materials on the basis of chromic acid tests.

Table 5.1. **Melting points of aldol condensation-dehydration products**

Aldehydes	Ketones			
	Acetone	Cyclo-pentanone	Cyclo-hexanone	4-Methylcyclo-hexanone
Benzaldehyde	113°C	189°C	118°C	98–99°C
4-Methylbenzaldehyde	175°C	235–236°C	170°C	133–135°C
4-Methoxybenzaldehyde	129–130°C	212°C	159°C	141–142°C
Cinnamaldehyde	144°C	225°C	180°C	163–164°C

4. Prepare the phenylhydrazone of the aldehyde and the 2,4-dinitrophenylhydrazone of the ketone, and determine their melting points.

5. Deduce the structure of your aldol condensation-dehydration product and your starting materials by evaluating the clues collected in steps 2–4.

5.1
Preparation of Aldol Condensation Product

Procedure

Techniques Selecting a Recrystallization Solvent: Technique 5.2

Microscale Recrystallization: Technique 5.5a

— **SAFETY INFORMATION** —

Wear gloves while performing the experiment and work in a hood, if possible.

The aldehydes and ketones used in this experiment are skin and eye irritants. Aqueous sodium hydroxide solutions are corrosive and cause burns.

Select (or your instructor will assign) a set of unknowns. The vials contain enough of each compound for the aldol condensation and the preparation of a derivative. The labels on the unknowns have a number followed by a volume, either 0.40 mL or 0.10 mL; for example, 29–0.40 mL and 29–0.10 mL.

The various possible condensation products form at different rates.

With a 1-mL graduated pipet, measure 0.40 mL of the unknown labeled "no. XX–0.40 mL" and place it in a 10-mL Erlenmeyer flask. With another 1-mL graduated pipet, measure 0.10 mL of the unknown labeled "no. XX–0.10 mL" and add that compound to the same Erlenmeyer flask. Add 2.0 mL of 95% ethanol and 1.5 mL of 2 M sodium hydroxide solution to the flask. Stir the solution with a magnetic stirrer for 15 min—or longer, if precipitate is still forming. If the solution is only cloudy or very little precipitate has formed after 15 min, heat the reaction mixture on a steam bath for 10–15 min. Cool the flask to room temperature.

When precipitation is complete, cool the flask in an ice-water bath for 10 min. While the flask is cooling, obtain 4 mL of 95% ethanol and transfer it to a test tube; repeat this procedure with 2 mL of 4% acetic acid in 95% ethanol (v/v). Chill the test tubes containing these solutions in the ice-water bath.

Collect the product by vacuum filtration on a Hirsch funnel [see Technique 5.5a]. Disconnect the vacuum and rinse the product with 2 mL of the ice-cold ethanol. Reconnect the vacuum and draw the liquid from the product. Repeat this washing procedure with the 2 mL of acetic acid/ethanol solution, and finally wash the crude product with the remaining 2 mL of ethanol.

Before recrystallizing the crude product, you need to find a suitable solvent. Test 95% ethanol and toluene according to the procedure described in Technique 5.2. The addition of a few drops of hexane may be necessary to promote crystallization from a toluene solution. If neither of these recrystallization solvents is satisfactory for your condensation product, test a 9:1 (v/v) mixture of 95% ethanol/acetone. Once you have selected a suitable solvent, carry out the recrystallization [see Technique 5.5a]. Leave the recrystallized product in your drawer to dry until the next laboratory period.

Cleanup: Place all filtrates in the container for flammable waste.

5.2
Characterization of the Aldol Product

Weigh your aldol product and determine its melting point. Obtain an NMR spectrum of your product, as directed by your instructor [see Spectrometric Method 2]. Put your aldol product in a vial, add a label with your name, the unknown number, your lab section, the amount of product, and the melting point; submit the product to your instructor.

Characterizing the Unknown Aldehyde and Ketone

Before you proceed with the preparation of derivatives, you need to ascertain which unknown is the ketone and which the aldehyde. As discussed in the introduction to this project, aldehydes undergo rapid oxidation with chromic acid, whereas ketones require a longer time to react.

--- SAFETY INFORMATION ---

Wear gloves while performing the following procedures and work in a hood.

The aldehydes and ketones used in this experiment are skin and eye irritants. Chromic acid solutions are toxic and corrosive. If you spill any on your skin, wash it off immediately with copious amounts of water.

Phenylhydrazine and 2,4-dinitrophenylhydrazine are toxic and possible carcinogens.

Chromic Acid Test

Perform this test on both unknown compounds. Add 1 drop of the unknown compound to a test tube and add 1 mL of acetone. Then add 1 drop of the chromic acid/sulfuric acid test reagent directly into the solution. Shake the mixture. A green or bluish green precipitate appears within 1 min if the unknown compound is an aldehyde. Aliphatic aldehydes give precipitates within 15 s, whereas aromatic aldehydes usually take 30–60 s. Acetone and other ketones oxidize in this solution only after 2–3 min.

Cleanup: Pour the test reaction mixture into the container for chromium or hazardous metal waste.

Phenylhydrazone of Unknown Aldehyde

Place 2 drops of the unknown aldehyde and 0.5 mL of ethanol in a 10×75 mm test tube. Add water dropwise, shaking after each drop, until the cloudiness produced just disappears. If too much water is used, add 1 or 2 drops of ethanol to give a clear solution.

Add 2 drops of phenylhydrazine and shake the test tube to mix the contents. A precipitate forms within a few minutes. Chill the test tube in an ice-water bath and collect the phenylhydrazone on a Hirsch funnel by vacuum filtration. Wash the precipitate three times with 0.5 mL of ice-cold ethanol.

Recrystallize the phenylhydrazone from ethanol / water mixed solvent by dissolving the product in ethanol, then adding water to the hot solution until it is just cloudy [see Technique 5.2]. Cool the solution to room temperature, then chill the recrystallized mixture in an ice-water bath before collecting the crystals on a Hirsch funnel by vacuum filtration. Allow the phenylhydrazone to dry and then determine its melting point. Compare the melting point of your derivative with those given in Table 5.2.

Cleanup: Pour the filtrate from the reaction mixture and from the recrystallization into the container for flammable (organic) waste.

2,4-Dinitrophenyl-hydrazone of Unknown Ketone

Thorough washing removes the sulfuric acid from the reagent solution. Any remaining acid catalyzes decomposition of the dinitrophenylhydrazone when you heat it, and the decomposition products lower the melting point and increase the melting-point range.

You should not confuse the melting point of unreacted 2,4-dinitrophenylhydrazine, 197–198°C, with that of a possible derivative.

Dissolve 3 drops of the unknown ketone in 0.5 mL ethanol in a 10 × 75 mm test tube. Add 1.5 mL of the 2,4-dinitrophenylhydrazine reagent solution. Shake the tube to mix the contents. A precipitate usually forms immediately. Let the mixture stand at room temperature for 15 min before collecting the solid product on a Hirsch funnel by vacuum filtration. Wash the crystals with three 1-mL portions of cold ethanol. To do this, remove the suction, add the ethanol, and stir the crystals gently to wash them completely. Then apply the suction again. Recrystallize the 2,4-dinitrophenylhydrazone from ethanol [see Technique 5.4]. Allow it to dry and determine its melting point. Compare the melting point of your derivative with those given in Table 5.3.

Cleanup: Pour the filtrate from the reaction mixture and the recrystallization into the container for flammable (organic) waste.

Report

Your laboratory notebook or report must include a brief synopsis of your reasoning leading to identification of all three compounds, chemical equations for the condensation-dehydration reactions,

Table 5.2. **Melting points of phenylhydrazone**

Aldehyde	Phenylhydrazone melting point, °C
Benzaldehyde	158
4-Methylbenzaldehyde	114
4-Methoxybenzaldehyde	120
Cinnamaldehyde	168

Table 5.3. **Melting points of 2,4-dinitrophenylhydrazone**

Ketone	2,4-Dinitrophenylhydrazone melting point, °C
Acetone	126
Cyclopentanone	142
Cyclohexanone	162
4-Methylcyclohexanone	130

the percent yield of your condensation product, and the reactions for the chromic acid test and the formation of both hydrazones. You should also include an analysis of the ^{1}H NMR spectrum of the product, with structural assignments for all signals, and a brief but complete summary of how this spectrum was useful in identifying your three compounds.

References

1. Hathaway, B. *J. Chem. Educ.* **1987,** *64*, 367–368.
2. Pickering, M. *J. Chem. Educ.* **1991,** *68*, 232–234.
3. Shriner, R. L.; Fuson, R. C.; Curtin, D. Y.; Morrill, T. C. *The Systematic Identification of Organic Compounds;* 6th ed.; Wiley: New York, 1980.

4. Rappoport, Z. *Handbook of Tables for Organic Compound Identification,* 3rd ed.; CRC Press: Boca Raton, 1967.

5. Pouchert, C. J.; Behnke, J. (Eds.) *Aldrich Library of ^{13}C and ^1H FT-NMR Spectra,* 3 vols.; Aldrich Chemical Co.: Milwaukee, WI, 1993.

Questions

1. Write the mechanism for the aldol condensation of propanal. Use hydroxide ion as the base and explain, using your mechanism, whether the base is a true catalyst.

2. Although acid is not needed to dehydrate the aldol products in the study you did, it certainly will enhance the rate of this elimination reaction. Write a mechanism for acid-promoted dehydration of the aldol product obtained in Question 1.

3. To undergo an aldol condensation with itself, an aldehyde or ketone must contain at least one α-hydrogen. Which of the following compounds have no α-hydrogens? (a) acetone; (b) 2-butanone; (c) diphenylketone; (d) cyclobutanone; (e) 2,2,4,4-tetramethylcyclobutanone

4. Formation of aldol products from carbonyl compounds is the result of an equilibrium reaction. Write the reaction for acetone, excluding dehydration, a case in which the equilibrium does not favor the aldol condensation. Discuss why product formation is more unfavorable than the condensation of acetaldehyde and suggest experimental approaches on how the equilibrium can be readjusted to produce a reasonable amount of aldol condensation product.

5. The crossed aldol condensation of acetaldehyde with benzaldehyde is aided by the fact that benzaldehyde bears no α-hydrogens. However, a possible competitive reaction is the "self-condensation" of two molecules of acetaldehyde. Slow addition of the acetaldehyde to a benzaldehyde solution containing the base will often result in less self-condensation. Explain.

6. There are four aldol condensation products possible when butanal and propanal are mixed in the presence of base. What are they and how are they formed?

7. Describe, using chemical reactions, how the chromic acid test and hydrazone formation can be used to characterize your aldol condensation-dehydration product.

8. Melting points of hydrazones can be complicated by the fact that hydrazones of aldehydes and unsymmetrical ketones can exist in both *syn* and *anti* forms. Using butanal and 2-butanone as examples, draw structures for the phenylhydrazones and explain the melting-point problem.

BENZOIN CHEMISTRY
Synthesis of an Anticonvulsant Drug Using
Biochemical Catalysis

In this project you will prepare an anticonvulsant drug, 5,5-diphenylhydantoin, by a three-step synthesis that uses benzaldehyde and urea, two simple molecules, as the building blocks.

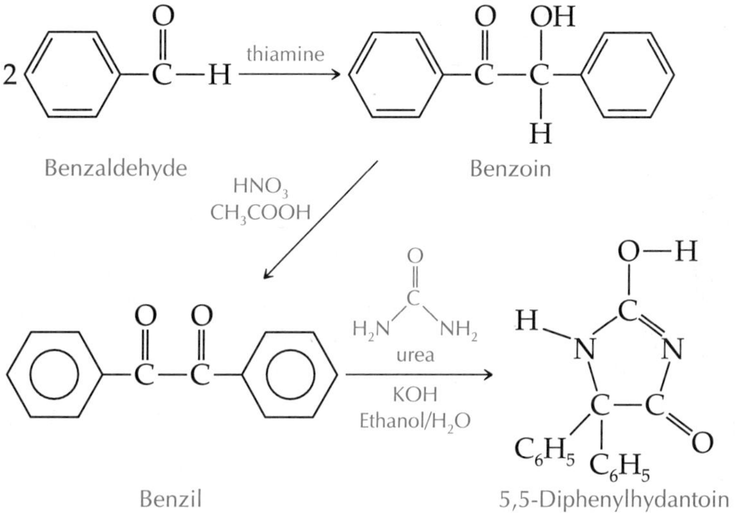

In the first step of this synthesis, benzaldehyde is converted to benzoin by a condensation reaction that is biochemically catalyzed by thiamine hydrochloride (vitamin B_1). Benzoin is oxidized by nitric acid to benzil in the second step. In the last step, benzil is combined with urea to form 5,5-diphenylhydantoin.

Thiamine-Catalyzed Benzoin Condensation

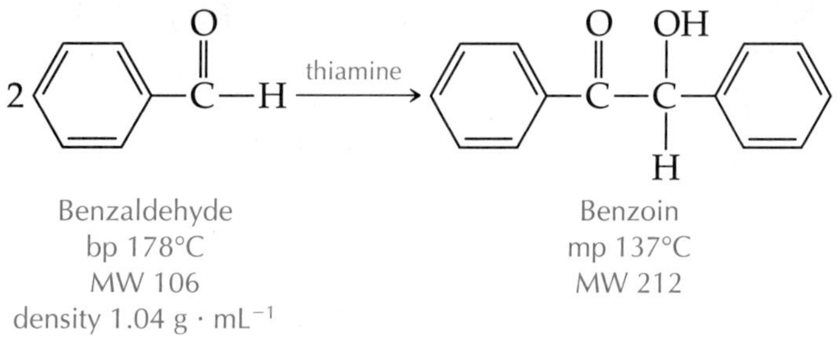

Benzaldehyde
bp 178°C
MW 106
density 1.04 g · mL⁻¹

Benzoin
mp 137°C
MW 212

The Benzoin Condensation

One of the greatest challenges in organic synthesis is the controlled formation of carbon-carbon bonds. The synthesis of a complex molecule from simple precursors involves the formation of many such bonds, and chemists have devoted their careers to the discovery of more efficient and specific synthetic methods. The benzoin condensation is a classic example of the specific catalysis of carbon-carbon bond formation.

The remarkably specific catalysis by cyanide ion in this reaction was first studied in the early years of this century. Cyanide is a highly specific catalyst because it can perform three functions: (1) it acts as a nucleophile; (2) it increases the acidity of the aldehydic proton; and (3) it can subsequently serve as a leaving group:

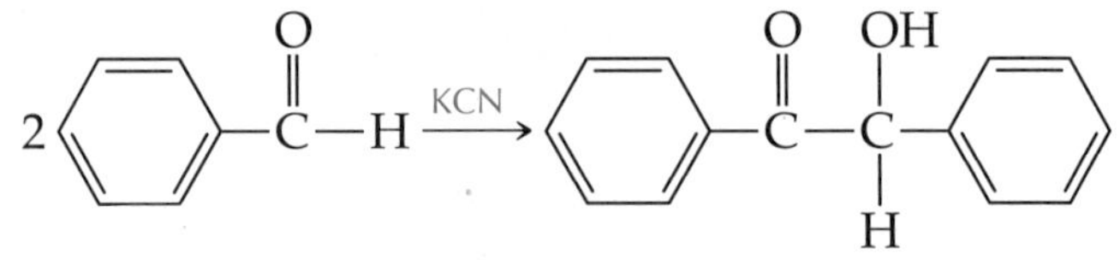

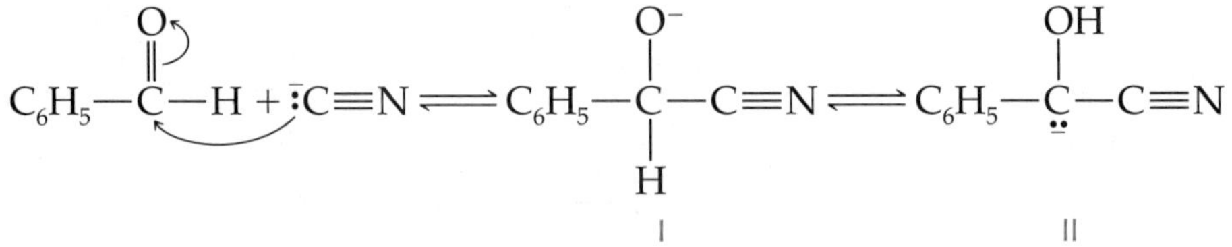

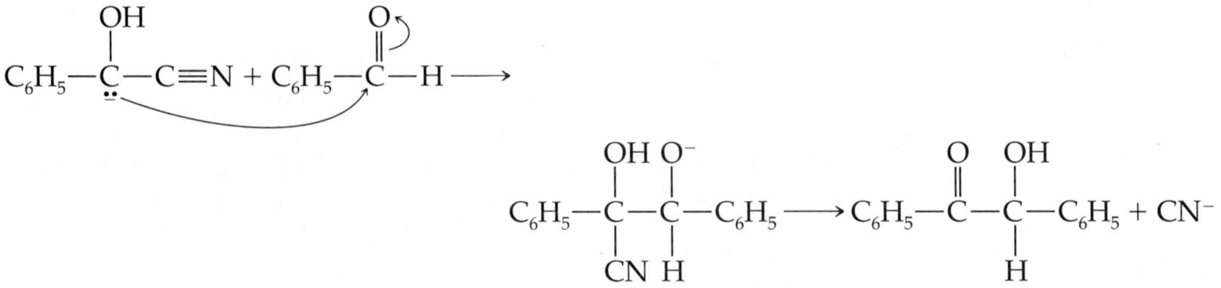

The electron-withdrawing cyano group gives a tremendous boost to the acidity of the α-proton in the cyanohydrin (I); removal of the proton by base then produces a carbon nucleophile (II). The carbon nucleophile reacts at the carbonyl-carbon atom of a second molecule of benzaldehyde, making a carbon-carbon bond. Loss of cyanide ion gives the product, benzoin; and the cyanide ion can recycle as a catalyst.

Although the cyanide-catalyzed benzoin condensation is an important reaction, it is not without danger. Potassium cyanide and hydrogen cyanide are violent poisons that can kill with little warning. Death may result from a few minutes exposure to 300 ppm of hydrogen cyanide in the atmosphere. Thiamine provides a far safer catalytic agent for the benzoin condensation.

Thiamine   Vitamins are organic compounds that are necessary for good health and growth, but cannot be synthesized efficiently in our bodies. A balanced diet contains sufficient quantities of the essential vitamins, but when people choose or are forced to restrict their diets, they may not ingest all the vitamins necessary for sound health. A deficiency of thiamine in the human diet can result in fatigue and depression. People who eat insufficient thiamine may get a disease of the nervous system called beriberi, which is characterized by partial paralysis of the extremities, emaciation, and anemia. After Casimir Funk isolated a dietary growth factor from rice polishings in 1911 and showed that this factor cured beriberi, he proposed that the term "vitamine" be used for the trace nutrients in foods whose lack produces disease. His term was accepted. When it was discovered that not all of these growth factors were amines, the final letter was dropped, giving us the word *vitamin.*

To describe why vitamins are necessary for good health, we need to discuss the high molecular weight proteins, or enzymes, that catalyze the chemical reactions in our bodies. An enzyme provides the structured active-site matrix for a biochemical reaction. Amino acid units at the active site of the enzyme provide a catalytic function at one or more steps of the reaction pathway. In all cases the enzyme performs a vital catalytic function by bringing the reactants together in the spatial relationships necessary for the specific metabolic reaction.

Many of the enzymes that catalyze and control our metabolic chemistry require additional small organic molecules as cocatalysts. These molecules are called *coenzymes.* In many biochemical reactions, much of the bond making and breaking takes place in the coenzymes and the substrates. In other words, a coenzyme is a chemical reagent necessary for the reaction. Many of these vital coenzymes are synthesized in our bodies from the vitamins that we eat with our food.

Thiamine pyrophosphate (TPP) is a coenzyme of universal occurrence in living systems (Figure 6.1). It is easily formed in human beings and other animals from the important nutritional factor thiamine.

One of the many metabolic reactions in which thiamine pyrophosphate is a necessary reagent is the decarboxylation of the α-keto acid pyruvic acid. Human beings, as well as many other species, produce pyruvate from carbohydrates. The Mg(II) ions

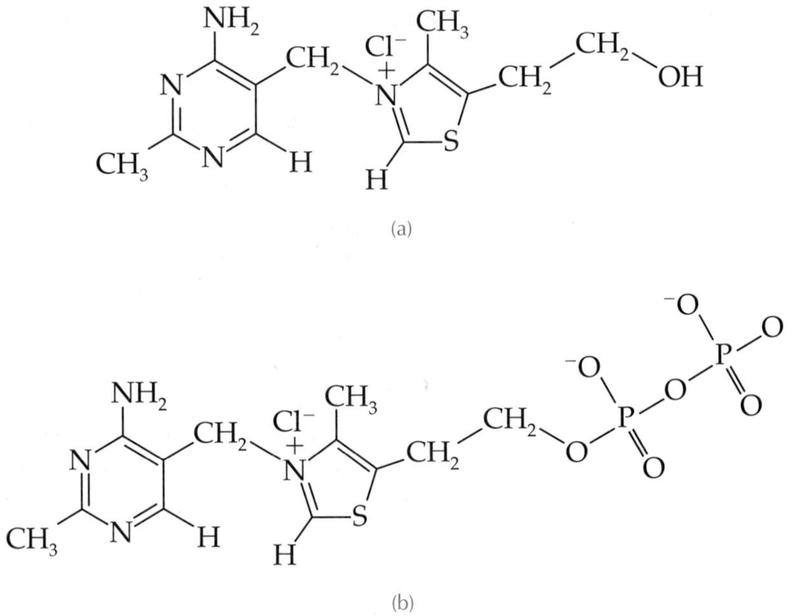

(a)

(b)

FIGURE 6.1 (a) Thiamine. (b) Thiamine pyrophosphate (TTP).

are required for binding the thiamine pyrophosphate to the enzyme's active site through charge attractions to the phosphate-oxygen atoms:

$$CH_3\overset{O}{\underset{\|}{C}}\overset{O}{\underset{\|}{C}}OH \xrightarrow[\substack{\text{thiamine pyrophosphate}\\ Mg^{2+}}]{\text{pyruvate decarboxylase}} CH_3\overset{O}{\underset{\|}{C}}H + CO_2$$

In human beings, acetaldehyde is not liberated from the enzyme surface but is oxidized to acetyl CoA as a prelude to further oxidation. Yeasts, for example, reduce the acetaldehyde to ethanol, which builds up to a significant concentration.

The mechanism of the biochemical action of TPP has been shown to involve the removal of a relatively acidic proton on the five-membered thiazolium ring component of thiamine by a basic group at the active site of the enzyme. This reaction produces a carbanion that is nucleophilic and can attack the carbonyl carbon of aldehydes and ketones in the same way a cyanide ion does. In this respect, thiamine is a biochemical analogue of cyanide ion.

The mechanism by which thiamine catalyzes the benzoin condensation is very similar to the mode of action of thiamine under physiological conditions, except that no enzyme is present in the experiment that you will do. To write out a complete mechanism for the thiamine catalysis of the benzoin condensation takes quite a bit of space, because the molecules involved are reasonably com-

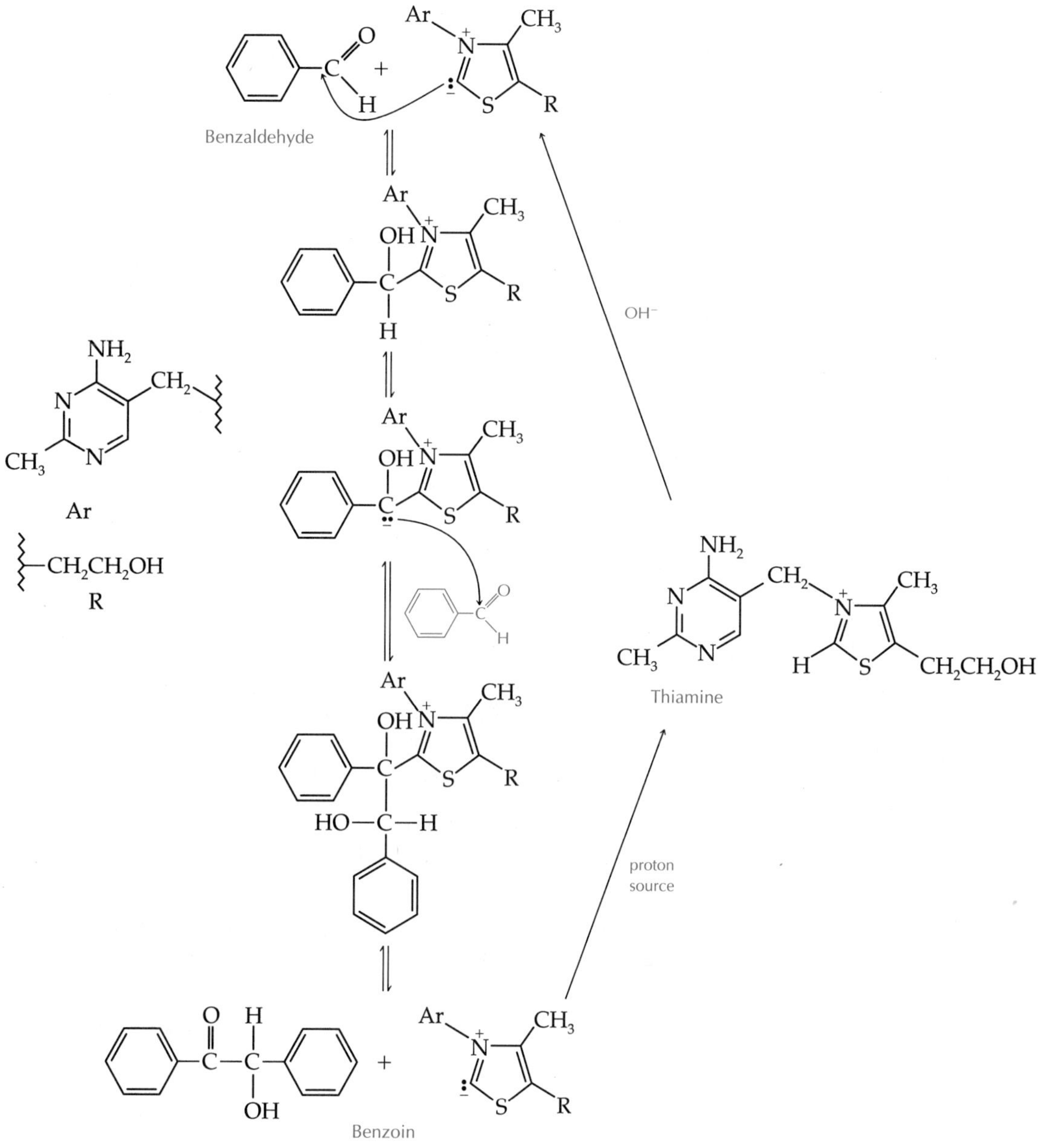

FIGURE 6.2 Thiamine catalysis of the benzoin condensation.

plex. Nevertheless, no new kinds of catalysis are present. The same laws of chemistry and physics are at work.

The most important steps in the mechanism are shown in Figure 6.2. Notice the similarity of the thiamine reactivity to that of cyanide shown earlier. The thiamine is able to serve the same three critical functions: (1) it acts as a nucleophile; (2) it boosts the acidity of the aldehydic proton of benzaldehyde; and (3) it can subsequently serve as a leaving group.

Reagents   The benzaldehyde used for this experiment must be free of benzoic acid. Benzaldehyde oxidizes in the presence of air, and crystals of benzoic acid are often visible at the bottom of an older bottle of benzaldehyde that has been opened several times in the past. Benzoic acid will change the pH of the reaction medium, so that the yield of benzoin is very low. *If you are unsure of the age or purity of your benzaldehyde, distill it* at atmospheric pressure before starting this experiment. Avoid exposing hot benzaldehyde to the air. A newly opened bottle of benzaldehyde works best for this experiment.

Thiamine hydrochloride is sensitive to heat and highly alkaline pH. It should be recently purchased for this experiment and stored in the refrigerator when not being used. It is important not to heat your reaction mixture any more than necessary and not to add any more sodium hydroxide than needed.

WWWMacroscale Procedure

Techniques   Recrystallization: Technique 5.3
IR Spectrometry: Spectrometric Method 1

SAFETY INFORMATION

Benzaldehyde is toxic and an irritant. Wear gloves and avoid contact with skin, eyes, and clothing.

Aqueous **sodium hydroxide** solutions are corrosive and cause burns. Solutions as dilute as 9% (2.5 M) can cause severe eye injury.

Weigh 1.1 g of recently purchased thiamine hydrochloride (vitamin B_1) and dissolve it in 2.5 mL of distilled water in a 50-mL round-bottomed flask. Add 10 mL of 95% ethanol and cool the resulting solution in an ice-water bath. While continuing to cool the thiamine hydrochloride solution in the ice-water bath, slowly add 2.2 mL of cold 3 M sodium hydroxide solution over a 5-min period. Gently swirl the thiamine solution during the entire addition to ensure thorough mixing. The solution will become yellow.

Measure 6.0 mL of benzaldehyde (freshly distilled or from a newly opened bottle) and add it to the reaction mixture. Fit the flask with a water-cooled condenser. Heat the reaction at 60°C for 90 min in a water bath *or* stopper the reaction and allow it to stand in your laboratory drawer until your next laboratory period (48 h or more).

If the reaction mixture was heated, allow it to cool to room temperature. Precipitate the benzoin by cooling the flask in an ice-water bath until the temperature is about 10°C. Benzoin appears as white crystals. If the benzaldehyde was not completely pure, some oil may also be floating on the surface of the reaction mixture. This oil is a mixture of benzaldehyde, benzoic acid, and some benzoin. It may be analyzed by thin-layer chromatography on silica gel plates, using a polar developing solvent [see Technique 10].

Collect the crude solid product by vacuum filtration. Wash it with 25 mL of cold water while it is still on the Buchner funnel.

Set aside 10–15 mg of crude benzoin for a melting-point determination before you recrystallize the rest from 95% ethanol [see Technique 5.3]. Approximately 8 mL of 95% ethanol are required per gram of crude benzoin. Be careful not to add more than the minimum amount of ethanol needed to dissolve the product.

Determine the melting points of both the crude and recrystallized benzoin. In some sources benzoin is reported to melt at 137°C; in others, at 133°–134°C. Calculate your percent yield. Typical yields are 40–50% if the recrystallization is done carefully. Determine the IR spectrum of your product. Allow your benzoin to dry overnight (or longer) before using it in the synthesis of benzil (Project 6.2). Your instructor may also ask you to obtain and analyze the NMR spectrum of your benzoin.

Cleanup: If the filtrate from the reaction mixture has an upper layer of oil, transfer it to a separatory funnel. Drain the lower aqueous layer into a beaker and neutralize the solution to pH 7, using 10% hydrochloric acid before washing it down the sink or pouring it into the container for aqueous inorganic waste. Pour the oil remaining in the separatory funnel into the container for flammable or organic waste. Pour the ethanol filtrate from the recrystallization into the container for flammable or organic waste.

Techniques   Microscale Recrystallization: Technique 5.5a
IR Spectrometry: Spectrometric Method 1

SAFETY INFORMATION

Benzaldehyde is toxic and an irritant. Wear gloves and avoid contact with skin, eyes, and clothing.

Aqueous **sodium hydroxide** solutions are corrosive and cause burns. Solutions as dilute as 9% (2.5 M) can cause severe eye injury.

Weigh 135 mg of thiamine hydrochloride (vitamin B_1) and dissolve it in 0.27 mL of distilled water in a 5-mL conical vial. Add 1.20 mL of 95% ethanol and cool the resulting solution in an ice-water bath. While continuing to cool the thiamine hydrochloride solution in the ice bath, slowly add 0.27 mL of cold 3 M sodium hydroxide solution dropwise over 2–3 min. Gently swirl the vial during the entire addition to ensure thorough mixing. The solution will become yellow.

Measure 0.75 mL of benzaldehyde from a newly opened bottle and add it to the reaction mixture. Fit the vial with a water-cooled condenser. Heat the reaction at 60°C for 75–80 min in a water bath *or* cap the reaction vial and allow it to stand in your laboratory drawer until your next laboratory period (48 h or more).

If the reaction mixture was heated, cool the vial to room temperature. Precipitate the benzoin by cooling the vial in an ice-water bath until the temperature is about 10°C. Benzoin appears as white crystals. If the benzaldehyde was not completely pure, some oil may also be floating on the water. This oil is a mixture of benzaldehyde, benzoic acid, and some benzoin. It may be analyzed by thin-layer chromatography on silica gel plates, using a polar developing solvent [see Technique 10].

Collect the crude solid product by vacuum filtration on a Hirsch funnel. Wash it with three 1-mL portions of cold water while it is still on the Hirsch funnel.

Set aside a few milligrams of crude benzoin for a melting point determination. Recrystallize your crude product from 95% ethanol in a 13 × 100 mm test tube, using a boiling stick to prevent bumping [see Technique 5.5a]. Be careful not to add more than the minimum amount of ethanol needed to dissolve the product. Cool the solution slowly to room temperature, then briefly in an ice-water bath. Collect the product by vacuum filtration on a Hirsch funnel.

Determine the melting points of both the crude and recrystallized products. In some sources benzoin is reported to melt at 137°C; in others, at 133–134°C. Calculate your percent yield. Determine the IR spectrum of your product. Your instructor may also ask you to obtain and analyze the NMR spectrum of your benzoin.

Cleanup: If the filtrate from the reaction mixture has an upper layer of oil, transfer it to a conical vial. Remove the lower aqueous layer with a Pasteur pipet and neutralize it to pH 7, using 10% hydrochloric acid before washing it down the sink or pouring it into the container for aqueous inorganic waste. Pour the oil remaining in the vial into the container for flammable or organic waste. Pour the ethanol filtrate from the recrystallization into the container for flammable or organic waste.

6.2 Oxidation of Benzoin

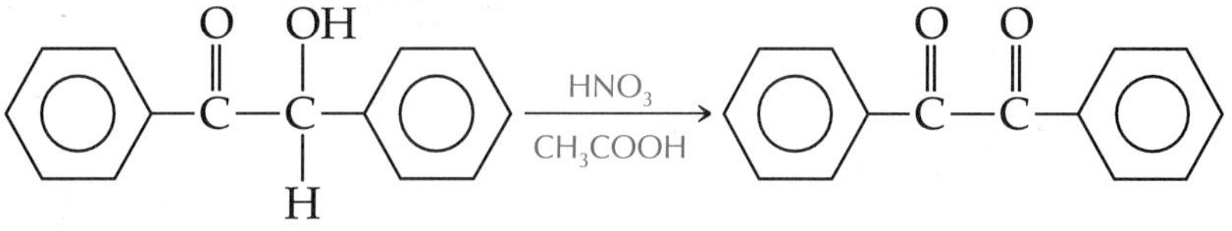

Benzoin
mp 137°C
MW 212

Benzil
mp 95°C
MW 210

A number of methods have been used to oxidize benzoin to the yellow diketone, benzil. Using nitric acid oxidation works well, and the progress of the reaction can easily be followed by thin-layer chromatography [see Technique 10]. While you cannot determine the exact composition of the reaction mixture on a thin-layer plate, you can readily detect changes from analyzing a series of thin-layer plates.

Monitoring a reaction by removing and analyzing small portions of the reaction mixture is a very useful practice. Without monitoring, the temptation is to heat the reaction mixture longer than necessary to make sure that the reaction is finished. This prolonged heating not only wastes time but often leads to decomposition of a desired product and results in a lower yield of a less pure product. Before beginning the experimental procedure, review the theory and techniques of TLC found in Technique 10.

/////Macroscale
Procedure

Techniques   Thin-Layer Chromatography: Technique 10
Recrystallization: Technique 5.3

SAFETY INFORMATION

Perform the entire experiment in a hood.

Both **glacial acetic acid** and concentrated **nitric acid** are corrosive. Wear gloves and avoid contact with skin, eyes, and clothing. If either acid gets on your skin, wash it off immediately with copious amounts of water.

Chloroform and **dichloromethane** are volatile and toxic.

Methanol is volatile, toxic, and flammable.

The quantities of reagents specified here are for 2.0 g of benzoin. If you are using the benzoin that you synthesized in Project 6.1, you should use all of your benzoin and proportionally adjust the quantities of the other reagents.

Before you start the reaction, analyze the 10% chloroform solutions of benzoin and benzil available in the laboratory, using silica gel thin-layer plates [see Technique 10]. Dichloromethane is a useful developing solvent for these compounds. Use an ultraviolet lamp to visualize the spots if you have fluorescent thin-layer plates; or use *p*-anisaldehyde visualization reagent if you are using glass- or aluminum-backed thin-layer plates [see Technique 10.5]. Calculate an R_f value for each substance [see Technique 10.1a]. Which one has the greater R_f?

Use a 100-mL round-bottomed flask fitted with a Claisen adapter, glass stopper, and reflux condenser for the reaction. The nitrogen dioxide gas passing out of the condenser during the reaction should be trapped in a water-ice mixture, in which NO_2 dissolves and decomposes. Attach a U-shaped piece of glass tubing to the top of the reflux condenser by means of a one-hole rubber stopper or a thermometer adapter. Place the open end of the U-tube **just above** the surface of about 50 mL of a water/ice mixture in a 125-mL filter flask, as shown in Figure 6.3.

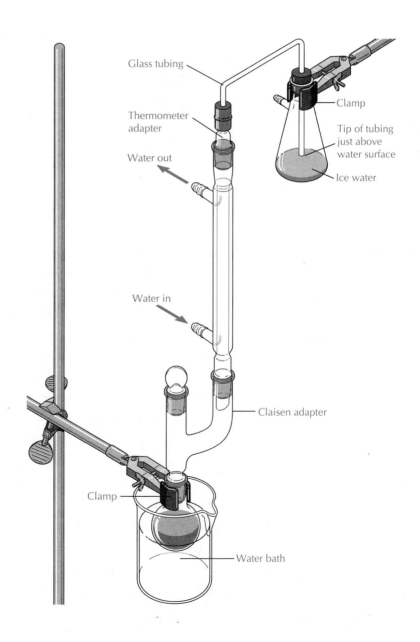

FIGURE 6.3 Apparatus for the oxidation of benzoin.

Brown, noxious nitrogen dioxide (NO$_2$) fumes escape from the reaction when you withdraw samples for TLC analysis during the heating period.

Place 2.0 g of benzoin in the 100-mL round-bottomed flask; add 10 mL of glacial acetic acid and 5 mL of concentrated nitric acid (70% HNO$_3$, 15 M). Assemble the apparatus as shown in Figure 6.3. Heat the reaction mixture in a water bath at 85°–95°C. After 15–20 min of heating, take out the glass stopper and remove about 0.2 mL of the reaction solution, using a Pasteur pipet. Quickly return the stopper to the Claisen adapter. Place the sample of reaction solution in a small test tube.

Spot a thin-layer plate with the reaction solution, using a micropipet [see Technique 10.3]. Allow the nitric and acetic acids to evaporate (in a hood), leaving the solids behind. Develop the plate with dichloromethane. Although benzil is yellow, the small amount present is not enough to allow you to see it. Visualize the spots and record your observations.

Continue taking 0.2-mL portions from the reaction mixture every 15–20 min until TLC analysis shows that no further oxidation is taking place. When the sizes and intensities of the visualized spots show no changes in two successive TLC analyses, the reaction can be considered complete. At that time, cool the reaction flask and add 40 mL of water containing some ice. Pale yellow crystals of benzil should form readily.

Filter the mixture by vacuum filtration and wash the crystals three times with 5-mL portions of cold water. Save about 10 mg of this crude product for a melting-point determination. Recrystallize your product from methanol, using a steam bath as the heat source [see Technique 5.3]. Allow crystallization to start at room temperature before cooling the mixture in an ice-water bath. Collect the product by vacuum filtration.

If you are not using your benzoin for Project 6.3 or will be using the reflux procedure for that step of the synthesis, allow the recrystallized product to dry overnight before determining the mass, the melting point, and the IR spectrum.

If you will be using your benzil to synthesize 5,5-diphenylhydantoin (Project 6.3) by the room-temperature method, draw air over the crystals for 20 min to complete the drying process. Weigh the product, then save enough crystals for a melting point and an IR spectrum. Before you leave the laboratory, set up the reaction for the synthesis of 5,5-diphenylhydantoin, using the rest of your benzil; store the reaction mixture for one week in your laboratory drawer.

Cleanup: Pour the aqueous filtrate from the oxidation of benzoin into a 600-mL beaker containing about 100 mL of water. Neutralize the solution with solid sodium carbonate **(Caution: Foaming.)** before washing it down the sink or pouring it into the container for aqueous inorganic waste. Pour the methanol filtrate from the recrystallization into the container for flammable or organic waste.

6.3 Synthesis of 5,5-Diphenylhydantoin, an Anticonvulsant Drug

Benzil	Urea	5,5-Diphenylhydantoin
MW 210	MW 60	MW 252
mp 137°C	mp 132.7°C	mp 295°–298°C

The sodium salt of 5,5-diphenylhydantoin (5,5-diphenyl-2,4-imidazolidinedione) is an anticonvulsant drug with significant water solubility, an aid in administration. It is sold under a variety of trade names including Dilantin, Dihydan, Dilabid, and Zantropil. It was originally developed as an antiepileptic and acts as an anticonvulsant to stabilize, but not decrease epileptic seizures (grand mal).

Diphenylhydantoin sodium
(Dilantin sodium)

You will synthesize diphenylhydantoin by treating benzil with urea in the presence of a strong base (KOH). The mechanism begins with reaction of a urea nitrogen atom with one of benzil's carbonyl groups. Loss of a proton produces a ketone containing an adjacent oxyanion. This anion undergoes phenyl migration in a benzilic acid-like rearrangement. Dehydration then gives rise to a heterocyclic compound that tautomerizes to the final product.

$+ H_2O$

Because this class of compounds has significant biological properties, it is important that special precautions be used in its handling, including the use of protective gloves.

∿∿Macroscale — **Procedure**

Technique   Recrystallization: Technique 5.3

SAFETY INFORMATION ———

Potassium hydroxide solution causes severe burns and eye damage. Handle it carefully.

The product of this reaction, **5,5-diphenylhydantoin**, is a potent anticonvulsant drug. Wear gloves during the entire workup procedure.

Room-temperature reaction method. Place 1.0 g of benzil (synthesized in Project 6.2) and 0.48 g of urea in a 50-mL Erlenmeyer flask. Add 25 mL of 95% ethanol and swirl the mixture to dissolve the solids. Add 2.8 mL of 9.4 M potassium hydroxide solution and warm the mixture on a steam bath for 5 min. The solution may turn brown and form a fine white residue on the bottom of the flask. Cork the flask tightly and store it in your drawer for one week.

Reflux reaction method. Alternatively, the reaction mixture may be heated under reflux for 2 h in a 50-mL round-bottomed flask fitted with a water-cooled condenser. Cool the reaction mixture, then proceed as directed below.

Workup of Reaction Mixture

Wear gloves while doing this procedure. If the reaction flask contains a precipitate, perform a vacuum filtration to separate it from the product solution.

Transfer the filtrate (or the reaction mixture, if no precipitate is present) to a 150-mL beaker and add 75 mL of water. Add 10% hydrochloric acid solution dropwise until the pH of the mixture reaches 4–5 when tested with pH test paper. The product, 5,5-diphenylhydantoin, will precipitate as the solution is neutralized. Chill the mixture in an ice-water bath for 10 min and collect the product by vacuum filtration.

The exact identity of the solid that precipitates from the reaction mixture is not known for certain, but it is suspected to be diphenylacetylene diureide (Ref. 1).

Recrystallize the product from ethanol [see Technique 5.3]. Allow the product to dry before determining the melting point, using a *sealed* capillary tube. After the sample is at the bottom of the capillary tube, heat the open end of the tube briefly in a Bunsen burner flame until the glass melts and closes the opening. Weigh your product, determine the percent yield, and submit it to your instructor in a labeled vial. Your instructor may also ask you to obtain and analyze the NMR spectrum of your 5,5-diphenylhydantoin.

Using a sealed capillary will prevent any vapors of diphenylhydantoin from escaping into the laboratory.

Cleanup: The aqueous filtrate from the reaction mixture can be washed down the sink or poured into the container for aqueous inorganic waste. Pour the ethanol filtrate from the recrystallization into the container for flammable (organic) waste.

Reference

1. Hayward, R. C. *J. Chem. Educ.* **1983,** *60,* 512.

Questions

1. How does a base, such as NaOH, help to catalyze the benzoin condensation when thiamine is present?
2. Suggest a reason why benzoic acid impurity in the benzaldehyde may drastically reduce the yield of the thiamine-promoted condensation of benzaldehyde.
3. You might have thought that adding more NaOH solution would neutralize the benzoic acid present if old benzaldehyde were used in the synthesis of benzoin. Unfortunately, too much base catalyzes the formation of noxious hydrogen sulfide. Where does the hydrogen sulfide come from?
4. What modifications in the reaction conditions would be necessary if an enzyme were present in the reaction?
5. Write a mechanism showing how thiamine pyrophosphate catalyzes the decarboxylation of pyruvic acid.
6. What major differences would be observed in the IR spectra of benzoin and benzil?
7. What structural features make benzil yellow and benzoin colorless (white)?
8. Explain differences in TLC R_f values for benzoin and benzil. Specifically, why does the compound with the lower R_f value cling to the stationary phase more tightly?
9. Do the properties of your product (color, physical appearance, crystal structure, etc.) match those described in *The Merck Index* for 5,5-diphenylhydantoin?
10. Write an equation that illustrates why 5,5-diphenylhydantoin precipitates when the reaction mixture is treated with 10% HCl.
11. Barbiturates are another example of biologically important heterocyclic compounds. The diethyl compound shown (Barbital) is a sedative and hypnotic. It is prepared by treatment of the diethyl ester of α,α-diethylmalonic acid with urea in the presence of an alkoxide base. Outline a mechanism for this reaction.

5,5-Diethylbarbituric acid
(barbital)

SYNTHESIS OF BENZOCAINE

In this project you will prepare compounds of pharmaceutical interest. In a two-step synthesis, *p*-nitrobenzoic acid is first reduced to *p*-aminobenzoic acid (PABA), whose derivatives are used in many sunscreens. An esterification reaction with ethanol then converts *p*-aminobenzoic acid to ethyl *p*-aminobenzoate (benzocaine), a local anesthetic.

p-Nitrobenzoic acid   *p*-Aminobenzoic acid   Ethyl *p*-Aminobenzoate (benzocaine)

The widely used local anesthetic benzocaine can be synthesized in a two-step sequence beginning with the reduction of *p*-nitrobenzoic acid. The product of the first step is *p*-aminobenzoic acid (PABA), whose derivatives are used as the active components of many sunscreens. PABA is also found as a structural unit of folic acid, a water-soluble vitamin.

Folic acid

Human beings must obtain folic acid from their diet, whereas bacteria and many plants can make their own folic acid, using PABA as one of the biosynthetic building blocks. Sulfa drugs, of which sulfanilamide is an example, are effective against bacteria because they block the biosynthetic incorporation of PABA into folic acid. The death rate due to pneumonia has been drastically reduced because of pharmaceuticals such as sulfanilamide.

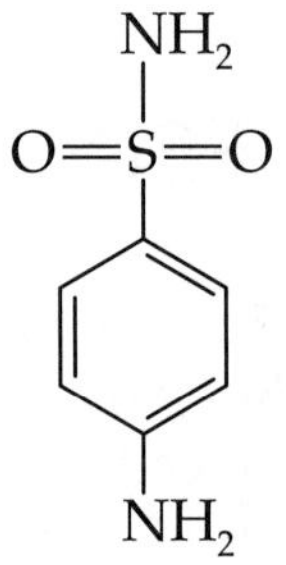

Sulfanilamide
(*p*-aminobenzenesulfonamide)

An interesting aside is the fact that the dye Prontosil has been found to have antibacterial properties. Microorganisms reduce Prontosil to sulfanilamide, which is actually the effective reagent.

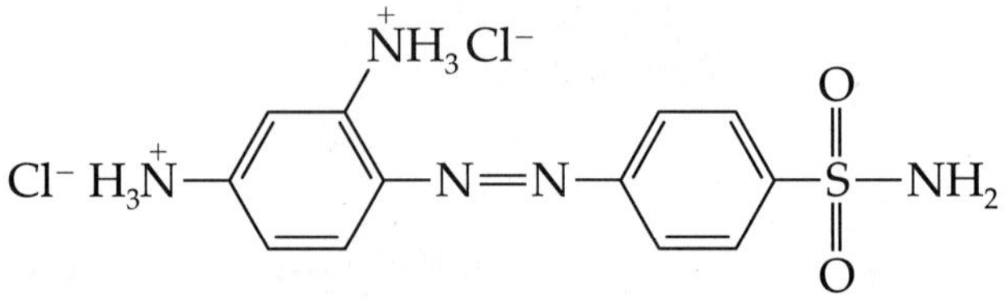

Prontosil
(2,4-diaminoazobenzene-4′-sulfonamide hydrochloride)

7.1
Synthesis of *p*-Aminobenzoic Acid

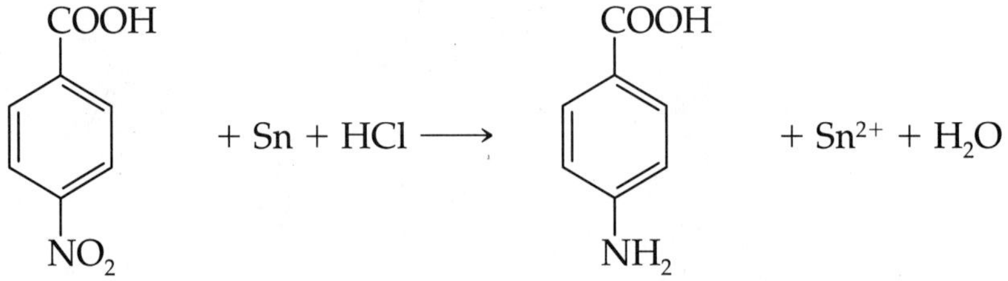

p-Nitrobenzoic acid
mp 242.2°C
MW 167

p-Aminobenzoic acid
mp 187°C
MW 137

The first step in the synthesis of benzocaine uses tin and hydrochloric acid to reduce the nitro group of *p*-nitrobenzoic acid to an amino group. The reduction mechanism is quite complex and not well understood. It is thought to proceed by a series of electron transfers from tin to the organic substrate. Sn is oxidized to Sn(II) during the reaction and is removed from the product mixture as the insoluble oxide, SnO.

<table><tr><td>

Macroscale Procedure

</td><td>

Technique   Recrystallization: Technique 5.3

</td></tr></table>

— SAFETY INFORMATION —

Conduct this experiment in a hood.

4-Nitrobenzoic acid is an irritant. Avoid breathing the dust. Avoid contact with skin, eyes, and clothing.

Concentrated hydrochloric acid is toxic, corrosive, and a severe irritant to mucous membranes. Concentrated **ammonium hydroxide** is corrosive and a lachrymator. Wear gloves and use both of these reagents in a hood. Avoid contact with skin, eyes, and clothing.

Glacial acetic acid is corrosive. Avoid contact with skin, eyes, and clothing.

Place 1.67 g of *p*-nitrobenzoic acid and 3.88 g of granular tin (20 mesh) in a 125-mL Erlenmeyer flask. Add 9.0 mL of concentrated (12 M) hydrochloric acid. Close the flask with a one-hole rubber stopper fitted with a drying tube containing loosely packed moistened (not dripping wet) cotton or glass wool.

— SAFETY PRECAUTION —

The cotton must not be so wet that it mats together and closes the opening at the bottom of the drying tube. You could be heating a closed system if this were the case!

The moistened cotton or glass wool absorbs HCl gas emitted as the reaction is heated.

Warm the flask on a hot plate (use the lowest setting). Bubbling will commence. Swirl the flask occasionally to loosen the thick suspension from the sides of the flask. When the yellowish solid has dissolved (10–15 min), heat the flask at a low to moderate setting for an additional 10 min until most of the residual tin dissolves.

Cool the flask to room temperature and add 3.0 mL of water. Slowly, with stirring, add 10.0 mL of concentrated (15 M) aqueous ammonia solution. Heat the suspension of SnO on the steam bath or in a boiling-water bath for 10 min. During this time, prepare a Celite filtration pad by making a slurry consisting of 4.0 g of Celite in 30 mL of water and pouring the mixture on a 3-in. Buchner funnel while the vacuum is turned on. Remove the aqueous filtrate from the filter flask before pouring the hot suspension of SnO onto the Celite pad. Rinse the Erlenmeyer flask with 10 mL of 20% (v/v) $NH_3(aq)$/water and pour the rinse solution onto the Celite pad.

When using any pH test paper, dip a stirring rod into the solution and touch the rod to the test paper, rather than putting the paper strip into the solution.

Transfer the filtrate, which contains your PABA as the ammonium salt, to a 250-mL beaker and, working in a hood, boil the solution down to 20–22 mL. Add glacial acetic acid dropwise to the hot solution until it is weakly acidic to litmus and your *p*-aminobenzoic acid has precipitated.

Cool the beaker in an ice-water bath until crystallization is complete. Collect the product by vacuum filtration, using a clean, but not necessarily dry, filter flask. A second crop of crystals may sometimes be recovered by treating the filtrate with 0.5 mL of glacial acetic acid and cooling the mixture again in an ice-water bath. Collect the second crop of crystals by vacuum filtration and recrystallize the combined crops of *p*-aminobenzoic acid, using a 1:8 (v/v) mixture of ethanol/water [see Technique 5.3]. After the crystals have thoroughly dried, determine the mass and the melting point. Calculate your percent yield.

Cleanup: Place the Celite containing SnO in the container for hazardous metal waste. Neutralize the aqueous filtrate from the crude product with sodium carbonate **(Caution: Foaming.)** before washing it down the sink or pouring it into the container for aqueous inorganic waste. The filtrate from the recrystallization may be washed down the sink. Rinse the cotton in the drying tube with sodium bicarbonate solution before placing it in the container for nonhazardous waste.

7.2

Synthesis of Benzocaine

p-Aminobenzoic acid   Ethanol   Ethyl p-Aminobenzoate
(Benzocaine)
mp 91°–92°C
MW 165

In the second synthetic step, the carboxylic acid portion of
p-aminobenzoic acid is converted to an ester by treatment with
ethanol in the presence of a catalytic amount of sulfuric acid. Acid-
catalyzed esterification reactions are equilibrium reactions that
favor the products and reactants to approximately the same
degree when a 1:1 molar mixture of the carboxylic acid and alcohol is used. Unless special conditions are employed, the equilibrium mixture contains roughly equal amounts of reactants and
products, thus effectively limiting the product yield. The yield
can, however, be greatly improved by taking advantage of the Le
Chatelier principle (the law of mass action). Use of a large excess
of one reactant can drive the reaction toward the product side of
the equilibrium. Another way of adjusting the equilibrium is by
continual removal of water, one of the products of the reaction.
For this synthesis, you will use an excess of ethanol to favor product formation.

Use the macroscale procedure if you have more than 0.3 g of
p-aminobenzoic acid from Project 7.1. Use the microscale procedure if you have less than 0.30 g of p-aminobenzoic acid from
Project 7.1. You will need to calculate the quantities of the reagents
for this synthesis on the basis of the amount of p-aminobenzoic
acid that you obtained in Project 7.1. Use the following informa-

tion to determine the amounts of ethanol and concentrated sulfuric acid to use for each 0.100 g of p-aminobenzoic acid:

0.100 g	p-aminobenzoic acid
1.3 mL	absolute (anhydrous) ethanol
0.1 mL	concentrated (18 M) H_2SO_4

Check your quantities with your instructor before beginning the synthesis.

MMMacroscale Procedure

Techniques Recrystallization from Mixed Solvent Pairs: Technique 5.2a

Recrystallization: Technique 5.3

Extraction: Technique 4.2

IR Spectrometry: Spectrometric Method 1

NMR Spectrometry: Spectrometric Method 2

SAFETY INFORMATION

Ethanol and **diethyl ether** are very flammable.

Concentrated sulfuric acid is corrosive and causes severe burns.

Place your *p*-aminobenzoic acid, a boiling stone, and the requisite volume of absolute ethanol in a dry 25- or 50-mL round-bottomed flask. Carefully add the concentrated sulfuric acid; some precipitate will form initially but will dissolve as the mixture is heated. Fit the flask with a water-cooled reflux condenser and heat the mixture under reflux for 75 min.

At the end of the reflux period, carefully pour the hot solution into a 150-mL beaker and cool it to room temperature. Slowly add small portions of 2 M sodium carbonate solution until the acid is neutralized and the pH reaches 9. **(Caution: Foaming will occur.)** At this point a precipitate will begin to form. Add 10 ml of water and cool the mixture in an ice-water bath for 15 min to complete the crystallization process. Collect the crude product by vacuum filtration on a Buchner funnel.

Recrystallize the crude product from an ethanol/water solution [see Technique 5.2a]. Add 95% ethanol in small portions until the crude product (which forms an oil as it heats) dissolves while the flask is being heated on a steam bath [see Technique 5.3]. Then add hot water dropwise until the solution becomes cloudy; the volume of water required may nearly equal the volume of ethanol. Stir the mixture while cooling it in an ice-water bath. Collect the crystals of benzocaine by vacuum filtration. Allow your product to dry thoroughly before determining the percent yield and the melting point. Obtain an IR or NMR spectrum of your dried product, as directed by your instructor.

Cleanup: The aqueous filtrate from the crude product may be washed down the sink or poured into the container for aqueous inorganic waste. Place the filtrate from the recrystallization in the container for flammable or organic waste.

microscale
Procedure

Techniques Recrystallization from Mixed Solvent Pairs: Technique 5.2a
Microscale Recrystallization: Technique 5.5a
Microscale Extraction: Technique 4.5c
IR Spectrometry: Spectrometric Method 1
NMR Spectrometry: Spectrometric Method 2

— **SAFETY INFORMATION** —

Ethanol and diethyl ether are very flammable.

Concentrated sulfuric acid is corrosive and causes severe burns.

Guidelines for reagent quantities are on p. 439.

Place your *p*-aminobenzoic acid, a boiling stone, and the requisite volume of absolute ethanol in a dry 10-mL round-bottomed flask. Carefully add the concentrated sulfuric acid; some precipitate will form initially but will dissolve as the mixture is heated. Fit the flask with a water-cooled reflux condenser and heat the mixture under reflux for 75 min.

At the end of the reflux period, carefully pour the hot solution into a 50-mL beaker and cool the solution to room temperature. Slowly add small portions of 2 M sodium carbonate solution until the acid is neutralized and the pH reaches 9. **(Caution: Foaming will occur.)** At this point a precipitate will begin to form. Add 3 mL of water and cool the mixture in an ice-water bath for 15 min to complete the crystallization process. Collect the crude product by vacuum filtration on a Hirsch funnel.

Recrystallize the crude product from an ethanol/water solution [see Technique 5.2a]. While the flask is being heated on a steam bath, add 95% ethanol dropwise until the crude product (which forms an oil as it heats) dissolves [see Technique 5.5a]. Then add hot water dropwise until the solution becomes cloudy; the volume of water required may nearly equal the volume of ethanol. Stir the mixture while cooling it in an ice-water bath. Collect the crystals of benzocaine by vacuum filtration on a Hirsch funnel. Allow your product to dry thoroughly before determining the percent yield and the melting point. Obtain an IR or NMR spectrum of your dried product, as directed by your instructor.

Cleanup: The aqueous filtrate from the crude product may be washed down the sink or poured into the container for aqueous inorganic waste. Place the filtrate from the recrystallization in the container for flammable or organic waste.

Questions

1. What is the role of the base (in the first step, the base is aqueous ammonia; in the second, it is sodium carbonate) in the last chemical step in each of these two synthetic reactions?
2. What is the reason for using anhydrous ethanol and concentrated (>99%) sulfuric acid rather than 95% ethanol or more dilute (say, 6 M) sulfuric acid in the esterification of *p*-aminobenzoic acid?
3. Should you be concerned if denatured alcohol (prepared by adding the poisonous compound methanol to 100% ethanol) is inadvertently used in the esterification step?

4. The starting nitro compound should show two strong IR bands in the range of 1550–1350 cm^{-1}. What are they due to and why should they not be expected to appear in the IR spectrum of *p*-aminobenzoic acid?

5. IR bands due to N—H stretching vibrations usually appear near 3300 cm^{-1}. They can be useful for characterizing amines, but *p*-aminobenzoic acid gives rise to a band that could readily interfere? What is it?

6. The IR band mentioned in Question 5 should largely disappear upon esterification? Why?

7. Would the number of ^{13}C NMR bands (proton-decoupled) be of use in differentiating (a) *p*-nitrobenzoic acid from *p*-aminobenzoic acid? (b) *p*-aminobenzoic acid from ethyl *p*-aminobenzoate?

8. The result of an ^{1}H NMR analysis of your final product (benzocaine) showed the following signals: δ 7.9, 2H, d; δ 6.25, 2H, d; δ 4.25, 2H, q; δ 4.0, 1.9H, s; δ 1.3, 3H, t (s = singlet, d = doublet, t = triplet, q = quartet). Assign all of the signals. Which one of the signals would be expected to show a significant concentration dependence? Why?

9. Predict the ^{1}H NMR spectra of ethyl *p*-nitrobenzoate and *p*-nitrobenzoic acid. Describe how each spectrum would differ from the spectrum of ethyl *p*-aminobenzoate.

Project 8
RESOLUTION OF (*R,S*)-BENZOYLAMINO ACIDS

In this project you will synthesize the benzoyl derivatives of the amino acids alanine or valine as racemic mixtures, then resolve, or separate, them into their optically active components by using papain, an enzyme isolated from papaya fruit. The optical purity of the resolved benzoylamino acid anilides will be determined by polarimetry.

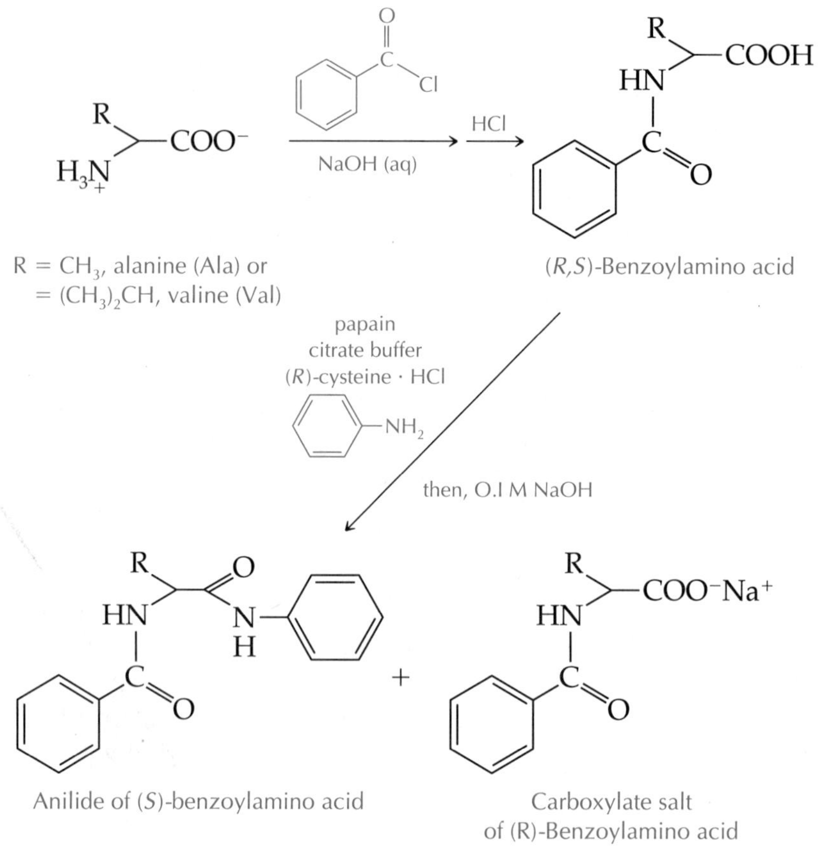

R = CH$_3$, alanine (Ala) or
= (CH$_3$)$_2$CH, valine (Val)

(R,S)-Benzoylamino acid

papain
citrate buffer
(R)-cysteine · HCl

then, O.I M NaOH

Anilide of (S)-benzoylamino acid

Carboxylate salt
of (R)-Benzoylamino acid

8.1
Papain

Relative to most proteins, the enzyme papain is rather small, with a molecular weight of only 23,350. Papain is one polypeptide chain made from 212 amino acid residues covalently linked by amide bonds. Its three-dimensional structure has been determined by x-ray crystallography (Figure 8.1).

Found in the juice of the papaya fruit, *Carica papaya*, papain constitutes 5% of the soluble protein in the milky papaya juice. The digestive action of papaya juice has been known and used for centuries. Most proteins are extensively degraded by the action of papain, which explains its use as a meat tenderizer. Papain is also used by brewers to prevent bottled beer from becoming cloudy when chilled. The cloudiness results from the precipitation of proteins, and papain cleaves them into smaller, more soluble peptide fragments. A variety of medical applications have been reported for papain, ranging from its use against intestinal worms to the

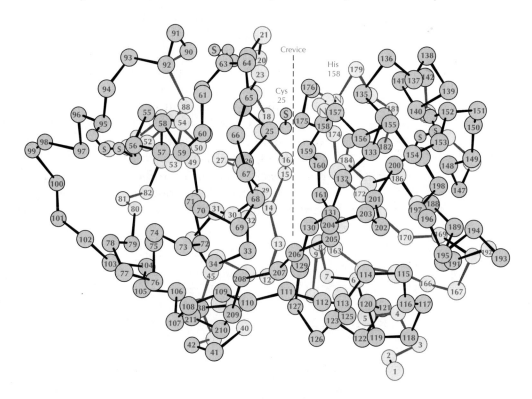

FIGURE 8.1 Three-dimensional structure of papain. The active-site crevice contains the essential cysteine thiol group. Each amino acid residue is depicted as a numbered disk.

treatment of slipped discs. Some hikers carry a small amount of papain-containing meat tenderizer on backpacking expeditions. A bit of wetted meat tenderizer applied to the site of a bee or wasp sting cleaves the protein linkages in the toxins of the insect venom.

Papain catalyzes a number of acyl addition-elimination reactions, including the hydrolysis of amides and esters. It acts on a wide variety of peptide bonds, and in dilute aqueous solution the equilibrium strongly favors the hydrolysis products:

$$\text{Polypeptide—NH—CH—}\overset{\displaystyle \overset{O}{\|}}{C}\text{—NH—Polypeptide} + H_2O \underset{}{\overset{papain}{\rightleftharpoons}}$$
$$\underset{\displaystyle R}{|}$$

$$\text{Polypeptide—NH—CH—COOH} + H_2N\text{—Polypeptide}$$
$$\underset{\displaystyle R}{|}$$

Studies on model compounds of low molecular weight show that the reaction is fastest at the carbonyl groups of arginine and lysine derivatives.

Papain is a chiral polymer that is composed of chiral amino acids, and the enzyme has a high specificity for the *S*-configuration of its substrates. You will use this property as the basis of an "asymmetric synthesis." When a benzoyl derivative of a racemic amino acid is mixed with a primary amine such as aniline (benzenamine), papain is able to catalyze the formation of an amide bond specifically with the *S*-enantiomer only. This anilide formation constitutes a resolution of the racemic mixture because the *R*-form is unreactive:

$$C_6H_5-\overset{\overset{\displaystyle O}{\|}}{C}-NH-\underset{\underset{\displaystyle R}{|}}{CH}-COOH + C_6H_5-NH_2 \underset{}{\overset{papain}{\rightleftharpoons}}$$

(*R,S*)-Benzoylamino acid

Aniline

$$C_6H_5-\overset{\overset{\displaystyle O}{\|}}{C}-NH-\underset{\underset{\displaystyle R}{|}}{CH}-\overset{\overset{\displaystyle O}{\|}}{C}-NH-C_6H_5 + C_6H_5-\overset{\overset{\displaystyle O}{\|}}{C}-NH-\underset{\underset{\displaystyle R}{|}}{CH}-COOH$$

(*S*)-Anilide of benzoylamino acid   (*R*)-Benzoylamino acid

The *S*-anilide (amide derivative of aniline) that forms is insoluble in the aqueous buffered reaction mixture and can be isolated by filtration. An excess of aniline and the insolubility of the anilide in the reaction mixture shifts the equilibrium of the reaction toward amide formation.

Just how does papain catalyze the formation and cleavage of amides? Does it employ the same kind of catalysis that low molecular weight catalysts do? The answer to the second question is a resounding "yes."

To answer the first question, we need to review briefly what is known about the mechanism of papain catalysis. First of all, papain reacts with amides, esters, and carboxylic acids to form a Michaelis-Menten complex. The enzyme then forms a covalent bond to the substrate in the form of a thiol ester:

$$\text{Enzyme}-CH_2SH + R-\overset{\overset{\displaystyle O}{\|}}{C}-NHR' \rightleftharpoons \text{Enzyme}-CH_2SH \cdot R-\overset{\overset{\displaystyle O}{\|}}{C}-NHR'$$

Michaelis-Menten complex

$$\text{Enzyme}-\text{CH}_2\text{SH} \cdot \text{R}-\overset{\overset{\displaystyle O}{\|}}{\text{C}}-\text{NHR}' \rightleftharpoons \text{Enzyme}-\text{CH}_2\text{S}-\overset{\overset{\displaystyle O}{\|}}{\text{C}}-\text{R} + \text{R}'\text{NH}_2$$

Acyl enzyme

The nucleophilic sulfur atom is known to be the thiol group of cysteine, which is amino acid 25 in the sequence of 212 amino acid residues that make up a molecule of papain (Figure 8.1).

It has been suggested that the thiol group is made even more nucleophilic by removal of its proton by the nearby imidazole ring of the amino acid histidine at position 159 in the protein chain. Reaction of this highly nucleophilic sulfur anion at the acyl-carbon atom forms the acyl-enzyme intermediate. In a few favorable cases, these covalent acyl enzyme intermediates are sufficiently stable to permit their direct examination by spectral and chemical methods.

The acyl papain is now quite reactive to nucleophilic attack because it is a thiol ester. The products of the hydrolysis reaction are a carboxylic acid and papain, which can then recycle to catalyze the hydrolysis of another amide bond:

$$\text{Enzyme}-\text{CH}_2\text{S}-\overset{\overset{\displaystyle O}{\|}}{\text{C}}-\text{R} + \text{H}_2\text{O} \rightleftharpoons \text{Enzyme}-\text{CH}_2\text{SH} + \text{RCOOH}$$

A word about stereochemical nomenclature is in order. There are three types of notation that can be used for amino acids and their derivatives. One type is the **R,S** notation for absolute configuration (based on the Cahn-Ingold-Prelog rules). Another notation refers to the direction of rotation of plane-polarized light, where the dextrorotatory form is denoted as (+), the levorotatory form as (−), and the racemic form, corresponding to a 1:1 mixture of the two enantiomeric forms, as (±). The third notation is D,L and is based on the configuration of the stereocenter relative to L-glyceraldehyde, the simplest carbohydrate. Naturally occurring amino acids normally have the L configuration. This is almost always identical to the *S* absolute configuration. It is almost impossible to predict whether an *S*-amino acid will be (+) or (−); experimental data must be used to tell.

(*S*)-Serine
(identical to L-serine
or (−)-serine)

$$\overset{\displaystyle \text{COO}^-}{\underset{\displaystyle \text{CH}_2\text{OH}}{^+\text{H}_3\text{N}\blacktriangleright\!\!-\!\!\blacktriangleleft\text{H}}}$$

In summary, R,S, ($\pm$) and D,L are all valid nomenclature methods. We have used the R,S notation in this project.

8.2 Resolution of (R,S)-Benzoylamino Acids

Papain is an excellent agent for separating racemic mixtures of benzoylamino acids. Derivatives of the amino acids alanine and valine work especially well, although several other amino acids can also be used. The products crystallize easily, and the success of the resolution can be determined by using polarimetry.

The synthesis of hippuric acid described in Experiment 26.1 serves as a model for this synthesis of the benzoyl derivatives of alanine and valine, and you may wish to read it to review the mechanism of the reaction:

R = CH₃, alanine (Ala) or
= (CH₃)₂CH, valine (Val)

(R,S)-Benzoylamino acid

The (R,S)-benzoylamino acids normally need no purification before their reaction with aniline. However, aniline becomes colored upon standing, as a result of oxidation. Any color should be removed by distilling the aniline before using it, and it should be stored in a dark bottle.

Commercial papain need not be highly purified for this asymmetric synthesis. It is remarkably stable in the presence of organic solvents and to changes of pH, ionic strength, and temperature. Papain may be kept at room temperature for a day or two, but it should be stored at 0–5°C for longer periods. Because the catalytic activity of papain depends on the reaction of a thiol (—SH) group on a cysteine residue, mild oxidizing agents deactivate it by converting two cysteine residues to their inactive disulfide form:

$$2\,R\text{—}S\text{—}H \longrightarrow R\text{—}S\text{—}S\text{—}R + [2H]$$

Reactivation occurs when thiol compounds are added to a solution of papain. These thiol compounds react with the disulfide to regenerate the essential thiol group of the enzyme. There are two common thiols that are used for this purpose, (*R*)-cysteine and 2-mercaptoethanol:

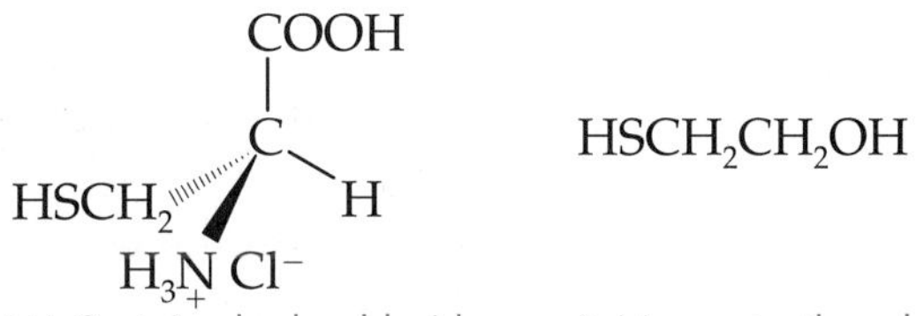

(*R*)-(+)-Cysteine hydrochloride   2-Mercaptoethanol

Both of these compounds work well, but each has a disadvantage. 2-Mercaptoethanol has a very disagreeable odor and should be handled only in the hood. (*R*)-cysteine is an optically active compound, and it would be nice to have no optically active compound in the reaction mixture other than papain. However, control experiments have shown that use of either (*R*)-cysteine or 2-mercaptoethanol gives exactly the same stereochemical result. We have chosen to use the less disagreeable (*R*)-cysteine to activate the papain.

8.3
Synthesis of the (*R,S*)-Benzoylamino Acid

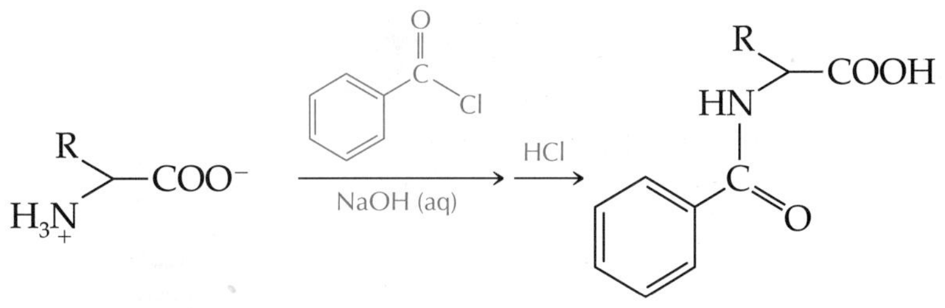

R = CH$_3$, alanine (Ala) or
= (CH$_3$)$_2$CH, valine (Val)

(*R,S*)-Benzoylamino acid

Macroscale Procedure

Techniques   Mixed Solvent Recrystallization: Technique 5.2a

Thin-Layer Chromatography: Technique 10

Follow the experimental procedure for the synthesis of hippuric acid given in Experiment 26.1, substituting either (R,S)-alanine or (R,S)-valine for glycine. Reaction of either racemic amino acid with benzoyl chloride in the presence of sodium hydroxide proceeds smoothly. Use 0.056 mol of either (R,S)-alanine or (R,S)-valine for the reaction and appropriate amounts of the other reagents such that the molar ratios remain constant. This will provide enough (R,S)-benzoylamino acid for the enzymic resolution step.

   Sometimes a gummy white product forms when the reaction mixture is poured into the cold hydrochloric acid. If this occurs in your reaction, cool the mixture in an ice water bath while you continue to stir it and rub the gummy material. A solid that can be filtered should result. After filtration and before the ether washing, the solid should be ground to a fine powder, using a mortar and pestle if it is very lumpy.

Normally, the (*R,S*)-benzoylamino acid does not need to be recrystallized. If you choose to recrystallize it, you may use a 1:1 ethanol/water mixture [see Technique 5.2a on recrystallization from a mixed solvent.] Compare the melting point of your (*R,S*)-benzoylamino acid with those listed below.

Compound	Melting point, °C
(*R,S*)-Benzoylalanine	164–166
(*R,S*)-Benzoylvaline	129–130

Thin-layer chromatography should be used to assay the purity of your product.

─── *SAFETY PRECAUTION* ───
Carry out the TLC in a hood and wear gloves.

A 1:1 methanol/chloroform mixture (by volume) works well for the TLC analysis on silica gel plates. Visualize the chromatograms with *p*-anisaldehyde reagent solution [see Technique 10.5]. Submit to your instructor that portion of your (*R,S*)-benzoylamino acid that you do not use in the next procedure.

Cleanup: Neutralize the aqueous filtrate from the crude product with solid sodium carbonate **(Caution: Foaming.)** before washing the solution down the sink or pouring it into the container for aqueous inorganic waste. Pour the ether filtrate and the aqueous ethanol filtrate from the recrystallization into the container for flammable (organic) waste. If you used decolorizing carbon in the recrystallization, place the filter paper containing the carbon in the container for nonhazardous solid waste. Pour the remaining TLC solvent into the waste container for halogenated materials.

8.4

Enzymatic Resolution: Synthesis of the Anilide of the (*S*)-Benzoylamino Acid

Anilide of (*S*)-benzoylamino acid

+

Carboxylate salt
of (*R*)-benzoylamino acid

ᴗᴗᴗ*Macroscale*
Procedure

Technique   Polarimetry: Technique 13

SAFETY INFORMATION

Aniline is toxic and absorbed through the skin. Wear gloves and pour it only in the hood.

Chloroform is volatile, toxic, and a suspected carcinogen. Wear gloves and use it only in a hood.

If the papain is from a bottle that has been stored for a long time, it may be necessary to double or even triple the amount used here.

Weigh out 0.02 mol of your (*R,S*)-benzoylamino acid and powder it with a mortar and pestle. Add the powdered material to 100 mL of citric acid/sodium citrate buffer (0.4 M, pH 5.0) in a 250- or 300-mL Erlenmeyer flask. Warm the mixture slightly on a steam bath if the solid does not dissolve with stirring; cool the solution

Note to instructor: One liter of the buffer may be made by dissolving 76.8 g (0.4 mol) of anhydrous citric acid in about 600 mL of distilled water and then adding enough 50% (19.1 M) sodium hydroxide solution to bring the pH to 5.0 (about 40 mL will be needed). Dilute the solution to 1.0 liter with distilled water and shake it to mix it thoroughly.

back to room temperature before adding 1.0 g of papain. Continue stirring until the papain is nearly dissolved. Weigh out 0.70 g (0.004 mol) of (*R*)-cysteine hydrochloride hydrate and add it to the reaction mixture.

Working in a hood, measure 4.6 mL (4.7 g, 0.050 mol) of colorless aniline, either from a new bottle or from a sample that has been recently distilled. Still using the hood, pour the aniline into the reaction mixture. Add an additional 150 mL of citrate buffer (0.4 M, pH 5.0) to the reaction flask, stopper it tightly with a solid rubber stopper, and swirl the flask to dissolve all the reactants. Bubble nitrogen through the mixture to remove dissolved air before corking the flask.

Allow the reaction to proceed in a laboratory bench drawer for 7 days. During the course of the reaction, the precipitated product can easily be seen. Reactions that are run for periods longer than 7 days develop a dark yellow color that does not easily wash out of the product. The color presumably results from the oxidation of aniline.

Filter the solid product by vacuum filtration. Remove the filtrate from the filter flask and set it aside for treatment later. (If you will be doing the optional experiment where you recover the unreacted (*R*)-benzoylamino acid, save this filtrate in a corked flask.) The following washings should be done in the Buchner funnel, applying suction only between each washing. Wash the product with 50 mL of cold water and then twice with 25-mL portions of a 1-M sodium hydroxide solution. Finally, wash the crystalline product with copious amounts of cold water.

Your instructor may ask you to prepare 1-M NaOH from 6-M NaOH solution.

Your product can be dried either at 90°C in an oven for 30 min or overnight at room temperature. Weigh the dried light tan product, determine the melting point, and calculate your percent yield. Normally it is not necessary to recrystallize the product to achieve satisfactory polarimetric measurements.

Optical Activity of (S)-Benzoylamino Acid Anilide

Ascertain the size of the polarimeter tube available in your laboratory. Prepare a solution of your (*S*)-benzoylamino acid anilide, using the quantities specified in the following table for that size of polarimeter tube.

Polarimeter tube	Amount of (S)-benzoylamino acid anilide, g	Amount of $CHCl_3$
1-dm (periscope type)	0.400	10.0
2-dm	1.00	25.0

Read Technique 13, Polarimetry, before you begin work with the polarimeter.

Weigh the specified amount of your completely dry (S)-benzoyl-amino acid anilide and transfer the solid to a dry volumetric flask (10- or 25-mL).

SAFETY PRECAUTION

Wear gloves and work in a hood while preparing your polarimetry solution and filling the polarimeter tube.

Fill the flask approximately two-thirds full with chloroform. Stopper the flask and swirl it until the (S)-benzoylamino acid anilide dissolves. Then add sufficient chloroform to bring the solution to the volume mark of the flask. Stopper the flask and shake it thoroughly to ensure a completely homogeneous solution.

If you can see any undissolved particles such as paper fibers or pieces of dust in the solution, filter the solution quickly by gravity through a short-stemmed funnel containing a small plug of glass wool or a cut-down piece of fluted filter paper. For a 2-dm tube, filter the solution into a 50-mL Erlenmeyer flask. Keep the flask tightly stoppered except while transferring it to the polarimeter tube to avoid evaporation of the solvent. If you are using a 1-dm periscope tube, filter the solution directly into the polarimeter tube and cork the polarimeter tube immediately.

dm = decimeter

Measure the optical activity of your chloroform solution [see Technique 13.4]. Calculate the specific rotation of your (S)-ben-zoyl-amino acid anilide and compare it with the specific rotation listed in the following table.

(S)-benzoylamino acid anilide	Melting point, °C	$[\alpha]_D^{25}$ found	Literature value of $[\alpha]_D^{25}$
Alanine	175–176	-74.6 ± 1.5 (CHCl$_3$, c = 0.040)	—
Valine	217–218	-82.6 ± 1.5 (CHCl$_3$, c = 0.041)	-80.6 ± 0.9 (CHCl$_3$, c = 0.045, 25°)

Calculate the percent of the (S)-enantiomer in your resolved product. Hand in that portion of your (S)-benzoylamino acid anilide that you did not use for polarimetric measurements.

Cleanup: **(Caution: Wear gloves while working with the reaction mixture filtrate.)** Adjust the pH of the filtrate from the reaction mixture to pH 9 with 1.0 M sodium hydroxide. Transfer the solution to a separatory funnel and extract the unreacted aniline with 15 mL of diethyl ether. Separate the layers and drain the lower aqueous phase into a beaker. Adjust the pH of the aqueous phase to 6–7 with 3 M HCl before washing the solution down the sink or pouring it into the container for aqueous inorganic waste. Pour the upper ether layer into the container for flammable (organic) waste. Adjust the pH of the filtrate collected from washing the product to pH 6–7 with 3 M HCl before washing the solution down the sink or pouring it into the container for aqueous inorganic waste. Pour the chloroform solution from the polarimetry into the waste container for halogenated materials.

Further Investigations

There are a number of experimental directions that you can take to extend your study of enzymatic resolution. These options make excellent projects for teams of two students.

- You can compare meat tenderizer and papain as the catalyst in the synthesis of (S)-benzoylamino acid anilides.
- You can carry out the resolution, using a different amino acid such as leucine, methionine, or phenylalanine, and compare the results from the different amino acids. (Note: (R,S)-Benzoylphenylalanine requires a citrate buffer of pH 6 for the reso-

lution step; and the anilide derivative must be dissolved in glacial acetic acid for the polarimetric measurement.)

- You can recover the unreacted (R)-benzoylamino acid and determine its optical activity; analyze and compare the NMR spectra of your racemic (R,S)-benzoylamino acid and the recovered R-benzoylamino acid.
- You can hydrolyze the anilide, recover the optically pure (S)-amino acid, and determine its optical activity.
- You can undertake a polarimetric study of the dependence of the specific rotation on solvent and concentration.
- Other possibilities are discussed in Reference 1 and in several references listed in that article.
- Obtain NMR spectra of your (R,S)-benzoylamino acid (in acetone-d_6) and the anilide of your (S)-benzoylamino acid (in $CDCl_3$). Analyze and compare these two spectra.

Reference

1. Mohrig, J. R.; Shapiro, S. M. *J. Chem. Educ.* **1976,** *53,* 586–589.

Questions

1. For what purpose was the solid product from the enzymatic resolution washed with 1.0 M sodium hydroxide solution?
2. Why is it necessary to control the pH of the reaction mixture when the (R,S)-benzoylamino acid undergoes reaction with aniline in the presence of papain?
3. Write a balanced equation showing the reaction whereby (S)-(+)-cysteine hydrochloride activates papain that has been oxidized.
4. Propose a method for converting an (S)-benzoylamino acid anilide to the (S)-amino acid.
5. Assuming that the acyl papain intermediate is formed by the usual acyl addition-elimination pathway, write a mechanism for its formation from the Michaelis-Menten complex.
6. Write a reasonable mechanism for the reaction of aniline with the acyl papain intermediate from (R,S)-benzoylalanine.

Project 9

SYNTHESIS OF PARA RED FROM ACETANILIDE

In the first part of this project, you will prepare *p*-nitroaniline, an aromatic amine, by the nitration of acetanilide and subsequent hydrolysis of the nitration product, *p*-nitroacetanilide. In the second part you will diazotize *p*-nitroaniline and use the diazonium salt in two reactions: the synthesis of *p*-iodonitrobenzene and the synthesis of a brilliant red dye called Para Red.

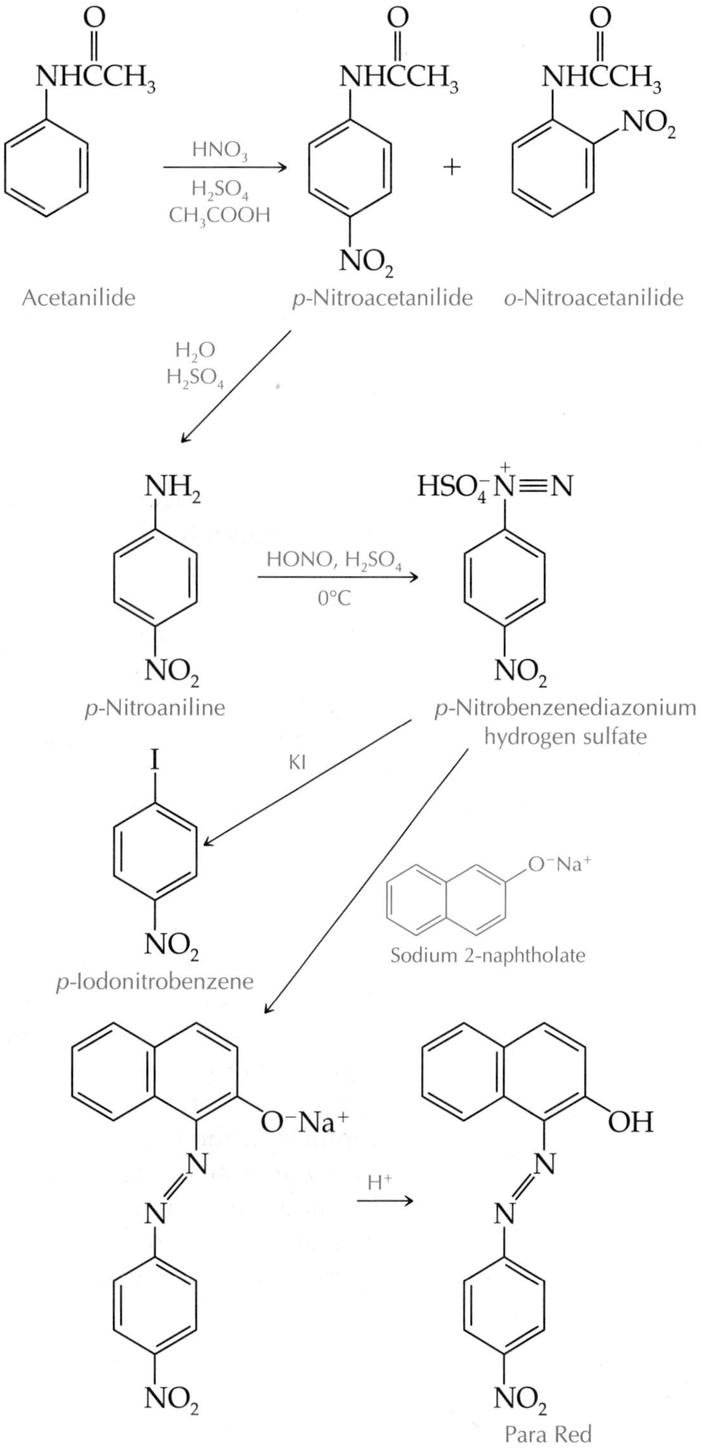

Acetanilide   *p*-Nitroacetanilide   *o*-Nitroacetanilide

p-Nitroaniline

p-Nitrobenzenediazonium
hydrogen sulfate

p-Iodonitrobenzene

Sodium 2-naphtholate

Para Red

Aromatic amines are of considerable interest because of their availability and their usefulness. Aniline, first isolated from coal tar, the thick liquid sludge formed during the heating of coal in the absence of oxygen to form coke, is the simplest aromatic amine.

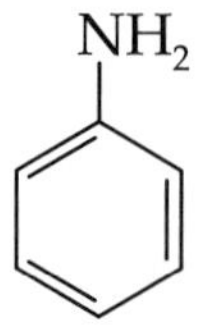

$$NH_2$$

Aniline
(benzenamine)

A major synthetic route to the aromatic amines is the reduction of aromatic nitro compounds, which in turn can easily be prepared by the nitration of a wide variety of aromatic compounds. Direct nitration of aniline with nitric acid, however, leads to tarlike oxidation by-products. This problem can be avoided by nitrating a less reactive derivative of aniline, as you will do in this project by using acetanilide.

$$Ar-H \xrightarrow[H_2SO_4]{HNO_3} Ar-NO_2 \xrightarrow{[H]} Ar-NH_2$$

Aromatic amines are important synthetic intermediates because they can easily be converted to other derivatives via diazonium salts.

$$Ar-NH_2 \xrightarrow{HONO} Ar-\overset{+}{N}\equiv N$$

Diazonium salts are useful in two major ways. Because molecular nitrogen is such a good leaving group, it can be displaced by a large variety of nucleophilic species, leading to aryl halides and phenols in particular. Diazonium salts are also effective electrophiles in the coupling of two aromatic compounds together as azo compounds. The words diazonium and azo both derive from the same word as the French word for nitrogen, which is *azote.* Derivatives of azo compounds are very important as dyes and pharmaceutical agents.

$$Ar'-H + Ar-\overset{+}{N}\equiv N \longrightarrow Ar'-N=N-Ar$$

Azo compound

Azo compounds find wide use as sulfa drugs, an example being salicylazosulfapyridine (Sulfasalazine):

Sulfasalazine

Sulfasalazine is an antibacterial compound that is used in the treatment of colitis.

Methyl orange is an azo compound that has found significant use as an acid-base indicator. At pHs greater than 3.5, it is in the form of the yellow unprotonated compound. When the pH is less than 3.5, methyl orange is protonated on an azo nitrogen atom, forming the red conjugate acid.

Methyl orange

pH > 3.5   pH < 3.5

Yellow form   Red form

9.1
Two-Step Synthesis of *p*-Nitroaniline

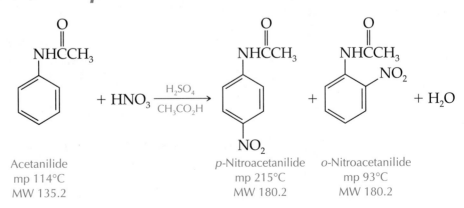

Acetanilide
mp 114°C
MW 135.2

p-Nitroacetanilide
mp 215°C
MW 180.2

o-Nitroacetanilide
mp 93°C
MW 180.2

$$p\text{-Nitroacetanilide} + H_2O \xrightarrow{\;H_2SO_4\;} p\text{-Nitroaniline} + CH_3COOH$$

p-Nitroacetanilide

p-Nitroaniline
mp 149°C
MW 138.1

The amide functional group of acetanilide is a strongly activating substituent for electrophilic aromatic substitution reactions, but it does not produce the colored, tarry oxidation products that form when aniline is nitrated directly. Like most activating groups, —NHCOCH$_3$ has an *ortho/para* directing influence. The *ortho* and *para* isomers of nitroacetanilide can easily be separated by fractional crystallization. Subsequent acid-catalyzed hydrolysis of the amide leads to *p*-nitroaniline.

꧁꧁꧁**Macroscale** Procedure

Technique   Recrystallization: Technique 5.4

— SAFETY INFORMATION —

Wear gloves while measuring the reagents and conducting the synthesis of both *p*-nitroacetanilide and *p*-nitroaniline.

Many aromatic amines are very toxic compounds and they can be carcinogenic. *p*-Nitroaniline is neither a known nor a suspected carcinogen, but it is highly toxic and can be absorbed through the skin. Wear gloves and avoid breathing it or getting it on your skin or clothing. In case of contact, immediately flush your skin with copious amounts of water.

Concentrated sulfuric acid and concentrated nitric acid are corrosive to all body tissues, and glacial acetic acid, especially when hot, can cause severe skin burns. Wear gloves. If any of these acids gets on your skin, wash it off with water immediately. Pour acetic acid only in the hood.

Sodium hydroxide solution (9 M) is extremely corrosive and causes severe burns. Solutions as dilute as 2.5 M can cause severe eye injury. Avoid contact with skin, eyes, and clothing.

Synthesis of
p-Nitroacetanilide

Add 12.0 g of acetanilide to 22 mL of glacial acetic acid (100% CH_3COOH) in a 125-mL Erlenmeyer flask. Carefully pour 22 mL of concentrated sulfuric acid (96% H_2SO_4) into the mixture while stirring it constantly. The mixture will become warm and the acetanilide will dissolve. Place the flask in a salt-ice bath and cool the contents to 5°–10°C.

Prepare a solution of 8 mL of concentrated nitric acid (70% HNO_3) and 5.3 mL of concentrated sulfuric acid. Carefully pour this acid solution into a separatory funnel supported over the flask containing the acetanilide solution. While stirring the reaction mixture, add the nitric acid solution dropwise. Control the rate of addition so that the temperature remains below 20°C. After all the nitrating solution has been added, let the reaction mixture stand for 20 min at room temperature. The reaction may continue to produce heat. If necessary, keep its temperature between 20° and 30°C by means of an ice-water bath.

Pour the reaction mixture into 100 mL of cold water mixed with 50 g of ice. The solid that precipitates is a mixture of *ortho*- and *para*-nitroacetanilide. Allow the aqueous mixture to stand for 5 min, stirring it occasionally.

Collect the solid on a Buchner funnel by vacuum filtration. Carefully pour the acidic filtrate into a 600-mL beaker labeled "acidic filtrate" and set it aside for treatment later. Wash the precipitate with 150-mL portions of cold water until the last portion of wash water is only slightly acidic (pH 5.0–6.0 with pH test paper such as pHydrion). Up to 1 L of cold water may be required. Save all these filtrates in a 1000-mL beaker; this solution contains most of the *ortho* product.

The solid remaining on the Buchner funnel is impure *p*-nitroacetanilide. Dry and weigh it. Calculate your percent yield. Recrystallize it from 95% ethanol [see Technique 5.4] and determine the melting point of the dried recrystallized product.

Cool the 1000-mL beaker containing the combined aqueous filtrates in an ice bath. Yellow, needlelike crystals of *o*-nitroacetanilide should precipitate. Collect the crystals on a Buchner funnel, dry and weigh them, and record their melting point. Calculate the percent yield of *o*-nitroacetanilide. Turn in this product to your instructor.

Cleanup: Neutralize the acidic filtrate in the 600-mL beaker with solid sodium carbonate **(Caution: Foaming.),** then wash the neutralized solution down the sink or pour it into the container for aqueous inorganic waste. If the aqueous filtrate from the *o*-nitroacetanilide is acidic, neutralize it with solid sodium carbonate before washing it down the sink. Pour the ethanolic filtrate from the recrystallization of *p*-nitroacetanilide into the container for flammable (organic) waste.

Synthesis of p-*Nitroaniline*

This procedure is written for 8.8 g of *p*-nitroaniline. If you have less than 8.8 g of *p*-nitroacetanilide after the previous step, you will need to proportionally adjust the quantities of all reagents in this procedure to your amount of *p*-nitroacetanilide.

Add 8.8 g of *p*-nitroacetanilide and 22 mL of water to a 100-mL round-bottomed flask. Slowly add 22 mL of concentrated sulfuric acid while swirling the flask. Reflux the mixture for 30 min. When the hydrolysis is complete, a 1-mL portion of the reaction mixture will be soluble in 5 mL of water.

Cool the reaction flask and pour the contents into a mixture of 50 g of crushed ice in 75 mL of water. Slowly add an excess of 9 M sodium hydroxide solution until the pH of the mixture reaches 4.0–5.0. During the addition of the sodium hydroxide solution, the mixture will get hot and should be well stirred. The precipitate is *p*-nitroaniline. Cool the mixture in an ice-water bath before collecting the *p*-nitroaniline by vacuum filtration. Wash the solid four times with 75-mL portions of water to remove any acid and inorganic salts that may be present.

Determine the melting point and recrystallize the product from ethanol if necessary. Calculate your percent yield. Save the product for use in Project 9.2.

Cleanup: Neutralize the aqueous filtrate from the crude product with solid sodium carbonate before washing the solution down the sink or pouring it into the container for aqueous inorganic waste. Pour the ethanolic filtrate from the recrystallization into the flammable (organic) waste container.

Preparation of Diazonium Salts

$$\underset{\substack{\text{NH}_2 \\ \\ \text{NO}_2 \\ \textit{p-Nitroaniline}}}{\bigcirc} + \text{HONO} + \text{H}_2\text{SO}_4 \xrightarrow{0°C} \underset{\substack{\text{HSO}_4^- \overset{+}{\text{N}} \equiv \text{N} \\ \\ \text{NO}_2 \\ \textit{p-Nitrobenzenediazonium} \\ \text{hydrogen sulfate}}}{\bigcirc} + 2\,\text{H}_2\text{O}$$

Diazonium salts are formed by the reaction of nitrous acid (HONO) with a primary amine in acidic solution. Because nitrous acid is unstable, it is formed in situ by the reaction of sodium nitrite and a strong acid such as hydrochloric or sulfuric acid.

$$\text{NaNO}_2 + \text{H}_2\text{SO}_4 \rightleftharpoons \text{HO}\overset{..}{\text{N}}\text{O} + \text{Na}^+ + \text{HSO}_4^-$$
$$\text{Nitrous acid}$$

The mechanism of the diazotization reaction is initiated by a nucleophilic attack of the primary amine at the nitrogen atom of nitrous acid. The intermediate nitrosated amine subsequently rearranges and loses a molecule of water under the acidic reaction conditions. One likely pathway is

$$\text{ArNH}_2 + \text{HONO} \longrightarrow \text{ArNHNO} + \text{H}_2\text{O}$$

$$\text{ArNHNO} \longrightarrow \text{Ar}-\text{N}=\text{N}-\text{OH}$$

$$\text{Ar}-\text{N}=\text{N}-\text{OH} + \text{H}^+ \longrightarrow \text{Ar}-\text{N}=\text{N}-\overset{+}{\text{O}}\text{H}_2 \longrightarrow \text{Ar}-\overset{+}{\text{N}}\equiv\text{N} + \text{H}_2\text{O}$$

The diazotization of aromatic primary amines is a useful synthetic procedure by which a variety of functional groups can be substituted onto an aromatic ring. You will use part of your *p*-nitrobenzenediazonium hydrogen sulfate in a reaction with potassium iodide in water solution. This nucleophilic displacement reaction results in the production of *p*-iodonitrobenzene and nitrogen gas:

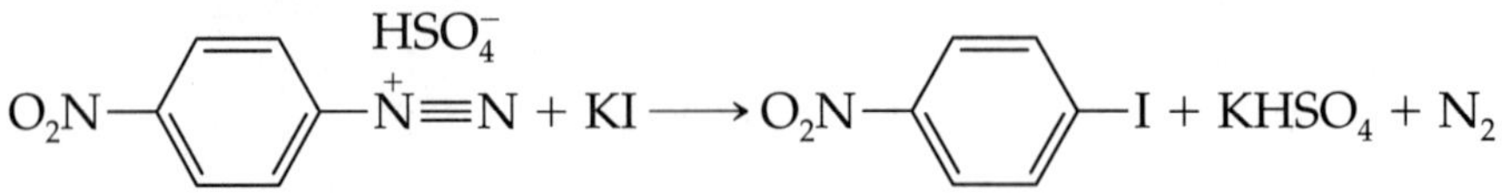

$$O_2N-\overset{\overset{HSO_4^-}{|}}{\underset{}{}}\overset{+}{N}{\equiv}N + KI \longrightarrow O_2N--I + KHSO_4 + N_2$$

p-Iodonitrobenzene
mp 171°C
MW 249

In the last part of the project you will prepare the azo dye Para Red by coupling *p*-nitrobenzenediazonium hydrogen sulfate with the sodium salt of 2-naphthol:

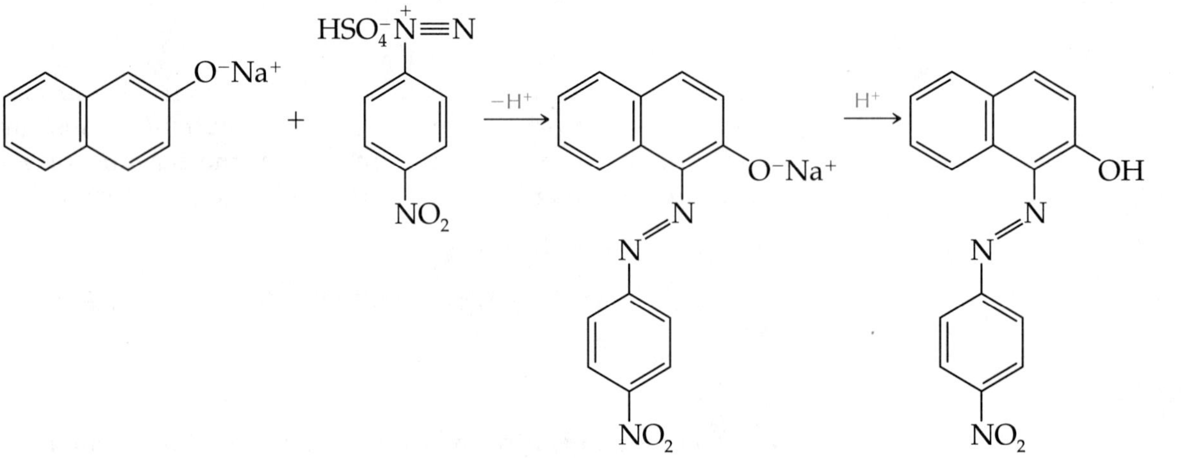

Para Red
MW 293.3

The oxygen function on the naphthalene ring strongly activates the 1-position toward electrophilic attack by the positively charged diazonium ion, thus resulting in the displacement of a proton.

It is important that the temperature of the reaction mixture be kept below 10°C during the formation and storage of the diazonium salt, so that the salt does not hydrolyze to *p*-nitrophenol:

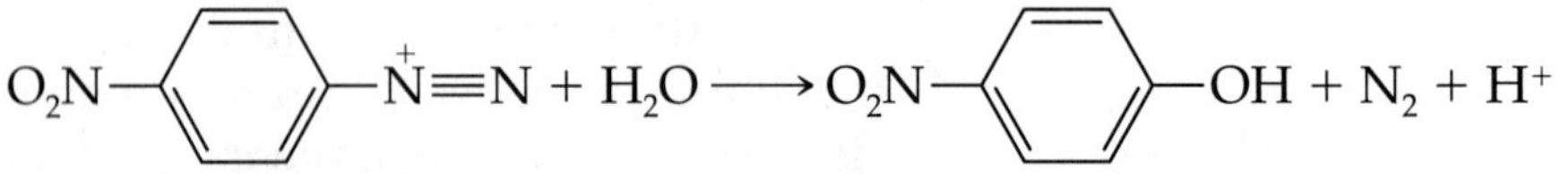

$$O_2N--\overset{+}{N}{\equiv}N + H_2O \longrightarrow O_2N--OH + N_2 + H^+$$

Many azo dyes will not bond well to fibers such as cotton. Nevertheless, these azo dyes can be used to dye cotton by synthesizing them inside the cloth fibers. The reactants can diffuse into the pores of the cloth, but the larger dye molecules are trapped inside the fibers.

Macroscale Procedure

Techniques Recrystallization: Technique 5.4

Thin-Layer Chromatography: Technique 10

— *SAFETY INFORMATION* —

Wear gloves while doing these procedures.

p-Nitroaniline is toxic and an irritant.

2-Naphthol is an irritant.

Diazonium salts are explosive if allowed to dry out, and they decompose if allowed to become warm. Keep the diazonium salt solution in an ice-water bath while you are doing these procedures.

Diazotization of p-Nitroaniline

Place a solution of 5 mL of concentrated sulfuric acid in 50 mL of water in a 250-mL Erlenmeyer flask. Stir the solution while adding 5.0 g (0.036 mol) of *p*-nitroaniline. Cool the mixture to 5°–10°C in a salt/ice bath.

While stirring the mixture, slowly pour in a solution of 2.5 g (0.036 mol) of sodium nitrite ($NaNO_2$) in 10 mL of water. Keep the temperature of the reaction under 10°C by adjusting the rate of addition of the sodium nitrite solution. Continue to keep the ice-water bath under 10°C so that the diazonium salt does not decompose before you are ready to use it in the following procedures.

Synthesis of p-Iodonitrobenzene

Foaming occurs as the nitrogen gas is evolved, hence the use of the large reaction container.

Slowly pour 25 mL of the cold diazonium salt solution into a solution of 4.2 g (0.025 mol) of potassium iodide in 20 mL of water contained in a 600-mL beaker. Use suction filtration to collect the *p*-iodonitrobenzene that precipitates. Recrystallize the product from 95% ethanol. Calculate the overall percent yield and determine the melting point.

Cleanup: Wash the filtrate from the reaction mixture down the sink or pour it into the container for aqueous inorganic waste.

Pour the filtrate from the recrystallization into the container for halogenated waste.

Dyeing Cloth Samples with Para Red

Place 0.5 g of 2-naphthol (pK_a 9.6) in a beaker containing 100 mL of hot water. Add 2.5 M NaOH solution dropwise while stirring the mixture until approximately three-fourths of the 2-naphthol dissolves. A little more than 1 mL of NaOH solution should suffice. It is important not to add too much alkali because cotton and woolen fabrics may disintegrate in strong alkali.

Dilute 5 mL of the cold *p*-nitrobenzenediazonium hydrogen sulfate solution to 20 mL with cold water. Thoroughly wet a sample of cotton cloth with the diluted solution for a few minutes, using a stirring rod to work it about. Remove the cloth with a forceps and drain off some of the excess diazonium salt solution by dragging the cloth along the inside of the beaker. Add the cloth sample to 50 mL of the alkaline 2-naphthol solution. After several minutes, remove the cloth sample and rinse it well with water. Investigate the dyeing of wool by treating a sample of woolen cloth in the same manner. Compare the results.

Synthesis of Para Red

Dissolve 2.7 g (0.019 mol) of 2-naphthol in 50 mL of 2.5 M sodium hydroxide solution. Cool the solution to 10°C by adding crushed ice. Carefully pour the solution into the flask containing the remainder of the cold *p*-nitrobenzenediazonium hydrogen sulfate solution while stirring the resulting mixture. Continue to stir the mixture vigorously for a few minutes, then acidify it with 1 M sulfuric acid, and filter the red precipitate. Wash the product with water under vacuum filtration and allow it to air dry. Because Para Red decomposes upon melting, you should assay its purity by thin-layer chromatography [see Technique 10].

Cleanup: Pour all the dye solutions into a waste bottle labeled "Waste Para Red Solutions."

Reference

1. Juster, N. J. *J. Chem. Educ.* **1962**, *39*, 596–601.

Questions

1. In the preparation of *p*-nitroaniline, why is acetanilide nitrated and hydrolyzed rather than aniline nitrated directly?
2. Why is *o*-nitroaniline soluble in acidic water solution?
3. If *p*-nitroacetanilide is hydrolyzed in acidic solution, why does hydrolysis not take place during the nitration with HNO_3 and H_2SO_4?
4. Why is the coupling reaction of the *p*-nitrodiazonium salt and 2-naphthol carried out in basic solution and the displacement reaction forming *p*-iodonitrobenzene done in acidic solution?
5. Would Para Red make a good acid-base indicator? How would you tell experimentally?

<hr />

Project 10

SYNTHESES STARTING WITH ALKYL HALIDES*

This project entails synthesizing three compounds, starting with an alkyl halide. You get to decide what directions your experimental work will take over the several weeks during which you will be working on the project. Your instructor will either assign an alkyl halide to you or allow you to select one from a list distributed in the laboratory.

Guidelines for the Project

Your task is to synthesize three different compounds from a total of 30 g of an alkyl halide. A list of suggestions for reactions, options for the three syntheses, and sources of experimental procedures are outlined below. Your instructor may require that a Grignard reaction be used in one of your three syntheses. The product of each synthesis must be characterized by one or more of the following methods: melting point or boiling point, ^1H NMR, IR, GC, or a chemical test.

Before you begin the experimental work, you will need to submit to your instructor a plan for your project that outlines the three syntheses that you intend to undertake. Your plan should include references for each synthesis, a copy of the synthesis, balanced equations, the modified quantities of reagents scaled to your actual starting amount of alkyl halide, the necessary safety

*This project is adapted from N. H. Potter and T. F. McGrath, *J. Chem. Educ.*, **1989**, *66*, 666–667.

considerations, and what by-products you will have left at the end of each synthesis. You also need to prepare a list of the chemicals and equipment required for each synthesis. Your instructor will provide specific deadlines for submitting this information.

─ **SAFETY INFORMATION** ─────────

You need to find the safety information pertaining to the reagents you will be using. Sources of safety information for chemicals are discussed in Technique 1.6.

Cleanup: Your instructor must approve your cleanup procedures before you begin the experiment. Prepare a list of all by-products remaining after each synthesis. Some examples of by-products include the aqueous layer from an extraction, used drying agent, solvent distilled from a product, and solutions prepared for GC analysis. Consult the cleanup procedures in experiments that you have done previously for ways of handling similar materials. Your instructor may also give you specific guidelines that pertain to disposal of by-products in your institution and community.

Suggestions for Syntheses

Suggestions for syntheses follow. You will find additional reactions of alkyl halides described in your class text. If there is a particular reaction that you would like to try, you should discuss it with your instructor.

Reactions of Primary Halides

1. $RX + Mg \longrightarrow RMgX$

2. $RX + R'OH \xrightarrow{NaOR'} ROR'$
or use a phase-transfer catalyzed (PTC) Williamson ether synthesis.

3. $RCH_2CH_2X \xrightarrow[\text{ethanol}]{\text{KOH}} RCH{=}CH_2$

4. $RBr \text{ (or Cl)} + NaI \xrightarrow{\text{acetone}} RI$

5. $RX + R'COO^-Na^+ \xrightarrow{\text{PTC}} R'COOR$

Reactions of Secondary and Tertiary Halides

1. Any of the Grignard reactions and sequential syntheses listed in Reaction 1 for primary halides.
2. A Friedel-Crafts synthesis with a tertiary halide or any secondary halide that cannot undergo rearrangement.
3. Avoid E2 eliminations with bases and S_N2 reactions with strong nucleophiles.

Synthesis Options

Option 1: You may select three single-step reactions, in each case starting with your alkyl halide. A particular type of reaction may not be repeated. We suggest using 10 g of substrate for each synthesis, but you may use more for one synthesis and then use a microscale reaction for another synthesis. This may be a good strategy if you carry out a macroscale synthesis that involves a distillation as the final step.

Option 2: You may select a two-step synthesis in which the starting material for the second reaction is the product of the first. Here we suggest that you use 20 g of alkyl halide for the first step. The third reaction would be a one-step synthesis, again starting with the alkyl halide.

Option 3: You may choose a three-step synthesis in which the second and third reactions start with the product of the previous reaction. We suggest that you use all your alkyl halide in the first step. You might need to use microscale techniques for the third reaction if your yields have been small in the two previous steps.

Grignard Reactions

Select a Grignard synthesis and use diethyl ether as the solvent. A suggested list of carbonyl compounds for reaction with your Grignard reagent follows. The alcohol produced in your reaction should have no more than eight or nine carbons to keep its boiling point below 220°C, so that a vacuum distillation of the final product is unnecessary. Alternatively, the Grignard reagent may be added to solid carbon dioxide (dry ice) to form a carboxylic acid.

Carbonyl compounds for reaction with Grignard reagents

Aldehydes	Ketones	Esters
acetaldehyde	acetone	ethyl formate
propanal	2-butanone	ethyl acetate
butanal	cyclopentanone	ethyl propionate
		diethyl carbonate

Sources of Synthetic Procedures

Sources of synthetic procedures include this text, other laboratory texts, and books that contain a large number of synthetic procedures such as *Organic Syntheses* and *Vogel's Textbook of Practical Organic Chemistry.* You may even want to go into the primary literature of organic chemistry (see Appendix B).

You may not find a procedure for the specific compound that you wish to make, but you should be able to find a preparation for an analogous compound. You will have to adapt the procedure to your compound *and* the amount of starting material that you are using for each reaction.

You also need to find the physical constants for your products. Sources for this information include *CRC Handbook of Chemistry and Physics, The Merck Index, Aldrich Catalog Handbook of Fine Chemicals,* and *The Dictionary of Organic Compounds.*

Product Characterization

The product of each synthesis must be characterized by one or more of the following methods (as specified by your instructor): melting point or boiling point, ^{1}H NMR, IR, GC, or a chemical test. A gas chromatographic analysis will help you assess the purity of your product. An IR or ^{1}H NMR spectrum will provide evidence that you have (or have not) synthesized the compound that you intended to make. It is vital that you characterize an intermediate product before trying to use it as a substrate for another reaction.

Chemical tests may be used to characterize the functional group in a product. Many of these are given in your class text. Directions for carrying out specific tests may be found in Part 3, Qualitative Organic Analysis, or in other laboratory texts containing a section on organic qualitative analysis.

Report

Your instructor will specify the report format for your project. You may also be asked to give an oral presentation or prepare a poster about your project.

DIELS-ALDER CYCLOADDITIONS OF VARIOUS DIENOPHILES TO 2,3-DIMETHYL-1,3-BUTADIENE

In this project you will be a member of a two-person team investigating Diels-Alder reactions between the diene 2,3-dimethyl-1,3-butadiene and various dienophiles in experiments of your own design.

In the first part of the project you will each prepare *cis*-4,5-dimethyl-1,2,3,6-tetrahydrophthalic anhydride by the Diels-Alder reaction of maleic anhydride with 2,3-dimethyl-1,3-butadiene. Although not all dienophiles are as reactive as maleic anhydride, this synthesis will familiarize you with Diels-Alder reactions. In the second part of the project, your two-person team will carry out experiments of your own design, using 2,3-dimethyl-1,3-butadiene and substituting other dienophiles for maleic anhydride.

11.1 Synthesis of cis-4,5-Dimethyl-1,2,3,6-Tetrahydrophthalic Anhydride

Refer to the introductory section of Experiment 16 for a discussion of Diels-Alder reactions.

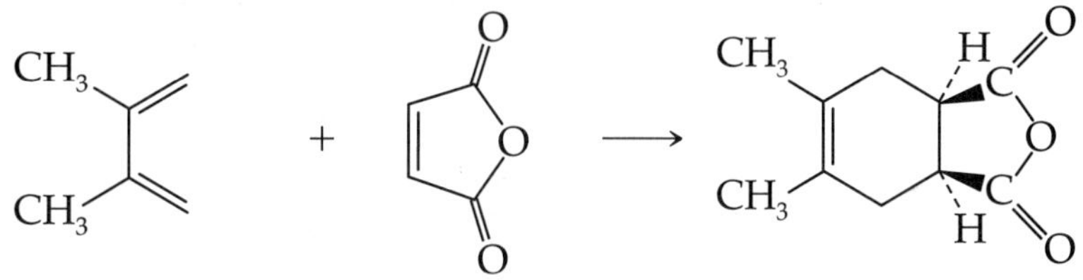

2,3-Dimethyl-1,3-butadiene
bp 69°C
MW 82.1
density 0.726 g · mL⁻¹

Maleic anhydride
mp 60°C
MW 98.1

cis-4,5-Dimethyl-1,2,3,6-tetrahydrophthalic anhydride
mp 78°–79°C
MW 180.1

mmmicroscale Procedure

Techniques IR Spectrometry: Spectrometric Method 1
NMR Spectrometry: Spectrometric Method 2

--- *SAFETY INFORMATION* ---

Maleic anhydride is corrosive and toxic. Wear gloves while handling it and grind it in a hood to avoid breathing the powder.

2,3-Dimethyl-1,3-butadiene is extremely volatile and flammable.

Hexane is also extremely volatile and flammable. Heat it only on a steam bath.

Weigh 0.40 g of powdered maleic anhydride; grind the maleic anhydride with a mortar and pestle before weighing it, if this has not been done previously. Transfer 0.44 mL of 2,3-dimethyl-1,3-butadiene with a graduated pipet into a 5-mL or 10-mL round-bottomed flask. Add the powdered maleic anhydride to the diene and fit the flask with a water-cooled reflux condenser [see Technique 3.3]. Within a few minutes, the reaction should begin spontaneously. Heat will be evolved as the reaction takes place, and the reaction mixture will boil vigorously.

When the reaction has ceased, allow the flask to cool to room temperature and the reaction mixture to solidify. It may take a few minutes before crystals appear. Add 2 mL of cold water and break up any large chunks of product with a microspatula; continue breaking the pieces of crude product until it consists of fine white crystals.

Allow the crystals to settle to the bottom of the flask. Using a Pasteur pipet fitted with a rubber bulb, expel the air from the rubber bulb and place the tip of the Pasteur pipet firmly against the bottom of the flask as shown in Figure 11.1. The pipet must be in a vertical position. Gently release the pressure on the rubber bulb. The water will be drawn into the pipet with few, if any, crystals. Remove the pipet and transfer the water to a 50-mL beaker. Continue this process until all of the water is removed from the crystals. Repeat the washing process with 2-mL portions of water until

Congo red paper turns blue in acidic solution.

a drop of the wash liquid placed on Congo Red test paper no longer gives an acid reaction. Three or four washes are usually

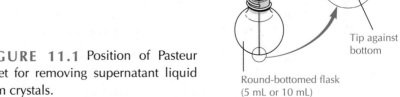

FIGURE 11.1 Position of Pasteur pipet for removing supernatant liquid from crystals.

Rubber bulb. Expel air *before* inserting in mixture.

Pasteur pipet

Tip against bottom

Round-bottomed flask (5 mL or 10 mL)

sufficient. Then collect the solid product on a small Buchner funnel, using vacuum filtration. Use the vacuum system to draw air over the crystals for 20 min.

Transfer the dry crystals to a 50-mL Erlenmeyer flask and recrystallize the product from hexane, using a steam bath as the heat source [see Technique 5.4]. Calculate the percent yield and obtain the product's melting point and infrared spectrum. The ^1H NMR spectrum of the product is also quite revealing. Obtain the ^1H NMR spectrum of your *cis*-4,5-dimethyl-1,2,3,6-tetrahydrophthalic anhydride, as directed by your instructor.

Cleanup: Combine the aqueous washes and the aqueous filtrate and neutralize the solution with sodium carbonate **(Caution: Foaming.)** before washing it down the sink or pouring it into the container for aqueous inorganic waste. Pour the hexane filtrate from the recrystallization into the container for flammable (organic) waste.

11.2
Addition of Various Dienophiles to 2,3-Dimethyl-1,3-Butadiene

You and your team member have the freedom to devise experiments that investigate the Diels-Alder reaction between 2,3-dimethyl-1,3-butadiene and one or two of the following dienophiles:

Diethyl fumarate (*trans*)   Diethyl maleate (*cis*)

Dimethyl maleate   Dimethyl acetylenecarboxylate   1,4-Benzoquinone

Suggestions for Experiments

The most useful assay technique for following the progress of these reactions is TLC [see Technique 10]. Reactions 1–3 should be done at reflux without any solvent.

1. Use a competitive reaction to compare the rates of cycloaddition of diethyl fumarate and diethyl maleate.

2. Compare the rates of cycloaddition for dimethyl maleate and dimethyl acetylenedicarboxylate.

3. Use dimethyl acetylenedicarboxylate and 2,3-dimethyl-1,3-butadiene, and subsequently convert the Diels-Alder adduct to an aromatic compound.

4. Carry out a competitive cycloaddition reaction between 2,3-dimethyl-1,3-butadiene and 1,3-cyclohexadiene with maleic anhydride. (Use refluxing diethyl ether or ethyl acetate for the reaction.)

5. Use different ratios of 1,4-benzoquinone and 2,3-dimethyl-1,3-butadiene, and prepare both the mono- and di-addition products. (Use refluxing ethanol for the reaction.)

6. You are not limited to the preceding suggestions. You may have another idea, such as using another dienophile or diene. Discuss it with your instructor.

Planning Your Experiments

Before you begin experimental work, you will need to submit to your instructor a plan for your project. Your plan should include quantities of reagents, reaction conditions and equipment, assay methods, balanced chemical equations, and safety considerations.

The products of each reaction must be characterized by one or more of the following methods: melting point, ^{1}H NMR, IR, or TLC. Thin-layer chromatography and melting points will indicate the purity of your products. NMR spectra will be very useful in determining whether or not you have synthesized the compounds that you intended to make. It will also be helpful to determine useful TLC developing conditions and R_f values for your substrates [see Technique 10], as well as to obtain and interpret their NMR spectra. It is vital that you characterize an intermediate product before trying to use it as a substrate for another reaction.

You need to consider the following questions before beginning your experimental work.

• How will you know when the addition reaction is complete? These reactions are not as rapid as the addition done in Project 11.1.

• If you are using a solvent for the reaction, how do you decide how much to use?

• How will you isolate your products? Adding water, as you did in Project 11.1, may not be appropriate for your reaction.

- What solvent will you use to recrystallize your products?
- If you are doing a competitive experiment, how will you determine the ratio of the two products?
- You may wish to consult one or more of the review articles on Diels-Alder reactions (Refs. 1–5) while you are planning your experiments. You need to find the melting point for any possible product. You may need to consult Beilstein's *Handbuch der organischen Chemie, Chemical Abstracts,* or the references listed in the review articles (Refs. 1–5) for this information.

Cleanup: Your instructor must approve your cleanup procedures before you begin your experiments. Prepare a list of all by-products from each reaction. Consult the cleanup procedures in experiments that you have done previously for ways of handling similar materials. Your instructor may also give you specific guidelines that pertain to disposal of your by-products in your institution and community.

Report   Your instructor will specify the report format for your project. You may also be asked to give an oral presentation or prepare a poster about your project.

References

1. Kloetzel, M. C. *Org. React.* **1948,** *4,* 1.
2. Holm, H. L. *Org. React.* **1948,** *4,* 60.
3. Butz, L. W.; Rytina, A. W. *Org. React.* **1949,** *5,* 136.
4. Wasserman, A. *Diels-Alder Reactions;* Elsevier: New York, 1965.
5. Larock, R. C. *Comprehensive Organic Transformations;* VCH Publishers: New York, 1989, pp. 263–272.

Questions

1. Propose a Diels-Adler reaction that would lead to each of the following:

2. Which of the following equations represents a 4+2 reaction and which a 2+2 reaction? Why?

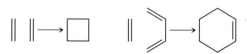

Project 12

MULTISTEP SYNTHESIS OF
N-PHENYLTETRAPHENYLPHTHALIMIDE*

In this project you will carry out a synthesis, using sequential reactions to convert two simple ketones to an imide. These reactions are an aldol condensation followed by dehydration, a Diels-Alder cycloaddition reaction accompanied by the loss of CO to form an aromatic diester, hydrolysis of the diester to a diacid, conversion of the diacid to an anhydride, and finally conversion of the anhydride to an imide.

*This project was developed by Rosemary G. Fowler of Cottey College, Nevada, MO.

The final phthalimide product is in a structural class that has been found in a number of biologically important molecules. A well-known example is the drug thalidomide. In 1962 this compound was made available in racemic form to pregnant women as an antinausea drug. It subsequently became clear that the (S)-$(-)$ form of this compound was a teratogen that induces a condition known as phocomelia, which causes deformities in the limbs of children born to these women. Apparently, although the (S) form causes the problem, the (R) form is benign in this respect. As a result of this and similar cases, the United States Food and Drug Administration (FDA) in 1988 began requiring submission of enantiomeric composition for all new drug applications.

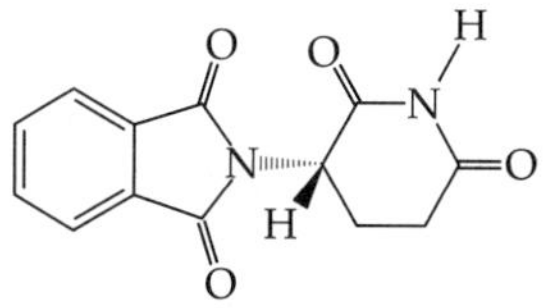

(S)-$(-)$-Thalidomide

The first two steps (Projects 12.1 and 12.2) have been presented earlier in this book: The aldol condensation reaction (here a "double-aldol" condensation) is described in Experiment 22 and the Diels-Alder reaction in Experiment 16.1. The details of the other three steps in this synthetic sequence follow.

The formation of phthalic anhydride from phthalic acid is so straightforward that the anhydride and the diacid are said to be synthetic equivalents. The intellectual process of breaking down the anhydride to its synthetic precursor (retro-synthetic analysis) can be depicted as follows:

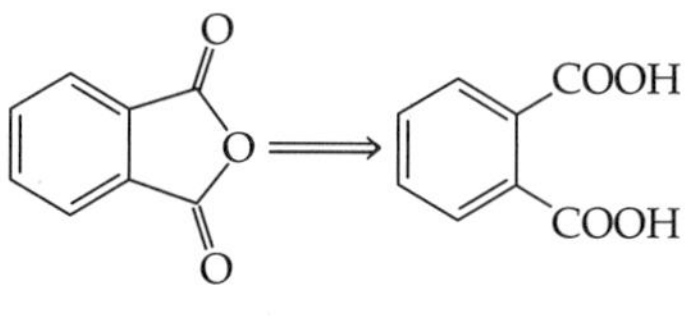

The concepts of retro-synthetic analysis have been extensively developed by E. J. Corey (Nobel prize in chemistry, 1990). In fact the entire reaction sequence used in this project can be visualized by the experienced organic chemist in a retrosynthetic sense:

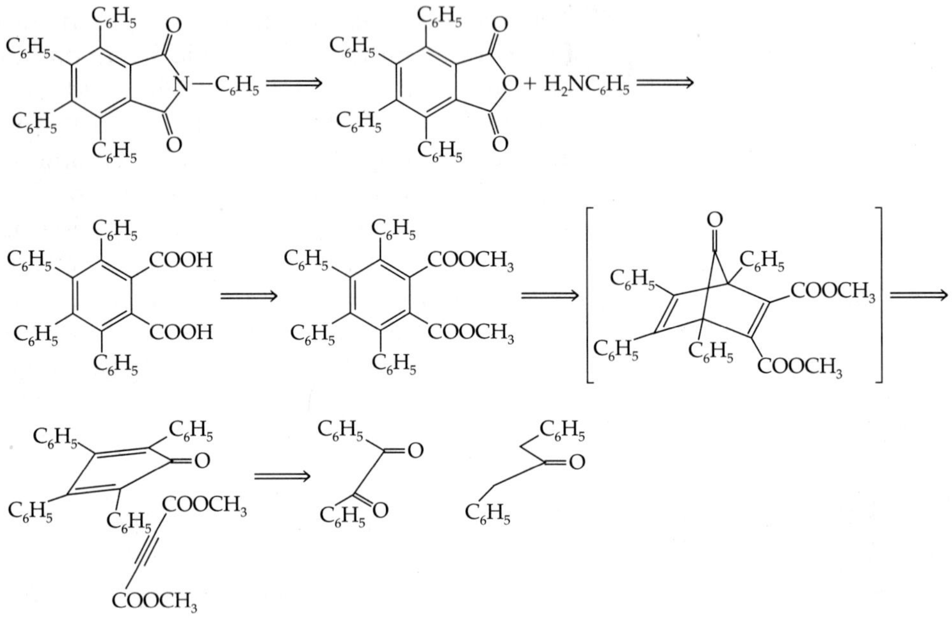

C_6H_5 C_6H_5 C_6H_5 C_6H_5 N—C_6H_5 $\Longrightarrow$ C_6H_5 C_6H_5 C_6H_5 C_6H_5 O + $H_2NC_6H_5$ $\Longrightarrow$

C_6H_5 C_6H_5 C_6H_5 C_6H_5 COOH COOH $\Longrightarrow$ C_6H_5 C_6H_5 C_6H_5 C_6H_5 COOCH$_3$ COOCH$_3$ $\Longrightarrow$ $\left[\; C_6H_5 \; C_6H_5 \; C_6H_5 \; C_6H_5 \; \text{COOCH}_3 \; \text{COOCH}_3 \;\right] \Longrightarrow$

C_6H_5 C_6H_5 C_6H_5 C_6H_5 COOCH$_3$ COOCH$_3$ $\Longrightarrow$ C_6H_5 C_6H_5 O O + C_6H_5 C_6H_5 O

12.1
Synthesis of Tetraphenylcyclopentadienone

1,3-Diphenyl-2-propanone
(dibenzyl ketone)
mp 34°–35°C
MW 210.3

1,2-Diphenyl-1,2-ethanedione
(benzil)
mp 96°C
MW 210.2

$—CH_2\overset{+}{N}(CH_3)_3\ OH^-$

triethylene glycol

Tetraphenylcyclopentadienone
mp 220°–221°C
MW 384.5

+ 2 H_2O

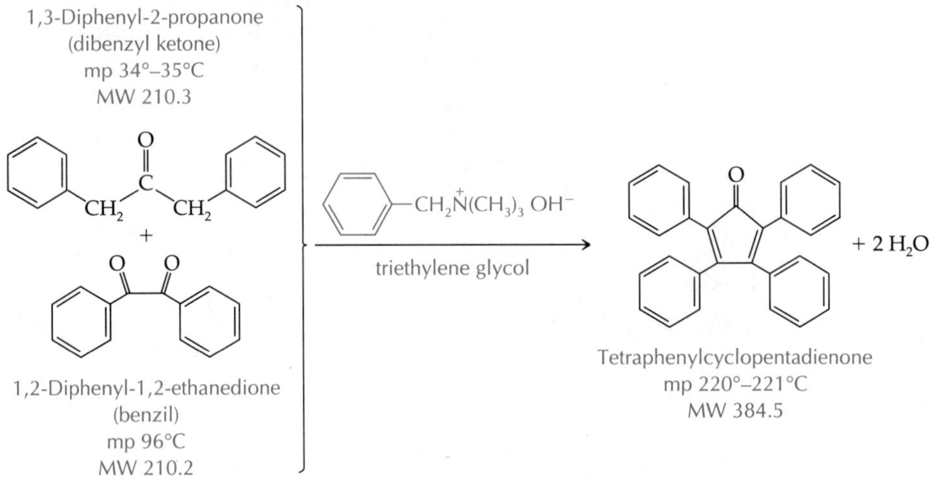

The first step in this multistep synthesis is the preparation of tetraphenylcyclopentadienone, described in Experiment 22. You will start with a scaled-down macroscale procedure.

<table>
<tr><td>

___WWW**Macroscale**___
Procedure

</td><td>

Follow the macroscale procedure given in Experiment 22, except begin with *0.500 g of benzil* and use *one-half* of the amounts listed for all other reagents. Carry out the reaction in a 13 × 100 mm test tube. The tetraphenylcyclopentadienone should be of sufficient purity to be used without further treatment in the synthesis of dimethyl tetraphenylphthalate (Project 12.2).

Weigh your product. Set aside about 40 mg of your tetraphenylcyclopentadienone to determine the melting point and ^{1}H NMR spectrum after the product dries thoroughly. Use the rest for the next step (Project 12.2).

</td></tr>
</table>

12.2
Synthesis of Dimethyl Tetraphenylphthalate

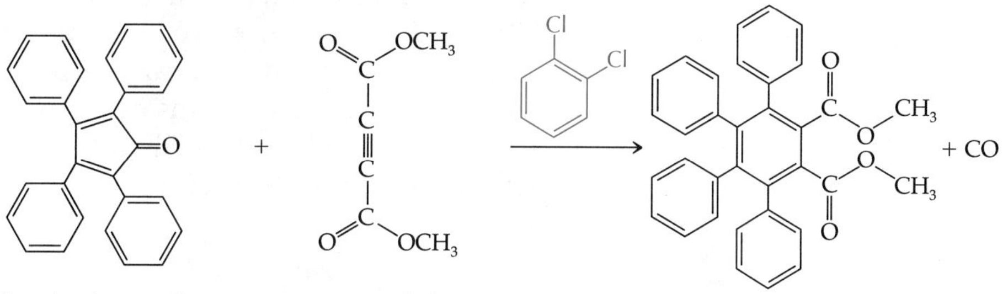

Tetraphenylcyclopentadienone	Dimethyl acetylenedicarboxylate	Dimethyl tetraphenylphthalate
mp 220°–221°C	bp 300°C	mp 258°C
MW 384.5	MW 142.1	MW 498

<table>
<tr><td>

___WWW**Macroscale**___
Procedure

</td><td>

Follow the macroscale synthesis of dimethyl tetraphenylphthalate described in Experiment 16.1. This procedure is written for 0.500 g of tetraphenylcyclopentadienone. You will need to proportionally adjust all reagent amounts for the amount of tetraphenylcyclopen-tadienone that you will use.

When you complete this step, store your product on a watch glass to dry until the next laboratory period. Set aside about 40 mg of your dried dimethyl tetraphenylphthalate to determine the melting point and ^{1}H NMR spectrum before you begin Project 12.3.

</td></tr>
</table>

12.3
Synthesis of Tetraphenylphthalic Acid

Dimethyl tetraphenylphthalate
mp 258°C
MW 498

Tetraphenylphthalic acid
mp 285°–287°C
MW 470

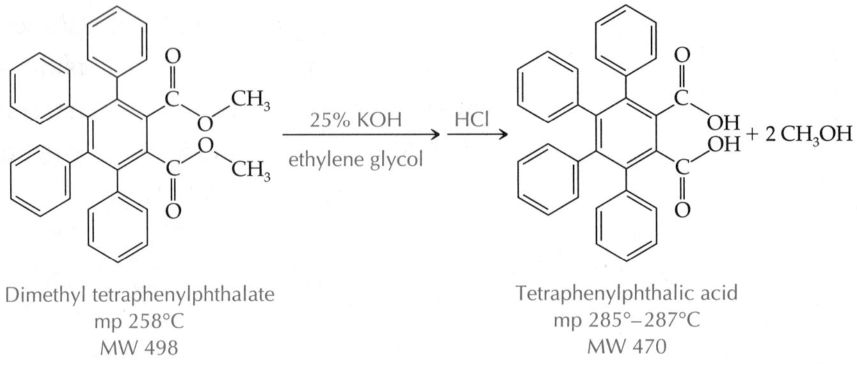

In this step you will hydrolyze the dimethyl ester that you synthesized in Project 12.2. We have chosen ethylene glycol, $HOCH_2CH_2OH$, as the solvent for this reaction because its low volatility easily accommodates the temperature of the reaction (225°C). The mechanism of this base-promoted hydrolysis reaction is well understood and is detailed here for one of the carbomethoxy groups (the other undergoes an identical reaction):

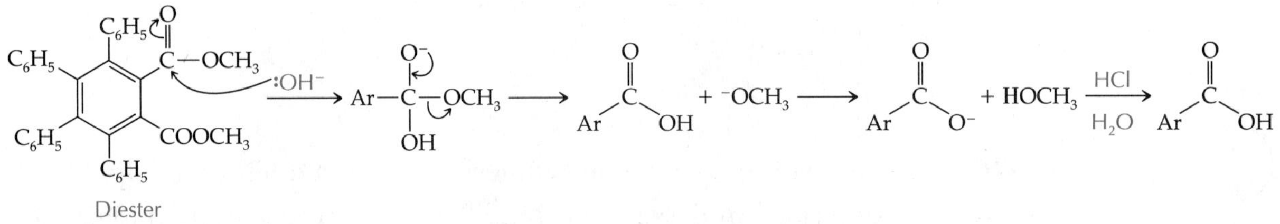

Diester

Neutralization with acid (HCl) is required because the hydrolysis product formed under alkaline conditions is a sodium salt, which is the conjugate base of the desired dicarboxylic acid.

Ester hydrolysis has been discussed elsewhere in this book: Project 1 and Experiment 4 provide a simple introduction. Insight into the process of ester hydrolysis can also be gained by reviewing ester formation reactions (Experiments 5 and 25).

<table>
<tr><td></td><td></td></tr>
</table>

mmicroscale
Procedure

Technique   Vacuum Sealed Capillary Tube:
Technique 6.3b

<table>
<tr><td>

— SAFETY INFORMATION —

The **ethylene glycol solution of potassium hydroxide** (25% by wt) is corrosive and causes severe burns.

Hydrochloric acid solution (6 M) is an irritant to the skin and mucous membranes. Avoid breathing the vapors.

For both these reagents, avoid contact with skin, eyes, and clothing. Should any spill on your skin, wash the area immediately with copious amounts of water.

</td></tr>
</table>

For this reaction, the standard taper joint on the microscale reflux condenser should be lightly greased because hot KOH solution can cause the joint to "freeze."

Begin heating an aluminum block (or sand bath) to 225°C. Place 320 mg of your dimethyl tetraphenylphthalate and 2.5 mL of 25% (wt) potassium hydroxide in ethylene glycol in a 10-mL round-bottomed flask. Add a small stirring bar to the flask. Lightly grease the bottom joint of a water-cooled reflux condenser before you fit the condenser to the flask.

Stir and heat the reaction mixture under reflux for 20 min. A precipitate will form as the reaction proceeds. At the end of the reflux period, let the reaction mixture cool to room temperature.

Wash the round-bottomed flask immediately so that the KOH solution does not etch the glass.

Transfer the reaction mixture to a 50-mL beaker, using, first, 3 mL of water, then 2 mL of 6 M HCl solution to rinse the flask. Add the rinses to the beaker containing the reaction mixture. Stir the mixture in the beaker and add 6 M HCl dropwise until the pH is below 2 when tested with pH test paper. Cool the beaker in an ice-water bath for 5 min. Collect the product by vacuum filtration on a Hirsch funnel. Wash the solid by covering it with water three times. Partially dry the solid acid by drawing air over it with the vacuum source turned on for 10 min. Transfer the acid (with the filter paper) to a watch glass and dry the acid under a heat lamp.

Weigh the product and determine the melting point in a sealed tube, prepared by the following directions. Seal the tip of a 20-cm Pasteur pipet in a Bunsen burner flame. Allow the glass to cool and introduce the usual amount of sample for a melting point through the top of the pipet. Tap the pipet on the tabletop to pack the sample. Attach the top of the pipet to a vacuum source and seal the stem about 8 cm above the sample with a burner flame [see Technique 6.3b].

The tetraphenylphthalic acid is pure enough to use in the next step without recrystallization. Determine the ^{1}H NMR spectrum of your acid, as directed by your instructor.

Cleanup: Pour the filtrate from the reaction mixture into about 50 mL of water in a 250-mL beaker. Neutralize the solution with solid sodium carbonate **(Caution: Foaming.)** before washing it down the sink or pouring it into the container for aqueous inorganic waste.

12.4 Synthesis of Tetraphenylphthalic Anhydride

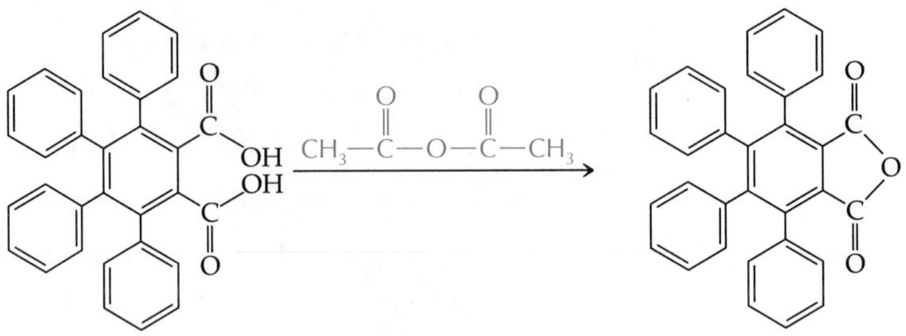

Tetraphenylphthalic acid
mp 285°–287°C
MW 470

Tetraphenylphthalic anhydride
mp 289°–290°C
MW 452

Treatment of tetraphenylphthalic acid, an *ortho*-dicarboxylic acid, with a dehydrating agent such as acetic anhydride at an elevated temperature leads to the bicyclic anhydride. Acetic anhydride is an effective dehydrating reagent because it readily consumes water to form acetic acid:

$$CH_3COCCH_3 + H_2O \longrightarrow 2\ CH_3COOH$$

For the reaction to be successful, it is important that the acetic anhydride and the phthalic acid substrate be dry.

It seems likely that the reaction proceeds through a mixed anhydride, a derivative vulnerable to ring closure, to the final product:

| Diacid | | Mixed anhydride | Tetraphenylphthalic anhydride |

microscale Procedure

SAFETY INFORMATION

Acetic anhydride is toxic, corrosive, and a lachrymator (causes tears). Measure it in a hood and avoid contact with skin, eyes, and clothing.

Hexane (or ligroin) is very volatile and flammable. Be sure that the heated hot plate is removed from your work area before you wash the product with hexane.

The tetraphenylphthalic acid prepared in Project 12.3 must be dry for this reaction to work successfully.

Begin heating an aluminum block or sand bath to 160°–170°C. Place 320 mg of your tetraphenylphthalic acid, 1.0 mL of acetic anhydride and a small stir bar in a 10-mL round-bottomed flask. Fit the flask with a water-cooled condenser and a calcium chloride-filled drying tube [see Technique 3.3a, Figure 3.6b]. Heat the reaction mixture under reflux for 30 min. Cool the flask and contents slowly to room temperature, then in an ice-water bath to induce crystallization of the anhydride. Collect the product by vacuum filtration on a dry Hirsch funnel. Wash the crystals with three 1-mL portions of cold hexane and partially dry the product by pulling air over it for 5 min while it is still in the funnel.

Transfer the anhydride (with the filter paper) to a watch glass and dry it under a heat lamp in a hood. Obtain the melting point and ^{1}H NMR spectrum of your product. This product is pure enough to use in the next step without recrystallization.

Cleanup: Pour the filtrate from the reaction mixture into the container for flammable (organic) waste.

12.5
Synthesis of *N*-Phenyl-3,4,5,6-Tetraphenylphthalimide

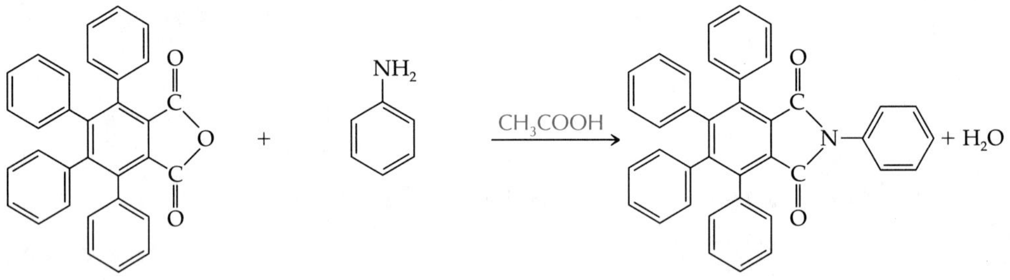

Tetraphenylphthalic anhydride	Aniline	N-Phenyltetraphenylphthalimide
mp 289°–290°C	bp 184°C	mp 358°C
MW 452	MW 93.1	MW 527
	density 1.002 g · mL⁻¹	

In the last step of this project you will convert tetraphenylphthalic anhydride to the corresponding phthalimide by heating the anhydride with aniline in acetic acid. Very likely, the intermediate is an amide that arises from nucleophilic attack by aniline on one of the carbonyl carbons of the anhydride:

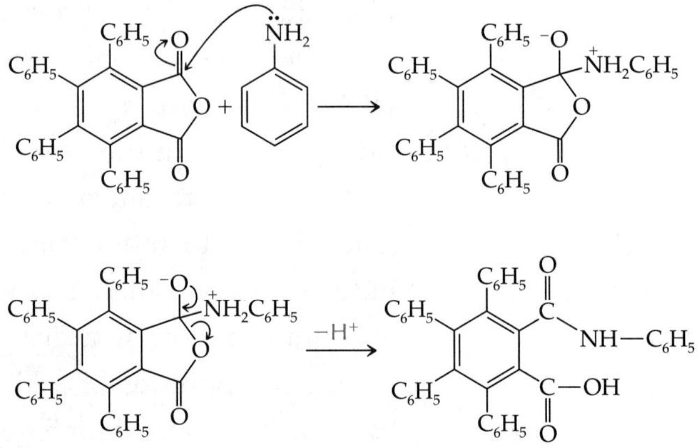

The intermediate amide is apparently nucleophilic enough to facilitate ring closure and imide formation. Most likely acetic acid catalyzes the reaction by serving as a proton donor.

Amide

N-Phenyltetraphenylphthalimide

ⅢⅢ*microscale* Procedure

--- **SAFETY INFORMATION** ---

Acetic acid is corrosive and causes burns. The vapor is extremely irritating to mucous membranes and the upper respiratory tract. Measure it in a hood and avoid contact with skin, eyes, and clothing.

Aniline is toxic and an irritant. Wear gloves and avoid contact with skin, eyes, and clothing.

Your tetraphenylphthalic anhydride must be dry for this reaction to work successfully.

The aniline used in this experiment should be colorless. Use a newly opened bottle or a sample that has been freshly distilled.

Heat an aluminum block or sand bath to 140°–150°C. Weigh 80 mg of aniline into a 10-mL round-bottomed flask, using a Pasteur pipet for the transfer. Add 2.0 mL of acetic acid, 225 mg of your tetraphenylphthalic anhydride, and a small stirring bar to the flask. Fit the flask with a water-cooled reflux condenser. Begin heating and stirring the mixture. After the anhydride dissolves, heat the reaction under reflux for 30 min. Then allow the reaction mixture to cool slowly to room temperature before cooling it in an ice-water bath. Collect the product by vacuum filtration on a Hirsch funnel. Use several 1-mL portions of ice-cold methanol to rinse the flask and pour these rinses over the crystals on the funnel. Dry the crystals by pulling air over them with the vacuum source for at least 15 min. Determine the melting point and percent yield of your product. Also obtain its ^{1}H NMR spectrum as directed by your instructor.

Cleanup: Pour the filtrate into a beaker containing 50 mL of water and neutralize the solution with solid sodium carbonate. **(Caution: Foaming.)** Wash the neutralized solution down the sink or pour it into the container for aqueous inorganic waste.

References

1. Vogel, A. I. *Vogel's Textbook of Practical Organic Chemistry;.* 3rd ed.; Longman: London, 1956, pp. 1062–1065.
2. Nicolet, B. H.; Bender, J. A. *Organic Syntheses;* Gilman, H.; Blatt, A. H., Eds.; Wiley: New York, 1941, Coll. Vol. I, pp. 410–411.
3. Fowler, R. G.; Caswell, L. R.; Sue, L. I. *J. Heterocycl. Chem.* **1973,** *10,* 407–408.

Questions

1. Suggest a mechanistic reason why base-promoted hydrolysis of esters is irreversible.
2. Suggest at least two reasons why the reagents must be scrupulously dry in the anhydride formation process.
3. If the product imide has both a very low melting point and a pungent, vinegarlike smell, what might be the problem? What simple manipulation could you do to correct the problem?
4. Explain why *N*-phenyltetraphenylphthalimide has such a high melting point (358°C)?

Project 13

SUGARS:
The Glucose Pentaacetates

In this project you will investigate the chemistry of carbohydrate interconversions, using glucose, a simple carbohydrate. You will prepare anomeric pentaacetates of α-D-glucose under two different reaction conditions and then design your own experiments to further probe the reaction.

Through the process of photosynthesis, green plants synthesize the carbohydrates that are vital foods for animals. Cellulose, the chief structural material of plants, is certainly one of the most abundant organic chemicals in all of nature; and although human beings cannot digest it, all grazing herbivores can. Starches, which are also carbohydrates, are our chief source of energy.

Carbohydrates are polyhydroxyaldehydes or ketones, usually containing five- or six-carbon units. The simplest carbohydrates, such as glucose and sucrose, are called sugars because of their sweetness. Ribose and deoxyribose are five-carbon sugars that form part of the structure of nucleic acids, the building blocks of DNA and RNA. Glucose, a six-carbon sugar, is an excellent fast-energy source. Our bodies store its polymeric form, glycogen, as a readily available energy supply. Glycogen is especially abundant in muscles and the liver. Both starch and cellulose are high molecular weight polymers of glucose. Because of their great importance in nature, carbohydrates have been studied extensively by organic chemists and biochemists.

Structure and Reactivity of Glucose

Glucose isolated from natural sources occurs as the optically active molecule (+)-glucose. Its formula is $C_6H_{12}O_6$, and it is a 2,3,4,5,6-pentahydroxyhexanal. (+)-Glucose is one of the 16 stereoisomers possible for this six-carbon pentahydroxyaldehyde, which is given the general name aldohexose. Among the other 15 stereoisomers of glucose are the C-2 diastereomer mannose and the C-4 diastereomer galactose, as well as (−)-glucose, the enantiomer of (+)-glucose.

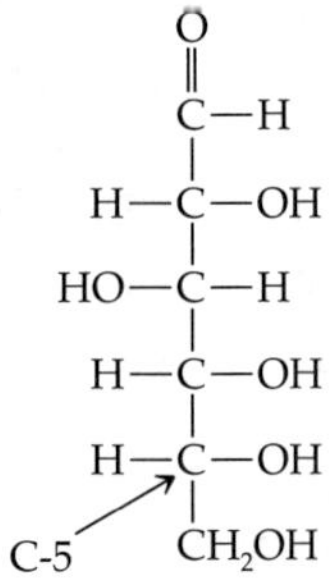

In a brilliant series of investigations, the German chemist Emil Fischer deduced the relative configuration of (+)-glucose in 1891. The beginnings of our modern understanding of carbohydrates date from this research. Fischer found that (+)-glucose has the structure shown in the margin. This structure establishes the relative configuration at each of the four stereocenters (chiral centers) of (+)-glucose. Fischer had no way to determine the absolute configuration of a glucose molecule, so he guessed at the configuration at C-5 and made the configurations at the other chiral centers relative to it. This configuration became known as D-(+)-glucose. When knowledge of the absolute configuration became available in 1951, through the x-ray studies of Bijvoet, Fischer's guess was found to be correct.

In water solution, glucose exists almost entirely as the cyclic hemiacetal formed by the addition of the hydroxyl group at C-5 to the carbonyl double bond of the aldehyde. This reaction produces a new stereocenter (C-1), so there are two cyclic diastereomers of D-glucose:

D-(+)Glucose

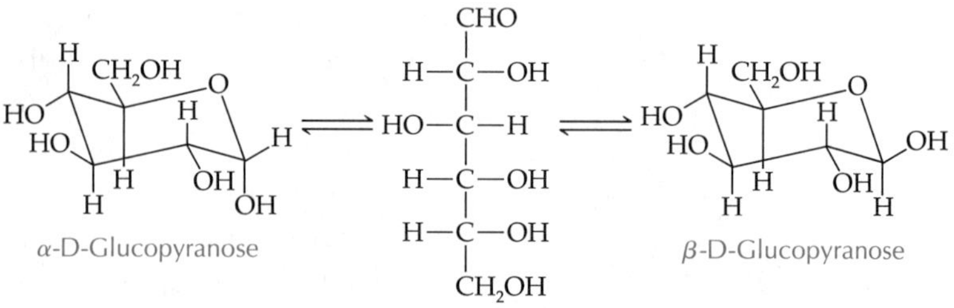

α-D-Glucopyranose   β-D-Glucopyranose

The two cyclic diastereomers are differentiated by the Greek letters α and β. These cyclic diastereomers of glucose are called glucopyranoses to indicate that they have six atoms in the ring (five carbons and one oxygen). Cyclic diastereomers of sugars that differ from each other in their configurations only at C-1 are called **anomers**. C-1 is at the far right in each of the rings drawn above. Both anomers of D-glucose have been isolated as pure compounds. β-D-Glucose has an optical rotation $[\alpha]_D^{20} = +18.7°$; the α anomer has $[\alpha]_D^{20} = +112°$. If either of them is dissolved in water, the optical rotation gradually changes until the equilibrium value of $[\alpha]_D^{20} = +52.7°$ is reached. Both acids and bases catalyze this equilibration. In water solution, the equilibrium mixture contains 63.6% of the β anomer and 36.4% of the α anomer.

The esterification of D-(+)-glucose with acetic anhydride produces the cyclic ester glucose pentaacetate, in which each of the five hydroxyl groups of the cyclic hemiacetal is acetylated. Either acidic or basic conditions may be used to catalyze the esterification, but different anomers are produced, using the different catalysts.

A commonly used acidic catalyst is anhydrous zinc chloride, a moderate Lewis acid. Reaction of D-(+)-glucose with acetic anhydride in the presence of zinc chloride leads primarily to the α anomer. This is an equilibrium reaction in which the α and β anomers of the pentaacetate are interconvertible. The more stable α isomer dominates in the equilibrium mixture.

The symbol Ac refers to an acetyl group, $CH_3\overset{\displaystyle O}{\overset{\|}{C}}-$.

Anhydrous sodium acetate is a common basic catalyst. Using basic catalysis, the β anomer is the favored product. Here the product that forms the fastest is trapped and is unable to equilibrate to a more stable structure. The β anomer is favored kinetically.

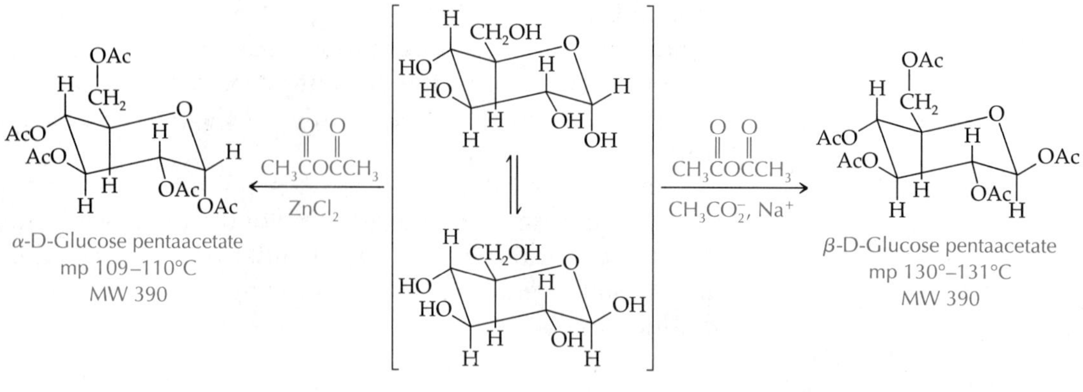

α-D-Glucose pentaacetate
mp 109–110°C
MW 390

β-D-Glucose pentaacetate
mp 130°–131°C
MW 390

¹H NMR Analysis of Carbohydrates

Historically, the assignment of the configuration at C-1 of the α and β anomers of carbohydrates presented a real problem. Both synthetic means and the magnitude of the optical rotation were used to assign structures to the diastereomers. It was a difficult problem and the wrong configurations were sometimes assigned.

Today, ¹H NMR spectrometry provides a convenient way to assign the configurations. The α- and β-D-glucose pentaacetates give complex ¹H NMR spectra as a result of the large number of protons with nearly identical chemical shifts. However, the one proton attached to C-1 of each compound appears furthest downfield because of the strong deshielding of two nearby electronegative oxygen atoms. C-1 is the only carbon atom in the pentaacetates that is bonded to two oxygen atoms.

The spin-spin splitting pattern of the anomeric proton at C-1 also shows a simple $n + 1$ splitting pattern. The protons at C-1 in both the α- and β-D-glucose pentaacetates have only one nearby proton neighbor, the one at C-2. Therefore, the proton at C-1 appears as a doublet in each anomer.

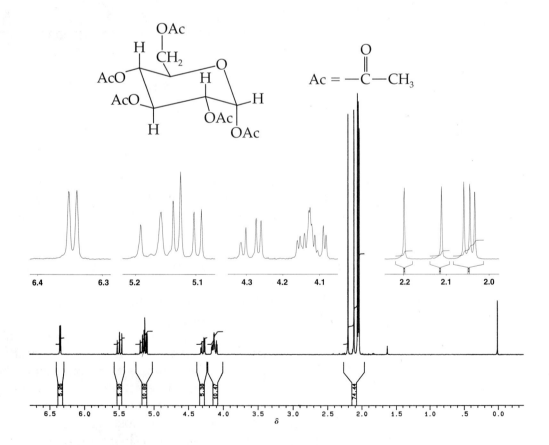

FIGURE 13.1 ¹H NMR spectrum (300 MHz) of α-D-glucose pentaacetate.

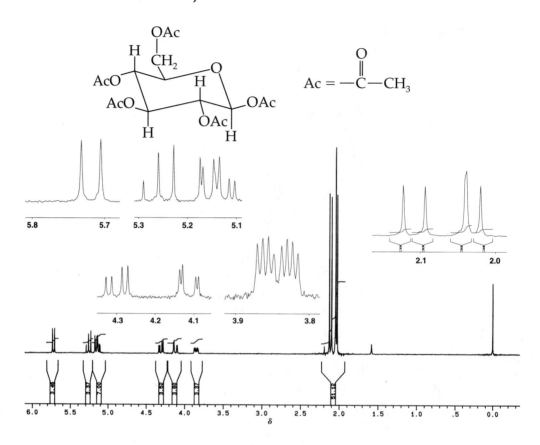

FIGURE 13.2 ^{1}H NMR spectrum (300 MHz) of β-D-glucose penta-acetate.

Careful examination of the spectra in Figures 13.1 and 13.2 shows subtle differences in the peaks due to the resonance of the anomeric protons. They differ both in chemical shift and in the magnitude of the coupling constant.

^{1}H NMR studies on a large number of cyclohexane derivatives have shown that equatorial protons generally appear about 0.5 ppm further downfield than do axial protons. This pattern also holds true for sugars in the pyranose ring form.

The sizes of the coupling constants also provide useful information about the configurations of the α and β anomers of the glucose pentaacetates. Analysis of spin-spin splitting from the ^{1}H NMR spectra of a large number of compounds shows that the efficiency of the coupling interaction of protons on adjacent carbon atoms depends on their conformational relationships. When the neighboring protons are eclipsed or **anti** to each other, their coupling constants are largest. When they are **gauche**, the coupling constants are far smaller (Figure 13.3).

$J \approx 8$–13 Hz
anti

$J \approx 3$–5 Hz
gauche

FIGURE 13.3 Magnitude of the NMR coupling constant, J, with different conformations.

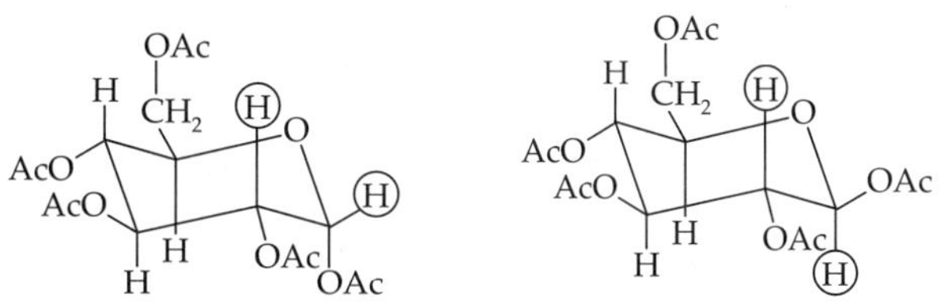

α-D-Glucose pentaacetate   β-D-Glucose pentaacetate

FIGURE 13.4 Conformations of the D-glucose pentaacetates, showing the angular relationships of the protons at C-1 and C-2.

The chair conformations shown in Figure 13.4 are more stable than any other conformation of the glucose pentaacetates. In effect, each of these compounds is frozen in a single chair conformation.

The two axial protons on β-glucose pentaacetate have a larger coupling constant than do the axial (C-2) and equatorial (C-1) protons on the α anomer.

13.1
Preparation of Anomeric D-Glucose Pentaacetates

\\\\\\\Macroscale
Procedure

Techniques   Polarimetry: Technique 13
                NMR Spectrometry: Spectrometric Method 2

--- SAFETY INFORMATION ---

Acetic anhydride is a strong irritant. Avoid contact with skin, eyes, or clothing. Pour acetic anhydride in the hood.

Anhydrous zinc chloride is corrosive. Avoid contact with skin, eyes, or clothing.

Chloroform is toxic and a suspected carcinogen. Wear gloves and use it only in a hood.

1. Preparation of α-D-Glucose Pentaacetate

Grind a small amount of anhydrous zinc chloride in a mortar *as rapidly as possible.* Anhydrous $ZnCl_2$ reacts rapidly with moisture in the atmosphere, which renders it useless as a catalyst, so it is important to work quickly.

The anhydrous zinc chloride must be recently purchased. Store it in a desiccator after opening.

Place 0.5 g of the powdered anhydrous zinc chloride, 12.5 mL (13.5 g) of acetic anhydride, and a magnetic stirring bar in a 100-mL round-bottomed flask equipped with a reflux condenser. Heat the mixture gently in a boiling water bath and stir until the $ZnCl_2$ dissolves. Slowly add 2.5 g of powdered D-(+)-glucose. Remove the condenser for each addition of glucose and stir briefly after each addition. The reaction may be vigorous if your zinc chloride is very dry, so care must be exercised in the addition process.

Start Procedure 2 at this point.

When all the glucose has been added, heat the mixture for 1 hour. Then pour the contents of the flask into 125 mL of ice and water. Stir the mixture at 0°C intermittently for 45 min to 1 h, during which time the oil that has separated will solidify.

Filter your crude product by vacuum filtration. Recrystallize it from a methanol/water solution made from one part methanol and two parts water by volume [see Technique 5.3]. Ten milliliters of this solvent mixture are required for each gram of α-D-glucose pentaacetate. Weigh your dried product and calculate the percent yield in the reaction. Save your product for analysis by polarimetry [see Technique 13] and ^{1}H NMR spectrometry.

Cleanup: Place the filtrate from the reaction mixture in the container for inorganic waste. Place the filtrate from the recrystallization in the container for flammable waste.

2. Preparation of β-D-Glucose Pentaacetate

Mix 2.5 g of powdered, dry D-(+)-glucose and 2.0 g of powdered, anhydrous sodium acetate in a 100-mL round-bottomed flask equipped with a reflux condenser. The glucose and sodium acetate can be powdered by grinding in a mortar. Add 12.5 mL (13.5 g) of acetic anhydride and heat the mixture in a boiling water bath for 1.5 h. Swirl the reaction mixture regularly during the heating period to maintain a clear solution.

As you stir 125 mL of ice and water in a beaker, slowly pour in the reaction mixture. Filter the product by vacuum filtration. Recrystallize it from a 1:2 methanol/water solution (by volume) [see Technique 5.3]. Approximately 10 mL of this solvent mixture are required for each gram of β-D-glucose pentaacetate. The crystals often have a beautiful, lustrous appearance.

Weigh your dried product and calculate the percent yield in the reaction. Analyze your product by polarimetry [see Technique 13] and ^{1}H NMR spectrometry.

Cleanup: Neutralize the filtrate from the reaction mixture with solid sodium carbonate **(Caution: Foaming.)** before washing it down the sink or pouring it into the container for aqueous inorganic waste. Place the filtrate from the recrystallization in the container for flammable (organic) waste.

13.2
^{1}H NMR Analysis of α- and β-D-Glucose Pentaacetates

Obtain the ^{1}H NMR spectra of your recrystallized α- and β-D-glucose pentaacetates, using $CDCl_3$ as the solvent. Analyze the ^{1}H NMR spectrum of each glucose pentaacetate, focusing on the chemical shift and coupling constant of the anomeric protons.

13.3
Optical Activity of Sugars

Read Technique 13, Polarimetry, before you begin work with the polarimeter.

There are three compounds in this experiment whose optical activity are of interest. The first is your starting material, D-(+)-glucose. It is not clear from its name alone whether the D-(+)-glucose is in the form of α-D-glucopyranose, β-D-glucopyranose, or a mixture of both. You will be able to determine quite simply the cyclic form of the D-(+)-glucose that you are using by measuring the optical activity of a sample. Calculation of its specific rotation tells you the ratio of α and β anomers of D-(+)-glucose.

Compound	$[\alpha]_D^{20°C}$	Solvent
α-D-Glucopyranose	+112°	H_2O
β-D-Glucopyranose	+18.7°	H_2O
α-D-Glucose pentaacetate	+101.6°	$CHCl_3$
β-D-Glucose pentaacetate	+4.2°	$CHCl_3$

The specific rotation of α-D-glucose pentaacetate has also been accurately measured and is very different from the rotation of β-D-glucose pentaacetate. Determine the specific rotations of D-(+)-glucose as well as your α- and β-glucose pentaacetates as described in Technique 13. Use water as the solvent for glucose and chloroform as the solvent for the pentaacetates.

SAFETY PRECAUTION ───────────────

Wear gloves and prepare the chloroform solution in a hood.

Use dry volumetric flasks to make up the chloroform solutions. After dissolving one of the sugars, stopper the volumetric flask and shake it numerous times to ensure a completely homogeneous solution. If you can see any undissolved particles such as paper fibers or pieces of dust in the solution, filter it by gravity through a small plug of glass wool in a short-stem funnel into a dry 50-mL Erlenmeyer flask. Keep the solution tightly stoppered during storage to avoid evaporation of the solvent. Use 5.0% solutions (0.050 g·mL^{-1}) of each sugar and pentaacetate in your measurements of optical activity.

Cleanup: Pour the chloroform solutions into the container for halogenated waste. The aqueous solutions can be washed down the sink.

Further Investigations

You can design a number of experiments with the glucose pentaacetates that you have synthesized in this project. Some suggestions follow. Monitor the progress of the reaction by TLC on silica gel thin-layer plates with cyclohexane/acetone (7:3 v/v) as the developing solvent. Visualize the chromatograms with *p*-anisaldehyde dipping reagent [see Technique 10.5]. Use polarimetry and ^{1}H NMR analysis to characterize your products.

1. Determine whether one pentaacetate isomerizes to the other in acetic anhydride with a catalytic amount of concentrated H_2SO_4.

2. Investigate kinetic versus equilibrium control in the pentaacetate formation using reaction temperatures between 75°C and 100°C. Investigate both acid and base catalysis. This investigation makes a good class project; different teams can use different temperatures.

References

1. Pearson, W. A.; Spessard, G. O. *J. Chem. Educ.* **1975,** *52*, 814–815.
2. Furniss, B. S.; Hannaford, A. J.; Smith, P. W. G.; Tatchell, A. R. *Vogel's Textbook of Practical Organic Chemistry;* 5th ed.; Longman: New York, 1989, pp. 642–646.

Questions

1. What other cyclic structures might D-(+)-glucose form? *Hint:* Look up furanose structures.
2. An alternative method for preparing the glucose pentaacetates substitutes acetyl chloride (CH_3COCl) for acetic anhydride. What advantages or disadvantages would this method have over the acetic anhydride method?
3. What structural factors freeze each of the glucose pentaacetates in a single conformation?
4. Sketch an energy-reaction progress diagram for the conversion of D-(+)-glucose to α- and β-D-glucose pentaacetates. The energy-reaction diagram should reflect the fact that the β product is the result of kinetic control and the α product is the result of thermodynamic control.
5. It may be surprising that α-D-glucose pentaacetate is more stable than the β isomer because the α isomer has its anomeric acetate group in the sterically less favorable axial position. This stability difference has been rationalized on the basis of the "anomeric" effect. That is, at the anomeric position, the C—O(Ac) bond of the β isomer has a dipole that is electronically unfavorable in its relation to the dipole of the ring oxygen atom. Explain this electronic unfavorability.

α-D-Glucose pentaacetate   β-D-Glucose pentaacetate

Organic
Qualitative Analysis

Organic chemists must regularly identify the compounds that are formed in chemical reactions or isolated from natural sources. Discovering the identity of an unknown organic compound requires finding which functional groups it contains and then determining its molecular and three-dimensional structure. Both chemical and spectrometric methods are used by practicing organic chemists. As you progress through this laboratory course, you will have the chance to learn and use the techniques of organic qualitative analysis while determining the identity of sample compounds whose identities are unknown to you (the test sample is designated the "unknown").

Because well over 15 million different carbon compounds have been discovered, synthesized, and characterized, the experimental techniques in Part 3 deal with a tremendous diversity of chemical substances. The tables in Appendix A contain over 500 compounds; thus, a bit of sleuthing is generally necessary to identify your unknown compound. Part of the challenge of organic qualitative analysis is that borderline cases and exceptions to the general rules are possible in many of the tests. If you work with an open, unprejudiced mind, ready to make, test, and reject hypotheses, you will have more fun and be much more successful in finding the identities of your unknowns. Your task is to reject every compound but the correct one, not to prove that your unknown is this or that.

The experiments are restricted to include only the following 16 functional-group classes.

Functional-Group Classes

Acyl chlorides	Aromatic hydrocarbons (arenes)
Alcohols	Aryl halides
Aldehydes	Carboxylic acids
Alkanes	Esters
Alkenes	Ethers
Alkyl halides	Ketones
Amides	Nitriles
Amines	Phenols

Organic qualitative analysis (structure identification) and the quantitative analysis of organic reaction mixtures is what most practicing research chemists make their living doing. The following material introduces you to the fundamentals and the excitement of organic analysis. But you learn how to do it well only from doing it over and over and over again.

You will have better success if you remember two fundamental concepts:

1. You learn nothing definitive from any test carried out on an impure compound.
2. You must be systematic—don't jump to conclusions.

Good luck in identifying your unknowns. It is like a puzzle—so treat it as one. You will be doing a lot of qualitative analysis if you become a professional chemist. Even if you become a physician, qualitative analysis and the thought processes of it are much like those of clinical diagnosis. We are sure you will enjoy it.

Qualitative Analysis 1
IDENTIFICATION OF UNKNOWNS

Six basic areas of experimental inquiry are useful for identifying an unknown compound. It is important that you understand what information can and cannot be obtained from each of them. The six areas are given in Table 1.1.

For each of these areas of inquiry, the importance of good experimental data cannot be stressed too much. When you are identifying an unknown compound, there is nothing worse than running tests that give ambiguous results. Given the diversity of organic compounds, it is often impossible to avoid all ambiguity, but careful and well-planned experimental work can greatly

Table 1.1. **Six areas of experimental inquiry for identifying unknowns**

Experimental areas of inquiry	Qualitative Analysis section presenting discussion of this topic
Physical properties	2
Classification of solubilities	3.1
Elemental analysis by sodium fusion	3.2
IR, NMR, and mass spectrometric analysis	4
Classification tests for functional groups	5
Synthesis of solid derivatives	6

restrict it. You cannot work directly with a mixture of substances—therefore, the first step in any analysis must be the purification of individual materials.

Every one of the areas of experimental inquiry in Table 1.1 depends on what can be called the structural theory of organic chemistry. By discovering how compounds act under certain conditions, a chemist can infer what their structures are. The specific areas of experimental inquiry in Table 1.1 have been found to be the most valuable for identifying organic compounds.

When you are asked to identify an unknown but pure compound, the first experiments you should do will require you to observe and determine the physical properties of the unknown; to determine (by sodium fusion) whether the compound contains nitrogen, sulfur, or halogen; to classify its solubility characteristics; and to examine its infrared (IR) and nuclear magnetic resonance (NMR) spectra. We have recommended a sequence of investigative steps, but the order in which these tests are done is not crucial. The determination and characterization of structure is usually more efficient when all these steps are carried out; however, in some cases your instructor may ask you to identify your unknown by using only a few of these strategies.

Familiarity with the spectral and solubility characteristics of your unknown compound, coupled with a knowledge of its physical properties, will help you decide what chemical classification tests you should use. The chemical reactions that a compound undergoes are determined by the functional groups it contains.

One of the major tools of organic qualitative analysis is the use of classification tests for specific functional groups. Depending on the results of your classification tests and evaluation of your spectrometric and physical data, you may want to analyze your compound for the presence of halogen or nitrogen atoms. The last step of the compound's identification is often the synthesis of a solid derivative whose melting point can be compared to the melting points of known compounds. (Reported melting points for derivatives can be found in the tables in Appendix A.)

You will find that once you have a large number of characteristics in hand, you can deduce the structure of a compound in much the same way that you put together a simple jigsaw puzzle. Often a major focus is provided with very few results: For example, carboxylic acids of reasonably high molecular weight are insoluble in water but soluble in alkali; phenols show somewhat different solubility behavior in basic solution. Infrared analysis and chemical tests both are used to determine the functional groups present; each of these two techniques should confirm the results of the other.

The compounds in the tables in Appendix A are either liquids or solids at room temperature. There are special handling techniques for each of these physical states. The purification of volatile liquids can often be achieved through short-path distillation [see Technique 7.3a] or microscale distillation [see Technique 7.3b]. The boiling point of a pure liquid compound is an important physical constant and can be accurately determined from distillation data or by a microscale boiling-point determination [see Technique 7.1a]. The purity of a liquid sample can be assessed through gas-liquid chromatography [see Technique 11]. However, improperly using a gas chromatograph to analyze samples of low volatility (particularly salts of organic acids or bases) may plug the chromatograph and may also give you misleading data caused by the appearance of decomposition products.

Handling small quantities of crystallizing solids is normally simple. For example, less than 100 mg of a solid can be recrystallized if a microscale apparatus such as a Craig tube is used [see Technique 5.5b]. An apparatus for microscale vacuum filtration is shown in Technique 5, Figure 5.7. Microscale procedures serve as an excellent background for organic qualitative analysis.

Spectrometric analysis and chemical classification tests are much the same for liquids and solids, except that solids may have to be dissolved in a solvent before the tests can be done.

When you are given an unknown compound, you may be told that it is pure. If not, you must assay its purity by using the techniques of gas chromatography (GC) or thin-layer chromatography (TLC), depending on its physical properties. (Gas chromatography is discussed in Technique 11 and thin-layer chromatography in Technique 10.) Ethyl acetate is a developing solvent of medium polarity and would be a good first choice if you are using silica gel thin-layer plates. If you wish to select your TLC developing solvent according to a less arbitrary method, see Technique 10.4.

It is best to distill a liquid before using GC analysis, because the GC conditions will depend in large part on the compound's volatility. As a general rule of thumb, you would wish to begin the search for proper column conditions with a column 30°–40°C below the compound's boiling point. See Technique 11.2 for the details of choosing a proper liquid stationary phase for your sample. If it proves to be necessary, purify your compound by distillation if it is a liquid or by recrystallization if it is a solid.

Purification is a crucial preliminary to identification. Chemical and spectrometric tests should only be done on pure compounds. Melting points and boiling points are rarely meaningful when obtained for impure compounds.

Qualitative Analysis 2
PHYSICAL PROPERTIES

The physical appearance of an unknown will be your first datum in the search to discover its identity. Simply knowing that the compound is a solid, rather than a liquid, at room temperature narrows the search considerably. A few solids have characteristic bright colors that may be of great significance in reaching a final answer. The color of a liquid sample must be considered much more cautiously. Many liquid compounds oxidize when they are stored for a long time. Often the oxidation products are intensely colored—yellow, green, red, brown, or black. In these cases the color of the liquid will tell you little if anything that will be useful. When such oxidation has occurred, distillation of the liquid often leads to a pure, colorless distillate.

The melting point or boiling point of an unknown compound is of critical importance, not only because these physical constants will definitely tell you whether your unknown is pure, but also because most of the tables of possible unknowns list compounds by functional group in order of increasing melting and boiling points. The tables of compounds in Appendix A also follow this pattern. Melting points tend to be more reliable than boiling points in this context.

Accurate physical constants are a must! For example, if you can trust a boiling point of a liquid alcohol that boils at 132 ± 2°C, you have narrowed the choice to only three or four possibilities from over 40 liquid alcohols in Table 2 in Appendix A. You should remember that the melting points listed in standard tables may be

2–3°C higher than your values, because the highest melting point of the purified compounds is normally listed. If you have doubts about how to obtain an accurate melting range, review Technique 6.3. Also, remember how important proper thermometer placement is for getting accurate boiling points during a distillation [see Technique 7.3, Figure 7.7].

One source of confusion with physical properties commonly occurs with compounds that melt just above room temperature when they are pure. Sometimes their melting points may be depressed by as much as 30°C because of impurities. In other words, a compound could be listed as a low-melting-point solid in our tables, yet it could be in the liquid state when you receive it. You can determine if an unknown liquid freezes near room temperature by putting about 2 mL of it in a small test tube. Place a thermometer in the test tube and immerse the test tube in a mixture of cracked ice and sodium chloride. If the liquid solidifies, remove the test tube from the ice bath and stir the mushy solid carefully and steadily. Note the temperature when the last crystals melt. This temperature is the melting point.

Most organic liquids are less dense than water, but alkyl bromides, alkyl iodides, polyhalogenated compounds, and many aryl halides are more dense. If an organic liquid is more dense than water, it probably contains halogen.

SAFETY PRECAUTIONS WITH UNKNOWNS

You should treat any unknown as if it were toxic. Wear gloves while handling it; be careful not to spill it on your skin; and do not breathe its vapor.

Some common but toxic compounds have been omitted from the Appendix A tables because we feel that especially dangerous compounds should not be given as unknowns. In the early days of organic chemistry, it was common not only to smell each new compound but also to taste a bit of it. We are much more alert to the dangers of toxic chemicals today, and therefore we **never smell (or taste) any unknown compound.** Only when an unknown compound has a penetrating, pronounced odor will the smell be useful to you in identification of the compound.

CLASSIFICATION BY SOLUBILITY AND ELEMENTAL ANALYSIS BY SODIUM FUSION

3.1
Classification by Solubility

Solubility tests should be performed on every general unknown, because they are quick and reliable and use only a small amount of sample. One can gather valuable information about possible functional groups through the use of the solubility classifications. If you already know what functional groups a compound has, a complete series of solubility tests is not necessary.

Five common reagents used for solubility tests are

> Water
>
> 2.5 M (9%) NaOH
>
> 0.6 M (5%) $NaHCO_3$
>
> 1.4 M (5%) HCl
>
> Concentrated (96%) H_2SO_4

Except in the case of water, solubility experiments probe the acid-base properties of organic compounds. If a compound is an acid, you can obtain a relative measure of its acid strength by testing it against the weak base sodium bicarbonate and the stronger base sodium hydroxide. By using hydrochloric acid and the powerful acid concentrated sulfuric acid, you can determine how basic the compound is if it is a base at all. Naturally *any organic compound that is soluble in water will probably also be soluble in 0.6 M $NaHCO_3$, 1.5 M HCl, and 2.5 M NaOH solutions.* These solutions are composed largely of water, so water-soluble compounds may also be soluble in them, revealing virtually nothing about the acid-base properties of the unknown.

Table 3.1 summarizes the solubility of each functional-group class considered in this section. Figure 3.1 uses a flow diagram to present the same kind of solubility data as that shown in Table 3.1. Use whichever is more convenient for you.

Table 3.1. **Solubility summary**[a]

	Water	2.5 M NaOH	0.6 M NaHCO$_3$	1.5 M HCl	Conc. H$_2$SO$_4$
Acyl chlorides	– (some react and dissolve)	+	+	–	+
Alcohols	– (except under C$_6$)	–	–	–	+
Aldehydes	– (except under C$_5$)	–	–	–	+
Alkanes	–	–	–	–	–
Alkenes	–	–	–	–	+
Alkyl halides	–	–	–	–	–
Amides	– (except under C$_6$)	–	–	–	+
Amines	– (except under C$_6$)	–	–	+	[b]
Arenes	–	–	–	–	– [d]
Aryl halides	–	–	–	–	–
Carboxylic acids	– (except under C$_6$)	+	+	–	+
Esters	– (except under C$_4$)	– [c]	–	–	+
Ethers	– (except under C$_5$)	–	–	–	+
Ketones	– (except under C$_5$)	–	–	–	+
Nitriles	– (except under C$_5$)	–	–	–	+
Phenols	– (except for a few)	+	– [d]	–	+

a. Compounds that are soluble in water may also be soluble in 2.5 M NaOH, 0.6 M NaHCO$_3$, and 1.5 M HCl solutions because these solutions are largely composed of water.
b. If HCl solubility and the presence of nitrogen indicate that the compound is an amine, do not test solubility in concentrated H$_2$SO$_4$, because a violent reaction can occur.
c. If cold.
d. Except for a few.

Solubility Test Procedure

— SAFETY INFORMATION —

Concentrated (96%) sulfuric acid is corrosive and causes severe burns. Wear gloves while conducting the solubility tests.

All solids must be *finely* powdered before weighing 15 mg on glassine weighing paper; use 1 drop (0.05 mL) for a liquid unknown. Place the unknown sample in a small test tube and add 0.5 mL of solvent. Calibrated disposable plastic pipets provide reasonably accurate volume measurements for the solvent.

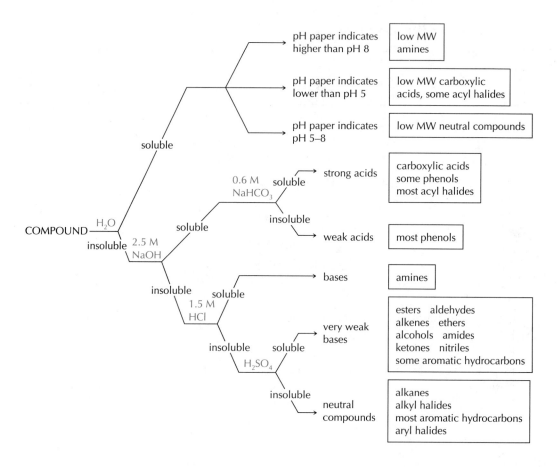

FIGURE 3.1 Solubility flow chart.

After you add the solvent, vigorously tap the test tube with your finger to ensure thorough mixing. Because of changes in the refractive index, mixing lines may appear when a liquid dissolves. If a liquid unknown does not dissolve, you will see the tiny droplets that were dispersed by your tapping coalesce into a larger drop at the top or the bottom of the solvent solution, depending on the relative densities of the solvent and unknown. Sulfuric acid mixtures may require stirring with a small glass rod to thoroughly mix the solvent and sample, because the acid is very dense and viscous.

Cleanup: Combine all aqueous test solutions (except sulfuric acid; see the next paragraph) from an unknown in a screw-capped centrifuge tube and adjust the pH with a few drops of dilute HCl or NaOH solution to a pH of 7 as determined by pH test paper. Add 1 mL of ether to the centrifuge tube, cap the tube, and shake it to mix the layers. Allow the layers to separate and remove the lower aqueous phase with a Pasteur pipet. Wash the aqueous phase down the sink or place it in the container for aqueous inorganic waste. Place the ether solution in the flammable (organic) waste container if you have determined that no halogen is present, or in the halogenated-waste container if you have found that your unknown contains a halogen.

Cautiously pour the concentrated sulfuric acid test mixture into a beaker containing 30 mL of water and neutralize the solution with solid sodium carbonate. Cool the solution to room temperature by adding a few ice chips, transfer the solution to a separatory funnel, and extract it with 4–5 mL of ether. Separate the layers. The lower aqueous phase may be washed down the sink or placed in the container for aqueous inorganic waste. Pour the ether solution into the flammable (organic) or halogenated-waste container as directed in the previous paragraph.

Sometimes it is difficult to decide whether two layers are present at the very top of a liquid level or whether you are seeing an optical illusion. You should try solubility tests with several known reference compounds until you learn to distinguish the appearance of two layers from that of a normal meniscus. Solids may take several minutes to dissolve. When a colored compound dissolves, normally the solution also will be colored.

We will use the following standard but arbitrary definition of solubility throughout this set of organic qualitative analysis experiments: *If a compound is soluble in a solvent to the extent of approximately 3–5%, that compound is declared to be "soluble" in that solvent.* Therefore, if 15 mg of an unknown dissolves in 0.5 mL of a solvent, the unknown is considered to be soluble in that solvent. (One could imagine many other definitions of solubility, but this is a convenient one. Of course, all compounds are soluble in all solvents to a small extent. If a drop of hexane were put into the Atlantic Ocean, it would surely dissolve; yet by our definition of solubility, hexane is insoluble in water.)

Solubility in Water   Most organic compounds are not soluble in distilled water, so if your unknown is insoluble in water, you have not learned much about it. However, if your unknown is water-soluble, you have learned a great deal. Solubility in water narrows the possibilities considerably, as shown in Table 3.1 and Figure 3.1.

If your compound dissolves in water, you should estimate the pH of the resulting solution by using indicating pH paper or litmus paper. It is best to wet a small stirring rod with the water solution and then touch the tip of it to the pH paper. In this way, the indicator dye of the pH paper does not dissolve, and a more accurate pH reading is possible. (See Figure 3.1 for the significance of a water solution whose pH is less than 5 or greater than 8.) Distilled water itself often has a pH of 5–6 as a result of the presence of dissolved carbon dioxide, which forms a dilute solution of carbonic acid.

Certain compounds—for example, acyl chlorides—react with water, forming a carboxylic acid, which may in turn be soluble in water. Heat is usually evolved in these hydrolyses.

$$R-\overset{\displaystyle O}{\overset{\displaystyle \|}{C}}-Cl + H_2O \longrightarrow R-\overset{\displaystyle O}{\overset{\displaystyle \|}{C}}-OH + HCl$$

Thus, if your compound dissolves because of a chemical reaction, you have learned something about its structure.

There is an indefinite borderline with respect to the solubility of carbon compounds in distilled water. For example, a few six-carbon alcohols, such as cyclohexanol, are soluble in water even though, according to Table 3.1, alcohols are water-soluble only if they have fewer than six carbons. Of course, polyhydroxy compounds, such as carbohydrates, are often very soluble in water. The same kind of borderline solubility is to be found among all functional-group classes whose low-molecular-weight members are soluble in water. Solubility depends on the exact structure of a compound, not only on the number of carbon atoms that it contains, and it is necessary to interpret solubility characteristics cautiously.

Solubility in 2.5 M (9%) NaOH Solution

Organic acids, which are insoluble in water, will normally dissolve in a 2.5 M (9%) NaOH solution. The pH of this solution is greater than 14, so any acid whose dissociation constant is greater than 10^{-12} ($pK_a < 12$) will be converted almost entirely to its conjugate base. The conjugate base of a carboxylic acid ($pK_a \sim 5$) is a carboxylate anion. The conjugate base of a phenol ($pK_a \sim 10$ is a phenoxide ion. Ionic carboxylate salts and phenoxide salts are quite soluble in water. Thus, water-insoluble carboxylic acids and phenols will dissolve in a 2.5 M sodium hydroxide solution.

$$R\overset{\displaystyle O}{\overset{\displaystyle \|}{C}}OH + OH^- \rightleftharpoons R\overset{\displaystyle O}{\overset{\displaystyle \|}{C}}O^- + H_2O$$

Water-
soluble

$$R\!\!-\!\!\underset{}{\underset{\text{Ar}}{\bigcirc}}\!\!-\!\!OH + OH^- \rightleftharpoons R\!\!-\!\!\underset{}{\underset{\text{Ar}}{\bigcirc}}\!\!-\!\!O^- + H_2O$$

Water-soluble

The hydrolysis of esters is strongly promoted by sodium hydroxide, but the process is too slow to allow an ester to dissolve in 2.5 M NaOH within a few minutes. However, heating the mixture could produce a soluble carboxylate salt. It is best not to use warm sodium hydroxide solution for this solubility classification. Most acyl chlorides hydrolyze quickly in sodium hydroxide solution, giving the water-soluble carboxylate salt.

$$\underset{}{\overset{O}{\overset{\|}{R\!C\!OR'}}} + OH^- \xrightarrow{\text{heat}} \underset{}{\overset{O}{\overset{\|}{R\!C\!O^-}}} + R'OH$$

Solubility in 0.6 M (5%) NaHCO$_3$ Solution

When a compound is insoluble in water but soluble in 2.5 M NaOH, a third solubility test using 0.6 M (5%) sodium bicarbonate is called for. Sodium bicarbonate is a weaker base than sodium hydroxide; a 0.6 M NaHCO$_3$ solution has a pH of approximately 9. It will dissolve a water-insoluble organic acid whose pK_a is less than 7.5 by converting it to a water-soluble salt. Whereas carboxylic acids (pK_a < 5) dissolve in a sodium bicarbonate solution (see equation), most phenols do not.

$$\underset{}{\overset{O}{\overset{\|}{R\!C\!OH}}} + HCO_3^- \rightleftharpoons \underset{}{\overset{O}{\overset{\|}{R\!C\!O^-}}} + H_2O + CO_2$$

So the use of a 2.5 M sodium hydroxide solution followed by the addition of 0.6 M NaHCO$_3$ to another sample of the compound can distinguish organic acids from other organic compounds and can differentiate carboxylic acids from phenols.

Only a few phenols are strong enough acids to completely dissolve in 0.6 M NaHCO$_3$ solution. Phenol derivatives with strongly electron-withdrawing substituents are much stronger acids than phenol itself. For example, 4-nitrophenol has a pK_a of 7.15; 4-nitrophenol will dissolve in a 0.6 M NaHCO$_3$ solution.

When an acidic compound dissolves in a sodium bicarbonate solution, carbon dioxide gas is released. Observation of effervescence in a liquid confirms that an acid-base reaction did indeed occur.

Solubility in 1.4 M (5%) HCl Solution

The only organic compounds that are insoluble in distilled water but soluble in dilute hydrochloric acid solution are amines, the major class of basic organic compounds. Nearly all amines react with HCl to produce ionic ammonium salts, which are almost always soluble in water. This behavior is the same for secondary (R_2NH), tertiary (R_3N), and primary amines (RNH_2). The reaction of a primary amine can be represented as follows:

$$R\ddot{N}H_2 + HCl \rightleftharpoons RNH_3^+ + Cl^-$$

Solubility in Concentrated H₂SO₄ (96%) Solution

SAFETY INFORMATION

Concentrated (96%) sulfuric acid is corrosive and causes severe burns. Wear gloves and avoid contact with skin, eyes, and clothing. **Immediately** wash any spill off the skin with a great deal of water.

Water-insoluble nitrogen compounds that are soluble in 1.4 M HCl should not be treated with concentrated H_2SO_4 because the heat produced may lead to a violent reaction.

Concentrated sulfuric acid will protonate all organic compounds that contain oxygen and/or nitrogen, as well as alkenes and a few aromatic hydrocarbons. These protonated organic compounds exist as ionic salts in sulfuric acid. Because sulfuric acid is a highly polar liquid, it dissolves these salts. The dissolution of compounds in H_2SO_4 may also produce large amounts of heat.

All organic compounds that contain nitrogen or oxygen are weak bases. In aqueous mineral acid solutions of moderate concentration, the conjugate acids of these compounds are present in modest amounts. In concentrated sulfuric acid, however, these conjugate acids are often stable enough to be present in sizable quantity:

$$ROH + H_2SO_4 \rightleftharpoons ROH_2^+ + HSO_4^-$$

$$\overset{O}{\underset{\|}{R C R}} + H_2SO_4 \rightleftharpoons \overset{OH^+}{\underset{\|}{R C R}} + HSO_4^-$$

$$RC{\equiv}N + H_2SO_4 \rightleftharpoons RC{\equiv}NH^+ + HSO_4^-$$

The solubility of alkenes in sulfuric acid results from protonation of the carbon-carbon double bond to form carbocations. (The carbocations in turn may react further, giving alkyl hydrogen sulfates—both of which are soluble in the sulfuric acid.) The carbocations may also react with unprotonated alkene to give insoluble brown polymers. A pronounced color change is thus a positive indication of a compound's solubility in sulfuric acid.

Some aromatic hydrocarbons and a few aryl halides are soluble in sulfuric acid because they are converted to sulfonic acids through electrophilic aromatic substitution. The polar sulfonic acids are very soluble in sulfuric acid. The possibility that aromatic hydrocarbons or aryl halides will dissolve in H_2SO_4 then depends on a fast sulfonation reaction. Alkyl substituents speed up electrophilic aromatic substitution and therefore enhance the solubility of aromatic hydrocarbons and halides in H_2SO_4. Sulfonic acid groups can be protonated and the resulting cationic structure would be very polar and thus soluble in the polar sulfuric acid solution.

You will notice in Table 3.1 that very few functional-group classes are insoluble in concentrated sulfuric acid. Thus, insolubility in H_2SO_4 is a more valuable piece of information than solubility. Alkanes, alkyl halides, aromatic hydrocarbons, and aryl halides are the only classes that are insoluble. These classes can easily be distinguished by using spectrometric analysis and functional-group classification tests.

3.2

Elemental Analysis by Sodium Fusion

Virtually all organic compounds contain carbon and hydrogen; many also contain oxygen. Carbon and hydrogen are usually detected by the ignition test (p. 541). The presence of oxygen can be inferred from solubility tests and spectrometric evidence but usually is not confirmed directly.

The halogens and nitrogen also are commonly found in the unknown compounds that you will encounter, and it can become quite important to know whether any of these elements are present in a compound as you seek to discover its identity. When you are working with compounds that have already been characterized and whose properties appear in reference books, it is not essential to determine the composition of a compound quantitatively. But it is important to know whether N, Cl, Br, or I are present or absent.

The simplest test for qualitative elemental analysis is the Beilstein test (see p. 535). This test can quickly indicate the presence or absence of halogen in a compound, but it can be so sensitive that trace impurities give false information about an unknown.

Fusing an unknown compound with sodium metal cleaves covalent bonds in an unselective manner. It leaches halide and sulfide and converts any carbon-bound nitrogen to cyanide ions:

$$(C,H,N,X,S) + excess\ Na \longrightarrow NaCN + NaX + Na_2S$$
$$(X = Cl,\ Br,\ I)$$

These anions can be detected by the classic techniques of inorganic qualitative analysis. Having a clean surface of hot sodium metal is required for accurate results in a sodium fusion test. Doing a sodium fusion test on an unknown compound without having done one or two sodium fusion tests on reference compounds usually leads to ambiguous results and is a waste of time. A good compound to use for testing your sodium fusion technique is 4-bromobenzenesulfonamide.

Run a sodium fusion test on one or two known reference compounds before you do one on an unknown.

Sodium Fusion

Sodium metal is stored under saturated hydrocarbon solvents to prevent its reaction with moisture or oxygen in the atmosphere.

— SAFETY INFORMATION —

Sodium metal is extremely caustic to all tissues. It also reacts violently with water, and the hydrogen gas that is evolved can explode. The heat produced when sodium reacts with water may set organic compounds on fire. Handle the metal carefully, using a pair of forceps. Excess sodium metal should be destroyed immediately by adding it *slowly* to a large quantity of ethanol.

Some organic compounds react violently with hot sodium. Always perform the sodium fusion test on small samples. Make sure that the tube containing the sodium fusion test mixture *does not* point directly at anyone nearby.

When sodium metal is stored, its surface can become covered with white flaky sodium oxide. This impurity must be cut away with a sharp knife before using the metal for the sodium fusion test. If you have doubts about the condition of the sodium in your laboratory, check with your instructor about using it before running the test.

Procedure. Place a small piece of clean sodium metal, the size of a small pea, in a dry 13 × 100 mm Pyrex test tube clamped in a vertical position. Use a clamp without rubber or plastic coating because these materials would fuse to the outside of the hot test tube during the analysis. Mix 50 mg of confectioner's (powdered) sugar with either 4 drops or about 80 mg of your unknown in another test tube or on a piece of weighing paper.

Heat the test tube containing the sodium with a Bunsen burner in a ventilating hood until the sodium melts and a shiny surface of clean, completely molten metal appears. Remove the flame and quickly (before cooling occurs), but carefully, add half of the powdered sugar/unknown mixture directly on top of the molten sodium; avoid getting the mixture on the walls of the test tube. Again heat the test tube until the sodium melts, remove the heat, and add the remaining sugar/unknown mixture. Heat the bottom of the tube to a red-hot state for 1–2 min and then cool the test tube to room temperature. Add 1 mL of methanol and break up the solid with a stirring rod to expose and completely destroy the excess sodium. Then cautiously add 4 mL of distilled water. Add a boiling stone to the test tube and heat the water solution to just below its boiling point while carefully agitating the mixture with a stirring rod.

Filter the solution, using folded filter paper in a small funnel. Portions of the filtrate will be used in the following tests for nitrogen and the halogens.

Elemental Analysis for Nitrogen

Procedure. To a small test tube, add 1 mL of a 1.5% solution of *p*-nitrobenzaldehyde in 2-methoxyethanol (methyl Cellosolve), 1 mL of 1.7% solution of *o*-dinitrobenzene in 2-methoxyethanol (methyl

Cellosolve), and 2 drops of aqueous 2% sodium hydroxide solution. Add 2 drops of the filtrate from the sodium fusion.

Positive test: A blue-violet colored solution indicates a positive test.

Cyanide ion converts the *p*-nitrobenzaldehyde to the conjugate base of a cyanohydrin, which then reduces *o*-nitrobenzene to a blue-violet dianion complex. Negative tests usually result in a tan or yellow color. The presence of halide or sulfide ions in the filtrate does not interfere with this test.

Cleanup: Pour the reaction mixture into the flammable (organic) waste container.

Elemental Analysis for Halogens

— **SAFETY INFORMATION** ——————

Carbon tetrachloride is toxic and a suspected carcinogen. **Concentrated (15.7 M) nitric acid** is very corrosive and causes severe burns. Wear gloves and work in a ventilating hood while using both of these reagents.

Procedure. Acidify 1 mL of the filtrate from the sodium fusion procedure by dropwise addition of 3 M nitric acid. If your elemental analysis for nitrogen was positive, boil the acidified solution in a hood for about 20 s to remove the hydrogen cyanide (extremely toxic), which interferes with the test for halogens. Then add 5 drops of 0.2 M silver nitrate solution.

Positive test: A heavy precipitate of silver halide indicates the presence of a halogen.

A *heavy* precipitate of silver halide indicates the presence of a halogen. A tiny amount of precipitate usually indicates a small amount of impurity. If the silver halide that forms is bright yellow, it is silver iodide. Silver bromide and silver chloride are white. The precipitate of silver halide may darken rapidly on exposure to light, because silver halides easily undergo a photochemical reduction to silver metal.

Iodine, bromine, and chlorine can be distinguished on the basis of differences in the ease with which they are oxidized by nitric acid.

Place 0.5 mL of sodium fusion filtrate in a 13 × 100 mm test tube. Add 0.5 mL of carbon tetrachloride and 3 drops of concentrated (15.7 M) nitric acid. Mix the two layers by drawing the mixture into a Pasteur pipet and expelling the liquid back into the test tube several times; allow the two phases to separate.

Positive test: The presence of iodine is indicated by a violet color in the lower carbon tetrachloride layer.

The presence of iodine is indicated by a violet color in the lower carbon tetrachloride layer.

If iodine is present, remove the carbon tetrachloride layer with a Pasteur pipet. Then add a fresh 0.5-mL portion of carbon tetrachloride to the aqueous solution remaining in the test tube. Repeat the mixing of the two phases, using a Pasteur pipet, and allow the layers to separate. If the violet color persists, remove the carbon tetrachloride layer and again extract the aqueous phase with a fresh 0.5-mL portion of CCl_4. Repeat these extractions until the CCl_4 layer is colorless. Then add 2 mL of concentrated (15.7 M) nitric acid to the colorless CCl_4 aqueous mixture test tube and again mix the phases with a Pasteur pipet. Allow the layers to separate.

Positive test: The presence of bromine is indicated by the appearance of a tan or tannish red color in the carbon tetrachloride layer.

The presence of bromine is indicated by the appearance of a tan or tannish red color in the carbon tetrachloride layer.

If bromine is present, carry out successive extractions with 0.5-mL portions of CCl_4, as directed above, until the CCl_4 layer again becomes colorless. Remove the CCl_4 layer and add five drops of 2% aqueous silver nitrate solution.

Positive test: The immediate formation of a white precipitate (AgCl) indicates the presence of chlorine.

The immediate formation of a white precipitate (AgCl) indicates the presence of chlorine. The chloride precipitate should be compared to that formed earlier in the first halogen test. The Beilstein test (p. 535) provides a cross-check for the presence of halide, although it does not identify the specific halogen.

Cleanup: Pour all carbon tetrachloride solutions into the halogenated-waste container. Neutralize all aqueous test solutions containing acid with sodium carbonate before washing them down the sink or placing them in the aqueous inorganic-waste container, as directed by your instructor.

Elemental Analysis for Sulfur

Procedure. Place 10 drops of the aqueous solution reserved from the fusion procedure in a small test tube. Add acetic acid dropwise until the solution becomes acidic (as indicated by litmus paper); then add 2 or 3 drops of 1% aqueous lead acetate solution.

Positive test: The formation of a black precipitate (PbS) indicates the presence of sulfur.

The formation of a black precipitate (PbS) indicates the presence of sulfur.

Cleanup: Pour the test solution and precipitate in the hazardous-metal/inorganic-waste container.

INFRARED, NUCLEAR MAGNETIC RESONANCE, AND MASS SPECTROMETRIC ANALYSIS

Spectrometric techniques are extremely powerful methods for the identification of organic compounds. These techniques can be so efficient and quick that chemists often use only a compound's spectra to determine its structure. Because this laboratory textbook has an extensive section on spectrometric methods, which includes chapters on infrared, NMR, and mass spectrometry, they will not be treated in detail here. Rather, we will discuss how to tackle the problem of determining the structure of an unknown compound by combining a variety of spectrometric data. This approach may also be applied in conjunction with tests relating to the sample's physical properties, solubility, and functional groups.

Nuclear magnetic resonance spectrometry (NMR) is a particularly powerful technique. It is usually the best, most direct route to the determination of organic structure. One of the most important pieces of information in an ^{1}H NMR spectrum is the chemical shift of the various kinds of protons in the sample. Usually, it is difficult or even impossible to decipher structural information from an ^{1}H NMR spectrum without paying close attention to the chemical shifts of the NMR peaks. Table 4.1 lists some of the chemical shift information for protons that is commonly needed. Carbon-13 (^{13}C) NMR also provides structural information from the chemical shifts, as shown in Table 4.2. Observed chemical shifts sometimes fall slightly outside these ranges, especially when different solvents or concentrations are used.

With NMR analysis, however, it can be useful to see if your conclusions fit with other observations, such as those from infrared and mass spectrometry, and chemical reactivity. For example, if your NMR spectrum suggests that the unknown may be an alcohol, chemical and infrared analyses can be used to confirm this conclusion. Solubility characteristics of a carboxylic acid in water and aqueous NaOH can be almost as important as the NMR spectrum.

Table 4.1. ¹H NMR chemical shifts of functional-group classes

Compound class	Functional group[a]	Chemical shift
Acyl chlorides	H̲C—C(=O)Cl	δ 1.9–3.0
Alcohols	R—OH̲	δ 0.5–4.0
	H̲C—COH	δ 3.0–4.0
Aldehydes	—C(=O)H̲	δ 9.0–10.0
	H̲C—C(=O)H	δ 1.9–3.0
Alkanes		δ 0.2–1.9
Alkenes	H̲C=C	δ 4.5–6.5
Alkyl halides	H̲CX	δ 3.0–4.0
Amides	—C(=O)NH̲	δ 5.0–8.7
	H̲C(=O)NH	δ 1.9–3.0
	—C(=O)NHCH̲	δ 2.9–3.8
Amines	—NH̲	δ 0.5–3.0
Arenes	H̲—Ar	δ 6.5–8.5
Aryl halides	H̲—ArX	δ 6.9–7.7
Carboxylic acids	RC(=O)OH̲	δ 10.0–13.5
	H̲CC(=O)OH	δ 1.9–3.0
Esters	H̲CC(=O)OR	δ 1.9–3.0
	—C(=O)OCH̲	δ 3.0–4.0
Ethers	H̲COR	δ 3.0–4.0
Ketones	H̲CC(=O)R	δ 1.9–3.0
Nitriles	H̲CC≡N	δ 2.15–3.9
Phenols	Ar—OH̲	δ 4.0–8.0
	intermolecular H-bond	δ 5.5–12.5
	ring	δ 6.8–7.1

a. Chemical shift is produced by underlined H atom.

The most obvious information from a compound's mass spectrum is its molecular mass. The molecular ion peak will often be the peak at the greatest mass in the spectrum. In practice, however, it can be tricky to rely on this generalization, because impurities in the spectrometer may produce small peaks of higher molecular mass. Then it becomes difficult to know exactly what the molecular mass of your compound really is. In some compounds, the molecular ion peak can be negligibly small because of its extensive fragmentation in the electron beam. With practice, however, one can use evidence for the presence of fragment ions to gain valuable insight into the structure of an organic compound.

Table 4.2. ^{13}C NMR chemical shifts of functional-group classes

Compound class	Functional group[a]	Chemical shift
Acyl chlorides	C̲(=O)X	δ 168–170
Alcohols	C̲OH	δ 50–75
Aldehydes	C̲(=O)H	δ 190–200
Alkanes		δ 0.0–39
Alkenes	HC̲=C	δ 100–150
Alkyl halides	HC̲X	δ 0–80
Amides	R—C̲(=O)NR$_2$	δ 160–173
Amines	R$_2$NC̲—	δ 25–60
Arenes	Ar—H	δ 95–165
Aryl halides	Ar—X	δ 95–165
Carboxylic acids	RC̲(=O)OH	δ 170–180
Esters	RC̲(=O)OR′	δ 1.9–3.0
Ethers	C̲OR	δ 65–75
Ketones	RC̲(=)OR′	δ 198–215
Nitriles	RC̲≡N	δ 117–120
Phenols[b]	C̲$_1$	δ 145
	C̲$_2$	δ 116
	C̲$_3$	δ 130
	C̲$_4$	δ 120

a. Chemical shift is produced by underlined C atom.
b. These values are for phenol. Substituted phenols will have shifts that vary from these as a result of the effect of the additional substituent(s).

Mass spectrometry is discussed in more detail in Spectrometric Method 3 in Part 4.

Because infrared (IR) spectra often exhibit many absorption bands, it may be very difficult to extract all the rich structural information from an infrared spectrum. Nevertheless, *infrared spectrometry is especially useful in identifying the functional groups in unknown compounds.* Review Spectrometric Method 1.4 in Part 4 as you prepare to analyze an infrared spectrum. The IR spectral correlations discussed in Spectrometric Method 1 and Table 1.1 (p. 590) will be useful as you work on the analysis of an unknown compound. Here we will indicate only a few principal features of the IR spectra for the functional groups considered in this chapter. As you would expect, the functional group region (4000–1300 cm^{-1}) is the major focus of the IR methods used in organic qualitative analysis.

A huge amount of effort has been focused on the infrared spectra of carbonyl compounds. The carbon-oxygen double bond appears in many interesting compounds, and this bond acts like a well-behaved localized vibration. Therefore, accurate conclusions can be drawn from the exact position of a carbonyl bond. Six of the functional group classes in the list on p. 499 contain the *carbonyl group,* $\diagup C = O$. The stretching absorption of this carbon-oxygen double bond is one of the most intense bands to appear on an infrared spectrum. The usual frequency ranges for these carbonyl stretching vibrations are

Acyl chlorides	1820–1750 cm^{-1}
Aldehydes	1730–1660 cm^{-1}
Amides	1700–1640 cm^{-1}
Carboxylic acids	1740–1680 cm^{-1}
Esters	1750–1700 cm^{-1}
Ketones	1740–1660 cm^{-1}

Unsaturated, conjugated carbonyl compounds have their carbon-oxygen double-bond stretching absorptions at the lower end of these frequency ranges. For example, whereas 2-butanone absorbs at 1715 cm^{-1}, the conjugated ketone, acetophenone, absorbs at 1686 cm^{-1}. The diaryl ketone, benzophenone, absorbs at about 1665 cm^{-1}.

Two other multiple-bond systems studied by IR spectrometric methods (see Table 1.1, p. 590) are also worthy of mention. The *carbon-carbon double bond* stretching absorption appears as a band of medium intensity, but a definitive one nonetheless, between 1670 cm^{-1} and 1600 cm^{-1}. The carbon-nitrogen triple bond of the *nitrile* functional group exhibits a moderately strong absorption at 2260–2200 cm^{-1}, a region where other absorptions rarely occur. Such a characteristic absorption is especially useful in the identification of an unknown compound. Weaker bands for unsymmetrical alkynes occur in virtually the same region, but symmetrical alkynes frequently show negligible or no bands in that region.

Another infrared region of great importance is between 4000 and 3000 cm^{-1}, where the *O—H* and *N—H stretching* vibrations occur.

Alcohols and phenols	3700–3100 cm^{-1}	(strong intensity)
Amides	3500–3100 cm^{-1}	(weak intensity)
Amines	3500–3100 cm^{-1}	(medium intensity)
Carboxylic acids	3300–2500 cm^{-1}	(strong intensity)

The O—H bond absorptions are very intense and often quite broad when the sample is a pure liquid or a mineral oil mull. This effect is largely due to hydrogen-bonding effects. The relative intensities of the sharp absorption band near 3600 cm^{-1}, which is due to a free O—H stretching vibration, and the broad band at lower frequencies, are dependent on the concentration of the infrared sample. The characteristically broad O—H stretching absorption of a carboxylic acid that is hydrogen bonded is a sure sign that the sample contains a carboxylic acid.

Although N—H stretching vibrations of amines are much weaker than O—H stretching bands, they are still useful—particularly in the identification of primary amines. These compounds show twin peaks at 3550–3420 cm^{-1} and 3450–3320 cm^{-1}, absorptions corresponding to the asymmetric and symmetric N—H stretching vibrations (see Spectrometric Method 1).

The characteristic and important *C—H stretching* frequencies of alkanes, alkenes, aldehydes, and aromatic compounds and the other diagnostic infrared bands for aromatic compounds are discussed in some detail in Spectrometric Method 1.

The fingerprint region (1400–1000 cm^{-1}) contains so many bands that it is often impractical to assign vibrations to them. This region does contain the C—O stretching frequency for alcohols and other compounds; they are normally intense and readily assigned.

Primary alcohols	1070–1010 cm^{-1}
Secondary alcohols	1170–1050 cm^{-1}
Tertiary alcohols	1230–1130 cm^{-1}

The *C—O stretching* vibration of esters is also intense, appearing in the 1300–1100 cm^{-1} region. Ethers also show C—O stretching vibrations in this region. The region below 1000 cm^{-1} often reveals strong bands that are useful for characterizing aromatic compounds (see Spectrometric Method 1).

4.1 Using Spectrometric Data

Determining the structure of an unknown organic compound by using only spectrometry can be a fascinating puzzle. Even though NMR, infrared, and mass spectrometry are discussed in detail elsewhere in this book, as well as in organic chemistry textbooks, it is instructive to include here two examples of overall spectrometric analyses in the context of organic qualitative analysis.

Example 1   In Figure 4.1 you see the NMR, infrared, and mass spectra of an unknown solid. Where to begin? The mass spectrum shows a prominent peak at 150 mass units, which is likely to be the molecular mass of the compound. The ^{1}H NMR spectrum has a large set

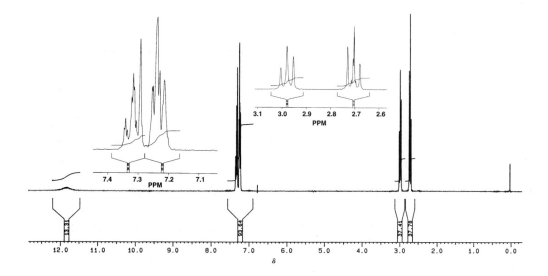

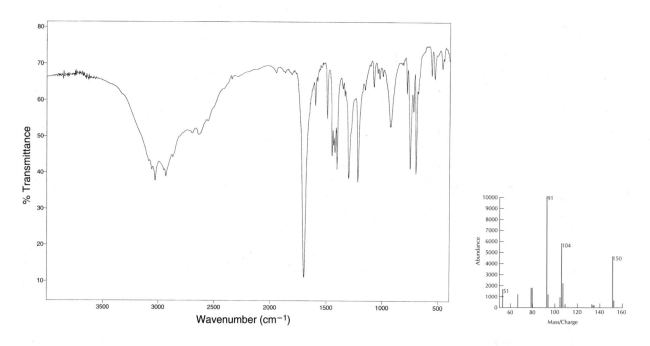

FIGURE 4.1 NMR, IR, and mass spectra for Example 1.

of peaks near δ 7.2, a finding that is a strong indication of a benzene ring. This 300-MHz ^{1}H spectrum, measured on a sample dissolved in chloroform, shows a relative integration of 5:2:2 for the three sets of peaks at δ 7.2, 2.9, and 2.5. Because there are five protons attached to the benzene ring, it seems to be a phenyl group, C_6H_5—. Now let us move to the infrared spectrum. The very broad, strong absorption in the region of 3000 cm^{-1} is characteristic of the strongly hydrogen-bonded O—H stretching vibration of a carboxylic acid. If so, there must also be an intense C=O stretch in the carbonyl region of the infrared spectrum. There it is at 1710 cm^{-1}!

Now we have used the simplest and most obvious aspects of all three spectra. They could have been used in any order up to this point. We have learned that it is likely that the compound is an aromatic carboxylic acid. Now let's return to the NMR spectrum. There are two other groups of peaks apparent. Where should peaks come for the protons α to the benzene ring and the carbonyl groups? Table 4.1 tells us that the carbonyl group deshields the α-protons, so they should appear in the δ 1.9–3.0 region—exactly where both sets of protons appear. Looking in the tables in the NMR chapter shows that benzyl protons α to the benzene ring also appear between δ 2 and 3.

It is time to make a proposal for the structure of the unknown compound. Let's say it is $C_6H_5CH_2CH_2COOH$. Now let's look at the coupling pattern in the NMR spectrum (Figure 4.1). Notice that each set of peaks between δ 2 and δ 3 is a triplet, showing two protons adjacent to each set of protons. This observation fits our proposal. What about the relative integration? This fits as well. The molecular ion at 150 is also consistent with the molecular mass of $C_9H_{10}O_2$. One could pay closer attention to the additional peaks in the infrared and mass spectra, such as the aromatic ring C—C stretch at 1610 cm^{-1} and the C—O stretch at 1290 cm^{-1}; however, the major pieces already fit. But wait a minute! Where is the carboxylic acid proton attached to oxygen? It doesn't appear anywhere in the NMR spectrum. That could be because strongly hydrogen-bonded protons appear at δ 10.0–13.5, beyond the region shown in the spectrum. It would be wise to do a chemical test at this point to confirm that the unknown is indeed a carboxylic acid. Solubility in 2.5 M NaOH solution and 0.6 M NaHCO$_3$ (and insolubility in H$_2$O) and a melting range of 46–47°C clinch the answer.

Example 2   In the second example (see Figure 4.2), we will follow the same procedure we did in the first, that is, we will look for the most useful information from all the spectra before analyzing any of them in detail. The NMR spectrum of this unknown liquid indicates five different kinds of protons. The two kinds farthest downfield, in the vicinity of δ 7, are likely to be due to aromatic protons. Their symmetry strongly suggests a disubstituted ring with the substituents in a para- or 1,4-relationship. The infrared spectrum shows a strong, broad absorption at about 3400 cm^{-1}. It is not nearly as broad nor at such a low frequency as the O—H stretching vibration of a carboxylic acid; it is the O—H stretch of an alco-

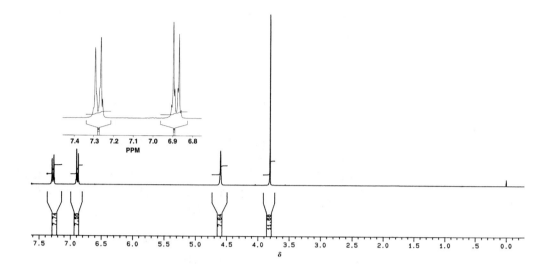

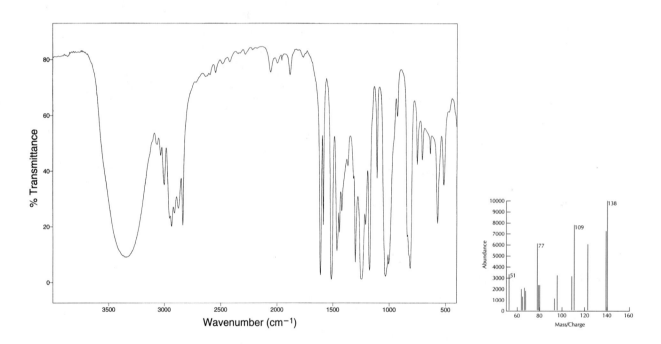

FIGURE 4.2 NMR, IR, and mass spectra for Example 2.

hol. Note also that there is no carbonyl absorption in the vicinity of 1700 cm^{-1}, as there was in the IR spectrum of the carboxylic acid in Example 1. If one were not confident at this point that the unknown is an alcohol, the chromic acid test for alcohols could be used to test the hypothesis. The mass spectrum shows a probable molecular ion with a mass of 138.

As before, let's return now to a more thorough analysis of the NMR spectrum. The relative integrations of the five sets of peaks are in a ratio of 2:2:2:3:1. In addition to the two aromatic protons each at δ 6.8 and 7.2, there are three additional kinds of protons in each molecule of the compound. The peak with relative integration of 1 is likely to be due to the OH proton, and that with relative integration of 3 is likely to be a methyl group. Now we must use chemical shifts to determine the molecular environments of the methyl and the remaining set of protons. The methyl singlet appears at about δ 3.8, which, according to Table 4.1, indicates a proton on a carbon atom α to an oxygen atom of an alcohol or an ether. The methyl group must be attached as a methyl ether. Therefore, there must be a methoxy (CH_3O—) group attached to the benzene ring. Across the ring, there is a substituent that must contain the —OH group of the alcohol indicated by the IR spectrum. If the —OH group were attached to a carbon that was also attached to the benzene ring, that —CH_2 group should be farther downfield than δ 4, because it is attached both to an oxygen atom and a phenyl ring.

A reasonable hypothesis for the structure of the unknown is 4-$CH_3OC_6H_4CH_2OH$, 4-methoxybenzyl alcohol. The NMR spectrum fits! Notice that there is even a hint of splitting between the —CH_2 and —OH groups, which means that the rate of exchange of the —OH proton must be much slower than usual so that the vicinal coupling becomes evident. In the IR spectrum, there are peaks at about 1620 cm^{-1}, due to the aromatic ring C—C stretch, and at about 1250 cm^{-1}, due to C—O stretching vibrations. The molecular mass of 138 also fits for $C_8H_{10}O_2$. To complete the determination, one could confirm that the primary alcohol is oxidized in the chromic acid test and that its boiling point is 259°C.

Qualitative Analysis 5
CLASSIFICATION TESTS FOR FUNCTIONAL GROUPS

If you have done everything recommended up to this point, you will have examined the physical properties of the unknown; determined whether the compound contains nitrogen, sulfur, or halogen; analyzed its solubility properties; and scrutinized the IR and NMR spectra. (Even if you were asked to do only part of this list, you should have a number of pieces of valuable information.) It is time to use a few of the classification tests for specific functional groups that you suspect may be in the compound. We have included only the simplest and most dependable of the many classification tests that have been proposed over the years. These are listed under the functional groups to which they apply.

You should plan carefully what chemical tests to use, based on the results you have obtained so far. You need not—indeed, should not—run all these classification tests. If your data suggest that your unknown may be a ketone with no other functional group likely, you can confirm this by using the aldehyde and ketone classification tests. You would not want to waste time doing a Lucas test (for alcohols) or a Hinsberg test (for amines), unless you suspect that the compound might also contain an alcohol or amino group. In other words, be selective with the tests that you run. Evaluate your hypotheses with crucial tests. If your early ideas need modification, then carefully reevaluate your data, make a new hypothesis and proceed to test it out. A "shotgun" approach is inefficient and frustrating.

Always try out an unfamiliar classification test on one or two reference compounds of known structure before using the test on your unknown compound. For example, if you suspect that your compound may be an alkene, you will want to test whether it can decolorize bromine. You should first try this test on known compounds such as cyclohexene and cyclohexane, so that you can observe directly both positive and negative results for this test. The inexperienced student may be frustrated easily by ambiguous results. This frustration can be minimized by doing control tests on reference compounds.

Functional-Group Tests

Alcohols
    Chromic acid oxidation
    Iodoform test
    Lucas test
Aldehydes and ketones
    Reaction with 2,4-dinitrophenylhydrazine
    Oxidation of aldehydes with chromic acid
    Iodoform test
    Tollens test
Alkenes
    Reaction with bromine
    Oxidation with potassium permanganate
Alkyl halides
    Beilstein test
    Reaction with ethanolic silver nitrate
    Reaction with sodium iodide in acetone
Amines
    Hinsberg test
Aromatic hydrocarbons (arenes)
    Ignition
    Reaction with aluminum chloride/chloroform

Aryl halides
  Ignition
  Beilstein test
  Reaction with aluminum chloride/chloroform
Carboxylic acids
  Neutralization equivalent
Carboxylic acid derivatives
  Acyl chlorides: Reaction with silver nitrate
  Esters: Ferric hydroxamate test
  Amides, acyl chlorides, esters, nitriles: Hydrolysis
Ethers
  Ferrox test
Phenols
  Reaction with bromine/water
  Reaction with ferric chloride

───── SAFETY INFORMATION ─────

Wear gloves while conducting any functional-group test.

5.1
Alcohols

*Chromic Acid Oxidation**

───── SAFETY INFORMATION ─────

Chromic acid solutions are very toxic and corrosive. Wear gloves and avoid contact with skin, eyes, and clothing. If you spill any on your skin, wash it off immediately with copious amounts of water.

Positive test: A primary or secondary alcohol will reduce the orange-red chromic acid/sulfuric acid reagent to an opaque green or blue suspension of Cr(III) salts in 2–5 s.

Procedure. Add 1 drop or 30 mg of the unknown or reference compound to 1 mL of reagent-grade acetone in a test tube. Then add a drop of the chromic acid/sulfuric acid reagent directly into the solution. Shake the mixture. A primary or secondary alcohol will reduce the orange-red chromic acid/sulfuric acid reagent to an opaque green or blue suspension of Cr(III) salts in 2–5 s. For future reference, try the oxidation test with 1-butanol, 2-butanol, and 2-methyl-2-propanol (*tert*-butyl alcohol) and record your results.

*Bordwell, F. G.; Wellman, K. M. *J. Chem. Educ.* **1962,** *39,* 308–310.

Cleanup: In some laboratories, the entire reaction mixture may be discarded in the chromium-waste container. Alternatively, the chromium-containing solution should be treated with 5 mL of 50% aqueous sodium bisulfite to reduce all the metal ion to green Cr(III). Chromic ion (Cr^{3+}) may be disposed of in a number of ways, and local regulations should of course be the basis for choice. Alternatives include simply placing the entire solution in the hazardous-waste container reserved for inorganic compounds. Another approach is to convert all the Cr^{3+} to $Cr(OH)_3$ by treatment with excess ammonium hydroxide. The metal hydroxide precipitate is collected on filter paper and placed in the waste container intended for inorganic or solid compounds. The filtrate from the latter procedure can be washed into the inorganic-waste container or down the drain, depending on local regulations.

Primary and secondary alcohols differ from tertiary alcohols in their reactions with oxidizing agents. Upon oxidation with Cr(VI), primary alcohols yield aldehydes, which are further oxidized to carboxylic acids:

Primary alcohol:

$$3RCH_2OH + 2CrO_3 + 3H_2SO_4 \xrightarrow{H_2O} 3R\overset{\overset{\displaystyle O}{\|}}{C}H + Cr_2(SO_4)_3 + 6H_2O$$
Green

Aldehyde:

$$3R\overset{\overset{\displaystyle O}{\|}}{C}H + 2CrO_3 + 3H_2SO_4 \xrightarrow{H_2O} 3R\overset{\overset{\displaystyle O}{\|}}{C}OH + Cr_2(SO_4)_3 + 3H_2O$$
Green

Secondary alcohols react with chromic acid to yield ketones, which do not oxidize further under these conditions:

$$3R_2CHOH + 2CrO_3 + 3H_2SO_4 \xrightarrow{H_2O} 3R\overset{\overset{\displaystyle O}{\|}}{C}R + Cr_2(SO_4)_3 + 6H_2O$$
Green

Tertiary alcohols are usually unreactive:

$$R_3COH + CrO_3 + H_2SO_4 \xrightarrow{H_2O} \text{no reaction}$$

With a few tertiary alcohols, the sulfuric acid may catalyze an elimination reaction. The resulting alkene will be oxidized, producing insoluble green Cr(III) salts.

To the instructor: The chromic acid reagent is prepared by slowly pouring, with stirring, 25 mL of concentrated sulfuric acid into a solution of 25 g of chromic acid (CrO_3) in 75 mL of water. (Safety Precaution: Prepare this solution in a hood. CrO_3 dust may be carcinogenic.)

Other functional groups can also be oxidized by the chromic acid reagent, for example, aldehydes, phenols, and many amines. Thus, other data must be used to discover whether or not these groups are present in your compound. For example, solubility tests will be extremely useful in detecting a phenol or an amine. And an aldehyde will give a solid 2,4-dinitrophenylhydrazone when reacted with the 2,4-dinitrophenylhydrazine reagent, whereas alcohols do not give a precipitate in that test. Thus a false positive test would be observed for these compounds with chromic acid.

Iodoform Test

See Section 5.2, Aldehydes and Ketones, for directions on how to carry out the iodoform test. Its main use is for the identification of ketones with a methyl group attached to the carbonyl group. However, any alcohol that can be oxidized to such a ketone will also give a positive iodoform test:

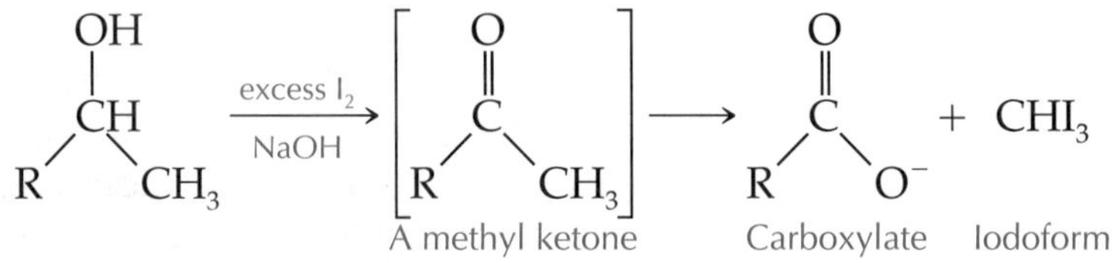

Lucas Test

SAFETY INFORMATION ──────────

The Lucas test reagent contains **concentrated hydrochloric acid,** which is corrosive and emits toxic HCl vapors. Wear gloves and dispense the reagent in a hood.

Procedure. Add 0.1 mL of the unknown or reference alcohol to a test tube and then add 1 mL of the hydrochloric acid/zinc chloride test reagent. Stir the whole mixture vigorously with a stirring rod to ensure complete mixing.

Alkyl chloride formation is noted by the formation of an insoluble layer or emulsion. Tertiary alcohols form the second layer in less than a minute, secondary alcohols require somewhat longer (5–10 min), whereas primary alcohols are essentially unreactive. The test is only applicable to alcohols that are soluble in water.

1-Butanol, 2-butanol, 2-methyl-2-propanol (*tert*-butyl alcohol), and 2-propene-1-ol (allyl alcohol) can be used as reference (control) compounds.

Positive test: Alkyl chloride formation is noted by the formation of an insoluble layer or emulsion. Tertiary alcohols form the second layer in less than a minute, secondary alcohols require somewhat longer (5–10 min), whereas primary alcohols are essentially unreactive.

Cleanup: Pour the reaction mixture into a 100-mL beaker containing 20 mL of water. Add solid sodium carbonate until the acid is neutralized and the solution exhibits a basic reaction when tested with pH paper. **(Caution: Foaming!)** Collect the resulting zinc hydroxide by vacuum filtration. Discard the zinc hydroxide in the inorganic-waste or solid-waste container and wash the filtrate down the sink or pour it into the container for aqueous inorganic waste.

Primary, secondary, and tertiary alcohols differ from one another in their reactions with hydrochloric acid as a result of their natural abilities to produce stable carbocations, which react further to yield alkyl chlorides. The alkyl chlorides are insoluble in aqueous HCl solution. Because tertiary carbocations form far more easily than secondary carbocations, which, in turn, form more easily than primary carbocations, reactivity of alcohols with the hydrochloric acid/zinc chloride reagent is in the order tertiary > secondary > primary. Of course, any alcohol that can produce a carbocation easily will produce an alkyl halide quickly. Primary alcohols, which can form carbocations stabilized by resonance, react more like secondary or tertiary alcohols, for example, benzyl alcohol or allyl alcohol. Zinc chloride is a good Lewis acid; it makes the reaction mixture even more acidic and helps to speed up the formation of carbocations.

To the instructor: The reagent is prepared by adding 136 g of anhydrous zinc chloride to 105 g of concentrated hydrochloric acid with cooling.

The reaction mechanism involves a carbocation intermediate:

$$R\text{—}OH + H^+ \longrightarrow R\text{—}OH_2^+ \xrightarrow{-H_2O} R^+ \xrightarrow{Cl^-} RCl + \text{alkene(s)}$$

The more stable the carbocation, the greater the driving force for this reaction.

5.2
Aldehydes and Ketones

Reaction with 2,4-Dinitrophenylhydrazine

— *SAFETY INFORMATION* —

2,4-Dinitrophenylhydrazine is an irritant and stains the skin. The reagent solution used in this procedure contains sulfuric acid, which is corrosive. Wear gloves while carrying out this procedure.

Procedure. Dissolve 20 mg of a solid unknown (1 or 2 drops of a liquid) in 0.5 mL of ethanol in a small test tube. Add 1 mL of the

Positive test: Formation of a large amount of yellow to red, insoluble 2,4-dinitrophenylhydrazone product indicates a positive test.

2,4-dinitrophenylhydrazine test reagent. Shake the test tube vigorously. Formation of a large amount of yellow to red, insoluble 2,4-dinitrophenylhydrazone product indicates a positive test. If no precipitate forms, heat the mixture to boiling for 30 s and shake the tube again. If there is still no precipitate, allow the tube to stand for 15 min. As a control, try the test on a known aldehyde and ketone. Note: the precipitate from this test may be collected with a Hirsch funnel and purified by the procedure described for the derivative preparation of 2,4-dinitrophenylhydrazones on p. 555. Hence, it serves as a derivative of the unknown compound.

To the instructor: Prepare the 2,4-dinitrophenylhydrazine reagent by carefully dissolving 8.0 g of 2,4-dinitrophenylhydrazine in 40 mL of concentrated sulfuric acid and adding 60 mL of water slowly while stirring the mixture to ensure complete solution. Add 200 mL of reagent-grade ethanol to this warm solution and filter the mixture if a solid precipitates.

Cleanup: Pour the test reaction mixture (or the filtrate, if you saved the precipitate) into the flammable-waste or organic-waste container.

The reaction of 2,4-dinitrophenylhydrazine with an aldehyde or ketone in an acidic solution is a dependable and sensitive test. Carbonyl compounds react with the phenylhydrazine derivative by nucleophilic attack at the carbonyl carbon atom followed by elimination of a molecule of water:

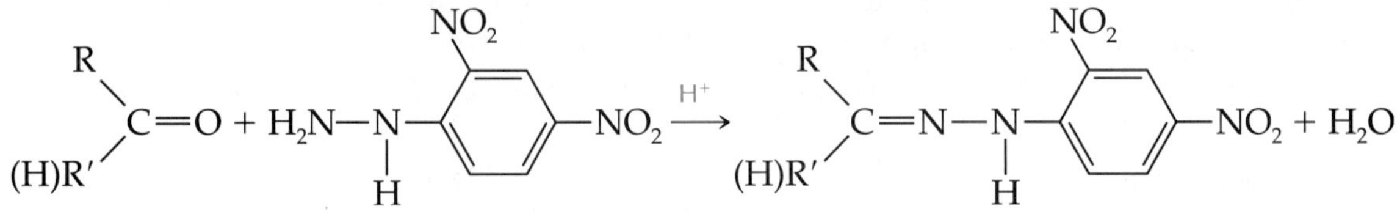

Most aromatic aldehydes and ketones produce red dinitrophenylhydrazones, whereas many nonaromatic aldehydes and ketones produce yellow products. If an orange precipitate forms, no definite conclusions can be drawn from the color.

The melting point of your derivative should not be confused with that of the starting reagent, 2,4-dinitrophenylhydrazine (mp 198°C, with decomposition).

Oxidation of Aldehydes with Chromic Acid*

— **SAFETY INFORMATION** ———————

Chromic acid solutions are very toxic and corrosive. Wear gloves and avoid contact with skin, eyes, and clothing. If you spill any on your skin, wash it off immediately with copious amounts of water.

---

*Morrison, J. D. *J. Chem. Educ.* **1965,** *42,* 554.

Positive test: A green or bluish green precipitate will appear within 1 min in the presence of an aldehyde.

Procedure. Add 1 drop or 30 mg of the unknown or reference compound to 1 mL of reagent-grade acetone in a test tube. Then add 1 drop of the chromic acid/sulfuric acid reagent directly into the solution. Shake the mixture. A green or bluish green precipitate will appear within 1 min in the presence of an aldehyde. Aliphatic aldehydes give precipitates within 15 s, whereas aromatic aldehydes take 30–45 s. Acetone and other ketones will oxidize in this solution, but only after 2–3 min. Try the test with an aliphatic aldehyde, an aromatic aldehyde, a ketone, and a primary or secondary alcohol as reference compounds.

*To the instructor: The chromic acid reagent is prepared by slowly pouring, with stirring, 25 mL of concentrated sulfuric acid into a solution of 25 g of chromic acid (CrO_3) in 75 mL of water. **(Safety Precaution: Prepare this solution in the ventilating hood. CrO_3 dust may be carcinogenic.)***

Cleanup: In some laboratories, the entire reaction mixture may be discarded in the chromium-waste container. Alternatively, the chromium-containing solution should be treated with 5 mL of 50% aqueous sodium bisulfite to reduce the metal ion to green Cr(III). Depending on local regulations, this solution can be placed in the inorganic-waste container, or otherwise suitably dispensed with (see p. 527).

A positive 2,4-dinitrophenylhydrazine test does not allow us to distinguish between an aldehyde and a ketone. This distinction can best be made by observing the rate of oxidation of the carbonyl compound with chromic acid. Aldehydes oxidize very easily, whereas ketones oxidize only slowly. Alcohols, phenols, and many amines can also be oxidized by the chromic acid reagent (see p. 526).

Iodoform Test

Procedure. Place 75 mg or 3 drops of the unknown in a 25-mL Erlenmeyer flask and add 1 mL of water; then, for a water-insoluble compound, add 1,2-dimethoxyethane dropwise until the unknown dissolves. Add 1 mL of 2.5 M sodium hydroxide and the iodine/potassium iodide stock solution dropwise until the red color of iodine persists after the reagents are thoroughly mixed. You have added enough when the reaction mixture is a dark red-brown color, almost as dark as the iodine/potassium iodide reagent itself.

Heat the mixture to 60°C in a water bath. Add more iodine/potassium iodide solution, if necessary, to maintain a definite red color for 2 min while heating at 60°C. Then discharge the excess iodine color by adding, with shaking, a few drops of 2.5 M sodium hydroxide and 5 mL of water. With a positive test, a yellow precipitate of iodoform appears within 15 min. Filter the product, wash it with water, dry it, and determine its melting point. Iodoform melts at 119–121°C. Try the test with acetone, 3-pentanone, and ethanol as reference compounds.

Positive test: A yellow precipitate of iodoform appears within 15 min.

Cleanup: Place all reaction mixtures in a beaker and add a few drops of acetone to consume any remaining iodine. (The solution should be colorless.) Collect any iodoform that forms by vacuum filtration; place the iodoform in the halogenated-waste container. Neutralize the aqueous filtrate to pH 7 before washing it down the sink or pouring it into the container for aqueous inorganic waste.

Those ketones and alcohols with a methyl group directly adjacent to a carbonyl group or to a carbon atom bearing a hydroxyl group react with an alkaline solution of iodine to produce an easily identifiable canary yellow solid, iodoform:

$$\underset{\overset{|}{\text{OH}}}{\text{RCHCH}_3} + \tfrac{1}{2}\text{I}_2 + 2\text{OH}^- \longrightarrow \underset{\overset{\|}{\text{O}}}{\text{RCCH}_3} + \text{I}^- + 2\text{H}_2\text{O}$$

$$\underset{\overset{\|}{\text{O}}}{\text{RCCH}_3} + 3\text{I}_2 + 4\text{OH}^- \longrightarrow \text{RCO}^- + \underset{\text{iodoform}}{\text{CHI}_3} + 3\text{I}^- + 3\text{H}_2\text{O}$$

Both these reactions involve the oxidation of the organic substrate by iodine. The mechanism of iodoform synthesis occurs through a series of enolate anions, which are iodinated. In the final steps, hydroxide displaces the CI_3^- anion through an addition-elimination pathway.

To the instructor: Prepare the iodine/potassium iodide stock solution by adding 50 g of potassium iodide and 25 g of iodine to 200 mL of distilled water. Stir the mixture until it is a homogeneous solution.

Only one aldehyde, acetaldehyde (ethanal), gives a positive iodoform test. Ethanol is the only primary alcohol that gives a positive iodoform test. Naturally, many ketones and secondary alcohols with the correct structural components give a positive test.

Tollens Test

Procedure. Fresh reagent should be prepared each time the test is conducted. The reagent is prepared by placing 2 mL of a 5% aqueous solution of silver nitrate in a small test tube. A drop of 10% NaOH is added, the result being a silver oxide precipitate. This precipitate is dissolved by adding just enough dilute (2%) ammonia to dissolve the silver oxide precipitate. The test is carried out by adding a few drops of this reagent to a small amount (2 drops, or a microspatula full of solid) of unknown. Frequently, a silver mirror or colloidial silver appear instantly. If not, gently warm the tube for a few minutes.

Positive test: A silver mirror or colloidal silver appears in the test tube.

Control reactions should be carried out on the following reference compounds: acetone, benzaldehyde, formalin, and glucose.

Cleanup: Combine all the reagents, whether used or unused, in a beaker. Add 6 M HCl until the precipitate stops forming. Collect the solid silver chloride by vacuum filtration and discard it in the inorganic-waste container. The filtrate should be neutralized with solid sodium carbonate before being washed down the drain or poured into the container for aqueous inorganic waste.

The preparation of Tollens reagent is based on the formation of a silver diamine complex that is water soluble in basic solution:

$$AgNO_3(aq) + NaOH(aq) \longrightarrow Ag_2O(s) \xrightarrow{\text{NH}_4\text{OH}} Ag(NH_3)_2^+ + OH^-$$

$$2Ag(NH_3)_2^+ + 2OH^- + \underset{R \quad H}{\overset{O}{\underset{\|}{C}}} \longrightarrow 2Ag(s) + \underset{R \quad O^-}{\overset{O}{\underset{\|}{C}}}$$

Tollens reagent gives a fast positive test with most simple aldehydes. Some α-hydroxyketones, such as fructose, also give a positive test. The reactions are sometimes characterized by a brief induction period, likely due to the limited water solubility of the organic unknown and to the necessity of silver metal formation as an autocatalytic reagent.

5.3
Alkenes

Reaction with Bromine

— SAFETY INFORMATION —

Bromine solutions cause burns and emit toxic bromine vapors. Wear gloves and use the solution only in a hood.

Procedure. Dissolve 30 mg or 2 drops of the unknown or reference alkene in 0.5 mL of dichloromethane. Add a 0.5 M solution of bromine in CH_2Cl_2 dropwise with shaking after each drop is

Positive test: Alkenes react with bromine and the characteristic red-brown color of bromine disappears. For most alkenes, this reaction occurs so rapidly that the reaction solution never acquires the red color of the bromine until the alkene is completely brominated.

added. Alkenes react with bromine and the characteristic red-brown color of bromine disappears. For most alkenes, this reaction occurs so rapidly that the reaction solution never acquires the red color of the bromine until the alkene is completely brominated. Try this test on cyclohexane, cyclohexene, and acetophenone as reference reactions.

Cleanup: The entire reaction mixture should be poured into the halogenated-waste container.

Almost all alkenes react quickly and smoothly with a dilute solution of Br_2 in CH_2Cl_2 to form dibromoalkanes:

$$\diagdown C=C \diagup + Br_2 \longrightarrow -\underset{Br}{\overset{Br}{\underset{|}{\overset{|}{C}}}}-\underset{}{\overset{}{C}}-$$

To the instructor: The Br_2/CH_2Cl_2 solution should be replenished periodically because it has a limited shelf life.

Very few other functional group classes interfere with this test. Although some compounds undergo free-radical or ionic substitution under these conditions, hydrogen bromide gas (HBr) is then evolved. You can test for the presence of HBr by blowing with your breath across the top of the test tube. If a fog is produced, then HBr is present. Phenols and some aldehydes and ketones may also be brominated under these conditions.

Oxidation with Potassium Permanganate

Positive test: Alkenes are oxidized, thereby causing the purple color of the permanganate solution to be replaced within 2–3 min by a brown precipitate of manganese dioxide.

Procedure A. Add 30 mg or 2 drops of an unknown or reference compound to 1 mL of acetone in a small test tube. Then add 1 drop of a 0.1 M aqueous solution of potassium permanganate and shake the test tube vigorously. Alkenes are oxidized, thereby causing the purple color of the permanganate solution to be replaced within 2–3 min by a brown precipitate of manganese dioxide. Try the test on an alkene, an aldehyde, and an alcohol.

Cleanup: Rinse the reaction mixtures down the sink, or into the inorganic-waste container.

Procedure B. Dissolve 800 mg of sodium chloride and 4 mg of potassium permanganate in 4 mL of distilled water contained in a 10-mL conical vial (or small flask). Add 4 mL of toluene and stir for a few minutes. Stop to allow separation of the two layers, and add 8 mg of tetrabutylammonium bromide to the mixture; stir until the toluene layer becomes a deep purple color as a result of the transfer of permanganate into this layer with a Pasteur pipet. Again stop and allow the layers to separate; then remove the toluene layer and place it in a clean conical vial. To this add 5–10 mg (or 1–2 drops) of unknown and reinitiate stirring. If the purple color turns brown within 5 min, this may be taken as a positive test result.

Cleanup: The entire reaction mixture should be placed in the flammable-waste or organic-waste container.

The permanganate test for double bonds is generally superior to the bromine test because the manganese reagent adds hydroxyl groups to both simple alkenes and alkenes containing strongly electron-withdrawing substituents:

$$3\ \diagdown C = C \diagup + 2MnO_4^- + 4H_2O \longrightarrow 3 - \underset{OH}{\overset{|}{C}} - \underset{OH}{\overset{|}{C}} - + 2MnO_2 + 2OH^-$$

The phase-transfer catalysis of procedure B (above) is superior to the reaction of procedure A in that the nonpolar layer (toluene in procedure B) normally contains the organic unknown and the catalyst greatly enhances transfer of the ionic permanganate to that layer.

Because some aldehydes and alcohols may also be oxidized by potassium permanganate, both a positive permanganate oxidation test and a positive bromine addition test are used to identify a compound as an alkene.

5.4 Alkyl Halides

Beilstein Test

Procedure. Bend a small loop in the end of a copper wire about 12 cm long. Hold the loop in a Bunsen burner flame until the flame is no longer green. Allow the wire to cool. Dip the loop in the unknown compound so that some of the compound sticks to the copper surface. Place the wire in the flame. A blue-green flame indicates the presence of halogen. Try the test on a reference compound containing halogen and one that doesn't contain halogen.

Positive test: A blue-green flame indicates the presence of halogen.

The Beilstein test is a quick preliminary check for halogens. Because the test is extremely sensitive to even trace impurities, a positive test should always be confirmed by the sodium fusion procedure (see p. 512). The blue-green color is due to the emission of light from excited states of copper halide that has vaporized in the burner flame.

Reaction with Silver Nitrate

Procedure. Add 1 drop of the unknown or reference compound to 1 mL of 0.1 M silver nitrate in ethanol. The formation of a precipitate (silver halide) is a positive indication of the presence of halide. If there is no reaction at room temperature within 5 min, heat the reaction mixture at its boiling point on a steam bath for 3–4 min. If a precipitate forms, note its color. Silver iodide is yellow, whereas silver chloride and silver bromide are white.

Add 2 drops of 1 M nitric acid. Some organic acids give insoluble silver salts that dissolve in the presence of nitric acid. Silver halides remain insoluble. Try this test on bromocyclohexane, bromobutane, 2-chloro-2-methylpropane (*tert*-butyl chloride), and chlorobenzene for reference.

Positive test: The formation of a precipitate (silver halide) is a positive indication of the presence of halide.

Cleanup: Add excess sodium chloride (about 5 mL) to precipitate unreacted silver nitrate as silver chloride, and place the entire mixture in the inorganic-waste container.

The reaction of halides with ethanolic silver nitrate can be useful for classification. Halides in group A—see Table 5.1—give a precipitate immediately at room temperature, those in group B react at high temperatures, and compounds in group C are essentially inert toward the reagent. The general reaction can be written

$$RX + Ag^+ \xrightarrow{\text{ethanol}} [R \cdots X \cdots Ag^+] \xrightarrow{S_N 1} R^+ + AgX(s) \xrightarrow{CH_3CH_2OH} ROCH_2CH_3 + \text{alkenes}$$

X = Cl, Br, I

Silver nitrate reacts instantaneously with any halide ion in solution or with compounds such as acyl chlorides that react quickly with ethanol to produce chloride ions. Other compounds in Table 5.1 react with silver nitrate because they easily form carbocations in an $S_N 1$ process.

The reaction is also catalyzed by Ag^+, a Lewis acid that can form complexes with the unshared electron pairs of the halogen. Any structural features that stabilize a carbocation intermediate will produce a faster precipitate of the silver halide. Thus al-

Table 5.1. **Reaction of various halides (X = Cl, Br, or I) with ethanolic silver nitrate**

Group A halides: Reaction at room temperature	Group B halides: Reaction at higher temperature	Group C halides: No reaction
$RCH=CHCH_2X$	RCH_2Cl	ArX
$RCHBrCH_2Br$	R_2CHCl	$RCH=CHX$
R_3CX	$RCHBr_2$	$HCCl_3$
RI		
RBr		
$ArCH_2X$		

lylic and benzylic compounds are quite reactive. The rate also depends on leaving group ability, where the order of reactivity is $I > Br > Cl$.

Reaction with Sodium Iodide in Acetone

Procedure. To a test tube containing 1 mL of a 15% (by weight) solution of sodium iodide in acetone, add 2 drops (or ~100 mg of solid in a minimum volume of acetone) of the unknown. Shake the test tube and allow the solution to stand at room temperature for 3 min. Note whether a precipitate forms (and whether the solution turns reddish brown as a result of the formation of free iodine). If there is no reaction, heat the solution to 50°C in a warm-water bath and, at the end of 6 min, allow the sample to cool to room temperature, noting whether a reaction has occurred. Try this test on 1-bromobutane, 2-bromobutane, 2-chloro-2-methyl-propane (*tert*-butyl chloride), and benzyl chloride for reference.

Positive test: The formation of an obvious precipitate (NaCl or NaBr) is a positive test.

Cleanup: Place entire mixture in the halogenated-waste container.

The reaction upon which this test is based is written

$$RX + NaI \xrightarrow[S_N2]{\text{acetone}} [\overset{\delta-}{I}\cdots\overset{\delta+}{R}\cdots\overset{\delta-}{X}] \longrightarrow I-R + NaX(s)$$

X = Cl, Br

The observation of a reasonable amount of precipitate (NaCl, NaBr) is a positive result. This test makes use of the high nucleophilic reactivity of sodium iodide in acetone, a reasonably polar solvent. Iodide thus directly displaces chloride or bromide. This test is an excellent complement to the ethanolic silver nitrate test described in the previous section. The NaI test is positive for compounds that undergo direct displacement (S_N2) reactions readily. Thus the reactivity order is primary > secondary > tertiary for simple alkyl halides. Benzylic and allylic compounds, however, give positive results with both NaI in acetone and ethanolic silver nitrate. The test relies on the fact that both NaCl and NaBr are much less soluble in acetone than NaI is. Primary substrates give a reaction at room temperature in 3 min; secondary substrates, in 6 min at 50°C; but the tertiary compounds, even at elevated temperatures, take days to react. Here, formation of molecular iodine (reddish brown color) indicates reaction; but this is a free-radical reaction rather than a direct displacement, and this test should not be taken as a positive result. In such cases you must rely on other tests to determine the nature of the unknown.

5.5
Amines

If an organic compound is a strong enough base to dissolve in 1.5 M HCl solution, we are justified in suspecting that it is an amine. Many amines also have characteristic, pervasive odors either like that of ammonia or resembling the smell of fish. Detection of nitrogen by the sodium fusion procedure (see p. 513) serves as excellent support for identification of an amine. The Hinsberg test is the best chemical test for classifying amines as primary, secondary, or tertiary. In practice, IR and NMR spectrometric methods are more efficient for this classification.

Hinsberg Test

— SAFETY INFORMATION —————————

Benzenesulfonyl chloride is a lachrymator (causes tears). Use it only in a hood.

Procedure. Add 6 drops or 200 mg of the unknown or reference amine to a test tube containing 3–4 mL of a 2.5 M potassium hydroxide solution. Then add 8 drops of benzenesulfonyl chloride. Stopper the tube and shake it intermittently for 3–5 min. At this point, the mixture should still be strongly basic. If this is not so, *add KOH solution dropwise until the mixture is again strongly basic.*

If the reaction mixture forms a homogeneous solution, a soluble primary benzenesulfonamide is present. Cautiously adjust the pH to approximately 4 with 1.5 M HCl solution. The primary sulfonamide will precipitate.

If the original reaction mixture has two layers, separate them and test the solubility of the organic layer in 1.5 M hydrochloric acid solution. Add the acid until the mixture is distinctly acidic, but avoid a large excess of acid. A tertiary amine will be soluble at this point. A secondary amine forms a benzenesulfonamide that is insoluble in hydrochloric acid.

High-molecular-weight or cyclic primary amines may form benzenesulfonamides that are not completely soluble in the alkaline reaction mixture. They remain insoluble after the hydrochloric acid is added. Therefore, they may give the same apparent result as you would obtain with a secondary amine. To distinguish these two possibilities, cautiously adjust the pH of the aqueous layer withdrawn from the original reaction mixture to pH 4 with 1.5 M HCl solution. A precipitate signifies that a primary benzenesulfonamide is present.

It is very important that you try the Hinsberg test on known amine samples before you use it to classify the functional groups in your unknown compound.

Cleanup: Place the entire reaction mixture in the flammable-waste or organic-waste container.

The Hinsberg test is based on the ability of benzenesulfonyl chloride to form sulfonamides:

Benzenesulfonyl chloride → Benzenesulfonamide

In principle, primary, secondary, and tertiary amines each exhibit observably different responses to alkaline benzenesulfonyl chloride. Simple primary amines give a homogeneous solution, because the sulfonamide initially formed is converted to its conjugate base, an anion that is frequently water-soluble. (Sometimes the anion is water-insoluble, especially if the R groups are exceptionally long.) Treatment of this anion with HCl reforms the sulfonamide as a precipitate:

N-Alkyl substituted
benzenesulfonamide

Conjugate base of
a benzenesulfonamide

Treatment of a secondary amine with alkaline benzenesulfonyl chloride results in the formation of a sulfonamide as an insoluble oil or solid. Because this sulfonamide does not bear a proton on its nitrogen, it is insoluble in alkali. Subsequent treatment with HCl normally gives little or no observable change:

N,N-disubstituted
benzenesulfonamide

Tertiary amines frequently give little or no indication of chemical change when treated with alkaline benzenesulfonyl chloride. This behavior may be due either to the fact that the tertiary amine does not undergo reaction with benzenesulfonyl chloride or to the fact that the reaction forms a water-soluble salt that is difficult to detect:

Quaternary
salt

In cases where the amine is water-soluble, the latter reaction is in effect "invisible"; moreover, often the quaternary salt formed in this way reverts to the original amine. Furthermore, because the quaternary salt bears no proton on the nitrogen, it does not undergo the type of reaction described earlier for sulfonamides formed from primary amines.

In the past, attempts have been carried out to broaden the utility of the Hinsberg test by inducing chemical reaction by increased heat. These approaches should be avoided because they have led to confusing results. Even though the reactions of amines with the Hinsberg reagent are complex (further details have been published*), careful interpretation accompanied by a reasonable amount of luck can provide chemical results that support conclusions drawn from IR and NMR spectrometry, solubility classification, and sodium fusion.

5.6
Aromatic Hydrocarbons (Arenes)

Ignition Test

Positive test: Many alkanes and their substituted derivatives burn with a clean yellow or bluish flame, whereas most aromatic compounds burn with a smoky flame.

Procedure. Place a small amount of the unknown (a few drops or a few milligrams) in a porcelain crucible cover. While holding the cover with tongs, bring it very near a Bunsen burner flame and allow the substance to ignite. Many alkanes and their substituted derivatives burn with a clean yellow or bluish flame, whereas most aromatic compounds burn with a smoky flame. Some halogen compounds also burn with a smoky flame, but elemental analysis easily distinguishes the two. Polyhalogenated compounds can be very difficult to ignite. Inorganic and organic salts always leave a residue upon burning or may not burn at all. The higher the oxygen content of a compound, the greater the propensity for the compound to burn with a colorless or blue flame.

Reaction with Aluminum Chloride/Chloroform

*Fanta, P. E.; Wang, C. S. *J. Chem. Educ.* **1964,** *41,* 280–281; Gambill, C. R.; Roberts, T. D.; Shechter, H. *J. Chem. Educ.* **1972,** *49,* 287–291.

Procedure. Place 1 mL of dry chloroform in a dry test tube; add 0.1 mL of a liquid, or 75 mg of a solid, unknown, or reference compound; mix thoroughly; and tilt the test tube to moisten the wall of the tube. Then add 0.25 g of anhydrous aluminum chloride so that some of the powder strikes the wetted side of the test tube. Note the color change of the powder and the solution. Try the test with toluene, chlorobenzene, and hexane as reference compounds.

Positive test: Note the color change of the powder and the solution.

Cleanup: Place the entire reaction mixture in the halogenated-waste container.

Aromatic compounds and their derivatives usually give characteristic colors when they come into contact with a mixture of aluminum chloride and chloroform. Generally, nonaromatic compounds do not produce a color on the $AlCl_3$. These color effects may be summarized as follows:

Compound Class	Color
Benzene derivatives	Orange to red
Naphthalene	Blue
Biphenyl or phenanthrene	Purple
Anthracene	Green

The colors in this classification test result from Friedel-Crafts reactions between chloroform and the aromatic compounds. Stable, highly colored carbocation salts form on the aluminum chloride surface through alkylation and hydride-transfer reactions:

5.7
Aryl Halides

Ignition   Aryl halides burn with a smoky, sooty flame in the ignition test discussed on p. 541.

Beilstein Test   Aryl halides give a blue-green flame in the Beilstein test, as described on p. 535.

Reaction with Aluminum Chloride/Chloroform   This test produces the same colors with aryl halides as it does with aromatic hydrocarbons (see p. 541).

5.8
Carboxylic Acids

Neutralization Equivalent   **Procedure.** Weigh at least two samples (three samples would be preferable to ensure more precise results) of an unknown or reference carboxylic acid to three significant figures on a balance of milligram sensitivity. The size of the samples should be in the 200–300 mg range. Place the acid in a 125-mL Erlenmeyer flask. If the acid is water-soluble, dissolve the sample in 50 mL of distilled water. If the acid is not water-soluble, add 25 mL of water to the sample, followed by enough 95% ethanol to dissolve the sample. Add 2–3 drops of phenolphthalein indicator solution (use bromothymol blue if the sample solution contains more than 50% ethanol). Titrate the acid solution with standardized 0.100 M sodium hydroxide to the indicator end point.

Cleanup: Pour the titrated solution into the inorganic-waste container or other waste container, as directed by your instructor.

The neutralization equivalent test is based on the ability of NaOH to react stoichiometrically with the carboxyl group:

$$RCOOR' + NaOH \longrightarrow RCOO^- Na^+ + R'OH$$

Identification of a particular carboxylic acid is aided by measuring the neutralization equivalent. *In the case of carboxylic acids containing one carboxyl group, the neutralization equivalent and the molecular weight are the same.* An acid that contains more than one carboxyl group gives a neutralization equivalent equal to the molecular weight divided by the number of acidic groups. Because it is desirable to know the neutralization equivalent of a carboxylic acid to *three significant figures*, it is important that you know the exact weight of the carboxylic acid sample that you titrate.

Calculate the neutralization equivalent (N.E.) of your acid:

$$\frac{\text{Milliequivalents base}}{\text{used in the titration}} = \text{milliequivalents acid present}$$

$$\text{Milliequivalents acid} = \text{(molarity of NaOH solution)} \times \text{(mL of NaOH used)}$$

$$\frac{\text{mg acid}}{\text{(molarity of NaOH)} \times \text{(mL of NaOH used)}} = \text{N.E.}$$

$$\text{(Number of acid groups)} \times \text{(N.E.)} = \text{molecular weight}$$

5.9
Carboxylic Acid Derivatives

5.9a Acyl Chlorides

Reaction with Silver Nitrate

Acyl chlorides react at room temperature with a 0.1 M solution of silver nitrate in ethanol to give a precipitate of silver chloride (see p. 536).

5.9b Esters

Hydroxamic Acid Test

Procedure. Place 50 mg or 2 drops of the unknown or reference compound along with 1 mL of 0.5 M hydroxylamine hydrochloride in methanol in a test tube. Add 2.5 M sodium hydroxide solution dropwise until the mixture is alkaline (use pH paper). Then add 3 drops more of the sodium hydroxide solution. Heat the reaction mixture just to boiling, cool the tube, and add 1.5 M hydrochloric acid dropwise with shaking until the pH of the mixture is 3. Add 2 or 3 mL more of methanol if a cloudy mixture results. Then add 1 drop of 0.7 M ferric chloride solution. A positive test is a blue-red color. Try the test on one or two esters and on phenol as reference compounds.

Positive test: A positive test is a blue-red color.

Cleanup: Place the entire reaction mixture in the inorganic-waste container.

Esters react with hydroxylamine in basic solution to form hydroxamic acids, which in turn react with ferric chloride in acidic solution to form bluish red ferric hydroxamates:

$$\underset{\substack{\parallel \\ \text{O}}}{\text{RCOR}'} + \text{NH}_2\text{OH} \longrightarrow \underset{\substack{\parallel \\ \text{O}}}{\text{RCNHOH}} + \text{R}'\text{OH}$$

$$\underset{\substack{\parallel \\ \text{O}}}{\text{RCNHOH}} + \text{Fe}^{3+} \longrightarrow \underset{\substack{\parallel \\ \text{O}}}{\text{RCNHOFe}^{2+}} + \text{H}^+$$

Ferric chloride will react directly with phenols to give products of much the same color, and some phenols may interfere with this test. However, the acidic solution used here will usually convert the ferric-phenolate complex into the free phenol, which is not deeply colored.

5.10
Amides, Acyl Chlorides, Esters, and Nitriles

Each of these carboxylic acid derivatives may be hydrolyzed to the parent carboxylic acid and, in the case of an amide or an ester, to the amine or alcohol component as well.

Amide Hydrolysis

Procedure. Reflux 1 g of the unknown or reference amide and 25 mL of 2.5 M sodium hydroxide solution for 1–2 h. Connect a trap containing a small volume of 3 M hydrochloric acid to the top of the condenser if a low-boiling-point amine component is expected to form (Figure 5.1). This setup traps the amine by converting it to a nonvolatile ammonium salt.

After cooling the hydrolysis mixture, perform an extraction procedure with two successive 20-mL portions of diethyl ether. The organic layer should now contain the amine unless a low-boiling-point amine was distilled into the hydrochloric acid trap; in this case, it can be extracted following neutralization of the acid with 2.5 M sodium hydroxide solution.

Make the aqueous solution acidic with dilute hydrochloric acid. If a carboxylic acid precipitates, filter it. If not, extract the solution with two 20-mL portions of diethyl ether. This extract should then contain the acid. Dry the diethyl ether extracts with anhydrous magnesium sulfate and distill the ether from each solution, leaving the amine and the carboxylic acid residues. Identify the amine and the carboxylic acid.

Cleanup: Pour the recovered ether solutions into the flammable (or organic) waste container. Neutralize the aqueous filtrate with

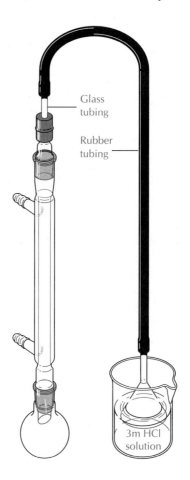

Glass
tubing

Rubber
tubing

3m HCl
solution

FIGURE 5.1 Apparatus for amide hydrolysis.

sodium carbonate before placing it in the container for aqueous inorganic waste or washing it down the sink.

Hydrolysis of an amide in sodium hydroxide solution produces the carboxylate salt of the parent acid and ammonia (for a 1° amide) or an amine (for a 2° or 3° amide). Addition of HCl converts the carboxylate salt to the parent carboxylic acid.

$$RCONH_2 + NaOH \longrightarrow RCOO^- \, Na^+ + NH_3$$

$$RCOO^- \, Na^+ + HCl \longrightarrow RCOOH + NaCl$$

Acyl Chloride Hydrolysis

Procedure. Dissolve 0.1 g or 3 drops of the unknown in 1 mL of water. Separate the carboxylic acid from the water solution, which has been acidified with 1.5 M HCl, by filtration or by extraction with diethyl ether. Dry the diethyl ether extract and distill off the ether, leaving the carboxylic acid as a residue. Identify the carboxylic acid.

Cleanup: Pour the recovered ether solutions into the container provided for flammable (or organic) waste. Neutralize the aqueous filtrate with sodium carbonate before placing it in the inorganic-waste container or washing it down the sink.

Many acyl chlorides react immediately with water at room temperature, forming the parent carboxylic acid. Those that do not hydrolyze at room temperature do so at elevated temperatures in alkaline solution.

$$RCOCl + H_2O \longrightarrow RCOOH + HCl$$

Ester Hydrolysis **Procedure.** Place 1 g of your unknown or reference ester and 10 mL of 2.5 M sodium hydroxide in a 50-mL round-bottomed flask equipped with a reflux condenser. Reflux the mixture for 1 h, then extract the cooled product with two 20-mL portions of diethyl ether. After drying the ether solution, distill it to obtain the alcohol. Identify the alcohol.

Make the aqueous sodium hydroxide solution distinctly acidic with 1.5 M HCl and cool well. If a precipitate forms, collect the solid by vacuum filtration. If no precipitate forms, extract the solution with two 20-mL portions of diethyl ether. Dry the ether solution and distill it. Identify the carboxylic acid residue.

Cleanup: Pour the recovered ether solutions into the container provided for flammable (or organic) waste. Neutralize the aqueous filtrate with sodium carbonate before placing it in the inorganic-waste container or washing it down the sink.

Basic hydrolysis converts an ester to the carboxylate salt of the parent acid and the alcohol from which the ester was formed. Acidification of the carboxylate salt solution with HCl leads to the recovery of the parent acid.

$$RCOOR' + NaOH \longrightarrow RCOO^- Na^+ + R'OH$$

$$RCOO^- Na^+ + HCl \longrightarrow RCOOH + NaCl$$

Often you will find that either the carboxylic acid or the alcohol formed from the hydrolysis of an ester is extremely soluble in water and does not appear in the ether extract. This situation occurs with ethyl benzoate and 1-phenylethyl acetate, for example. Whereas the benzoic acid is easy to isolate, most of the ethanol remains in the water solution. Similarly, the 1-phenylethanol can be readily extracted with ether, but this is not so for the acetic acid. In these cases, you may have to rely on a definite identification of the water-insoluble component and combine this evidence with the physical and spectral data for the unknown ester in order to reach a final answer.

Nitrile Hydrolysis

SAFETY INFORMATION

Aqueous sulfuric acid (8 M) solution is corrosive and causes burns. Wear gloves while measuring it.

Procedure. Reflux 1 g of your unknown or reference nitrile for 30–45 min with 10 mL of 8 M sulfuric acid solution. If you expect a volatile carboxylic acid product, add 5 mL of water to the cooled solution and distill off 5–6 mL of the liquid. The distillate contains the parent carboxylic acid of the nitrile.

If you expect a nonvolatile carboxylic acid to be formed, cool the reaction solution in an ice bath, add 20 mL of water, and neutralize the sulfuric acid *carefully* with 2.5 M sodium hydroxide solution. When it reaches pH 1–2, cool the solution again. If the carboxylic acid has precipitated, filter it. If not, extract it with two 20-mL portions of diethyl ether. Dry the ether solution and distill the ether from it. Identify the carboxylic acid residue.

Cleanup: Pour the recovered ether solutions into the container provided for flammable (or organic) waste. Neutralize the aqueous filtrate with sodium carbonate before placing it in the inorganic-waste container or washing it down the sink.

A nitrile can be hydrolyzed to its parent carboxylic acid by aqueous sulfuric acid.

$$RC \equiv N + H_2SO_4(aq) \longrightarrow RCOOH + NH_4Cl$$

5.11
Ethers

Ferrox Test

Procedure. Use a stirring rod to grind together a crystal of ferric ammonium sulfate and a crystal of potassium thiocyanate in a dry test tube. A colored mass of ferric hexathiocyanatoferrate(III) will adhere to the stirring rod.

Positive test: A red to reddish purple color appears if the unknown contains oxygen.

Place 3–4 drops of a liquid unknown or reference compound in a clean test tube and stir it, using the stirring rod with the solid mass attached. The ferric hexathiocyanatoferrate(III) will dissolve to give a red to reddish purple color if the unknown contains oxygen.

Solid compounds can be tested by using a saturated solution of the compound in 0.5 mL of carbon tetrachloride. Warm the mixture briefly on a steam bath in a hood and transfer the solution to the test tube containing the colored mass of ferric hexathiocyanatoferrate(III) with a Pasteur pipet. Stir the mixture with the stirring rod that has the solid mass of ferric complex attached.

Cleanup: For liquid compounds, place the residue left from the reaction in the inorganic-waste container; for solid compounds, discard the reaction mixture in the halogenated-waste container.

The Ferrox test distinguishes ethers from the hydrocarbons that are also soluble in concentrated sulfuric acid. Some high-molecular-weight ethers do not give positive tests.

5.12
Phenols

Reaction with Bromine/Water

Procedure. Add a saturated solution of bromine in water dropwise to 0.1 g of the unknown or reference phenol in 10 mL of water, shaking until the bromine color is no longer discharged.

Disappearance of the orange bromine color, accompanied by a precipitate, is a positive test. If your suspected phenol is insoluble in water, the test can be done in an ethanol/water mixture. Try the test again with phenol and cyclohexene as reference compounds.

Cleanup: Place the mixture in the halogenated-waste container.

Phenols are extremely reactive toward electrophilic substitution. For example, solid tribromophenol is formed almost immediately when bromine water is added to a solution of the phenol in water:

Aniline and some of its derivatives may interfere with this test

because they can also react rapidly with bromine in water to produce insoluble precipitates.

Reaction with Ferric Chloride

Positive test: Most phenols react with ferric chloride to form red to blue ferric phenolate complexes.

Procedure. Suspend 30 mg or 1 drop of the suspected phenol in 1 mL of chloroform. Add 1 drop of pyridine (in the hood) and 3 drops of a 0.06 M solution of anhydrous ferric chloride in chloroform. *Most phenols react with ferric chloride to form red to blue ferric phenolate complexes.* Try the test on phenol and ethyl acetate as reference compounds.

Cleanup: Place the entire mixture in the halogenated-waste container.

Because all phenols do not produce colored complexes under the test conditions, a negative test must be considered as an ambiguous result.

$$3ArOH + 3py + FeCl_3 \longrightarrow Fe(OAr)_3 + 3pyH^+ Cl^-$$
(py = pyridine)

Qualitative Analysis 6
SYNTHESIS OF SOLID DERIVATIVES

The last step in the identification of an unknown compound, the experiment that eliminates any remaining doubt, is the synthesis of a solid derivative whose melting point can be compared with the melting points of known compounds. The tables in Appendix A, as well as the references listed at the end of Part 3, Organic Qualitative Analysis, contain lists of a large number of organic compounds and the melting points of their derivatives. Most of the physical constants of such organic compounds were compiled before the advent of infrared and NMR spectrometry, and modern spectrometric techniques have replaced their use in large part. In some cases, the physical properties of organic compounds and their derivatives can still be useful. For example, it would be difficult to be confident that an unknown compound was cholesterol even if you possessed the infrared and NMR spectra of the compound, unless a sample of cholesterol or reference spectra were also available. Knowing that the phenylurethane of the unknown melted at 167–168°C would remove a good deal of the uncertainty attaching to its identity.

For some functional groups, the synthesis of a derivative is straightforward. These are the ones that we have included. For others, the synthesis of a suitable derivative is quite involved and difficult. We have omitted them. In these cases, you will have to rely on more detailed spectral analyses, functional-group classifications, solubility data, and elemental analyses.

Your derivatives will probably melt a few degrees (2–5°C) lower than those recorded in the literature, because the highest melting point obtained after recrystallization is usually reported. The suggested recrystallization solvent for many of the derivatives is a mixture of ethanol and water. It is impossible to state an exact ratio of ethanol to water in each recrystallization, because the compounds to be purified are so diverse in structure and solubility. Like most organic compounds, however, the derivatives are more soluble in ethanol than in water. The general procedure for recrystallizing a solid from a mixed solvent pair is discussed in

Technique 5.2a. It works well for the recrystallizations of derivatives from ethanol/water mixtures.

Derivatives have great importance in certain situations. The isomers *m*-methylaniline (*m*-toluidine) and *p*-methylaniline (*p*-toluidine) have extremely similar IR and NMR spectra. Thus the preparation of derivatives with widely differing melting points is helpful. When choosing the specific derivative, very often the difference in melting points is the deciding factor. For example, if your list of possible unknowns includes both 1-propanol (bp 97°C) and 2-butanol (bp 99°C), you should choose the 1-naphthylurethane rather than the 3,5-dinitrobenzoate. The melting points are, respectively, 80°C and 97°C for the naphthylurethane, and 74°C and 75°C for the 3,5-dinitrobenzoate. You have virtually no hope of discerning such a small melting-point difference as 1°.

Useful Solid Derivatives

Alcohols
   3,5-Dinitrobenzoates
   1-Naphthylurethanes
   Phenylurethanes
Aldehydes and ketones
   2,4-Dinitrophenylhydrazones
   Semicarbazones
Amines
   Benzamides
   Benzenesulfonamides
   Picrates

Carboxylic acids
   Amides
   Anilides
   Toluidides
Phenols
   Bromo derivatives
   1-Naphthylurethanes
   3,5-Dinitrobenzoates

6.1 Alcohols

3,5-Dinitrobenzoates

SAFETY INFORMATION

3,5-Dinitrobenzoyl chloride is a strong irritant; wear gloves and avoid contact with skin, eyes, and clothing. This reagent is moisture-sensitive; recap the bottle quickly.

Pyridine is toxic and has an extremely unpleasant odor; dispense it in a hood.

Procedure. Place 800 mg of 3,5-dinitrobenzoyl chloride, 0.20 mL (or 200 mg) of the unknown alcohol, 2.0 mL of dry pyridine, and a boiling chip in a 5-mL conical vial. Attach a water-cooled condenser and heat the mixture at reflux on an aluminum block at 140°C for 15–20 min. Put 5 mL of water in a 10-mL Erlenmeyer flask and pour the reaction mixture slowly into the water. Cool the solution in an ice bath until the product precipitates. Scratching the walls of the flask with a stirring rod may be necessary to induce crystallization. Collect the crystals by vacuum filtration. Transfer the crystals to a small beaker and stir them thoroughly with 3 mL of 5% sodium carbonate solution to remove any 3,5-dinitrobenzoic acid that may be present. Again collect the solid by vacuum filtration in a Hirsch funnel and wash it with 1 mL of cold water. Recrystallize the derivative from a mixture of ethanol and water.

Cleanup: Acidify the sodium carbonate filtrate. Use vacuum filtration to collect any 3,5-dinitrobenzoic acid that forms as a precipitate; place it in the solid organic waste container. Neutralize the filtrate with solid sodium carbonate and place the resulting solution in the inorganic-waste container or wash it down the sink, as directed by your instructor. Place the filtrate from the recrystallization in the container provided for flammable (or organic) wastes.

Alcohols react with carboxylic acids and some of their derivatives to form esters. An example is the reaction of an acyl chloride with an alcohol:

$$\text{ROH} + \text{R}'\overset{\overset{\displaystyle O}{\displaystyle \|}}{\text{C}}\text{Cl} \longrightarrow \text{R}'\overset{\overset{\displaystyle O}{\displaystyle \|}}{\text{C}}\text{OR} + \text{HCl}$$

A useful derivative of an unknown alcohol is the 3,5-dinitrobenzoate ester, which is prepared through the reaction of the alcohol with 3,5-dinitrobenzoyl chloride:

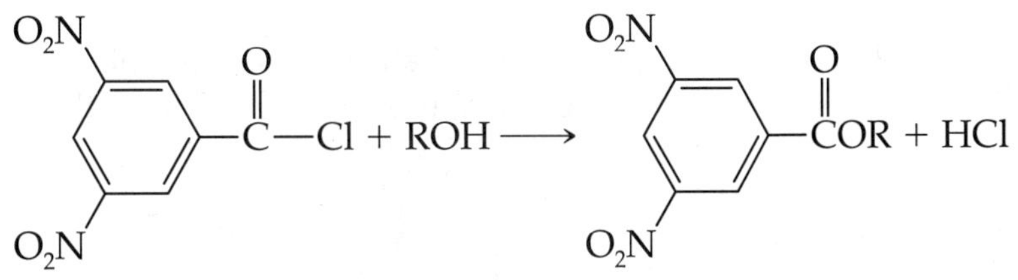

The hydrogen chloride produced immediately reacts with the tertiary amine pyridine. Hindered tertiary alcohols sometimes react quite slowly with 3,5-dinitrobenzoyl chloride. In such cases, it is advisable to reflux the reaction mixture for 1–2 h.

1-Naphthylurethanes

Procedure. This reaction requires a dry alcohol. Place 0.5 mL of a liquid alcohol in a small test tube and dry it for 10 min with anhydrous Na_2SO_4.

Place 0.2 g (or 5 drops) of dry unknown alcohol in a dry 4-in. test tube. Add 0.20 mL of 1-naphthyl isocyanate. A precipitate usually forms spontaneously. If the reaction is not spontaneous, place a plug of cotton in the top of the test tube and warm the mixture on a steam bath for up to 15 min. When the reaction is complete, cool the test tube and add 3 mL of hexane (petroleum ether or ligroin may be substituted). Place a boiling stick in the test tube and heat the mixture carefully in a 70°C water bath to dissolve the product. Gravity filter the hot solution through a small fluted filter paper to remove the undissolved di-1-naphthylurea (mp = 297°C) that usually forms as a by-product. Cool the solution in an ice-water bath and collect the product by vacuum filtration on a Hirsch funnel. Further recrystallization is not usually necessary.

Cleanup: Pour the filtrate into the container provided for flammable (or organic) wastes. Place the filter paper containing dinaphthylurea in the container provided for organic solid wastes.

Most anhydrous alcohols react vigorously with 1-naphthyl isocyanate to form solid 1-naphthylurethanes:

Phenylurethanes

Procedure. Follow the procedure described earlier for the preparation of 1-naphthylurethanes, except use 0.20 mL of phenyl isocyanate. (The melting point of the by-product, diphenylurea, is 240°C.)

Phenyl isocyanate reacts more slowly with alcohols than 1-naphthyl isocyanate does. Phenylurethanes may not form spontaneously if bulky substituents are present near the hydroxyl group of the alcohol.

6.2
Aldehydes and Ketones

2,4-Dinitrophenyl-hydrazones

Procedure. Dissolve 100 mg (or, with liquids, 3 drops) of the unknown carbonyl compound in 2 mL of reagent-grade ethanol and add 3 mL of the 2,4-dinitrophenylhydrazine reagent (see p. 529). A large quantity of crystals usually forms immediately; however, heating the reaction mixture in a water bath at 50°C for

A thorough washing is necessary to remove all the sulfuric acid that may be clinging to the crystals. Any remaining acid will catalyze decomposition of the dinitrophenylhydrazone when you heat it, and the decomposition products lower and broaden the melting range.

2 min may be necessary to produce a derivative. Let the mixture stand at room temperature for 15–20 min before collecting the solid product by vacuum filtration on a Hirsch funnel. Wash the crystals with two or three 1-mL portions of cold ethanol. To do this, discontinue the suction, add the ethanol, and stir the crystals gently to wash them completely. Then apply suction again. After thorough washing, 2,4-dinitrophenylhydrazones often need no recrystallization. Determine the melting point. If recrystallization proves to be necessary, use ethanol as the solvent.

Cleanup: Pour the filtrate into the container provided for flammable (or organic) wastes.

The reaction between a carbonyl compound and 2,4-dinitrophenylhydrazine is written:

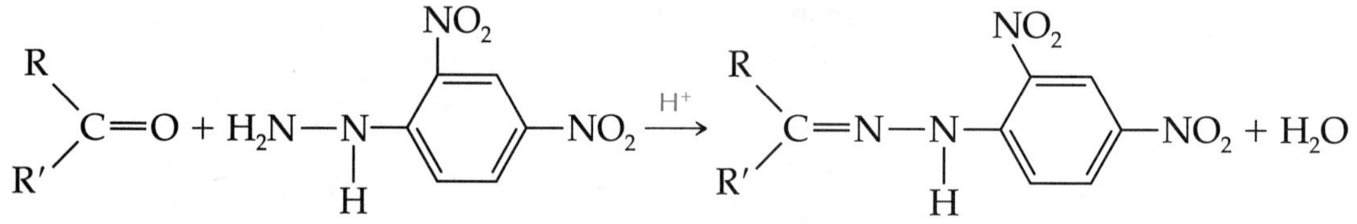

Semicarbazones   **Procedure.** Dissolve 0.5 g of semicarbazide hydrochloride in a mixture of 4.0 mL of water and 7.0 mL of ethanol in a 25-mL round-bottomed flask. Add 0.80 g of sodium acetate trihydrate or 0.50 g of anhydrous sodium acetate to provide the proper pH for the formation of the semicarbazone. Add 250 mg of the unknown carbonyl compound and a boiling chip; fit the flask with a water-cooled condenser. Reflux the mixture on a steam bath for 30 min. Allow the solution to cool for a few minutes, add 3.0 mL of cold water, and cool the mixture in an ice-water bath. Scratch the insides of the flask with a stirring rod to induce crystallization of the semicarbazone. If crystallization still does not occur, evaporate at least half of the solvent on a steam bath and repeat the cooling process. Collect the crystals on a Hirsch funnel by vacuum filtration and recrystallize the product from an ethanol/water solvent.

Cleanup: Dilute the filtrate from the reaction mixture with water and add enough 3 M HCl dropwise to make the solution slightly acidic (pH ~ 5) and then wash the solution down the sink or place it in the container for aqueous organic waste. Pour the filtrate from the recrystallization into the container provided for flammable (or organic) wastes.

Semicarbazones form readily from most aldehydes and ketones as highly crystalline solids.

$$\underset{\text{NH}_2\text{CNHNH}_2}{\overset{\overset{\displaystyle O}{\|}}{}} + \underset{(\text{H})\text{R}}{\overset{\displaystyle R'}{\diagdown}}\text{C}{=}\text{O} \xrightarrow{\text{H}^+} \underset{}{\overset{\overset{\displaystyle O}{\|}}{\text{NH}_2\text{CNHN}}}{=}\underset{\text{R(H)}}{\overset{\displaystyle R'}{\text{C}\diagup}} + \text{H}_2\text{O}$$

6.3 Amines

Benzamides

Procedure. Dissolve 0.15 g of the amine in a solution of 1.5 mL of pyridine and 3.0 mL of toluene in a 10-mL round-bottomed flask. Add 0.30 mL of benzoyl chloride to the flask. Attach an air condenser fitted with a drying tube to the flask. Heat the reaction mixture on a steam bath for 20 min and then pour it into 20 mL of water. Transfer half of the mixture to a 15-mL screw-capped centrifuge tube; remove the lower aqueous phase with a Pasteur filter pipet, leaving the toluene layer in the centrifuge tube. Add the rest of the water/toluene mixture and remove the lower aqueous phase. Wash the toluene layer with 5 mL of 0.5 M Na_2CO_3 solution. Dry the organic layer with anhydrous Na_2SO_4 and remove the solvent by distillation. Recrystallize the crude benzamide from ethanol or an ethanol/water mixture.

Cleanup: The aqueous residues from the extractions can be washed down the sink or poured into the container for aqueous inorganic waste. Pour the recovered toluene and the recrystallization filtrate into the container provided for flammable (or organic) wastes.

Primary and secondary amines form solid benzamides upon reaction with benzoyl chloride:

$$\text{C}_6\text{H}_5\overset{\text{O}}{\underset{\|}{\text{C}}}\text{Cl} + \text{RNH}_2 \longrightarrow \text{C}_6\text{H}_5\overset{\text{O}}{\underset{\|}{\text{C}}}\text{NHR} + \text{HCl}$$

Benzenesulfonamides

--- *SAFETY INFORMATION* ---

Benzenesulfonyl chloride is corrosive and a lachrymator. Wear gloves and use it only in a hood.

Procedure. Place 0.3 g of the amine in a dry test tube, then add 0.3 mL of benzenesulfonyl chloride and 6 mL of 2.5 M NaOH solution. Close the test tube with a rubber stopper and shake it intermittently for 3–5 min. Cautiously acidify the reaction mixture with 6 M HCl solution to pH 5. Collect the crystals by vacuum filtration in a Hirsch funnel and wash them thoroughly with water. Recrystallize the product from an ethanol/water mixture.

Cleanup: If the unknown is a primary or secondary amine, the filtrate from the reaction mixture may be diluted and washed down the sink or placed in the container for aqueous inorganic waste. For a tertiary amine, add dilute NaOH to the filtrate until the solution is basic. Extract the unreacted amine with 5 mL of hexane (petroleum ether). Place the hexane solution in the container provided for flammable (or organic) wastes and wash the aqueous solution down the sink or pour it into the container for aqueous inorganic waste. Place the recrystallization filtrate in the container provided for flammable (or organic) wastes (or the halogenated-waste container, if the unknown contained a halogen).

Primary and secondary amines react with benzenesulfonyl chloride, as described in the discussion of the Hinsberg functional-group test (p. 538).

Picrates

SAFETY INFORMATION

Anhydrous **picric acid** (2,4,6-trinitrophenol) is explosive when it is rapidly heated or subjected to a sharp blow. This procedure uses a saturated ethanolic solution of picric acid prepared from the moist solid, which can be handled safely.

Procedure. Dissolve 0.15 g of the unknown amine in 5 mL of ethanol; if the amine does not dissolve entirely, use the saturated solution for the preparation of the derivative. Prepare 5 mL of saturated picric acid solution by adding small amounts of *moist* picric acid to 5 mL of ethanol until the solution is saturated. Add the amine solution to the saturated picric acid solution and heat the reaction mixture to boiling on a steam bath. Cool the solution slowly and collect the precipitate by vacuum filtration. Wash the crystals with 1–2 mL of ice-cold ethanol. Recrystallization is not usually necessary.

Cleanup: Pour the filtrate from the reaction mixture into the container provided for flammable (or organic) wastes.

Picric acid reacts with amines (here, a tertiary amine is shown) to form an ammonium picrate:

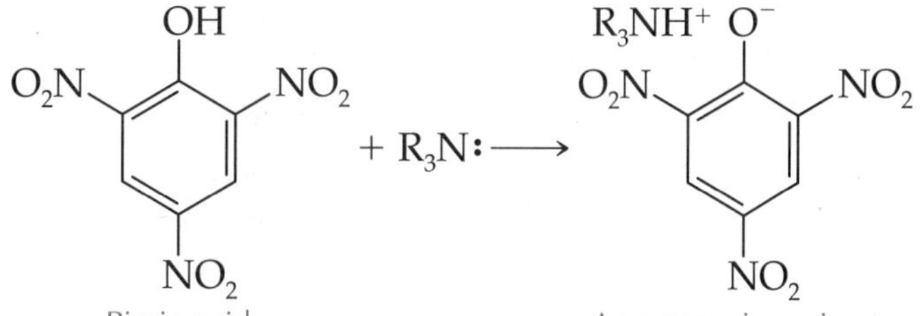

Picric acid
2,4,6-trinitrophenol

An ammonium picrate

Picric acid is a strong acid ($pK_a = 0.25$) that will form crystalline compounds with primary, secondary, and tertiary amines. The picrate salts have convenient, sharp melting points.

6.4 Carboxylic Acids: Amides, Anilides, Toluidides

Step 1: Preparation of the Acyl Chloride

The thionyl chloride must be relatively pure; otherwise the procedure does not work well.

— SAFETY INFORMATION —

If thionyl chloride is spilled on the skin, serious burns can result. It is volatile and reacts with moisture in the atmosphere, forming hydrogen chloride and sulfur dioxide. Wear gloves. It should be dispensed in a hood and used in a hood, if possible; and the bottle should be kept tightly capped.

Piperidine is a cyclic secondary amine that serves as a catalyst.

Procedure. Place 250 mg of the unknown carboxylic acid, 0.30 mL of thionyl chloride and 1 drop of piperidine in a 5-mL conical vial fitted with a water-cooled condenser and a drying tube filled with anhydrous calcium chloride. Heat the reaction mixture in a water bath at 50–55°C for 25 min. Cool the solution in a bath of cold tap water. The acyl chloride solution is now ready for addition to the selected base, either ammonia, aniline, or 4-toluidine.

Synthesis of an amide, an anilide, or a toluidide begins with the preparation of the acyl chloride from the unknown carboxylic acid, using thionyl chloride:

$$R-C\overset{\displaystyle O}{\underset{\displaystyle OH}{\diagdown}} + SOCl_2 \longrightarrow R-C\overset{\displaystyle O}{\underset{\displaystyle Cl}{\diagdown}} + SO_2 + HCl$$

Step 2: Preparation of Amides

— SAFETY INFORMATION —

The addition of an acyl chloride solution to cold concentrated ammonia can cause a violent reaction when the thionyl chloride hydrolyzes to hydrochloric acid. Carry out the addition dropwise with stirring. Work in a hood while doing this procedure.

Procedure. Place 3 mL of concentrated ammonia solution in a 30-mL beaker and chill the solution in an ice-water bath. While stirring the ammonia solution with a glass rod, add the acyl chloride solution dropwise, using a Pasteur pipet. Stir the mixture until the reaction is complete. Collect the solid amide by vacuum filtration in a Hirsch funnel and recrystallize it from an ethanol/water mixture.

This procedure does not work well for an unknown carboxylic acid that is water-soluble, because the amide is likely to be water-soluble also.

Cleanup: Dilute the filtrate with water and neutralize it with 6 M hydrochloric acid before flushing it down the sink or pouring it into the container for aqueous inorganic waste. Pour the filtrate from the recrystallization into the container provided for flammable (or organic) wastes.

Amides are the most common solid derivatives of carboxylic acids and are prepared by the reaction of acyl chlorides with ammonia:

$$R-C \begin{matrix} O \\ \diagdown \\ Cl \end{matrix} + 2NH_3 \longrightarrow R-C \begin{matrix} O \\ \diagdown \\ NH_2 \end{matrix} + NH_4Cl$$

Amide of unknown acid

Step 2: Preparation of Anilides or Toluidides

─ **SAFETY INFORMATION** ─

Aniline and **4-toluidine** are very toxic and can be absorbed through the skin. Measure aniline in a hood. Wear gloves while using either of these reagents.

Procedure. Place 5 mL of dichloromethane and 0.8 mL of aniline (or 0.9 g of 4-toluidine) in a 25-mL Erlenmeyer flask. While stirring the dichloromethane solution, add the acyl chloride dropwise, using a Pasteur pipet. Rinse the reaction vial with a few drops of dichloromethane and add this rinse to the Erlenmeyer flask. The reaction is quite vigorous and a thick slurry should form. Allow the mixture to stand for 5 min, then add 7 mL of ice-cold water and stir thoroughly to mix the layers; the solid should dissolve. Pour the two-phase mixture in a 15-mL centrifuge tube. Transfer the lower organic phase to another 15-mL centrifuge tube, using a Pasteur pipet. Wash the organic layer successively with 2 mL of water, 2 mL of 10% HCl (to remove the excess amine), then 2 mL of water [see Technique 4.5b]. Dry the organic layer with anhydrous calcium chloride. Transfer the dried solution to a 10-mL

Erlenmeyer flask. Working in a hood, evaporate the dichloromethane on a steam bath or with a stream of nitrogen. Recrystallize the product from ethanol or an ethanol/water solution.

Cleanup: Combine the aqueous phases from the extractions and adjust the pH to 10 with 5% NaOH. Transfer the solution to a screw-capped centrifuge tube and extract the aqueous phase with 2 mL of ether. Remove the lower aqueous phase and adjust the pH to 7 with 5% HCl before washing the solution down the sink or pouring it into the container for aqueous inorganic waste. Pour the ether solution and the filtrate from the recrystallization into the container provided for flammable (or organic) wastes. Place the calcium chloride pellets in the inorganic-waste container.

Anilides are formed by the reaction of acyl chlorides with aniline:

Aniline Anilide of unknown acid (Ar = aryl)
(phenylamide)

Toluidides are analogous compounds containing a methyl group in the para position of the aromatic ring.

4-Toluidine Toluidide of unknown acid
(*N*-arylamide)

6.5 Phenols

Bromo Derivatives **Procedure.** Dissolve 0.1 g of the unknown phenol in 1 mL of methanol in a 50-mL Erlenmeyer flask and add 1 mL of water. Slowly add a saturated solution of bromine in water, a few milliliters at a time, until the bromine color persists. Shake the flask after each addition. Filter the solid product and wash it well with water. Recrystallize it from a methanol/water solvent mixture.

A typical case is the bromination of phenol:

Phenol   2,4,6-Tribromophenol

1-Naphthylurethanes **Procedure.** Follow the same synthetic procedure as that used for alcohols (p. 554), *except add 1 drop of pyridine to the reaction mixture.*

The reaction of phenol and α-naphthyl isocyanate is:

Phenol α-Naphthyl isocyanate   Phenyl α-naphthylurethane

3,5-Dinitrobenzoates **Procedure.** Follow the same synthetic procedure as that used for alcohols (p. 552).

A phenyl dinitrobenzoate, for example, forms by this reaction:

Phenol   3,5-Dinitrobenzoyl chloride   Phenyl 3,5-dinitrobenzoate

Qualitative Analysis 7
SEPARATION OF MIXTURES

The practicing chemist is usually faced with a mixture of products and unreacted starting materials, rather than a single easily purified unknown. In fact, you have already separated many mixtures as you isolated and purified products from the experiments you have done.

When faced with a mixture of compounds whose separate identities you want to know, you need to discover first of all how many compounds there are in the mixture. Then you must develop a method for their separation before you can begin to use the identification tests that we have already discussed.

The science of mixture separation continues to be improved every day. When an organic chemist considers a separation, the first question to be addressed is whether a preparative or an analytical separation is necessary. For an analytical separation, merely enough of each compound (just a milligram or so) to determine the presence and identity of all components is often sufficient. The use of analysis as a means of determining a mixture's composition is not a viable alternative unless the chemist already has a good idea about the identity of all the components (after using many of the techniques described in this section). The analysis can be carried out more readily when a pure sample of each component is in hand. Where the preparative approach is necessary, it is normally appropriate to separate much larger amounts of compound, and it is usually desirable to isolate between 100 mg and a gram of each component, if possible.

There are many weapons in the arsenal of separations. It is clear, however, that only some of these weapons are available in the educational laboratory. Conversely, it is worthwhile to learn about the procedures of the research scientist at the same time. The techniques we will use in this text are

Thin-layer chromatography (TLC)
Extraction
Distillation
Column chromatography
High-performance liquid chromatography (HPLC)
Gas chromatography (GC, or GLC, or GLPC)
Mass spectrometry (MS)

We will summarize the strengths and weaknesses of each of these seven techniques to familiarize you with those situations to which they may best lend themselves. Full descriptions appear in Part 5, Techniques, as the cross references indicate.

Thin-Layer Chromatography (TLC) [Technique 10] TLC is the technique most frequently used by the organic chemist. It requires only small amounts of compounds, and with practice it is easily modified (by varying the choice of solvent) to resolve even very complex mixtures. Silica gel is a relatively gentle absorbent; thus TLC does not harm most sensitive compounds. Elevated temperatures are not required, as is the case for distillation or gas chromatography. TLC is also a companion technique: For example, a TLC procedure is often carried out before examining a sample by means of GC in order to be sure that the compound is sufficiently pure and will not foul the GC columns. Moreover, TLC analysis is frequently run on fractions taken from column chromatography in order to determine the number of components in a sample and often the identity of these components. "Thick"-layer chromatography can be carried out on substances when a preparative-scale TLC separation is desired. Special large chromatographic plates are used for this procedure, thereby ensuring the separation of enough material to allow the recovery of individual compounds.

Extraction [Technique 4] Because extractions are frequently used to purify the products of organic synthesis, the equipment (separatory funnels and so on) is commonly available. Extraction requires fairly large amounts of compound (at least several hundred milligrams). Specific solubility properties such as the polarity characteristics and the acid-base chemistry of the organic compounds of interest must be considered. Extraction does not harm most organic compounds, as long as they can tolerate 5–10% solutions of strong acids or bases. Elevated temperatures are not required. TLC procedures can be performed before and after subjecting a sample to acid extraction. Mechanical losses, however, can be fairly substantial. The details of a typical extraction carried out on a mixture of simple organic compounds are described in Organic Qualitative Analysis 7.3.

Distillation [Technique 7] Because distillation is frequently used to purify the products of organic synthesis, the equipment (condensers, adapters, and so on) is normally readily available. Distillation requires fairly large amounts of compound (at least a few grams for a macro distillation). All of the components of the mixture must be reasonably stable with respect to heat. Vacuum distillation can be used for thermally sensitive compounds of modest thermal stability, but it is very inefficient. Mechanical losses can be fairly substantial in any distillation. One can run TLC procedures before distillation and on each fraction obtained from the distillation.

Column Chromatography [Technique 12] and Thin-Layer Chromatography (TLC) [Technique 10]   Column chromatography is frequently used when the sample is thermally sensitive enough to prohibit preparative-scale separation by either distillation or gas chromatography. TLC is often used to identify useful solvent systems for column chromatography, although the separations are not normally as good on a column as they are on thin-layer plates. Columns require substantial amounts of compound (100 mg, and often much more), and TLC must be used to monitor the identity of the samples as they come off the column. Compounds with close retardation factor (R_f) values are frequently not easily separable by means of column chromatography. Because silica gel is a gentle absorbent, compound stability is not normally an issue. Elevated temperatures are not required. Special techniques such as flash [Technique 12.8] and dry column chromatography [Technique 12.7] use pressure-driven liquids forced through very small particles of adsorbent to carry out improved separations.

High-Performance Liquid Chromatography (HPLC) [Technique 12.9]
HPLC is similar to flash chromatography in that pressure is used to drive mixtures through adsorbents of an especially small particle size. HPLC is used when high speed is needed and especially when the sample is thermally sensitive. TLC is often used to decide what solvent system might work as the mobile phase. Although HPLC separations are frequently excellent as an analytical technique, an expensive, elaborate instrument is required, especially when doing preparative-scale separations. Sample sizes akin to that used for TLC (milligrams) or even smaller amounts are usually sufficient for analytical work. Elevated temperatures are not required.

Gas Chromatography (GC, or GLC, or GLPC) [Technique 11]   In all the preceding techniques, the samples are in the liquid phase. In gas chromatography, the sample is in the gaseous phase and components are separated by adhesion to a liquid adsorbent. Thus, significant thermal stability and vapor pressure is required of all components of the mixture. Sample sizes akin to those used for TLC and HPLC (milligrams or microliters) or even smaller amounts are usually sufficient. Preparative-scale procedures can be carried out on specially adapted instruments. The analytical procedure itself is usually quite fast (a matter of minutes).

Mass Spectrometry [Spectrometric Method 3]   One particularly powerful method allows you to obtain mass spectra of compounds separated by gas chromatography in a simple and straightforward manner. The instrument used in this method links a gas chromatograph directly to a mass spectrometer; the method is thus designated "GC-MS." To obtain a mass spectrum of each component of a mixture, a tiny amount of the mixture is dissolved in a solvent such as methanol or ether and the solution is injected into the gas chromatograph. A time delay before the mass spectrometer begins to operate is built into the procedure so that the solvent can exit the gas chromatograph before MS data is collected. This delay means that low-boiling-point organic compounds will not be observed. A mass spectrum is recorded for each GC peak; so the molecular ion and fragmentation peaks can be interpreted by the methods discussed in Spectrometric Method 3.

In the early part of your experimental work, obtain an infrared spectrum of the test mixture. This can be a big help in deciding what procedures to use for separating the compounds. Initially, try your separation methods on small amounts of the unknown mixture before you commit the bulk of your sample. Separation of components from a mixture is, in part, a trial and error process.

7.1
Assaying the Number of Compounds

Solids If you are given a mixture of solids, run thin-layer chromatograms on the mixture. Ethyl acetate is a developing solvent of medium polarity and would be a good initial choice if you are using silica gel thin-layer plates. [See Technique 10.4 for a discussion of a general method of selecting a TLC developing solvent.]

Liquid-Solid Samples With a mixture that is part liquid and part solid, suction filtration is a natural first step [see Technique 5 and Figure 5.6]. Gas-liquid chromatography is the method of choice for analyzing the number of compounds in a liquid; however, there may be solids dissolved in the liquid portion of a liquid-solid mixture. These will plug up the GC column and never reach the detector. Most solids are not volatile enough to pass through standard GC columns at the column temperatures that you will normally be using. TLC analysis at least gives a rough idea of the purity of a sample that you are considering for GC analysis.

To determine whether the liquid portion of your mixture contains any solids in solution, place a few drops of the liquid on a watch glass in the hood. Allow the liquid to evaporate and observe whether a solid residue remains. The only ambiguity in this analysis occurs with liquid amines, which often react with carbon dioxide from the air to form solid carbonates and ureas.

Do not put any liquid that contains dissolved solids through a gas-liquid chromatograph. First, find a method for separating out the dissolved solids. Only when a solid compound has already been identified and is known to be volatile can an exception be made to this rule.

Liquids   If you are given a homogeneous liquid mixture, place a few drops of it on a watch glass in the hood. Allow the liquid to evaporate and observe whether a solid residue remains. When a solid is present, treat the mixture as a liquid-solid mixture.

If no solid is present, analyze the liquid mixture by gas-liquid chromatography. It is easiest to distill a small amount of the liquid first in order to gain some idea of its volatility; see the discussion on microscale distillation in Technique 7.3b. This preliminary distillation allows you to choose a GC column and working temperature more exactly. Initially, choose a column temperature from 20°C to 75°C below the point at which the liquid boils. If it boils over a wide temperature range, you may have to do GLC analyses at several different temperatures. [You may at this point review GC techniques in Technique 11.]

7.2 Separating the Mixture

Filtration, distillation, extraction, and chromatography are the primary separation techniques. We have already discussed the use of filtration for separating solids from liquids. For separating liquids from one another, distillation can be an effective method. If this method looks like a feasible approach for your unknown, try it out on a small amount (1–2 mL) of the sample. Remember that it is difficult to separate liquids that boil within 50°C of one another by using simple fractionating columns. In addition, you must be aware of the possibility that two liquids will react with each other at the elevated temperatures of a distillation. This reaction is especially likely if you heat an alcohol and a carboxylic acid, or an amine and a carboxylic acid or one of its derivatives. If no solids are present in solution, check the purity of the mixture by GLC both before and after the distillation. If a distillation has produced an 80:20 mixture from a 50:50 mixture of two volatile liquids, a second and even a third distillation are probably warranted.

Extraction techniques provide one of the best ways to separate a mixture of compounds. Extraction can take advantage of the chemical nature of the compounds to be separated. Both the polarity and the acid-base properties of the test compounds can be exploited for this purpose.

Because most organic compounds are insoluble in water, it is unlikely that an entire mixture will be soluble. However, there is a chance that one of its components may be soluble in water, especially if O—H stretching vibrations appear in the infrared spectrum. When one of the components is soluble in water, this behavior is the basis of an excellent separation method.

Determining the water solubility of one component of a mixture is not always easy. Even with a positive result, there will still be two layers present. Use a 10-mL graduated cylinder to test the solubility of the component. Add 0.5 mL of the mixture and then a few milliliters of water. Shake the liquids together and allow them to separate again; then carefully check the volume of the organic layer. You should be able to see whether more than 0.1 mL of the mixture has dissolved in the water. If in doubt, add another measured quantity of the mixture and check to see how much has been extracted into the water layer overall.

If one component is soluble in water and the others insoluble, extraction of the mixture by means of a separatory funnel will allow you to separate out the water-soluble one. Separating that component from the water will require distillation. Remember, however, that some water-soluble compounds form azeotropic mixtures with water. If this is the case, you may have to add a low-boiling-point solvent, such as dichloromethane or diethyl ether, to the distillate and dry the solution with anhydrous magnesium sulfate. If a drying agent is added directly to a small volume of liquid, a large percentage of that liquid may be lost as it clings to the solid particles. After filtration and distillation of the low-boiling-point solvent, the water-soluble component that remains can be identified by the usual methods. [See Technique 4.6 for the techniques of drying a solution.]

One of the best methods for separating a mixture of compounds is extraction with acid or base. As discussed in Organic Qualitative Analysis 3, strong acids and bases can be converted to water-soluble salts by neutralization with 2.5 M NaOH and 1.5 M HCl, respectively. In this way, a carboxylic acid can be separated from an ester and an alcohol, a phenol from an aryl halide, an amine from an amide, and so forth.

Take the case of a water-soluble carboxylic acid and an ester. If you shake a few milliliters of 2.5 M NaOH solution with your unknown mixture, the carboxylic acid dissolves in the aqueous layer as the carboxylate salt. After separation of the organic and aqueous layers, the aqueous layer can be acidified with 6 M HCl solution. A solid carboxylic acid precipitates and can be filtered. A liquid carboxylic acid can be distilled or extracted with diethyl ether and isolated as discussed earlier.

A water-insoluble amine can be extracted with 1.5 M HCl solution. After separation of the layers, the aqueous mixture can be made strongly alkaline with 2.5 M NaOH solution. The amine can be recovered as a residue by extraction with ether, drying, and distillation of the solvent. (Solubility data relating to various classes of compounds are given in Figure 7.1 in the form of a flow chart.)

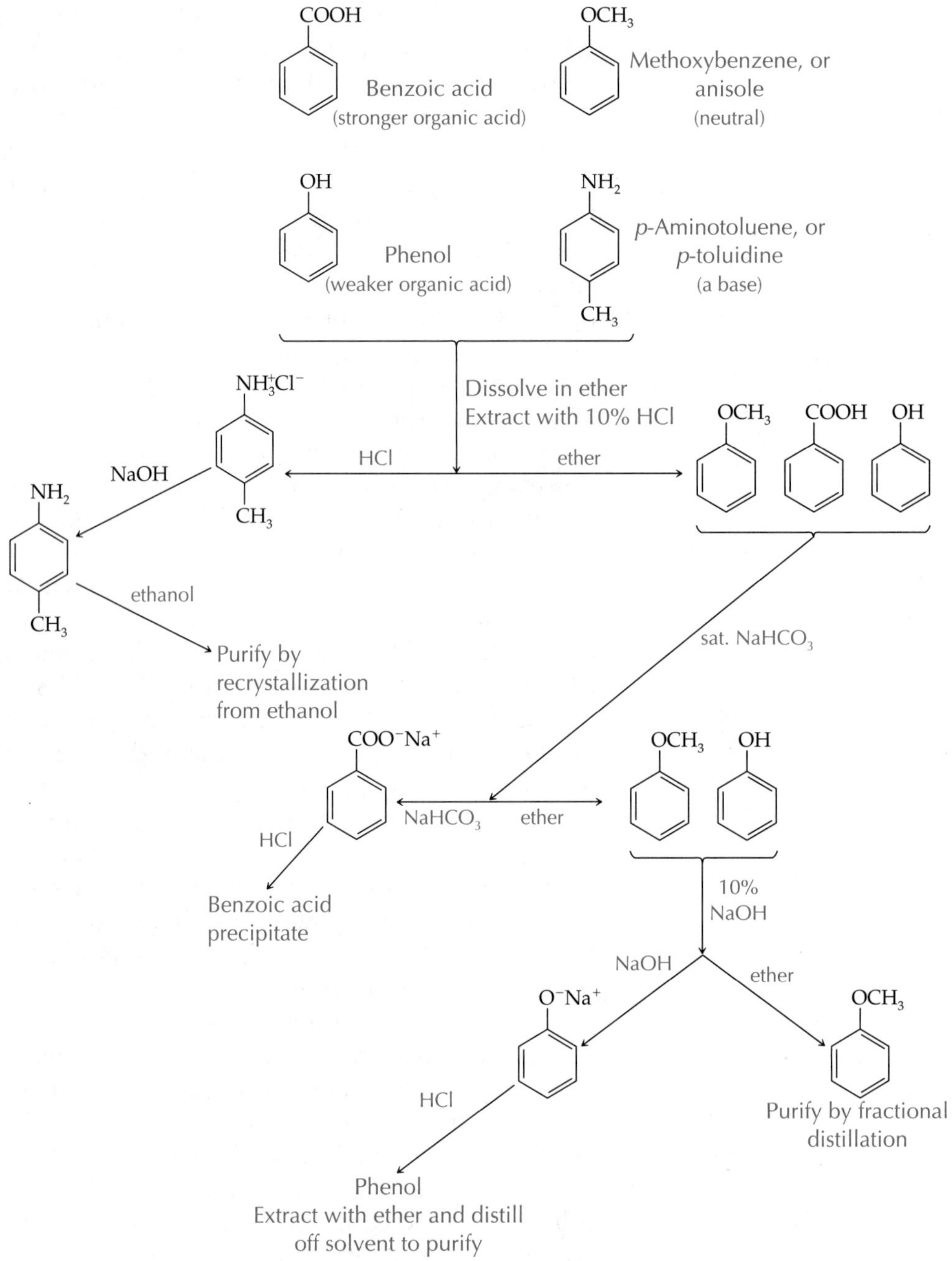

FIGURE 7.1 Extraction of organic compounds with acids and bases.

7.3
Model Extraction Procedure

In the next paragraph, we describe a typical extraction procedure that can be used to separate four classes of organic compounds: a stronger organic acid (a carboxylic acid), a weaker acid (a phenol), a base (an aniline), and a neutral compound.

Assume that a mixture (total weight 5–20 g) is composed of benzoic acid, phenol, *p*-methylaniline (*p*-toluidine), and methoxybenzene (anisole). The task of analyzing a mixture of, respectively, a strong organic acid, a weak acid, a base, and a neutral compound represents a typical situation for extraction techniques.

At this point, follow the procedure outlined in the flow diagram shown in Figure 7.1. Dissolve the entire mixture in 100 mL of ether (filter off any solids). In a separatory funnel, extract [Technique 4.2] this solution with 2 × 35-mL portions of 10% HCl. The aqueous layer contains *p*-methylanilinium chloride, which can be treated with ammonia (until the solution is no longer acidic as determined with pH paper); as a result, the amine will precipitate out. The solid amine can be recrystallized from ethanol for purification. Extract the remaining ether layer twice with 35-mL portions of saturated sodium bicarbonate solution. By this means the stronger acid is removed from the ether. Treat the sodium benzoate in the water layer with 10% HCl solution until the water layer is no longer basic; this step will precipitate the benzoic acid (mp 121°C), which can be recrystallized from hot water. The ether solution now contains only phenol (a weak acid) and methoxybenzene, which is neutral. Extract this solution with 2 × 35 mL portions of 10% NaOH; and treat the water layer, which now contains sodium phenoxide, with 10% HCl until it is no longer basic as tested with pH paper.

The result is that phenol is regenerated, and it can be extracted with 2 × 35-mL portions of ether; recover the phenol by ether solvent evaporation. The ether layer now contains only methoxybenzene, which can be recovered by fractional distillation, after first removing the ether (bp 35°C) and then the methoxybenzene (bp 154°C).

Remember that infrared spectrometry, as well as a number of the procedures for making derivatives *requires dry reagents*. After you have done aqueous extractions, always remove any water in the organic phase with a small amount of a drying agent.

Numerous chromatographic techniques may be helpful in separating a mixture of unknowns. Column chromatography [Technique 12] is the most important one. If you have a particularly difficult separation, in which distillation and extraction methods are of no avail, consider column chromatography. This is a powerful technique, but it is often very time-consuming to develop a chromatographic separation from scratch. Recall that TLC is a good way to determine the mobile phase (solvent) and stationary phase (adsorbent) that could work with column chromatography. Gravity columns, however, will not yield as clean a separation as suggested by a simple TLC test. Try to maximize the TLC separation (in other words, seek a large ΔR_f, namely, the difference between the R_f values of the components of interest) before performing column chromatography.

After separating a mixture, identify the individual components. Treat the individual compounds as separate unknowns. Use methods based on solubility, physical constants, and spectral properties; perform classification tests and elemental analyses, and prepare derivatives to identify the individual compounds in your mixture.

Qualitative Analysis 8
SUMMARY OF IDENTIFICATION PROCEDURE FOR AN UNKNOWN COMPOUND

1. **Purity**

   Estimate purity: color (one color? off-color? decomposition?), appearance (crystallinity of solid? solid and liquid?, mp of solids, bp of liquids), TLC, GC

   Ignition test: residue? soot? flame color?

   Improve purity? distillation, recrystallization, chromatography

2. **Physical properties**

   Solid or liquid? color? bp or mp? odor **(hazardous)**

3. **Solubility**

   Determine classification (Qualitative Analysis 3)

   Suggest functional-group classes

   Make list of compounds from tables by functional group (mp/bp)

4. **Elemental analysis by sodium fusion**

   Sulfur? Nitrogen? Halogen?

   Refine list of compounds

5. **IR and NMR analysis**

   Refine list of compounds (identify compound?)

   Assign IR and NMR absorptions

6. **Chemical tests for functional groups**

   Choose tests found in Qualitative Analysis 5.

7. **Derivative**

   Choose product that differentiates proposed structures

Qualitative Analysis 9
SUMMARY OF THE ANALYSIS OF A MIXTURE OF UNKNOWN COMPOUNDS

1. **Preliminary procedures**

   Note color, appearance, pH

   Take advantage of easy separations (filter, etc.)

2. **Assay the number of compounds in mixture**

   TLC, GC (only with instructor's permission), IR, and NMR of mixture

3. **Carry out separation**

   Decision based on conclusions from 1 and 2

   Distillation, chromatography, or extraction

4. **Monitor success**

   Use TLC (or GC, with permission) to determine whether fractions contain one compound

5. **Identification**

   Use TLC (GC?), mp, bp results already determined

   Use procedures of previous summary on each component

References

1. Shriner, R. L.; Fuson, R. C.; Curtin, D. Y.; Morrill, T. C. *The Systematic Identification of Organic Compounds*; 6th ed.; Wiley: New York, 1980.
2. Cheronis, N. D.; Entrikin, J. B.; Hodnett, E. M. *Semimicro Qualitative Analysis*; 3rd ed.; Interscience Publishers: New York, 1965.

3. Rappoport, Z. *Handbook of Tables for Organic Compound Identification;* 3rd ed.; CRC Press: Cleveland, OH, 1967.

4. Kemp, W. *Qualitative Organic Analysis;* 2nd ed.; McGraw-Hill: New York, 1986.

5. Furniss, B. S.; Hannaford, A. J.; Smith, P. W. G.; Tatchell, A. R. *Vogel's Textbook of Practical Organic Chemistry;* 5th ed.; Longman: New York, 1989.

6. Lide, D. R.; Milne, G. W. A. *CRC Handbook of Data for Organic Compounds;* 2nd ed.; CRC Press: Boca Raton, FL, 1994.

7. Codagan, J. I. G.; Buckingham, J.; Macdonald, F. *Dictionary of Organic Compounds;* 6th ed.; Chapman and Hall: New York, 1996; 9 volumes.

8. Pasto, D. J.; Johnson, C. R. *Organic Structure Determination;* Prentice-Hall: Englewood Cliffs, NJ, 1969.

9. Utermark, W.; Schicke, W. *Melting Point Tables of Organic Compounds;* Interscience Publishers: New York, 1963.

10. Siverstein, R. M.; Bassler, G. C.; Morrill, T. C. *Spectrometric Identification of Organic Compounds;* 5th ed.; Wiley: New York, 1991.

11. Grasselli, J. G.; Ritchey, W. M. *CRC Atlas of Spectral Data and Physical Constants for Organic Compounds;* 2nd ed.; CRC Press: Cleveland, OH, 1975; 6 volumes.

12. Criddle, W. H. and Ellis, G. P. *Spectral and Chemical Characterization of Organic Compounds: A Laboratory Handbook;* 3rd ed.; Wiley: New York, 1990.

13. Pouchert, C. J. *The Aldrich Library of FT IR Spectra;* Aldrich Chemical Co.: Milwaukee, WI, 1985; 2 volumes.

14. Pouchert, C. J. *The Aldrich Library of ^{13}C and ^1H NMR Spectra;* Aldrich Chemical Co.: Milwaukee, WI, 1993; 3 volumes.

Questions

1. Caproic acid (molecular formula $C_6H_{12}O_2$) is only slightly soluble in water but is readily soluble (with stirring or shaking) in both 0.6 M $NaHCO_3$ and 2.5 M NaOH. Conversely, formic acid is readily soluble in both water and the two alkaline reagents. Explain.

2. Picric acid (2,4,6-trinitrophenol) is slightly soluble in water but is readily soluble (with stirring or shaking) in both 0.6 M $NaHCO_3$ and 2.5 M NaOH. Conversely, *o*-ethylphenol is insoluble in both water and 0.6 M $NaHCO_3$, but soluble in 2.5 M NaOH. Explain.

3. Both *p*-ethylaniline and triethylamine are only slightly soluble in water but both are readily soluble with stirring in 1.5 M HCl. In contrast, triphenylamine shows virtually no solubility in either water or 1.5 M HCl. Explain.

4. Match each one of the five compounds (1–5 on the right) with one of the following sets of IR bands (one of a–e on the left) selected from the IR spectra of the five compounds. All except the last two

bands are strong, and the band at 3373 cm^{-1} is exceptionally broad.

a. 1715 cm^{-1}	1. acetophenone
b. 3373 cm^{-1}	2. 2-butanone
c. 3008,2940,741 cm^{-1}	3. octylamine
d. 1685 cm^{-1}	4. phenol
e. 3372,3290 cm^{-1}	5. *o*-xylene

5. The ^{1}H NMR spectrum of a compound of molecular formula $C_3H_6Cl_2$ shows only two multiplets, a triplet at δ 3.72 and a quintet at δ 2.23. The high-field:low-field integration ratio is 2:1. Deduce the structure of the organic halide and explain the NMR signals.

6. A compound of bp 204°C is water-insoluble and is insoluble in 2.5 M NaOH and 1.5 M HCl as well. The compound does dissolve in concentrated H_2SO_4. The original compound gives rise to both a positive iodoform test and a positive 2,4-dinitrophenylhydrazine test. Deduce the structure and name of the compound and write reactions for all positive chemical tests.

7. A compound of bp 130°C is water-soluble and an aqueous solution of this compound is basic as tested with pH paper. Elemental analysis by sodium fusion indicates the presence of only nitrogen. The original compound gives rise to a precipitate when treated with benzenesulfonyl chloride and this precipitate remains insoluble in aqueous alkali. The original compound forms a benzene sulfonamide which melts at 118°C. Deduce the structure and name of the compound and write reactions for all positive chemical tests.

8. A compound of bp 152–153°C is water-insoluble and insoluble in 2.5 M NaOH, 1.5 M HCl, and concentrated H_2SO_4. The original compound reacts negatively to a 2,4-dinitrophenylhydrazine test. The ^{1}H NMR spectrum gives rise to the following signals: δ 1.25, 6H, d; δ 2.9, 1H, septet; δ 7.1–7.4, m, 5H. The ^{13}C NMR spectrum in the completely decoupled proton mode shows singlets at δ 22, 34.5, 126, 127, 129, and 149. The IR spectrum shows four peaks between 2980–2800 cm^{-1} and three peaks in the 3100–3000 cm^{-1} region. Deduce the structure and name of the compound and assign all spectral absorptions.

9. A compound of mp 150°C is water-insoluble and also insoluble in 1.5 M HCl. The compound does dissolve in 2.5 M NaOH, in 0.6 M NaHCO$_3$, and in concentrated H_2SO_4. Elemental analysis by sodium fusion indicates the presence of only bromine. The original compound gives a negative 2,4-dinitrophenylhydrazine test. Treatment with aluminum chloride gives rise to an intense color. When treated with thionyl chloride, followed by ammonia, the compound forms an amide that melts at 154°C. Deduce the structure and name of the compound and write out all chemical reactions.

10. A compound of mp 150°C is water-insoluble and also insoluble in 1.5 M HCl. The compound does dissolve in 2.5 M NaOH, in 0.6 M NaHCO$_3$, and in concentrated H$_2$SO$_4$. Elemental analysis by sodium fusion reveals no elements. The original compound is acidic as tested with pH paper and gives a positive 2,4-dinitrophenylhydrazine test. Treatment with aluminum chloride gives rise to an intense color. When treated with thionyl chloride, followed by ammonia, the compound forms an amide derivative that melts at 166°C. Deduce the structure and name of the compound and write out all chemical reactions.

11. Carbohydrates are normally water-soluble (for example, sucrose) but ether-insoluble. Briefly explain.

12. Actually the compound on which the tests in Question 6 are carried out gives rise to 2,4-dinitrophenylhydrazones of different melting points (240°C, 250°C), depending on the conditions of the derivatization procedure. Suggest a reason for this behavior.

Spectrometric Methods

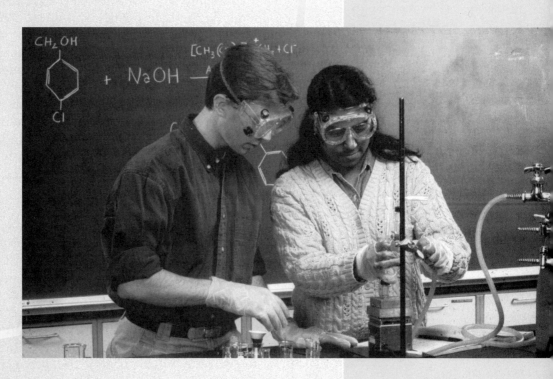

Part 4 introduces the four spectrometric methods most frequently used today by chemists to determine the structure of organic compounds: infrared spectrometry (IR), nuclear magnetic resonance spectrometry (NMR), mass spectrometry (MS) and ultraviolet-visible spectrometry (UV-VIS). *Infrared spectrometry* provides information on the functional groups (for example, —OH, —COOH, —(C=O)NH$_2$, benzene ring or —C=O) present in a compound. *Nuclear magnetic resonance spectrometry* reveals many details of an organic compound's structure, particularly the interrelated connectivity of its hydrogen and carbon atoms. *Mass spectrometry* allows chemists to determine the molecular weight of a compound, and the fragmentation pattern provides data that can lead to identification of the compound. *Ultraviolet-visible spectrometry* is chiefly used for characterization of compounds containing conjugated double bonds or one (or more) pairs of nonbonded electrons.

Integrating the data obtained from the four different spectrometric methods discussed in Part 4 is an important part of using spectrometric techniques in the characterization of an organic compound. One spectral method may reveal different features about a compound that may not be clear from another method, or in some cases one spectral method may confirm the existence of a structural unit suggested by another method. For example, the skeleton of a carbonyl-containing compound is usually readily determined by the use of ^{1}H and ^{13}C NMR, but the nature of the carbonyl-containing functional group (ketone, ester, acid, amide, etc.) is often yielded by the IR spectrum.

The last section of Part 4 contains *integrated problems* with data about a compound obtained by several spectrometric techniques. Solving these problems requires you to consider how the data from one spectrometric technique assist in interpreting the data obtained by another spectrometric method. The mass spectrometric data is usually a good starting point because it gives the molecular weight of the compound. Next, consider the IR spectrum, which provides data for the identification of the functional groups present in the compound. Interpretation of the ^{1}H and ^{13}C NMR spectra allows you to define the carbon and hydrogen skeleton of the compound. For compounds with systems of conjugated double bonds, UV-VIS spectral data can supply valuable confirmatory evidence.

Spectrometric Method 1
INFRARED SPECTROMETRY

Throughout the study of organic chemistry, we are asked to think in terms of structure—the experienced organic chemist can anticipate many of the physical and chemical properties of various compounds by a quick glance at the structure. Fifty or more years ago, structure was determined largely by chemical methods. But the chemical methods are very time consuming.

Fortunately there are now much quicker ways of deducing structures. These new techniques are based on spectrometric (less correctly, spectroscopic) methods of analysis and usually involve the absorption of "light," or radiation representing various portions of the electromagnetic spectrum (Figure 1.1). In effect, they provide "snapshots" of all or a portion of a molecule. One of the most useful types of spectrometry is *infrared (IR) spectrometry*, which has also been called *vibrational spectroscopy* because the

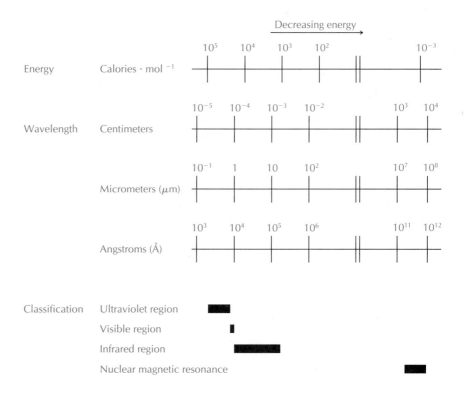

FIGURE 1.1 Comparison of wavelengths and energies in various spectral regions.

amount of IR radiation absorbed by a compound corresponds to the energy it takes to stretch or bend bonds in a given compound. IR absorption is generally used to identify organic functional groups. The detailed analysis of the carbon skeleton (and any attached hydrogens) is usually carried out by other techniques (especially nuclear magnetic resonance spectrometry; see Spectrometric Method 2).

1.1
Theory of Molecular Vibration

Molecules selectively absorb specific frequencies of IR radiation that correspond to the frequencies of the vibrational oscillations of the atoms connected by covalent bonds. When IR radiation is absorbed by a molecule, the amplitudes of these vibrations increase. The absorption corresponding to these oscillations appears in certain definite wavelength (or frequency) regions of the spectrum, regardless of the particular compound in which the group of atoms is contained. For example, the stretching regions of the oxygen-hydrogen bonds in all alcohols appear at nearly the same frequency, and absorptions of the O—H bonds can easily be detected in the IR spectra of all alcohols.

Absorption positions in IR are given in wavenumbers (cm⁻¹).

The frequency at which a characteristic IR absorption occurs depends on the mass of the atoms involved in the vibration and the strength of the bond connecting them. Consider, for example, the molecule HCl. The wave number for stretching the H—Cl bond is 2886 cm⁻¹, and when hydrogen chloride absorbs IR radiation with the wavenumber 2886 cm⁻¹, the motion within the molecule becomes more vigorous and the average displacement (amplitude) of the atoms from the equilibrium bond length becomes greater.

Water has three fundamental vibrational frequencies. They are drawn in Figure 1.2. The vibrations (a) and (b) correspond to the stretching of the oxygen-hydrogen bonds. Figure 1.2a shows the symmetric vibration in which the hydrogen atoms stretch in and out together. In the asymmetric vibration in Figure 1.2b, as one hydrogen atom springs out, the other springs in, whereas the oxy-

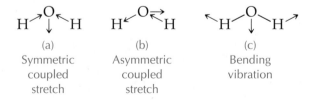

(a) | (b) | (c)
Symmetric coupled stretch | Asymmetric coupled stretch | Bending vibration

FIGURE 1.2 Fundamental vibrations of water. The oxygen moves to maintain a constant center of gravity.

gen atom, moving slowly because of its large mass, shifts a little bit to one side. The last of the three vibrations (Figure 1.2c) involves the bending of the O—H bonds, a kind of scissoring motion in which the H—O—H bond angle changes.

1.2
IR Spectrum

When IR radiation is absorbed by a sample, the changes are picked up electrically (see later section on instruments) and are passed on to a detector. The raw data are run through a computer and eventually recorded as an off-null (by convention, usually downward from the baseline, or null) absorption. Thus, a two-dimensional recording—called an **IR spectrum**—is obtained, with the position of the absorption on the abscissa (energy, wavenumber, or wavelength), and the amount of intensity of absorption on the ordinate. Figures 1.3 and 1.4 are examples of such spectra.

The equation that relates the energy (E) of the absorbed radiation to its frequency (ν) is

$$E = h\nu$$

where h = Planck's constant. Frequency is proportional to the inverse of the wavelength (λ), so

$$E = hc(1/\lambda)$$

In other words, frequency may be equated to the product of the wavenumber ($\bar{\nu} = 1/\lambda$) and the speed of light (c); thus

$$E = hc\bar{\nu}$$

An **IR band** is normally reported as a position of maximum absorption in wavenumbers ($\bar{\nu}$), in units of reciprocal centimeters (cm^{-1}). Both Figures 1.3 and 1.4 show the most intense band for cyclopentanone between 1700 and 1800 cm^{-1}.

Organic compounds, which are composed of a number of different covalent bonds, usually give rise to a number of IR bands; thus the spectrum of such a compound can be quite detailed. The IR spectra of cyclopentanone shown in Figures 1.3 and 1.4 reveal the same peak positions, but their shapes are different because they have been measured on different types of spectrometers. An IR spectrum normally encompasses the region 4000–400 cm^{-1}, although the lower limit may vary from one instrument to another. If a molecule has 10 or 20 atoms, its IR spectrum will be complex. The large number of fundamental vibrations, their overtones, and combinations of vibrations make it far too difficult to

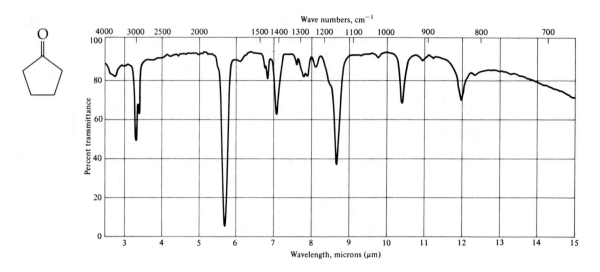

FIGURE 1.3 IR spectrum of cyclopentanone. The sample was prepared as a liquid film ("neat"), and the spectrum was recorded on a dispersive instrument with prism optics. Note that the wavelength scale is linear; consequently, the wavenumber scale is not.

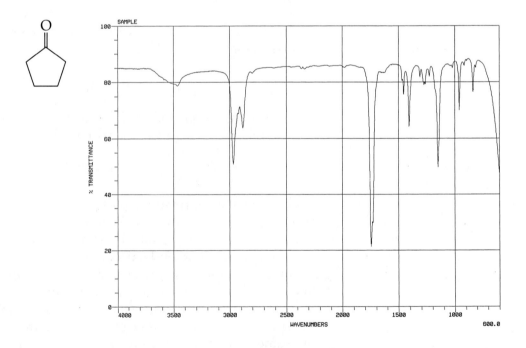

FIGURE 1.4 IR spectrum of cyclopentanone recorded from a capillary film between salt plates on an FT IR instrument (thus "neat"). In this spectrum, the linear scale is in wavenumber, not wavelength. Reproduced from "A Spectrum of Spectra," by R. A. Tomasi, Sunbelt R&T Pub., Tulsa, OK, 1992, with permission.

quantitatively describe the IR spectra of many organic compounds, but nonetheless spectra can easily yield a great deal of qualitative information.

The *intensity* (peak size) of an IR absorption can be reported in terms of either transmittance (T) or absorbance (A). **Transmittance** is the ratio of the intensity of the absorbed beam (I) divided by the intensity of the incident beam (I_0):

$$T = I/I_0$$

Peak intensity can also be reported as absorbance (A), where **absorbance** is the log of the inverse of transmittance:

$$A = \log(1/T)$$

In practice, peak intensities are reported in a less quantitative fashion, and it is desirable to record a spectrum in which the most intense peak nearly fills the vertical height of the chart. Peaks of that order of magnitude are termed strong (s) peaks; smaller peaks are called either medium (m) or weak (w). Peaks can also be described as broad or sharp (narrow). It is important that the most intense peak be completely on the chart so that all peak maxima can be observed.

It is common to calibrate IR spectra by using a polystyrene standard. Often a peak at 1600 cm^{-1} (separate from the main spectrum) reveals that this has been done.

What determines the position and intensity of IR bands? A variety of factors contribute to the position of an IR band: bond order, bond length, and the electronegativity difference between the two atoms or groups in a bond. **Bond order** is simply the amount of bonding between two atoms (here, between carbons); bond order increases from one to two to three for ethane (CH_3—CH_3), ethene (ethylene, CH_2=CH_2), and ethyne (acetylene, $HC{\equiv}CH$), respectively. In general, the higher the bond order, the higher the frequency (and the wavenumber) of the position of IR absorption:

$C{\equiv}C, C{\equiv}N$	2300–2000 cm^{-1}
$C{=}C, C{=}O, C{=}N$	1900–1500 cm^{-1}
$C{-}C, C{-}O, C{-}N$	1300–800 cm^{-1}
$C{-}H, O{-}H, N{-}H$	3800–2700 cm^{-1}

Of course, factors other than bond order also influence the positions of these bonds.

Polarity greatly influences the intensity of IR bands. If a vibration (stretch or bend) induces a significant change in molecular dipole movement, an intense IR band will result. Thus it is not surprising to find that stretching $C{-}O, C{=}O, O{-}H$ bonds produce intense bands. In each of the bonds listed earlier, there are large electronegativity differences between the atoms, so stretching polar bonds like these will cause a large change in molecular dipole and, usually, intense IR bands.

1.3 IR Instrumentation

There are two major classes of instruments used to measure IR absorption: dispersive and Fourier Transform (FT) spectrometers. **Dispersive spectrometers** contain either a prism or a grating, the purpose of which is to break up the IR light into its component wavelengths. Thus the compound is subjected to incrementally varied wavenumbers of light, and recording the compound's ability to absorb at each wavelength gives rise to the spectrum. The spectrum in Figure 1.3 is a dispersive spectrum (prism optics). Note that the wavelength scale is linear (μm, micrometers).

In *FT IR spectrometers,* radiation containing all IR wavenumbers (for example, $5000-400 \text{ cm}^{-1}$) is used in a single burst. After this radiation has passed through the sample, Fourier transform mathematics is used to sort out the components of the spectrum, so each burst provides a complete IR spectrum.

A Fourier Transform Infrared (FT IR) spectrum recorded by an FT IR instrument has a linear wavenumber scale in cm^{-1} (Figure 1.4). In recent years, most laboratories have converted to FT IR instruments, but there is still a large body of literature that lists

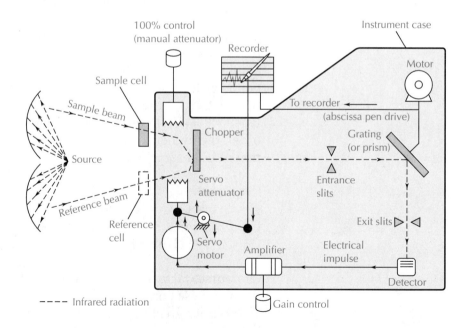

FIGURE 1.5 Schematic diagram of a dispersive (double-beam) infrared spectrometer.

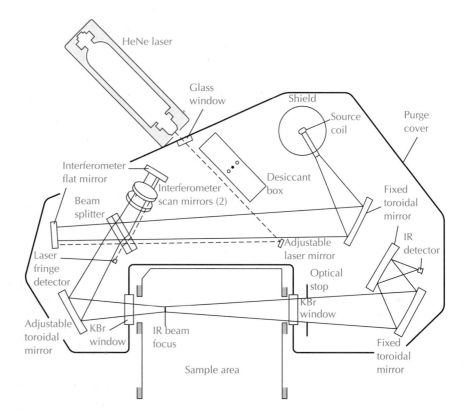

FIGURE 1.6 Schematic for a Model 1600 Perkin-Elmer FT IR (Fourier Transform Infrared) Spectrometer. Reproduced with permission of Perkin-Elmer Corp., Norwalk, CT.

data in wavelength units. A formula that allows interconversion of wavenumber and wavelength is

$$\bar{\nu} = 10{,}000/\lambda$$

where $\bar{\nu}$ is wavenumber in reciprocal centimeters (cm^{-1}) and λ is wavelength in micrometers (μm, 10^{-6} meters—called simply μ or microns in the older literature).

Figures 1.3 and 1.4 show different IR spectra of the same compound. Because the linear scales are different (they have an inverse relationship), the peak shapes are quite different. However, the actual positions of absorption should be exactly the same. We will find that at higher wavenumber (for example, in the 3500 to 2500 cm^{-1} region) FT IR spectra have more well-defined peaks with better resolution.

We can appreciate the differences in IR instrumental analysis by examining schematic diagrams of the two classes of instruments: dispersive (Figure 1.5) and FT IR (Figure 1.6). For the dispersive instrument, a source (for example, a heated filament) provides a beam of IR radiation that is directed by mirrors through both sample and reference cells. The reference cell contains pure solvent (the same solvent that is used to prepare the sample solution). If no solvent is used, the reference sample is merely air. The two beams are directed by a series of mirrors (not shown) to the chopper, a mechanical device that alternately blocks the beams, first that coming through the sample solution and then that coming through the reference cell. Alternators serve to control the amount of radiation passing through. Mirrors now focus this combined beam into the dispersion device (either a prism or a grating). The dispersion device incrementally separates the beam into the various energies that are available from the source. This process eventually allows only a narrow region of energy through the exit slits to the detector. Collecting scans over the entire region gives the complete spectrum. A reference cell allows the instrument to produce IR signals that represent the difference between absorptions due to the solution and those of the pure solvent.

When the sample is "neat" (that is, simply a thin film of the pure liquid as shown in Figure 1.3c), the reference is merely air (no reference cell is used).

A schematic for an FT IR (Fourier Transform Infrared) spectrometer is shown in Figure 1.6. We can follow the light path in the schematic. Irradiation is focused from a source coil by a fixed toroidal mirror onto a beam splitter. The splitter allows radiation to directly hit either a fixed-position mirror (labeled Interferome-

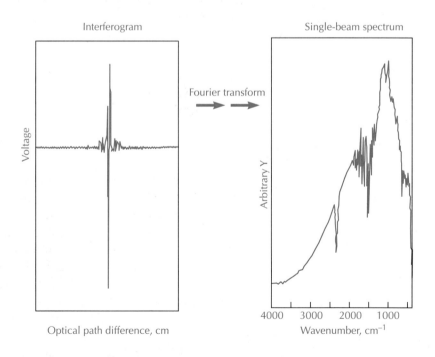

FIGURE 1.7 Illustration of how an interferogram is Fourier transformed to generate a single-beam infrared spectrum.

ter flat mirror) or scan mirrors whose position can be varied. The position of the movable mirror is incrementally changed by a laser beam, and the splitter allows combination of the fixed and variable path beams. The relationships between the distance moved by adjustable scan mirrors and the resulting interferogram are well known. The detailed description of an interferogram is beyond the scope of this book. It has, however, become routine to use variations in position of the movable mirror and Fourier mathematics to transform an interferogram into an IR spectrum (see Figure 1.7). Moreover, this approach offers the advantage of providing the entire scan within one burst (although additional bursts are often used to intensify the signal). Also, a reference cell is not necessary in this single-beam approach.

There are a number of advantages to the FT IR method. Because a monochromator (dispersion procedure) is not used, the entire range of radiation is passed through the sample simultaneously, with a great savings in time. Thus a signal with a high signal to noise ratio and excellent resolution is rapidly obtained. Results of a number of scans are combined to average out random artifacts, and excellent spectra from very small samples can be

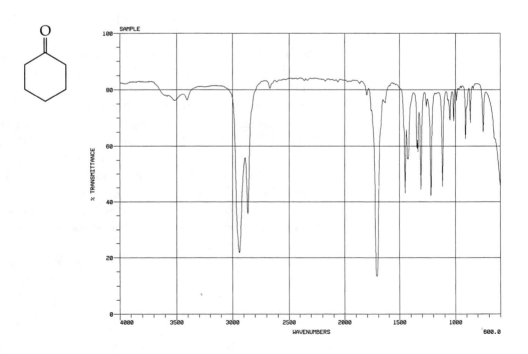

FIGURE 1.8 IR spectrum of cyclohexanone recorded from a capillary film between salt plates on an FT IR instrument. Reproduced from "A Spectrum of Spectra," by R. A. Tomasi, Sunbelt R&T Pub., Tulsa, OK, 1992, with permission.

obtained. Moreover, because the data undergo analog-to-digital (A/D) conversion, the IR spectrum is easily manipulated by a computer, or spectra of pure solvents or of common impurities can be subtracted from the spectrum of a mixture. In addition, printouts can be done in numerical fashion or can be graphically linear in either wavenumber or wavelength. FT IR scans may show undesirable peaks due to atmospheric water and CO_2. Thus computer subtraction of an atmospheric scan can often greatly improve the quality of a spectrum.

Instrument makers provide spectrometers that interface to various types of chromatographs (GC, LC, HPLC), and spectra can be obtained from nanogram amounts of vaporous samples.

A comparison of Figures 1.8 (cyclohexanone) and 1.4 (cyclopentanone) shows that these IR spectra are very similar, a result that is not surprising, because they are both cycloalkanones (that is, they both contain a carbonyl group) and they differ in ring size by only one carbon atom. Careful scrutiny of these two figures, however, indicates that the two spectra have differences in their fine structure. That is, the two compounds have different "fingerprints" (see Section 1.5, Interpreting IR Spectra).

1.4
Techniques of Sample Preparation

A number of considerations influence the method of choice for IR sample preparation. The sample must be evenly dispersed within the IR cell; thus it might appear that simply using a thin film of any liquid sample would be appropriate and straightforward. A problem associated with this method is the fact that such a sample is very highly concentrated, so running the analysis on a solution of the compound in, for example, a chlorinated alkane solvent such as chloroform often is preferable. A problem with solvents, however, is the fact that they often interfere by absorbing in the IR region. Thus a mulling compound such as potassium bromide might be desirable because KBr does not absorb in the 4000–400 cm^{-1} region. Mulled samples do, however, require some skill in their preparation. It is therefore desirable to become acquainted with the details of various sampling methods.

Solid or liquid compounds can be dissolved in a solvent and analyzed in the solution cell shown in Figure 1.9a. This type of solution cell can be used in most IR spectrometers. A solvent should be selected from Table 1.1. It is, of course, desirable to use a solvent that has a minimum of IR absorptions in regions of interest for the primary sample. When solvents do absorb, we should pick those that are not in regions important to our sample. Dispersive instruments (Figure 1.5) are double-beam instruments; thus the use of a solution cell requires a matched reference cell. Here

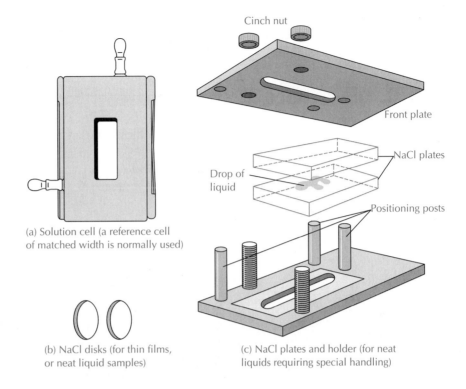

(a) Solution cell (a reference cell of matched width is normally used)

(b) NaCl disks (for thin films, or neat liquid samples)

(c) NaCl plates and holder (for neat liquids requiring special handling)

FIGURE 1.9   Three sample holders.

the reference cell should have exactly the same width as the sample cell and should be filled with pure solvent. The solvent peaks will be subtracted and the resulting spectrum should show only peaks due to the sample of interest. In practice, although well-matched cells may remove the solvent peaks, sample peaks (especially weak peaks) that occur in the region where the solvent absorbs will be difficult to find.

If the compound of interest is a liquid, IR spectra are often obtained by using one of the devices shown in Figure 1.9b,c. Thus films of so-called neat samples (pure liquids, no added solvents) can be analyzed. When the sample is fairly viscous, the sample can be smeared between the two disks shown in Figure 1.9b; the disks are then placed in the sample beam. When the sample has a low viscosity, it requires special handling, and the holder shown in Figure 1.9c can be used.

Naturally, the sample cells themselves must be transparent to IR radiation. Unfortunately, glass absorbs far too much IR radiation to be a useful cell material. Most sample cells for IR spectrometry are made from alkali halides; polished sodium chloride is by far the most popular. Because the cells used in IR spectrometry are usually made of sodium chloride disks or plates, *all of your samples*

Table 1.1. **Region of absorption of common solvents and mulling compounds**

Carrier	Absorption region (in cm^{-1})
Solvents	
Carbon tetrachloride[a]	1590–1565
	835–625
Carbon disulfide[b]	2450–2100
	1650–1440
	880–850
Chloroform[c]	3125–2940
	1250–1190
	835–625
Dichloromethane[d]	3300–2850
	2450–2300
	1550–1150
	<930
Mulling compounds	
Fluorolube	1300–1080
	1000–920
	910–870
	<670
Hexachlorobutadiene[e]	1600–1500
	1270–1230
	1050–760
Nujol®	3000–2700
	1500–1450
	1430–1360
	720–750
Potassium bromide	transparent

a. Toxicity hazard.
b. Toxicity hazard, highly flammable.
c. Toxicity hazard.
d. Methylene chloride, toxicity hazard.
e. Toxicity hazard.

must be completely dry. Water makes salt disks cloudy and virtually useless.

The sample cell is filled by using a syringe or a disposable pipet. In the student laboratory, use of carbon tetrachloride solvent should be avoided. After recording a spectrum, the cells must be carefully cleaned with pure carbon tetrachloride. Infrared cells can be dried either by passing dry nitrogen gas through them or by drawing dry air into them by applying vacuum suction to one of the outlets.

— SAFETY PRECAUTION —————

Carbon tetrachloride is toxic. Use it only with the instructor's permission and use a hood and disposable gloves.

Carbon tetrachloride and chloroform are simple, symmetrical molecules with few intense IR absorptions, and therein lies their attraction as IR solvents. However, both of them have intense absorptions due to the carbon-chlorine stretching vibrations. Chloroform also has absorptions due to carbon-hydrogen stretching and bending. Such a small amount of IR radiation passes through CCl_4 and $CHCl_3$ in these frequency regions that the spectrometer may simply not respond to any other absorptions that the sample may have there. These absorbance regions are given in Table 1.1.

Solvents other than carbon tetrachloride or chloroform are also used as IR solvents, the most common of which is carbon disulfide (CS_2). This solvent is most useful for study of the $835-625$ cm^{-1} region, where chlorinated solvents show strong absorptions.

— SAFETY PRECAUTION —————

Carbon disulfide is very toxic and highly flammable; we recommend that it not be used where other methods suffice.

Solids that are insoluble in carbon tetrachloride, chloroform, or carbon disulfide pose another problem. The most satisfactory general sampling technique for solids is to press them into a transparent disk made from *potassium bromide.* This is done by carefully grinding a small quantity of the solid sample (0.5–2.0 mg) together with 100 mg of completely dry potassium bromide (KBr). The grinding operation is usually carried out with a polished mortar and pestle made of agate or some other nonporous material or by vibrating the mixture in a small ball mill similar to that used by dentists in mixing amalgam fillings. The mixture is pressed into transparent disks with special dies (Figure 1.10).

This mixture is subjected either to 14,000–16,000 psi in a high-pressure disk die or to hand pressure in the minipress shown in

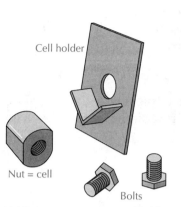

Cell holder

Nut = cell

Bolts

(a) Minipress components and holder

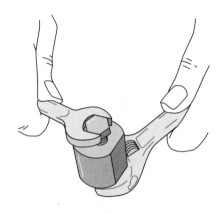

(b) Squeezing the KBr pellet in the press

FIGURE 1.10 Aids in the preparation KBr pellets.

Figure 1.10b. Although KBr disks are potentially excellent methods for IR analysis (no solvent interference), their preparation is challenging and requires great care. For example, the smallest traces of water in the disk can disrupt homogeneous sample preparation and also produce spurious O—H peaks in the spectrum. Modern FT IR instruments, as mentioned earlier, often allow analysis of a sample simply dispersed on a KBr bed.

Many modern FT IR instruments preclude the necessity of making KBr disks. Instruments that are now available will detect 1 mg of sample simply sprinkled on the surface of dry KBr.

Other common techniques for obtaining IR spectra also use mulls. *Mulls* are not true solutions but are dispersions of organic compounds either in a solid such as anhydrous potassium bromide (KBr) or in liquids such as Nujol (an oily mixture of long alkanes), Fluorolube (fluorinated, chlorinated alkanes), and hexachlorobutadiene. The halogenated mulling substances are used for more polar compounds. Mulls are normally used for solid compounds; these should be ground in a mortar and pestle until the sample is highly dispersed within the mortar and has a caked, glassy appearance. To a sample of 15–20 mg of ground solid is added 1–2 drops of mulling liquid. This mixture is ground for a few minutes to make a margarine-like paste. This paste is transferred to plates such as those shown in Figure 1.9b or c, and these are gently pressed together and put in a sample holder (see V-device of Figure 1.10a), which is placed in the beam of an IR spectrometer. The disk is simply placed on the V such that the hole through which the IR beam passes is covered.

Routinely, Nujol and Fluorolube display bands that, of course, cover peaks due to the compound dispersed in these liquids.

1.5
Interpreting Spectra

Confirming the identity of a compound is one of the most important uses of IR spectrometry. A compound thought to have a certain structure will have an IR spectrum *superimposable* on that of an authentic sample of the known compound. No two different compounds are known to have identical IR spectra. However, all too often, one cannot count on knowing enough about the identity of a compound to choose an authentic sample for comparison. Therefore, it becomes necessary to interpret the spectrum.

Interpreting IR spectra is an acquired skill. If the spectrum is adequately resolved, the peaks of the proper intensity, and the spectrometer calibrated so that the position of the peaks is recorded accurately, the spectrum can be enormously useful in assigning the compound's proper structure. A precise analysis of all the vibrations and modes of motion is not usually feasible, except in the simplest compounds; therefore, the spectra are analyzed empirically.

IR absorption frequencies are characterized according to the functional group absorptions given in Figure 1.11. Because the absorptions of some groups vary a good deal and are frequently weak intensity, this absorption frequency table is useful only within a certain restricted set of guidelines. One has to learn which regions of the IR spectrum are most useful and interpret the spectral information accordingly.

The IR spectrum can be broken down into three regions, whose titles are nearly self-explanatory:

Functional group region	$4000-1300$ cm^{-1}
Fingerprint region	$1300-900$ cm^{-1}
Aromatic region	$900-400$ cm^{-1}

The functional group region ($4000-1300$ cm^{-1}) provides unimpeded, reasonably strong bands for most major functional groups. The fingerprint region ($1300-900$ cm^{-1}) is normally so complex that it can be used for no more than fingerprint identification. That is, the identity of an unknown compound can be established by matching the fine structures of its IR spectrum with the spectrum of a known compound. The aromatic region ($900-400$ cm^{-1}) provides useful information for most benzenoid and nonbenzenoid aromatic compounds.

Keep in mind that the absence of IR bands is also useful information. The absence of substantial bands in the $900-400$ cm^{-1} region rules out aromatic compounds. The absence of appropriate IR bands in the $4000-1300$ cm^{-1} region argues against alkenes; alkynes; nitriles, amines, and amides with at least one N—H

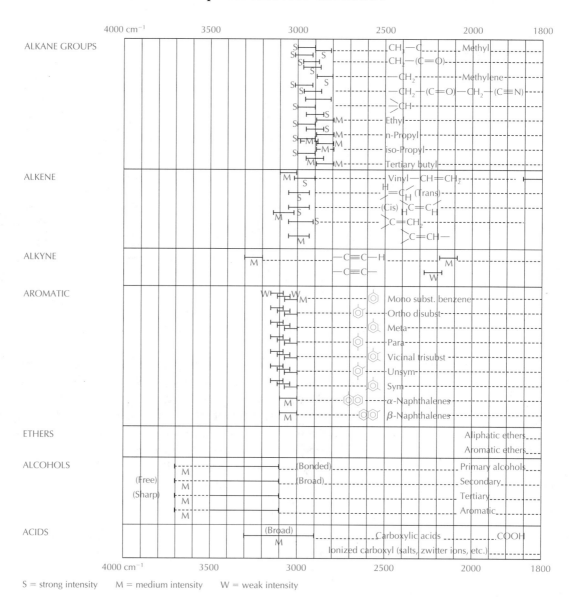

FIGURE 1.11 Characteristic infrared absorption peaks of the functional groups.

bond; alcohols and phenols; carboxylic acids; aldehydes; and ketones. This conclusion is justified because $C=C$, $C≡C$, $C≡N$, $N—H$, $O—H$, and $C=O$ bonds always show IR bands in this region.

A definitive region of the IR spectrum is that between 4000 and 3000 cm^{-1}. The stretching frequencies of $O—H$ and $N—H$ bonds

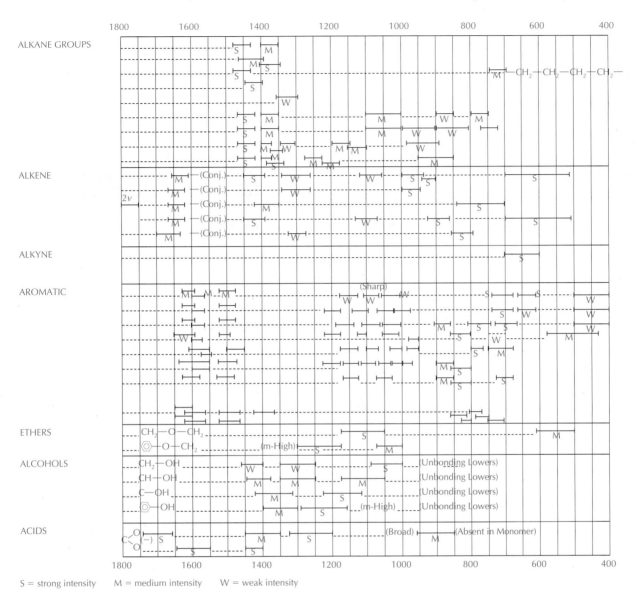

S = strong intensity   M = medium intensity   W = weak intensity

FIGURE 1.11 (continued) Characteristic infrared absorption peaks of the functional groups.

occur in this region. There are discernable IR bands for an oxygen-hydrogen bond in an alcohol, phenol, or carboxylic acid, or nitrogen-hydrogen bonds in primary or secondary amines. Thus, each of these compounds shows some absorption above 3000 cm^{-1}. The form of the absorption is highly varied, being sharp and strong in phenol, sharp and weak in diethylamine, and very broad in the

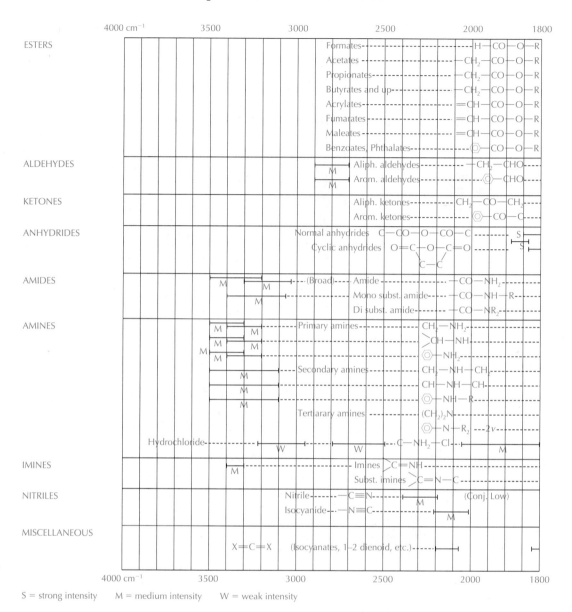

FIGURE 1.11 (continued) Characteristic infrared absorption peaks of the functional groups.

carboxylic acids. In fact, with carboxylic acids, the O—H stretching absorption may tail all the way to 2500 cm^{-1}. The reason for this variation is that compounds containing O—H and N—H bonds are capable of hydrogen bonding. The higher the concentration of the compound and the more capable the compound is of hydrogen bonding, the broader is the absorption. Hydrogen bonding also causes the O—H and N—H stretching absorptions to appear at lower wavenumbers.

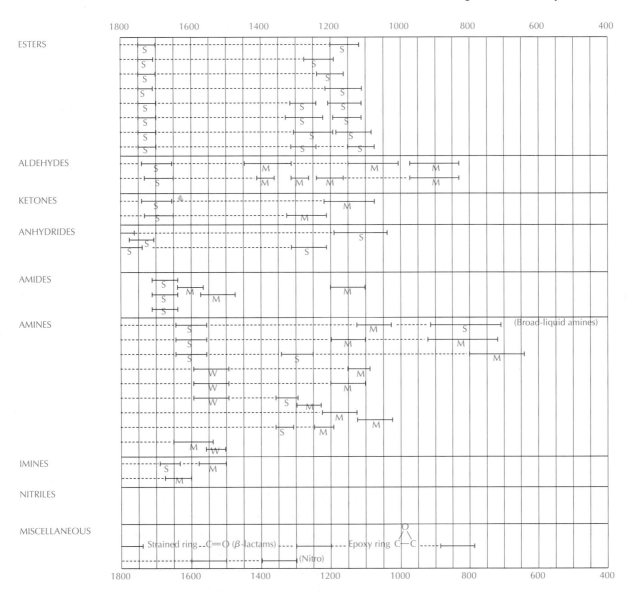

FIGURE 1.11 (continued) Characteristic infrared absorption peaks of the functional groups.

Many other functional groups also show specific absorptions in the IR spectrum. Nitriles (acetonitrile, for example) have strong absorptions around 2240 cm^{-1}. An absorption found in that region, therefore, is likely to be that of a nitrile, because few other functional groups absorb in this region. Alkynes absorb around 2150 cm^{-1}, although their absorptions are likely to be quite weak.

To see how IR spectra can be used to characterize organic compounds, we will survey a cross section of compounds containing

$CH_3CH_2CH_2CH_2CH_3$

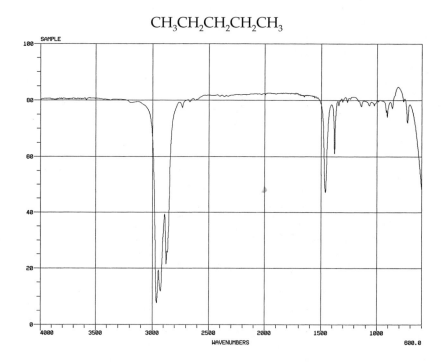

FIGURE 1.12 IR spectrum of pentane (FT IR mode). Capillary film between salt plates. Reproduced from "A Spectrum of Spectra," by R. A. Tomasi, Sunbelt R&T Pub., Tulsa, OK, 1992, with permission.

standard functional groups and comment on the appearance of the appropriate portions of the spectra. In the IR spectrum of pentane (Figure 1.12), we see only absorptions due to C—H bonds. There is some detail in the C—H stretching region (3000–2800 cm^{-1}) because there are both methyl and methylene C—H bonds, and there also are coupled interactions within these groups. The methyl shows both symmetric and asymmetric coupling: **symmetric coupling** means that the three hydrogens all move away from the carbon simultaneously; in **asymmetric coupling,** while two hydrogens move out, one moves in (Figure 1.13). Coupled vibrations like these help identify functional groups including amines, amides, anhydrides, nitro compounds, and sulfones.

The rest of the spectrum of pentane is made up of absorptions due to various C—H bending modes. The IR spectrum of 2,2,4-trimethylpentane (isooctane) is quite similar to that for pentane. A feature here is the absorption pattern in the 1400–1340 cm^{-1} region. When two methyl groups are on the same carbon (called a *gem*-dimethyl group, from *gemini*, meaning "twins"), a pair of peaks is expected: The isopropyl group shows one peak in the 1385–1380 cm^{-1} region and the other in the 1370–1365 cm^{-1}

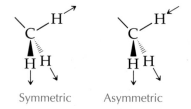

Symmetric   Asymmetric

FIGURE 1.13 Coupled vibrations of the methyl group.

region; the *t*-butyl group shows one peak in the 1380 cm^{-1} region and the other in the 1370 cm^{-1} region. These first two spectra typify "aliphatic" IR spectra and have features (C—H stretching near 3000 cm^{-1} and bending near 1300 cm^{-1}, etc.) that we expect for most organic compounds bearing alkyl groups.

Unsaturated hydrocarbons have IR spectra that include the usual alkyl bands just as described and also absorptions indicative of double or triple bonds. Alkenes (olefins) show medium to weak peaks between 1667 and 1640 cm^{-1} that correspond to C=C stretch. If the C=C unit bears a proton, its stretching vibration normally gives a medium intensity band near 3000 cm^{-1}. Alkynes (acetylenes) show C≡C stretch between 2260 and 2100 cm^{-1}, and for terminal alkenes, ≡C—H stretch usually causes a strong peak at approximately 3300 cm^{-1}.

In the IR spectrum of toluene (Figure 1.14), we find that the C—H stretching absorptions distribute themselves on both sides of the 3000 cm^{-1} line. Those above 3000 cm^{-1} (3150–3000 cm^{-1}) correspond to the stretching of C—H bonds on the *sp*^{2} carbon hybrids in the benzene ring; those below 3000 cm^{-1} (3000–2800 cm^{-1}) correspond to the stretching of C—H bonds on the *sp*^{3} carbon hybrids on the alkyl side group. A series of four small

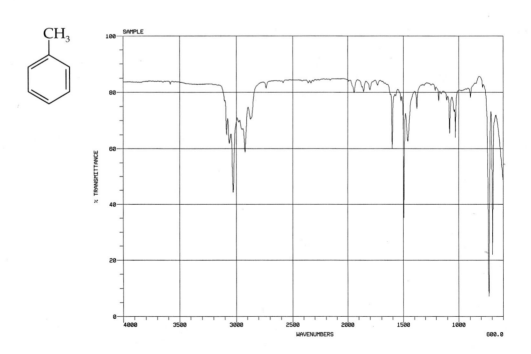

FIGURE 1.14 IR spectrum of toluene (FT IR mode). Reproduced from "A Spectrum of Spectra," by R. A. Tomasi, Sunbelt R&T Pub., Tulsa, OK, 1992, with permission.

peaks in the 2000–1650 cm^{-1} region are overtone bands and are typical of monosubstituted benzenes. The substantially stronger peaks in the 1700–1400 cm^{-1} region are due to the stretching of ring C$\doteq$C bonds; there are a number of bands in this region due to the various possibilities of symmetric and asymmetric coupling between these bonds. There are two strong aromatic bands between 700 and 750 cm^{-1}, indicating a monosubstituted benzene ring. Although it is enticing to expect great utility for both the 2000–1650 cm^{-1} (overtone) and 900–600 cm^{-1} (aromatic) region, there are severe limitations. In the former, it is frequently the case that a strong band (say, a carbonyl stretch near 1700 cm^{-1}) obliterates the weak overtone bands, and the aromatic pattern below 900 cm^{-1} is unreliable whenever there are polar functional groups on the benzene ring.

Ethanol (Figure 1.15) shows standard aliphatic features: C—H stretching at 3000–2800 cm^{-1} (C$_{sp3}$—H stretch) and various C—H bending bands at lower wavenumbers. The polar hydroxyl functional group causes strong IR bands. For example, the broad strong band from 3600 to 3100 cm^{-1} is due to O—H stretch. The position and breadth is due to self-association by hydrogen bonding. If the spectrum of an alcohol is recorded from a dilute solu-

CH_3CH_2OH

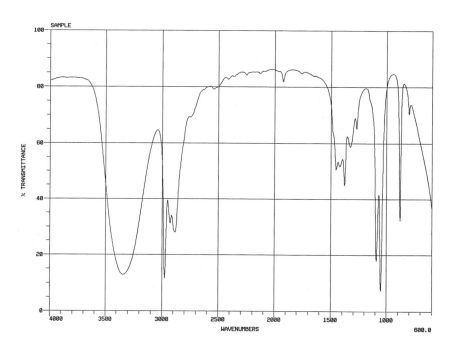

FIGURE 1.15 IR spectrum of ethanol (FT IR mode). Reproduced from "A Spectrum of Spectra," by R. A. Tomasi, Sunbelt R&T Pub., Tulsa, OK, 1992, with permission.

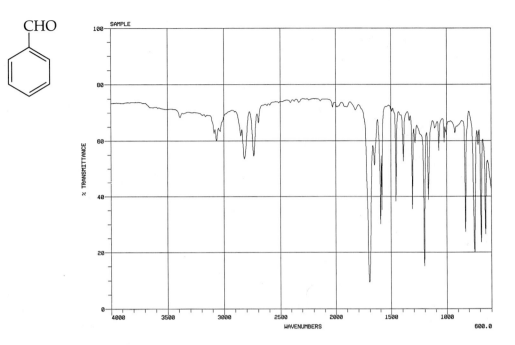

FIGURE 1.16 IR spectrum of benzaldehyde (FT IR mode). Reproduced from "A Spectrum of Spectra," by R. A. Tomasi, Sunbelt R&T Pub., Tulsa, OK, 1992, with permission.

tion or a gas-phase sample, there is simply one sharp peak at approximately 3650 cm^{-1}. A strong peak for C—O stretch is usual for almost any class of organic compounds, and here it occurs at 1050 cm^{-1}. Diethyl ether also shows standard aliphatic features: C—H stretching (3000–2800 cm^{-1}, Csp^3—H stretch) and various C—H bending bands (1500–1300 cm^{-1}). The strong peak at 1130 cm^{-1} clearly indicates that this compound is an ether; it is the asymmetric C—O—C stretch. Symmetric C—O—C stretching of ethers yields only weak bands that are not useful. Pure and dry diethyl ether should show no O—H bands.

In the IR spectrum of benzaldehyde (Figure 1.16), we find the ring C—H stretching absorptions at 3150–3000 cm^{-1}. The aldehyde functional group is revealed in two regions, the C=O (carbonyl) stretch at 1700 cm^{-1} and the aldehydic C—H stretching at 2800 and 2720 cm^{-1}. The spectra of ketones should show no such aldehydic stretching. The stronger peaks in the 1650–1200 cm^{-1} region of benzaldehyde are due to the stretching of ring C$\dot{=}$C bonds. The spectra of ketones should show no such aldehydic stretching.

Acetic acid (Figure 1.17) reveals the features of both its carbonyl and hydroxyl groups: a strong carbonyl (C=O) stretch at

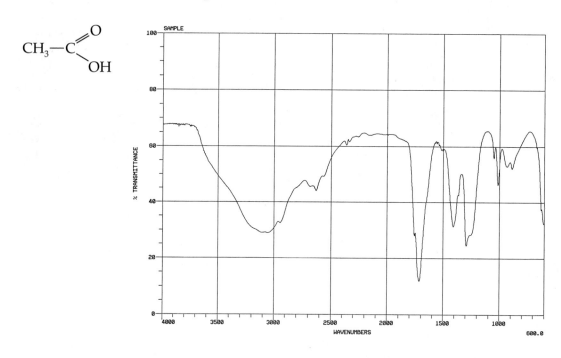

FIGURE 1.17 IR spectrum of acetic acid (FT IR mode). Reproduced from "A Spectrum of Spectra," by R. A. Tomasi, Sunbelt R&T Pub., Tulsa, OK, 1992, with permission.

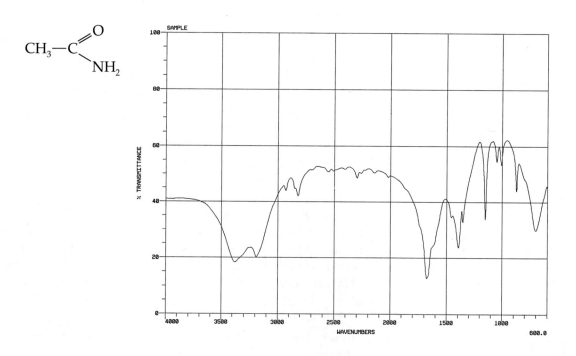

FIGURE 1.18 IR spectrum of acetamide (FT IR mode). Reproduced from "A Spectrum of Spectra," by R. A. Tomasi, Sunbelt R&T Pub., Tulsa, OK, 1992, with permission.

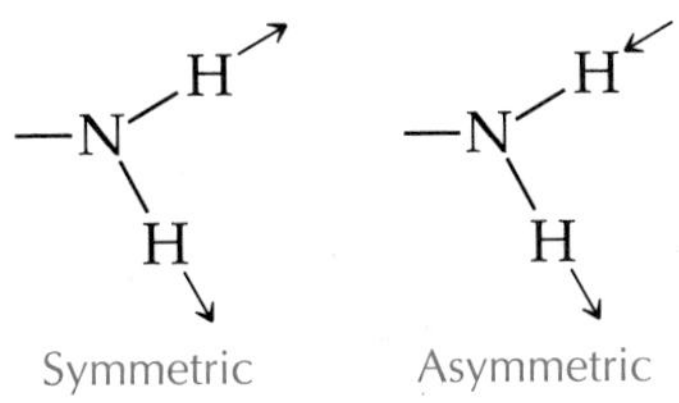

Symmetric   Asymmetric

FIGURE 1.19 Coupled vibrations of the amine group.

1710 cm^{-1} and a broad, strong O—H band (hydrogen-bonded O—H stretch) from 3700 to 2400 cm^{-1}. This can be compared to an ester such as ethyl acetate. Although ethyl acetate shows no O—H stretch, it does show a strong carbonyl stretch (1735 cm^{-1}) and two strong C—O stretches: an "acid" C—(C=O)—O stretch at 1240 cm^{-1} and an alcohol O—C—C stretch at 1050 cm^{-1}.

As we might expect, N—H stretching bands are in essentially the same region as O—H stretch. The IR spectrum of acetamide (Figure 1.18) shows two N—H stretching bands, one each at 3400 and 3180 cm^{-1} due to, respectively, asymmetrically and symmetrically coupled N—H stretches (Figure 1.19).

The carbonyl band (sometimes called the "amide I band") is at 1694 cm^{-1}, and an N—H band occurs at 1620 cm^{-1} as a shoulder on the amide I band. N—H bands can reveal whether an amine (or amide) is primary, secondary, or tertiary. A primary amine (1-aminobutane, or butylamine) shows two N—H bands, one at 3350 cm^{-1} (asymmetric coupling) and one at 3290 cm^{-1} (symmetric coupling). The secondary amine diethylamine shows one N—H band at 3300 cm^{-1}. Finally, the tertiary amine triethylamine shows no peak in the N—H (3500–3000 cm^{-1}) region.

In reality, determining the class of amine by IR can be misleading. Tertiary amines (and quaternary ammonium salts) are often hygroscopic and thus may show water peaks in the 3300 cm^{-1} region. Primary amines may show only one peak if the resolution is inadequate (for example, because of dissolved water), and, finally, primary amines sometimes show three or more N—H peaks that are due to various degrees of hydrogen bonding or other molecular environment variations.

In summary, IR analysis is a way to identify organic functional groups. In addition, some structural information can be obtained from IR, although, as we will see later, NMR is usually a better way of obtaining structural information. Finally, IR spectra are useful fingerprints for confirming the identity of organic compounds. Comparing IR spectra obtained in the laboratory to the spectra available in the Aldrich collection is an excellent way to identify a compound.

References

1. Pouchert, C. J. (Ed.) *The Aldrich Catalog of Infrared Spectra*; Aldrich Chemical Company: Milwaukee, 1975.
2. Bellamy, L. J. *Advances in Infrared Group Frequencies*; Methuen: London, 1968.
3. Bellamy, L. J. *The Infrared Spectra of Complex Molecules*, 3rd ed.; Halsted-Wiley: New York, 1975.
4. Colthup, N. B.; Daly, L. H.; Wiberly, S. E. *Infrared Absorption Spectroscopy—Practical*, 3rd. ed.; Academic: New York and London, 1990.

Questions

1. It is common to use a polystyrene window to calibrate an IR spectrometer. Some important IR peaks for polystyrene are at 2924, 1601, 1493, 757, 698 cm^{-1}. Assign these bands. It might be useful to review the discussions of Figures 1.12 and 1.14 (IR spectra, respectively, of pentane and toluene).

Polystyrene

2. In each of the sets below, match the proper compound with the appropriate set of IR bands.
   a. dodecane, 1-decene, 1-hexyne, *o*-xylene
      3311(s), 2961(s), 2119(m) cm^{-1}
      3020(s), 2940(s), 1606(s), 1495(s), 741(s) cm^{-1}
      3049(w), 2951(m), 1642(m) cm^{-1}
      2924(s), 1467(m) cm^{-1}
   b. phenol, benzyl alcohol, anisole
      3060(m), 2835(m), 1498(s), 1247(s), 1040(s) cm^{-1}
      3370(s), 3045(m), 1595(s), 1224(s) cm^{-1}
      3330(bs), 3030(m), 2980(m), 1454(m), 1023(s) cm^{-1}

   bs = broad, strong band

   c. 2-pentanone, acetophenone, 2-phenylpropanal, heptanoic acid, 2-methylpropanamide, propanoic anhydride, phenyl acetate, octyl amine
      3070(m), 2978(m), 2825(s), 2720(m) 1724(s) cm^{-1}
      3372(m), 3290(m), 2925(s) cm^{-1}
      2987(m), 1818(s), 1751(s), 1041(s) cm^{-1}
      3070(w), 1765(s), 1215(s), 1193(s) cm^{-1}
      3300–2500(bs), 2950(m), 1711(s) cm^{-1}
      3060(m), 2985(w), 1685(s) cm^{-1}
      3352(s), 3170(s), 2960(m), 1640(s) cm^{-1}
      2964(s), 1717(s) cm^{-1}

3. Consider the IR spectra shown in Figure 1.20 (Compounds A through H) and match them to the following compounds: biphenyl, 4-isopropylmethylbenzene, 1-butanol, phenol, 4-methylbenzaldehyde, ethyl propanoate, benzophenone, acetamide.

4. Review the following and comment on how the sample preparation techniques can affect the appearance of the high wavenumber end of the IR spectrum.
   a. The IR spectrum of a thin film of 1-pentanol reveals a single broad and strong band centered at 3300 cm^{-1}, whereas the spectrum of a dilute chloroform solution of the same compound shows both the same broad band and a sharp spike at 3650 cm^{-1}.

Compound A

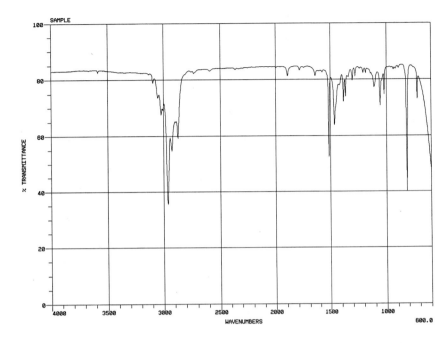

Compound B

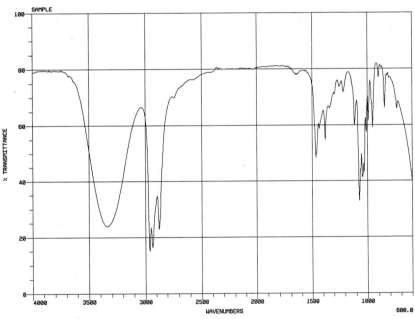

FIGURE 1.20 IR spectra for Question 3. Reproduced from "A Spectrum of Spectra," by R. A. Tomasi, Sunbelt R & T Pub., Tulsa, OK, 1992, with permission.

b. IR analysis of *o*-hydroxyacetophenone shows the same appearance in a wide range of concentrations in chloroform: a broad band centered at 3080 cm^{-1}.

c. When simple alcohols are subjected to analysis in any kind of mulling compound (Nujol, KBr, Fluorolube), the IR spectrum

Compound C

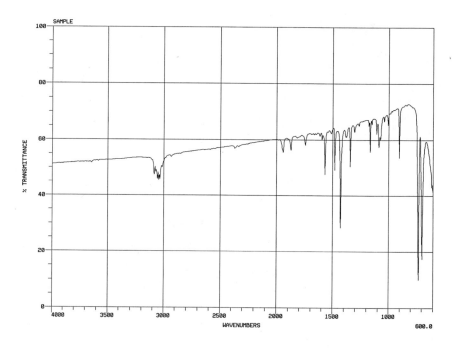

Compound D

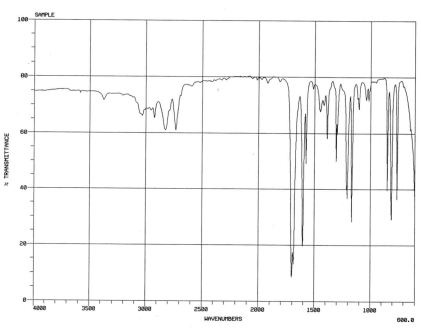

FIGURE 1.20 (continued) IR spectra for Question 3.

shows essentially the same broad strong band centered at approximately 3300 cm^{-1}, whereas dilute solutions in a chlorinated hydrocarbon solvent might show the same 3300 cm^{-1} band, and very often show a new, sharp peak at approximately 3650 cm^{-1}.

Compound E

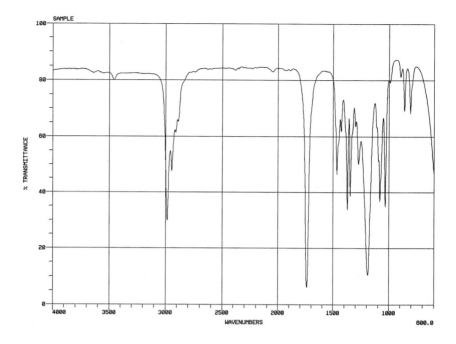

Compound F

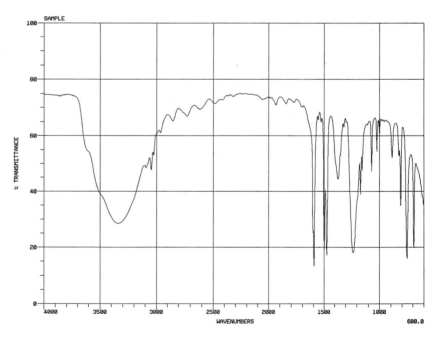

FIGURE 1.20 (continued) IR spectra for Question 3.

5. Treatment of cyclohexanone with sodium borohydride results in a product that can be separated by distillation. The IR spectrum of this product is shown in Figure 1.21. Identify the product and assign the major IR bands.

Compound G

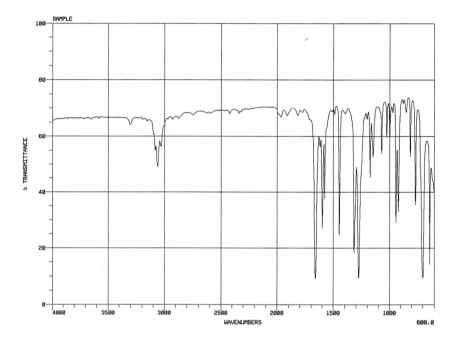

Compound H

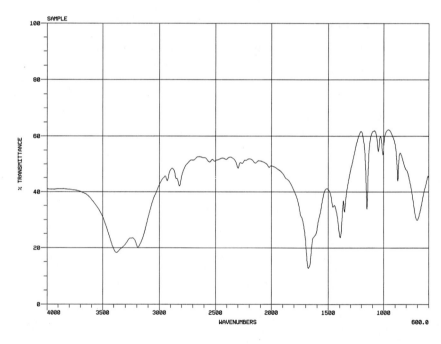

FIGURE 1.20 (continued) IR spectra for Question 3.

6. Heating cyclohexanol with a trace of sulfuric acid, followed by distillation, results in a product of substantially greater volatility than cyclohexanol. The IR spectrum of this product is shown in Figure 1.22. Identify the product and assign a few major IR bands.

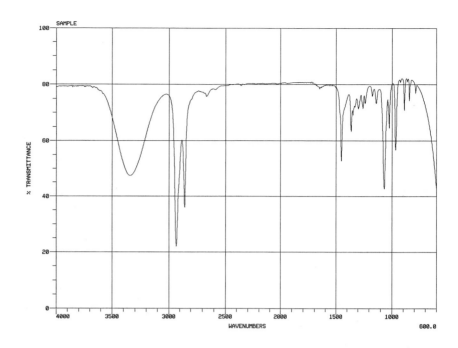

FIGURE 1.21 IR spectrum of product for Question 5. Reproduced from "A Spectrum of Spectra," by R. A. Tomasi, Sunbelt R & T Pub., Tulsa, OK, 1992, with permission.

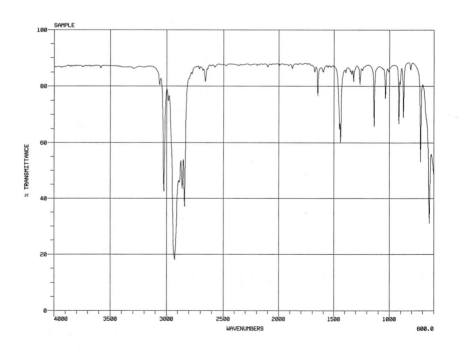

FIGURE 1.22 IR spectrum of product for Question 6. Reproduced from "A Spectrum of Spectra," by R. A. Tomasi, Sunbelt R & T Pub., Tulsa, OK, 1992, with permission.

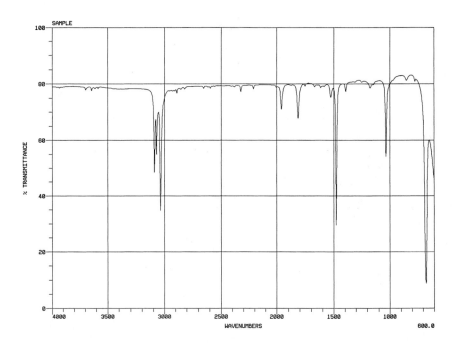

FIGURE 1.23 IR spectrum of product boiling at 80°C for Question 7. Reproduced from "A Spectrum of Spectra," by R. A. Tomasi, Sunbelt R & T Pub., Tulsa, OK, 1992, with permission.

7. When benzene is treated, in the presence of aluminum chloride, with chloroethane (ethyl chloride), the product is expected to be ethylbenzene (bp 136°C). During the isolation of this product by distillation, some liquid of bp 80°C was obtained. Identify this product, using its boiling point and the IR spectrum in Figure 1.23.

8. In principle, matched solution cells should remove IR bands due to the solvent and reveal peaks due to solute (sample) in the region where solvents absorb (Table 1.1). In practice, although solvent band removal is easily achieved, it is often difficult to discern the peaks due to the sample. Explain.

Spectrometric Method 2

NUCLEAR MAGNETIC RESONANCE SPECTROMETRY

Nuclear magnetic resonance (NMR) spectrometry is one of the most important modern instrumental techniques used in the determination of molecular structure. The first measurements of the absorption of energy by atomic nuclei in a varying magnetic field were reported independently by the American physicists

Felix Bloch and Edward Purcell in 1946. In 1952 they shared the Nobel prize in physics for this work. Since that time, nuclear magnetic resonance has been in the forefront of the spectrometric techniques that have completely revolutionized organic structure determination. Like other spectrometric techniques, NMR depends on the quantized energy changes that can be induced in simple molecules when they are struck by electromagnetic radiation. The energy requirements of NMR (10^{-6} kJ·mol^{-1}) are low relative to those of other spectrometric techniques (infrared radiation is in the 10^{-4} kJ·mol^{-1} range and ultraviolet in the 160–1300 kJ·mol^{-1} range.)

Any isotope whose nucleus has a nonzero magnetic moment is detectable by NMR spectrometry. These include ^{1}H, ^{2}H, ^{13}C, ^{15}N, ^{19}F, and ^{31}P, among others. Here we will focus on ^{1}H and ^{13}C NMR.

The physical model that explains the behavior of a spinning nucleus in an applied field comes directly from quantum mechanics (Figure 2.1). The precession of a child's top about a vertical axis can be used as a mechanical model for this process, and this precession will describe the upper circle in Figure 2.1a. The vertical axis corresponds to the instrument's magnetic field, and nuclei with nonzero magnetic moments will precess about this axis. In Figure 2.1 the vector at a 45° upward angle is aligned with the applied magnetic field B_0. The orientation of the 45° vector in Figure 2.1b is aligned against the field.

Both H_0 and B_0 have been used for instrument or applied magnetic field. In this book we use B_0.

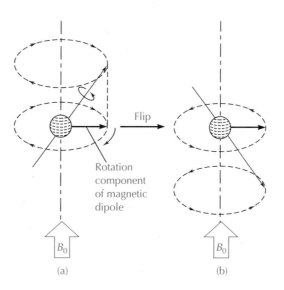

(a) (b)

Rotation component of magnetic dipole

B_0 B_0

Flip

FIGURE 2.1 Resonance condition. Adapted from D. C. Neckers and M. P. Doyle, *Organic Chemistry,* J. Wiley and Sons, New York, 1977.

Precession of a nucleus about the applied magnetic field is a quantized phenomenon, and the absorption of the excitation frequency of radio waves can cause a transition to a higher energy level. If the applied frequency (ν) is precisely tuned to the precessional angular frequency (the rotational frequency of the vector that describes the dashed-lined circles in Figure 2.1), the system is said to be in resonance and the nucleus will absorb energy and flip to a position opposed to the magnetic field (Figure 2.1b). The energy of this transition is extremely small by chemical standards, being only about $10^{-6}\,\text{kJ}\cdot\text{mol}^{-1}$. NMR signals correspond to measurement of this energy of transition. Nuclei give rise to various signals at different frequencies because of their structural environment in an organic compound.

2.1
NMR Instrumentation

The first NMR instruments included a radio-frequency transmitter used to continuously sweep a sample with incrementally varied energy until a match was obtained between this energy and the magnitude of the energy difference between nuclei in the two states shown in Figure 2.1. When an exact match was obtained, resonance caused a signal to occur at a frequency indicative of the environment of the nucleus (usually ^{1}H or ^{13}C) of interest. A radio-frequency receiver was used to absorb the energy caused by this spin state change. Because the transmitter signals changed incrementally (over a small range of energies), we refer to this as continuous wave (cw) NMR.

Modern instruments use the technique known as pulsed-Fourier Transform NMR (PFT NMR). In this technique, a broad pulse of electromagnetic radiation excites all the nuclei simultaneously and this results in a continuously decreasing oscillation (Figure 2.2a) caused by the decay of the excited nuclei back to a stable (or Boltzmann's) energy distribution. This stable distribution contains a slightly greater population of the lower energy state, that is, the state to the left in Figure 2.1. The oscillating, or decaying sine curve is called a *free-induction decay.* The frequency of the simple FID shown in Figure 2.2a may be read by simply counting the number of times the curve crosses an imaginary midpoint axis (horizontally centered in the oscillation) during complete decay. Imagine the axis to be horizontal and centered in the decaying sine curve. One cycle requires two crossings of this axis, and the FID for iodomethane (methyl iodide) shown in Figure 2.2a undergoes 92 crossings (46 cycles) over the 2 seconds required for complete decay; thus we have a signal that occurs at $(46/2) = 23$ cycles per second (23 Hertz, Hz; Figure 2.2b). Most organic compounds are much more complex than iodomethane, and a computer using Fourier mathematics converts the FID (a

FID-free induction decay

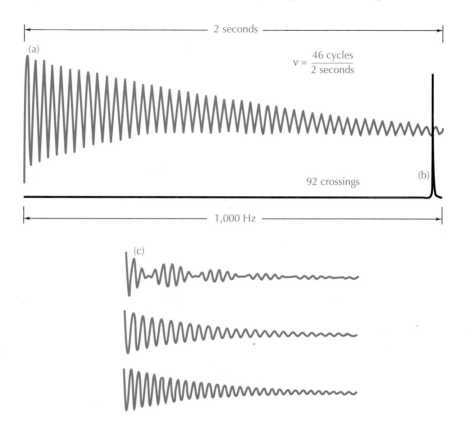

$$\nu = \frac{46 \text{ cycles}}{2 \text{ seconds}}$$

FIGURE 2.2 (a) Time domain—free induction decay (FID). (b) Frequency domain—spectrum (CH_3I, ^1H decoupled). (c) The middle FID is for a chloroform sample and the bottom FID is for a benzene sample. The top FID is for a sample of a mixture of 85% chloroform and 15% benzene.

time domain signal) to the "normal" (or frequency domain) signal shown in Figure 2.2b. In Figure 2.2c, the FID for chloroform also shows a simple pattern because there is only one type of carbon in this compound. The same degree of simplicity is seen in the FID for benzene (Figure 2.2c).

When two carbon environments are combined in one molecule, however, a more complex signal arises for the two carbons: constructive combinations of the two signals give rise to positive signals and destructive combinations give little or no signal. One organic structure alone can often provide many different kinds of carbons (or protons). Thus the FID of a compound containing many nuclei is very complex. Also in practice we find that the acquisition of the signal from more than one pulse (or "scan") is necessary to obtain NMR signals of substantial magnitude, especially when the size of the sample is small. Thus another use of the NMR computer is programming multiple pulses and collecting

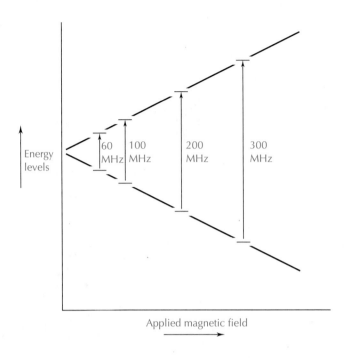

Energy levels

60 MHz   100 MHz   200 MHz   300 MHz

Applied magnetic field

FIGURE 2.3 Plot of the magnitude of splitting of nuclei of spin $\frac{1}{2}$ (for example, ^1H, ^{13}C) versus the applied or instrument magnetic field.

the data from these pulses. The computer is an indispensable part of the NMR spectrometer.

There are many different kinds of NMR spectrometers. In earlier instruments, a frequency of 60 megahertz (MHz) was used for ^1H NMR, thus requiring a magnetic field (B_0) of 1.4092 Tesla. Now it is common to use NMR instruments of substantially higher field strengths. Research instruments routinely have 200 to 300 MHz (7.046 Tesla) fields, and a number of centers have instruments with 400, 500, 600, and even 750 MHz fields. The benefit of higher field strength is suggested by Figure 2.3; higher field means larger energy differences between nuclei and thus greater instrument resolution, or signal separation. Moreover, higher field strengths mean higher sensitivity, which translates into a stronger signal relative to background noise. Higher field strengths have been achieved by the application of electrical superconductivity at very low temperatures. Magnets cooled by liquid helium readily obtain the desired high field strengths.

2.2 Chemical Shift

A nuclear magnetic resonance spectrum is a plot of the intensity of various signals versus magnetic field, or frequency. Because it is difficult to reproduce magnetic fields exact enough for NMR on an

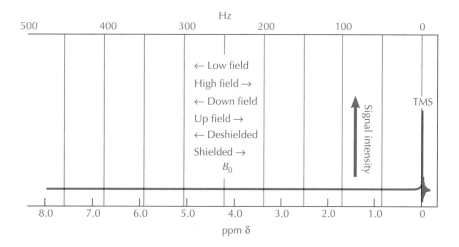

IGURE 2.4 ^{1}H NMR spectrum of tetramethylsilane (TMS, 60 MHz instrument) and terms used in NMR.

absolute basis, an internal standard is used. The position of the absorption, called the ***chemical shift,*** is measured relative to the absorption of a reference standard material. The standard is tetramethylsilane [$(CH_3)_4Si$, or TMS]. Nuclei that are chemically equivalent, such as the 12 protons in tetramethylsilane, the four protons in methane, or the two protons in dichloromethane, show only one absorption in the NMR spectrum of each. However, protons that are not chemically equivalent absorb at different frequencies. Terms that are used to describe chemical shift phenomena are displayed in Figure 2.4. Chemical shifts can be measured at a frequency (Hz) corresponding to the sample signal's position relative to TMS. It is conventional, however, to convert frequency to a value of δ (ppm) by dividing frequency by the field strength of the spectrometer:

Upfield is to the right in an NMR spectrum (lower δ value). Downfield is to the left.

$$\delta \text{ (ppm)} = \frac{\text{frequency of signal of interest (in Hz, from TMS)}}{\text{applied frequency (instrument field strength, MHz)}} \times 10^6$$

Thus a signal at 60 Hz on a 60 MHz spectrometer occurs at 1.00 ppm:

$$\delta = \frac{60\text{ Hz}}{60\text{ MHz}} \times 10^6 = 1.00$$

This signal would also be at 1.00 on 200, 300, or 500 MHz spectrometers. In other words, on the 300 MHz spectrometer, the signal would occur at a frequency of 300 Hz. This relationship makes clear the value of high field strength for resolution (see Figure 2.5): Two well-defined signals separated on a 60 MHz instrument by

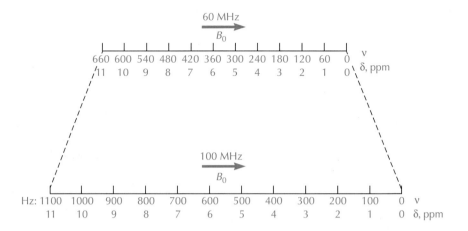

FIGURE 2.5 ^{1}H NMR scale at 60 MHz and 100 MHz. Adapted from R. M. Silverstein, G. C. Bassler, and T. C. Morrill, *Spectrometric Identification of Organic Compounds,* 5th ed., J. Wiley and Sons, New York, 1991.

10 Hz will be separated by $(100/60)10 = 16.67$ Hz on a 100 MHz instrument, and by $(300/60)(10)$, or 50 Hz on a 300 MHz instrument. Increases in peak separation such as these mean enhanced signal resolution with instruments of higher field strength.

TMS is arbitrarily said to have nuclei that absorb at zero ($\delta = 0.00$). In practice, FT NMR spectra are calibrated by "locking" on a solvent nucleus (typically, deuterium in heavy water, or $CDCl_3$ or other labeled solvent); the instrument "scan," however, is still scaled to set TMS at $\delta = 0.00$, even though TMS is normally no longer added to the sample.

There are two absorption peaks (in addition to the TMS standard) in the NMR spectrum of *tert*-butyl acetate (Figure 2.6). The nine equivalent methyl protons (a) in the *tert*-butyl group absorb at δ 1.45, whereas the three methyl protons (b) appear at δ 1.97; and these numerical values of δ apply no matter what the field strength of the instrument.

The chemical shift of a proton is strongly influenced by its atomic neighbors because the chemical shift of a proton depends on the electron density around the nucleus. Circulating electrons in the electron cloud of an atom close to the proton induce a magnetic field, in a direction opposite to that of the applied field. Thus, the actual magnetic field that the proton feels is less than the applied field, and the electron cloud is said to shield the nucleus from the magnetic field (Figure 2.7).

Decreasing the electron density around a particular nucleus has the effect of deshielding that nucleus from the applied magnetic field (B_0). With deshielding, a slightly lower applied field will cause NMR absorption. For instance, in substituted methane

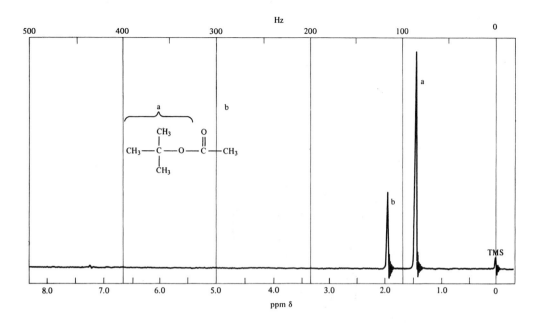

FIGURE 2.6 ^{1}H NMR spectrum of tert-butyl acetate measured at 60 MHz. The relative areas (9 to 3, or 3 to 1) of peaks a and b correspond to the ratio of protons (9 to 3) causing these peaks.

derivatives (Figure 2.8) such as methyl chloride, dichloromethane, and chloroform, the chlorine atoms withdraw electrons from the carbon atom to which the protons are attached. The electronegative chlorine atoms, therefore, have the overall effect of decreasing the electron density around the protons in the molecule, and the

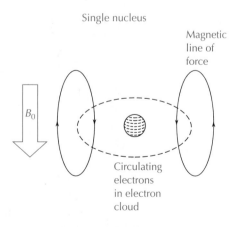

FIGURE 2.7 Magnetic lines induced in opposition to the applied field (B_0) due to shielding of the nucleus by circulating electrons.

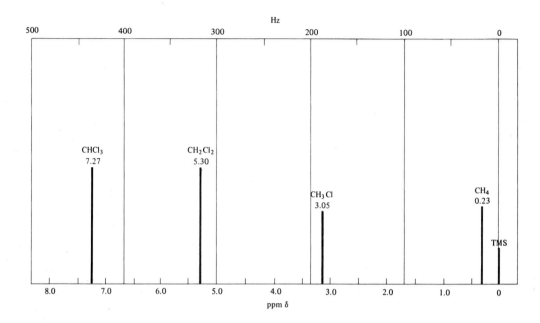

FIGURE 2.8 ^{1}H NMR chemical shifts for methane and chlorinated methanes. The lines mark the position of the centers of the signals.

protons are said to be deshielded. This effect increases with an increase in the number of attached chlorines.

Absorption of the four equivalent hydrogens of methane occurs at δ 0.23 (downfield, or to the left of the absorption of TMS at δ 0.00). Halogen deshielding causes the ^{1}H NMR absorptions of methyl chloride (CH_3Cl), dichloromethane (methylene chloride, CH_2Cl_2), and chloroform ($CHCl_3$) to occur at successively lower fields (Figure 2.8). The cumulative effect of the chlorine atoms is nearly additive, as demonstrated by the fact that each additional halogen induces a deshielding effect of approximately 2 ppm. For example, the chemical shift difference between $CHCl_3$ and CH_2Cl_2 is δ 7.27 − δ 5.30, or 1.97.

Examples of chemical shift ranges are illustrated in Figure 2.9. A more complete list of chemical shifts is found in Tables 2.1 and 2.2. In general, all the tabulated substituents are deshielding relative to the silicon of TMS, so the protons of virtually all compounds absorb at lower field than does TMS itself. Table 2.2 can be used to anticipate the deshielding effect of one or two functional groups on aliphatic protons.

You may notice in Table 2.1 that the protons attached to benzene rings absorb at an especially low field position. This effect is not accounted for simply by the localized deshielding that has been discussed but results from interactions of the π-electrons

FIGURE 2.9 Regions of chemical shifts of different kinds of protons. Adapted from R. M. Silverstein, G. C. Bassler, and T. C. Morrill, *Spectrometric Identification of Organic Compounds,* 5th ed., J. Wiley and Sons, New York, 1991.

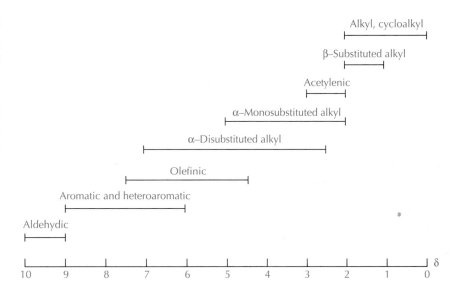

Table 2.1. Characteristic ^{1}H NMR chemical shifts

Compound	Chemical shift, ppm
TMS	0.00
Alkane[a]	0.20–1.90
Halides[a] (C$\underline{H}$X)[b]	2.10–4.10
Alkene (olefinic, C$\underline{H}$=C)	4.50–6.50
Aromatic (Ar—$\underline{H}$)	6.50–8.50
Aldehydic (C$\underline{H}$O)	9.0–10.0
Alcoholic (O$\underline{H}$)	0.5–4.0
	(4.2–6.2)[c]
Alkyne (C≡C—$\underline{H}$)	2.2–2.8
Alcohols, esters, ethers ($\underline{H}$CO)	3.0–4.0
Carboxylic acids, esters, anhydrides ($\underline{H}$CC=O)	1.9–3.0
Phenols (ArO$\underline{H}$)	4.0–7.5
Carboxylic acids (RCOO$\underline{H}$)	10.0–13.5
Amines (N$\underline{H}$)	0.5–3.0
Amides (N$\underline{H}$)	5.0–8.7
Anilines (N$\underline{H}$)	2.0–4.0

a. Tertiary hydrogens are at lower field than secondary hydrogens, which are at lower field than primary hydrogens. Table 2.2 shows more detail.
b. Halogens of higher electronegativity are more deshielding.
c. In DMSO.

Table 2.2. Chemical shifts (δ, ppm) of protons in various environments near functional groups

Group (Y)	CH$_3$Y	RCH$_2$Y	R$_2$CHY	YCH$_2$Y'[a]
—H	0.23	0.9	1.3	—
—CH=CH$_2$	1.71	2.0	1.7	1.32
—C≡CH	1.80	2.1	2.6	1.44
—C$_6$H$_5$	2.35	2.6	2.9	1.85
—F	4.27	4.4	4.8	—
—Cl	3.06	3.5	4.1	2.53
—Br	2.69	3.4	4.2	2.33
—I	2.16	3.2	4.2	1.82
—OH	3.39	3.5	3.9	2.56
—OR	3.24	3.3	3.6	2.36
$\overset{\text{O}}{\overset{\|}{-\text{OCCH}_3}}$	3.67	4.0	4.9	3.13
—CH=O	2.18	2.4	2.4	—
—(C=O)CH$_3$	2.09	2.4	2.5	1.70
—COOH	2.08	2.3	2.6	—
—NH$_2$	2.47	2.7	3.1	1.57
—NH(C=O)CH$_3$	2.71	3.2	4.0	2.27
—SH	2.00	2.5	3.2	1.64
—CN	1.98	2.3	2.7	1.70
—NO$_2$	4.29	4.3	4.4	—

a. This column provides data that can be used to calculate the expected chemical shifts for a methylene group bearing two functional groups. For example, we can use δ 0.23 for the shift of the bare methylene group, and the effect of attached chloro and phenyl groups would be to add, respectively, 2.53 and 1.85 ppm to that shift, causing the methylene group of benzyl chloride (C$_6$H$_5$—CH$_2$—Cl) to be expected at δ (0.23 + 2.53 + 1.85) or at δ 4.61.

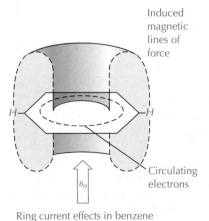

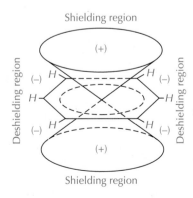

FIGURE 2.10 Ring current effects in aromatic compounds.

Ring current effects in benzene

(a) Induced magnetic fields

(b) Shielding/Deshielding areas

with the applied field. The protons are substantially deshielded because of the electron circulation set up in the aromatic ring by the applied field. This ring current induces a magnetic field, which deshields the aromatic protons (Figure 2.10). A similar effect takes place with alkenes (olefins), aldehydes, and carboxylic acids. In fact, the protons of aldehydes and carboxylic acids absorb even further downfield than do aromatic protons (Table 2.1). Protons attached to carbon-carbon double bonds absorb downfield from alkyl hydrogens, but not as far downfield as the corresponding aromatic protons. One can often determine at a glance whether a compound is aromatic, whether it is an alkene, whether it contains a CHO group, or whether it is a carboxylic acid from the regions of absorption in the NMR spectrum.

2.3
Spin-Spin Splitting

A simple ^{1}H NMR spectrum, that of chloroethane (ethyl chloride), is shown in Figure 2.11. The two signals of chloroethane show splitting due to spin-spin interactions. Two major absorptions are seen: a quartet centered at δ 3.57 and a triplet at δ 1.55. These multiplets are called *first-order* because they can be interpreted by a simple $N + 1$ rule. That is, we find $N + 1$ lines for a signal coupled by N equivalent nuclei. Thus the methylene protons (b) are

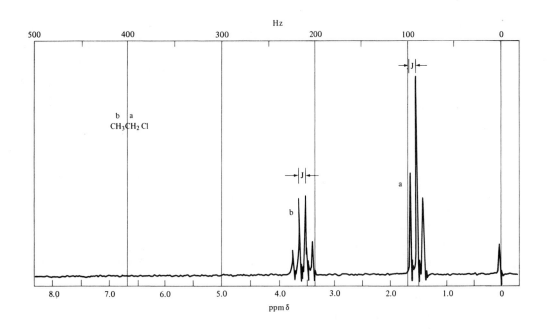

FIGURE 2.11 ^{1}H NMR (60 MHz) spectrum of chloroethane. The splitting (J) is equal for both signals a and b.

split by the three adjacent methyl protons ($N = 3$) to form a quartet ($N + 1 = 4$). Compounds such as this follow the simple $N + 1$ rule because the difference in chemical shifts ($\delta\,3.57 - \delta\,1.55$, or about 2 ppm, or 120 Hz) is large relative to the magnitude of the splitting (about 7 Hz, see Figure 2.11). The splitting occurs because of the fact that the signal of one set of protons is slightly affected by the spin of the protons in the adjacent set of protons. This effect is called ***spin-spin coupling*** (*J*).

Abbreviations to describe multiplets are used:
s = singlet, d = doublet,
t = triplet, q = quartet.

Now let's examine the fine structure of each of the two large signals in the ^{1}H NMR spectrum of chloroethane (Figure 2.11). The quartet at $\delta\,3.57$ shows splitting among the four peaks equal in magnitude to the splitting among the lines of the upfield triplet at $\delta\,1.55$.

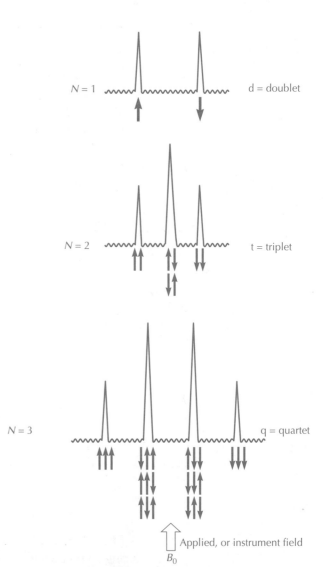

FIGURE 2.12 Spin-spin interaction of a proton with (from top to bottom) one, two, or three equivalent protons. Thus a triplet (middle) is caused when a proton is coupled by two equivalent protons, where the two can align against the instrument (or applied) field, the two can align with the applied field, or the effect of the two cancel, the last in two different ways. Thus a 1:2:1 triplet is the result.

There are two different ways that a pair of adjacent, nonequivalent protons can affect each other: They can be magnetically aligned either with each other or against each other (Figure 2.12, top). If they are aligned with each other, there is a slight deshielding factor; if they are aligned against each other, there is a slight, but equal, shielding factor. Thus the signal of one proton, when coupled by just one other proton, is split into a doublet, the two lines of which are of equal area. Therefore two adjacent and nonequivalent protons

$$\left(\begin{array}{cc} \text{H} & \text{H} \\ | & | \\ -\text{C} - \text{C} - \\ | & | \end{array} \,' \text{called vicinal protons} \right)$$

will appear as doublets with equal splitting.

This idea may be extrapolated to cases in which the splitting is caused by more than one proton, as in chloroethane. As long as the interaction is first order, the degree of splitting, as well as the relative intensities of lines may be predicted by the statistical combination of alignments of adjacent protons. For example, if there are two equivalent protons (Figure 2.12, $N = 2$) adjacent to a single proton of interest, the two protons could both align with the single proton, or against that proton, or the two could align one

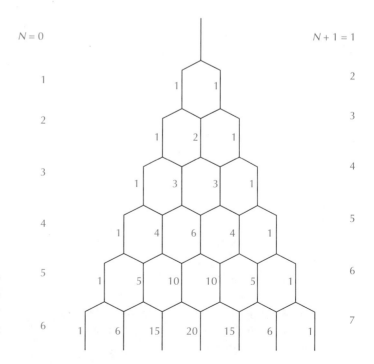

FIGURE 2.13 Pascal's triangle can be used to predict the multiplicity and relative intensities of the signal of any magnetically active nucleus attached to N equivalent nuclei of spin $\frac{1}{2}$ (applies to both ^{1}H and ^{13}C).

against and one with that proton (thus canceling the effects); the last takes place in two different ways. Thus the proton of interest would be a triplet of intensity 1:2:1 (Figure 2.12, middle). Fortunately, there is a shorter way to determine the number of components (or lines) in a multiplet. These numbers are determined by the coefficients of an expanded polynomial, and we can readily obtain them from a simple device known as Pascal's triangle (Figure 2.13). From this triangle, we find that a triplet should be composed of lines with relative areas of 1:2:1. Similarly, we find that a quartet will have relative areas of 1:3:3:1; these predictions are supported experimentally by the spectrum of chloroethane in Fig. 2.11.

If the coupling protons are not equivalent, they will occur at different chemical shifts and produce a complex spectrum. We must therefore distinguish between chemical shift equivalent protons and protons that are not equivalent when interpreting NMR spectra.

2.4
Detecting Chemical Equivalence

Chemically nonequivalent protons have different NMR spectral parameters. They have different chemical shifts and different coupling with nearby magnetic nuclei. To determine whether one proton is chemically equivalent with another, simply replace each of the protons in question by another group, say X, and then name the two new compounds. If the two protons replaced by the new functional group, X, are not chemically equivalent, two different compounds will result. This difference will be indicated by different names. On the other hand, if replacement of the two original protons by the new functional group results in two compounds that have the same name (or are mirror images), then the protons in the original molecule were chemically equivalent.

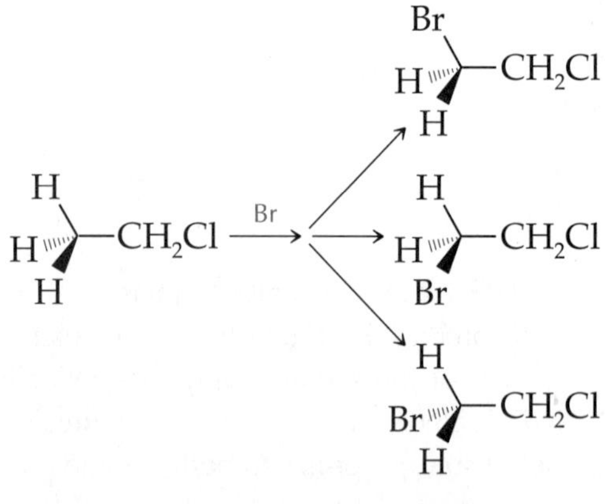

FIGURE 2.14 Substitution of bromine for each of the three hydrogens in the methyl group of chloroethane. The three product structures are identical.

Consider some examples. Chloroethane has the structure shown in Figure 2.14. We can replace each of the three methyl protons with bromine and obtain three new structures that are identical in every respect. Thus we can conclude that the three methyl protons are chemical shift equivalent; that is, the three hydrogens undergo chemical substitution with equal ease and they have identical chemical shifts. Moreover, chemical shift equivalence means that, although the methyl protons of chloroethane will couple with the methylene protons (CH_2), they do not couple to each other because all three CH_3 protons are chemical shift equivalent.

Contrast the situation for chloroethane with that for vinyl chloride (chloroethene, Figure 2.15). All three of the protons in the vinyl group are chemical shift nonequivalent because the rigidity of the system locks the positions of the four groups attached to a double bond. Thus H_a (*geminal*, or on the same carbon as the chlorine atom), H_b (*cis* to Cl), and H_c (*trans* to Cl) have three different environments. This conclusion may easily be confirmed by making an artificial substitution of X for each of the three hydrogens. Figure 2.15 reveals that three different compounds would arise. Thus we can expect that the three protons, which are chemical shift nonequivalent, would occur at different shifts, and they would be expected to couple to one another. In fact, their shifts are $H_a = \delta\ 6.28$, $H_b = \delta\ 5.47$ and $H_c = \delta\ 5.31$, and the coupling constants are in the order $J_{ab} > J_{ac} > J_{bc}$.

It is important to realize that chemical shift equivalent nuclei (such as the methyl protons of chloroethane) will all couple equally to other nuclei (such as the methylene protons of chloroethane). In such cases, simple multiplets predicted by the $N + 1$ rule will occur. When a proton is coupled by nonequivalent protons (as is the case for vinyl chloride), then the signals are more complex and often cannot be predicted by the $N + 1$ rule.

As you encounter more NMR spectra, you will be able to anticipate when the $N + 1$ rule does not apply and how to deal with those spectra. More complex molecules are described in the references listed at the end of this chapter.

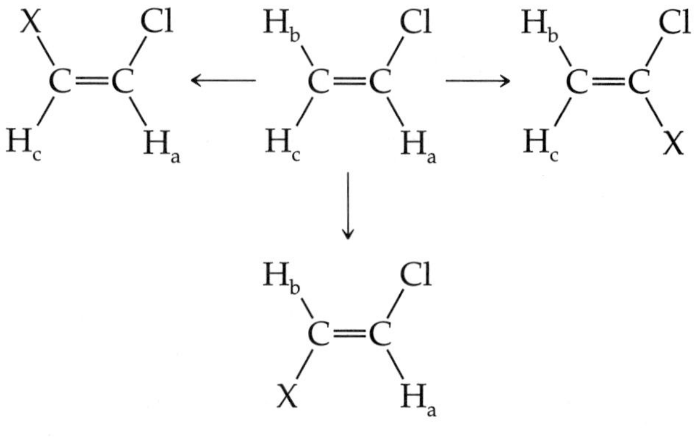

FIGURE 2.15 Chemical nonequivalence in vinyl chloride (chloroethene).

2.5
Integration

Chemical shifts and spin-spin splitting patterns give a great deal of information about the environments of protons in molecules. But there is still another kind of information that you can easily obtain from an NMR spectrum. Peak *integration* allows you to count to the *relative number* of each group of nonequivalent protons in a molecule. In an ^{1}H NMR spectrum, the area under each peak is proportional to the number of protons generating that peak. Integration of a spectrum gives a line that becomes higher in steps as it traces through the absorption regions of the compound. Areas can be used in the integrated spectrum of bromoethane (ethyl bromide, Figure 2.16). The height of the integral line (which is proportional to the area of the multiplet below it) does *not* give absolute numbers of hydrogens, but it does give relative numbers. Modern spectrometers routinely provide integrations in digital format also. ^{1}H NMR spectra provide integration-based proton counts, but ^{13}C NMR spectra often yield signals that are strongly dependent on more than simply the number of equivalent nuclei giving rise to that signal.

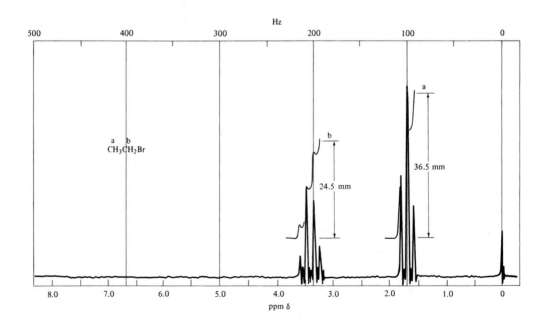

FIGURE 2.16 Integrated NMR spectrum of bromoethane (ethyl bromide). Modern instruments provide number integration as well as step curve integrals.

Usually it is not too difficult to combine NMR integration data with spin-spin splitting patterns to deduce the actual numbers of protons in a simple molecule. The splitting pattern of a triplet and a doublet for bromoethane immediately tells us that an ethyl group is present in the molecule (Figure 4.16). The triplet must be due to the absorption of the CH_3 group, and the quartet must come from the CH_2 group, because spin-spin splittings reflect the number of neighboring protons. The CH_3 pattern at δ 1.68 should have a relative area of 3, compared with 2 for the CH_2 absorption at δ 3.46. The actual heights (which are proportional to the peak areas) are found to be 36.5 mm and 24.5 mm (Figure 2.16). Thus the ratio of peak areas is

$$\frac{36.5}{24.5} = \frac{1.49}{1}$$

If we assume that the integrals have some error (it can be as much as 10%), we round to 1.5 as the relative number of non-equivalent protons. Because it would be impossible for a compound to have 1.5 protons of a certain type, the ratio must be 3:2 (or 6:4, etc.). In the case of bromoethane, three methyl protons to two methylene protons is just what we expected.

2.6

Preparing the NMR Sample

NMR analysis of solids requires a special probe; thus most NMR analysis in teaching and research laboratories is done on solutions. It is common to dissolve liquids as well as solids in a solvent. Only about 10–20 mg of compound dissolved in approximately 0.5 mL of solvent is necessary for modern FT NMR analysis. The NMR spectra of solids and viscous liquids are usually determined in solution, because, if these materials are not in solution, the nuclear-spin energy is efficiently transferred from one nucleus to another and quickly spreads out among many nuclei. In this case, the spectrometer sees many different magnetic environments for the protons and very broad, usually unhelpful, absorptions result. Table 2.3 lists standard NMR solvents. Water in the solvents may be a problem, so the shifts for the water protons have been provided also (Table 2.4).

We provide tube preparation guidelines here, but you should consult with your instructor, because instruments vary.

Put your solution into an NMR tube. Only a small part of an NMR tube is in the effective probe area of the instrument. About 0.5 mL of solution in the tube gives the best results (Figure 2.17). Often a clear plastic cylinder is available for checking the solution height for a given NMR spectrometer. A disposable glass pipet works well for transferring your solution to the sample tube.

To obtain spectra with sharp, well-defined peaks, the tube sometimes is spun to average the magnetic environments

Table 2.3. Solvents for ^{1}H and ^{13}C NMR

Solvent	Structure	Nucleus	δ_H or $\delta_C{}^a$	J_{HD} or J_{CH}, Hz
Acetone-d$_6$	CD$_3$(C=O)CD$_3$	^{13}C	29.8(septet)	20
			206.5(septet)^b	<1
Acetone-d$_6$	CD$_3$(C=O)CD$_3$	^{1}H	2.04(quintet)	2.2
Chloroform-d	CDCl$_3$	^{13}C	77(t)	32
Chloroform-d	CDCl$_3$	^{1}H	7.27(s)	
Deuterium oxide	D$_2$O	^{1}H	4.65(bs)	
Dimethyl sulfoxide-d$_6$	CD$_3$SOCD$_3$	^{13}C	39.7(quintet)	21
Dimethyl sulfoxide-d$_6$	CD$_3$SOCD$_3$	^{1}H	2.49(quintet)	1.7
Methanol-d$_4$	CD$_3$OD	^{13}C	49.0(septet)	21.5
Methanol-d$_4$	CD$_3$OD	^{1}H	3.30(quintet)	1.7
			4.78(s)	
Trifluoroacetic acid-d	CF$_3$CO$_2$D	^{1}H	11.50(s)	

a. s, singlet; d, doublet; t, triplet; q, quartet.
b. Long-range coupling.

throughout the sample. In fact, NMR tubes are finely machined for uniform wall thickness and minimum wobble. Too much sample in the tube tends to make it top-heavy and can give poor spinning performance and, thus, poor quality spectra.

The cw NMR method is not inherently sensitive. Very concentrated solutions are thus necessary to obtain good spectra. If you

Table 2.4. Chemical shifts of dissolved water in NMR solvent

Solvent	δ
CDCl$_3$	1.5
(suspended water)	($\approx$4.7)
C$_6$D$_6$	0.4
Acetone-d$_6$	2.75
CD$_2$Cl$_2$	1.55
D(C=O)(ND$_3$)$_2$ (DMF-d$_7$)	3.0
Pyridine-d$_5$	5.0
CD$_3$OD	4.9
CD$_3$C≡N	2.1
DMSO-d$_6$	3.35
D$_2$O	$\approx$4.75 (HOD)

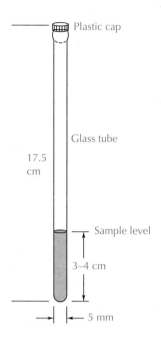

FIGURE 2.17 NMR sample tube filled to the correct height.

are using 0.5 mL of solvent, you will need about 0.15 mL of a liquid sample, or 150 mg of a solid sample.

Most of the sample in an NMR tube is solvent, so ideally we want a chemically unreactive solvent that does not absorb energy in the magnetic field. This means a solvent with no protons. Deuterochloroform ($CDCl_3$) has become very popular as an NMR solvent because it dissolves a wide range of organic compounds. Commercial $CDCl_3$ contains 99.8% (or more) deuterium and only 0.2% protium in its molecules, but this is still enough to give a small peak ($CHCl_3$) at 7.27. You may notice this in Figure 2.6.

Some NMR samples are not soluble in $CDCl_3$. Among these are polar compounds such as carboxylic acids and polyhydroxyl compounds. In most cases these compounds are soluble in water. Deuterium oxide (D_2O, heavy water) should then be used as the NMR solvent. If a carboxylic acid is not soluble in D_2O, it will probably be soluble in D_2O containing sodium hydroxide. Adding a drop or two of concentrated sodium hydroxide to the sample in D_2O is usually enough.

The use of deuterium oxide presents some problems. There will always be a broad peak at approximately $\delta\,4.5$ due to a small amount of HOD (and/or H_2O) present in the original D_2O solvent. This solvent peak can cover important peaks from your compound. Also, D_2O may exchange with protons in your sample compound, producing HOD. Consider, for example, what happens when a carboxylic acid or alcohol dissolves in D_2O:

$$R-\overset{\overset{\displaystyle O}{\|}}{C}-O-H + D_2O \rightleftharpoons R-\overset{\overset{\displaystyle O}{\|}}{C}-O-D + H-OD$$

Carboxylic acid

$$R-O-H + D_2O \rightleftharpoons R-O-D + H-OD$$

Alcohol

Deuterium nuclei are "invisible" in ^1H NMR spectra, and in an NMR solution there are many more molecules of solvent D_2O than of the sample's acid or alcohol protons. The equilibrium positions lie well to the right, and the carboxylic acid proton and the proton on the alcohol-oxygen atom do not show in the NMR spectrum. They are simply not present in high enough concentration. Exchangeability also causes OH and NH signals to occur anywhere over wide ranges in ^1H NMR spectra (Table 2.1).

Before using a deuterated solvent (expensive!) for an NMR analysis, first be sure that your compound dissolves in the deuterium-free solvent, $CHCl_3$ or H_2O. Other deuterated solvents are also available commercially (Table 2.3). Deuterated dimethyl sulfoxide (DMSO-d_6) is a very useful solvent of polarity intermediate between that of $CDCl_3$ and D_2O.

We discussed in Section 2.2 the need for a standard reference substance. Tetramethylsilane [TMS, or $(CH_3)_4Si$] was the choice because all its protons are equivalent, and they absorb at a magnetic field where very few other protons in carbon compounds absorb. TMS is also chemically inert and is soluble in most organic solvents. When used as an internal standard, 1–2% TMS is dissolved in the solution being analyzed.

Tetramethylsilane is not soluble in D_2O, so it cannot be used as a standard with this solvent. The reference substance that has been used for water solutions is sodium 2,2-dimethyl-2-silapentane-5-sulfonate (DSS), $(CH_3)_3SiCH_2CH_2CH_2SO_3^- Na^+$. Its major peak appears at very nearly the same position as the TMS absorption.

As mentioned earlier, modern FT NMR instruments do not actually require TMS in the solution but merely lock on deuterium atoms of a known reference compound (D_2O, or $CDCl_3$). The spectrum is nevertheless calibrated such that TMS is at δ 0.00. Keep in mind that a glance at the position of the $CHCl_3$ impurity (δ 7.27) in $CDCl_3$ can serve as a rough spectrum calibration.

NMR is a nondestructive analytical technique. Because none of the sample is destroyed, it can easily be recovered. A solid can easily be recovered by evaporating $CDCl_3$ solvent. A liquid sample can be recovered by distillation or chromatography, if evaporation does not work.

2.7
Recording an NMR Spectrum

In most cases, the instructor or an experienced laboratory assistant either will obtain the NMR spectrum of your sample or will supervise you as you operate the instrument. Instruction in the operation of expensive NMR spectrometers is *crucial* for good spectra and trouble-free instrument operation. The controls of NMR spectrometers vary widely, so no attempt will be made to discuss them here.

2.8
Interpreting ¹H NMR Spectra

The NMR spectra of organic compounds are important because they give a great deal of information about molecular structure. In this section we will look at the spectra of some simple organic molecules and show how the information derived from the spectra can help us determine their molecular structures.

Four pieces of information are used in the preliminary interpretation of an ¹H NMR spectrum: (1) the number of signals, which tell us how many kinds of nonequivalent protons are in the molecule; (2) the chemical shifts of the protons doing the absorbing; (3) the spin-spin splitting patterns (multiplicity) of the protons; and (4) the integrations of the areas under the peaks, which indicate the relative number of protons giving rise to the absorptions. Also important are the values of the coupling constants for the various interactions of protons within the molecule, although we will use this information infrequently. None of the information stands alone, however, and in each case the NMR spectrum is most useful in interpreting molecular structure when used in conjunction with other analytical data, including infrared spectrometry (Spectrometric Method 1), elemental analysis, and the other chemical evidence discussed in Part 3.

We have already indicated, in Figure 2.9 and Table 2.1, the regions of chemical shifts of some of the protons in organic compounds. In addition to the chemical shift data, it is helpful to keep in mind the use of spin multiplicity information. Some spin-spin splittings give rise to regular patterns, making the group producing the multiplet immediately obvious from the spectrum. The multiplet pattern of the ethyl group, for example, is seen in the ^{1}H NMR spectra of 2-butanone, ethanol, and ethyl acetate (respectively, Figures 2.18–2.20). Such an NMR pattern requires the $-CH_2CH_3$ in the molecule. The chemical shift positions of the

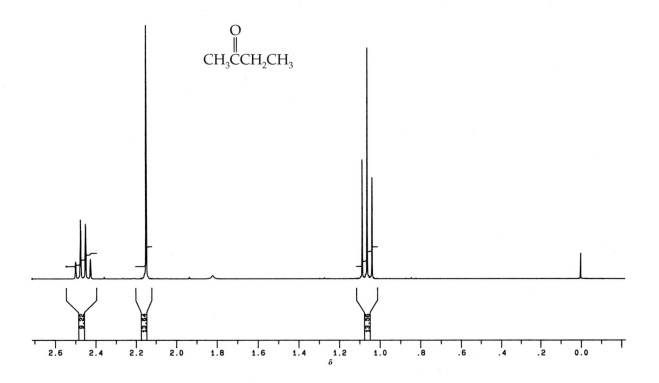

FIGURE 2.18 ^{1}H NMR spectrum (300 MHz) of 2-butanone. The vertical numbers provide the integrated areas.

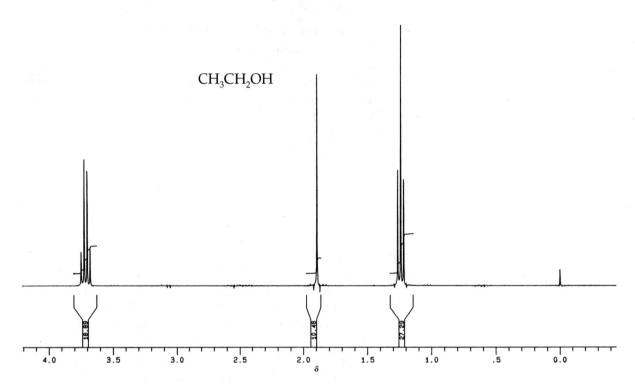

FIGURE 2.19 ¹H NMR spectrum (300 MHz) of ethanol. The chemical shift of the ¹H signal at δ 1.9 shows a strong concentration dependence.

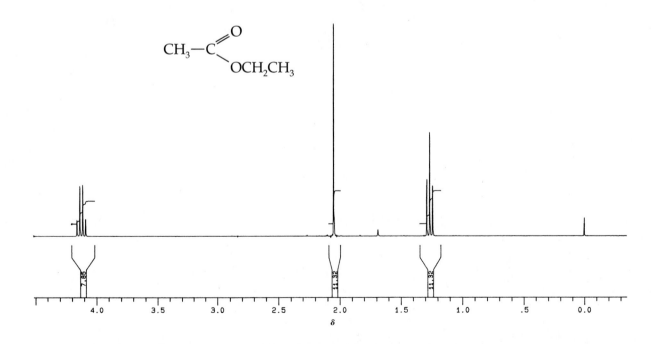

FIGURE 2.20 ¹H NMR spectrum (300 MHz) of ethyl acetate.

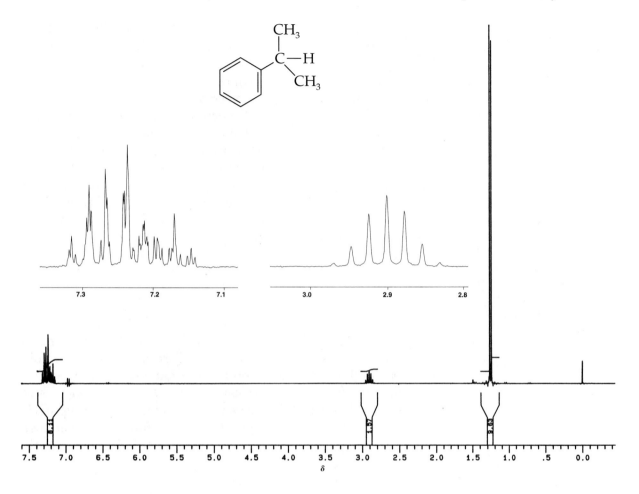

FIGURE 2.21 ¹H NMR spectrum (300 MHz) of isopropylbenzene.

methylene group quartets in Figures 2.19 and 2.20 also strongly indicate that an oxygen atom is attached to the carbon (see Table 2.1). Similarly, the pattern of the isopropyl group (Figure 2.21) is the same whether the compound is isopropyl cyanide, isopropyl mercaptan, or isopropylbenzene. Notice in the spectrum of isopropylbenzene (Figure 2.21) the intense absorption in the region (δ 7.2, 5H) where highly deshielded aromatic protons absorb. This peak, taken with the integration data of the spectrum, clearly indicates a mono-substituted benzene compound.

The methyl singlets in the ¹H NMR spectra of toluene (3H singlet at $\approx \delta$ 2.4) and *t*-butyl alcohol (9H singlet at $\approx \delta$ 1.25) are not split because these protons are isolated in each structure. There are other standard patterns that you should recognize. The isolated isopropyl group of isopropylbenzene (Figure 2.21) shows a 6H doublet at $\approx \delta$ 1.25 (the methyl hydrogens of the isopropyl group) and a septet at $\approx \delta$ 2.9 (the isopropyl methine hydrogen). Note that the septet had to be enlarged to observe the entire multiplet. Because the methine signal represents only one proton and because this proton is split by six equivalent protons, the relatively weak outer lines (one-twentieth the size of the center line, see Pascal's triangle, Figure 2.13) are too small to be seen easily.

Return to the spectrum of ethanol, Figure 2.19. The hydroxyl proton is at δ 2.2; its chemical shift and the fact that it is a singlet (and not a triplet as might be expected as a result of coupling by the protons in the attached methylene) are both due to the rapid exchange experienced by hydroxyl protons: This exchange is normally promoted by minute traces of acid in the $CDCl_3$ NMR solvent (in DMSO-d_6, this exchange is usually slower, and a triplet is usually observed for the hydroxyl proton).

2.9
^{13}C NMR

We can appreciate the use of carbon (^{13}C) NMR by building on what we have learned about ^{1}H NMR. Part of our ability to build is based on the fact that carbon has the same spin multiplicity as does the proton (^{1}H); thus we can use the $N + 1$ rule much as we have earlier. For example, the carbon of a methylene group (CH_2) can be split into a triplet by the attached protons. Moreover, there is a crude parallel between the chemical shifts for carbons (Tables

Table 2.5. **Characteristic ^{13}C NMR chemical shifts**

Compound	Chemical shift, ppm
TMS	0.00
Alkane	0.0–39
Halides ($\underline{C}HX$)	0–80
Alkene (olefinic, $\underline{C}{=}C$)	100–150
Aromatic	110–135
Aldehydic, ketonic ($\underline{C}{=}O$)	190–215
Alcoholic ($\underline{C}{-}OH$)	50–75
Alkyne ($\underline{C}{\equiv}C$)	75–95
Nitriles ($\underline{C}{\equiv}N$)	117–120
Alcohols, esters, ethers ($\underline{C}{-}O$)	50–70
Carboxylic acids, esters, anhydrides, amides ($\underline{C}{=}O$)	160–180

Table 2.6. Chemical shifts of carbons on monosubstituted benzene rings

Compound	C-1	C-2	C-3	C-4
C_6H_6	128.5	128.5	128.5	128.5
C_6H_5R	129–142	127–130	128	131
$C_6H_5CH=CH_2$	138	126	129	128
$C_6H_5C\equiv CH$	123	135	129	129
C_6H_5OH	155	116	130	121
$C_6H_5(C=O)G^a$	137	130	128	133
$C_6H_5COO)R^b$	131	130	129	133
$C_6H_5NH_2$	148	115	130	119
$C_6H_5NH(C=O)CH_3$	140	119	129	123
$C_6H_5NO_2$	148	123	129	134.5
C_6H_5F	164	114	129	124
C_6H_5Cl	135	129	130	126.5
C_6H_5Br	123	132	131	127.5
C_6H_5I	96	138	131	121

a. G = Cl, OH, OR, $\overset{O}{\overset{\|}{C}}R$.
b. R = alkyl, aryl.

2.5 and 2.6 and Figure 2.22) and those for protons: for example, alkanes appear upfield (lower magnitude of δ), and nearby electronegative atoms deshield both carbons and protons (moving signals downfield), and aromatic carbons and protons are both at low field. Carbon NMR, however, opens up new areas for analysis: For example, the shifts of carbonyl carbons are useful. Carbon shifts occur over a 200-ppm range compared with the usual 10-ppm range for protons. The simplest and most routine ^{13}C spectrum is that in which the protons are completely decoupled from the carbons; in that case, all carbon signals will be singlets. It is a very rare event for nonequivalent carbons to give coincidentally equivalent signals in such a ^{13}C NMR spectrum.

Samples are prepared for ^{13}C NMR much as they are for ^1H NMR. Table 2.3 lists solvents suitable for ^{13}C NMR. The compound $CDCl_3$ is again the solvent of choice because it dissolves most organic compounds.

There are two reasons why the ^{13}C NMR experiment has taken longer to evolve than the ^1H NMR experiment has. One is more obvious: We are dealing with a minor isotope of carbon (^{13}C is only 1.1% as abundant as ^{12}C). Second, the ^{13}C nucleus is

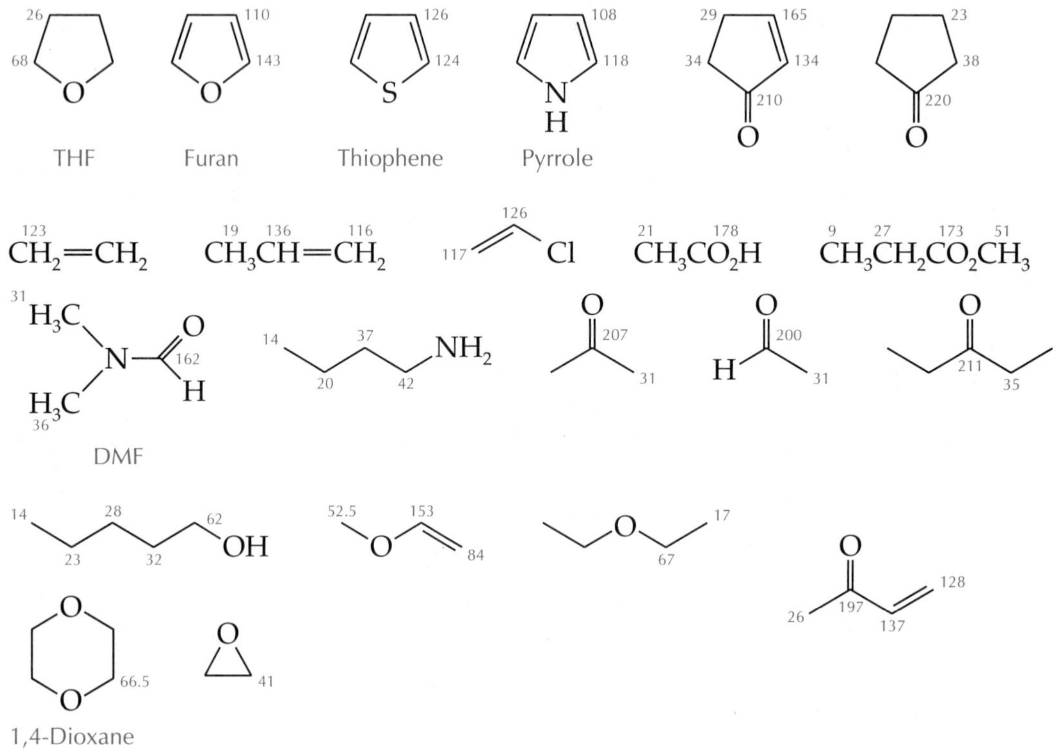

FIGURE 2.22 Chemical shifts for carbons of various organic compounds.

inherently more resistant to the NMR experiment, as suggested by its lower magnetogyric ratio (^{13}C is about one-fourth that of ^{1}H). This 1:4 ratio describes another aspect of the proton-carbon comparison; an instrument built to analyze protons at 300 MHz will be at essentially one-fourth that field strength, or 75 MHz, when ^{13}C nuclei are analyzed.

Table 2.5 and Figure 2.22 display carbon shifts for a variety of organic compounds. Carbon shifts for substituted benzenes are shown in Table 2.6.

In Figures 2.23 and 2.24, we see ^{13}C NMR spectra of bromoethane (ethyl bromide) done in two different ways. Figure 2.23 shows carbon signals that reveal the sizable coupling of protons directly attached to these two carbons. We can use the $N + 1$ rule to rationalize these multiplicities: The upfield methyl carbon is a quartet because it has three attached protons, and the downfield methylene group is a triplet because it has two attached protons. Note also that the lower field position for the methylene carbon

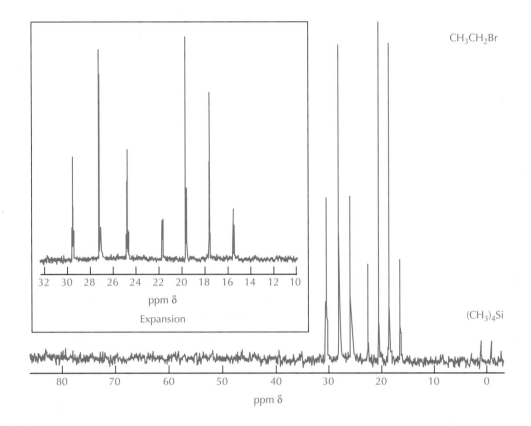

FIGURE 2.23 ^{13}C NMR spectrum of bromoethane (instrument field 62.8 MHz, or 5.875 Tesla). The upfield (δ 18.3 ppm) quartet is due to the methyl group, and the downfield triplet is due to the methylene. The coupling due to direct proton attachment is 118 MHz (methyl) and 151 MHz (methylene). The tetramethylsilane (TMS) signal is a quartet whose outer lines have nearly been lost. The inset shows fine structure due to long-range coupling of low (2–5 Hz) magnitude.

resonance is consistent with patterns we have learned for protons; that is, the methylene carbon is more deshielded by the directly attached bromine. In common practice, the decoupled spectrum shown in Figure 2.24 is the more straightforward place to begin ^{13}C NMR analysis. This 200-ppm wide spectrum shows singlets for each carbon because all the protons were decoupled by irradiation with a broad band of energy. Carbon-carbon coupling is not observed because there is a vanishingly small probability that two attached carbons are both ^{13}C.

Numerical values of ^{13}C chemical shifts are another area in which our earlier study of ^1H NMR is helpful. Table 2.5 reveals

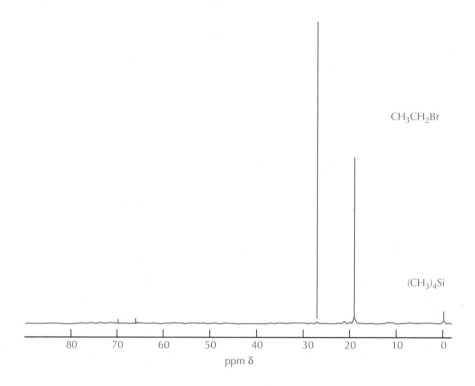

FIGURE 2.24 Decoupled ^{13}C NMR spectrum of bromoethane (instrument field 62.8 MHz). This sample has been irradiated by a broad-band decoupling signal centered at 250 MHz to remove all proton coupling. All three types of carbons appear as singlets.

trends that we have seen before. Carbons on electronegative atoms such as oxygen occur at much lower field than do the corresponding carbon of alkanes; we saw this to be the case earlier from protons on these carbons. Moreover, olefinic and aromatic carbons are downfield (recall we saw that the same was true for protons on these sp^2 carbons). Finally, carbonyl carbons show very low field shifts, the position of which is dependent on the type of functional group (for example, ketone versus ester) containing this carbon.

There are some shielding effects unique to ^{13}C NMR. First, it is clear that carbon is quite strongly deshielding. Specifically, it is known that carbons both α and β to a given (*) carbon will deshield that carbon by about 9 ppm (that is, both the α and β carbons of the —C*$C_\alpha C_\beta$— structural unit will increase the chemical shift of the C* carbon by 9 compared with the shift of methane). Also, although strongly electronegative halogens deshield carbon (the carbon of fluoromethane occurs at δ 75.4), large halogens

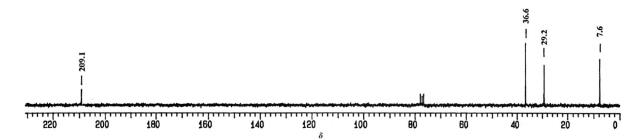

FIGURE 2.25 ^{13}C NMR spectrum (50 MHz) of 2-butanone. The 1:1:1 triplet near δ 77 is due to CDC13 solvent. Reproduced from "A Spectrum of Spectra," by R. A. Tomasi, Sunbelt R&T Pub., Tulsa, OK, 1992, with permission.

sometimes shield carbons (iodomethane has a carbon chemical shift of − δ 20.7, that is over 20 ppm upfield of TMS). This has been assigned to a "steric compression" effect. In such cases, steric buttressing apparently causes the electrons in the orbitals of the carbon to be compacted, thus more highly shielding the carbon nucleus.

The ^{13}C NMR spectra of 2-butanone, ethanol, and ethyl acetate are also instructive (Figures 2.25–2.27). First, the number of carbon signals (disregard TMS and solvent signals) provide a count of the number of nonequivalent carbons in each compound (2-butanone, 4; ethanol, 2; and ethyl acetate, 4). Second, the methylene carbon (δ 58) of ethanol is more downfield of the methyl carbon (δ 18) because it is closer to the deshielding oxygen atom. Third, the different shifts of the downfield signals of 2-butanone (at nearly δ 205 ppm) and ethyl acetate (δ 171 ppm) clearly indicate different types of carbonyl carbons.

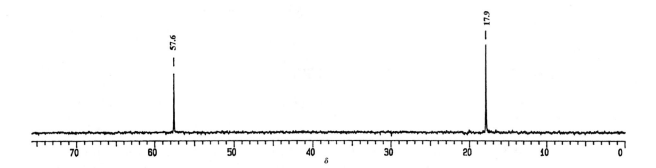

FIGURE 2.26 ^{13}C NMR spectrum (50 MHz) of ethanol. Reproduced from "A Spectrum of Spectra," by R. A. Tomasi, Sunbelt R&T Pub., Tulsa, OK, 1992, with permission.

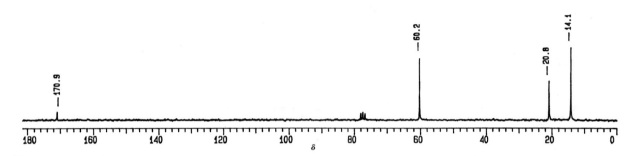

FIGURE 2.27 ^{13}C NMR spectrum (50 MHz) of ethyl acetate. Reproduced from "A Spectrum of Spectra," by R. A. Tomasi, Sunbelt R&T Pub., Tulsa, OK, 1992, with permission.

Frequently ^{13}C NMR spectra are used to determine the number of types of carbons by simply counting the number of carbon singlets in a proton decoupled spectrum and comparing that number of carbons in the compound's molecular formula. Obtaining the quantitative count of a given type of carbon by using the peak area is, however, far less straightforward because the area of a signal is caused not only by the number of carbon atoms causing the signal, but by variations in the so-called spin-lattice relaxation times (T_1's) and nuclear Overhauser effect (NOE), factors not elaborated in this book.

A comment on the signal just upfield of δ 80 in Figure 2.25a is worthwhile. This signal is due to the carbon of $CDCl_3$, which is split into a 1:1:1 triplet by the deuterium.

Finally, we can interrelate the use of ^1H and ^{13}C NMR spectra. In ethanol, for example, the ^1H NMR quartet at δ 3.72 and the —CH_2OH at δ 58 (Figure 2.26) are, respectively, downfield of the methyl protons and carbons, and both shifts reveal the direct attachment of the strongly deshielding oxygen.

2.10
An Approach to Solving NMR Problems

There are four main types of information provided by NMR spectra:

Chemical shift Suggests the environment of the proton or carbon. Downfield signals (larger δ values) suggest nearby deshielding structural units such as oxygen atoms, halogens, or π-systems.
Integration Determined by the number of nonequivalent protons ^1H NMR. Thus this information reveals collections of equivalent protons such as the nine methyl protons of a *t*-butyl group.

(a)

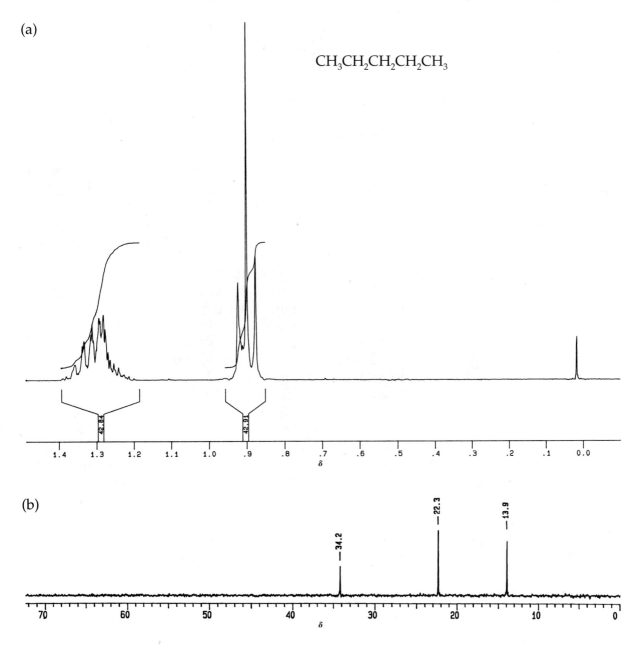

$CH_3CH_2CH_2CH_2CH_3$

(b)

FIGURE 2.28 (a) ^1H NMR spectrum (300 MHz) of pentane. (b) ^{13}C NMR spectrum (50 MHz) of pentane. Reproduced from "A Spectrum of Spectra," by R. A. Tomasi, Sunbelt R&T Pub., Tulsa, OK, 1992, with permission.

Splitting Caused by spin-spin coupling of protons to other protons or to carbons. Reveals positions of nearby protons causing the coupling. Proton coupling is often removed from ^{13}C spectra to simplify them.

Number of signals Depends on the symmetry of the structure. Often used with proton decoupled ^{13}C spectra. The number of singlets identifies the number of nonequivalent carbons.

The proton and carbon NMR spectra for pentane are shown in Figure 2.28. The ^1H NMR spectrum shows a well-resolved triplet at $\approx \delta$ 0.9 due to the equivalent terminal methyl groups. The poorly resolved methylene protons extend from $\approx \delta$ 1.2 to 1.4. Because each of the δ 0.9 and δ 1.2–1.4 signals are caused by the same number of protons (here 6H each), their areas are equal. The chemical shift range (rather than center) and instrument field strength should always be reported for such poorly resolved signals. The carbon spectrum shows the symmetry of the compound by displaying three signals indicative of the three different carbons: C_1 and C_5, C_2 and C_4, and C_3.

The ^{13}C NMR spectrum of 1-bromobutane (Figure 2.29b) simply shows that there are only four types of carbons in the structure. If we examine the ^1H NMR spectrum for 1-bromobutane (Figure 2.29a) the deshielding effect of the bromine is clearly revealed. We not only find that bromine has highly deshielded the protons on the attached carbon, we also find that to a gradually lessening extent it has also deshielded all the other protons on the chains. We are now able to find a well-defined signal for each set of shift-equivalent protons in the compound:

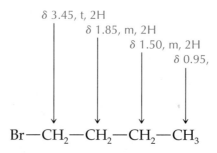

The simple first-order patterns for 1-bromobutane should be contrasted with the somewhat more complex situation for the butylbenzene spectra shown in Figure 2.30a,b. The ^1H NMR spectrum (Figure 2.30a), if examined superficially, seems to reveal a simple first-order pattern for its butyl group. That is, at first glance, the signals are simple multiplets that seem to fit the $N + 1$

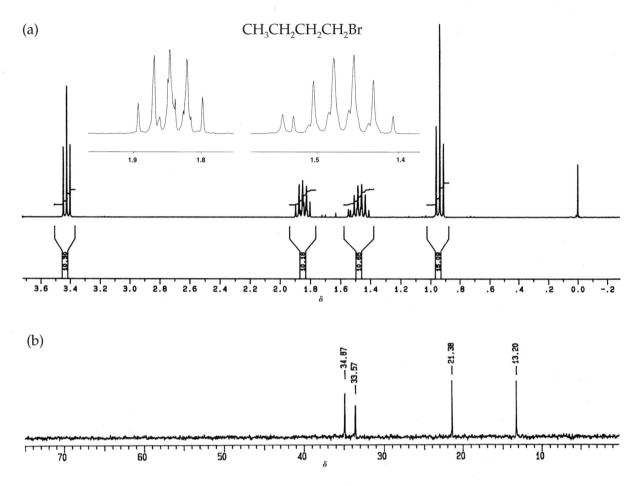

FIGURE 2.29 (a) ^{1}H NMR spectrum (300 MHz) of 1-bromobutane. (b) ^{13}C NMR spectrum (50 MHz) of 1-bromobutane. Reproduced from "A Spectrum of Spectra," by R. A. Tomasi, Sunbelt R&T Pub., Tulsa, OK, 1992, with permission.

rule. If, however, we examine the expanded signals for the methylenes at C-2 and C-3, more detail is revealed. Note that these multiplets are more complex than the simple quintet and sextet that might be expected from simple $N + 1$ analysis. These two signals show more complexity because the interset coupling constants from each side of the two methylene groups are not equal. Specifically, the coupling from the protons on C-1 to those on C-2 and from the protons on C-3 to those on C-2 are not quite equal; thus the signal at 1.6 (protons at C-2) is not a simple quintet (see expanded inset in Figure 2.30a). The standard multiplet at 7.2 is due to the five aromatic protons.

(a)

$CH_2CH_2CH_2CH_3$

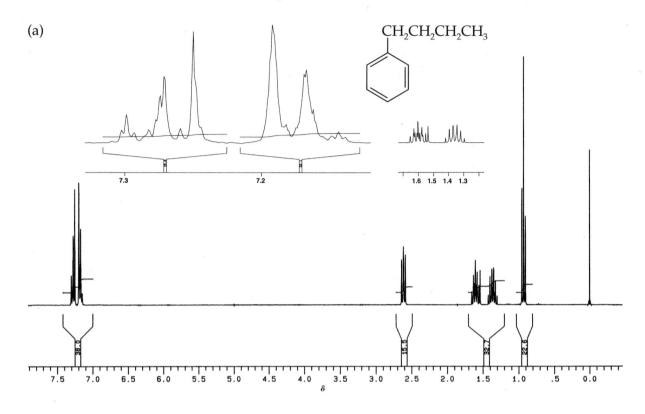

(b)

FIGURE 2.30 (a) ^{1}H NMR spectrum (300 MHz) of butylbenzene. (b) ^{13}C NMR spectrum (50 MHz) of butylbenzene. Reproduced from "A Spectrum of Spectra," by R. A. Tomasi, Sunbelt R&T Pub., Tulsa, OK, 1992, with permission.

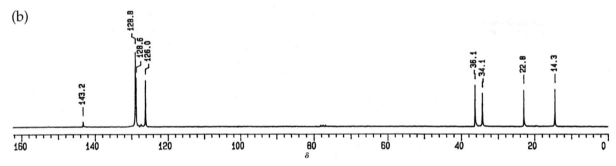

δ 141.7 — $CH_2CH_2CH_2CH_3$

An expanded inset is also necessary to reveal all the pertinent details for the ^{13}C NMR spectrum (Figure 2.30b) of butylbenzene because there are eight nonequivalent carbons in this molecule (four aromatic carbons and four aliphatic). Moreover, the noteworthy weak intensity of the carbon at 143.2 is caused by the fact that it has no attached hydrogens.

Simple first-order patterns occur for the ^{1}H NMR spectrum (Figure 2.31a) of 2-butylbenzene (*sec*-butylbenzene), and these

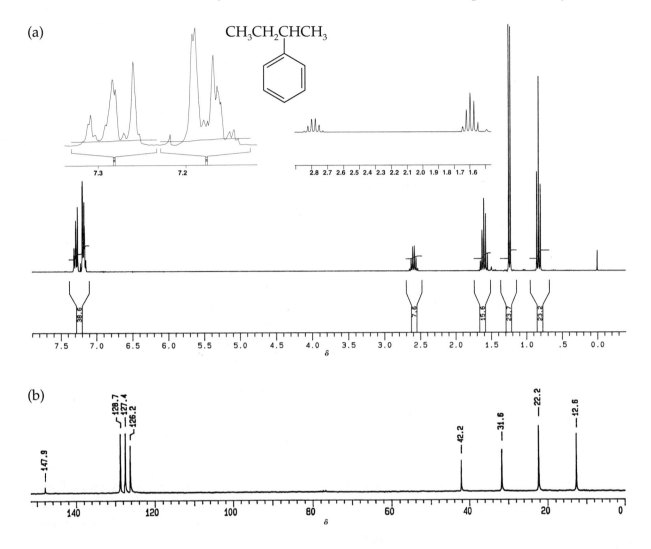

FIGURE 2.31 (a) ¹H NMR spectrum (300 MHz) of 2-butylbenzene. (b) ¹³C NMR spectrum (50 MHz) of 2-butylbenzene. Reproduced from "A Spectrum of Spectra," by R. A. Tomasi, Sunbelt R&T Pub., Tulsa, OK, 1992, with permission.

should be contrasted to the somewhat more complex situation for the butylbenzene proton spectrum discussed earlier. The ¹H NMR spectrum reveals a simple first-order pattern for its butyl group. If we examine the signals for the butyl group, the branched nature of this side chain is revealed. Specifically, the ¹H distorted sextet is consistent with the chemical shift of a proton on a carbon attached to the benzene ring (see arrow on following structure). Thus we can immediately propose the following partial structure:

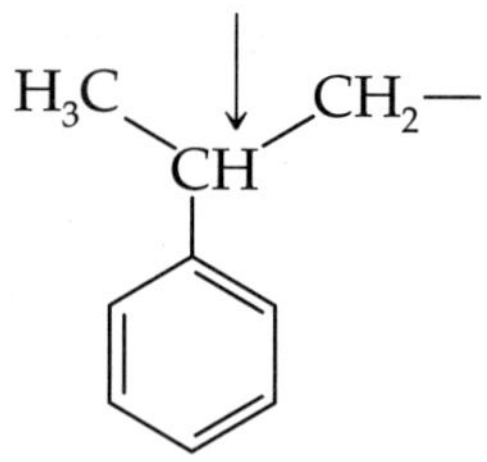

The open valence site must be filled by a methyl group to complete the butyl chain. Thus the following structure and its assignments confirm the identity of the compound.

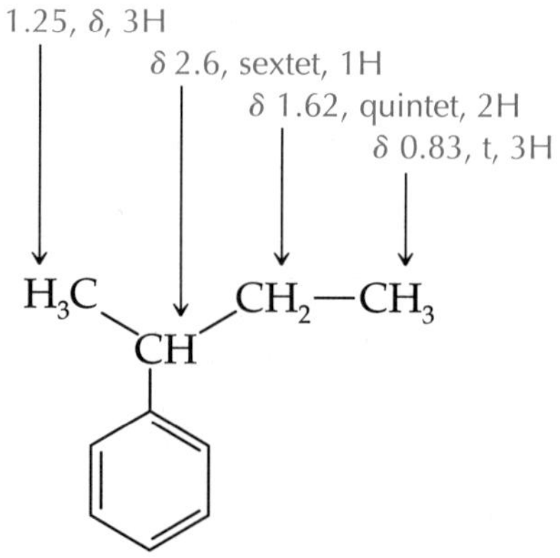

d = doublet, t = triplet

The details of the ^{13}C NMR spectrum of 2-butylbenzene (Figure 2.31b) confirm its identity as a member of the butylbenzene family but are of no value in differentiating this compound from other butylbenzenes.

Suppose that the labels were lost from the bottles of *ortho-*, *meta-*, and *para-*dimethylbenzene (xylenes). We can use NMR to reidentify these compounds.

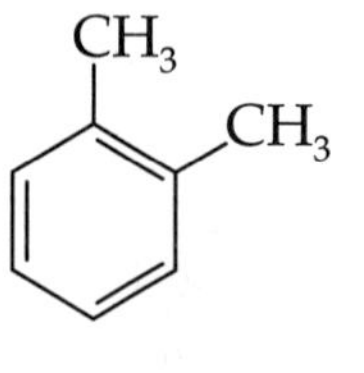

ortho isomer
(1,2-)

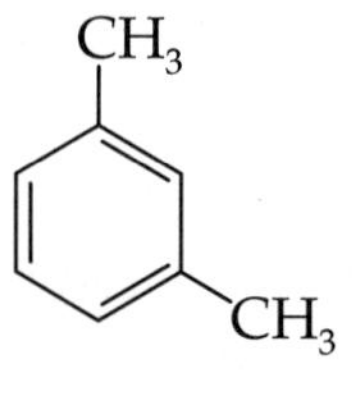

meta isomer
(1,3-)

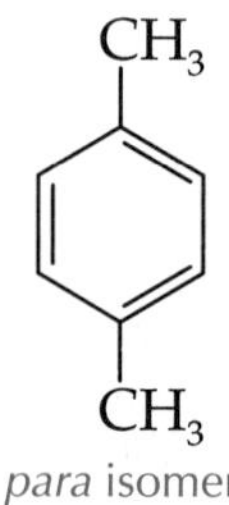

para isomer
(1,4-)

Because the greatest symmetry occurs for the *para* isomer, it is the easiest to identify. We expect only two kinds of protons (one each of methyl and ring protons) and three kinds of carbon (methyl and two kinds of ring carbons) and this clearly is shown in the spectra of the 1,4-isomer. The other two are identified with only slightly more effort. The *ortho* compound should show four kinds of carbon and the *meta*, five, that is, four benzene ring carbons and an upfield aliphatic carbon. At this field strength, differentiation of the *ortho* and *meta* isomers by proton NMR would be very difficult. The other hope would be to use a fingerprinting technique; that is, they could be identified by comparing these spectra with published proton spectra (at the same field strength, if at all possible) for the 1,2- and 1,3-isomers.

The proton NMR spectrum of butanal is shown in Figure 2.32. Even at an instrument field strength of only 60 MHZ, many details are revealed. Specifically, the aldehydic proton at δ 9.83 is a triplet, a pattern indicating the presence of the two protons on the C-2 methylene at c (δ 2.4). The straight-chain nature of the compound is revealed because the methyl at δ 1.0 (a) is a triplet and is thus connected to a methylene (b). Finally, note that the closer the protons are to the carbonyl group, the more they are deshielded.

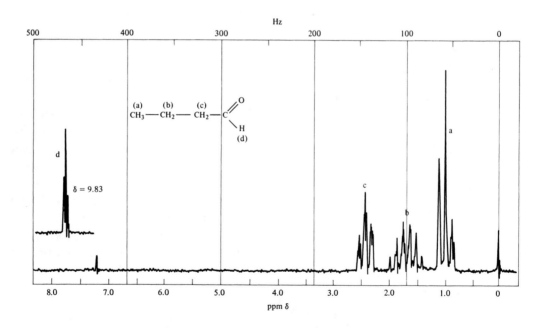

FIGURE 2.32 NMR spectrum of butanal (60 MHz). The spin-spin splitting follows the $N+1$ analysis. The aldehydic proton, with its large chemical shift, is visible at δ 9.83. Protons on the α-carbon are deshielded by the electronegative carbonyl group.

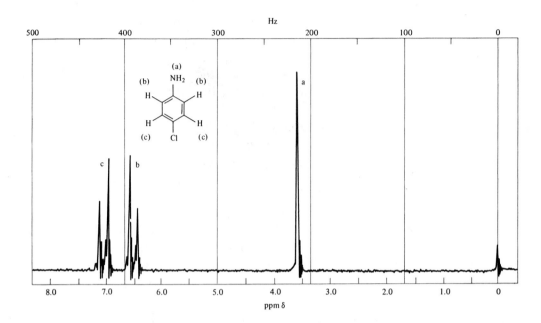

FIGURE 2.33 NMR spectrum of 4-chloroaniline (60 MHz). Note the symmetry of the aromatic pattern (δ 6.4–7.3).

The typical proton pattern (signals b and c) of a *para*-substituted compound is shown in the ^{1}H NMR spectrum of *p*-chloroaniline in Figure 2.33a. Moreover, because the amino protons (signal a) are exchangeable, they can be confirmed by treating the sample with D_2O.

Three NMR spectra for benzaldehyde are shown in Figure 2.34s. Figure 2.34a and b clearly illustrates the effect of increased magnetic field. Both spectra show a simple singlet at δ 10.00 for the isolated aldehydic proton. The aromatic protons reveal more at the higher field strength (Figure 2.34b); specifically, the doublet due to the *ortho* protons is clear at δ 7.87. This signal is lost in a high-order jumble in Figure 2.34a. The ^{13}C NMR (Figure 2.34c) shows (1) five kinds of carbon, (2) a carbonyl carbon at δ 188 (deshielded by the positive nature of this carbon), and (3) a weak

FIGURE 2.34 (opposite page) (a) Monosubstituted benzenes, like benzaldehyde, shown here, may also have a complicated aromatic spectral region. The *ortho*-protons are more deshielded than the *meta*- and *para*-protons. 60 MHz. (b) ^{1}H NMR spectrum (300 MHz) of benzaldehyde. (c) ^{13}C NMR spectrum (50 MHz) of benzaldehyde. Reproduced from "A Spectrum of Spectra," by R. A. Tomasi, Sunbelt R&T Pub., Tulsa, OK, 1992, with permission.

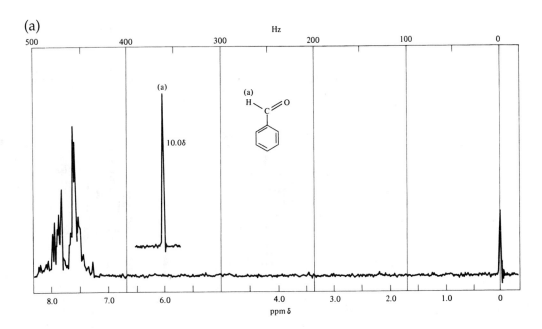

(a)

Hz

10.0δ

(a)

ppm δ

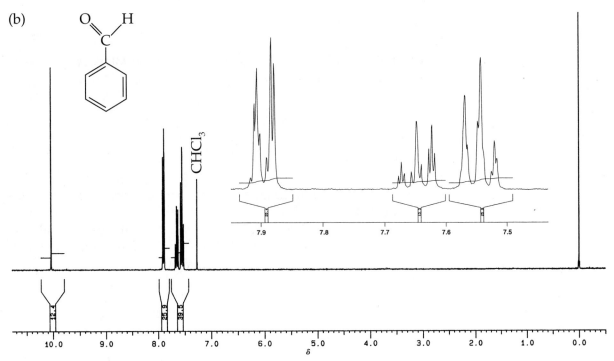

(b)

CHCl₃

δ

(c)

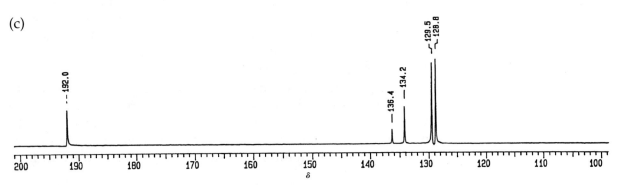

192.0

135.4

134.2

129.5
128.8

δ

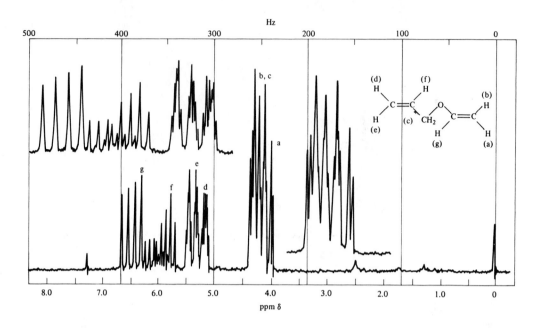

FIGURE 2.35 Spectrum of allyl vinyl ether (60 MHz). A spectrum at 300 or 500 MHz may be simpler.

signal at δ 136.4 due to an aromatic carbon bearing the aldehyde substituent.

Although NMR is a very powerful tool, it does not always work. The ¹H NMR spectrum of the unsaturated ether shown in Figure 2.35 defies straightforward analysis.

2.11
Two-Dimensional NMR and Related Techniques (Special Topic)

We have included two-dimensional NMR here, fully aware that many lab courses may not cover it. It is, however, an important aspect of NMR, and you should at least know something about it.

We have already discussed the advantage of both high-field NMR and high-speed computers, techniques that greatly expand the use of NMR. More recently, it has become possible to modify pulse sequences such that a given NMR spectrum can be made to yield even more information. Although a large number of techniques have been developed, we will only discuss a few to give you a sense of how they are obtained and where they are useful.

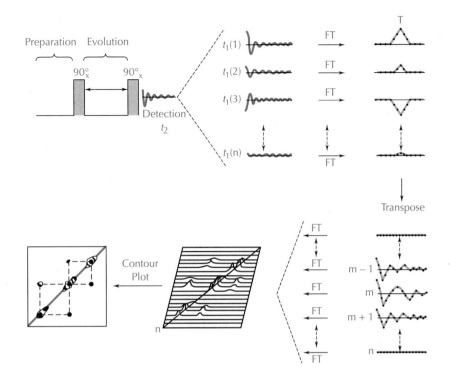

FIGURE 2.36 Generating two-dimensional spectra. A series of 90°–90° pulse sequences (over increasing times t_1) generate a series of one-dimensional FID's $t_1(1)$, $t_1(2)$, . . . , $t_1(n)$, which when transformed (FT) give rise to one-dimensional frequency domain spectra (top right-hand corner), which are "transposed" by plotting the maxima and minima, etc., to give decaying sine curves that can then be transformed in the normal FT way to provide the stacked two-dimensional grid (which is finally displayed as a contour plot).

Figure 2.36 introduces the general concept of two-dimensional NMR. With this technique, standard (one-dimensional) NMR spectra are evolved over time in a second dimension. In Figure 2.36, the pulse sequence (after preparation) is a $(90°_x)$ pulse followed by an incrementally increased time, t_1, followed by a sampling (second $90°_x$) pulse, leading to an FID (free induction decay) that is collected over time t_2. Thus we obtain a series of FID plots collected from time $t_1(1)$ to $t_1(n)$. Fourier transformation of each of these FIDs results in a series of one-dimensional (frequency domain) spectra that undulate. The signal intensity changes over time, actually giving rise to both positive and negative peaks. If we trace over the changing maxima and minima of these peaks, we find that a decaying sine curve is traced out (see the transpose operation in Figure 2.36). This decay curve is much like the original FIDs giving rise to the one-dimensional spectrum. If we then Fourier transform these traces, we obtain a two-dimen-

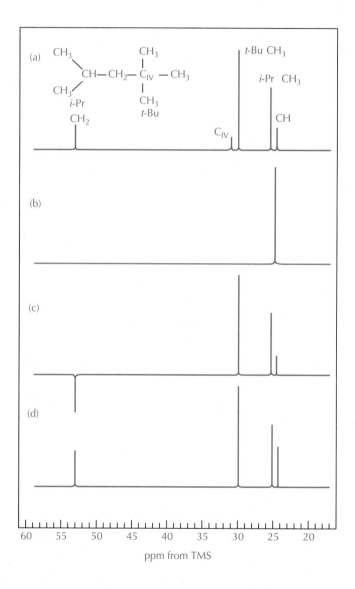

FIGURE 2.37 INEPT ^{13}C NMR spectrum, 50 MHz, C_6D_6 40% solution of isooctane (2,2,4-trimethylpentane) in benzene-d_6. (a) Normal (proton-decoupled) ^{13}C NMR spectrum (all carbons appear). (b) Pulse delay causes only the CH (methine) carbon to appear. (c) New pulse delay causes the CH (methine) and CH_3 carbons to appear and the CH_2 carbon is inverted. (d) New pulse delay causes only the quaternary C carbon to be absent.

sional array of signals, the off-diagonal elements of which reveal information (for example, spin-spin coupling) related to the one-dimensional signal (shown here along the diagonal of the stacked plot—one-dimensional spectra are often displayed along the vertical and horizontal axes of the stacked plot). Finally we see that the stacked plot may be alternatively displayed as a contour plot, a display that is merely a bird's-eye view of the stacked plot, revealing only the significantly sized peaks, but not the grid of lines, of the stacked plot. Such a procedure reveals information that would otherwise be hidden in the one-dimensional spectrum. For example, in a so-called H/H COSY (defined later) spectrum, the off-diagonal peaks reveal spin coupling between protons in the structure. This approach is especially useful where multiplets overlap and produce complex spectra that resist simple coupling analysis.

A quantitative measure of the proton to carbon coupling can be obtained by this technique, but that topic is not of interest here. It is possible to translate this coupling information to peak height information by an appropriately chosen pulse sequence, and the resulting spectra, although not rigorously two-dimensional spectra, are useful because they reveal the number of protons attached to a carbon. In Figure 2.37 we see such spectra, a so-called INEPT (Insensitive Nuclei Enhanced by Polarization Transfer) spectra. More than one experiment is necessary to obtain the entire story. We find a standard carbon spectrum (a, all the spectra in Figure 2.37 are completely proton decoupled), followed by b–d in which different pulse delays (which are related to the coupling proton-carbon coupling constant) are used to reveal structural clues. Spectrum b shows only the methine (CH) resonance. In spectrum c, the CH_3 and CH resonances show normal positive signals and the CH_2 resonance is an inverted signal (methyl, methylene, and methine resonances are now clearly identifiable by comparison of spectrum c to a and b). Finally, spectrum d shows the methyl, methylene, and methine resonances as upright signals, but the weak quaternary carbon signal has disappeared. Other pulse sequences reveal the same kinds of information: the DEPT (Distortionless Enhancement of Polarization Transfer) sequence, the SEFT (Spin Echo Fourier Transform) sequence, and (perhaps the kindest acronym) the APT (Attached Proton Test) sequence.

The most generally useful type of two-dimensional NMR spectrum is the H/H COSY (Correlation Spectroscopy spectrum; see Figure 2.38). H/H COSY is sometimes called HOMCOR; the acronyms vary from one instrument manufacturer to another. In our example, there is a one-dimensional spectrum both along the top, and, in contour fashion, on the diagonal. The off-diagonal elements in "box-fashion" identify spin-spin coupling interactions.

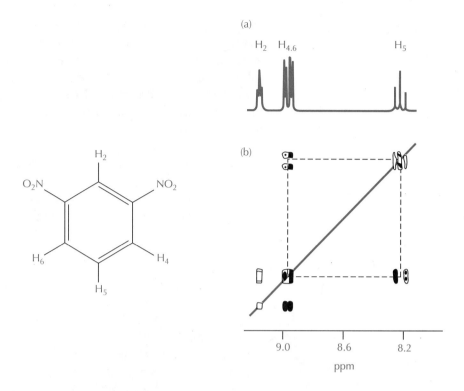

FIGURE 2.38 H/H COSY (¹H NMR, 250 MHz) of *m*-dinitrobenzene. Both *ortho* ($J_{24} = J_{26}$) is indicated by the off-diagonal "box" corners. Coupling between *para* protons was not detected.

Specifically, the spots show that there is coupling between the proton at C-5 (δ 8.25) and the two *ortho* protons at C-4 and C-6 (δ 8.85). The C-5 proton appears as a triplet, and the two pairs of off-diagonal spots (forming a "square box") indicate that the source of the splitting is the δ 8.85 signal. Moreover, with a little imagination we realize that, for more complex compounds, both on-diagonal elements of the box could overlap with other signals. Thus, even with an instrument field strength of 750 MHz, we might not see the triplet nature of the δ 9.25 signal. As an aside, we note the longer range coupling between H_2 and $H_{4,6}$ by the off-diagonal spots (it is not uncommon to see *meta* coupling and, occasionally, *para* coupling in benzene rings).

Another commonly used type of two-dimensional NMR spectrum is the H/C COSY spectrum. H/C COSY is sometimes called HETCOR. In our example (see Figure 2.39), there is a one-dimensional carbon spectrum along the bottom and a one-dimensional proton spectrum along the right side. The two-dimensional array

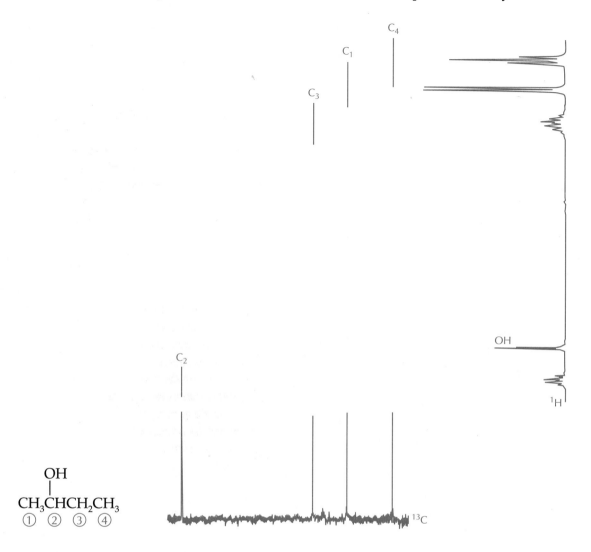

FIGURE 2.39 H/C COSY (HETCOR) ^{13}C NMR (2D) spectrum, 62.9 MHz, 2-butanol. The most downfield ^{13}C singlet is connected to a proton, showing the most downfield signal (sextet) in the ^{1}H NMR spectrum. This is certainly the carbon and its connected proton at C_2.

of four elongated spots identify proton-carbon connectivity. Now we can jump between the two types of spectra, using the more revealing spectrum (that is, the spectrum that is easier to interpret). By linking this to the other spectrum, we can probe and identify features in that other spectrum. For 2-butanol, the ^{1}H signal associated with the OH proton is shown as being a proton on a hetero atom because it does not correlate with a carbon. Overlap in compounds more complex than 2-butanol could be a problem. Specifically, we find that the correlated carbon C-4 corresponds to one for which an APT test has already revealed three attached protons. Again, a little information reveals other possibilities.

A useful concept of two-dimensional NMR is NOESY (<u>N</u>uclear <u>O</u>verhauser <u>E</u>nhancement <u>S</u>pectroscop<u>y</u>) spectroscopy. The two-dimensional array of a NOESY spectrum gives rise to information about spatial relationships of nuclei in organic structures. NOESY analysis is based on the nuclear Overhauser (NOE) effect, so it

A reasonable NOE is expected when two protons are 4Å (or less) apart.

Camphor

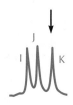

FIGURE 2.40 (a) ¹H NMR (2D) spectrum, 400 MHz, camphor. A "normal" spectrum (no NOE-inducing irradiation). (b) Difference NOE spectrum. This spectrum is the result of irradiation upfield at signal K (C-8 methyl protons). Irradiation results in NOE enhancement at A (3-*exo*-proton) and at B (proton 4). All the other protons receive negligible or slightly negative enhancements. This result is consistent with the 3-*exo* and 4- protons being the closest to the protons on C-8.

reveals distances between nuclei, because it is well known that the enhancement of signals greatly increases as the distance between a properly irradiated nucleus and the nucleus of interest decreases. We can identify nuclei through space distances with NOESY, but not with the other two two-dimensional methods, where we are normally dealing with interactions through bonds. In practice, NOESY analysis is difficult, and a superior procedure is difference NOE spectroscopy (which, although related to two-dimensional methods, is not rigorously a two-dimensional method). Examination of Figure 2.40 shows what difference NOE can reveal. Spectrum a (bottom trace) of Figure 2.40 is a simple one-dimensional spectrum of camphor. Spectrum b arises as the result of irradiation of the most upfield signal (K) and subtraction of spectrum a from the result. Irradiation (see arrow) at signal K is irradiation at methyl group 8, which is clearly closest to the 3-*exo* hydrogen (A) and to H_4(B). All the other signals show either no enhancement or negative enhancement. However, complete assignment of the protons of camphor relies on a wide array of NMR methods, and difference NOE is only part of the approach.

Finally, we will briefly mention a two-dimensional technique called 2D INADEQUATE (Incredible Natural Abundance Double Quantum Transfer Experiment). In simple terms, this technique allows us to connect carbon signals in a way that corresponds to the bonded skeleton for an organic structure. Thus, in principle, carbon skeletons can be obtained for any compound. In practice, it is a very demanding analysis and uses extensive instrument computer time. Thus it is often used only as a last resort.

References

1. Atta-ur-Rahman. *Nuclear Magnetic Resonance: Basic Principles;* Springer-Verlag: New York, 1986.
2. Bovey, F. A. *Nuclear Magnetic Resonance Spectroscopy,* 2nd ed.; Academic Press: New York, 1988.
3. Breitmaier, E.; Voelter, W. *Carbon-13 NMR Spectroscopy,* 3rd ed.; VCH: New York, 1987.
4. Derome, A. *Modern NMR Techniques for Chemistry Research;* Pergamon Press: Oxford, 1987.
5. Duddeck, H.; Dietrich, W. *Structure Elucidation by Modern NMR;* Springer-Verlag: New York, 1989.
6. Pouchert, C. J.; Behnke, J. (Eds.) *Aldrich Library of ^{13}C and ^1H FT-NMR Spectra,* 3 vols.; Aldrich Chemical Co.: Milwaukee, WI, 1993.
7. Sanders, J. K.; Hunter, B. K. *Modern NMR Spectroscopy,* 2nd ed.; Oxford University Press: Oxford, 1993.

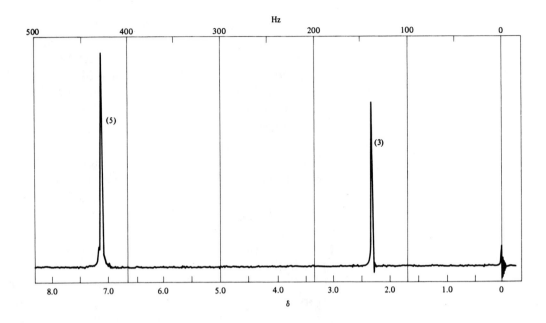

(b)

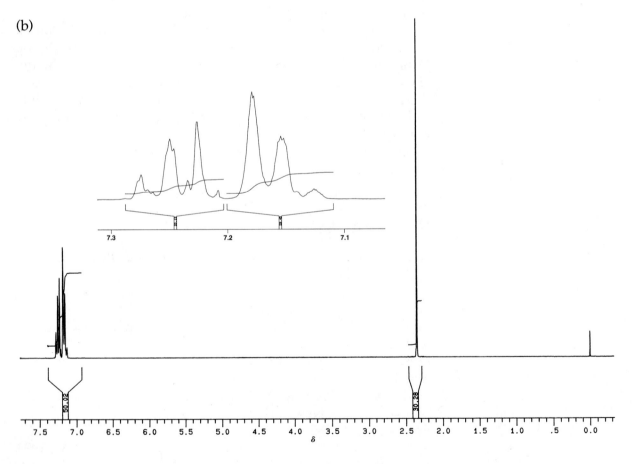

FIGURE 2.41 ^{1}H NMR spectra of toluene for Question 1. (a) 60 MHz. (b) 300 MHz.

Questions

1. Evaluate the ^{1}H NMR spectra of toluene recorded in Figure 2.41 at two substantially different field strengths (60 and 200 MHz). Assign the signals and explain the differences in resolution in the signals at δ 7.2.

2. The ^{1}H and ^{13}C NMR spectra of an unknown compound of molecular formula C_4H_9Cl as shown in Figure 2.42. Deduce the structure of this compound and assign all NMR signals.

3. Both the ^{1}H and the ^{13}C NMR spectra for *N,N*-dimethyl acetamide, $(CH_3)_2N(C{=}O)CH_3$, show two signals for the methyl groups on nitrogen. Use resonance theory to explain what this must mean about the bonding in this amide.

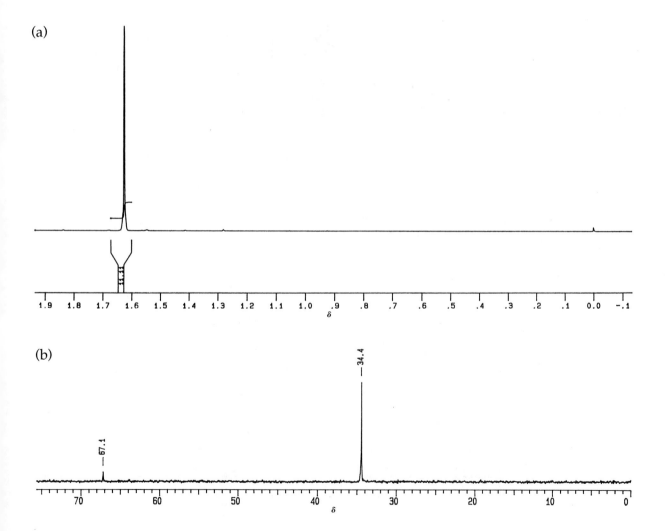

FIGURE 2.42 NMR spectra of C_4H_9Cl for Question 2. (a) ^{1}H, 200 MHz. (b) ^{13}C, 50 MHz.

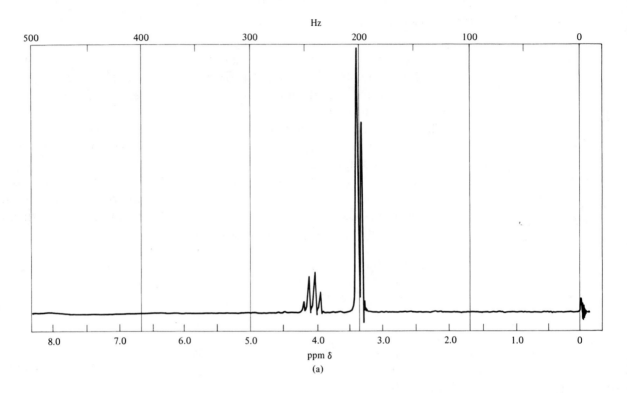

FIGURE 2.43 ¹H NMR spectrum of acid-free CH₃OH, 60 MHz, for Question 4.

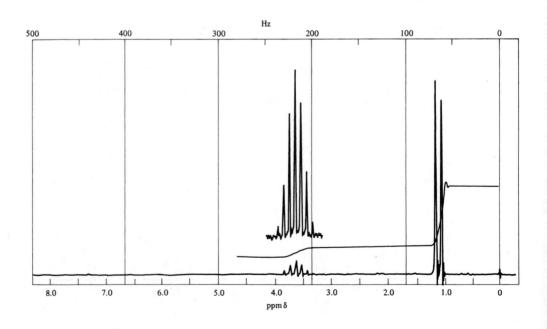

FIGURE 2.44 ¹H NMR spectrum of a dibromoether, 60 MHz, for Question 5.

4. The ^1H NMR spectrum of carefully purified methanol (with care to remove traces of acid) in Figure 2.43 shows multiplets that can be analyzed by first-order ($N + 1$) rules. If, however, even a small amount of acid is present (as is usually the case in even good grades of commercial $CDCl_3$), the doublet and quartet both coalesce to singlets. How can you explain this, using the concept of chemical exchange?

5. A compound is known to be either 1,1'-dibromoethyl ether or 2,2'-dibromoethyl ether (α,α'-dibromoethyl ether or β,β'-dibromoethyl ether). From the ^1H NMR spectrum in Figure 2.44, decide which one and explain your choice.

6. A compound of molecular formula C_4H_9Br gives rise to the ^1H NMR and ^{13}C NMR spectra shown in Figure 2.45. Deduce the structure of the compound and assign all NMR signals.

(a)

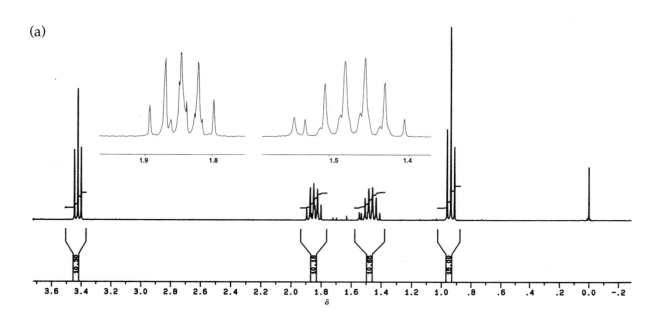

(b)

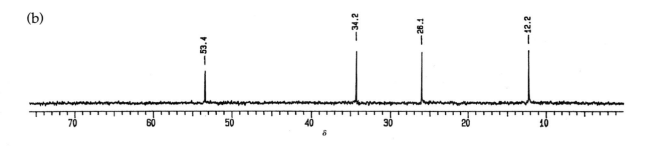

FIGURE 2.45 NMR spectra of C_4H_9Br for Question 6. (a) ^1H, 300 MHz. (b) ^{13}C, 50 MHz.

7. A compound of molecular formula $C_{10}H_{14}$ gives rise to the ^{1}H NMR spectrum shown in Figure 2.46. Using the facts that the ^{13}C NMR reveals seven singlets and those at δ 145.2 and 135.4 are very weak, deduce the structure of the compound and assign all NMR signals.

(a)

(b)

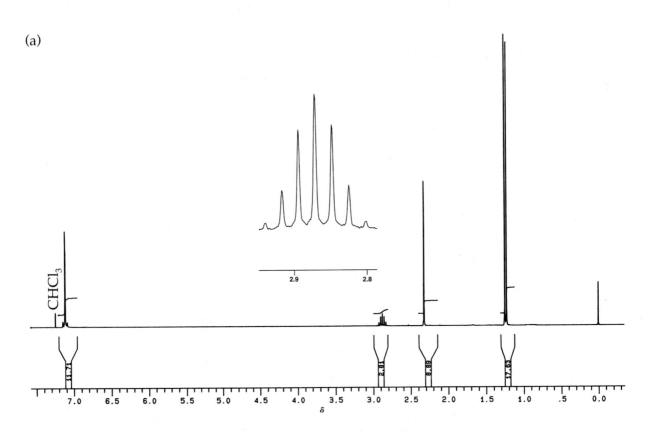

FIGURE 2.46 NMR spectra of $C_{10}H_{14}$ for Question 7. (a) ^{1}H, 300 MHz. (b) ^{13}C, 50 MHz.

8. A compound of molecular formula C_3H_8O gives rise to the ^1H NMR and ^{13}C NMR spectra shown in Figure 2.47. In addition, when this compound is treated with D_2O, the ^1H NMR signal at 2.0 disappears (another signal appears downfield). Moreover, when this compound is highly purified and care is taken to remove all traces of acid from the NMR solvent, the singlet at δ 2.0

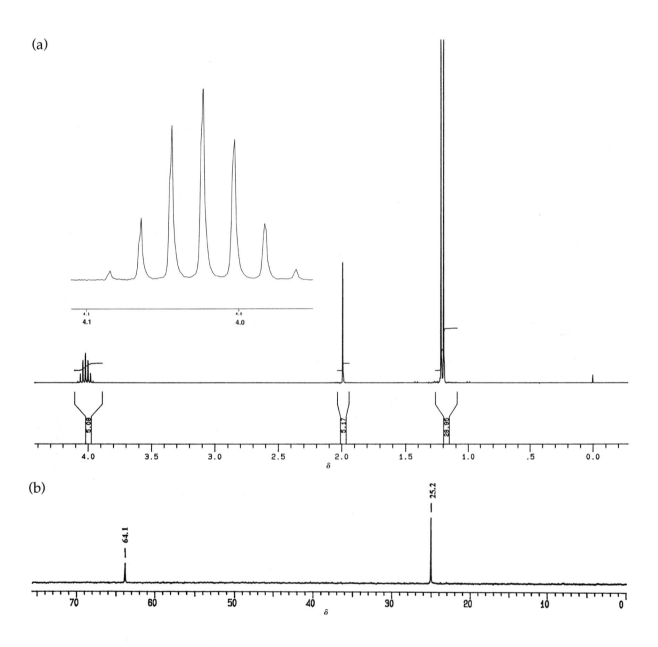

FIGURE 2.47 NMR spectra of C_3H_8O for Question 8. (a) ^1H, 300. (b) ^{13}C, 50 MHz.

is replaced by a doublet. Finally, it is found that the chemical shift of the δ 2.0 signal is highly concentration dependent: An increase in the concentration of this compound results in a downfield shift of this signal. Deduce the structure of this compound, assign all NMR signals, and explain the changes observed for the δ 2.0 signal.

Spectrometric Method 3

MASS SPECTROMETRY

Most spectrometric techniques used by organic chemists are based on the ability of a carbon compound to absorb light, or other radiation, of various energies. This absorption yields structural information. Mass spectrometry, however, is exceptional in that it does

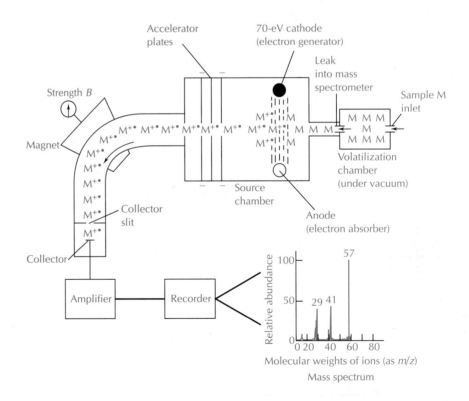

FIGURE 3.1 Schematic representation of a mass spectrometer that uses electron impact ionization. The flow of molecular ion species is from the upper right to the lower left. (From Vollhardt, K. P. C.; Schore, N. E. *Organic Chemistry,* 2nd ed.; W. H. Freeman and Company: New York, 1994; p. 797.)

not involve absorption of light. Mass spectrometry yields structural information because a molecule, when bombarded with a stream of electrons is converted to a radical cation. (See Figure 3.1 for an instrument schematic.) The cation formed in this way from an intact molecule is called a *molecular ion* ($M^{+\cdot}$). Such an approach is called electron impact (EI) mass spectrometry.

$$
M \xrightarrow[\text{electron stream}]{e^-} M \longrightarrow M^{+}\cdot
$$

M = molecule

Radical cation
form of molecule

In the foregoing reaction, the molecule loses one of its own electrons when struck by an external electron and becomes a molecular species ($M^{+}\cdot$) bearing a single positive charge, a molecular ion.

Because the molecular ion, once formed, can divide to form both uncharged and charged fragments, the reaction sequence for specific organic compounds is usually more complex, as illustrated for methane:

$$
CH_4 \xrightarrow{e^-} CH_4^{+}\cdot \xrightarrow{-H\cdot} CH_3^{+} \xrightarrow{-H\cdot} CH_2^{+}\cdot \longrightarrow \text{etc.}
$$

As shown in Figure 3.1, we can take advantage of the radical cation's charge by applying a magnetic field perpendicular to the ion pathway (in Figure 3.1, this field would be perpendicular to the page). The magnetic field causes the pathway for each cation to be curved. We can use this phenomenon for mass analysis because variations in the radius of curvature depend on the mass to charge (m/z) ratio for each ion. Because each ion usually bears a single positive charge, m/z simplifies to m, the mass of the ion.

Normally, a molecule is converted to its gaseous form for simple mass spectrometry. Thus this EI technique is limited to compounds with significant vapor pressures at room temperature. Special ionization techniques can, however, be used to directly ionize samples in the solid state. Thus many samples (such as polymers of biological or synthetic significance) with high molecular weights—and very low vapor pressures—can be studied with mass spectrometry.

Fragmentation   A molecular ion (sometimes called a parent ion) often fragments, so the resulting spectrum can be complex because any one of a number of covalent bonds might be broken during fragmentation. For example, as shown earlier, methane (CH_4) fragments to give CH_3^{+}, $CH_2^{+}\cdot$, and $:CH^{+}$ cations. Because a large number of peaks

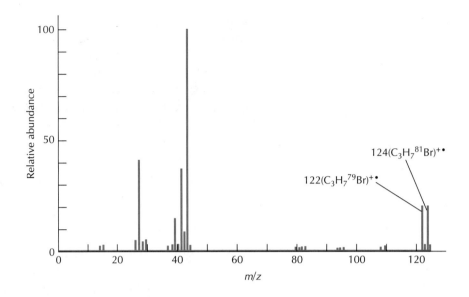

FIGURE 3.2 Mass spectrum of 1-bromopropane. Note the nearly equal heights of the peaks at $m/z = 122$ and 124, due to the almost equal abundances of the two bromine isotopes. (From Vollhardt, K. P. C.; Schore, N. E. *Organic Chemistry*, 2nd ed.; W. H. Freeman and Company: New York, 1994; p. 800.)

do arise for a given compound, a unique fingerprint for each compound can be produced. And, in many cases, the pathways for molecular cleavage can be predicted by using expected fragmentation patterns. In fact, rules of fragmentation, a topic too wide in scope to be covered in this book, have been established. The mass spectrum for 1-bromopropane is shown in Figure 3.2, and the most intense peak, called the base peak, is at m/z 43 ($C_3H_7^+$). It is common to set the abundance of the base peak at 100% and report the abundance of all other peaks as percentages of base, as has been done in Figure 3.2.

An important aspect of Figure 3.2 is the existence of, and the intensity ratio (approximately 1:1) for, the two peaks at m/z 122 and 124. The peak at m/z 122 is by convention defined as the molecular ion peak (M) because it has the formula C_3H_7Br. In this case, the bromine atom has a mass of 79. Although we don't, we could also call the m/z 124 peak a molecular ion peak, because it corresponds to an intact molecule of +1 charge. Even though the mass 124 species has an atomic composition corresponding to the molecular formula (C_3H_7Br), mass spectrometrically it is referred to as the M+2 peak because the bromine atom is a ^{81}Br species. Naturally occurring isotopes often make sizable contributions to M+2 peaks. For instance, it is known that natural bromine exists as a mixture of ^{79}Br and ^{81}Br in very close to a 1:1 ratio. Moreover, isotopic contributions of heavy carbon (^{13}C, ^{14}C) and heavy hydrogen (^{2}H, ^{3}H) to the M and M+2 peaks are negligibly small; thus the M and M+2 peaks also occur in very close to a 1:1 ratio. This ratio clearly shows that the molecule contains one bromine atom. Tiny contributions of the heavy isotopes of carbon and hydrogen can be observed, and in Figure 3.2 these are revealed as a small but observable m/z 123 peak (M+1 peak). Figure 3.2 shows clearly that the base peak need not be the molecular ion peak.

Carbon, nitrogen, hydrogen, and oxygen all make small contributions to the M+1 and M+2 peaks, and the resulting intensities have been used to reveal the molecular formulas of organic compounds. Figure 3.3 is the mass spectrum of methane. The relative intensities of the m/z 16 (M) and m/z 17 (M+1) peaks can be used to reveal the molecular formula (CH_4). The m/z 17 peak is due to molecules with formulas such as $^{13}C^{1}H_4$, $^{12}C^{1}H_3{}^{2}H$, and so on. In other words, the ratio of the intensity of the M+1 peak to that of the M peak is due to the isotopic abundances of the atoms in a molecule and the number of each type of atom.

Tables listing molecular formulas and the M/M+1/M+2 ratios expected for these formulas have been constructed. The data of Figure 3.3 suggests that the use of these expected ratios is reasonable. The m/z 17 peak of methane is expected (from such tables) to be 1.15% of the M peak (set at 100%), and the table of data in Figure 3.3 reveals that the M+1 peak was found to be 1.1% for this analysis of methane. Practical experience, however, has shown that for many C, H, N, and O compounds, the expected

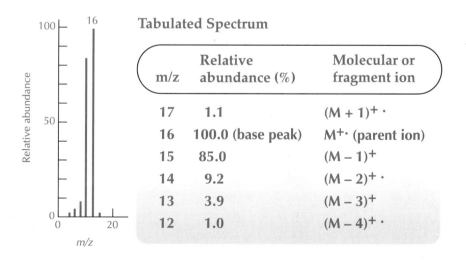

Tabulated Spectrum

m/z	Relative abundance (%)	Molecular or fragment ion
17	1.1	$(M + 1)^{+}\cdot$
16	100.0 (base peak)	$M^{+}\cdot$ (parent ion)
15	85.0	$(M - 1)^{+}$
14	9.2	$(M - 2)^{+}\cdot$
13	3.9	$(M - 3)^{+}$
12	1.0	$(M - 4)^{+}\cdot$

FIGURE 3.3 Mass Spectrum (bar graph form) and the corresponding tabular data for methane. (From Vollhardt, K. P. C.; Schore, N. E. *Organic Chemistry,* 2nd ed.; W. H. Freeman and Company: New York, 1994; p. 798.)

ratios are often in error. Errors can occur for many reasons. For example, the sample may undergo ion-molecule collisions that provide extensive and undesirable sources of the M+1 peak:

$$CH_4^+ \cdot + CH_4 \longrightarrow CH_5^+ + CH_3 \cdot$$
$$ m/z\ 16 m/z\ 17$$

Impurities also can cause the same types of reactions to occur. Even when the sample is pure, it is often necessary to carry out the time-consuming task of obtaining the entire spectrum several times and averaging the M+1 values.

Despite the limitations just described, the intensities of the M+1 and M+2 peaks occasionally assist in formula determinations for organic molecules. We can use the following equations to calculate the expected intensities of M+1 and M+2 peaks (relative to the M peak set arbitrarily at 100%). These formulas can work when the organic compounds contain nothing other than C, H, N, O, F, P, and I.

Comparison of observed M+1 and M+2 intensities to those calculated can support a molecular formula.

$$\% (M+1) = 100\left[\frac{M+1}{M}\right]$$
$$= 1.1\ (\text{no. of C atoms}) + 0.36\ (\text{no. of N atoms})$$

$$\% (M+2) = 100\left[\frac{M+2}{M}\right]$$
$$= \frac{1.1\ (\text{no. of C atoms})^2}{200} + 0.20\ (\text{no. of O atoms})$$

Table 3.1. Relative isotope abundances of common elements

Elements	Isotope	Relative abundance	Isotope	Relative abundance	Isotope	Relative abundance
Carbon	^{12}C	100	^{13}C	1.11		
Hydrogen	^{1}H	100	^{2}H	0.016		
Nitrogen	^{14}N	100	^{15}N	0.38		
Oxygen	^{16}O	100	^{17}O	0.04	^{18}O	0.20
Fluorine	^{19}F	100				
Silicon	^{28}Si	100	^{29}Si	5.10	^{30}Si	3.35
Phosphorus	^{31}P	100				
Sulfur	^{32}S	100	^{33}S	0.78	^{34}S	4.40
Chlorine	^{35}Cl	100			^{37}Cl	32.5
Bromine	^{79}Br	100			^{31}Br	98.0
Iodine	^{127}I	100				
Boron	^{10}B	24.8	^{11}B	100		

Table 3.1 provides information we can use to analyze certain compounds in much the same way we analyzed 1-bromopropane earlier (Figure 3.2). From this table, we can identify elements that will make substantial contributions to the M+2 peak. These include Si, S, Cl, and Br; and these contributions are all certainly much larger than those made by C, H, N, or O. For example, a monochloro compound is expected to have an M+2 peak that is 32.5% as intense as the M peak. A sulfide would have an M+2 peak 4.4% as intense as the M peak, and a disulfide, 8.8% as intense. These values ignore the negligible contributions made by C, H, N, and O. When we move on to polyhalo compounds, the absolute values are intensity

Atom content	M+2, M+4, M+6*
Br_2	195.0, 95.5
Br_3	293.0, 286.0, 93.4
Cl_2	65.3, 10.6
Cl_3	97.8, 31.9, 3.47
BrCl	130.0, 31.9
Br_2Cl	228.0, 159.0, 31.2
Cl_2Br	163.0, 74.4, 10.4

*Percentages relative to M = 100%.

High-Resolution Mass Spectroscopy

A superior approach to determining formulas is high-resolution mass spectrometry. High-resolution instruments use both electric and magnetic fields for focusing the ion pathways. Such double-resolution instruments yield masses that contain four figures beyond the decimal place. Table 3.2 provides the masses that should be used for this approach. Mass spectrometry can be described as a microscopic approach to molecular weight analysis, whereas simple laboratory measurements on a mole basis can be described as macroscopic procedures. On this basis, we can appreciate how to use Table 3.2. If we were weighing out a certain number of grams of a compound for a chemical reaction (a macroscopic procedure), we would use the atomic weights that reflect the natural isotopic composition (protium mixed with a small amount of deuterium, ^2H = D, and an even smaller amount of tritium, ^3H = T). Therefore we would use 1.00794 as shown in Table 3.1. But the molecular weight of a compound used for mass spectrometry (which we recall is a microscopic procedure) requires the use of the exact mass of the most abundant nuclide (here protium, mass 1.00783). Thus the molecular weight determined for mass spectrometric purposes will differ slightly from that used for macroscopic purposes. For example, if we were interested in carrying out the oxidation of a mole of methane (a macroscopic operation), we could use a molecular weight

Table 3.2. **Exact masses of isotopes**

Element	Atomic weight	Nuclide	Mass
Hydrogen	1.00794	^{1}H	1.00783
		D(^{2}H)	2.01410
Carbon	12.01115	^{12}C	12.00000 (std)
		^{13}C	13.00336
Nitrogen	14.0067	^{14}N	14.0031
		^{15}N	15.0001
Oxygen	15.9994	^{16}O	15.9949
		^{17}O	16.9991
		^{18}O	17.9992
Fluorine	18.9984	^{19}F	18.9984
Silicon	28.0855	^{28}Si	27.9769
		^{29}Si	28.9765
		^{30}Si	29.9738
Phosphorus	30.9738	^{31}P	30.9738
Sulfur	32.066	^{32}S	31.9721
		^{33}S	32.9715
		^{34}S	33.9679
Chlorine	35.4527	^{35}Cl	34.9689
		^{37}Cl	36.9659
Bromine	79.9094	^{79}Br	78.9183
		^{81}Br	80.9163
Iodine	126.9045	^{127}I	126.9045
Boron	10.811	^{10}B	10.0129
		^{11}B	11.0093

of $12.00115 + 4(1.00794) = 16.03291$. But the molecular weight for high-resolution mass spectrometry would be $[12.0000 + 4(1.00783)] = 16.03132$.

Problem Solving   Let's consider high-resolution mass spectrometry in a problem-solving context. Suppose that a compound with a nominal, or integral, mass of 98 was believed to have a molecular formula of either C_7H_{14} or $C_6H_{10}O$. The exact masses (using the exact masses of the most abundant isotopes of each atom) of these two compounds are, respectively, 98.1096 and 98.0732. A high-resolution result of 98.1082 is clearly more consistent with the formula C_7H_{14}.

For both high-resolution and unit-resolution mass spectrometry, fragmentation is a complicating but useful aspect. At first, the overwhelming number of peaks in a mass spectrum that are due to fragmentation is daunting. The spectrum of 1-bromopropane shown in Figure 3.2 contains 27 peaks, only four of which (m/z 122, 123, 124, and 125) correspond to molecular ion species. The peak of mass 122 corresponds to C_3H_7Br; that at 123 corresponds to a C_3H_7Br molecule containing an isotope one mass unit greater than the corresponding atom in the mass 122 molecule; the peak at 124 corresponds either to a molecule containing two atoms each with isotopes one mass unit higher, or one with an isotope two mass units higher (such as ^{81}Br); and so on. The mass spectrum of the female sex hormone estrone ($C_{18}H_{22}O_2$) contains over 100 peaks, only two of which are in the molecular ion cluster. As mentioned earlier, the array of fragmentation peaks constitutes a fingerprint that can be used for identification. In fact, modern spectrometers are routinely equipped with computer libraries containing tens of thousands of such fingerprints for matching purposes; and these fingerprints are very useful for identifications. An understanding of such fragmentation processes certainly enhances appreciation of the value of mass spectrometry.

Mechanisms depicting fragmentation are enhanced by the efficient use of arrows. The fishhooks (arrows with half-heads) represent the migration of a single electron; this notation is similar to that used in free radical or photochemical processes:

$$M \longrightarrow \underbrace{M^{+\cdot} = [f' \overset{\frown}{\rule{1em}{0pt}} f]^{+\cdot}}_{\substack{\text{Molecular} \\ \text{cation radical}}} \longrightarrow \underset{\text{Fragments}}{f'^{\cdot} + f^{+}}$$

Forces that contribute to the ease with which fragmentation processes occur include the strength of bonds in the molecule (for example, the f'—f bond) and the stability of the free radicals ($f'\cdot$) and carbocations (f^{+}) produced by fragmentation. Although these fragments are formed in the gas phase, we can still apply our "chemical intuition," which is generally based on reactions in solution. Dehydration is a common laboratory reaction carried out in solution; thus it is reassuring to find that mass spectrometric fragmentation also often involves the loss of simple molecules such as water, H_2S, CO, and NH_3. But it is less reassuring to find that these losses may occur in either one step or two steps. Furthermore, we may find fragments at m/z values corresponding to M−H, M−OH, and M−H_2O. Even more upsetting is the occasional observation of the loss of 3H or the appearance of a fragment that must have been formed in a very large number of steps involving exceedingly complex rearrangements.

Despite the aforementioned complexities, there are a large number of fragmentation patterns that occur in an orderly fashion. Benzylic, allylic, and ring carbons are good positions for cleavage (in each of the following reactions, the first structure corresponds to a molecular ion species; G represents various groups):

Benzylic cleavage:

$$\left[\begin{array}{c} \text{Ph–CH}_2\text{–G} \end{array}\right]^{+\cdot} \begin{array}{c} \nearrow \;\; \overset{+}{\text{CH}}_2\text{–Ph} + \text{G}\cdot \\ \\ \searrow \;\; \overset{\cdot}{\text{CH}}_2\text{–Ph} + \text{G}^+ \end{array}$$

Allylic cleavage:

$$\left[\begin{array}{c} \text{CH}_2\text{=CH–CH}_2\text{–G} \end{array}\right]^{+\cdot} \begin{array}{c} \nearrow \;\; \overset{+}{\text{CH}}_2 + \text{G}\cdot \\ \\ \searrow \;\; \overset{\cdot}{\text{CH}}_2 + \text{G}^+ \end{array}$$

Cleavage at a ring:

$$\left[(\text{CH}_2)_n\,\text{CH}{-}{-}\text{R}\right]^{+\cdot} \begin{array}{c} \nearrow \;\; (\text{CH}_2)_n\,\text{CH}^+ + \text{R}\cdot \\ \\ \searrow \;\; (\text{CH}_2)_n\,\text{CH}\cdot + \text{R}^+ \end{array}$$

The carbonyl group can be a fragmentation site because the resulting charged fragment can be stabilized by resonance:

$$\left[\begin{array}{c} \overset{\text{O}}{\underset{\text{G}\;\;\;\;\text{G}}{\overset{\|}{\text{C}}}} \end{array}\right]^{+\cdot} \longrightarrow \text{G–C=}\overset{..}{\underset{..}{\text{O}}} + \text{G}\cdot$$

Aldehydes, ketones, esters

$$[\text{G–}\overset{+}{\text{C}}\text{=}\overset{..}{\underset{..}{\text{O}}} \xleftrightarrow{\text{resonance}} \text{G–C}\equiv\text{O}{:}^+]$$

We can expect analogous C=N groups to facilitate the same type of cleavage.

Hetero atoms can provide the driving force for other types of cleavage. Primary alcohols show an intense m/z 31 peak, which arises as a result of a cleavage in which the electron pair on the oxygen atom stabilizes the fragment by resonance:

$$CH_3\!\!-\!\!CH_2\!\!-\!\!\overset{+\cdot}{\underset{\cdot\cdot}{O}}H \longrightarrow [CH_2\!\!=\!\!\overset{+}{\underset{\cdot\cdot}{O}}H \longleftrightarrow \overset{+}{C}H_2\!\!-\!\!\overset{\cdot\cdot}{O}H]$$

Mass 31

Appropriately structured amines, ethers, and sulfur compounds undergo similar cleavage:

$$R\!\!-\!\!\overset{+\cdot}{C}H_2\!\!-\!\!Y \xrightarrow{\ -R\cdot\ } [CH_2\!\!=\!\!\overset{+}{Y} \longleftrightarrow \overset{+}{C}H_2\!\!-\!\!\overset{\cdot\cdot}{Y}]$$

$Y = NH_2$, NHR, NR_2,
OR, SH, or SR

The fundamental nitrogen rule states that a compound (containing nothing other than C, H, N, or O) whose molecular weight is an even number must either contain no nitrogen atoms or an even number of nitrogens. A compound whose molecular weight is an odd number must contain an odd number of nitrogens. The following compounds support the nitrogen rule:

Pyridine	Triethylamine	Benzylamine	Aniline
MW 79	MW 101	MW 107	MW 93

$(CH_3CH_2)_3N$

There is a corollary to this rule that applies to fragmentation. The corollary states that simple fragmentation (cleavage of just one covalent bond) of a compound with an even molecular weight gives a fragment whose mass is an odd number, and cleavage of a molecule whose molecular weight is odd gives a fragment with a mass that is an even number:

$$\left[H_3C\!\!-\!\!CH_3\right]^{+\cdot} \longrightarrow CH_3^+ + CH_3\cdot$$

MW 30 Mass 15

$$(CH_3CH_2)_2\overset{+\cdot}{N}\!\!-\!\!CH_2\!\!-\!\!CH_3 \longrightarrow (CH_3CH_2)_2\overset{+}{N}\!\!=\!\!CH_2 + CH_3\cdot$$

MW 101 Mass 86

Although there are exceptions to these rules, in general they are useful in the analysis of mass spectrometric data.

In summary, mass spectrometry can be used to identify organic compounds. Molecular weights and molecular formulas can be obtained for many organic compounds, both of which are frequently significant clues to the identity and the structure of an organic compound. Moreover, the basic mechanisms for molecular fragmentation have been extensively studied; the detailed patterns provided by a compound's fragmentation profile can very often establish the identity of a compound.

References

1. Beynon, J. H.; Saunders, R. A.; Williams, A. E. *The Mass Spectrometry of Organic Molecules;* Elsevier: New York, 1968.
2. Budzikiewicz, H.; Djerassi, C.; Williams, D. H. *Mass Spectrometry of Organic Compounds;* Holden-Day: San Francisco, 1967.
3. Budzikiewicz, H.; Djerassi, C.; Williams, D. H. *Structure Elucidation of Natural Products by Mass Spectrometry;* Holden-Day: San Francisco, 1964; 2 vols.
4. McLafferty, F. W. *Interpretation of Mass Spectra,* 3rd ed.; University Scientific Books: Mill Valley, CA. 1980.
5. Stenhagen, E.; Abrahamsson, S.; McLafferty, F.; Eds. *Atlas of Mass Spectral Data;* Interscience-Wiley: New York, 1969; Vol. 1: Molecular Weight 16.0313-187.1572; Vol. 2: Molecular Weight 142.0185-213.2456; Vol. 3: Molecular Weight 213.8629-702.7981. The three volumes have complete EI data for about 6000 compounds.
6. Stenhagen, E.; Abrahamsson, S.; McLafferty, F.; Eds. *Registry of Mass Spectral Data;* Wiley-Interscience: New York, 1974; Vol. I: MW 16.0313-187.1572, Formula CH_4-$C_{10}H_{21}NO_2$; Vol. II: MW 187.1572-270.1256, Formula $C_{10}H_{21}NO_2$-$C_{17}H_{18}O_3$; Vol. III: MW 270.2671-402.3305, Formula $C_{16}H_{34}N_2O$-$C_{25}H_{13}O_3B$; Vol. IV: MW 40.3305-1519.806, Formula $C_{25}H_3O_3B$-$(C_9H_{18}D_3O_3Al)n$. The four volumes contain bar graphs of 18,806 compounds. Volume IV also contains the index for all four volumes.

Questions

1. Match the following compounds (azobenzene, ethanol, pyridine) to their molecular weights (46, 79, 182). How does the fact that in one case the molecular weight is odd and in the other two cases the molecular weight is even help in the selection process?
2. If the mass spectrum of a compound of molecular weight 93 shows a peak m/z 46.5, what does this suggest about the ionization process?
3. The mass spectrum of 1-bromopropane is shown in Figure 3.2. Propose a source for the m/z 43 peak.
4. 1-Pentanol has its base peak at m/z 31, whereas that for 2-pentanol is at m/z 45. Explain briefly.
5. Similar types of cleavages give rise to two peaks for 4-chlorobenzophenone, one at m/z 105 (the base peak) and one at m/z 139

(70% of base). A clue to their identities is the fact that, whereas the 139 peak is accompanied by a peak at 141 that is about one-third the intensity of the 139 peak, the 105 peak has no such partner. What structures correspond to these peaks?

Spectrometric Method 4

ULTRAVIOLET AND VISIBLE SPECTROMETRY

Electronic transitions in organic compounds often lead to absorption of ultraviolet radiation having wavelengths between 200 and 400 nanometers (nm, 10^{-9} meters) and visible light between 400 and 800 nm. An electronic transition corresponds to the promotion of an electron either from a bonding (for example, a σ or π) orbital or from a nonbonding (n) orbital to an antibonding orbital (σ^* or π^*). We will be concerned only with $\pi \rightarrow \pi^*$ and n $\rightarrow \pi^*$ transitions, and from those classes, largely only the transitions for conjugated organic compounds because these transitions are likely to fall in the 200–400 nm region. Conjugated compounds (Figure 4.1) have either one or more pairs of alternating double bonds, as seen in 1,3-butadiene, or n–π conjugation, as seen in methyl vinyl ether. Thus the following conjugated compounds are ones that readily lend themselves to characterization by UV:

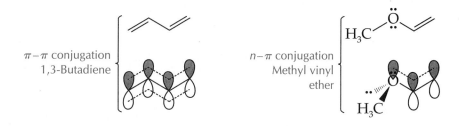

FIGURE 4.1 Structures illustrating π–π and n–π conjugation.

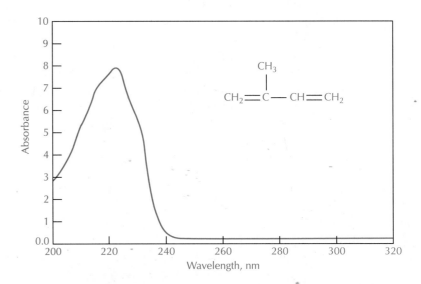

1,3,5-Hexatriene   Benzene   Styrene   Anisole
(methoxybenzene)

The wavelength of maximum absorbance is denoted λ_{max}.

Molar absorptivity (ε) is a molecular property of the absorbing species that depends on the structure of the compound and on the wavelength of irradiation used.

Figure 4.2 is the ultraviolet spectrum (trace) of 2-methyl-1,3-butadiene (isoprene). This compound has a λ_{max} of 222.5 nm and a molar absorptivity (ε) of 10,800.

The expressions that describe the magnitude of the energy gap (ΔE) in terms of the wavelength of absorbed radiation (measured in nanometers) for a transition are

$$\Delta E = h\nu = h(c/\lambda)$$

where h = the Planck constant and c = the speed of light. Thus the energy gap has an inverse dependence on the wavelength of absorbed light; therefore, the smaller the gap, the longer the wavelength of light.

Numerical determinations of the energy for electronic transitions can be done, using the following form of the energy equation:

FIGURE 4.2 UV spectrum of 2-methyl-1,3-butadiene in methanol. There is a λ_{max} at 222.5 nm (ε_{max} = 10,800) and shoulders on both sides of the maximum. Adapted from Vollhardt, K. P. C.; Schore, N. *Organic Chemistry*, 2nd ed.; W. H. Freeman: New York, 1994.

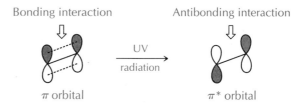

Bonding interaction   Antibonding interaction

π orbital π^* orbital

FIGURE 4.3 Bonding and antibonding molecular orbitals of the ethylene (ethene) π system.

$$\Delta E = h(c/\lambda) = (1.20 \times 10^5 \text{ kJ} \cdot \text{mol}^{-1})/(\lambda \text{ in nm})$$

It is common to obtain the UV spectrum over a range of wavelengths (λ) equal to 200 to 400 nm. The preceding equation shows that this wavelength range corresponds to an energy range of $600-300$ kJ·mol^{-1}.

We can readily visualize the $\pi \rightarrow \pi^*$ transition for ethene (Figure 4.3). When light having an energy corresponding to a wavelength of 171 nm (a value that falls below the practical region for UV analysis) is absorbed, the π bonding orbital of ethylene is excited to a π^* orbital. Support for this bonding change comes from the fact that UV radiation can be used to isomerize substituted double bonds. Thus *cis-* and *trans-2*-butene can be interconverted by treatment with ultraviolet radiation:

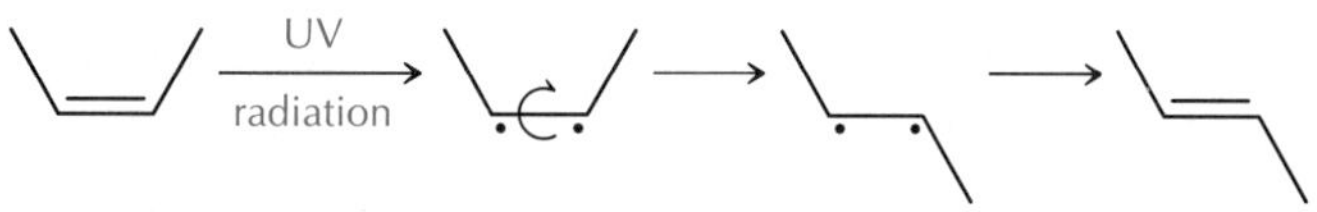

The energy gap for the $\pi \rightarrow \pi^*$ is shown in Figure 4.4. When radiation of an energy corresponding to the difference between

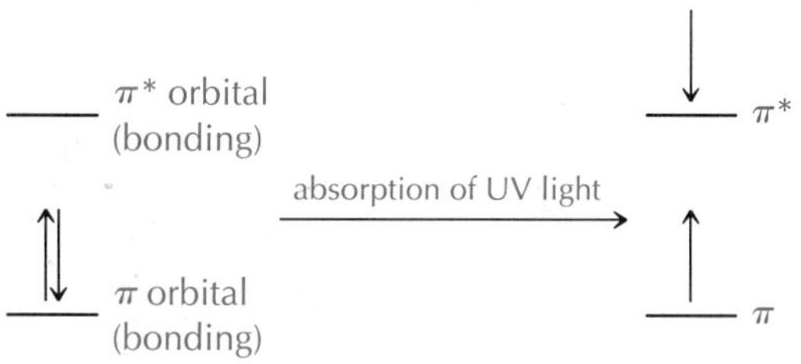

π^* orbital
(bonding)

absorption of UV light

π orbital
(bonding)

π^*

π

FIGURE 4.4 Orbital energy levels for ethylene, illustrating promotion of an electron from the π (bonding) level to the π^* (antibonding) level.

the π and π^* levels is directed on the compound, an electron is promoted from the π to π^* level and UV radiation is absorbed.

When reporting UV results, it is conventional to report the solvent used, the wavelength of maximum absorption (λ_{max}), and the intensity of absorption (ε = molar absorptivity) at the λ_{max} (thus ε_{max}). The molar absorptivity is the ratio of the absorbance (A) to the molar concentration of the sample (C):

$$\varepsilon = A/C$$

The dependence of UV patterns on the basic conjugated units of a compound is made clear by comparing the absorbances of mesityl oxide and cholest-4-en-3-one. Despite the fact that their overall structures are drastically different, the UV spectra for these

Table 4.1. **UV data for various organic compounds**

Compound	Name	λ_{max} [a]	ε_{max}
$CH_2{=}CH_2$	Ethene (ethylene)	171	15,500
	1,4-Pentadiene	178	
	1,3-Butadiene	217	21,000
	2-Methyl-1,3-butadiene (isoprene)	222.5	10,800
	1,3-Cyclopentadiene	239	4200
	2,5-Dimethyl-2,4-hexadiene	241.5	13,100
	1,3-Cyclohexadiene	259	10,000
	1,3,5-Hexatriene	268	36,300

a. All transitions are $\pi \rightarrow \pi^*$ except acetone and the longer wavelength band of methyl vinyl ketone, which are n $\rightarrow \pi^*$ transitions.
b. These are log(ε_{max}) values.
c. An orange precursor of vitamin A.

FIGURE 4.5 Electronic transitions between energy levels, illustrating the $\pi \rightarrow \pi^*$ transition and the lower energy $n \rightarrow \pi^*$ transition.

two compounds are virtually identical as a result of the conjugated enone chromaphore (C=C—C=O) common to both.

Mesityl oxide

Cholest-4-en-3-one

There are differences between $\pi \rightarrow \pi^*$ and $n \rightarrow \pi^*$ transitions. The horizontal lines in Figure 4.5 correspond to orbital energy

Table 4.1. UV data for various organic compounds *(continued)*

Compound	Name	λ_{max}^{a}	ε_{max}
	Acetone	279	13
	3-Butene-2-one (methyl vinyl ketone)	212.5 320	3.85[b] 1.32[b]
	5,7,9(11)-Cholestratriene	324	
	1,3,5,7-Octatetraene	330	
	β-Carotene[c]	497	133,000

a. All transitions are $\pi \rightarrow \pi^*$ except acetone and the longer wavelength band of methyl vinyl ketone, which are $n \rightarrow \pi^*$ transitions.
b. These are $\log(\varepsilon_{max})$ values.
c. An orange precursor of vitamin A.

levels. You will find that usually these levels exist at increasing energy positions, going from bonding (for example, π) to nonbonding (n) up to antibonding (here π^*). Thus $\pi \rightarrow \pi^*$ transitions are usually observed at shorter wavelengths than $n \rightarrow \pi^*$ transitions because of the larger energy gap for $\pi \rightarrow \pi^*$ transitions and the inverse relationship between energy and wavelength. Also you will likely find that $\pi \rightarrow \pi^*$ transitions are more intense (ε = in the thousands or more), whereas $n \rightarrow \pi^*$ transitions are weaker (ε = 10–500). For example, acrolein ($CH_2=CH-CH=O$) shows a $\pi \rightarrow \pi^*$ transition at 210 nm (ε = 11,500) and an $n \rightarrow \pi^*$ transition at 315 nm (ε = 14). The reason for these differences is the fact that $\pi \rightarrow \pi^*$ transitions are "allowed" in the quantum mechanical sense and $n \rightarrow \pi^*$ transitions are "forbidden."

Table 4.1 provides the λ_{max} and the ε values for a variety of conjugated hydrocarbons. Table 4.2 provides the cutoff wavelength for standard UV solvents. For example, cyclohexane can be used as a solvent from 400 nm down to 210 nm, whereas chloroform is not useful below 255 nm.

Another aspect of UV spectra that is readily predicted on the basis of structural theory is the effect of increased conjugation on the position of the λ_{max}. You will see (Figure 4.6) that connecting the ends of two ethylene systems (λ_{max} = 171 nm) to form 1,3-butadiene causes a substantial reduction in the energy gap for the major $\pi \rightarrow \pi^*$ transition. As a result, the λ_{max} for 1,3-butadiene

Table 4.2. **Low end wavelength cutoffs for standard UV solvents**[a]

Solvent[a]	Low end cutoff (nm)
Acetonitrile	195
Chloroform	255
Cyclohexane	210
1,4-Dioxane	220
Ethanol	210
Hexane	200
Methanol	210
Isooctane	200
Water	200

a. All solvents are useful from the table value up to 400 nm.

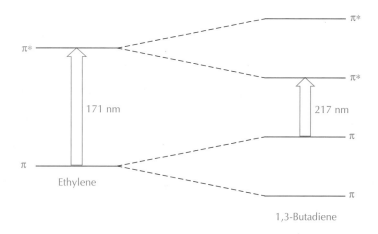

FIGURE 4.6 The split of π energy levels upon changing from ethylene to 1,3-butadiene.

occurs at a substantially longer wavelength (217 nm) because the structural change from ethylene to butadiene increases the total number of orbital energy levels from two to four, and the four butadiene levels now include two bonding levels (π) and two antibonding (π^*) levels. Finally the gap between the lowest antibonding level and the highest bonding level is significantly less than the corresponding gap for the ethylene orbitals.

Visible Light Absorption Colored compounds are structural examples that illustrate the relationship between UV and visible (VIS) absorption. Compounds with eight or more conjugated double bonds absorb in the visible region (400–800 nm). One example is the orange compound β-carotene (Table 4.1, λ_{max} = 497 nm), which has 11 conjugated double bonds. Another is lycopene (λ_{max} = 505 nm), which also has 11 conjugated double bonds. Lycopene is isolated from fresh, ripe tomatoes and is one of the major reasons for their red color.

Lycopene

There is a complementary relationship between the color of a compound and the color (or wavelength) of the light it absorbs. For example, lycopene absorbs light with a wavelength of 505 nm.

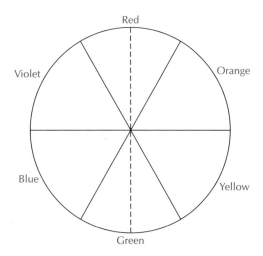

FIGURE 4.7 Color wheel. A compound that is (transmits) red will absorb green light.

Therefore, the color of the light absorbed is green. The color wheel in Figure 4.7 reveals the fact that lycopene, which absorbs green light, is expected to display (transmit) the color on the opposite side of the wheel, that is, red.

Using UV Spectrometry

A practical context for the application of UV spectrometry would be the determination of the structure for the isomeric pentadienes (A through C) of molecular formula C_5H_8. Exhaustive hydrogenation (metal catalyst) of A–C gives pentane in all three cases, a result establishing the straight-chain nature of the carbon skeleton. Compound A has a λ_{max} at 178 nm, compound B at 211 nm, and compound C at 215 nm. The λ_{max} (178 nm) for 1-pentene is a useful reference value. We can conclude that A is 1,4-pentadiene, because the double bonds are not conjugated. Thus the 1,4-isomer would be expected to have a λ_{max} similar to that of 1-pentene. Because the UV spectrum of B and C shows the λ_{max} values at longer wavelengths, we know that their double bonds are conjugated. Differentiating between B and C would require another technique, such as ^{1}H NMR.

References

1. Jaffe, H. H.; Orchin, M. *Theory and Application of Ultraviolet Spectroscopy;* Wiley: New York, 1962
2. Scott, A. I. *Interpretation of the Ultraviolet Spectra of Natural Products;* Pergamon (Macmillan): New York, 1964
3. Stern, E. S.; Timmons, T. C. J. *Electronic Absorption Spectroscopy in Organic Chemistry;* St. Martin's: New York, 1971.

Questions

1. Acetaldehyde shows two UV bands, one with a λ_{max} of 289 nm (ε_{max} = 12.5) and one with a λ_{max} of 182 nm (ε_{max} = 10,000). Which is the $n \rightarrow \pi^*$ transition and which is the $\pi \rightarrow \pi^*$ transition, and why?

2. It should not be surprising to find that cyclohexane and ethanol are reasonable UV solvents, whereas toluene is not. Why?

3. Benzene shows more than one UV maximum. Use the energy levels below (obtained for benzene from simple Hückel MO calculations) to explain this.

$$\underline{\quad} \pi_6^*$$

$$\pi_4^* \underline{\quad} \qquad \underline{\quad} \pi_5^*$$

$$\pi_2 \underline{\uparrow\downarrow} \qquad \underline{\uparrow\downarrow} \pi_3$$

$$\underline{\uparrow\downarrow} \pi_1$$

4. 1-Pentene and 1,4-pentadiene cannot be differentiated on the basis of their λ_{max} values. They can, however, be identified on the basis of their molar absorptivities (ε_{max}). Explain.

INTEGRATED PROBLEMS

1. A compound of molecular formula C_4H_8O shows a peak of substantial size in its mass spectrum at m/z 43. This compound gives rise to an infrared spectrum that shows, among other absorptions, four peaks in the 2999–2850 cm^{-1} range and a strong peak at 1715 cm^{-1}. There are no IR peaks of substance at greater than 3000 cm^{-1}. The ^1H NMR spectrum contains a triplet (t) at $\delta 1.08$ (3H), a singlet (s, 3H) at $\delta 2.15$, and a quartet (q) at $\delta 2.45$ (2H). The magnitudes of the splitting of both the quartet and triplet are identical.

The proton-decoupled ^{13}C NMR spectrum reveals singlets at δ209, 37, 29, and 8.0. DEPT analysis shows these signals to be due respectively to quaternary (C), methylene, methyl, and methyl carbons. Deduce the structure of this compound and assign all peaks.

2. a. A compound shows a molecular ion peak in its mass spectrum at m/z 92 and a satellite peak at m/z 94 that is 36% the intensity of the m/z 92 peak. The ^{1}H NMR spectrum contains only one signal, a singlet at δ1.65. The proton-decoupled ^{13}C NMR spectrum reveals a strong singlet at δ34.6 and a weak singlet at δ67.2. DEPT analysis shows these signals to be due respectively to methyl and quaternary (C) carbons. Deduce the structure of this compound and assign all peaks.

 b. A compound shows a molecular ion peak in its mass spectrum at m/z 112 and strong satellite peaks at m/z 114 (66% the intensity of the m/z 112 peak) and at m/z 116 (11% the intensity of the m/z 112) peak. The ^{1}H NMR spectrum contains a 3H doublet (d) at δ1.63, two doublets of doublets—one at δ3.58 (J = 9, 7 Hz) and one at δ3.78 (J = 7, 5 Hz)—and a 1H complex multiplet at δ4.16. The proton-decoupled ^{13}C NMR spectrum reveals singlets at δ22.6, 49.3 and 56. DEPT analysis shows these signals to be due respectively to methyl, methylene and methine carbons. Deduce the structure of this compound and assign all peaks.

3. A compound of molecular formula $C_7H_{16}O$ gives rise to an infrared spectrum that shows among other absorptions a strong, very broad peak at 3300 cm^{-1} and a reasonably strong peak at 1080 cm^{-1}. The ^{1}H NMR spectrum contains a distorted triplet at δ0.9 (3H), a somewhat broadened singlet at δ1.22–1.35 (8H), and a distorted quintet (2H) at just above δ1.4, a broadened singlet at δ2.3 (1H) that has a chemical shift that is strongly concentration dependent, and a 2H triplet at δ3.65. The proton-decoupled ^{13}C NMR spectrum reveals singlets at δ63, 33, 29, 26, 23, and 14. DEPT analysis shows all of these signals to be due to methylene groups except the δ14 signal, which is due to a methyl group. Deduce the structure of this compound and assign all peaks.

4. A compound whose mass spectrum reveals its molecular ion peak at m/z 108 gives rise to an infrared spectrum that shows, among other absorptions, a strong, very broad peak at 3300 cm^{-1}, a strong peak at 1030 cm^{-1}, three weak peaks in the 3100–3000 cm^{-1} range, and two weak peaks in the 2950–2850 cm^{-1} range. The ^{1}H NMR spectrum contains a somewhat broadened singlet at δ7.2–7.5 (5H), a singlet (2H) at δ4.65, and a broadened singlet at δ2.5 (1H) that has a chemical shift that is strongly concentration dependent. The proton-decoupled ^{13}C NMR spectrum reveals singlets at δ140.5,

129, 128, 127, and 65. DEPT analysis shows these signals to be due to methine (CH) groups except the δ65 signal, which is due to a methylene group, and the δ140.5 signal, which is due to a quaternary carbon (C). Deduce the structure of this compound and assign all peaks.

5. A compound shows a molecular ion in its mass spectrum at m/z 108. This compound gives rise to an infrared spectrum that shows, among other absorptions, a strong, broad peak at 3350 cm^{-1}, a weak peak at 3020 cm^{-1}, a strong peak at 1230 cm^{-1}, and weak peaks at 2900 and 2850 cm^{-1}. The ^{1}H NMR spectrum contains doublets (d) at δ6.7 and 7.05 (2H each), a broadened singlet (s, 1H) at δ4.7, and a singlet (s) at δ2.3 (3H). The magnitude of the splitting of the two doublets is identical. The proton-decoupled ^{13}C NMR spectrum reveals singlets at δ153, 130, 118, and 20.5. DEPT analysis shows these signals to be due respectively to quaternary (C), methine (CH), methine (CH), and methyl carbons. Deduce the structure of this compound and assign all peaks.

5

Techniques of the
Organic Laboratory

The chapters in Part 5 describe the techniques and operations that organic chemists use to carry out chemical reactions, separate products from their reaction mixtures and purify them, and analyze the results. These techniques include the following:

Heating methods and how to assemble glassware for doing reactions

Extraction, recrystallization, distillation, sublimation, and column chromatography for separating and purifying products

Melting points, thin-layer and gas-liquid chromatography, polarimetry, refactrometry for analyzing products

You will use these techniques repeatedly as you engage in experimental organic chemistry, so we have grouped them together to simplify locating the technique needed. Spectrometric methods are so important in the analysis of your results that we have treated them separately in Part 4.

The first technique in Part 5 discusses laboratory safety, because safety is of fundamental concern in any chemistry laboratory. The safe performance of every task and safe handling practices for the chemicals you use must always be uppermost in your mind while you are working in the laboratory. An important goal of your laboratory work is to learn how to handle chemicals safely. We urge you to consider this chapter required reading before you begin your study of organic chemistry in the laboratory.

Extensive cross-listing occurs within the experiments referring you to the techniques you will use. For example, if you are directed to perform a microscale extraction, you will find the notation "[see Technique 4.2]." Technique 4.2 discusses and illustrates the steps in performing an extraction. If you are doing a microscale recrystallization, the notation "[see Technique 5.5b]" directs you to the procedure for this operation. Many techniques of the organic laboratory involve multiple steps with equipment that may be unfamiliar to you when you first need to use it. Read the introductory information about technique and the specific directions about the operation the first time you use it. Continue to utilize the cross-listings to look up the technique whenever you are in doubt about how to do a specific operation until you acquire enough experience with the technique that it becomes part of your repertoire of laboratory skills.

SAFETY

The organic chemistry laboratory is a place where accidents can occur and where safety is everyone's business. As a consumer, you are protected by laws. In the laboratory, you are protected by the experiments we choose and by the design of the laboratory. Your laboratory is designed to safeguard you from most routine hazards, but it *cannot* protect you from the worst hazard—your own or your neighbors' carelessness. In this chapter we will familiarize you with basic laboratory safety information and some common problems relating to safety.

In all likelihood, your chemistry department and your college or university have safety committees, which concern themselves with your safety in the organic chemistry laboratory. You are probably better protected from toxic chemicals in your chemistry laboratory than you are in your automobile—or from your friend's cigarettes, or from a drunken driver. As consumers, however, it should be obvious that some substances are dangerous; it is equally clear that they can be handled safely if proper care is taken. The same thing is true in the laboratory. Part of learning to do organic chemistry in the laboratory is learning how to do it safely.

1.1
Causes of Laboratory Accidents

Laboratory accidents are of three general types: those involving fires and explosions; those producing cuts or burns; and those occurring from inhalation, ingestion, or absorption through the skin of hazardous materials.

Fires and Explosions   Fire is the chemical union of a fuel with an oxidizing agent, usually oxygen, and is accompanied by the evolution of heat and a flame. Most fires are those of ordinary combustible materials, which are hydrocarbons or their derivatives.

These fires are *extinguished* by removing oxygen or the combustible material or by decreasing the heat of the burning fire. Fires are *prevented* by keeping flammable materials away from a flame source or from oxygen (obviously, the former is easier).

There are four sources of ignition in the organic laboratory: *open flames*, *hot surfaces* such as hot plates or hot heating mantles, *faulty electrical connections*, and *chemical fires*. The most obvious way to prevent a fire is to prevent ignition.

Open flame ignition is easily prevented: **Never bring a lighted Bunsen burner or a match near a low-boiling-point flammable**

FIGURE 1.1
Heating devices.

Ceramic heating mantle

Hot plate

liquid. Furthermore, because vapors from organic liquids travel over long distances toward the floor (they are heavier than air), an open flame within 10 ft of diethyl ether, pentane, or other low-boiling-point (and therefore volatile) organic solvents is an unsafe practice. In fact, the use of a Bunsen burner or any other flame in an organic laboratory should be a rare occurrence and should be done only with the permission of your instructor.

Hot surfaces, such as a hot plate or heating mantle, present a trickier problem (Figure 1.1). An organic solvent spilled or heated

FIGURE 1.2 Fire hazards
with a hot plate.

(a) Flammable liquid spilled on a hot plate

(b) Flammable vapors emitted when
heating a volatile solvent in an
open container

recklessly on a hot plate surface may burst into flames (Figure 1.2). The thermostat on most hot plates is not sealed, and it sparks when it cycles on and off. This spark can ignite flammable vapors from an open container. It is also possible for the vapors from a volatile organic solvent to be ignited by the hot surface of a hot plate or a heating mantle, even if you are not actually heating a container of the substance with the heating device. Remove any hot heating mantle or hot plate from the vicinity before pouring a volatile organic liquid.

Electrical fires pose the same danger in the laboratory as they do in the home; do not use appliances with frayed or damaged electrical cords.

Chemical reactions sometimes produce enough heat to cause a fire and explosion. Consider, for example, the reaction of metallic sodium with water. The hydrogen gas that forms as a product of this reaction can explode and then ignite a volatile solvent that happens to be nearby.

Cuts and Injuries

Cuts and mechanical injuries are hazards anywhere. You have been taught to be careful of broken glass since you were a child. Be just as careful now.

When you break a glass rod or a glass tube, do it correctly. Score (scratch) a small line on one side of the tube with a file. Then, holding the tube on both sides with a towel, with the scored part away from you, quickly snap it by pulling the ends toward you (Figure 1.3).

Do not carelessly insert glass tubes or thermometers into cork or rubber stoppers. Do it the correct way. First, lubricate the end of the tube with a drop of water or glycerin; then, while holding it with a towel close to the lubricated end, insert the tube slowly by firmly rotating it into the stopper. Never hold the glass tube by the end far away from the stopper—it will break every time!

Check the upper rim of beakers, flasks, and other glassware for chips. Discard any piece of glassware that is chipped, because you could be cut very easily by the sharp edge of the chip while handling the item.

FIGURE 1.3 Breaking a glass rod properly.

Inhalation, Ingestion, and Skin Absorption

The hoods in your laboratory protect you from noxious fumes, hazardous vapors, or dust from finely powdered materials. A hood is an enclosed area where such materials can be used safely. Because a great many compounds are at least potentially dangerous, the best practice is to run every experiment in a hood. But, in some instances, this is not possible. Your instructor will tell you when an experiment should be carried out in a hood, and we will tell you if a hood is required for one of our experiments. Make sure that the hood you use is turned on before you use it. Your hood probably has a door or sash; it is there for a reason. When using really obnoxious or dangerous substances, the sash must be pulled down. Remember, hoods are there to protect you; use them.

Ingestion of chemicals by mouth is easily prevented. **Never taste any substance or pipet any liquid by mouth.** Wash your hands with soap and water before you leave the laboratory. No food or drink of any sort should be brought into a laboratory nor eaten there.

Many organic compounds are absorbed through the skin. Wear gloves while handling reagents and reaction mixtures. If you spill any substance on your skin, wash the affected area thoroughly with water for 10–15 min.

1.2
Safety Features of Your Laboratory

Like the jet airplane, your organic laboratory contains many safety features for your protection and comfort. It is unlikely that you will have to use them, but in the event that you do, you must know what and where they are and how they operate.

Fire extinguishers. *Note:* Colleges and universities all have standard policies regarding the handling of fires. You should consult your instructor as to whether evacuation or fire extinguisher use takes priority. **Learn where the exits from your laboratory are located.**

Fire extinguishers are strategically located in your laboratory. There are several kinds and your teacher should demonstrate their use.

Water is the simplest fire extinguisher, and it is found in every organic lab. However, in the organic lab, water as a fire extinguisher has a disadvantage. Because many flammable liquids are lighter than water, such a burning liquid will rise to the surface. When water is splashed on this type of fire, it will actually spread it. Your lab, therefore, is equipped with chemical fire extinguishers. These fire extinguishers, found in various positions around the lab, usually produce a CO_2 foam or a dry chemical dust.

Most older portable fire extinguishers contain liquid carbon dioxide. The CO_2 expands when the valves of the extinguishers are released, laying CO_2 gas over the fire and smothering it by preventing the necessary oxygen from reaching it.

Many laboratories are also equipped with dry chemical fire extinguishers, which can be used at greater distances and for larger fires. The dry chemical, which simply smothers the fire, is usually sodium bicarbonate, potassium bicarbonate, or ammonium phosphate.

Some industrial laboratories and chemical storage areas are equipped with automatic fire extinguishing systems. With these, a fire sensor releases a heavier than air, nonflammable halocarbon, which floods an entire area.

Fire blankets. Fire blankets are used for one thing and one thing only—to smother a fire involving a person's clothing. Fire blankets will be available in most labs. If a person's clothing catches fire, tightly wrap the blanket around the person.

Safety showers. Safety showers are for acid burns and any other spill of corrosive, irritating, or toxic chemicals on the skin or clothing. If a shower is nearby, it can also be used when a person's clothes are ablaze. The typical safety shower dumps a huge volume of water in a short period of time and thus is also effective in the case of acid spills, when speed is of the essence. **Do not use the safety shower routinely, but do not hesitate to use it in an emergency.**

Eye wash stations. You should always wear safety goggles while working in a laboratory, but if you accidentally splash something in your eyes, immediately rinse them with copious quantities of water. Do this at an eye wash station. Learn the location of the eye wash stations in your laboratory and examine the instructions on the eye wash stations during the first (check-in) lab session. You need to keep your eyes open while using the eye wash. Because this may be difficult, assistance may be required. Do not hesitate to call for help!

First aid kits. A typical lab first aid kit will contain most of the following items:

adhesive bandages, 3/4 in.
sterile pads, large
sterile pads, medium
bandages, 2 in. × 126 in.
bandage, 4 in. × 126 in.
first aid tape, 1/2 in. × 180 in.
cleansing wipes
ammonia inhalants, 0.33 mL/min
triangular bandage

nonadhering dressings
tube of first aid cream, 8 oz
eye irrigating solution
eyecup
tourniquet
pair of scissors
pair of tweezers
first aid guide booklet

Your instructor will send you to the college health service, accompanied by another person, after any accident, except for the most trivial. However, some immediate first aid will almost always be required. Your instructor will show you the first aid station and instruct you in its use.

1.3
Preventing Accidents

Accidents can largely be prevented by common sense and a few simple safety rules:

1. Think about what you are doing while you are in the laboratory. Read your experiment ahead of the laboratory time and perform laboratory operations with careful forethought.

2. It is a state law in many states and common sense in the remainder to wear safety glasses or goggles. **Do it.**

3. Watch what your neighbors are doing. Many accidents and injuries in the laboratory are caused by other people. Often the person hurt worst in an accident is the one standing next to the place where the accident occurred.

4. Never work alone in the laboratory. Being alone in a situation in which you may be helpless is dangerous.

5. Make yourself aware of the procedures that should be followed in case of any accident. (See Technique 1.4, What to Do If an Accident Occurs.)

6. Never taste any chemical unless specifically instructed to do so. Always use the hood when working with toxic or obnoxious materials. Treat no chemical carelessly, and remember that many chemicals can enter the body through the skin, as well as through the mouth or lungs.

7. Disposable gloves are available in all laboratories. Wear them to prevent chemicals from coming into contact with your skin unnecessarily. A lab coat should be worn when working with hazardous chemicals; cotton is the preferred fabric because synthetic fabrics could melt in a fire or undergo a reaction that causes the fabric to adhere to the skin. If a chemical is spilled on the skin, wash it off immediately with soap and water. If you are handling a particularly hazardous compound, find out how to remove it from your skin before you begin the experiment. Consult your instructor if in doubt as to the toxicity of the chemical. **Always wash your hands at the end of the laboratory period.**

8. Wear clothing that protects your body. Shorts, tank tops, and sandals (or bare feet) are not suitable attire for the laboratory. Avoid loose clothing and long hair, which are a fire hazard or could become entangled in apparatus.

9. Keep your laboratory space clean. Aside from your own bench area, the balance and chemical dispensing areas should be left clean and orderly. If you spill anything while measuring out your materials, clean it up immediately. Do not leave a mess that other students must work around. After weighing your chemicals, replace the caps on the containers. Although moisture in the air does not greatly affect results in experiments involving aqueous systems, many chemicals used in organic preparations are rendered useless when exposed to moist air for even fairly short periods.

10. **Never heat a closed system! Never completely close off an assembly of apparatus in which a gas is being evolved.** Always provide a vent in order to prevent an explosion.

11. Diethyl ether is particularly dangerous. Flammable solvents with boiling points of less than 100°C such as ether, methanol, pentane, hexane, ethanol, and acetone should be distilled, heated, or evaporated on a steam bath or heating mantle, **never** with a Bunsen burner. Keep flammable solvents in flasks and not in open beakers.

12. Most laboratories have explicit directions for waste disposal, and you should obtain details from your lab instructor. We have provided cleanup procedures with the experiments in this book, but local regulations take precedence over waste disposal methods suggested here.

13. Whenever you insert a glass tube or a thermometer into a cork or rubber stopper, follow three simple rules. First, lubricate the tube or thermometer with glycerin or water so that it slips into the hole easily. Second, protect your hands with a towel while you rotate the tube slowly and gently into the hole. Third, hold the tube close to the lubricated end.

14. Remember that hot glass looks just like cold glass. When heating glass, do not touch the hot spot. Do not lay hot glass on a bench where someone else might pick it up.

15. Keep gas and water valves closed whenever they are not in use.

16. Steam and boiling water cause severe burns. Turn off the steam source before removing containers from the top of a steam bath or steam cone. Handle containers of boiling water very carefully.

17. Never eat or drink in the lab; you might accidentally contaminate your food with laboratory chemicals.

18. Never apply cosmetics in the laboratory.

19. Floors become very slippery if water is spilled; wipe up any spill immediately.

1.4

What to Do If an Accident Occurs

Act quickly, but think first. The first few seconds after an accident may be crucial. Acquaint yourself with the instructions below so that you can be of immediate assistance.

Fire. In case of fire, get out of danger and then immediately notify your instructor. If possible, remove any containers of flammable solvents from the fire area.

You should consult your laboratory instructor about the proper response to a fire. **It is important to know the policy of your institution concerning when to evacuate and when to use an extinguisher.**

Know the location of the fire extinguishers and how they operate. A fire extinguisher *will always be available.* Aim low and direct its nozzle first toward the edge of the fire and then toward the middle. Water is not always useful for extinguishing chemical fires and can actually make some fires worse, so always use the fire extinguisher.

If a person's clothes catch on fire, smother the flames with either a lab coat, a fire blanket, or a water shower. Be sure you know where the fire blanket and safety shower are located. **If your clothing is on fire, do not run.** Rapid movement fans flames.

Cuts and burns. Be aware of the location of the first aid kit and of the materials it contains for the treatment of burns and cuts. The first thing to do for any chemical wound, unless you have been specifically told otherwise, is to wash it well with water. This treatment will rinse away the excess chemical reagent. For acids, alkalies, and toxic chemicals, quick dousing with water will save pain later. Specific treatments for chemical burns are available in *The Merck Index.* Notify your instructor immediately if you are cut or burned.

If chemicals get into your eye, use the eye wash fountain and wash your eye immediately with a large amount of water. Hold your head in the shower or under the tap. **Hold your eyes open** to allow the water to flush the eyeballs. Do not use very cold water, however, because it can also damage the eyeball. Then seek medical treatment immediately.

General policy. Always inform the instructor of any accident that has happened to you or your neighbors. **Let your instructor decide whether a physician's attention is needed.** If a physician's attention is necessary, the injured person should always be accompanied to the medical facility, even when he/she protests that he/she is fine. The injury may be more serious than it initially appears.

1.5
Chemical Toxicology

The toxicity of a compound generally refers to its ability to produce injury once it reaches a susceptible site in the body. The toxicity of a substance is related to its probability of causing injury and is a species-dependent term. What is toxic for people may not be toxic for animals and vice versa.

A substance is *acutely toxic* if it causes a toxic effect in a short time; it is *chronically toxic* if it causes toxic effects with repeated small exposures over a long duration. A major concern in chemical toxicology is quantity or dosage. A *poison* is a substance that causes harm to living tissues when applied in small doses. *Poison* is a word that comes to us through many centuries of usage; it has meaning but is not a scientifically useful term. Remember the dose makes the poison.

Most compounds in their pure form are toxic or poisonous in one way or another. For example, oxygen is strongly implicated in many forms of cell aging and cell destruction and is also important in cancer cell growth. Oxygen at a partial pressure of 2 atm is toxic; for this reason, scuba tanks are filled, not with O_2, but with compressed air. Thus oxygen, although essential for life, is also toxic to people—but its good features outweigh its hazards. In the organic laboratory, most compounds you use are more toxic than oxygen, however, and you must be careful with them.

How Toxic Substances Enter the Body

There are three main routes into the body for any toxic substance: through the *lungs,* through the *mouth* and *gastrointestinal tract,* and through the *skin.* Through one of these routes, a toxic substance can reach the bloodstream, where it can be carried to other body tissues.

Many organic solvents are absorbed through the skin. Wearing gloves while working in the organic laboratory protects the hands. The common industrial practice of using grease-removing solvents to wash one's hands is a mistake and can cause dermatitis.

Many organic solvents are also absorbed through the lungs, because most of them have an appreciable vapor pressure at room temperature. The dangerous glue-sniffing craze takes advantage of the vapor pressure of the esters, ketones, or chlorinated solvents present in airplane glue to induce a "kick" or "high." In fact, it is really a poisoning of the glue sniffer that gives the apparent euphoria.

Be particularly careful when using the solvent benzene. This important and highly toxic compound is known to cause leukemia, and its use must be carefully controlled. If you need to use benzene, work with it only in a hood, and wear gloves. We have designed experiments in this book that do not require the use of benzene.

Fortunately, not all poisons that accidentally enter the body get to a site where they can be deleterious. Even though a toxic substance is absorbed, it is often excreted rapidly. Our body protects us with various devices: the nose, scavenger cells, metabolism, cell replacement, and rapid exchange of good air for bad. Most toxic substances are only detectably toxic because they are present in our environment over long periods of time. The majority of all foreign substances are discharged from the body immediately.

The action of toxic substances varies from individual to individual. Although many substances are toxic to the entire system (arsenic, for example), many others act on specific sites; that is, they are site specific. Carbon monoxide, for example, forms a complex with blood hemoglobin, destroying the blood's ability to absorb and release oxygen. Acute benzene toxicity seems to affect only the platelet-producing bone marrow.

In certain instances (e.g., methanol poisoning), the metabolites (in this case, formaldehyde and formic acid) are more poisonous than the starting material; the formic acid affects the optic nerve, causing blindness. Impaired health can be the result of a substance acting at some body site to cause a structural change; as far as the body is concerned, it does not matter whether the poison is the original substance or a metabolic product of it.

Toxicity Testing and Reporting

Consumers are protected by a series of laws that define toxicity, the legal limits and dosages of toxic materials, and the procedures for measuring these toxicities.

Acute oral toxicity is measured in terms of LD_{50} (LD stands for "lethal dose"). The LD_{50} represents the dose, in milligrams per kilogram of body weight, that will be fatal to 50% of a certain population of experimental animals (mice, rats, etc.). Other tests include dermal toxicity (skin sensitization) and irritation of mucous membranes (eyes, nose).

There are huge tables of toxicities of almost every known organic chemical, and every year many more are added to the list. A wall chart of toxicities may be hanging in your lab or near your stockroom. A useful and handy reference is *The Merck Index*, and a page from the twelfth edition (1996) is reproduced in Figure 1.4.

1.6 Where to Find Chemical Safety Information

All laboratories must make available the Material Safety Data Sheets (MSDS) for every chemical used. Figure 1.5 shows the MSDS for dichloromethane (methylene chloride), detailing the hazards, safe handling practices, first aid information, storage, and disposal of dichloromethane. MSDS information can also be obtained on a CD-ROM (Aldrich Chemical Co.).

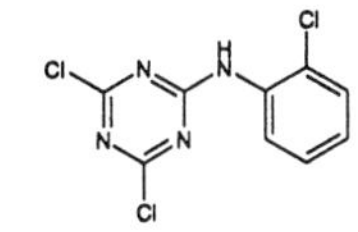

Anilinephthalein 698

White to tan crystals, mp 159-160°. Insol in water. Soly at 30° (g/100 ml): toluene, 5; xylene, 4; acetone, 10. Subject to hydrolysis; not compatible with oils and alkaline materials. LD$_{50}$ orally in rats: > 5000 mg/kg, Mobay Technical Information Sheet, Jan. 1979.

USE: Fungicide.

695. Anileridine. *1-[2-(4-Aminophenyl)ethyl]-4-phenyl-4-piperidinecarboxylic acid ethyl ester; 1-(p-aminophenethyl)-4-phenylisonipecotic acid ethyl ester; ethyl 1-(4-aminophenethyl)-4-phenylisonipecotate; N-[β-(p-aminophenyl)ethyl]-4-phenyl-4-carbethoxypiperidine; N-β-(p-aminophenyl)ethylnormeperidine; Leritine; Nipecotan; Alidine; Apodol.* C$_{22}$H$_{28}$N$_2$O$_2$; mol wt 352.48. C 74.97%, H 8.01%, N 7.95%, O 9.08%. Synthesis: Weijlard *et al.*, *J. Am. Chem. Soc.* 78, 2342 (1956); U.S. pat. 2,966,490 (1960 to Merck & Co.).

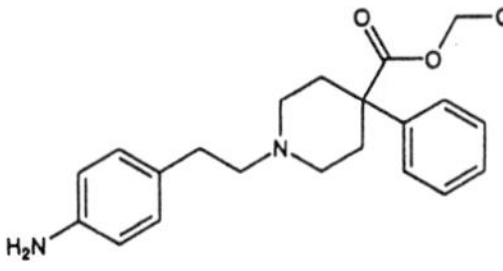

mp 83°.

Dihydrochloride, C$_{22}$H$_{28}$N$_2$O$_2$.2HCl, crystals from methanol + ether, mp 280-287° (dec). Freely sol in water, methanol. Solubility in ethanol: 8 mg/g. pH of aq solns 2.0 to 2.5. Solns are stable at pH 3.5 and above. At pH 4 and higher the insol free base is precipitated. uv max (pH 7 in 90% methanol contg phosphate buffer): 235, 289 nm (A$_{1 cm}^{1\%}$ 293, 34.5). Distribution coefficient (water, pH 3.6/n-butanol): 0.9.

Note: This is a controlled substance (opiate) listed in the U.S. Code of Federal Regulations, Title 21 Part 1308.12 (1995).

THERAP CAT: Analgesic (narcotic).

696. Aniline. *Benzenamine; aniline oil; phenylamine; aminobenzene; aminophen; kyanol.* C$_6$H$_7$N; mol wt 93.13. C 77.38%, H 7.58%, N 15.04%. First obtained in 1826 by Unverdorben from dry distillation of indigo. Runge found it in coal tar in 1834. Fritzsche, in 1841, prepared it from indigo and potash and gave it the name aniline. Manuf from nitrobenzene or chlorobenzene: Faith, Keyes & Clark's *Industrial Chemicals*, F. A. Lowenheim, M. K. Moran, Eds. (Wiley-Interscience, New York, 4th ed., 1975) pp 109-116. Procedures: A. I. Vogel, *Practical Organic Chemistry* (Longmans, London, 3rd ed./ 1959) p 564; Gattermann-Wieland, *Praxis des organischen Chemikers* (de Gruyter, Berlin, 40th ed., 1961) p 148. Brochure "*Aniline*" by Allied Chemical's National Aniline Division (New York, 1964) 109 pp, gives reactions and uses of aniline (877 references). Toxicity study: K. H. Jacobson, *Toxicol. Appl. Pharmacol.* 22, 153 (1972).

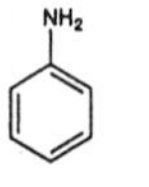

Oily liquid; colorless when freshly distilled, darkens on exposure to air and light. *Poisonous!* Characteristic odor and burning taste; combustible; volatile with steam. d$_4^{20}$ 1.022. bp 184-186°. Solidif —6°. Flash pt, closed cup: 169°F (76°C). n$_D^{20}$ 1.5863. pKb 9.30. pH of 0.2 molar aq

soln 8.1. One gram dissolves in 28.6 ml water, 15.7 ml boil. water; misc with alcohol, benzene, chloroform, and most other organic solvents. Combines with acids to form salts. It dissolves alkali or alkaline earth metals with evolution of hydrogen and formation of anilides, e.g., C$_6$H$_5$NHNa. *Keep well closed and protected from light.* Incompat. Oxidizers, albumin, solns of Fe, Zn, Al, acids, and alkalies. LD$_{50}$ orally in rats: 0.44 g/kg (Jacobson).

Hydrobromide, C$_6$H$_7$N.HBr, white to slightly reddish, crystalline powder, mp 286°. Darkens in air and light. Sol in water, alc. *Protect from light.*

Hydrochloride, C$_6$H$_7$N.HCl, crystals, mp 198°. d 1.222. Darkens in air and light. Sol in about 1 part water; freely sol in alc. *Protect from light.*

Hydrofluoride, C$_6$H$_7$N.HF, crystalline powder. Turns gray on standing. Freely sol in water; slightly sol in cold, freely in hot alc.

Nitrate, C$_6$H$_7$N.HNO$_3$, crystals, dec about 190°. d 1.36. Discolors in air and light. Sol in water, alc. *Protect from light.*

Hemisulfate, C$_6$H$_7$N.½H$_2$SO$_4$, crystalline powder. d 1.38. Darkens on exposure to air and light. One gram dissolves in about 15 ml water; slightly sol in alc. Practically insol in ether. *Protect from light.*

Acetate, C$_6$H$_5$NH$_2$.HOOCCH$_3$. Prepd from aniline and acetic acid: Vignon, Evieux, *Bull. Soc. Chim. France* [4] 3, 1012 (1908). Colorless liquid. d 1.070-1.072. Darkens with age; gradually converted to acetanilide on standing. Misc with water, alc.

Oxalate, C$_6$H$_5$NH$_2$.HOOCCOOH.H$_2$NC$_6$H$_5$. Prepd from aniline and oxalic acid in alc soln: Hofmann, *Ann.* 47, 37 (1843). Triclinic rods from water, mp 174-175°. Readily sol in water; sparingly sol in abs alc. Practically insol in ether.

Caution: Intoxication may occur from inhalation, ingestion, or cutaneous absorption. *Acute Toxicity:* cyanosis, methemoglobinemia, vertigo, headache, mental confusion. *Chronic Toxicity:* anemia, anorexia, wt loss, cutaneous lesions, *Clinical Toxicology of Commercial Products*, R. E. Gosselin *et al.*, Eds. (Williams & Wilkins, Baltimore, 4th ed., 1976) Section III, pp 29-35.

USE: Manuf dyes, medicinals, resins, varnishes, perfumes, shoe blacks; vulcanizing rubber; as solvent. Hydrochloride used in manuf of intermediates, aniline black and other dyes, in dyeing fabrics or wood black.

697. Aniline Mustard. *N,N-Bis(2-chloroethyl)benzenamine; N,N-bis(2-chloroethyl)aniline; phenylbis[2-chloroethylamine]; β,β'-dichlorodiethylaniline; Lymphochin; Lymphocin; Lymphoquin.* C$_{10}$H$_{13}$Cl$_2$N; mol wt 218.13. C 55.06%, H 6.01%, Cl 32.51%, N 6.42%. Prepd by the action of phosphorus pentachloride on *N,N-bis-[2-hydroxyethyl]-aniline (phenyldiethanolamine):* Robinson, Watt, *J. Chem. Soc.* 1934, 1538; Korshak, Strepikheev, *J. Gen. Chem. USSR* 14, 312 (1944).

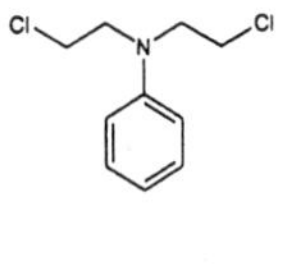

Stout prisms from methanol, mp 45°. bp$_{14}$ 164°; bp$_{0.5}$ 110°. Sol in hot methanol, ethanol. Very slightly sol in ether.

Hydrochloride, C$_{10}$H$_{14}$Cl$_3$N, crystals. *Vesicant.* Freely sol in water. Sol in alcohol.

USE: In cancer research.

698. Anilinephthalein. *3,3-Bis(4-aminophenyl)-1(3H)-isobenzofuranone; 3,3-bis(p-aminophenyl)phthalide.* C$_{20}$H$_{16}$N$_2$O$_2$; mol wt 316.36. C 75.93%, H 5.10%, N 8.85%, O 10.11%. Prepn: Hubacher, *J. Am. Chem. Soc.* 73, 5885 (1951).

Consult the Name Index before using this section.   Page 111

FIGURE 1.4 Page from *The Merck Index*, 12th ed. (Reprinted with permission from Merck and Co., Inc., Whitehouse Station, NJ.)

An extensive collection of safety information is found in the three-volume set *The Sigma-Aldrich Library of Regulatory and Safety Data* (Aldrich Chemical Co., 1993). This set reports the Toxic Substances Control Act (TSCA) status for a large number of the chemicals available from the Aldrich Chemical Company. TSCA requires a listing for any chemical used for commercial purposes.

Sigma Chemical Co.   Aldrich Chemical Co., Inc.   Fluka Chemical Corp.
P.O. Box 14508   1001 West St. Paul   980 South Second St.
St. Louis, MO 63178   Milwaukee, WI 53233   Ronkonkoma, NY 11779
Phone: 314-771-5765   Phone: 414-273-3850   Phone: 516-467-0980
    Emergency Phone: 516-467-3535

SECTION 1. - - - - - - - - - CHEMICAL IDENTIFICATION- - - - - - - - - - -
   CATALOG #:   D65100
   NAME: DICHLOROMETHANE, 99.6%, A.C.S. REAGENT
SECTION 2. - - - - - COMPOSITION/INFORMATION ON INGREDIENTS - - - - - -
   CAS #:   75-09-2
   MF: CH2CL2
   EC NO: 200-838-9
   SYNONYMS
    AEROTHENE MM * CHLORURE DE METHYLENE (FRENCH) * DICHLOROMETHANE (DOT:
    OSHA) * METHANE DICHLORIDE * METHYLENE BICHLORIDE * METHYLENE
    CHLORIDE (ACGIH:DOT:OSHA) * METHYLENE DICHLORIDE * METYLENU CHLOREK
    (POLISH) * NARKOTIL * NCI-C50102 * R 30 * R30 (REFRIGERANT) * RCRA
    WASTE NUMBER U080 * SOLAESTHIN * SOLMETHINE * UN1593 (DOT) *
SECTION 3. - - - - - - - - - - - HAZARDS IDENTIFICATION - - - - - - - - -
   LABEL PRECAUTIONARY STATEMENTS
    TOXIC (USA)
    HARMFUL (EU)
    HARMFUL BY INHALATION, IN CONTACT WITH SKIN AND IF SWALLOWED.
    IRRITATING TO EYES, RESPIRATORY SYSTEM AND SKIN.
    POSSIBLE RISK OF IRREVERSIBLE EFFECTS.
    CALIF. PROP. 65 CARCINOGEN.
    POSSIBLE CARCINOGEN/MUTAGEN.
    NEUROLOGICAL HAZARD.
    READILY ABSORBED THROUGH SKIN.
    TARGET ORGAN(S):
    LIVER, PANCREAS
    IN CASE OF ACCIDENT OR IF YOU FEEL UNWELL, SEEK MEDICAL ADVICE
    IMMEDIATELY (SHOW THE LABEL WHERE POSSIBLE).
    IN CASE OF CONTACT WITH EYES, RINSE IMMEDIATELY WITH PLENTY OF
    WATER AND SEEK MEDICAL ADVICE.
    AFTER CONTACT WITH SKIN, WASH IMMEDIATELY WITH PLENTY OF WATER.
    WEAR SUITABLE PROTECTIVE CLOTHING, GLOVES AND EYE/FACE
    PROTECTION.
    HANDLE AND STORE UNDER NITROGEN.
SECTION 4. - - - - - - - - - - FIRST-AID MEASURES- - - - - - - - - - -
    IN CASE OF CONTACT, IMMEDIATELY WASH SKIN WITH SOAP AND COPIOUS
    AMOUNTS OF WATER.
    CONTAMINATION OF THE EYES SHOULD BE TREATED BY IMMEDIATE AND PROLONGED
    IRRIGATION WITH COPIOUS AMOUNTS OF WATER.
    ASSURE ADEQUATE FLUSHING OF THE EYES BY SEPARATING THE EYELIDS
    WITH FINGERS.
    IF INHALED, REMOVE TO FRESH AIR. IF NOT BREATHING GIVE ARTIFICIAL
    RESPIRATION. IF BREATHING IS DIFFICULT, GIVE OXYGEN.
    IF SWALLOWED, WASH OUT MOUTH WITH WATER PROVIDED PERSON IS CONSCIOUS.
    CALL A PHYSICIAN.
    WASH CONTAMINATED CLOTHING BEFORE REUSE.

SECTION 5. - - - - - - - - FIRE FIGHTING MEASURES - - - - - - - - - -
   EXTINGUISHING MEDIA
    NONCOMBUSTIBLE.
    USE EXTINGUISHING MEDIA APPROPRIATE TO SURROUNDING FIRE CONDITIONS.
   SPECIAL FIREFIGHTING PROCEDURES
    WEAR SELF-CONTAINED BREATHING APPARATUS AND PROTECTIVE CLOTHING TO
    PREVENT CONTACT WITH SKIN AND EYES.
   UNUSUAL FIRE AND EXPLOSIONS HAZARDS
    EMITS TOXIC FUMES UNDER FIRE CONDITIONS.
SECTION 6. - - - - - - - - - - ACCIDENTAL RELEASE MEASURES- - - - - - - - -
    EVACUATE AREA.
    WEAR SELF-CONTAINED BREATHING APPARATUS, RUBBER BOOTS AND HEAVY
    RUBBER GLOVES.
    ABSORB ON SAND OR VERMICULITE AND PLACE IN CLOSED CONTAINERS FOR
    DISPOSAL.
    VENTILATE AREA AND WASH SPILL SITE AFTER MATERIAL PICKUP IS COMPLETE.
SECTION 7. - - - - - - - - - - HANDLING AND STORAGE- - - - - - - - - - -
    REFER TO SECTION 8.
SECTION 8. - - - - - - EXPOSURE CONTROLS/PERSONAL PROTECTION- - - - -
    WEAR APPROPRIATE NIOSH/MSHA-APPROVED RESPIRATOR, CHEMICAL-RESISTANT
    GLOVES, SAFETY GOGGLES, OTHER PROTECTIVE CLOTHING.
    USE ONLY IN A CHEMICAL FUME HOOD.
    SAFETY SHOWER AND EYE BATH.
    DO NOT BREATHE VAPOR.
    AVOID CONTACT WITH EYES, SKIN AND CLOTHING.
    AVOID PROLONGED OR REPEATED EXPOSURE.
    READILY ABSORBED THROUGH SKIN.
    WASH THOROUGHLY AFTER HANDLING.
    TOXIC.
    POSSIBLE CARCINOGEN.
    IRRITANT.
    NEUROLOGICAL HAZARD.
    POSSIBLE MUTAGEN.
    KEEP TIGHTLY CLOSED.
    KEEP AWAY FROM HEAT AND OPEN FLAME.
    STORE IN A COOL DRY PLACE.
SECTION 9. - - - - - - - PHYSICAL AND CHEMICAL PROPERTIES - - - - - - -
   APPEARANCE AND ODOR
    COLORLESS LIQUID
   PHYSICAL PROPERTIES
    BOILING POINT: 39.8 C TO 40 C
    MELTING POINT: -97 C
    FLASHPOINT   NONE
    EXPLOSION LIMITS IN AIR:
     UPPER   22%
     LOWER   14%
    AUTOIGNITION TEMPERATURE: 1223 F   661C
    VAPOR PRESSURE: 6.83PSI 20 C   24.48PSI 55 C
    SPECIFIC GRAVITY: 1.325
    VAPOR DENSITY: 2.9
SECTION 10. - - - - - - - - -STABILITY AND REACTIVITY - - - - - - - -
   INCOMPATIBILITIES
    ALKALI METALS
    ALUMINUM

    HEAT
   HAZARDOUS COMBUSTION OR DECOMPOSITION PRODUCTS
    CARBON MONOXIDE, CARBON DIOXIDE
    HYDROGEN CHLORIDE GAS
    PHOSGENE GAS
SECTION 11. - - - - - - - - - TOXICOLOGICAL INFORMATION - - - - - - - -
   ACUTE EFFECTS
    HARMFUL IF SWALLOWED, INHALED, OR ABSORBED THROUGH SKIN.
    VAPOR OR MIST IS IRRITATING TO THE EYES, MUCOUS MEMBRANES AND UPPER
    RESPIRATORY TRACT.
    CAUSES SKIN IRRITATION.
    DICHLOROMETHANE IS METABOLIZED IN THE BODY PRODUCING CARBON MONOXIDE
    WHICH INCREASES AND SUSTAINS CARBOXYHEMOGLOBIN LEVELS IN THE BLOOD,
    REDUCING THE OXYGEN-CARRYING CAPACITY OF THE BLOOD.
    EXPOSURE CAN CAUSE:
    NAUSEA, DIZZINESS AND HEADACHE
    MAY CAUSE NERVOUS SYSTEM DISTURBANCES.
   CHRONIC EFFECTS
    POSSIBLE CARCINOGEN.
    LABORATORY EXPERIMENTS HAVE SHOWN MUTAGENIC EFFECTS.
    TARGET ORGAN(S):
    LIVER
    PANCREAS
    NERVES
    CARDIOVASCULAR SYSTEM
    TO THE BEST OF OUR KNOWLEDGE, THE CHEMICAL, PHYSICAL, AND
    TOXICOLOGICAL PROPERTIES HAVE NOT BEEN THOROUGHLY INVESTIGATED.
   RTECS #: PA8050000
    METHANE, DICHLORO-
   IRRITATION DATA
    SKN-RBT 810 MG/24H SEV   EJTXAZ 9,171,76
    SKN-RBT 100 MG/24H MOD   85JCAE -,88,86
    EYE-RBT 162 MG MOD   EJTXAZ 9,171,76
    EYE-RBT 10 MG MLD   TXCYAC 6,173,76
    EYE-RBT 500 MG/24H MLD   85JCAE -,88,86
   TOXICITY DATA
    ORL-HMN LDLO:357 MG/KG   34ZIAG -,390,69
    ORL-RAT LD50:1600 MG/KG   FAONAU 48A,94,70
    IHL-RAT LC50:52 GM/M3   TPKVAL 15,64,79
    IPR-RAT LD50:916 MG/KG   ENVRAL 40,411,86
    UNR-RAT LD50:5350 MG/KG   GISAAA 53(6),78,88
    IHL-MUS LC50:14400 PPM/7H   NIHBAZ 191,1,49
    IPR-MUS LD50:437 MG/KG   AGGHAR 18,109,60
    SCU-MUS LD50:6460 MG/KG   TXAPA9 4,354,62
    UNR-MUS LD50:4770 MG/KG   ESKGA2 28,P31,82
    UNR-RBT LD50:1225 MG/KG   GISAAA 53(6),78,88
   TARGET ORGAN DATA
    PERIPHERAL NERVE AND SENSATION (PARESTHESIA)
    BEHAVIORAL (ALTERED SLEEP TIME)
    BEHAVIORAL (EUPHORIA)
    BEHAVIORAL (SOMNOLENCE)
    BEHAVIORAL (CONVULSIONS OR EFFECT ON SEIZURE THRESHOLD)
    BEHAVIORAL (ATAXIA)
    CARDIAC (CHANGE IN RATE)

FIGURE 1.5 Material Safety Data Sheet (MSDS) for dichloromethane. (Reprinted with permission from Aldrich Chemical Co., Inc., Milwaukee, WI.)

PRODUCT #: D65100   NAME: DICHLOROMETHANE, 99.6%, A.C.S. REAGENT
MATERIAL SAFETY DATA SHEET, Valid 11/96 - 1/97
Printed Thursday, November 21, 1996 4:38PM

LUNGS, THORAX OR RESPIRATION (TUMORS)
LIVER (LIVER FUNCTION TESTS IMPAIRED)
SPECIFIC DEVELOPMENTAL ABNORMALITIES (MUSCULOSKELETAL SYSTEM)
SPECIFIC DEVELOPMENTAL ABNORMALITIES (UROGENITAL SYSTEM)
TUMORIGENIC (CARCINOGENIC BY RTECS CRITERIA)
ONLY SELECTED REGISTRY OF TOXIC EFFECTS OF CHEMICAL SUBSTANCES
(RTECS) DATA IS PRESENTED HERE. SEE ACTUAL ENTRY IN RTECS FOR
COMPLETE INFORMATION.
SECTION 12. - - - - - - - - - - ECOLOGICAL INFORMATION - - - - - - - - -
DATA NOT YET AVAILABLE.
SECTION 13. - - - - - - - - - DISPOSAL CONSIDERATIONS - - - - - - - - -
DISSOLVE OR MIX THE MATERIAL WITH A COMBUSTIBLE SOLVENT AND BURN IN A
CHEMICAL INCINERATOR EQUIPPED WITH AN AFTERBURNER AND SCRUBBER.
OBSERVE ALL FEDERAL, STATE AND LOCAL ENVIRONMENTAL REGULATIONS.
SECTION 14. - - - - - - - - - - TRANSPORT INFORMATION - - - - - - - - -
CONTACT ALDRICH CHEMICAL COMPANY FOR TRANSPORTATION INFORMATION.
SECTION 15. - - - - - - - - - REGULATORY INFORMATION - - - - - - - - -
EUROPEAN INFORMATION
EC INDEX NO: 602-004-00-3
HARMFUL
R 40
POSSIBLE RISK OF IRREVERSIBLE EFFECTS.
S 23
DO NOT BREATHE VAPOR.
S 24/25
AVOID CONTACT WITH SKIN AND EYES.
S 36/37
WEAR SUITABLE PROTECTIVE CLOTHING AND GLOVES.
REVIEWS, STANDARDS, AND REGULATIONS
OEL=MAK
ACGIH TLV-TWA 50 PPM   85INA8 6,981,91
ACGIH TLV-SUSPECTED HUMAN CARCINOGEN   85INA8 6,981,91
IARC CANCER REVIEW:ANIMAL SUFFICIENT EVIDENCE IMEMDT 41,43,86
IARC CANCER REVIEW:HUMAN INADEQUATE EVIDENCE IMEMDT 41,43,86
IARC CANCER REVIEW:GROUP 2B   IMSUDL 7,194,87
EPA FIFRA 1988 PESTICIDE SUBJECT TO REGISTRATION OR RE-REGISTRATION
FEREAC 54,7740,89
MSHA STANDARD-AIR:TWA 500 PPM (1750 MG/M3)
DTLVS* 3,171,71
OSHA PEL (GEN INDU):8H TWA 500 PPM;CL 1000 PPM;PK 2000 PPM/5M/2H
CFRGBR 29,1910.1000,94
OSHA PEL (CONSTRUC):SEE 56 FR 57036
CFRGBR 29,1926.55,94
OSHA PEL (SHIPYARD):8H TWA 500 PPM (1740 MG/M3
CFRGBR 29,1915.1000,93
OSHA PEL (FED CONT):8H TWA 500 PPM (1740 MG/M3)
CFRGBR 41,50-204.50,94
OEL-AUSTRALIA:TWA 100 PPM (350 MG/M3);CARCINOGEN JAN93
OEL-AUSTRIA:TWA 100 PPM (360 MG/M3) JAN93
OEL-BELGIUM:TWA 50 PPM (174 MG/M3);CARCINOGEN JAN93
OEL-DENMARK:TWA 50 PPM (175 MG/M3);SKIN;CARCINOGEN JAN93
OEL-FINLAND:TWA 100 PPM (350 MG/M3);STEL 250 PPM (870 MG/M3) JAN93
OEL-FRANCE:TWA 100 PPM (360 MG/M3);STEL 500 PPM (1800 MG/M3) JAN93
OEL-GERMANY:TWA 100 PPM (360 MG/M3);CARCINOGEN JAN93

Page 4

PRODUCT #: D65100   NAME: DICHLOROMETHANE, 99.6%, A.C.S. REAGENT
MATERIAL SAFETY DATA SHEET, Valid 11/96 - 1/97
Printed Thursday, November 21, 1996 4:38PM

OEL-HUNGARY:STEL 10 MG/M3;CARCINOGEN JAN93
OEL-JAPAN:TWA 100 PPM (350 MG/M3) JAN93
OEL-THE NETHERLANDS:TWA 100 PPM (350 MG/M3);STEL 500 PPM JAN93
OEL-THE PHILIPINES:TWA 500 PPM (1740 MG/M3) JAN93
OEL-POLAND:TWA 50 MG/M3 JAN93
OEL-RUSSIA:TWA 100 PPM;STEL 50 MG/M3 JAN93
OEL-SWEDEN:TWA 35 PPM (120 MG/M3);STEL 70 PPM (250 MG/M3);SKIN JAN93
OEL-SWITZERLAND:TWA 100 PPM (360 MG/M3);STEL 500 PPM JAN93
OEL-THAILAND:TWA 500 MG/M3;CARCINOGEN JAN93
OEL-THAILAND:TWA 500 MG/M3;NOS MG/M3 JAN93
OEL-TURKEY:TWA 500 PPM (1740 MG/M3) JAN93
OEL-UNITED KINGDOM:TWA 100 PPM (350 MG/M3);STEL 250 PPM JAN93
OEL IN BULGARIA, COLOMBIA, JORDAN, KOREA CHECK ACGIH TLV
OEL IN NEW ZEALAND, SINGAPORE, VIETNAM CHECK ACGIH TLV
NIOSH REL TO METHYLENE CHLORIDE-AIR:CA LOWEST FEASIBLE CONCENTRATION
NIOSH* DHHS #92-100,92
NOHS 1974: HZD 47270; NIS 374; TNF 89025; NOS 192; TNE 975696
NOES 1983: HZD 47270; NIS 363; TNF 87086; NOS 212; TNE 1438196; TFE
352536
ATSDR TOXICOLOGY PROFILE (NTIS** PB/89/194468/AS)
EPA GENETOX PROGRAM 1988, POSITIVE: CELL TRANSFORM.-RLV F344 RAT EMBRYO
EPA GENETOX PROGRAM 1988, POSITIVE: HISTIDINE REVERSION-AMES TEST
EPA GENETOX PROGRAM 1988, POSITIVE: S CEREVISIAE GENE CONVERSION; S
CEREVISIAE-HOMOZYGOSIS
EPA GENETOX PROGRAM 1988, POSITIVE: S CEREVISIAE-REVERSION
EPA GENETOX PROGRAM 1988, NEGATIVE: D MELANOGASTER SEX-LINKED LETHAL
EPA TSCA SECTION 8(B) CHEMICAL INVENTORY
EPA TSCA 8(A) PRELIMINARY ASSESSMENT INFORMATION, FINAL RULE
FEREAC 47,26992,82
EPA TSCA SECTION 8(D) UNPUBLISHED HEALTH/SAFETY STUDIES
ON EPA IRIS DATABASE
EPA TSCA TEST SUBMISSION (TSCATS) DATA BASE, JULY 1996
NIOSH CURRENT INTELLIGENCE BULLETIN 46, 1986
NIOSH ANALYTICAL METHOD, 1994: METHYLENE CHLORIDE, 1005
NTP CARCINOGENESIS STUDIES (INHALATION);CLEAR EVIDENCE:MOUSE,RAT
NTPTR* NTP-TR-306,86
NTP 7TH ANNUAL REPORT ON CARCINOGENS, 1992 : ANTICIPATED TO BE
CARCINOGEN
OSHA ANALYTICAL METHOD #ID-59
U.S. INFORMATION
THIS PRODUCT IS SUBJECT TO SARA SECTION 313 REPORTING REQUIREMENTS.
SECTION 16. - - - - - - - - - OTHER INFORMATION- - - - - - - - - - - -
THE ABOVE INFORMATION IS BELIEVED TO BE CORRECT BUT DOES NOT PURPORT TO
BE ALL INCLUSIVE AND SHALL BE USED ONLY AS A GUIDE. SIGMA, ALDRICH,
FLUKA SHALL NOT BE HELD LIABLE FOR ANY DAMAGE RESULTING FROM HANDLING
OR FROM CONTACT WITH THE ABOVE PRODUCT. SEE REVERSE SIDE OF INVOICE OR
PACKING SLIP FOR ADDITIONAL TERMS AND CONDITIONS OF SALE.
COPYRIGHT 1996 SIGMA CHEMICAL CO., ALDRICH CHEMICAL CO., INC.,
FLUKA CHEMIE AG
LICENSE GRANTED TO MAKE UNLIMITED PAPER COPIES FOR INTERNAL USE ONLY

Page 5

The three-volume set also lists the SARA 302 and 313 status of many Aldrich compounds. SARA is the Superfund Amendments and Reauthorization Act. SARA 313 lists compounds that must be reported if released into the environment. SARA 302 lists extremely hazardous compounds.

Many suppliers provide a code for all chemicals, describing hazards associated with the compounds they sell. For example, J. T. Baker, on the basis of the guidelines of the National Fire Protection Association, gives a 1 (slight) rating to the health hazards of acetone, a 3 (severe) rating to acetone's flammability, a 2 (moderate) rating to its reactivity, and a 1 (slight) to its contact hazard. In comparison, nitric acid has a 3 (severe) rating for its health hazard, a 0 (none) rating for its flammability, a 3 (severe) rating for its reactivity, and a 4 (extreme) rating for its contact hazard. Sigma Chemical Company and Aldrich Chemical Company print pictograms, which have been approved by the European Economic Community (EEC), on the labels of their chemicals that have hazards (Figure 1.6).

Explosive

Oxidizing

Highly flammable or
extremely flammable

Toxic or
very toxic

Harmful or
irritant

Corrosive

Biohazard

Dangerous for
the environment

FIGURE 1.6 Pictograms indicating chemical hazards.

We have designed the experiments in this book for the smallest amount of reagents that will lead to a satisfactory yield with the equipment found in a typical undergraduate laboratory. Moreover, we have included many microscale experiments, which minimize even further the degree of exposure to hazardous chemicals by using only milligram amounts of chemicals. At the beginning of every experimental procedure, we have provided a brief summary of the relevant safety and handling information for most of the chemicals that we use. Figure 1.7 shows the format for this safety information.

FIGURE 1.7 Example of Safety Information, which appears at the beginning of each experiment.

SAFETY INFORMATION

1-Chlorobutane is harmful if inhaled, ingested, or absorbed through the skin. Wear gloves and use it in a hood, if possible.

2,2,4-Trimethylpentane is flammable and a skin irritant. Use it in a hood, if possible.

Sodium hypochlorite solution, when acidified, emits chlorine gas, which is toxic and an eye and respiratory irritant. Use it in a hood, if possible.

Cleanup: The aqueous phrase separated from the reaction mixture may be washed down the sink or poured into the container for aqueous inorganic waste. The calcium chloride should be placed in the hazardous solid waste container, because it is coated with halogenated hydrocarbons. After performing the GC analysis, the product mixture should be poured into the halogenated waste container.

FIGURE 1.8 Example of Cleanup directions, which appear at the end of each experiment.

1.7
Protecting the Environment

Protection in the laboratory does not stop with the persons who work there. Every student and industrial worker must also be aware of the impact of what he does on others outside his immediate sphere of influence.

Before disposing of anything in the lab, therefore, we must be aware of the impact this disposal will have on the environment. Although zero waste is impossible, minimum waste is essential. More and more, industries are being required to account for almost everything they put into the environment, either as gases, liquids, or solid wastes. In the college laboratory, we must do likewise. We have designed experiments in this book to minimize waste by using the smallest possible quantities for the macroscale procedures and by providing microscale procedures for a large number of experiments. Every experiment includes a cleanup procedure listing what to do with each by-product produced (Figure 1.8). Sometimes you are directed to neutralize an acidic or basic aqueous solution before washing it down the sink, or you are instructed to use an appropriate waste container for remaining solvents, filtrates, or solids.

Constructing an Environmental Profile

You may be required to produce an environmental profile for some experiments you do in this laboratory. Such an environmental profile provides a quantitative accounting for as many of the chemicals as possible.

An environmental profile accounts, as nearly as possible, for every material used: What goes in must be accounted for by what comes out. You cannot simply throw things down the drain, because that isn't throwing things away—it is just putting them in another place.

Let's construct an environmental profile: Consider a common organic laboratory situation, the preparation of methyl benzoate from methanol and benzoic acid. The equation is

$$C_6H_5COOH + CH_3OH \xrightarrow{H_2SO_4} C_6H_5COOCH_3 + H_2O$$

mol. wt. = 122   mol. wt = 32   mol. wt = 136   mol. wt. = 18

The experiment is conducted by refluxing a mixture of 12 g of benzoic acid and 25 mL of methanol with a sulfuric acid catalyst. We know that the density of methanol is $0.79 \ g \cdot mL^{-1}$, so we can determine that 0.10 mol of benzoic acid and 0.62 mol of methanol are present. The methanol is in substantial excess, because benzoic acid and methanol react in a 1:1 mole ratio according to the balanced chemical equation.

Ignoring heat and water, the input statement is

Input	Amount	Moles
Benzoic acid	12 g	0.10
Methanol (25 mL)	20 g	0.62
Sulfuric acid	(approx. 1 mL)	
Dichloromethane	25 mL	

After heating the mixture under reflux for 1 h, the solution is cooled, poured into water, and extracted with dichloromethane (25 mL).

After this step, the sulfuric acid is neutralized, and we isolate, not only the product methyl benzoate, but also any unused benzoic acid and all of the dichloromethane, the solvent used for extraction. Because it is water soluble, the excess methanol is lost when the reaction mixture is dumped into water.

The output statement might look like this:

Output	Amount	Moles
Benzoic acid (recovered)	3.7 g	0.033
Methyl benzoate (formed)	8.3 g	0.06
Dichloromethane (recovered)	20 mL	

Lost in the experiment were substantial amounts of methanol, 0.007 mole of benzoic acid, 5 mL of dichloromethane, and about 1 mL of sulfuric acid.

Where did it go? The answer is, into the waste waters, into waste crocks, or into the air. And in that "going," it contributed to environmental pollution.

We can prevent some of this waste by being more careful, although much of the loss cannot be prevented on such a small scale. On an industrial scale, the loss we indicated would be too large, both for economic and for environmental reasons, and the experimental procedures would be changed to prevent it.

We can greatly minimize losses in the organic laboratory by using microscale procedures. Because we use hundreds of milligrams, instead of tens of grams, the benefits for the environment are clear.

References

1. Lewis, Sr., R. J. *Sax's Dangerous Properties of Industrial Materials*, 9th ed.; Van Nostrand-Reinhold: New York, 1996.

2. The Manufacturing Chemists Association, *Guide for Safety in the Chemistry Laboratory*; Van Nostrand-Reinhold: New York, 1972.

3. Furr, A. K. (Ed.) *CRC Handbook of Laboratory Safety*; 3rd ed.; CRC Press: Boca Raton, FL, 1990.

4. The Manufacturing Chemists Association, Chemical Safety Data Sheets; Washington, DC.

5. U.S. Department of Labor, *Occupational Exposure to Hazardous Chemicals in Laboratories*; no. OSHA 95-33; Government Printing Office: Washington, DC, 1995.

6. U.S. Department of Health, Education, and Welfare, NIOSH, *Suspected Carcinogens*; 2nd ed.; Government Printing Office: Washington, DC, 1976.

7. Budavari, S.; O'Neil, M. J.; Smith, A.; Heckelman, P. E.; Kinneary, J. F. (Eds.) *The Merck Index: An Encyclopedia of Chemicals, Drugs, and Biologicals*; 12th ed.; Merck & Co., Inc.: Whitehouse Station, NJ, 1996.

8. Sax, N. I.; Lewis, Sr., R. J. *Rapid Guide to Hazardous Chemicals in the Workplace*; 3rd ed.; Van Nostrand-Reinhold: New York, 1986.

9. Lenga, R. E.; Votoupal, K. L. (Eds.) *The Sigma-Aldrich Library of Regulatory and Safety Data*; Aldrich Chemical Co.: Milwaukee, WI, 1993; 3 volumes.

10. Hajian, H. G. *Working Safely in the Chemical Laboratory*; American Chemical Society: Washington, DC, 1994.

11. National Research Council *Prudent Practices in the Laboratory, Handling and Disposal of Chemicals*; National Academy Press: Washington, DC, 1995.

12. American Chemical Society *Less is Better: Laboratory Chemical Management for Waste Reduction*; 2nd ed., American Chemical Society: Washington, DC, 1993.

13. American Chemical Society *Safety in Academic Chemistry Laboratories*; 5th ed.; American Chemical Society: Washington, DC, 1990.

Technique 2
LABORATORY GLASSWARE

In Technique 2, we discuss the glassware used in the organic laboratory. For most of the experiments in this book, you will be using carefully constructed and polished glassware that has ground glass joints designed to fit together tightly. This glassware is called *standard taper glassware,* and it comes in a variety of sizes, some suitable for macroscale experiments and some used only for microscale experiments. The last section of this technique describes the preparation of a Pasteur filter pipet, needed for many microscale experiments.

SAFETY PRECAUTION

Before you use any glassware in an experiment, check it carefully for cracks or chips. Glassware with spherical surfaces, such as round-bottomed flasks, can develop small star-shaped cracks. **Replace damaged glassware.** When cracked glassware is heated, it can break, thereby ruining your experiment and possibly causing a serious spill or fire.

2.1
Standard Taper Glassware

Standard taper glassware is designated by the abbreviation ⵧ. All of the joints in standard taper glassware have been carefully ground so that they are exactly the same size, and all the pieces fit together interchangeably. We recommend the use of ⵧ 19/22 glassware for macroscale experiments. You may use ⵧ 14/20 or ⵧ 24/40 glassware in your laboratory. The numbers represent the diameter, in millimeters, and the length, also in millimeters, of all the joints on the glassware (Figure 2.1). Figure 2.2 shows a typical set of ⵧ 19/22 glassware found in introductory organic laboratories.

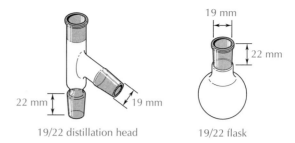

19/22 distillation head   19/22 flask

FIGURE 2.1 Standard taper distillation head and round-bottomed flask.

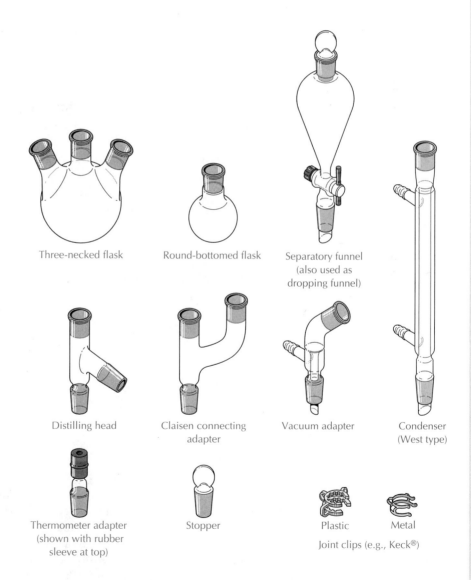

Three-necked flask   Round-bottomed flask   Separatory funnel (also used as dropping funnel)

Distilling head   Claisen connecting adapter   Vacuum adapter   Condenser (West type)

Thermometer adapter (shown with rubber sleeve at top)   Stopper   Plastic   Metal

Joint clips (e.g., Keck®)

FIGURE 2.2 Standard taper glassware for macroscale experiments.

Greasing Joints   Although standard taper joints fit together tightly, they are never put together dry. You must grease them with the lubricating grease provided in the laboratory (Figure 2.3). If ground glass joints are not greased, the glass may react with the chemicals used in the experiment, thus causing the joints to "freeze," or stick together. Taking apart stuck joints, while not impossible, is often not an easy task, and standard taper glassware frequently gets broken in this way. So, an ounce of prevention is worth a pound of cure. In other words, **always grease ground glass joints.**

A variety of greases are commercially available. For general purposes in an undergraduate laboratory, a hydrocarbon grease, such as Lubriseal, is preferred because it can be removed easily. Silicone greases are intended for sealing a system that will be under vacuum. They are nearly impossible to remove completely because they do not dissolve in detergents or organic solvents.

Using excess grease is also bad practice. It is messy; but worse, it may contaminate your reaction or coat the inside of your flasks, thus making them difficult to clean. If grease oozes above the top or below the bottom of the joint, you have used too much. Take the joint apart, wipe off the excess grease with a towel or tissue, and assemble the pieces again.

When you have finished your experiment, clean the grease from the joints by using a brush, detergent, and hot water. If this scrubbing does not remove all the grease, dry the joint and clean it with a towel moistened with toluene or hexane.

— **SAFETY PRECAUTION** ——————

Toluene or hexane pose a fire hazard; work in a hood.

Grease

FIGURE 2.3 Assembling standard taper joint on macroscale glassware. Apply two *thin* strips of grease that run the entire length of the inner joint and are about 180° apart. Insert the inner joint into the outer joint and rotate one of the pieces. The joint should rotate easily and show a clear band of grease on its entire surface.

**Assembling Standard
Taper Glassware**

The interchangeable pieces of standard taper glassware can be assembled in a variety of ways, depending on the operation being performed. Figure 2.4 shows a simple distillation apparatus, a typical example of how standard taper glassware is assembled for a laboratory operation.

In the lab you must clamp the glassware firmly to a ring stand or metal rack. Many of the procedures that call for standard taper glassware involve a round-bottomed flask. **When you set up the apparatus, you should always start by firmly clamping the neck of this flask to a ring stand.** The rest of the apparatus builds from this flask. The use of joint clips greatly reduces the breakage that sometimes occurs while assembling a distillation apparatus. More

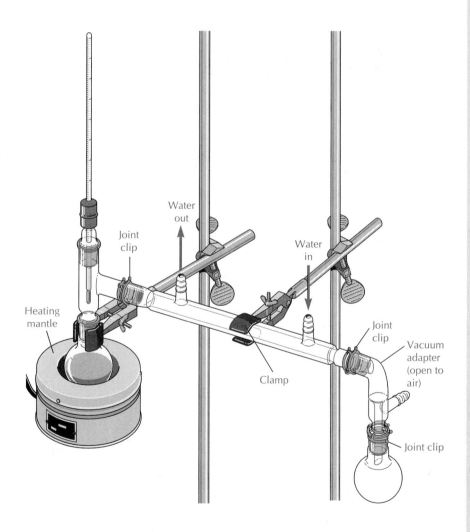

FIGURE 2.4 Simple distillation apparatus.

glassware is broken as a result of the action of gravity than for any other reason. (The bond formed by grease on the joints between the still head and the condenser or the condenser and the vacuum adapter is not sufficiently strong to overcome gravity if these joints are not secured by clips.)

2.2 Microscale Glassware

Many of the experiments in this book use milligram quantities of reagents; we call these procedures microscale experiments. Recovering any product from an operation of this scale would be almost impossible if you were using 19/22 standard taper glassware; most of the material would be lost on the glassware surfaces. The various pieces of glassware shown in Figure 2.5 are used for the microscale experiments in this book.

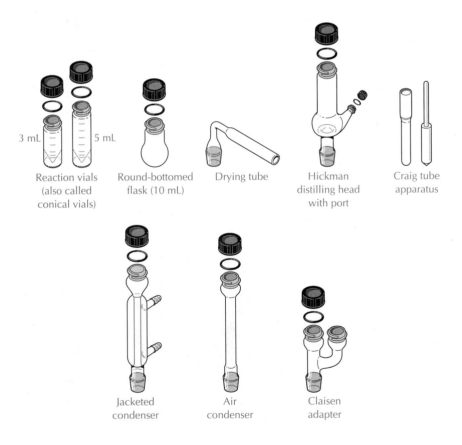

FIGURE 2.5 Microscale glassware.

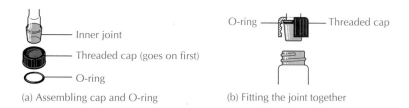

(a) Assembling cap and O-ring   (b) Fitting the joint together

FIGURE 2.6 Assembling standard taper joint on microscale glassware.

The microscale glassware fits together with 14/10 standard taper joints. **Grease is NOT used with microscale glassware** because its presence could cause significant contamination of the reaction mixture. Instead, a threaded cap and O-ring ensure a tight seal and hold the pieces together, thus eliminating the use of clamps or clips. Place the threaded cap over the inner joint, then slip the O-ring over the tapered portion. Fit the inner joint inside the outer joint and screw the threaded cap tightly onto the outer joint (Figure 2.6). This type of connection effectively prevents the escape of vapors and is vacuum tight.

2.3
Cleaning Your Glassware

Part of careful laboratory technique includes cleaning your glassware before you leave the laboratory. Clean glassware is essential for any organic reaction, but in many instances it also needs to be dry. Try not to have to wash something immediately prior to using it, because then you will waste precious time while it dries in the oven.

Strong detergents, hot water, and elbow grease are the ingredients that will clean most glassware used for organic reactions. Scrubbing with a paste made from water and scouring powder, such as Ajax, will remove many organic residues from glassware. Organic solvents, such as acetone or hexane, will help dissolve the polymeric tars that sometimes coat the inside of a flask during a distillation.

SAFETY PRECAUTION
These solvents are flammable; use them in a hood and dispose of them in the flammable (organic) waste container.

A solution of alcoholic sodium hydroxide* is our favorite cleanser for removing organic residues from flasks or other vessels.

> — **SAFETY PRECAUTION** —————————————
>
> Strong alkali can cause serious burns.

A final rinse with distilled water prevents water spots. After your lab is over, there is absolutely no reason why good quality glassware should look one bit different from the way it looked when you began. Take care of it. It's expensive.

2.4
Pasteur Filter Pipets

Microscale laboratory work often requires the transfer of small volumes (1–5 mL) of liquids from one container to another. The Pasteur pipet serves as the tool of choice for these transfers. Volatile liquids, however, will frequently drip out of a Pasteur pipet during such transfers because the solvent vapors build up in the rubber bulb and create increased pressure that forces the liquid out of the pipet. Losses of even a few drops need to be avoided in microscale work. Placing a small plug of cotton in the tip of the Pasteur pipet provides enough resistance to this increased pressure to prevent dripping during the transfer of a volatile liquid. The cotton plug also allows the solution to be drawn into or expelled from the pipet at a slower rate so that better control can be maintained, especially during microscale extractions (discussed in Technique 4).

Airborne particles, such as dust and lint, also present problems when working with microscale volumes of liquids. The cotton plug serves to filter the solution each time it is transferred with a filter tip pipet. The filter pipet also removes a solution very efficiently from a drying agent (discussed in Technique 4.7).

Filter pipets are made by using a piece of wire that has a diameter slightly less than the inside of the capillary portion of the pipet to push a tiny piece of cotton into the tip of a Pasteur pipet (Figure 2.7). A piece of cotton of the appropriate size should offer only slight resistance to being pushed by the wire. If there is so

---

*Made by dissolving 120 g of NaOH in 120 mL of water and diluting to 1 L with 95% ethanol.

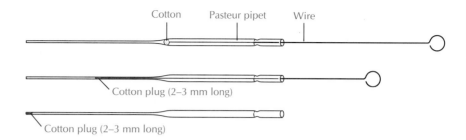

FIGURE 2.7 Preparing Pasteur filter pipet.

much resistance that the cotton cannot be pushed to the tip of the pipet, then the piece is too large. Remove the wire and insert it through the tip to push the cotton back into the upper part of the pipet, and take a bit off the piece of cotton before putting it back into the pipet. The finished cotton plug should be 2–3 mm long and should fit snugly but not too tightly. If the cotton is packed too tightly in the tip, liquid will not flow through it; if it fits too loosely, it may be expelled with the liquid. With a little practice, you should be able to prepare a filter pipet easily.

When using a Pasteur filter pipet (or any other pipet with a bulb attached), you should set the pipet in a test tube or Erlenmeyer flask to keep it upright. Laying the pipet on the bench top or other horizontal surface allows the rubber bulb to be contaminated with solvent or compounds dissolved in the solvent. (Figure 2.8).

FIGURE 2.8 Correct ways to temporarily store Pasteur pipet when you are not using it to transfer liquid.

Technique 3
HEATING METHODS

Most organic reactions do not occur spontaneously when the reactants are mixed together but require a period of heating to reach completion. The use of Bunsen burners in the organic laboratory can pose an **extreme fire hazard** because of the presence of vapors of volatile organic compounds. We recommend using electrical heating devices or a steam bath for the experiments in this book. Using a Bunsen burner or other flame should be a rare event.

3.1
Boiling Stones

Before describing various heating methods, let's consider the role of boiling stones (also called boiling chips) or boiling sticks. Liquids heated in laboratory glassware tend to boil by forming large bubbles of superheated vapor, a process called "bumping." The inside surface of the glass is so smooth that no tiny crevices exist where air bubbles can be trapped, unlike the surfaces of metal pans used for cooking. The addition of boiling stones supplies a porous surface. The air in the pores of the boiling stone expands and escapes as the liquid is heated, carrying vapor with it and promoting the formation of many small bubbles instead of a few large ones. The boiling stones commonly used in the laboratory consist of small pieces of carborundum, a chemically inert compound of carbon and silicon. (Their black color makes them easy to identify and remove from your product, if you have not removed them earlier by filtration.) Boiling sticks are short pieces of wooden applicator sticks. Boiling sticks should not be used in reaction mixtures, nor with any solvent that might react with wood, nor in a solution containing an acid.

One or two boiling stones suffice for smooth boiling of most liquids. **Boiling stones should always be added before heating the liquid.** Adding boiling stones to a hot liquid may cause the liquid to boil violently out of the flask because superheated vapor trapped in the liquid is released all at once. If you forget to add boiling stones before heating, **the liquid must be cooled well below the boiling point before putting boiling stones into it.**

If a liquid you have boiled requires cooling and reheating, additional boiling stones should be added **before** reheating commences. As the boiling stones cool, their pores fill with liquid. This liquid does not escape from the pores as readily as air when the boiling stone is reheated, rendering the boiling stone less effective in promoting smooth boiling.

Mechanical or magnetic stirring agitates a boiling liquid enough to make boiling stones unnecessary, and stirring is used for some experiments in this book. You should **always** add boiling stones or a boiling stick to any unstirred liquid before boiling it— unless you are instructed to the contrary.

3.2
Heating Devices

Heating Mantles   Many reactions and other operations are carried out in round-bottomed flasks. These flasks are heated with electric heating mantles, shaped to fit the bottom on the flask. Several types of heating mantles are available commercially. One type consists of woven fiberglass, with the heating element embedded between the layers of fabric. Fiberglass heating mantles come in a variety of sizes to fit specific sizes of round-bottomed flasks; one sized for a 100-mL flask will not work well with a flask of any other size. Another type of heating mantle, called a Thermowell, has a metal housing and a ceramic well covering the heating element. Thermowell heating mantles can be used with flasks smaller than the designated size of the mantle because of radiant heating from the surface of the well. Both types of heating mantles have no controls and **must** be plugged into a variable transformer or other variable controller to adjust the rate of heating (Figure 3.1). The variable transformer is then plugged into a wall outlet.

Heating mantles are supported underneath a round-bottomed flask by an iron ring or lab jack. Fiberglass heating mantles should not be used on wooden surfaces, because the bottom of the heating mantle may become hot enough to char the wood.

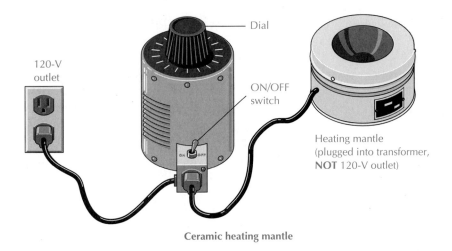

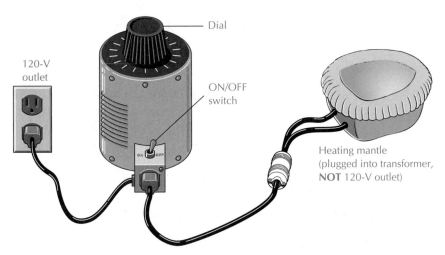

FIGURE 3.1 Heating mantle and variable transformer. (Note: The transformer dial is usually calibrated in percentage of line voltage, *not* in degrees.)

Hot Plates   Hot plates work well for heating flat-bottomed containers such as beakers, Erlenmeyer flasks, and crystallizing dishes used as water baths. We also utilize hot plates to heat the aluminum blocks used with microscale glassware. Figure 3.2a shows a typical reaction apparatus with a conical vial and an air condenser. Several types of aluminum blocks are available commercially; these have holes sized to fit the 14/10 standard taper microscale glassware and conical (reaction) vials described in Technique 2.2. The blocks also have a hole designed to hold a thermometer, so that the tempera-

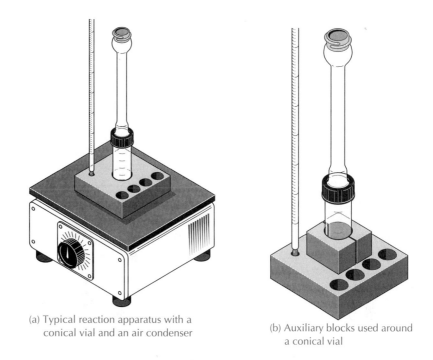

(a) Typical reaction apparatus with a
conical vial and an air condenser

(b) Auxiliary blocks used around
a conical vial

FIGURE 3.2 Heating aluminum blocks used with microscale glassware.

ture of the block can be ascertained. The microscale experiments in this book specify the temperature to which the aluminum block should be heated. Auxiliary blocks designed in two sections to fit around a conical vial provide extra heat to the upper portion of the vial and sit on top of the aluminum block, as shown in Figure 3.2b.

Safety precautions pertaining to all electrical heating devices:
1. The hot surface on the inside of a heating mantle or the heated top of a hot plate poses a fire hazard in the presence of volatile, flammable solvents. An organic solvent spilled on the hot surface can burst into flames if its autoignition temperature is exceeded. It is a prudent laboratory practice to remove a hot heating mantle or hot plate from your work area before pouring any flammable liquids.
2. Heating flammable solvents in open containers on a hot plate can cause a buildup of flammable vapors around the hot plate and should **never** be done. The thermostat on most laboratory hot plates is not sealed and arcs each time it cycles on and off, providing an ignition source for flammable vapors. Steam baths, oil baths, or heating mantles are safer choices if explosion-proof hot plates are not available (reference 1).

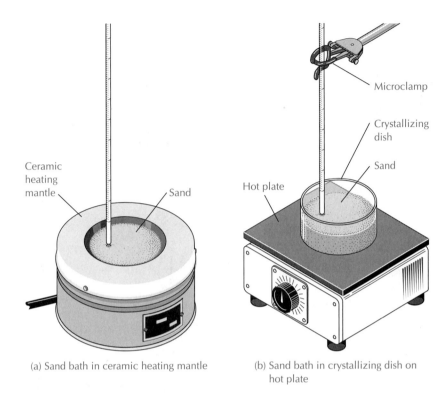

(a) Sand bath in ceramic heating mantle

(b) Sand bath in crystallizing dish on hot plate

FIGURE 3.3 Preparing a sand bath.

Sand Baths   A sand bath provides another method for heating microscale reactions. Sand is a poor conductor of heat, so a temperature gradient exists along the various depths of the sand, with the highest temperature occurring at the bottom of the sand and the lowest temperature near the top surface of the sand.

One method of preparing a sand bath uses a ceramic heating mantle, such as a Thermowell, about two-thirds full of washed sand (Figure 3.3a). A second method employs a crystallizing dish, heated on a hot plate, containing 1–1.5 cm of washed sand (Figure 3.3b). A thermometer is inserted in the sand with the thermometer bulb completely submerged. The rate of heating experienced by the reaction vessel can be closely controlled by raising or lowering the vessel to a different depth in the sand, as well as by changing the heat supplied by the hot plate or heating mantle.

Steam Baths   Steam baths or steam cones provide a safe and efficient way of heating low-boiling-point flammable organic liquids (Figure 3.4) and are extensively used in the organic laboratory for heating liq-

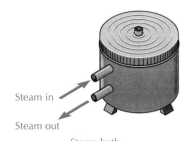

Steam in

Steam out

Steam bath

Steam in

Drain

Steam cone

FIGURE 3.4 Devices for heating liquids below 100°C.

uids below 100°C, where precise temperature control is not required. The concentric rings on the top of the steam bath can be removed to accommodate containers of various sizes. A round-bottomed flask should be positioned so that the rings cover the flask to the level of the liquid it contains. For an Erlenmeyer flask, remove only enough rings to create an opening that is slightly larger than one-half of the bottom diameter of the flask.

Steam baths operate at only one temperature, 100°C. Increasing the rate of steam flow will not raise the temperature but only produce clouds of moisture within the laboratory or hood, and in your sample! Adjust the steam valve for a slow to moderate rate of steam flow when using a steam bath.

A steam bath does have two disadvantages. It cannot be used to boil any liquid with a boiling point above 100°C. Water vapor from the steam may contaminate moisture-sensitive compounds being heated on the steam bath unless special precautions are taken to exclude moisture.

— *SAFETY PRECAUTION* —

Steam is nearly invisible and causes severe burns. Turn off the steam before placing a flask on the steam bath or removing it. Grasp the neck of the hot flask with flask tongs.

3.3
Refluxing a Reaction Mixture

As we stated earlier, most organic reactions do not occur spontaneously but require a period of heating. If this heating were done with the reaction in an open container, the solvent and other liquids in the system would soon evaporate; if the system were closed, it would probably explode when heated. Chemists have developed a simple method of heating a reaction mixture for extended time periods; they call it *refluxing.* Refluxing simply means boiling a liquid while continually cooling the vapor and returning it to the flask as a liquid. A condenser mounted vertically above the reaction flask provides the means of cooling the vapor so that it condenses and flows back into the reaction flask (Figure 3.5).

With a reflux apparatus, the rate of heating is not critical as long as the liquid is boiling at a moderate rate. With more heat, faster boiling occurs but the temperature of the liquid in the flask

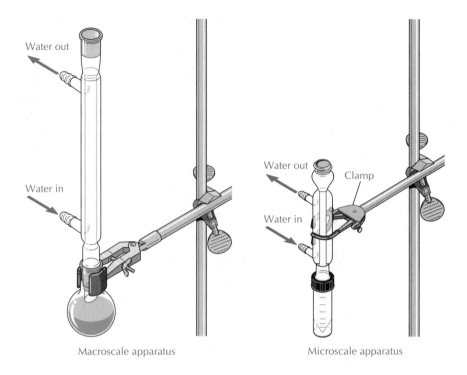

Macroscale apparatus   Microscale apparatus

FIGURE 3.5 Two setups for simple refluxing.

cannot rise above the boiling point of the solvent or solution. If the system is boiling too rapidly, the capacity of the condenser to cool the vapors may be exceeded and solvent will be lost from the top of the condenser. Also, boiling the liquid faster than necessary wastes energy.

(3.3a) Refluxing a Reaction Mixture Under Anhydrous Conditions

Sometimes it is necessary to prevent atmospheric moisture from entering the reaction chamber during refluxing. In this case, a drying tube filled with a suitable drying agent, often anhydrous calcium chloride, is placed at the top of the condenser. For standard taper glassware, a thermometer adapter serves to hold the drying tube (Figure 3.6a). Note that a small piece of cotton is placed at the bottom of the drying tube to prevent the desiccant particles from plugging the outlet of the tube; cotton is also placed over the desiccant at the top of the drying tube to keep the desiccant from spilling. The microscale drying tube has a standard taper 14/10 ground glass joint that fits into the corresponding joint at the top of the condenser and seals with a screw cap (Figure 3.6b).

When it is also necessary to add reagents during the reflux period, a separatory funnel is used as a dropping funnel. If the round-bottomed flask has only one neck, a Claisen adapter provides a second opening into the flask, as shown in Figure 3.7a. For a three-necked flask, the third neck is closed with a ground glass stopper, as shown in Figure 3.7b. Both the condenser and the separatory funnel are fitted with drying tubes filled with a suitable desiccant.

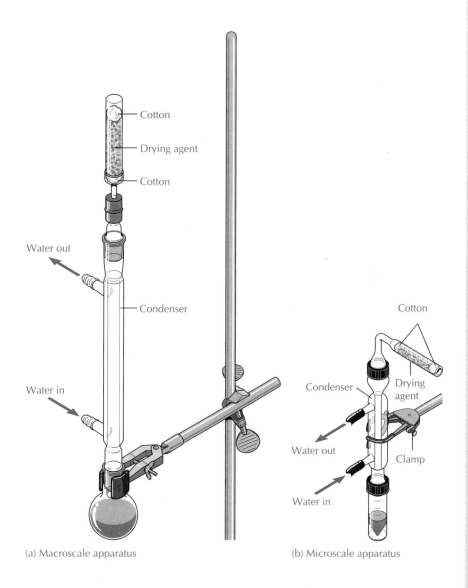

(a) Macroscale apparatus   (b) Microscale apparatus

FIGURE 3.6 Refluxing under anhydrous conditions.

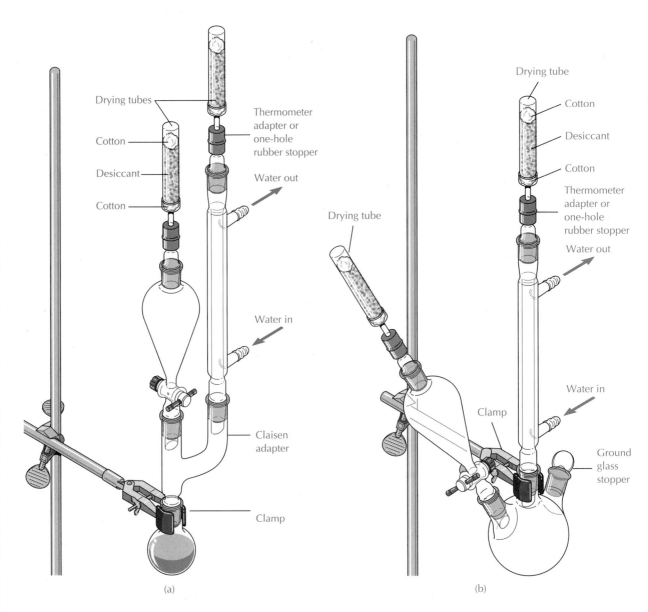

FIGURE 3.7 Maintaining anhydrous conditions while adding reagents to a reaction heated under reflux (a) in a one-necked reaction flask; (b) in a three-necked flask.

The addition of reagents to a microscale reaction being carried out under anhydrous conditions is done with a syringe, as shown in Figure 3.8. The Claisen adapter provides two openings in the system. The opening used for the syringe can be capped either with a screw cap and Teflon septum or with a fold-over rubber septum.

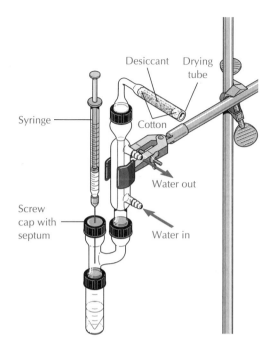

FIGURE 3.8 Using a syringe to add reagents to a microscale reaction being carried out under anhydrous conditions.

— *SAFETY PRECAUTION* —

Syringe needles are puncture hazards. They must be handled carefully. Keep them capped except when actually in use. Their disposal must be done appropriately (as specified by your instructor), in a manner that does not create a hazard to lab occupants or housekeeping staff.

References

1. James A. Kaufman & Associates. *The Laboratory Safety Newsletter* **1993**, *1*(4), 15. (Address: 101 Oak Street, Wellesley, MA 02181.)
2. Lodwig, S. N. *J. Chem. Educ.* **1989**, *66*, 77–84.

Questions

1. Explain why a reaction mixture heated in a reflux apparatus for an extended period of time does not lose solvent.
2. A student continually raises the temperature of a flask with a heating mantle until it appears that the liquid in the flask is very hot, yet no reflux action has begun. The student's neighbor mentions

that either a boiling stick or boiling stones are needed. What is the *first* thing that the student should do to correct the situation?

3. Bubbling an air stream continuously through a solution while refluxing it can serve as an alternative to the use of boiling stones. However, the use of air sometimes poses a serious problem. Explain. (*Hint:* Use of nitrogen, or even better, helium would be superior to air.) Give two reasons why we use boiling stones instead of helium.

4. While allowing a reaction mixture to reflux in apparatus in which the condenser is fitted with a drying tube containing a drying agent such as calcium chloride, a student overheats the liquid in the round-bottomed flask and drives vapors into the drying tube. The vapors condense on the drying agent and drip back into the condenser and flask. Suggest why this may disrupt the desired reaction.

5. Steam causes more severe burns than boiling water, even though both have a temperature of 100°C. Explain.

6. Explain why it takes longer to carry out a reaction on a steam bath in Denver, Colorado, than in New York City.

Technique 4
EXTRACTION AND DRYING AGENTS

Extraction is a technique used for separating a compound from a mixture. It is often used to separate a compound from the other materials present in a natural product, for example, removing caffeine from coffee. Liquid-liquid extraction involves the distribution of a compound, called a **solute**, between two immiscible, or nonsoluble, liquid **solvents.** Generally, although not always, one of the solvents in an extraction is water and the other is a much less polar organic solvent such as dichloromethane, ether, ethyl acetate, or hexane. By taking advantage of the differing solubilities of a solute in a pair of solvents, compounds can be selectively dissolved in one member of the pair. For example, inorganic salts, water-soluble acids and bases, and water-soluble organic compounds can be conveniently separated from a water-insoluble organic compound by extracting them into water.

In the extraction procedure, an aqueous solution and an immiscible organic solvent are usually shaken in a separatory funnel (Figure 4.1). The various solutes distribute themselves between the water layer and the organic layer, according to their solubilities. Inorganic salts generally prefer the aqueous layer,

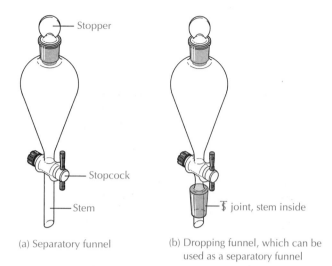

(a) Separatory funnel

(b) Dropping funnel, which can be used as a separatory funnel

FIGURE 4.1 Funnels.

whereas most organic substances dissolve more readily in the organic solvent layer. Thus, two or three extractions of a mixture of materials will usually suffice to draw a nonpolar organic compound such as a hydrocarbon or a halocarbon into an organic solvent. Alcohols or other more polar organic compounds may require additional extractions or a different approach.

When a given organic compound is partitioned between an organic solvent and water, the ratio of the solute concentration in the organic solvent to its concentration in the water is equal to the ratio of its solubilities in the two solvents. This distribution of an organic solute, either liquid or solid, can be expressed by Equation (1):

$$K = \frac{\text{g compound per mL organic solvent}}{\text{g compound per mL water (solvent)}} \tag{1}$$

where *K* is the ***distribution coefficient***, or ***partition coefficient.***

Any organic compound with a distribution coefficient greater than 1 can be separated efficiently from water by extractions with a water-insoluble organic solvent. As you will soon see, working through the mathematics of Equation (1) shows that a series of extractions using small volumes of solvents is more efficient than one single, large-volume extraction. A volume of solvent about one-third the volume of the aqueous phase is sufficient for most extractions. Commonly used extraction solvents are indicated in Table 4.1.

Table 4.1. **Extraction solvents**

Solvent	Boiling point, °C	Solubility in water, $g \cdot (100 \text{ cc})^{-1}$	Hazard	Density, $g \cdot mL^{-1}$	Fire hazard[b]
Diethyl ether	35	6	Inhalation, fire	0.71	++++
Pentane	36	0.04	Inhalation, fire	0.62	++++
Petroleum ether[c]	40–60	low	Inhalation, fire	0.64	++++
Dichloromethane	40	2	LD_{50},[a] $1.6 \text{ mL} \cdot kg^{-1}$	1.32	+
Petroleum ether[c]	60–80	low	Inhalation, fire	0.65	++++
Chloroform	61	0.5	Banned for use in drugs and cosmetics	1.48	–
Hexane	69	0.02	Fire, inhalation	0.66	++++
Toluene	111	0.06	Inhalation, fire; less toxic than benzene	0.87	++

a. LD_{50}, lethal dose orally in young rats.
b. Scale: extreme fire hazard = ++++.
c. A mixture of hydrocarbons.

4.1 Extraction: The Theory

If the distribution coefficient, K, between water and an organic solvent is large (>100) for a solute, a single extraction will suffice to extract the compound from water into the organic solvent. Most often, however, the distribution coefficient is between 1 and 10, thereby making multiple extractions necessary.

Suppose we consider a simple case of extraction from water into ether, and we assume a distribution coefficient of 5 for the organic compound being extracted. Just for illustration, let's use 5.0 g of compound in 500 mL of water being extracted into 150 mL of ether. If C_1 = final concentration in ether = 5 parts, C_2 = final concentration in water = 1 part, and x = amount of solute remaining in the water layer after extraction, then

$$5 = K = \frac{C_1}{C_2} = \frac{\dfrac{5.0 \text{ g} - x \text{ g}}{150 \text{ mL}}}{\dfrac{x \text{ g}}{500 \text{ mL}}} = \frac{(5.0 - x)500}{150x}$$

$$x = 2.0 \text{ g}$$

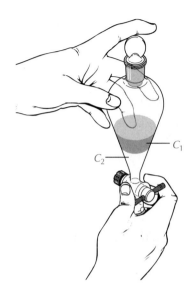

So 2.0 g remain in the water layer after extraction, and 5.0 g − x g = 3.0 g of solute are in the ether layer.

Using these two quantities of solute, we can calculate the final concentration of solute in each of the layers.

$$C_1 \text{ in ether} = \frac{5.0 \text{ g} - x \text{ g}}{150 \text{ mL}} = \frac{3.0 \text{ g}}{150 \text{ mL}} = 0.020 \text{ g} \cdot \text{mL}^{-1}$$

$$C_2 \text{ in water} = \frac{x \text{ g}}{500 \text{ mL}} = \frac{2.0 \text{ g}}{500 \text{ mL}} = 0.0040 \text{ g} \cdot \text{mL}^{-1}$$

A second extraction will increase the overall efficiency of the extraction procedure. Thus, extracting the remaining water solution with a fresh 150-mL portion of ether will remove more of the organic substance that still remains in the water.

Again, C_1 = final concentration in ether, C_2 = final concentration in water, and x = amount of solute remaining in the water layer after extraction. As before,

$$5 = K = \frac{C_1}{C_2} = \frac{\dfrac{2.0 \text{ g} - x \text{ g}}{150 \text{ mL}}}{\dfrac{x \text{ g}}{500 \text{ mL}}} = \frac{(2.0 - x)500}{150x}$$

$$x = 0.80 \text{ g}$$

So, 0.80 g solute remain in the water layer after the second extraction, and 2.00 g − x g = 1.20 g have been extracted into the ether layer.

A third extraction of the residual water with a 150-mL portion of ether will remove another 0.48 g of the organic compound from the water layer. In the three ether extractions, 3.0 g + 1.2 g + 0.48 g, for a total of 4.68 g of the organic substance, have been extracted from the water into the organic solvent. Only 6.4% of the organic substance remains behind in the residual water. Most of this could be extracted into another 150 mL of ether.

Suppose that, instead of extracting with three 150-mL portions of ether, we had extracted the original solution all at once with a 450-mL portion of ether. Compared with the total of 4.68 g extracted by the three 150-mL portions, the single 450-mL extraction is less efficient. The single extraction leaves 18% of the organic solute behind in the water layer.

Let x, C_1, and C_2 be defined as before, and $K = 5$, as before. Then,

$$5 = \frac{\dfrac{5.0 \text{ g} - x\,\text{g}}{450 \text{ mL}}}{\dfrac{x\,\text{g}}{500 \text{ mL}}} = \frac{(5.0 - x)500}{450x}$$

$$x = 0.91 \text{ g}$$

After one large-volume extraction, 0.91 g of solute remains in the water layer, and 4.09 g of solute is now in the ether layer.

In general, the fraction of the solute remaining in the original water solvent is given by

$$\frac{C_{2,\,\text{final}}}{C_{2,\,\text{initial}}} = \left(\frac{V_2}{V_2 + V_1 K}\right)^n$$

where V_1 = volume of ether in each extraction; V_2 = original volume of water; n = number of extractions; and K = distribution coefficient.

4.2
Macroscale Extractions

Place a separatory funnel big enough to hold two to four times the total solution volume in a metal ring firmly clamped to a ring stand or metal rack (Figure 4.2a). Make sure that the stopcock is closed. Pour the cooled aqueous solution to be extracted into the separatory funnel. Add a volume of organic solvent equal to approximately one-third the total volume of the aqueous solution (Figure 4.2b), and put the stopper in place. Remove the funnel from the ring, hold the stopper firmly in place with your index finger (Figure 4.2c), invert the funnel, and **open the stopcock immediately** to release the pressure from solvent vapors (Figure 4.2d). Close the stopcock, and shake the mixture by inverting the separatory funnel four or five times before releasing the pressure by opening the stopcock; continue this shaking and venting process for five or six repetitions to ensure complete mixing of the two phases.

Pour the top layer out of the top of the funnel so that it is not contaminated by the residual bottom layer adhering to the stopcock and tip.

Place the separatory funnel in the ring once more and wait until the layers have completely separated (Figure 4.3a). Remove the stopper and draw off the bottom layer into an Erlenmeyer flask (Figure 4.3b). Pour the remaining layer out of the funnel through the top (Figure 4.3c).

Do this entire procedure each time an extraction is carried out. When extracting a solution several times with a solvent more

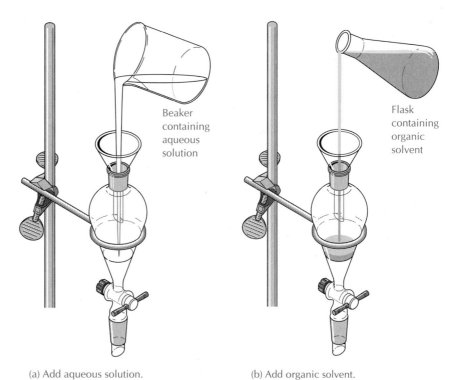

(a) Add aqueous solution.　　　　(b) Add organic solvent.

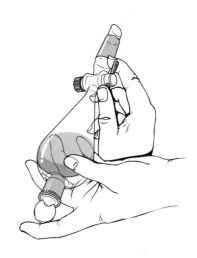

(c) Insert stopper and hold
    stopper with your finger.

(d) Invert funnel and immediately
    open stopcock to release pressure.

FIGURE 4.2 Using a separatory funnel.

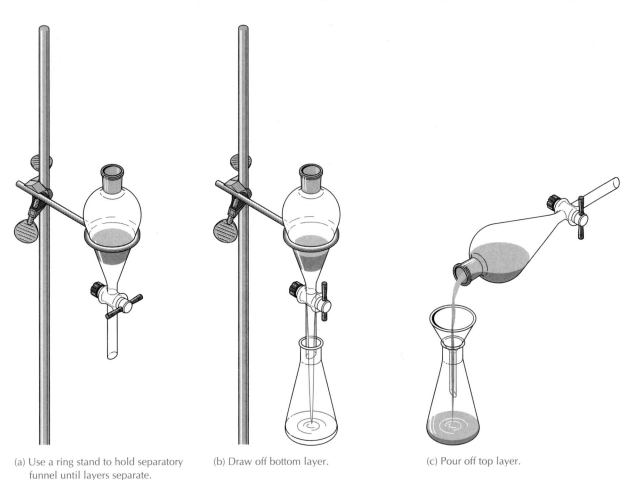

(a) Use a ring stand to hold separatory funnel until layers separate.

(b) Draw off bottom layer.

(c) Pour off top layer.

FIGURE 4.3 Using a separatory funnel (continued).

dense than water, it is not necessary, of course, to pour the top layer out of the separatory funnel until all of the extractions are done.

Finally, **do not discard any solution** until you have completed the entire extraction procedure and are certain which flask contains your desired product.

4.3 Additional Information about Extractions

Densities   Before you begin any extraction, look up the density of the organic solvent in a handbook to determine whether the solvent you are using for the extraction is *more dense* or *less dense* than water. The denser layer will *always* be on the bottom. Solvents that are less dense than water form the upper layer in the separatory funnel

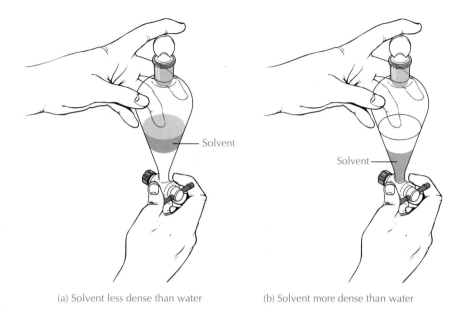

(a) Solvent less dense than water   (b) Solvent more dense than water

FIGURE 4.4 Solvent densities.

(Figure 4.4a), whereas those that are more dense form the lower layer (Figure 4.4b).

If you cannot be sure whether the top or bottom layer is the organic one, you can add a few drops of the layer in question to a milliliter or two of water in a test tube and observe whether it dissolves or not.

Temperature of the Extraction Mixture

Be sure that the extraction solution is at room temperature or slightly cooler before you add the organic extraction solvent. Most solvents used for extractions have low boiling points and will boil if added to a warm solution. A few pieces of ice can be added to cool the solution.

Venting the Funnel

Be sure that you vent an extraction mixture by carefully inverting the funnel and **immediately** opening the stopcock (Figure 4.2d) before you begin the shaking process. If you do not do this, the stopper may pop from the funnel and liquids as well as gases will be released, as shown in Figure 4.5. Pressure buildup within the separatory funnel is always a problem when using low-boiling-point solvents such as diethyl ether, pentane, and dichloromethane for extractions.

Venting extraction mixtures is especially important when you use a dilute sodium carbonate or bicarbonate solution to wash an organic phase containing traces of an acid. Carbon dioxide gas is

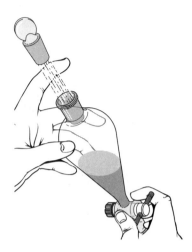

FIGURE 4.5 Failure to vent the separatory funnel when washing with sodium carbonate or bicarbonate solution can cause the stopper to blow out.

given off in the neutralization process. The CO_2 pressure buildup can easily force the stopper out of the funnel and cause losses of your solutions (and product!). When using sodium carbonate or bicarbonate to extract or wash acidic contaminants from an organic solution, **vent the extraction mixture immediately** after the first inversion and subsequently after every three or four inversions.

Washing the Organic Phase

After the extraction is completed and the two immiscible liquids are separated, the organic layer is often extracted or *washed* with water or perhaps a dilute aqueous solution of an acid or a base. For example, a chemical reaction involving alkaline reagents often yields an organic extract that still contains some alkaline material. The alkaline material can be removed by washing the extract (organic phase) with a 5% solution of hydrochloric acid. Similarly, an organic extract obtained from an acidic solution should be washed with 5% solutions of sodium hydroxide, sodium carbonate, or sodium bicarbonate (see previous subsection on venting). The salts that are formed in these extractions are very soluble in water but not in typical organic extraction solvents, so they are easily removed in the aqueous phase. If acid or base washes are required, they are done in the same manner as any other extraction and are usually followed by a water wash.

Salting Out

If the distribution coefficient for a substance to be extracted from water into an organic solvent is close to or lower than 1.0, a simple extraction procedure will not be very effective. In this case, a ***salting out*** procedure can help. Salting out is done by adding a saturated solution of NaCl (sometimes called brine) or Na_2SO_4, or the salt crystals themselves to the aqueous layer. The presence of a salt in the water layer decreases the solubility of the organic compound in the aqueous phase. Therefore, the distribution coefficient increases because more of the organic compound is transferred from the aqueous phase to the organic layer. Salting out can also help to separate a homogeneous solution of water and a water-soluble organic compound into two phases, making it easier to extract the organic compound into an organic solvent.

Avoiding Emulsions

The formation of an *emulsion* between the organic and aqueous layers is sometimes encountered while doing extractions. When an emulsion forms, the entire mixture has a milky appearance, with no clear separation between the immiscible layers; or there may be a third milky layer between the aqueous and the organic phases. Emulsions are not usually formed during diethyl ether extractions but frequently occur when aromatic or chlorinated organic solvents are used. An emulsion often disperses if the separatory funnel and its contents are allowed to sit in a ring stand for a few minutes. However, when you use aromatic or chlorinated solvents to extract organic compounds from aqueous solutions, be gentle; preventing emulsions is simpler than dealing with them. Instead of shaking the mixture vigorously, invert the separatory funnel and gently swirl the two layers together for 2–3 min. However, use of this swirling technique may mean that you need to extract an aqueous solution with three or four portions of organic solvent rather than with only two or three.

Prepare a Celite pad.

Should an emulsion occur, it can be dispersed by vacuum filtration through a pad of the filter aid Celite. Prepare the Celite pad by pouring a slurry of Celite and water onto a filter paper in a Buchner funnel. Remove the water from the filter flask before pouring the emulsion through the Buchner funnel. Return the filtrate to the separatory funnel and separate the two phases. Another method, useful when the organic phase is the lower layer, involves filtering the organic phase through a silicone impregnated filter paper. For microscale extractions (discussed in Technique 4.5), centrifugation of the emulsified mixture effectively separates the phases.

Cleaning and Storing the Separatory Funnel

When the entire operation is complete, clean up. Stopcocks and stoppers stick easily, especially when the funnel has been used with alkaline solutions. Cleaning the funnel immediately and regreasing the stopcock will prevent "frozen" stopcocks later. Grease is not necessary with Teflon stopcocks, but these may also freeze if not loosened during storage.

4.4
Summary of Macroscale Extraction
From Water to an Organic Solution

1. Place the aqueous mixture in a separatory funnel with a capacity at least twice as great as the amount of mixture.

2. Add a volume of immiscible organic solvent approximately one-third that of the aqueous phase. **You must know the density of the organic solvent.**

3. Invert the funnel with one hand on the stopper and neck of the funnel, and open the stopcock to release any pressure buildup.

4. Shake the funnel vigorously two or three times (however, see precautions about emulsions on page 733); vent the funnel.

5. Allow the two phases to separate.

6. **For an organic solvent less dense than water,** draw off the lower phase into a **labeled** Erlenmeyer flask; pour the organic phase out of the top of the funnel into a **labeled** Erlenmeyer flask; return the aqueous phase to the separatory funnel for step 7. **For an organic solvent more dense than water,** draw off the lower organic phase into a **labeled** Erlenmeyer flask; the upper aqueous phase remains in the separatory funnel ready for step 7.

7. Extract the original aqueous mixture once or twice more with fresh solvent.

8. Combine the organic solvent extracts in the same Erlenmeyer flask and pour this solution into the separatory funnel. Extract this organic solution with dilute acid or base if necessary, to neutralize any bases or acids remaining from the reaction.

9. Wash the organic phase with water or saturated NaCl.

10. Dry the organic phase with an anhydrous drying agent (see Technique 4.6).

4.5
Microscale Extractions

The small volumes of liquids used in microscale reactions cannot be handled in a separatory funnel because most of the material would be lost on the glassware. Instead, a conical vial or a centrifuge tube holds the two-phase system, and a Pasteur pipet serves as the tool for separating one phase from the other (Figure 4.6). The conical bottom of the vial or centrifuge tube enhances the visibility of the interface between the two phases in the same way that the conical shape of the separatory funnel enhances the visibility of the interface just above the stopcock. Conical vials, with a capacity of 5 mL, work well for extractions in which the total vol-

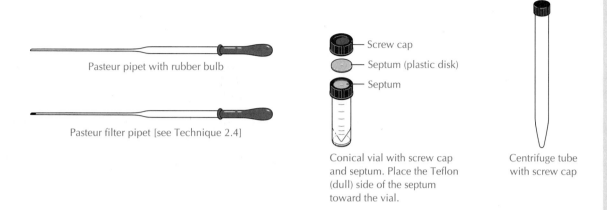

FIGURE 4.6 Equipment for microscale extractions.

ume of both phases does not exceed 4 mL; screw-capped centrifuge tubes with a 15-mL capacity serve for extractions involving a total volume of 4 to 12 mL. Centrifuge tubes are particularly useful for combinations of organic and aqueous phases that form emulsions. The tubes can be spun in a centrifuge to produce a clean separation of the two phases.

Techniques Common to All Microscale Extractions

Before discussing specific types of extractions, we should consider the techniques common to all microscale extractions.

1. Conical vials tip over very easily. **Always place the vial in a 30- or 50-mL beaker.** Set centrifuge tubes in a test tube rack.

2. The plastic septum used with the screw cap on a conical vial has a chemically inert coating of Teflon on one side. The Teflon looks dull and should be placed against the top of the vial. (*Note:* The shiny side of the septum is not inert to all organic solvents.)

3. As with extractions performed in a separatory funnel, thorough mixing of the two phases is essential for complete transfer of the solute from one phase to the other. Mix the two phases by capping the vial or centrifuge tube and shaking it vigorously.

4. Slowly loosen the screw cap to vent the vial or centrifuge tube.

5. A Pasteur pipet with a cotton plug in the tip provides better control than one without the cotton, particularly when transferring volatile solvents such as dichloromethane or ether.

6. **Expel the air from the rubber bulb before inserting the pipet** to the bottom of the vial or centrifuge tube.

7. Slowly release the pressure on the bulb and draw the lower layer into the pipet. Maintain a steady pressure on the rubber bulb while transferring the liquid to another container.

8. Hold the receiving container, usually another conical vial or centrifuge tube, close to the extraction vial so that the transfer can be accomplished quickly without any loss of liquid.

9. Carefully label all containers holding aqueous or organic layers. Do not discard any solution until you have completed the entire extraction procedure and know which vessel contains your product.

(4.5a)   Extracting an Aqueous Solution with an Organic Solvent More Dense Than Water

The microscale extraction of an aqueous solution with a solvent, such as dichloromethane (CH_2Cl_2), that is more dense than water is an example of this type of extraction. The product is transferred

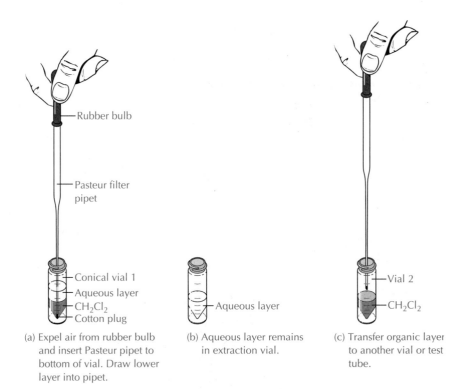

(a) Expel air from rubber bulb and insert Pasteur pipet to bottom of vial. Draw lower layer into pipet.

(b) Aqueous layer remains in extraction vial.

(c) Transfer organic layer to another vial or test tube.

FIGURE 4.7 Extracting an aqueous solution with a solvent more dense than water.

from the aqueous phase to the organic phase and the lower organic phase (dichloromethane) needs to be removed from the vial in order to separate the layers.

Set conical vials in 30- or 50-mL beakers to prevent tipping.

Place the aqueous solution and the specified amount of solvent in a conical vial (or centrifuge tube) labeled vial 1 (Figure 4.7). Tightly cap the vial and shake the mixture thoroughly. Unscrew the cap slightly to release the pressure and allow the layers to separate completely.

Hold vial 2 close to vial 1 so that the transfer can be accomplished without loss of liquid.

Prepare a Pasteur filter pipet [see Technique 2.4]. Press the air from the rubber bulb and put the pipet into the vial with the tip touching the bottom of the cone. Partially release the pressure on the bulb and draw the lower layer into the pipet until the interface between the two layers is exactly at the bottom of the vial. Maintain a steady pressure on the rubber bulb while transferring the pipet to another vial (vial 2 in Figure 4.7c) or test tube. The aqueous layer remains in the extraction vial and can be extracted again with a second portion of CH_2Cl_2. The second portion of CH_2Cl_2 is added to vial 2 after the separation.

(4.5b) Washing an Organic Liquid That Is More Dense Than Water

When an organic phase (either a liquid product or a solution of organic solvent and product) more dense than the aqueous phase is being washed, the aqueous phase needs to be removed. There are two methods for doing this operation. Method A is preferable when a liquid organic product is not dissolved in a solvent, because losses on the glassware are minimized. Method B is easier to perform because the lower phase is always removed, but it requires several more transfers than Method A, plus two or three additional conical vials or centrifuge tubes.

Tipping the vial helps you to see the interface.

Remember, as in any extraction, no material should be discarded until you are certain that you have the desired product in hand.

Method A. After mixing the layers thoroughly and allowing them to separate, carefully remove the upper aqueous layer with a Pasteur filter pipet (Figure 4.8). Press the rubber bulb to expel an amount of air approximately equal to the volume of the aqueous phase. Hold the tip of the pipet in the aqueous phase 1–2 mm above the interface and slowly draw the aqueous solution into the pipet. Draw all of the aqueous layer, plus a small amount of the lower organic phase into the pipet. Maintain a steady pressure on the rubber bulb and hold the pipet in a vertical position just above the liquid surface. The phases will separate in the pipet and the lower layer can be gently expelled back into the extraction vial. When the interface is exactly at the cotton plug, transfer the aqueous layer to a test tube. When used in this fashion, the Pasteur pipet becomes a micro-separatory funnel.

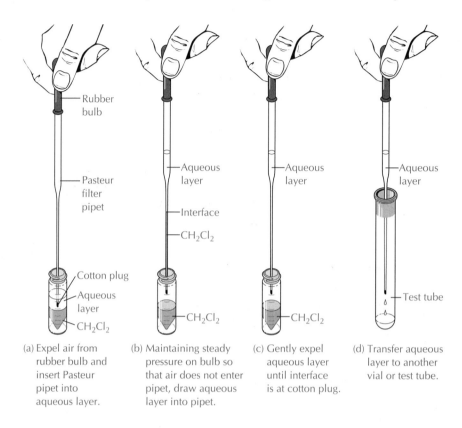

(a) Expel air from rubber bulb and insert Pasteur pipet into aqueous layer.

(b) Maintaining steady pressure on bulb so that air does not enter pipet, draw aqueous layer into pipet.

(c) Gently expel aqueous layer until interface is at cotton plug.

(d) Transfer aqueous layer to another vial or test tube.

FIGURE 4.8 Method A for washing an organic liquid that is more dense than water.

Method B. Two conical vials or centrifuge tubes are needed for this procedure; a third is useful (Figure 4.9). The lower organic phase is transferred to vial 2 in the first separation. Pour the aqueous phase remaining in vial 1 into a labeled test tube. Clean and dry vial 1 (not necessary if a third vial is available). Add the second portion of wash liquid (e.g., water, NaOH solution). Proceed with the second washing of the organic phase in vial 2 by capping the vial and shaking the two phases together. Allow the phases to separate and again remove the lower organic phase with a Pasteur filter pipet, placing the organic phase in the clean vial 1 (or vial 3). Repeat the procedure if additional washings are required. After the last washing, the organic phase will be in a clean vial, ready for the next step of the experimental procedure. Treat the aqueous washes as directed in the cleanup procedure.

Remember, as in any extraction, no material should be discarded until you are certain that you have the desired product in hand.

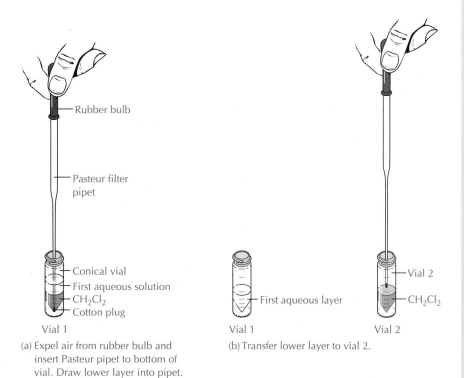

Vial 1

(a) Expel air from rubber bulb and insert Pasteur pipet to bottom of vial. Draw lower layer into pipet.

(b) Transfer lower layer to vial 2.

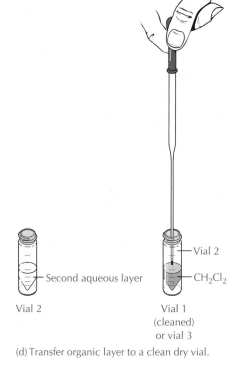

Vial 2

(c) After washing with second aqueous solution, draw lower organic layer into Pasteur pipet.

(d) Transfer organic layer to a clean dry vial.

FIGURE 4.9 Method B for washing an organic liquid that is more dense than water.

FIGURE 4.10 Extracting an aqueous solution with an organic solvent less dense than water.

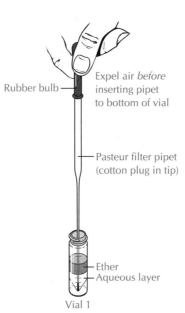

Rubber bulb

Expel air *before* inserting pipet to bottom of vial

Pasteur filter pipet (cotton plug in tip)

Ether
Aqueous layer

Vial 1

(a) Remove lower aqueous phase with Pasteur pipet.

Ether
Aqueous layer

Vial 1   Vial 2

(b) Transfer aqueous phase to vial 2.

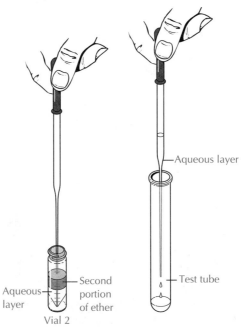

Aqueous layer
Second portion of ether

Aqueous layer

Vial 2

Test tube

(c) Remove aqueous phase and transfer to a test tube.

Vial 1   Vial 2

(d) Combine ether solution from vial 1 with ether solution in vial 2.

(4.5c) Extraction With an Organic Solvent Less Dense Than Water

The microscale extraction of an aqueous solution with an organic solvent less dense than water, for example, diethyl ether, exemplifies this type of extraction. In this case, the lower aqueous phase can be removed from the vial or centrifuge tube, leaving the upper organic phase in the container.

Two conical vials or centrifuge tubes are needed for this procedure; a third is helpful (Figure 4.10). Place the aqueous solution in conical vial or centrifuge tube 1. Add the specified amount of organic solvent; we will use diethyl ether in this example. Cap the vial and shake it to mix the layers. Vent the vial by slowly releasing the cap and allow the phases to separate. Expel the air from the bulb on a Pasteur filter pipet. Insert the tip of the pipet to the bottom of the conical vial. Slowly draw the aqueous layer into the pipet until the interface between the ether and aqueous solution is at the bottom of the vial. Maintain a steady pressure on the rubber bulb and transfer the aqueous solution to vial 2. The ether solution remains in vial 1. Add a second portion of ether to vial 2, cap the vial, and shake it to mix the phases. After the phases separate, again remove the lower aqueous layer with the filter pipet and place it in a test tube. Using a Pasteur pipet, add the ether solution in vial 1 to the ether solution in vial 2. Figure 4.10 illustrates the steps of this procedure. If the experiment specifies washing the ether solution, continue with the procedure described in Technique 4.5d.

Remember, as in any extraction, no material should be discarded until you are certain that you have the desired product in hand.

(4.5d) Washing an Organic Liquid That Is Less Dense Than Water

When washing an organic solution or product less dense than water, the lower or aqueous layer needs to be removed from the vial. This operation is similar to the procedure in Section 4.5c for extracting an aqueous solution with a solvent less dense than water. The only difference is that the organic phase remains in the same vial for the entire procedure and only the aqueous solutions are removed (Figure 4.11). Add the specified amount of water or aqueous solution, cap the vial, and shake it to mix the phases. Open the cap to vent the vial and allow the layers to separate. Expel the air from the rubber bulb of a Pasteur filter pipet and insert the pipet into the bottom of the conical vial. Draw the aqueous layer into the pipet and transfer it to a test tube. The upper organic phase remains in the extraction vial, ready for the next step, which may be another washing or drying with an anhydrous salt.

Remember, as in any extraction, no material should be discarded until you are certain that you have the desired product in hand.

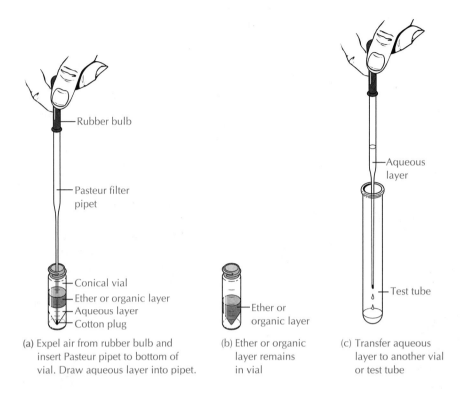

(a) Expel air from rubber bulb and
    insert Pasteur pipet to bottom of
    vial. Draw aqueous layer into pipet.

(b) Ether or organic
    layer remains
    in vial

(c) Transfer aqueous
    layer to another vial
    or test tube

FIGURE 4.11 Washing an organic liquid that is less dense than water.

4.6
Drying Organic Liquids

Most organic separations involve extractions from an aqueous solution; therefore, no matter how careful you are, some water will usually be transferred along with the organic solvent. This transfer happens either because a small amount of water is dissolved in the nonaqueous solvent (see Table 4.1) or because the physical separation of the layers in the extraction process is incomplete. As a result, the nonaqueous (organic) layer usually needs to be dried before performing additional operations on it.

There are two ways to dry an organic solvent. One way is to add an anhydrous chemical drying agent that binds with the water. The other is to distill the water from the solvent [see Technique 7]. The first method is generally easier and most often used.

Chemical drying agents react with water in solution or suspension to form organic-phase insoluble hydrates.

$$n\text{H}_2\text{O} + \text{drying agent} \longrightarrow (\text{drying agent}) \cdot n\text{H}_2\text{O}$$

Usually, anhydrous inorganic salts are used as drying agents. Table 4.2 lists several common drying agents. Several things need

Table 4.2. **Common anhydrous chemical drying agents**

Drying agent	Acid-base properties	Capacity value[a]	Capacity	Efficiency[b] mg·L^{-1}	Speed
$MgSO_4$	neutral	7	high	2.8	fairly rapid
$CaCl_2$	neutral	6	high	1.5	fairly slow
Na_2SO_4	neutral	10	high	25	slow
K_2CO_3	basic	2	low	moderate	fairly rapid
$CaSO_4$	neutral	0.5	low	0.004	fast
KOH	basic	high	high	0.1	fast
H_2SO_4	acidic	very high	high	0.003	fast

a. Capacity value = (maximum number moles of H_2O)/(moles of drying agent).
b. Efficiency = equilibrium water pressure [mg·(L air)$^{-1}$] at 25°C.

to be considered in selecting a drying agent, namely, the capacity, the efficiency, the speed, and chemical inertness. The capacity refers to the number of moles of water bound in the hydrated form of the salt, and the efficiency expresses how well the hydrate holds water. The speed with which the hydrate forms determines how long the drying agent needs to be in contact with the organic solution. Drying agents must be unreactive toward both the organic solvent and any organic compound dissolved in the solvent. Table 4.3 gives the type of drying agent to use with various classes of organic compounds. A base, such as K_2CO_3, is not suitable for drying an acidic organic compound because it would undergo a chemical reaction with the acid.

Table 4.3. **Preferred drying agents**

Class of compounds	Recommended drying agents
Alkanes and alkyl halides	$MgSO_4$, $CaCl_2$, $CaSO_4$, H_2SO_4
Hydrocarbons and ethers	$CaCl_2$, $MgSO_4$, $CaSO_4$
Aldehydes, ketones, and esters	Na_2SO_4, $MgSO_4$, $CaSO_4$
Alcohols	$MgSO_4$, K_2CO_3, $CaSO_4$
Amines	KOH, K_2CO_3
Acidic compounds	Na_2SO_4, $MgSO_4$, $CaSO_4$

FIGURE 4.12 (a) Adding powdered drying agent to solution. (b) Drying agent clumped on the bottom of the flask.

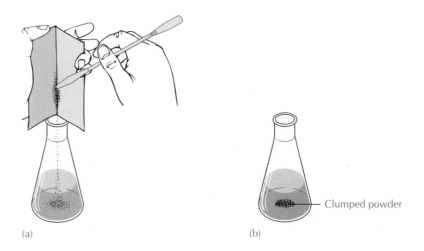

(a)

(b)

Clumped powder

As you can see from Table 4.2, some drying agents have a high capacity but leave water in the organic solution. Na_2SO_4 is a good example. It is particularly useful as a preliminary drying agent, but it is also widely used as a general purpose drying agent because it is inexpensive and can be used with many types of compounds. $CaSO_4$, on the other hand, is very efficient but has a low capacity, which means that it works better after a preliminary drying of the liquid with, for example, Na_2SO_4 or $MgSO_4$. Also some drying agents, such as Na_2SO_4 or $CaCl_2$, do not react quickly with water, so they must be given time to form the hydrate. Of the drying agents listed in Tables 4.2 and 4.3, magnesium sulfate is used most often because it very rarely reacts with solutes or solvents, it has a good capacity for water, and it is quite efficient. However, its exothermic reaction with the water in the solution being dried sometimes causes the solvent to boil if the drying agent is added too rapidly. Slow addition of the drying agent prevents this problem.

To remove water from an organic liquid, add a small quantity of the powdered drying agent directly to the solution to be dried (Figure 4.12a). If an organic solution has water in it, the first bit of drying agent that you add will clump together (Figure 4.12b). When some of the drying agent can easily move about in the mixture while gently swirling the flask, you have added enough. The solution may be stirred with a magnetic stirring bar or simply swirled occasionally by hand to ensure as much contact with the surface of the drying agent as possible. Usually, the best drying efficiency is obtained if the sample is left over the drying agent for at least 10 min. Often a preliminary drying period of 30–60 s, followed by filtration, then by the addition of a second portion of drying agent that stands in the liquid for 10 min removes the water more completely than the use of a single portion of drying agent.

Always place the organic liquid being treated with drying agent in an Erlenmeyer flask closed with a cork. This practice prevents evaporation losses.

4.7
Methods for Separating Drying Agents and Organic Liquids

After the drying agent has absorbed the water present in the organic liquid, it must be separated from the organic liquid. The container receiving the liquid should be clean and dry and have a volume twice the volume of the organic liquid.

Macroscale Separation of Drying Agents

If the solvent is going to be evaporated, place fluted filter paper [see Technique 5.3a, Figure 5.5] in a small funnel and set the funnel in an Erlenmeyer flask (Figure 4.13). If the solvent is being distilled, use a round-bottomed flask, set on a cork ring, as the receiver. Decant the solution slowly into the filter paper, leaving most of the drying agent in the flask. Rinse the drying agent in the flask with 2–3 mL of solvent and also pour this rinse into the filter paper. The extraction procedure is thus complete, and the organic liquid is now ready for the removal of the solvent.

In some extraction procedures, the organic liquid is not dissolved in a solvent. In this situation, you must minimize the loss of liquid (product!) during the removal of the drying agent. Instead of filter paper, a tightly packed cotton plug about 5–6 mm in diameter put in the outlet of a funnel traps the drying agent and absorbs only a small amount of the organic liquid (Figure 4.14).

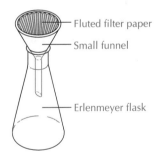

FIGURE 4.13 Filtration of the drying agent from a solution when the solvent will be evaporated. (Replace the Erlenmeyer flask with a round-bottomed flask if the solvent is to be removed by distillation or with a rotary evaporator.)

Fluted filter paper
Small funnel
Erlenmeyer flask

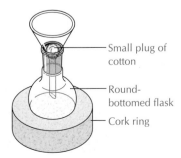

FIGURE 4.14 Filtration of drying agent from an organic liquid when no solvent is present.

Small plug of cotton
Round-bottomed flask
Cork ring

FIGURE 4.15 Using a Pasteur filter pipet (a) to separate an organic liquid from the drying agent and (b) as a funnel.

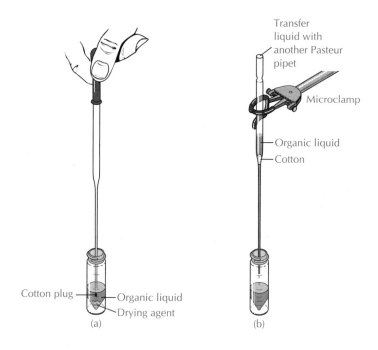

Slowly pour the liquid from the drying agent. If the drying agent is granular or chunky, the cotton plug can be omitted and the liquid carefully decanted into the funnel, leaving all of the drying agent in the original flask. The drying agent may or may not be rinsed with a few milliliters of solvent in this procedure. The organic liquid is ready for the final distillation or analysis.

Liquids cannot be decanted cleanly from MgSO₄ because it is finely powdered.

Microscale Separation of Drying Agents

In microscale extractions, the organic liquid is usually dried in a conical vial or centrifuge tube. A Pasteur filter pipet [see Technique 2.4] is used to remove the liquid from the drying agent and to transfer it to a clean dry container (Figure 4.15a). An alternative procedure, useful for a powdered drying agent such as magnesium sulfate, uses one Pasteur pipet as a funnel and a second Pasteur pipet to transfer the liquid. Here the cotton is packed tightly at the top of the tip (Figure 4.15b).

4.8

Recovery of the Organic Product from the Dried Extraction Solution

Once the extraction solution has been dried, it is necessary to remove the solvent in order to recover the desired organic product. In some of the experiments in which the amount of solvent is small (less than 25 mL), we have directed you to remove the solvent by evaporation on a steam bath in a hood or by blowing off the solvent with a stream of nitrogen in a hood. Your instructor will advise you if these methods for removing solvents are followed in your laboratory. Concern for the environment and environmental laws now limit, or even prohibit, this practice. Removing solvents by distillation or with a rotary evaporator are alternatives to evaporation, and both methods allow the solvents to be recovered.

Evaporation

Place a boiling stick or boiling stone in the Erlenmeyer flask containing the solution to be evaporated and heat the flask on a steam bath in a hood. Your product will be the liquid or solid residue left in the flask when the boiling ceases.

Distillation

Assemble the simple distillation apparatus shown in Technique 7, Figure 7.7. If the solvent is ether, pentane, or hexane, then use a steam bath as a heat source to eliminate the fire hazard an electric heating mantle poses with the very flammable vapors from these solvents. Continue the distillation until the solvent has completely distilled, an end point indicated by a drop in the temperature read on the thermometer. The drop in temperature occurs because there is no longer enough hot vapor surrounding the thermometer bulb. Your product remains in the boiling flask.

Using a Rotary Evaporator

Read Technique 7.7 on vacuum distillation before using a rotary evaporator.

A rotary evaporator is an apparatus for removing solvents rapidly in a vacuum (Figure 4.16). No boiling stones or bubbler are necessary because the rotation of the flask minimizes bumping. Rotary evaporation is done in a round-bottomed flask that is no more than half-filled with the solution being evaporated.

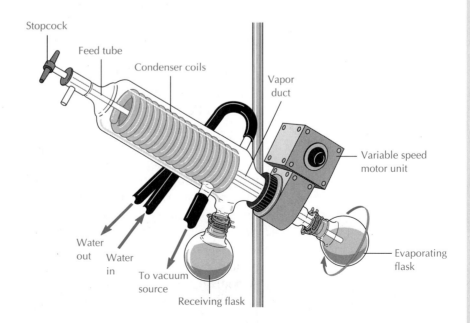

FIGURE 4.16
Schematic diagram of a rotary evaporator.

The following is a generalized outline of the steps in using a rotary evaporator, but you should consult your instructor about the exact operation of the rotary evaporator in your laboratory. Select a round-bottomed flask that will be half-full or less with the solution undergoing evaporation. Connect the flask to the rotary evaporator with a joint clip. Use an empty receiving flask and be sure that it is also clipped tightly to the condenser. Adjust the apparatus so that the flask containing the solution is approximately one-third submerged in a room-temperature water bath. Turn on the water to the condenser, then turn on the vacuum source. As the vacuum develops, turn on the motor that rotates the evaporating flask. When the vacuum stabilizes at 20–30 mmHg or better, begin to heat the water bath. A water bath temperature of 50°–60°C will quickly evaporate solvents with boiling points under 100°C.

When solvent no longer condenses in the receiving flask, the distillation is complete. Stop the rotation of the flask and raise it from the heating bath. Open the stopcock and allow air to bleed into the system. Hold the flask with one hand and remove the clip holding it to the evaporator. Turn off the vacuum source and the condenser water. Disconnect and empty the solvent in the receiving flask into the appropriate waste container or recovered solvent container.

Questions

1. An extraction procedure specified that the aqueous solution containing dissolved organic material be extracted twice with 10-mL portions of diethyl ether. A student removed the lower layer after the first extraction and added the second 10-mL portion of ether to the upper layer remaining in the separatory funnel. After shaking the funnel, the student observed only one phase, with no interface. Explain.
2. A crude nonacidic product mixture dissolved in diethyl ether contains acetic acid. Describe an extraction procedure that could be used to remove the acetic acid.
3. It is usually desirable to minimize the number of transfer steps while conducting an extraction. For extracting an organic compound from an aqueous solution to an organic solvent, would a solvent more dense or less dense than water require fewer transfers? Explain.
4. What precautions need to be observed when using aqueous sodium carbonate to extract an organic solution containing traces of acid?
5. A poorly prepared student forgets to record the densities of water and ether in his/her lab notebook. When the two layers form dur-

ing an ether/water extraction, what would be a convenient, easy way to tell which layer is which?

6. You have 75 mL of a solution of benzoic acid in water, estimated to contain about 2.5 g of the acid. The distribution coefficient of benzoic acid in diethyl ether and water is approximately 10. Calculate the amount of acid that would be left in the water solution after four 25-mL extractions with ether. Do the same calculation, using one 100-mL extraction with ether, to determine which method is more efficient.

Technique 5
RECRYSTALLIZATION

A *pure organic compound* is one in which there are no detectable impurities and which has consistent physical properties. Because experimental work requires an immense number of molecules (Avogadro's number per mole), it is not necessarily true that 100% of the molecules of a given compound with consistent physical properties are identical to one another. Seldom will a pure compound be purer than 99.99%. Even if it were that pure, 1 mol would still contain 10^{19} molecules of other compounds. Nevertheless, we want to work with compounds that are as pure as possible; therefore, we must have ways to purify impure materials.

5.1
Theory of Recrystallization

Recrystallization is a common method for separating a relatively pure, homogeneous solid from impurities. Most solids absorb heat when they dissolve in a solvent. The technique of recrystallization depends on the increasing solubility of a compound in a solvent as the solvent is heated. The important point is that a saturated solution at a higher temperature normally contains more solute than the same solute-solvent pair at a lower temperature. Therefore, the solute precipitates when a warm saturated solution cools. Recrystallization is a laboratory operation whereby a crystalline material dissolves in a solvent, then returns to a crystalline solid by recrystallization. Because the total concentration of impurities in the solid of interest is usually significantly lower than the concentration of the substance being purified, as the mixture cools the impurities remain in solution while the highly concentrated product crystallizes.

Crystal formation of a solute from a solution is a selective process. When a solid crystallizes at the right speed under the appropriate conditions of concentration and solvent, an almost perfectly pure crystalline material can result, because only molecules of the right shape fit into the crystal lattice. In recrystallization, dissolution of the impure solid in a suitable hot solvent destroys the impure crystal lattice, and crystallization from the cold solvent selectively produces a new, purer crystal lattice. Slow cooling of the saturated solution promotes the formation of pure crystals because the molecules of the impurities do not fit properly into the newly forming lattice. Crystals that form slowly are larger and purer than the ones that form quickly. Indeed, rapid crystal formation often traps the impurities again, because the lattice grows so rapidly that the impurities are simply surrounded by the crystallizing solute as the crystal forms.

The most crucial aspect of a recrystallization procedure is the choice of solvent, because *the solute should have a maximum solubility in the hot solvent and a minimum solubility in the cold solvent.* Table 5.1 lists common recrystallization solvents.

In general, a solvent with a structure similar to that of the solute being dissolved is a better recrystallization solvent than solvents with dissimilar structures. Although the appropriate choice of solvent is a trial and error process, a relationship exists between the solvent's molecular structure and the solubility of the solute. This relationship is best described as *like dissolves like*. Nonionic

Table 5.1. **Common recrystallization solvents**

Solvent	Formula	Boiling point, °C	Freezing point, °C
diethyl ether	$(C_2H_5)_2O$	34.6	−116
acetone	$(CH_3)_2CO$	56	−95
petroleum ether[b]	—	60–80	—
chloroform	$CHCl_3$	61	−63
methanol	CH_3OH	65	−98
hexane	C_6H_{14}	69	−94
ethyl acetate	$CH_3CO_2C_2H_5$	77	−84
ethanol	C_2H_5OH	78.5	−117
water	H_2O	100	0
toluene	$C_6H_5CH_3$	110.6	−95
acetic acid	CH_3CO_2H	118	16

a. Scale: Extreme fire hazard = ++++.
b. Petroleum ether (or ligroin) is a mixture of isomeric alkanes. The term "ether" refers to volatility, not the presence of oxygen.

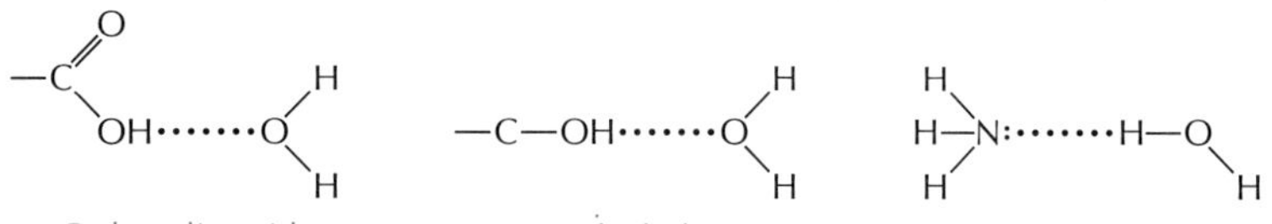

FIGURE 5.1 Compounds that associate with water by hydrogen bonds (dotted lines).

compounds dissolve in water only when they can associate with the water molecules through hydrogen bonding. Thus, all the hydrocarbons and the alkyl halides are virtually insoluble in water, whereas carboxylic acids, which are very polar, are often recrystallized from water solution. Molecules that associate with water through hydrogen bonds include carboxylic acids, alcohols, and amines (Figure 5.1). Carboxylic acids hydrogen bond to a lone pair of electrons on water through the acidic proton; alcohols do likewise. Amines hydrogen bond through the lone pair of nitrogen to the water hydrogen.

Polarity of the solvent is a crucial factor in solubility. One mea-

Misci-bility in water	Solvent polarity	Dielectric constant (ϵ)	Fire[a] hazard	Inha-lation toxicity
−	intermediate	4.3	+ + + +	—
+	intermediate	20.7	+ + +	—
−	nonpolar	ca. 2	+ + + +	—
−	intermediate	4.8	0	high
+	polar	32.6	+ +	high
−	nonpolar	1.9	+ + + +	—
−	intermediate	6.0	+ +	—
+	polar	24.3	+ +	—
	very polar	80	0	—
−	nonpolar	2.4	+ +	high
+	intermediate	6.15	+	high

sure of polarity is the dielectric constant, ϵ (see Table 5.1). A rough expectation is that solvents with higher dielectric constants should more readily dissolve polar (or ionic) compounds. The first four or five members (carbon content of C_4 to C_5) of a homologous series of carboxylic acids, alcohols, or amines are water soluble. But as the molecular weight increases, the water solubility of the species decreases, because the hydrocarbon, or nonpolar, portion of the molecule begins to dominate its physical behavior.

The salts of low molecular weight carboxylic acids are quite water soluble. For example, potassium acetate, CH_3COOK, and sodium propionate, CH_3CH_2COONa, are largely ionic compounds with solubility characteristics similar to sodium chloride or potassium bromide.

Nonpolar solvents such as ether ($\epsilon = 4.3$) dissolve most nonionic organic compounds with ease. The solution process is one of molecular mixing. Polar organic compounds dissolve in nonpolar solvents if the ratio of polar functional groups per carbon atom is not too high (for example, approaching 1). Thus, monofunctional lower members of all homologous series are soluble in nonpolar solvents. For example, methanol ($\epsilon = 32.6$) is extremely soluble in ether.

Among the relatively nonpolar solvents, diethyl ether appears to provide the best solvent properties, although its extreme flammability and low boiling point (34.5°C) require careful attention to safety when using it. Ether in combination with hexane, methanol, or dichloromethane also has excellent solvent properties for recrystallizations. Hexane is even more nonpolar than ether and has the advantage of a higher boiling point (69°C), yet it is still easy to remove from the recrystallized solid.

SAFETY PRECAUTION

Both ether and hexane are very flammable and should not be heated with a flame or on a hot plate, but with a steam bath.

Among the more polar liquids, methanol and ethanol are both commonly used as recrystallization solvents, because they evaporate easily, possess water solubility, and dissolve a wide range of both polar and nonpolar compounds.

5.2 Selecting a Proper Recrystallization Solvent

To select a solvent for recrystallization, take a small sample (20–30 mg) of the compound to be recrystallized, place it in a test tube, and add 5–10 drops of a trial solvent. Shake the tube to mix the materials. If the compound dissolves immediately, it is probably too soluble for a recrystallization to be effective in that solvent. If no solubility is observed, heat the solvent to its boiling point. If solubility is observed then, cool the solution to observe crystallization. The formation of crystals in 10–20 min suggests that you have a good recrystallization solvent at hand.

(5.2a) Recrystallization from Mixed Solvent Pairs

When no single solvent seems to work, a pair of miscible solvents can sometimes be used. Usually mixed solvent pairs include one solvent in which a particular solute is very soluble and another in which solubility is marginal. Often such pairs consist of a polar solvent (with, for example, a high dielectric constant) mixed with a nonpolar solvent.

To recrystallize a solid from a mixed solvent pair, first dissolve the solute in the solvent in which it is more soluble; warm the solvent before adding it to the solute. Then warm the solution nearly to the boiling point and add the other solvent dropwise, until a slight cloudiness appears (indicating that the hot solution is saturated in the solute). Add some of the first solvent again until the cloudiness just disappears and then add a few drops of the first solvent to ensure an excess (more if you are working with a large volume of solution). Let the solution cool slowly to initiate crystallization. Typical mixed solvent pairs are listed in Table 5.2.

Another technique uses a mixture of miscible solvents (for example, ethanol and water) to dissolve the impure solid. This method sometimes requires testing several different proportions to find the optimum ratio of the two solvents. In this case, if the

Table 5.2 Solvent pairs for mixed solvent recrystallizations[a]

Solvent 1	Solvent 2	Solvent 1	Solvent 2
Ethanol	Acetone	Ethyl acetate	Hexane
Ethanol	Petroleum ether	Methanol	Dichloromethane
Ethanol	Water	Methanol	Diethyl ether
Acetone	Water	Methanol	Water
Chloroform	Petroleum ether	Diethyl ether	Hexane (or petroleum ether)

a. Properties of these solvents are given in Table 5.1.

solid is more soluble in the solvent with the lower boiling point, the excess solvent can simply be boiled away until cloudiness is reached. Then the solution is cooled.

Usually laboratory experiments in this book specify the preferred recrystallization solvent in the experimental directions, the choice being based on experience.

(5.3) Macroscale Procedure for Recrystallizing a Solid

Dissolving the Solid

— SAFETY PRECAUTION —

Most organic solvents used for recrystallization are volatile and flammable. Therefore, they should be heated on a steam bath, not on a hot plate.

Place the solid to be recrystallized on a creased weighing paper and carefully pour it into an Erlenmeyer flask (Figure 5.2a).

— SAFETY PRECAUTION —

Lift a hot Erlenmeyer flask with flask tongs.

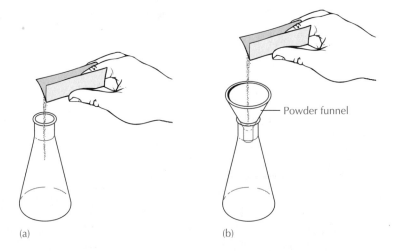

(a)   (b)

Powder funnel

FIGURE 5.2 Two ways to add the solid to an Erlenmeyer flask for recrystallization.

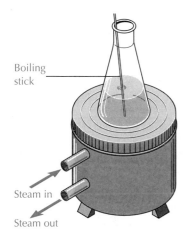

Boiling
stick

Steam in

Steam out

FIGURE 5.3 Heating a solution on a steam bath.

Alternatively, a plastic powder funnel may be set in the neck of the Erlenmeyer flask to prevent spillage (Figure 5.2b). Add one or two boiling chips or a boiling stick. Heat an appropriate volume of the solvent in another Erlenmeyer flask. Then add small portions of hot (just below boiling) solvent to the solid being recrystallized. Begin heating the solid-solvent mixture, allowing it to boil briefly between additions, until the solid dissolves; then add a little excess solvent. Remember that some impurities may be completely insoluble, so do not add too much solvent in trying to dissolve the last bit of solid. Bring the solution to a boil on a steam bath (Figure 5.3).

With particularly volatile organic solvents such as ether or hexane, it is often easier to add a small amount of cold solvent, then heat the mixture nearly to boiling. Slowly add more cold solvent to the heated mixture until the solid just dissolves when the solution is boiling; then add a slight excess.

Cooling the Solution

Filtration of the hot solution is sometimes necessary to remove insoluble impurities [see Technique 4.3a]. However, if there are no insoluble substances in your hot recrystallization mixture, allow the solution to cool slowly. The size of the crystals obtained will depend on the rate at which the solution cools. Allowing the hot solution to stand on the bench top until crystal formation begins and the flask reaches room temperature, followed by final cooling in an ice-water bath, usually produces crystals of a reasonable and intermediate size. The whole cooling process should take at least 15 min; occasionally, it may take much longer (>30 min) before crystals appear.

Collecting the Recrystallized Solid

To complete the recrystallization procedure after all the crystals appear to have formed, collect the crystals by vacuum filtration, using a Buchner funnel, filter flask, and trap bottle or flask. The trap flask keeps backflow from a water aspirator out of the filter flask; with a house vacuum system, the trap flask keeps any overflow of the filter flask out of the vacuum line (Figure 5.4).

Choose the correct size of filter paper, one that will fit flat on the bottom of the funnel and just cover all the holes. Turn on the vacuum source and wet the paper with the solvent to pull it tight over the holes in the funnel. Pour the slurry of crystals and solvent into the funnel. Wash the crystals on the Buchner funnel with a small amount of *cold* solvent (1 to 5 mL, depending on the amount of crystals) to remove any supernatant liquid adhering to them. To do this, allow air to enter the filtration system by removing the rubber tubing from the water aspirator nipple before turning off the water (to prevent backup of water into the system), or turn off

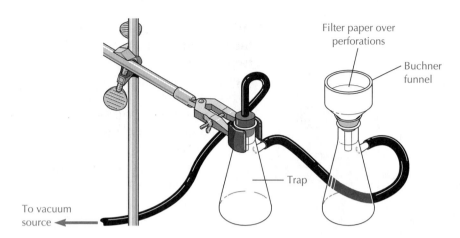

FIGURE 5.4 Apparatus for vacuum filtration. The second filter flask serves as a backflow trap.

Chemists sometimes refer to the supernatant liquid from a recrystallization as the "mother liquor."

the vacuum line and loosen the rubber adapter connecting the Buchner funnel to the filter flask. Then cover the crystals with the cold solvent, reconnect the vacuum, and draw the liquid off the crystals. Initiate the crystal drying process by pulling air through them for a few minutes. Then, again disconnect the vacuum as described earlier. You will probably need to leave your crystals open to the air in your desk for a time in order for them to dry completely; place the crystals on a tared (weighed) watch glass. Remove any boiling chips or sticks before you weigh your crystals. Use this mass to calculate the recrystallization yield.

A second "crop" of crystals can sometimes be obtained by evaporating about half of the solvent from the filtrate and again cooling the solution. This crop of crystals should be kept separate from the first crop of crystals until its melting point [see Technique 6] has been determined. If the melting points of both crops are the same, indicating that the purity is the same, they may be combined. Usually the second crop has a slightly lower melting point, indicating that some impurities crystallized with the desired product.

Careful attention to detail and slow cooling of the hot solution often results in the formation of beautiful pure crystals. Beautiful crystals are to the organic chemist what a home run is to a baseball player!

5.3a Removing Impurities from the Recrystallization Solution

Filtering Insoluble Impurities

A gravity filtration is necessary before cooling your recrystallization solution, if you have some insoluble material that needs to be removed—things like dust, bits of filter paper, or other insoluble impurities. If your recrystallization solution needs filtration, pre-

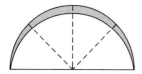

(a) Crease filter paper.

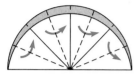

(b) Fold each quarter inward.

(c) Fluted filter paper.

FIGURE 5.5 Fluting filter paper.

pare a fluted filter paper or obtain prefolded filter paper from the supply in the laboratory. Fluted filter paper provides a larger surface area than the usual filter paper cone, making for a faster filtration. Suction filtration under reduced pressure does not work well, because the solution cools rapidly during this process and premature crystallization can occur. Also, small particles may pass through the filter paper when suction is used. To make a fluted filter, crease a regular filter paper in half four times (Figure 5.5a). Then fold each of the eight sections of the filter paper inward, so that it looks like an accordion (Figure 5.5b). Finally, the paper is opened to form a fluted cone, as illustrated in Figure 5.5c. Alternatively, filter paper already folded in this manner is commercially available.

Place the fluted filter paper in a short-stemmed funnel (a plastic powder funnel works well) and put the funnel in a second clean Erlenmeyer flask. Add a small amount (1–3 mL) of the recrystallization solvent to the second flask and heat the flask, funnel, and solvent on your steam bath. The boiling solvent warms the funnel and helps prevent premature crystallization of the solute during filtration (Figure 5.6a). If your steam bath is large enough, keep both flasks hot during the filtration process; if it is too small for both, keep the unfiltered solution hot and set the receiving flask on the bench top. Next, pour the hot recrystallization solution through the fluted filter (Figure 5.6b).

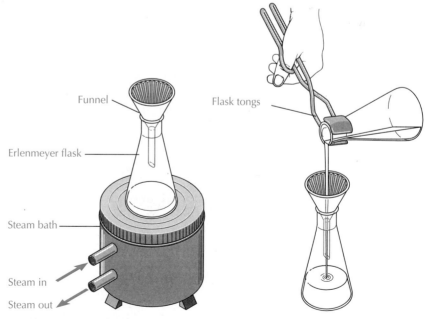

Funnel

Erlenmeyer flask

Steam bath

Steam in

Steam out

Flask tongs

(a) Heat receiving flask and funnel.

(b) Pour hot solution through fluted filler paper

FIGURE 5.6 Filtering recrystallization solution.

Be sure that the hot solution is added in small quantities to the fluted filter paper, because cooling at this stage may cause crystallization in the filter paper. Keep the unfiltered solution hot during this step so that all of the solid remains in solution. If you have difficulty in keeping the solution from crystallizing on the filter paper, add extra hot solvent to the flask containing the unfiltered solution and reheat it to the boiling point before continuing the filtration. Then after the filtration is complete, boil away the extra solvent that you have added.

When all the hot solution has filtered through the paper, check to see whether any crystallization occurred in the Erlenmeyer flask during the rapid cooling of the filtration step. If it has, reheat the mixture to dissolve the solid. Then allow the solution to cool slowly. While the solution is cooling, the Erlenmeyer flask should be loosely stoppered or covered.

Using Activated Charcoal to Remove Colored Impurities

If the solution is deeply colored after the solid dissolves in the recrystallization solvent, treatment with activated charcoal (Norit, Darco, etc.) may remove the colored impurities. Particles of activated charcoal have a large surface area and a strong affinity for adsorbing highly conjugated and thus colored compounds, but too much charcoal will also adsorb the compound that you are purifying. Add 20–30 mg of the activated carbon to the hot, *but not boiling*, solution.

Now heat the solution to boiling for a few minutes. While the solution is still very hot, gravity filter it according to the procedure described in the previous section.

5.4
Summary of the Macroscale Recrystallization Procedure

1. Dissolve the solid sample in a minimum volume of hot solvent with a boiling chip or boiling stick present.
2. If the color of the solution reveals impurities, add a small amount of adsorbing charcoal to the hot, but not boiling, solution (optional).
3. If insoluble impurities are present or charcoal treatment is used, gravity filter the hot solution through a fluted filter paper.
4. Cool the solution slowly to room temperature and then in an ice-water bath to induce crystallization.
5. Remove the crystals from the solvent by vacuum filtration.
6. Wash the crystals with a small amount of cold solvent.
7. Allow the crystals to air dry completely on a watch glass before weighing them and determining their melting point.

5.5
Microscale Recrystallization

Microscale methods are used for recrystallizations of less than 500 mg of solid. If the amount of solid ranges between 150 and 500 mg, we use smaller versions of the equipment specified in the macroscale procedure summarized in Technique 5.6. Recrystallizations of 150 mg or less are done in a Craig tube, a device that serves as both recrystallization vessel and filtration apparatus.

5.5a Recrystallizing 150 to 500 Milligrams of a Solid

Recrystallization of 150 to 500 mg of material generally follows the same steps outlined in Technique 5.3 using smaller equipment. A test tube or 10-mL Erlenmeyer flask holds the recrystallization solution and a Hirsch funnel replaces the Buchner funnel for collecting the crystals. The following steps outline the procedure for recrystallizing 150 to 500 mg of solid.

1. Place the solid in a 13 × 100 mm test tube or a 10-mL Erlenmeyer flask; add a boiling stick. With a Pasteur pipet, add enough solvent to just cover the crystals.

2. Boil the contents of the test tube or flask, then add additional solvent dropwise, allowing the mixture to boil briefly after each addition. Continue this process until just enough solvent has been added to dissolve the solid.

3. If colored impurities are present, cool the mixture slightly and add 10 mg of Norit carbon-decolorizing pellets (about 10 pellets). Boil the mixture briefly. If the color is not removed after 1–2 min, add a few more Norit pellets and boil briefly. Prepare a Pasteur filter pipet [see Technique 2.4]. Warm the Pasteur pipet by immersing it in a test tube of hot solvent and drawing the hot solvent into it several times; use the heated pipet to separate the hot recrystallization solution from the Norit pellets and transfer it to another test tube or flask. If crystallization begins in the solution with the carbon pellets during this process, add a few drops of solvent and warm the mixture to boiling to redissolve the crystals before completing the transfer.

4. If the recrystallization mixture contains insoluble impurities, follow the hot filtration process outlined in step 3 to separate the solution from the insoluble impurities.

5. Allow the solution to cool slowly to room temperature, then chill it in an ice-water bath to complete crystallization.

6. Collect the crystals by vacuum filtration, using a Hirsch funnel (Figure 5.7).

7. Allow the crystals to air dry completely on a watch glass before weighing them and determining their melting point.

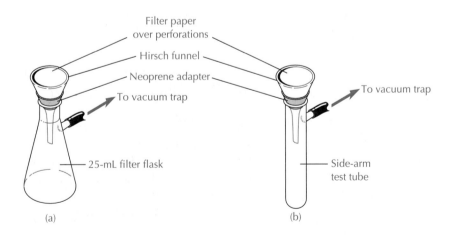

FIGURE 5.7 Vacuum filtration, using a Hirsch funnel.

$\left(\boxed{5.5b}\right)$ **Recrystallizing Less Than 150 Milligrams of a Solid in a Craig Tube**

Recrystallizations of 150 mg or less are done in a Craig tube, a small device that serves as both recrystallization vessel and filtration apparatus. The Craig tube eliminates the losses associated with separation of the crystals from the solution by vacuum filtration. It also allows multiple recrystallizations to be performed without removing the crystals from the tube, thus preventing significant losses of material on the glassware.

The Craig tube consists of a glass tube that has a band with a rough surface at the point where the tube widens and a separate plug made of glass or Teflon (Figure 5.8). If the plug is glass, the large end has a rough surface. When a plug is inserted into the tube and the apparatus inverted as described later, the rough glass surface of the tube forms an incomplete seal with the plug. This imperfect seal allows the crystallization solution (mother liquor) to flow out during centrifugation, leaving the crystals in the tube. This process separates the crystals from the recrystallization solution.

Heat for a Craig tube recrystallization is supplied by an aluminum block heated to 10°C above the boiling point of the solvent; one of the holes in the block is sized for the Craig tube. Place the material to be recrystallized in the outer tube of the Craig assembly. Bumping and boiling over occurs *very* quickly with a Craig tube; prevent this by using a boiling stick or by stirring rapidly by rolling a microspatula, inserted in the Craig tube, between your fingers while the mixture is heating. Boiling chips should not be used with a Craig tube because their subsequent removal is difficult.

If the amount of solvent is not specified, cover the crystals with a few drops of solvent and begin heating the mixture on the

Chemists sometimes refer to the supermutant liquid from a recrystallization as the "mother liquor."

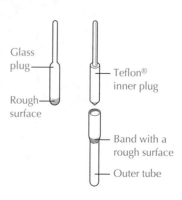

FIGURE 5.8 Craig tube.

FIGURE 5.9 Apparatus for removing solvent by centrifugation.

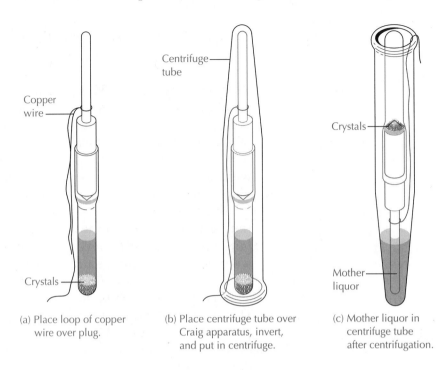

(a) Place loop of copper wire over plug.

(b) Place centrifuge tube over Craig apparatus, invert, and put in centrifuge.

(c) Mother liquor in centrifuge tube after centrifugation.

aluminum block; continue adding solvent dropwise and boiling the mixture gently between additions until the solid dissolves completely. When dissolution is complete, insert the plug into the tube and set the assembly inside a 25-mL Erlenmeyer flask to cool slowly and undisturbed. When crystallization appears to be complete, cool the tube in an ice-water bath for 5 min.

Slow cooling promotes the formation of the larger and purer crystals essential for good recovery.

Recover the crystals in the following manner: First, obtain a centrifuge tube (Figure 5.9). If the centrifuge tube and the Craig tube plug are glass, place a small piece of cotton in the bottom of the centrifuge tube; the cotton is unnecessary with either a plastic centrifuge tube or a Teflon plug. Slip a loop of thin copper wire over the stem of the plug. Invert the centrifuge tube over the Craig apparatus, then turn the centrifuge tube upright. Bend the end of the wire over the lip of the centrifuge tube (Figure 5.9).

During centrifugation, the recrystallization solvent flows through the small openings in the rough glass surface of the Craig tube, while the crystals remain on the end of the inner plug.

Set the tube in a centrifuge and balance its weight in the opposite hole with another centrifuge tube that is two-thirds to three-fourths full of water; centrifuge for 1–2 min. (If you are having difficulty balancing the centrifuge, weigh the centrifuge tube containing the Craig apparatus in a small beaker, then adjust the amount of water in the other centrifuge tube until the mass approximately equals that of the tube with the Craig apparatus.) Remove the tube from the centrifuge and carefully remove the Craig tube from the centrifuge tube by pulling on the copper wire. Turn the Craig apparatus upright. Remove the plug from the outer tube and set the tube upright in your desk to allow the crystals to dry. Carefully scrape any crystals adhering to the plug onto a small watch glass and store them too until the crystals in the Craig tube are dry. The loss of crystals adhering to the inside of the tube is less if the crystals have dried before they are removed from the tube. The solvent in the bottom of the centrifuge tube should be treated as directed in the experimental cleanup procedure.

5.6
Summary of the Microscale Recrystallization Procedures

Recrystallizing 150 to 500 Milligrams of a Solid

1. Dissolve the solid in a minimum volume of hot solvent in a small test tube or 10-mL Erlenmeyer flask; use a boiling stick or boiling chip to prevent bumping.

2. If colored impurities are present, boil the mixture briefly with 8–10 Norit pellets.

3. If insoluble impurities are present or Norit pellets were used, transfer the hot recrystallization solution to another test tube or flask, using a heated Pasteur filter pipet.

4. Cool the solution slowly to room temperature to induce crystallization, then complete cooling in an ice-water bath.

5. Collect the crystals by vacuum filtration on a Hirsch funnel.

6. Allow the crystals to air dry completely on a watch glass before weighing them.

Recrystallizing up to 150 Milligrams in a Craig Tube

1. Dissolve the solid in a minimum volume of hot solvent in the outer tube of the Craig apparatus; use a boiling stick or continuously roll a stirring rod between your fingers to prevent bumping.

2. Insert the plug into the tube and set the assembly inside a 25-mL Erlenmeyer flask to cool undisturbed to room temperature. When crystallization is complete, cool the tube in an ice-water bath.

3. Place an inverted centrifuge tube over the Craig apparatus, then turn the centrifuge upright. Separate the solvent from the crystals by spinning the centrifuge tube in a centrifuge.

4. Allow the crystals to dry in the outer tube before removing them for weighing.

5.7
What to Do If No Crystals Appear in the Cooled Solution

In many instances, recrystallization fails because too much solvent was used in the process. In these cases, you need to boil off a bit of the solvent and try the recrystallization again.

If recrystallization still does not occur from the supersaturated solution, or if oils form, there are several tricks that can be used to induce crystallization. *Always save a few crude crystals, as seeds, in the event recrystallization does not occur.* Adding a few of these to the cooled solution may induce crystallization by providing nuclei around which crystals can grow. Another way to promote crystal formation is to vigorously scratch the inside of the bottom of the flask with a stirring rod. The tiny particles of glass scratched from the flask serve as centers for crystallization to begin. It is said that Louis Pasteur accidentally initiated a significant recrystallization with dandruff from his beard, but we do not recommend this technique.

The formation of oils is probably the most frustrating outcome of an attempted recrystallization. The presence of impurities lowers the melting point of substances, making "oiling out" especially prevalent during recrystallization of a solute with a melting point below the boiling point of the solvent being used. Compounds with low melting points often form oils during recrystallization. Impurities distribute themselves in the two liquids, the solvent and the oil, before crystallization occurs, so the impurities are trapped in the oil when it cools and hardens into a viscous glasslike substance. If you have an oil rather than crystals, you can add more solvent so that the compound does not come out of solution at so high a temperature, because the point of supersaturation may have been reached when the solvent temperature was still above the freezing (melting) point of the compound. It may also help to switch to a solvent with a lower boiling point (consult Table 5.1).

5.8 Pointers for Successful Recrystallizations

The pointers listed here apply to both macroscale and microscale recrystallizations. When you are recrystallizing a product, attention to these details will increase the purity and amount of your product.

Many students recover a smaller amount of product from a recrystallization than they should because of mechanical losses. Losses occur because (1) too much solvent is added, (2) too much charcoal is added to decolorize colored solutions, (3) premature crystallization occurs during a gravity filtration, or (4) the crystals are filtered or centrifuged before recrystallization is complete.

When higher boiling-point solvents such as ethyl alcohol, water, or toluene are used, the recrystallized product may not dry completely for a rather long time and should be allowed to dry at least overnight before determining its mass and melting point.

If time permits, collecting a second crop of crystals by boiling off some of the solvent from the filtrate for the first crop will increase the yield. However, the purity of this second crop is likely to be lower than that of the first crop; determine the melting points of both crops before combining them.

Questions

1. Describe the characteristics of a good recrystallization solvent.
2. The solubility of a compound is 59 g per 100 mL in boiling methanol and 30 g per 100 mL in cold methanol, while its solubility in water is 7.2 g per 100 mL at 95°C and 0.22 g per 100 mL at 2°C. Which solvent would be better for recrystallization of the compound? Explain.
3. Explain how the rate of crystal growth can affect the purity of a recrystallized compound.
4. Under what circumstances is it necessary to filter a hot recrystallization solution?
5. Why should a hot recrystallization solution be filtered by gravity rather than by vacuum filtration?
6. An organic compound is quite polar and is thus much more soluble in methanol than in pentane (bp 36°C). Why would this be an awkward solvent pair for recrystallization? Consult Table 5.1 to assist you in deciding how to change this solvent pair so that recrystallization will proceed smoothly.
7. For a variety of reasons, *N,N*-dimethylformamide (DMF) is not usually utilized as a recrystallization solvent even though it is often used as a reaction solvent and it has a moderate dielectric constant ($\epsilon = 38$). Included among its properties are an unpleasant odor, a high boiling point (153°C), and the fact that it is highly hygroscopic. Refer to Table 5.1 and discuss what other solvent(s) might serve as an alternative.

Technique 6
MELTING POINTS
AND MELTING RANGES

Molecules in a crystal are arranged in a regular pattern. Melting occurs when this fixed array of molecules rearranges to the more random, freely moving liquid state. Such a transition requires energy in the form of heat to break down the crystal lattice, and the *melting point* of a solid is the temperature at which this transition occurs. It is the temperature at which the solid phase and the liquid phase of a pure substance exist in equilibrium. The melting point offers a particularly useful and quick criterion for the identification and the determination of purity of crystalline organic compounds.

6.1
Melting Point Theory

A solid at any temperature has a finite vapor pressure. As the temperature of the solid is increased by heating, the vapor pressure exerted by the solid increases as well. Both the solid and the liquid are in equilibrium with the vapor, and, at the melting point, are in equilibrium with each other (Figure 6.1). Figure 6.2 is a vapor pressure–temperature diagram for both the solid and liquid phases of a chemical compound.

The temperature at which a compound melts is a physical characteristic of that substance and, for pure compounds, is generally reproducible. The presence of even a small quantity of impurity, however, usually depresses the melting point a few degrees and causes melting to occur over a relatively wide temperature range. Because the melting point is the temperature at which the vapor pressure of the pure liquid and the pure solid are equal, the presence of an impurity that is soluble in the liquid can change the temperature at which this equilibrium state occurs.

Consider, for example, the behavior of a solid containing 80% of A and 20% of B (Figure 6.3). The component with the lower melting point, A, begins to melt while solid B dissolves in the liquid. The vapor pressure of the liquid solution of A and B is lower than that of pure liquid A at the melting point, whereas the vapor pressure of solid A at a particular temperature is virtually unchanged by the impurity B, because the solids do not mix together intimately. This means that the temperature at which solid A melts is lower when B is present.

There is a limit, however, to how far the melting point can be lowered. This limit is reached when the liquid solution becomes saturated in B, a condition causing some of solid B to remain after all of A has melted. In Figure 6.3, up to point E all of B dissolves in the melting A. After point E, when all of A is melted, a portion of solid B remains. Point E defines the composition of a saturated solution of B in liquid A and is called the eutectic composition. A

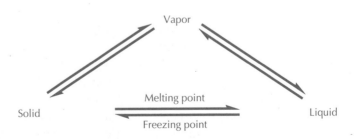

FIGURE 6.1 Solid and liquid in equilibrium.

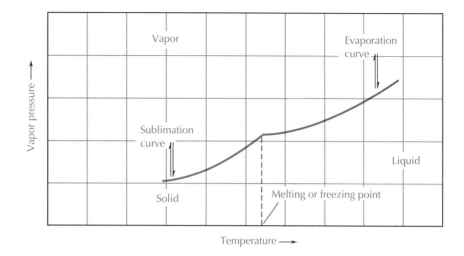

FIGURE 6.2 Vapor pressure–temperature diagram for solid and liquid phases of a chemical compound.

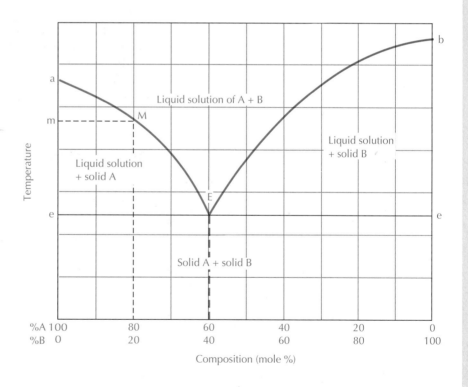

FIGURE 6.3 Melting-point composition diagram for the binary mixture A + B. The presence of an impurity in a solid not only decreases the melting point but also produces melting over a relatively wide temperature range. In this diagram, a is the melting point of pure solid A, b of pure solid B, and e of eutectic mixture E. The temperature range e–m is the melting range of a solid containing 80% A and 20% B.

solid mixture with the eutectic composition (60% A–40% B) will melt sharply at the eutectic temperature, e.

Again consider the melting of a solid mixture composed of 80% A and 20% B. As heat is applied, the temperature of the solid mixture rises. When it reaches temperature e, A and B will melt together at a constant ratio (the eutectic composition) and the temperature will remain constant. When the minor component, B, is completely liquified and more heat is applied, solid A, in equilibrium with the eutectic composition of liquid A and B, continues to melt as the temperature increases. Because the vapor pressure of liquid A increases as the mole fraction of A in the liquid increases, the temperature required to melt A also rises. Melting will occur along curve EM in Figure 6.3, giving an observed temperature range of e–m.

It is possible for a binary mixture, like A + B, to have a more complex melting-point composition diagram than the one shown in Figure 6.3. There can be two eutectic points when the two components interact to form a molecular compound of definite composition. Discussions of more complex melting phenomena are found in the reference given at the end of the chapter.

Relatively pure compounds normally melt sharply over a temperature range of 0.5° to 2.0°, whereas impure substances often melt over a much larger range. Until the advent of thin-layer chromatography, the melting point was the primary index of purity for an organic solid. A melting point is still used as an effective, quickly determined, preliminary indication of purity.

6.2

Apparatus for Determining Melting Ranges

Several types of electrically heated melting-point devices are commercially available. Many undergraduate laboratories use the Mel-Temp apparatus shown in Figure 6.4. A thin-walled glass capillary tube holds the sample. The capillary tube fits into one of three sample chambers in the heating block; multiple chambers allow simultaneous determinations of several melting points. A cylindrical cavity in the top of the heating block holds the thermometer, a light illuminates the sample chamber, and an eyepiece containing a small magnifying lens facilitates observation of the sample. A rheostat controls the rate of heating by allowing continuous adjustment of the voltage. The higher the rheostat setting, the faster the rate of heating. Figure 6.5 shows graphically how the rate of heating changes at different rheostat settings. Heating at any particular setting occurs more rapidly at the start, then levels off as the temperature increases. The decreasing rate of heating at the higher temperatures allows for the slower rate of heating needed as one approaches the melting point.

The numbers on the rheostat are voltages, not temperatures.

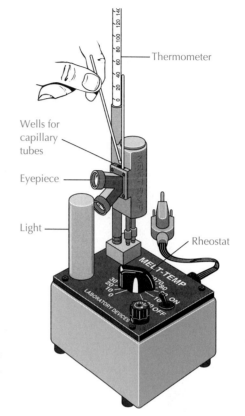

FIGURE 6.4 The Mel-Temp apparatus. (Courtesy of Laboratory Devices, Inc., Holliston, MA.)

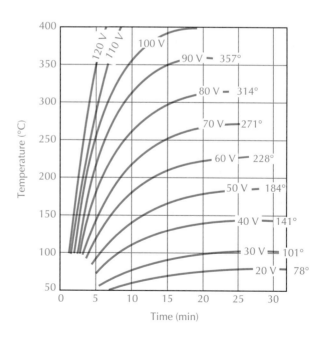

FIGURE 6.5 Heating rate curves for a Mel-Temp apparatus at various rheostat settings (indicated in volts). (Courtesy of Laboratory Devices, Inc., Holliston, MA.)

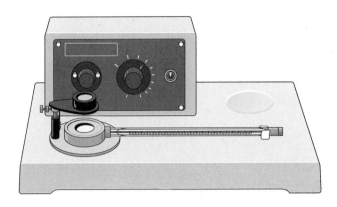

FIGURE 6.6 The Fisher-Johns hot stage melting-point apparatus. (Courtesy of Fisher Scientific, Pittsburgh, PA.)

The Fisher-Johns hot stage apparatus represents a second type of apparatus for the determination of melting points (Figure 6.6). The sample is crushed between thin circular microscope coverslips instead of being placed in a capillary tube. The coverslips fit in a depression in the metal block surface. A rheostat controls the rate of heating, and the lighted sample area is viewed through a small magnifying glass.

── *SAFETY PRECAUTION* ──

If the heater is not turned off after the sample melts, the high heat may ruin the thermometer calibration or even break the thermometer.

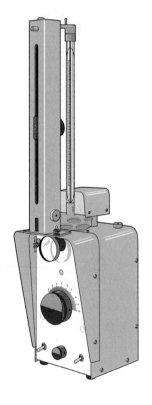

The Thomas-Hoover melting point apparatus represents a third type of melting-point apparatus. In this instrument, capillary tubes holding the samples are submerged in an electrically heated oil bath. An illuminated chamber holds the oil bath, and a magnifying glass allows you to observe the samples (Figure 6.7).

── *SAFETY PRECAUTION* ──

A chamber with fouled oil (usually caused by breakage of melting-point tubes) should not be used because toxic fumes may be driven out by the heat.

FIGURE 6.7 The Thomas-Hoover melting-point apparatus. (Courtesy of Thomas Scientific, Swedesboro, NJ.)

6.3
Determining Melting Ranges

Sample Preparation The melting range of an organic solid can be determined by introducing a small amount of the substance into a capillary tube that has one end sealed. Such tubes are commercially available. Place a few milligrams of the dry solid on a piece of smooth-surfaced paper and crush it to a fine powder with a spatula. Introduce the solid into the capillary tube (approximately 1 mm in diameter) by tapping the open end of the tube in the solid substance. A small amount of material will stick in the open end. Invert the tube (the sealed end is now "down"), hold it very near the sealed end, and with quick motions tap the sealed end of the tube against the bench top. The solid will fall to the bottom of the tube. If the solid is still wet from a recrystallization, it will not fall to the bottom of the tube, but will stick to the capillary wall. This failure to behave properly is probably a good thing, because *melting points of wet solids are nearly worthless.* If your sample is still wet, allow it to dry completely before continuing with the melting-range determination. The amount of solid in the tube should be about 1–2 mm in height. Melting-point determinations made with too much material will produce a large melting range, because more time is required to melt the complete sample.

SAFETY PRECAUTION

Care must be taken while tapping a capillary tube against the bench top: it could snap off and cut you.

An alternative method for getting the solid to the bottom of the capillary tube is to drop it down a piece of glass tubing about a meter in length or down the inside tube of your condenser, the bottom end of which is resting on the lab bench. After a few trips down the tube, the solid will have fallen to the bottom of the capillary tube.

Samples for the Fisher-Johns apparatus also need to be finely powdered. Put a few grains of the powdered sample on one coverslip and set the coverslip in the metal heating block. Place a second coverslip over the sample and gently flatten the powder until the two glass surfaces just touch each other; this contact ensures good heat transfer.

Heating the Sample to the Melting Point

Always prepare a new sample for each melting-point determination.

Electrical melting-point devices have a significant time lag between a change in the rheostat setting and achievement of the new rate of heating.

The melting-point apparatus can be heated rapidly until the temperature is about 20° below the expected melting point. Then decrease the rate of heating so that the temperature rises only one to two degrees per minute and the sample has time to melt before the temperature rises above the true melting point. If you do not know the approximate melting point of a solid sample, you can take a quick preliminary reading by heating the sample rapidly and watching for the temperature at which melting begins. In a more accurate second determination, you can carefully control the temperature rise to one to two degrees per minute when you get to within 15°–20° of the expected melting point. Always prepare a fresh sample for each melting-point determination, because many organic compounds decompose at the melting point, thus making reuse of the solidified sample invalid. Moreover, many low-melting-point samples (mp 25°–80°C) do not resolidify upon cooling. When taking successive melting points, remember that the apparatus needs to cool at least 20° below the expected melting point before it can be used for the next determination.

There are other sources of error in a melting-point determination. Heating faster than one to two degrees per minute may lead to an observed melting range that is higher than the correct value. And, if the rate of heating is extremely rapid (>10° per minute), you may instead observe thermometer lag, a condition caused by the failure of the mercury in the thermometer to rise quickly enough to accurately show the temperature of the metal heating block. This error causes the observed melting range to be lower than it actually is. Determining accurate melting points requires patience.

Remember, unless you have an extraordinarily pure compound in hand, you will always observe and report a *melting range*—from the temperature where the first drop of liquid appears to the temperature where the solid is melted and only a clear liquid is present. This melting range is usually one to two degrees or slightly more. For example, the salicylic acid synthesized in Experiment 4 may have a melting range of 156° to 159°C. An extremely pure sample of salicylic acid, melting over less than a one-degree range (for example, 160.0°–160.5°C) may have 160°C listed as its melting point.

When you heat a sample for a melting-point determination, you may see some strange and wonderful things happen before the first drop of liquid actually appears. The compound may soften and shrivel up as a result of changes in crystal structure. It may "sweat out" some solvent of crystallization. It may decompose, changing color as it does so. None of these should be called melting. Only a small drop of liquid indicates the onset of true melting. It can be difficult to distinguish exactly when melting does start. In fact, even with careful heating, two people may disagree by as much as a degree or two.

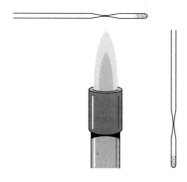

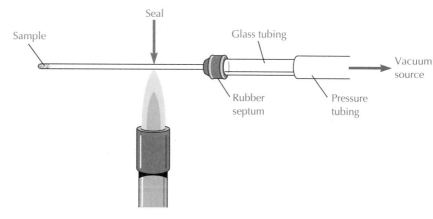

FIGURE 6.8 Left: Sealing a capillary tube with a Bunsen burner. Right: Evacuating and sealing a capillary tube.

Another possible complication in melting-point determinations occurs if the sample sublimes. Sublimation is the change that occurs when a solid is transformed directly to a gas, without passing through the liquid phase [see Technique 8]. In such a case, the sample in the capillary tube will sublime and disappear as heat is applied. Many common substances such as camphor and caffeine sublime, but you can determine their melting points by sealing the open end of the capillary tube in a Bunsen burner flame before it is placed in the melting-point apparatus (Figure 6.8).

Some compounds decompose as they melt, a behavior usually indicated by a change in color of the sample to dark red or brown. You will find the melting point of such a compound reported in the literature with the symbol "d" after the temperature, for example, 186°C d, meaning that the compound melts at 186°C with decomposition. Sometimes the decomposition occurs as a result of a reaction between the compound and oxygen in the air. If the capillary tube is evacuated and sealed, the melting point can be determined without this type of decomposition (Figure 6.8). Place the sample in the capillary tube as directed earlier. Punch a hole in a rubber septum with a small nail and **insert the sealed end of the capillary tube through the septum, from the inside.** Fit the septum over a piece of glass tubing that is connected to the vacuum line. Turn on the vacuum source, hold and pull on the sample end of the capillary, while heating the upper portion until it seals.

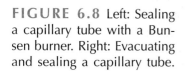

6.4
Mixture Melting Points

We have already discussed how impurities can lower the melting point of a compound. This behavior can be useful not only in evaluating a compound's purity but also in helping to identify it. Let us assume that two compounds have virtually identical melting ranges. Are the compounds identical? Possibly, but not necessarily, because the identical melting ranges may be simply a coincidence.

If roughly equal amounts of the two compounds are finely ground together with a spatula, the melting point of the mixture can provide useful information. If there is a melting-point depression or if the melting range is expanded by a number of degrees, it is reasonably safe to conclude that the two compounds are not identical. One of them has acted as an impurity toward the other by lowering the melting range. If there is no lowering of the mixture's melting range relative to the melting range of each compound, the two are very likely the same compound.

Sometimes only a modest melting-point depression is observed. To know whether this change is significant, the mixture melting point and the melting point of one of the two compounds should be determined simultaneously in separate capillary tubes. This experiment allows simultaneous identity and purity checks. Infrequently, a eutectic point (E, Figure 6.3) can be equal to the melting point of the pure compound of interest. In a case where you have accidentally used the eutectic mixture, purity would be incorrectly suggested by a mixture melting point. This error can be discerned by testing various mixtures other than a 1:1 composition.

Other ways of determining the identity of a compound that you have synthesized involve spectrometric methods and thin-layer chromatography [see Technique 10].

6.5
Thermometer Calibration

The accuracy of your melting-point determinations can be no better than the accuracy of your thermometer. Often one simply assumes that the thermometer has been accurately calibrated. Although frequently this is the case, it is not always true. Thermometers can give high or low temperature readings of one or two degrees or more.

A thermometer can be calibrated with a series of compounds that are readily available in the pure state and whose melting points are easy to reproduce. A useful series of such compounds is given in Table 6.1.

The melting point of ice can be determined by simply measuring the temperature of a beaker of ice water; the others are done in the usual way in capillary tubes. The boiling points of acetone (56°C) and water (100°C) are also useful calibration reference points [see Technique 7.1]. You may want to record the temperature deviation of your thermometer at a number of these points and make a graph of thermometer corrections. Plot observed temperature against temperature correction and interpolate to correct the future determinations. Usually these plots are linear.

Table 6.1. **Melting point standards**

Compound	Melting point, °C
water	0
benzoic acid	122
salicylic acid	160
3,5-dinitrobenzoic acid	205

6.6
Summary of the Melting-Point Determination

1. Introduce the powdered, dry solid sample into a capillary tube that is sealed at one end.

2. Place the capillary tube in the melting-point apparatus.

3. Adjust the rate of heating so that the temperature rises at a moderate rate. This can be a faster rate if, for example, the melting point is 170°C rather than 70°C.

4. When a temperature 15°–20° below the expected melting point is reached, decrease the rate of heating so that the temperature rises only one to two degrees per minute. *Note:* There is a time lag on electrically heated devices before the rate of heating changes.

5. If the temperature is rising more than one to two degrees per minute at the time of melting, retake the melting point, using a new sample.

6. Record the melting range as the range of temperatures that begins with the onset of melting and ends with the temperature at which only liquid remains in the tube.

Reference

Skau, E. L.; Arthur, J. C., Jr. In *Physical Methods of Chemistry*, A. Weissberger and B. W. Rossiter, eds.; Wiley-Interscience: New York, 1971, Vol. 1, Part V.

Questions

1. A white crystalline compound melts at 111–112°C and the melting-point capillary is set aside to cool. Repeating the melting-point analysis, using the same capillary, reveals a much higher melting

point of 140°C. Yet repeated recrystallization of the original sample yields sharp melting points no higher than 114°C. Explain the behavior of the sample that was cooled and remelted.

2. A student performed two melting-point determinations on a crystalline product. In one determination, the capillary tube contained a sample about 1–2 mm in height and the melting point was found to be 141°–142°C. In the other determination, the sample height was 4–5 mm and the melting point was found to be 141–145°C. Explain the broader melting-point range observed for the second sample. The reported melting point for the compound is 143°C.

3. Another student reported a melting point of 136–138°C for the melting point of the unknown in Question 2 and mentioned in his/her notebook that the rate of heating was about 12 degrees per minute. NMR analysis of this student's product did not reveal any impurities. Explain the low melting point.

4. A compound melts at 120–122°C on one apparatus and at 128–129°C on another. Unfortunately, neither apparatus is calibrated. How might you check the identity of your sample without calibrating either apparatus?

Technique 7
BOILING POINTS, DISTILLATION, AND AZEOTROPES

Distillation is a method for separating two or more liquid compounds on the basis of boiling point differences. Unlike the liquid-liquid and solid-liquid separation techniques of extraction and crystallization, distillation is a liquid-gas separation. In distillation vapor pressure differences are used to separate materials.

A liquid at any temperature exerts a pressure on its environment. This pressure, the *vapor pressure,* results from molecules leaving the surface of the liquid to become vapor and occurs because the molecules are in constant motion.

$$(\text{molecules})_{\text{liquid}} \rightleftharpoons (\text{molecules})_{\text{vapor}}$$

As a liquid is heated, its kinetic energy increases; the equilibrium shifts and more molecules move into the gaseous state, thereby increasing the vapor pressure. This relationship is shown for benzene, water, and *tert*-butylbenzene in Figure 7.1.

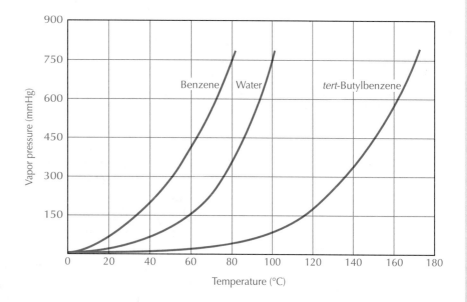

FIGURE 7.1 Dependence of vapor pressure on temperature.

Boiling Points

The boiling point of a pure liquid is defined as the temperature at which the vapor pressure of the liquid exactly equals the pressure exerted on it by the atmosphere. For example, at an external pressure of 1 atm (760 mmHg), the boiling point is reached when the vapor pressure equals 1 atm.

The boiling point of a liquid depends on the atmospheric pressure. Table 7.1 gives boiling points of several common solvents in

Table 7.1. **Boiling points of common compounds at different elevations (pressures)**

Compound	Death Valley $P = 1.10$ atm	New York City $P = 1.00$ atm	Laramie, Wyoming $P = 0.75$ atm
Water	100.3	100.0	93
Diethyl ether	35.0	34.6	27
Benzene	80.8	80.2	73
Acetic acid	119.0	118.8	110

Pentane
bp = 36.1°C

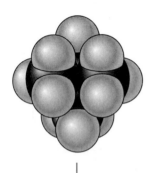

2,2-Dimethylpropane
(Neopentane)
bp = 9.5°C

Laramie, Wyoming (elevation 7520 ft), at the experimental station of the U.S. Department of Agriculture in Death Valley, California (elevation −280 ft), and in New York City (elevation 0 ft). Because a liquid's boiling point is dependent on the atmospheric pressure, both the atmospheric pressure and the boiling-point temperature are always recorded when a distillation is done.

Every stable organic compound has a characteristic boiling point at atmospheric pressure. The boiling point of an organic compound reflects its molecular structure, specifically the types of intermolecular interaction that bind the molecules together in the liquid state. Polar compounds have higher boiling points than corresponding nonpolar compounds of approximately the same molecular weight. Increased molecular weight usually leads to a higher boiling point, if compound polarities remain constant. Molecular shape also affects the boiling point of a liquid; a spherical molecule will boil at a lower temperature than one of its isomers that is cigar shaped because the spherical molecules have fewer intermolecular interactions. Two isomers of C_5H_{12} illustrate this decrease in boiling point as the shape of the molecules changes from cylindrical to spherical. The cylindrical molecules align in a manner that allows stronger intermolecular interactions than occur with spherical molecules. Thus the greater energy needed to overcome the intermolecular interactions is reflected in the higher boiling point.

(7.1a) Determination of Boiling Points

Macroscale Determination of Boiling Points

The boiling point of 5 mL or more of a pure liquid compound can be determined by a simple distillation. The procedure for setting up a simple distillation is described in Technique 7.3. When distillate is condensing steadily and the temperature stabilizes, the boiling point of the substance has been reached.

Rather than use all the sample for a distillation, the microscale procedure described next can be used to determine the boiling point of any pure liquid and requires only 0.3 mL of the liquid.

Microscale Determination of Boiling Points

Place 0.3 mL of the liquid and a boiling stone in the outer tube of a Craig apparatus. Set the tube in the appropriate sized hole of an aluminum heating block [see Technique 3.2]. (Alternatively, heat may be supplied by a sand bath, in which case, the tube and the thermometer need to be held by separate clamps [see Technique 3.2].) Clamp a thermometer so that the bottom of the bulb is about 0.5 cm above the surface of the liquid, being sure that the thermometer does not touch the wall of the tube (Figure 7.2). Gradually heat the sample to boiling and continue to increase the rate of heating slowly until the ring of condensate is 1–2 cm above the

FIGURE 7.2 Apparatus for microscale boiling-point determination.

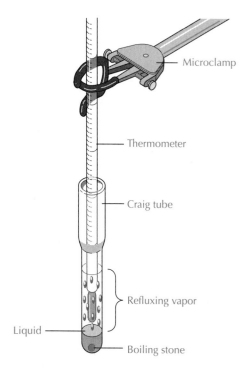

— Microclamp

— Thermometer

— Craig tube

Refluxing vapor

Liquid —

— Boiling stone

top of the mercury bulb. When the temperature reaches a maximum and stabilizes for at least 1 min, you have reached the boiling point of the liquid. Rapid or excessive heating of the tube can lead to superheating of the vapor, causing the observed boiling point to be too high.

7.2 Boiling Points and Separation of Mixtures

The boiling point of a mixture is a function of the vapor pressures of the various components in the mixture. Impurities either raise or lower the observed boiling point of a sample, depending on how the impurity interacts with the compound for which the boiling point is being measured. In fact, just as the melting range of a solid indicates its degree of purity, the boiling range of a liquid is a good criterion of its purity. Although there are more sensitive methods for assaying the purity of a liquid, a substance that boils over a range of several degrees in temperature is *usually* not pure. Consider, for example, the boiling characteristics of a mixture of pentane and hexane.

Pentane and hexane are mutually soluble, and their molecules interact with one another only by weak van der Waals forces. A solution composed of both pentane and hexane will boil at temperatures intermediate between the boiling points of pentane

(36°C) and hexane (69°C). If pentane alone were present, the vapor pressure above the liquid would be due only to pentane. However, with pentane as only a fraction of the solution, the vapor pressure exerted by pentane (P) will be equal to only a fraction of the vapor pressure of pure pentane at the same temperature ($P°$), where X is the *fraction of molecules of pentane in solution*, called the mole fraction of pentane.

$$P_{pentane} = P°_{pentane} X_{pentane} \qquad (1)$$

The same is true for the hexane component:

$$P_{hexane} = P°_{hexane} X_{hexane} \qquad (2)$$

Raoult's law applies only to liquids miscible in one another.

These relationships are strictly valid only for ideal liquids in the same way that the ideal gas law strictly applies only to ideal gases. Equations 1 and 2 are applications of Raoult's law, named after the French chemist François Raoult, who studied the vapor pressures of solutions in the late nineteenth century.

Using Dalton's law of partial pressures, we can now calculate the total vapor pressure of the solution, which is the sum of the partial vapor pressures of the individual components:

$$P_{total} = P_{pentane} + P_{hexane} \qquad (3)$$

Figure 7.3 shows the pentane/hexane system, using Raoult's and Dalton's laws. The boiling point of a pentane/hexane mixture is the temperature at which the individual vapor pressures of both pentane and hexane add up to the total pressure exerted on the liquid by its surroundings.

Being able to calculate the total vapor pressure of a solution can be extremely useful to a chemist; knowing the composition of the vapor above a solution can be just as important. Qualitatively, it is not hard to see that the vapor above a 50:50 pentane/hexane solution will be richer in pentane as a result of its greater volatility and vapor pressure. Quantitatively, we can predict the composition of the vapor above a solution, for which Raoult's law is valid, simply by knowing the vapor pressures of the volatile compounds and the composition of the liquid solution.

Here is an illustration of how it is done: Applying the ideal gas law to the mixture of gases above a solution of pentane and hexane, we have equation 4. Y is the *fraction of pentane molecules in the vapor* above the solution.

$$Y_{pentane} = \frac{P_{pentane}}{P_{total}} \qquad (4)$$

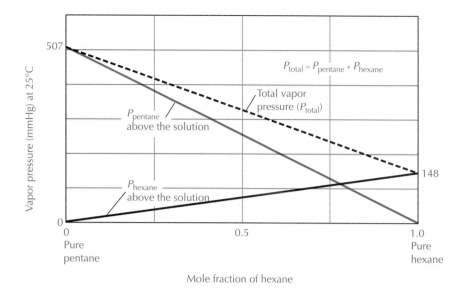

FIGURE 7.3 Vapor pressure–mole fraction diagram for pentane/hexane solutions at 25°C.

A single expression for the total vapor pressure (equation 5) can be derived easily from equations 1, 2, and 3, because $X_{hexane} = 1.0 - X_{pentane}$.

$$P_{total} = X_{pentane}\left(P^{\circ}_{pentane} - P^{\circ}_{hexane}\right) + P^{\circ}_{hexane} \qquad (5)$$

Finally, the combination of equations 4 and 5, plus Raoult's law, allows the calculation of the mole fraction of pentane in the vapor state.

$$Y_{pentane} = \frac{P^{\circ}_{pentane}X_{pentane}}{X_{pentane}\left(P^{\circ}_{pentane} - P^{\circ}_{hexane}\right) + P^{\circ}_{hexane}} \qquad (6)$$

So, if you know the vapor pressures of pure pentane and pure hexane at various temperatures and the composition of the liquid, you can calculate the fraction of pentane in the vapor above the solution. This kind of calculation can be used to construct a temperature-composition diagram (sometimes called a phase diagram) like the one shown in Figure 7.4. This diagram can also be constructed directly from experimental data.

It is useful to follow the dotted line in Figure 7.4, moving from L_1 to V_1 to L_2, etc. Point L_1 indicates a boiling point of 44°C at atmospheric pressure for a solution containing a 1:1 molar ratio of pentane to hexane. Upon removing a sample of the vapor, we find

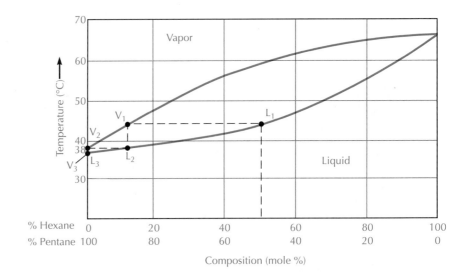

FIGURE 7.4 Estimated temperature-composition diagram for pentane/hexane solutions at 1.0 atm pressure.

that it has a molar composition of 87% pentane and 13% hexane, as indicated by point V_1. *The mole fraction of the component with the lower boiling point is greater in the vapor than in the liquid.* If the vapor at V_1 condenses, the liquid that collects (L_2) will have the same composition as the vapor (V_1). Now if the condensed liquid (L_2) is revaporized, the new vapor will be even richer in pentane (V_2). Repeating the boiling and condensing several more times allows us to obtain pure pentane, uncontaminated by hexane.

As pentane is removed in the vapor, the composition of the liquid, originally at L_1, becomes richer in hexane, the component with the higher boiling point. As the mole fraction of hexane in the liquid increases, the boiling point of the liquid also increases until the boiling point of pure hexane, 69°C, is reached. In this way pure hexane can also be separated. This process of repeated vaporizations and condensations, called *fractional distillation,* allows us to separate components of a mixture because of differences in the vapor pressures of the components.

We use a *fractionating column* in the distillation apparatus to provide the large surface area over which a number of separate liquid-vapor equilibria can occur. As vapor travels up a column, it cools, condenses into a liquid, revaporizes as more heat reaches it, and repeats the process many times. Each successive equilibrium enriches the condensate returning to the boiling flask in the component with the higher boiling point. If the fractionating column is efficient, the vapor that reaches the distilling head at the top of the column will be composed entirely of the component with the

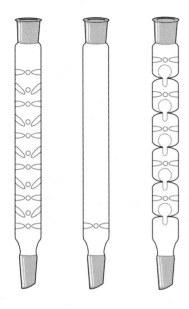

FIGURE 7.5 Examples of fractionating columns.

lower boiling point. Thus we have carried out the process denoted by the dotted lines in Figure 7.4.

The efficiency of a fractionating column is given in terms of theoretical plates. It is simplest to define this term by referring back to Figure 7.4. Let us assume that the original solution being distilled has a 1:1 molar ratio of pentane to hexane. A column would have one theoretical plate if the liquid that distills from the top of the fractionation column has the composition L_2. In other words, a column has one theoretical plate if one complete vaporization of the original solution followed by recondensation of the vapor occurs in the column. The column would have two theoretical plates if the liquid that distills has the composition L_3; notice that L_3 is already 98% pentane and only 2% hexane. Starting with a 1:1 solution, a column with three theoretical plates would seem sufficient to separate pure pentane, V_3, from hexane. However, as the distillation progresses, the residue becomes richer in hexane, so more theoretical plates are required for complete separation of the two compounds.

Fractionating columns that can be used to separate two liquids boiling at least 25°C apart are shown in Figure 7.5. The larger the surface area on which liquid-vapor equilibria can occur, the more efficient the column will be. The fractionating columns shown in Figure 7.5 have from two to eight theoretical plates. A fractionating column with two theoretical plates can be used to separate liquids with boiling points differing by about 70°C; an eight theoretical plate column can be used to separate liquids boiling 25°C apart.

More efficient columns can be made by packing a simple fractionating column with a wire spiral, glass helixes, metal sponge, or thin metal strips. These packings provide additional surface area on which liquid-vapor equilibria can occur. Some care must be used with metal packings, because they can become involved in chemical reactions with the hot liquids in the column.

Some of the most efficient fractionating columns are those with helical bands of Teflon mesh that spin at thousands of rotations per minute. These spinning band columns can have more than 150 theoretical plates and can be used to separate liquids having a boiling-point difference of only one or two degrees. Further discussion on the use of packings for increasing the efficiency of a fractionating column can be found in the references at the end of this chapter.

7.3 Simple Distillation

In a *simple distillation,* only one vaporization and condensation occurs, corresponding to points L_1 and V_1 in Figure 7.4. This process would not effectively separate a mixture such as pentane and hexane. If a 1:1 solution of pentane and hexane undergoes a simple distillation, the first vapor that condenses has a molar composition of 87% pentane and 13% hexane. The molar composition of the remaining liquid now contains more hexane and less pentane than originally; consequently, the boiling point of the mixture will increase. As additional vapor condenses into the receiving flask, the boiling point of the remaining mixture continues to increase. We can represent this graphically by a distillation curve showing vapor temperature versus volume of distillate for the simple distillation of our pentane/hexane mixture (Figure 7.6). The initial distillate is collected at a temperature above the boiling point of pure pentane and the final distillate never reaches the boiling point of pure hexane, a result indicating a poor separation of the two compounds.

Even though simple distillation does not effectively separate a mixture of liquids whose boiling points differ by less than 60–70°C, organic chemists use simple distillations in two commonly encountered situations. The last step in the purification of a liquid compound usually involves a simple distillation to obtain the pure product and determine its boiling point. Simple distillation is also used to remove the solvent when recovering an organic compound with a high boiling point from a solution.

The distilling flask should be about twice as large as the volume of liquid being distilled. If it is too full, liquid can easily bump over into the condenser. If it is nearly empty, a substantial fraction of the material will be needed just to fill the flask and distilling head with vapor and form a thin liquid film on the glass surfaces. When the desired liquid is dissolved in a large quantity of a solvent with a lower boiling point, the distillation should be

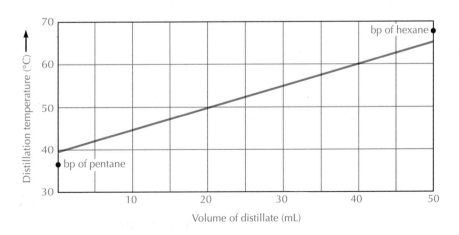

FIGURE 7.6 Distillation curve for a simple distillation of a 1:1 molar solution of pentane and hexane.

interrupted after the solvent has distilled, and the liquids with higher boiling points should be poured into a smaller flask before continuing the distillation. Figure 7.7 shows the apparatus for a simple distillation.

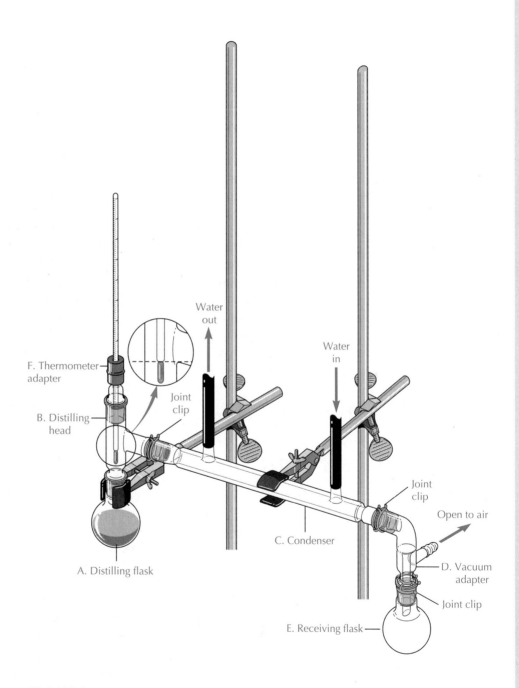

FIGURE 7.7 Simple distillation apparatus: (A) distilling flask, (B) distilling head, (C) condenser, (D) vacuum adapter, (E) receiving flask, (F) thermometer adapter. The detail shows the correct placement of the thermometer bulb.

Steps in Assembling
a Simple Distillation
Apparatus

A funnel keeps the
ground-glass joint from
being coated with the liquid
and prevents spills on the
outside of the flask.

1. Select a round-bottomed flask (A in Figure 7.7) of a size that will be one-third to one-half filled with the liquid being distilled. **Place a clamp firmly on the neck of the flask** and attach the clamp to a ring stand or rack. Using a conical funnel, pour the liquid into the flask. Add one or two boiling stones.

--- SAFETY PRECAUTION ---

Boiling stones should **never** be added to a hot liquid because this may cause the liquid to boil violently out of the flask.

2. Grease the bottom joint and the side-arm joint [see Technique 2.1a] on the distilling head (B). Fit the distilling head to the round-bottomed flask and twist the joint to achieve a tight seal. Finish assembling the rest of the apparatus *before* inserting the thermometer adapter and thermometer.

3. Attach rubber tubing to the outlets on the condenser jacket. Wire hose clamps are often used to prevent water hoses from being blown off the outlets by a surge in water pressure. Grease the inner joint on the bottom of the condenser, attach the vacuum adapter (D) and, while the pieces are lying on the desk top, place a joint clip over the joint. Clamp the condenser (C) to another ring stand or upright bar on the racking as shown in Figure 7.7. Fit the upper joint of the condenser to the distilling head, twist to spread the grease, and place a joint clip over the joint.

The use of a joint clip ensures
that ground glass joints do
not come apart.

4. Figure 7.7 shows a round-bottomed flask (E) serving as the receiving vessel. Depending on the particular procedure being carried out, an Erlenmeyer flask or a graduated cylinder may be substituted for the round-bottomed flask. A beaker should not be used as the receiving vessel, because its wide opening readily allows vapor to escape from your product. It is usually necessary to have at least two receiving vessels at hand; the first container is for collecting the initial distillate that consists of impurities with low boiling points before the expected boiling point of the main fraction is attained.

The position of the
thermometer bulb is crucial in
obtaining an accurate boiling
point. If the thermometer bulb
is too high, it is not completely
surrounded by vapor and the
observed boiling point will
be lower than the true
boiling point.

5. Gently push the thermometer through the rubber sleeve on the thermometer adapter. Grease the joint on the thermometer adapter and fit it into the top joint of the distilling head. *Adjust the position of the thermometer bulb until the top of the thermometer bulb is aligned with the bottom of the side arm on the distilling head* (see detail in Figure 7.7). Alternatively, a thermometer with a

standard taper fitting may be used instead of the thermometer and rubber sleeved adapter.

6. Check to ensure that the rubber tubing is tightly attached and then start water flowing through the condenser jacket.

7. Place a heating mantle under the distillation flask, using an iron ring or lab jack to support the mantle, and begin heating the flask.

A slow to moderate flow rate for the condenser water is usually sufficient and lessens the chance of blowing the rubber tubing off the condenser.

Carrying Out the Distillation

The expected boiling point of the liquid being distilled determines the rate of heating, controlled by a variable transformer [see Technique 3.2]; a liquid with a high boiling point requires more heat to vaporize than a liquid with a low boiling point. Heat the liquid slowly to a gentle boil. A ring of condensate will begin to move up the inside of the flask and then up the distilling head. The temperature observed on the thermometer will not rise until the ring of vapor reaches the thermometer bulb, because it is measuring the vapor temperature and not the temperature of the boiling liquid in the distilling flask. If the ring of vapor stops moving before reaching the thermometer, increase the rate of heating slightly.

When the vapor reaches the thermometer, the temperature reading increases rapidly. Collect any liquid that condenses below the expected boiling point as the first fraction, then change to a second receiving vessel to collect the main fraction when the temperature stabilizes at or slightly below the expected boiling point of the liquid. Record the temperature at which you begin to collect the main fraction. Adjust the rate of heating to maintain a distillation rate of one drop every one or two seconds. It may be necessary to increase the rate of heating during the distillation if the distillation rate decreases. **Stop the distillation by lowering the heating mantle before the distillation flask reaches dryness** or when the temperature either begins to climb above the expected range or begins to drop. Record the temperature at which the last drop of distillate is collected; the initial and final temperatures for the main fraction are the boiling range.

--- *SAFETY PRECAUTION* ---

A distillation flask should **never** be allowed to reach dryness. By leaving a small amount of boiling residue in the flask, you will not overheat the flask and break it, nor will you char the last drops of residue, which causes a difficult cleaning problem. Some compounds, such as ethers, secondary alcohols, and alkenes, form peroxides by air oxidation. If a distillation involving one of these compounds is carried to dryness, the peroxides could explode.

(**7.3a**) **Short-Path Distillation**

Some experiments in this book involve the distillation of only 4–6 mL of liquid. For distillation of these small volumes, we modify a simple distillation apparatus by omitting the condenser. Figure 7.8 illustrates such a short-path distillation apparatus. The short path reduces the amount of space that must be filled with vapor. A beaker or crystallizing dish of water surrounding the

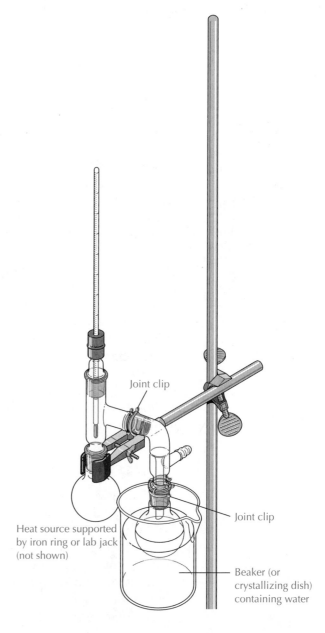

Joint clip

Heat source supported by iron ring or lab jack (not shown)

Joint clip

Beaker (or crystallizing dish) containing water

FIGURE 7.8 Short-path distillation apparatus.

receiving flask replaces the condenser. If the liquid boils between 50°C and 100°C, the beaker should contain an ice-water mixture; in this case, it may be necessary to attach a drying tube to the side arm of the vacuum adapter to prevent moisture from condensing inside the receiving flask. If the liquid boils above 100°C, tap water provides sufficient cooling. For liquids that boil above 150°C, air cooling of the receiving flask will suffice.

Carry out the distillation as described in Technique 7.3 for a simple distillation, but do the short-path distillation at a rate of less than one drop per second. If the receiving flask is being cooled by a water bath, it may be necessary to stop the distillation by removing the heat source while changing receiving flasks.

(*7.3b*) Microscale Distillation

When the volume of liquid to be distilled is only a few milliliters, a significant part of the sample would be used just to fill the short-path system described in Technique 7.3a with vapor (called the *holdup volume*). For these small volumes, use a microscale distillation apparatus consisting of a 3-mL or 5-mL conical vial or a 10-mL round-bottomed flask and a Hickman distilling head (Figure 7.9). The Hickman distilling head serves as both condenser and receiving vessel, an arrangement that considerably reduces the holdup volume. Vapors condense on the upper portion of the Hickman still and drain into the bulbous collection well. One version of the Hickman still has a port at the side for easy removal of the condensate (see Figure 7.9a).

To carry out a microscale distillation, select a conical vial or 10-mL round-bottomed flask appropriate for the volume of liquid to be distilled; the vessel should be no more than two-thirds full. Place the liquid in the vial and add a magnetic spin vane (for a 10-mL flask, use a "flea" bar magnet) or a boiling stone. Attach the Hickman distilling head to the vial with a screw cap and O-ring. Usually an air condenser or a water-cooled condenser (for particularly volatile liquids) is placed above the Hickman distilling head to minimize the loss of vapor (see Figure 7.10).

Grease is not used on the ground glass joints of microscale glassware because its presence could contaminate your product.

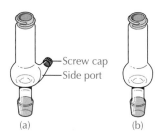

FIGURE 7.9 Hickman distilling heads (a) with a side port and (b) without a side port. The condensate collects in the well at the bottom of the head in both versions.

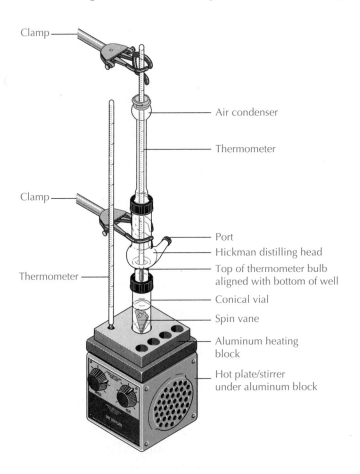

FIGURE 7.10 Apparatus for a microscale distillation using a Hickman distilling head with a side port.

Clamp the assembled apparatus at the Hickman distilling head, with the vial placed in an aluminum heating block. Turn on the stirrer motor and begin heating the aluminum block slowly to a temperature 20–30°C above the boiling point of the liquid being distilled. Position a thermometer inside the condenser and the Hickman distilling head, with the top of the thermometer bulb aligned with the bottom of the head's collection well, as shown in Figure 7.10; clamp the thermometer firmly above the condenser. (*Note:* The inside diameter of some water-jacketed condensers is too small to accommodate a thermometer; in this case, you can ascertain the boiling point of the distillate by a microscale boiling-point determination on the distillate [see Technique 7.1a].)

After the liquid in the vial boils, you will notice a ring of condensate slowly moving up the vial and into the Hickman still. The temperature observed on the thermometer will rise as the vapor reaches the thermometer bulb. You may also see the inside walls of the Hickman still above the collection well become wet and shiny as the vapor condenses and begins to fill the well. The distillation should be done at a rate slow enough to allow the vapor to condense and not evaporate out of the system.

The collection well has a capacity of about 1 mL, so the distillate may need to be removed once or twice during a distillation. Open the port and remove the distillate with a clean Pasteur pipet or a syringe inserted through the plastic septum in the screw cap of the port.

7.4 Fractional Distillation

In a fractional distillation, the use of a fractionating column allows repeated vaporizations and condensations to occur, as discussed in Technique 7.2. The composition of the vapor phase at the top of the column as it encounters the thermometer bulb and moves into the side arm determines the composition of the liquid that forms in the condenser and collects in the receiving flask. Let us illustrate this with a fractional distillation of a 1:1 molar solution of pentane and hexane. If the fractionating column has enough theoretical plates, the initial condensate will appear when the temperature is very close to 36°C, the boiling point of pure pentane. The observed boiling point will remain essentially constant while all the pentane distills. Then the boiling point will rise rapidly to 69°C, the boiling point of hexane. Figure 7.11 shows a distillation curve for the fractional distillation of pentane and hexane.

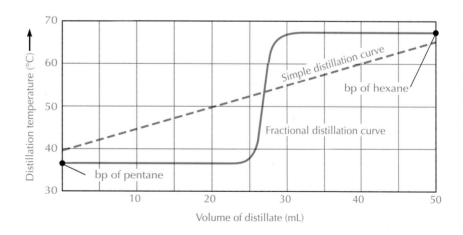

FIGURE 7.11 Distillation curve for the fractional distillation of a 1:1 molar solution of pentane and hexane. The dotted line represents the distillation curve for a simple distillation of the same solution. The abrupt temperature increase from the bp of pentane to that of hexane demonstrates the greater efficiency of fractional distillation.

Assembling a Fractional Distillation Apparatus

The distilling flask should be about twice as large as the volume of liquid being distilled. When the desired material is contained in a large quantity of a solvent with a lower boiling point, the distillation should be interrupted after the solvent has distilled, and the liquids with higher boiling points should be poured into a smaller flask before continuing the distillation.

Figure 7.12 shows the apparatus for a fractional distillation. Follow the steps listed in Technique 7.3 for assembling a simple

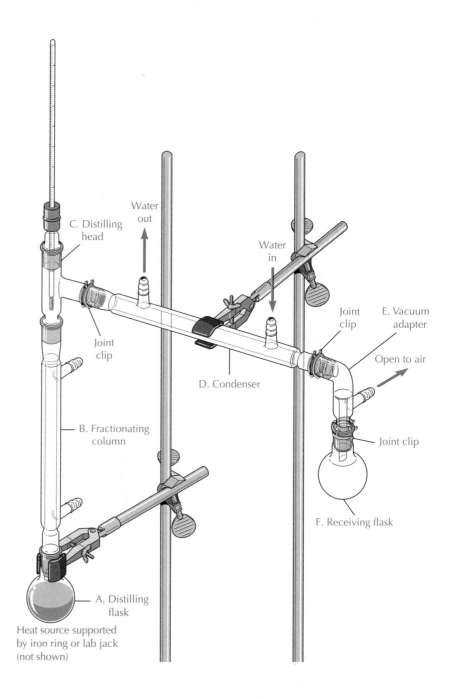

FIGURE 7.12 Fractional distillation apparatus: (A) distilling flask, (B) fractionating column, (C) distilling head, (D) condenser, (E) vacuum adapter, (F) receiving flask.

distillation apparatus with the addition of the fractionating column (B) between the boiling flask (A) and the distilling head (C). Be sure that you have added one or two boiling stones to the distilling flask and that the thermometer is placed correctly, as shown in the detail in Figure 7.7.

Carrying Out a Fractional Distillation

Heat the distilling flask slowly. Control of heating in a fractional distillation is extremely important. The rate of heating needs to be increased gradually as the distillation proceeds. However, applying too much heat causes the distillation to occur so quickly that the repeated liquid-vapor equilibria required to bring about separation on the surfaces of the fractionating column cannot occur. If too little heat is applied, the column may lose heat faster than it can be warmed by the vapor, thus preventing the vapor from reaching the top of the column. Therefore, too little heat during the distillation causes the thermometer reading to drop below the boiling point of the liquid, simply because vapor is no longer reaching the top of the column. The thermometer temperature may also drop during a fractional distillation when a compound with a lower boiling point has completely distilled and not enough heat is being supplied to force the vapor of a compound with a higher boiling point up to the top of the column. The addition of more heat will soon correct this situation.

The rate of distillation is always a compromise between the speed and the efficiency of the fractionation. For an easy separation, one or two drops per second can be collected. Generally a slow, steady distillation where one drop is collected every two or three seconds is a more reasonable rate. Difficult separations (when the boiling points of the distilling compounds are close together) require a slower rate of distillation as well as more efficient fractionating columns. The distillation rate can be increased during collection of the last fraction, when all of the compounds with lower boiling points have already distilled.

Collecting the Fractions

You will need a labeled receiving vessel (round-bottomed flask or Erlenmeyer flask) for each fraction you plan to collect. The cutoff points for the fractions are the boiling points (at atmospheric pressure) of the substances being separated. For example, in a fractional distillation of the 1:1 solution of pentane (bp 36.1°C) and hexane (bp 68.7°C) described in Figure 7.4, the first fraction would be collected when the temperature at the distilling head reaches 36°C. The temperature will stay at 36°C for a period of time while the pentane distills. Eventually the temperature will either rise or drop several degrees, the latter change indicating that there is no longer enough pentane vapor to maintain the boiling-point temperature. When this happens, increase the rate of heating and change to the second receiving flask. Liquid will begin to distill again; leave the second receiver in place until the temperature reaches 69°C, the boiling point of hexane, then change to the third receiving flask. Continue collecting fraction 3 (hexane) until only 1–2 mL of liquid remain in the boiling flask.

—— *SAFETY PRECAUTION* ——

A distillation flask should **never** be allowed to boil dry.

Summary of Fractional Distillation Procedure

1. Use a round-bottomed flask that has a capacity of about twice the volume of the liquid mixture you wish to distill. Clamp the flask to a ring stand or rack upright. Pour the liquid into the flask and add one or two boiling chips.

2. Set up the rest of the apparatus, as shown in Figure 7.12.

3. Heat the mixture to boiling and collect the distilling liquid in fractions based on the boiling points of the individual components in the mixture.

7.5
Azeotropic Distillation

The systems described up to this point are solutions whose components interact only slightly with one another and thus approximate the behavior of an ideal solution. As discussed earlier, the behavior of such solutions follows Raoult's law:

$$P = P°X$$

Most liquid solutions, however, deviate from this ideality. Such deviations result from intermolecular interactions in the liquid state. In the distillation of some solutions, mixtures that boil at a constant temperature are produced. These *constant boiling mixtures* cannot be further purified by distillation and are called *azeotropes.*

One of the best-known binary mixtures that forms an azeotropic mixture during distillation is the ethanol/water system, shown in Figure 7.13. The azeotrope boils at 78.2°C and is composed of 95.6% ethanol and 4.4% water by weight. Liquid of this composition in a fractionating column will vaporize to a gas having exactly the same composition, because the liquid and vapor curves of Figure 7.13 intersect at this point. No matter how

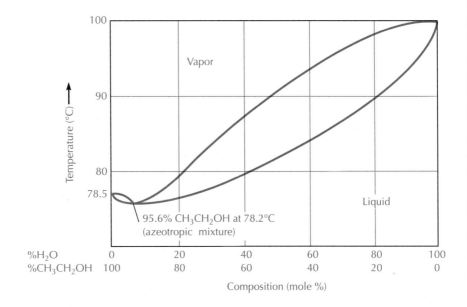

FIGURE 7.13 Temperature/composition diagram for ethanol/water solutions at 1.0 atm pressure. The mixture of 95.6% ethanol and 4.4% water is an azeotrope with a boiling-point minimum.

many more liquid-vapor equilibria take place as the materials travel up the column, no further separation occurs. Continued distillation will never yield a liquid that is higher than 95.6% in ethanol. Pure ethanol must be obtained by other means.

More detailed discussion about the formation of azeotropic distillation mixtures from nonideal solutions can be found in the references at the end of the chapter. Extensive tables of azeotropic data are available in references such as the *Handbook of Chemistry and Physics.* Table 7.2 lists a few azeotropes formed by common solvents.

Table 7.2. Azeotropes formed by common solvents

Component x (bp)	% by wt	Component y (bp)	% by wt	Azeotrope bp
Water (100)	13.5	Toluene (110.7)	86.5	84.1
Water (100)	1.4	*n*-Pentane (36.1)	98.6	34.6
Methanol (64.7)	12	Acetone (56.15)	55.5	55.5
Methanol (64.7)	72.5	Toluene (110.7)	27.5	63.5
Ethanol (78.3)	68	Toluene (110.7)	32	76.7

7.6
Steam Distillation

Many organic compounds decompose near their boiling points, so codistillation with water is useful because it prevents decomposition of the organic compounds. Codistillation with water is called *steam distillation,* because water vapor (steam) is present in the distillation apparatus. It can be thought of as a special kind of azeotropic distillation. Steam distillation is especially useful for separating volatile organic compounds from nonvolatile inorganic salts or from the leaves and seeds of plants. Indeed, the process has found wide application in the flavors and fragrance industries as a means of separating the essence or flavor oil from the plant material.

Steam distillation depends on the *mutual insolubility or immiscibility of many organic compounds with water.* In such two-phase systems, at any given temperature, each component exerts its own vapor pressure independently of the other compounds present. The total vapor pressure above the two-phase mixture is equal to the sum of the vapor pressures of the pure components. Consider the codistillation of iodobenzene (bp 188.5°C) and water (bp 100°C). The vapor pressures ($P°$) of both substances increase with temperature, but the vapor pressure of water will always be higher than that of iodobenzene, because water is more volatile. At 98.25°C,

$$P°_{\text{iodobenzene}} = 46 \text{ mmHg}$$

$$P°_{\text{water}} = 714 \text{ mmHg}$$

$$P°_{\text{iodobenzene}} + P°_{\text{water}} = 760 \text{ mmHg}$$

Therefore, a mixture of iodobenzene and water codistills at 98.25°C. An ideal gas law calculation shows that the mole fraction of iodobenzene in the vapor at the distillating head is 0.06 (46/760), and therefore the mole fraction of water in the vapor is 0.94. However, because iodobenzene has a much higher molecular weight than water (204 g·mol^{-1} versus 18 g·mol^{-1}), its weight percentage in the vapor is much larger than 0.06, as the following calculation shows:

$$\frac{\text{moles}_{\text{iodobenzene}}}{\text{moles}_{\text{water}}} = \frac{P°_{\text{iodobenzene}}}{P°_{\text{water}}}$$

$$\frac{g_{\text{iodobenzene}}/\text{MW}_{\text{iodobenzene}}}{g_{\text{water}}/\text{MW}_{\text{water}}} = \frac{P°_{\text{iodobenzene}}}{P°_{\text{water}}}$$

Rearranging the previous expression and substituting the molecular weights and vapor pressures allows us to calculate the weight ratio of iodobenzene to water:

$$\frac{g_{iodobenzene}}{g_{water}} = \frac{0.73 \; g_{iodobenzene}}{g_{water}}$$

In other words, the distilling liquid contains 42% iodobenzene and 58% water by weight. In any steam distillation, a large excess of water is used in the distilling flask so that virtually all the organic compound (iodobenzene in this example) can be distilled from the mixture and at a temperature well below the boiling point of the pure compound.

The temperature in any steam distillation of a reasonably volatile organic compound will never rise above 100°C, the boiling point of water, unless your laboratory is below sea level. The steam distillation of most compounds occurs between 80°C and 100°C. For example, at 1 atm, octane (bp 126°C) steam distills at 90°C, and 1-octanol (bp 195°C) steam distills at 99.4°C.

Apparatus for Steam Distillation

Add an excess of water (> 100%) to the organic mixture being distilled and select a distilling flask that will be no more than half filled with this organic/water mixture. Add one or two boiling stones to the flask. For a steam distillation, modify a simple distillation apparatus by adding a Claisen connecting tube or adapter between the boiling flask and the distilling head. This adapter provides a second opening into the system to accommodate a source of steam or the addition of water. Figure 7.14 shows the apparatus for a steam distillation that uses externally generated steam, such as a steam line or a flask of boiling water. The bent adapter collects any water that condenses from the steam; opening the pinch clamp drains the water out of the adapter. A piece of glass tubing conducts the steam to the bottom of the distillating flask.

Steam also can be generated simply by boiling a large amount of water with the mixture in the distillation flask (Figure 7.15). If the codistillation of an organic compound with low volatility requires a large volume of steam, a separatory funnel placed in the second opening of the Claisen connecting tube provides a way of adding more water to the system without stopping the distillation.

Steps in a Steam Distillation

1. Set up the apparatus as shown in Figure 7.14 or 7.15.
2. Add the organic mixture and an excess of water to a firmly clamped distilling flask at least twice as large as the combined organic-water volume. Add one or two boiling chips.

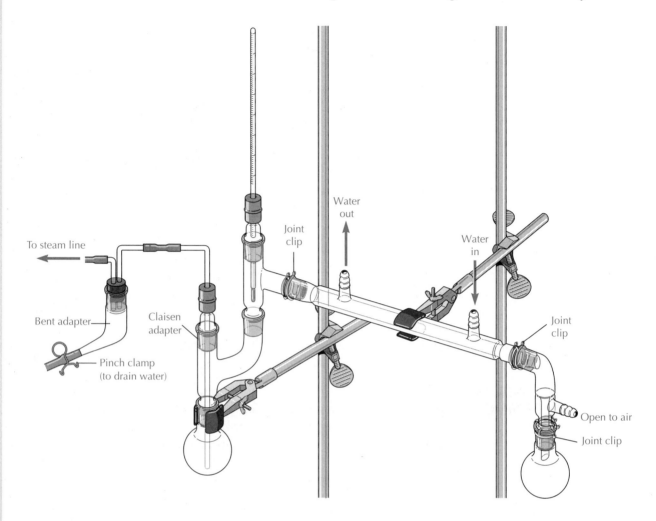

FIGURE 7.14 Steam distillation apparatus for use with an external steam source.

To check whether or not the organic material has completely distilled, collect a few milliliters of distillate in a clean receiving flask and see whether oily droplets are still present in the water.

3. Heat the mixture until all the organic layer has distilled into the receiving flask. Sometimes it is worthwhile to collect an additional 10–15 mL of water after organic material is no longer detected to ensure complete recovery.

4. Separate the organic and water layers of the distillate according to the procedure specified in the experiment.

7.7
Vacuum Distillation

Distillation at reduced pressure, called *vacuum distillation,* takes advantage of the fact that the boiling point of a liquid is a function

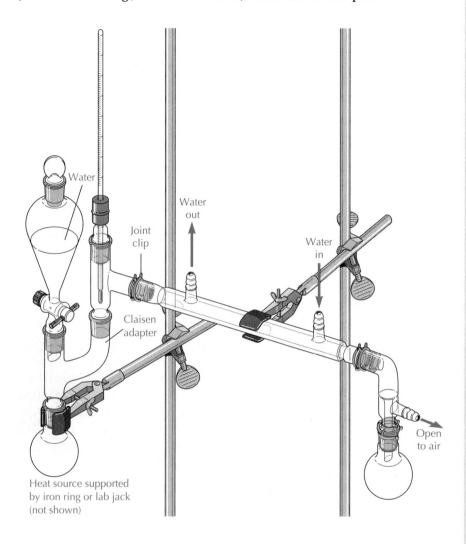

FIGURE 7.15 Steam distillation apparatus for internal generation of steam.

of the pressure under which the liquid is contained [see Technique 7.1]. Vacuum distillation involves reducing the pressure over the liquid so that distillation can be carried out at a lower temperature.

Many organic compounds with high boiling points decompose at temperatures below their atmospheric boiling points and cannot be steam distilled because they undergo chemical reactions with hot water. These compounds, and others whose boiling points are inconveniently high, distill at temperatures lower than their atmospheric boiling points when a partial vacuum is produced in the distillation apparatus. Vacuum distillation is most

Table 7.3. Boiling points (°C) at reduced pressures

Pressure (mmHg)	Water	Benzaldehyde	Diphenyl ether
760	100	179	258
100	51	112	179
40	34	90	150
20	22	75	131

1 mmHg = 1 Torr

useful in the final purification of these liquids. It should be noted that the separation of liquids is not nearly as good under vacuum distillation as it is in fractional distillation at atmospheric pressure.

A partial vacuum can be obtained in the laboratory with either a vacuum pump or a water aspirator. Vacuum pumps can easily produce pressures of less than 0.5 mmHg. The pressure obtained with a water aspirator can be no lower than the vapor pressure of water, which is 13 mmHg at 15°C and sea level. In practice, a water aspirator produces a partial vacuum of 15–25 mmHg.

The boiling point of a compound at any given pressure other than 760 mmHg is difficult to calculate exactly. As a rough estimate, a drop in pressure by one-half lowers the boiling point of an organic liquid 15–20°C. Below 25 mmHg, reducing the pressure by one-half lowers the boiling point approximately 10°C (Table 7.3).

A nomograph provides another way of estimating the boiling points of relatively nonpolar compounds at either reduced or atmospheric pressure (see Figure 7.16). The graph gives a less accurate estimate of boiling points for polar compounds that associate strongly in the liquid phase. For example, if the boiling point of a compound at 760 mmHg is 240°C and the vacuum distillation is being done at 20 mmHg, the approximate boiling point is found by aligning a straight edge on 240 in column B with 20 in column C; the straight edge intersects column A at 125°C, the approximate boiling point of the compound at 20 mmHg. Similarly, one can estimate the boiling point at atmospheric pressure if the boiling point at a reduced pressure is known. By aligning the boiling point in column A with the pressure in column C, the straight edge intersects column B at the approximate atmospheric boiling point.

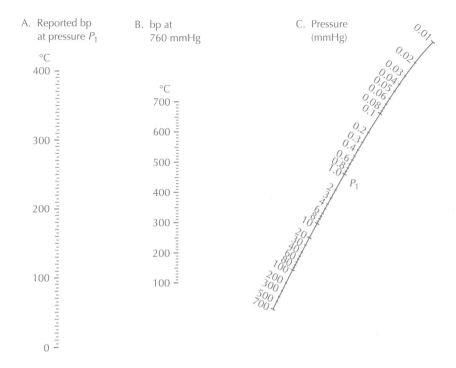

FIGURE 7.16 Nomograph for estimating boiling points at different pressures.

Apparatus for Vacuum Distillation

The vacuum distillation apparatus shown in Figure 7.17 works adequately for most vacuum distillations, although a fractionating column must sometimes be included to provide satisfactory separation of some mixtures. Because liquids often boil violently at reduced pressures, a Claisen connecting adapter is always used in a vacuum distillation to lessen the possibility of liquid bumping up into the condenser. If undistilled material jumps through the Claisen adapter into the condenser, you must begin the distillation again. Uncontrolled bumping during a vacuum distillation can be lessened by using a large distillation flask and by adding small pieces of wood splints in place of boiling chips or by magnetic stirring.

Instead of wood splints or boiling chips, a very finely drawn out capillary tube can provide a steady stream of very small bubbles (Figure 7.18). The bottom of the capillary tube bubbler should be just above the bottom surface of the distilling flask and must always be *below* the liquid's surface. Do not use wood splints or boiling chips when you use a capillary bubbler; their violent motions may break the fragile tip of the bubbler, rendering it useless. A capillary bubbler should not be used with air-sensitive compounds unless an inert gas is used in place of air.

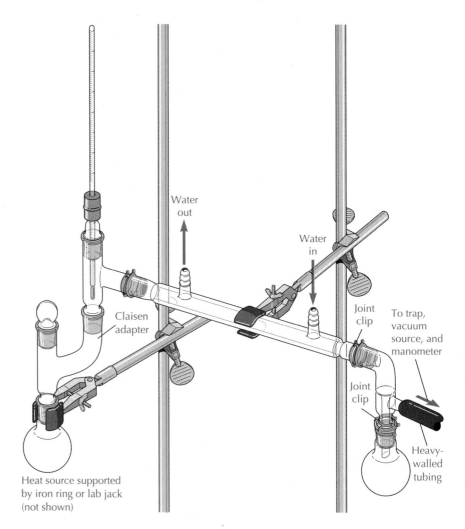

Water
out

Water
in

Claisen
adapter

Joint
clip

To trap,
vacuum
source, and
manometer

Joint
clip

Heavy-
walled
tubing

Heat source supported
by iron ring or lab jack
(not shown)

FIGURE 7.17 Vacuum distillation apparatus for use with a water aspirator-generated vacuum.

If a satisfactory vacuum is to be maintained, each connecting surface must be completely greased with *high-vacuum silicone grease,* and the rubber tubing to the aspirator or vacuum pump should be thick-walled so that it does not collapse. If the partial vacuum is not as low as expected, carefully check all connections for possible leaks. The pressure can be continuously monitored with a manometer (Figure 7.19). When a vacuum pump is used, two traps kept at isopropyl alcohol/dry ice temperature ($-77°C$) or lower must be placed between the distillation system and the pump. These traps collect any volatile materials that could get into the pump oil and cause a rise in the vapor pressure of the oil, thereby raising the lowest partial vacuum that the pump can supply, and possibly damaging the pump.

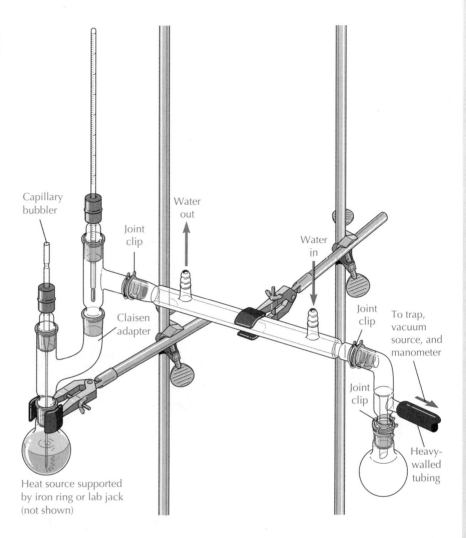

FIGURE 7.18 Vacuum distillation apparatus fitted with a capillary bubbler.

Capillary bubbler

Joint clip

Water out

Water in

Joint clip

To trap, vacuum source, and manometer

Claisen adapter

Joint clip

Joint clip

Heavy-walled tubing

Heat source supported by iron ring or lab jack (not shown)

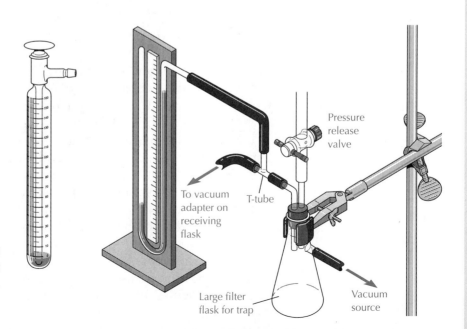

FIGURE 7.19 Two types of closed-end manometers used in vacuum distillation.

Pressure release valve

To vacuum adapter on receiving flask

T-tube

Large filter flask for trap

Vacuum source

Steps in a Vacuum Distillation

1. Add the liquid to be distilled to a round-bottomed flask sized so that it will be less than half filled, then add some wood splints or boiling chips or a magnetic stirrer (if you use the apparatus in Figure 7.17) and set up the apparatus as shown in Figure 7.17 or 7.18.

2. Totally close the apparatus and turn on the vacuum.

3. Heat the distilling flask cautiously to obtain a moderate distillation rate.

4. When the distillation is complete, remove the heat source and allow the apparatus to cool nearly to room temperature before breaking the vacuum. **Turn off the aspirator or vacuum pump after the vacuum is broken.**

References

1. Lide, D. R., Ed.; *Handbook of Chemistry and Physics;* 77th ed. CRC Press: Boca Raton, FL, 1997.

2. Weissberger, A. *Techniques of Organic Chemistry;* 2nd ed.; Wiley-Interscience: New York, 1951, Vol. IV, Distillation.

Questions

1. Explain why the observed boiling point for the first drops of distillate collected in the simple distillation of a 1:1 molar solution of pentane and hexane illustrated in Figure 7.6 was above the boiling point of pentane.

2. A mixture contains 80% hexane and 20% pentane. Use the phase diagram shown in Figure 7.4 to estimate the composition of the vapor over this liquid. This vapor is condensed and the resulting liquid is heated. What is the composition of the vapor over this liquid?

3. Arrange the following compounds in order of decreasing boiling point: pentane, octane, hexane, 2-methylpentane.

4. A student carried out a simple distillation on a compound known to boil at 124°C and reported an observed boiling point of 116–117°C. Gas chromatographic analysis of the product showed that the compound was pure, and a calibration of the thermometer indicated that it was accurate. What procedural error did the student make in setting up the distillation apparatus?

5. The directions in an experimental procedure specify that the solvent, diethyl ether, be removed from the product by using a simple distillation. Why should the heat for this distillation be supplied by a steam bath and not an electrical heating mantle?

6. Azeotropes can be used to assist chemical reactions. Treatment of 1-butanol with acetic acid in the presence of a nonvolatile acid cat-

alyst results in formation of the ester (butyl acetate) and water. The mixture of 1-butanol/butyl acetate/water forms a ternary azeotrope that boils at 90.7°C. This azeotrope separates into two layers, the upper being largely ester and the lower, water. The ester forms by an equilibrium reaction that does not especially favor product formation. Describe an apparatus that you could use to take advantage of azeotrope formation to drive the equilibrium toward the products, thus maximizing the yield of ester.

7. In the past, chemists sometimes used benzene as the solvent for esterifications because water is a side product of the reaction and benzene forms an azeotrope with water that boils at a convenient temperature (69.3°C). The reaction could be monitored by collecting the distillate in a vessel with calibrated volume markings, and the reaction could be stopped when the volume of water observed (the azeotrope separates upon condensation) corresponds to that expected from the reaction's stoichiometry. Why is it hazardous and thus improper to carry out the esterification this way?

8. A compound has a boiling point of 300°C at atmospheric pressure. Use a nomograph (Figure 7.16) to determine a pressure at which the compound would boil at about 200°C.

Technique 8
SUBLIMATION

The evaporation of most solid organic compounds first requires melting, a process that usually takes a reasonably high temperature. However, some substances, such as iodine, camphor, and 1,4-dichlorobenzene (*para*-dichlorobenzene, or moth balls), exhibit appreciable vapor pressures below their melting points. You have probably already seen iodine crystals evaporate to a purple gas during gentle heating and smelled the characteristic odors of camphor or moth balls. These substances all change directly from the solid phase to the gas phase without forming the intermediate liquid phase by a phenomenon called **sublimation.**

Sublimation requires two conditions, namely, that the vapor pressure of the solid equals the pressure above the solid and that the temperature of the solid remains below its melting point. In other words, the only source of the gas phase is the vapor pressure of the solid. The process of sublimation seems somewhat unusual, in that, unlike normal phase changes from solid to liquid to gas, no liquid phase forms between the solid and gas phases. The interconversion of the solid and gaseous forms of carbon dioxide (solid = "dry ice") may be the best-known example of sublimation. Carbon dioxide does not have a melting point at atmospheric pressure. More than 5 atm of pressure are necessary before dry ice will melt (at −57°C), while the sublimation point for CO_2 at atmospheric pressure is −78°C, well below room temperature.

In the laboratory we can use sublimation as a purification method for an organic compound if it is stable enough to vaporize without melting, if the vapor condenses to the solid, and if the impurities present do not also sublime. Many organic compounds that do not sublime at atmospheric pressure will sublime readily at reduced pressure, thus enabling their purification by sublimation. Use of reduced pressure, supplied by a vacuum source, also makes decomposition and melting less likely to occur during the sublimation.

8.1

Assembling the Apparatus for a Sublimation

The apparatus for a sublimation consists of an outer vessel, connected to a vacuum source, that holds the sample being purified. An inner container, sometimes called a "cold finger," provides a cold surface upon which the vaporized compound can condense as a solid.

Two simple arrangements for sublimation under reduced pressure are shown in Figure 8.1. The inner test tube, which contains cold water or ice and water, serves as a condensation site for the sublimed solid. The outer vessel, a side-arm test tube or a filter

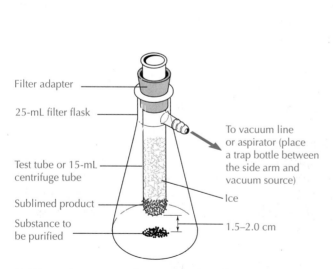

Filter adapter

25-mL filter flask

Test tube or 15-mL centrifuge tube

Sublimed product

Substance to be purified

To vacuum line or aspirator (place a trap bottle between the side arm and vacuum source)

Ice

1.5–2.0 cm

(a) Microscale or macroscale apparatus, depending on the sizes of the flask and the test tube

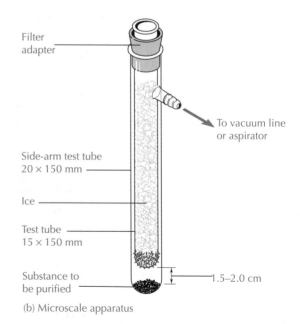

Filter adapter

Side-arm test tube 20 × 150 mm

Ice

Test tube 15 × 150 mm

Substance to be purified

To vacuum line or aspirator

1.5–2.0 cm

(b) Microscale apparatus

FIGURE 8.1 Two simple apparatuses for sublimation.

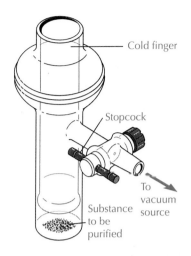

- Cold finger
- Stopcock
- To vacuum source
- Substance to be purified

FIGURE 8.2 Commercially available sublimation apparatus for gram quantities of material.

flask, holds the substance being purified, and the side arm provides a connection to the vacuum source. The inner and outer vessels are sealed together by a rubber filter adapter. The distance between the bottom surfaces of the inner tube and the outer tube or filter flask should be 0.5- to 1.0 cm. If the vapor has to travel a long distance, a higher temperature is needed to keep it above the freezing point, and decomposition of the sample may very well occur. If the surfaces are too close, impurities can spatter up to contaminate the condensed solid on the surface of the inner tube. Connect the side arm of the test tube or filter flask to a water aspirator or vacuum line, using a safety trap flask between the aspirator and the sublimation apparatus.

The side-arm test tube apparatus serves well for 10–150 mg of material. The filter flask apparatus can be sized to suit the amount of material being purified. For example, microscale quantities of 10–150 mg can be sublimed in a 25-mL filter flask; whereas 1 g of material would require a 125-mL filter flask with a correspondingly larger test tube for the cold finger. The apparatus shown in Figure 8.2 is a commercially available sublimation apparatus used for gram quantities of material.

8.2
Carrying Out a Sublimation

SAFETY PRECAUTION

The lip of the inner test tube must have a lip large enough to prevent it from being pushed through the bottom of the filter adapter by the difference in pressure created by the vacuum. Slippage of the inner test tube could cause both vessels to shatter as the inner test tube hits the outer test tube or flask. Placing a microclamp on the test tube above the filter adapter will help to keep the test tube from moving once it is positioned in the filter adapter.

*The ice and water are added **after** the vacuum is applied so that moisture from the air will not condense on the inner tube before sublimation takes place.*

Place the sample (10–150 mg) being sublimed in a side-arm test tube, and turn on the water aspirator or vacuum line. After a good vacuum has been achieved, fill the inner test tube with ice and water, then proceed to heat the sublimation tube **gently** using an aluminum block on a hot plate or a sand bath [see Technique 3.2]. If a filter flask (25 mL for 10–150 mg, 125 mL for 0.25–1.0 g) is being used as the outer container, heat it **gently** on a hot plate or with a sand bath.

During sublimation, you will notice material disappearing from the bottom of the outer vessel and reappearing on the cool outside surface of the inner test tube. If the sample begins to melt, briefly withdraw the apparatus from the heat source. If all of the ice melts, remove the water from the inner test tube with a Pasteur pipet, then add additional ice. After sublimation is complete, remove the heat source, slowly let air back into the system by gradually removing the rubber tubing from the water aspirator and then turn off the water flow in the aspirator, or turn off the vacuum source and slowly disconnect the rubber tubing from the side arm. Carefully remove the inner test tube and scrape off the purified solid onto a tared weighing paper. After weighing the sublimed solid, store it in a tightly closed vial.

Questions

1. Which of the following three compounds would most likely be amenable to purification by sublimation: polyethylene, menthol, or benzoic acid?
2. A solid compound has a vapor pressure of 65 mmHg at its melting point of 112°C. Give a procedure for purifying this compound by sublimation.
3. Hexachloroethane has a vapor pressure of 780 mmHg at its melting point of 186°C. Describe how the solid would behave while carrying out a melting-point determination at atmospheric pressure (760 mmHg) in a capillary tube open at the top.

Technique 9
REFRACTOMETRY

A beam of light traveling from a gas (air, for example) into a liquid undergoes a decrease in its velocity, and the beam bends downward as it passes from the gas into the liquid. Application of this phenomenon allows the determination of a physical property known as the *refractive index*, a measure of how much the light is bent, or *refracted*, as it enters the liquid. The refractive index can be determined quite accurately to four decimal places, thus making this physical property useful for assessing the purity of liquid compounds. The closer the experimental value approaches the value reported in the literature, the purer the sample. Even trace amounts of impurities (including water) change the measured value, so unless the compound has been extensively purified, the experimentally determined refractive index may not agree with the literature value past the second decimal place.

9.1
Theory of Refractometry

The bending of a light beam as it passes from one medium to another is known as refraction. The refractive index, n, represents the ratio of the velocity of light in a vacuum or in air to the velocity of light in the liquid being studied. The velocities of light in both mediums are related to the angles that the incident and the refracted beams make with a theoretical line drawn vertical to the liquid surface (Figure 9.1).

$$n = \frac{V_{vac}}{V_{liq}} = \frac{\sin \theta}{\sin \theta'}$$

where

$$V_{vac} = \text{velocity of light in a vacuum (or air)}$$
$$V_{liq} = \text{velocity of light in the liquid}$$
$$q = \text{angle of light in a vacuum (or air)}$$
$$q' = \text{angle of light in the liquid}$$

The velocity of light in a liquid sample is always less than that of light in a vacuum (or air), so refractive index values are numerically greater than one. Making measurements of light velocity in a vacuum creates many experimental difficulties; thus, in practice, refractive index measurements use the velocity of light in air.

The variables of temperature and the wavelength of the light being refracted influence the refractive index for any substance. The temperature of the sample affects its density; and the density change, in turn, affects the velocity of the light beam as it passes through the sample. Therefore, the temperature (20°C in the following example) at which the refractive index was determined is always specified by a superscript in the notation of n:

$$n_D^{20} = 1.3910$$

The wavelength of the light used also affects the refractive index, because light of differing wavelengths bends at different angles. The bright yellow line of the sodium spectrum at 589 nm, commonly called the sodium D line, usually serves as the standard wavelength for refractive index measurements. This standard is noted by the D in the subscript of the symbol n. If light of some other wavelength is used, the specific wavelength in nanometers appears in the subscript.

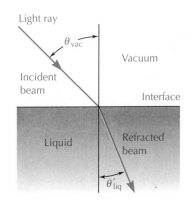

Light ray

θ_{vac}

Vacuum

Incident beam

Interface

Liquid

Refracted beam

θ'_{liq}

FIGURE 9.1 Refraction of light in a liquid.

9.2
The Refractometer

We measure the refractive index of a compound with an instrument called a *refractometer.* Figure 9.2 shows an Abbe refractometer, an example of the type commonly found in undergraduate organic laboratories. The instrument includes a built-in thermometer for measuring the temperature at the time of the measurement and a system for circulating water at a constant temperature around the sample holder. This type of refractometer uses a white light source instead of a sodium lamp and contains a series of compensating prisms that give a refractive index equal to that obtained with the D line of sodium at 589 nm.

A few drops of sample are introduced between a pair of hinged prisms (Figure 9.3). The light passes through the sample and is reflected by an adjustable mirror. When the mirror is properly aligned, the light is reflected through the compensating prisms and, finally, through a lens with crosshairs to the eyepiece.

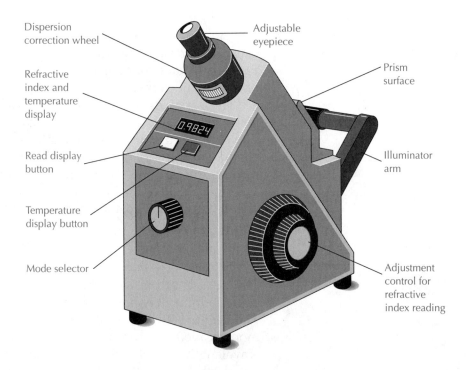

FIGURE 9.2 Abbe refractometer. (Courtesy of Leica, Inc., Optical Products Division, P.O. Box 123, Buffalo, NY 14240.)

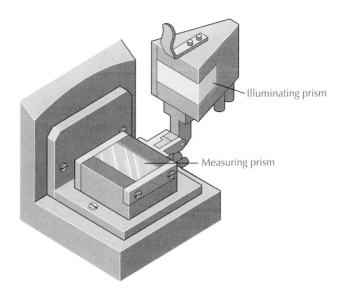

Illuminating prism

Measuring prism

FIGURE 9.3 Hinged prisms of the Abbe refractometer, shown in the open position. (Courtesy of Leica, Inc.)

9.3
Steps in Determining a Refractive Index

*Do **not** use acetone to clean prisms because it dissolves the adhesive holding the prism.*

1. Check the surface of the prisms for residue from the previous determination. If the prisms need cleaning, place a few drops of methanol on the surfaces and blot **(do not rub)** the surfaces with lens paper. Allow the residual methanol to evaporate completely before placing the sample on the lower prism.

2. With a Pasteur pipet held 1–2 cm above the prism, place 4–5 drops of the sample on the measuring (lower) prism. **Do not touch the prism with the tip of the dropper, because the highly polished surface scratches very easily, and scratches ruin the instrument.** Lower the illuminating (upper) prism carefully so that the liquid spreads evenly between the prisms.

3. Rotate the instrument wheel until the dark and light fields are centered on the crosshairs in the eyepiece. (Figure 9.4). If color (usually red or blue) appears as a horizontal band at the interface of the dark and light fields, rotate the chromatic adjustment, or dispersion correction wheel, until the interface is sharp and uncolored (achromatic). Occasionally the sample evaporates from the prisms, making it impossible to produce a sharp achromatic interface between the light and dark fields. If this happens, apply more sample to the prism and repeat the adjustment procedure.

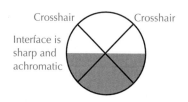

FIGURE 9.4 View through the eyepiece when the refractometer is adjusted correctly.

4. Press the display button and record the refractive index in your notebook. Then press the temperature button and record the temperature.

5. Open the prisms, blot up the sample with lens paper, and follow the cleaning procedure with methanol outlined in step 1.

9.4
Temperature Correction

Values reported in the literature are often determined at a number of different temperatures, although 20°C has become the standard. To compare an experimental refractive index with a reported value at a different temperature, a correction factor must first be calculated for the experimental refractive index. The refractive index increases by 4.5×10^{-4} for each one-degree decrease in temperature. Refractive indexes vary inversely with temperature because the density of a liquid almost always decreases as the temperature increases. This decrease in density produces an increase in the velocity of light in the liquid, causing a corresponding decrease in the refractive index.

To compare an experimental refractive index measured at 25°C to a reported value at 20°C; a temperature correction needs to be calculated:

$$\Delta n = 4.5 \times 10^{-4} \times (T_1 - T_2)$$

where

T_1 = the observed temperature in degrees Celsius

T_2 = the temperature reported in the literature in degrees Celsius

Add the correction factor, *including its sign,* to your experimentally determined refractive index. For example, if your experimental refractive index is 1.3888 at 25°C, then you obtain a corrected value at 20°C of 1.3910 by adding the correction factor of 0.0022 to the experimental refractive index.

$$\Delta n = [4.5 \times 10^{-4} \times (25 - 20)] = 0.00225 \quad \text{(rounds to 0.0022)}$$

$$n^{20} = n^{25} + 0.0022 = 1.3888 + 0.0022 = 1.3910$$

This correction needs to be applied before comparing the experimental value to a literature value reported at 20°C.

For an experimental refractive index determined at a temperature lower than that of the literature value to which it is being compared, the correction has a negative sign and the corrected refractive index would be lower than the experimental value. For example, if the observed refractive index at 19°C is 1.4179 and the reported value was determined at 25°C, the following correction applies to the observed value:

$$\Delta n = [4.5 \times 10^{-4} \times (19 - 25)] = -0.0027 \quad \text{(rounds to } -0.0032)$$

The corrected refractive index (n^{25}) would be 1.4147, the sum of 1.4179 (n^{19}) and −0.0027.

Questions

1. A compound has a refractive index of 1.3191 at 20.1°C. Calculate its refractive index at 25.0°C.
2. It is usually recommended that the glass surfaces of a refractometer be cleaned with ethanol or methanol, and **not** with acetone. What is the reason for this?
3. In the past (before the availability of IR and NMR), hydrocarbons were often identified by refractive index. Explain why at that time chemists frequently had to use refractive index to identify hydrocarbons, while, for example, alcohols, aldehydes, or ketones could be identified by other methods.

Technique 10
THIN-LAYER CHROMATOGRAPHY

Few experimental techniques are either as versatile or as useful in separating complex mixtures of compounds as chromatography is. First used in the early 1900s by the Russian botanist Mikhail Semenovich Tsvet, *chromatography* got its name because it was used to separate substances of different colors.

Although the development of chromatography was very slow until the 1930s, progress accelerated once chemists realized that chromatography could be used to separate colorless substances as well as colored ones. Three types of chromatography are used extensively in organic chemistry and will be discussed in detail in this book: *thin-layer chromatography (TLC), gas-liquid chromatography (GC),* and *column chromatography. Flash chromatography* and *high-pressure liquid chromatography (HPLC)* will be discussed briefly. There are other chromatographic methods, too, such as *gel permeation chromatography (GPC)* and *ion-exchange chromatography,* that are beyond the scope of this book and more frequently are used for separations in other areas of the physical and biological sciences, such as molecular biology and biochemistry.

Thin-layer chromatography, which appeared in the late 1950s, has rapidly become one of the most widely used analytical techniques. It is simple, inexpensive, fast, sensitive, efficient, and requires only milligram quantities of material. Thin-layer chromatography, or simply TLC, is especially useful for determining the number of components in a mixture, for possibly establishing whether or not two compounds are identical, and for following the course of a reaction.

Every type of chromatographic analysis depends on the distribution of the substances being separated between two phases, a mobile phase and a stationary phase. The **mobile phase** consists of a liquid or gas, which carries the sample through the solid or liquid that forms the **stationary phase.** For example, in both thin-layer and column chromatography, a finely ground solid forms the stationary phase. A liquid solvent provides the mobile phase. The separation of the compounds in the sample being analyzed occurs because polar compounds, which adsorb more tightly than nonpolar ones, stay on the stationary phase longer and need a greater amount of solvent (mobile phase) to carry them through the stationary phase. The compounds being separated adsorb onto and desorb from the stationary phase many, many times as the solvent passes through the stationary phase; hence this kind of chromatography is called **adsorption chromatography.**

It is also possible to coat a solid phase with a liquid that does not dissolve off the solid particles when a solvent passes over the particles. In this instance, the stationary phase is actually the coating liquid and not the solid support. For example, water and methanol may bind so tightly to a polar solid surface that they stay put. In such a case, the compounds to be separated partition themselves between this stationary liquid phase and the solvent traveling through the stationary phase, as the solute partitions itself between the two solvents used for an extraction [see Technique 4]. This kind of chromatography is called **partition chromatography.**

In both adsorption and partition chromatography, a dynamic equilibrium exists, and the compounds being separated move slowly along the adsorbent (stationary phase) in the direction of the liquid (mobile phase) flow. The compounds separate because of the differences in their adsorptivities on the stationary phase and their solubilities in the mobile phase.

10.1
Principles of Thin-Layer Chromatography

Thin-layer chromatography uses glass, metal, or plastic plates coated with a thin layer of adsorbent as the stationary phase. Silica gel ($SiO_2 \cdot xH_2O$) and aluminum oxide (Al_2O_3) are the most common solid *adsorbents* for thin-layer plates. Most nonvolatile solid organic compounds can be analyzed by thin-layer chromatography. However, the method does not work well for liquid compounds, because their volatility can lead to loss of the sample by evaporation from the thin-layer plate during the analysis.

A small amount of the mixture being separated is applied or spotted on the adsorbent near one end of the plate. Then the thin-layer plate is placed in a closed chamber, with the lower edge (near the applied spot) immersed in a shallow layer of *developing solvent,* the mobile phase (Figure 10.1). The solvent rises through the stationary phase by capillary action. This process is called *developing the chromatogram.*

As the solvent ascends the plate, the sample is distributed between the mobile liquid phase and the stationary solid phase. The separation during the development process occurs as a result of the many equilibrations taking place between the mobile and stationary phases and the compounds being separated. *The more tightly a compound binds to the absorbent, the more slowly it moves on the thin-layer plate.* The developing solvent moves nonpolar substances up the plate most rapidly. Polar substances travel up the plate slowly, or possibly not at all, as the solvent ascends by capillary action.

The thin-layer plate is removed from the developing chamber when the *solvent front is about 1 cm from the top of the plate.* The position of the solvent front is marked immediately with a pencil, before the solvent evaporates. The plate should then be placed in a ventilating hood to dry.

Several methods are available to *visualize* the spots. If the thin-layer plate is impregnated with a fluorescent indicator, the plate can be illuminated by exposure to ultraviolet radiation. Alternatively, placing the plate in a jar containing a few iodine crystals will color the spots with the brown of iodine. In another method, the thin-layer plate is dipped in a reagent that produces a

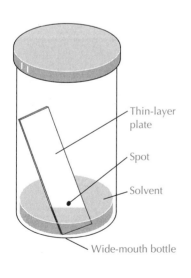

Thin-layer plate

Spot

Solvent

Wide-mouth bottle

FIGURE 10.1 Developing chamber containing a thin-layer plate.

colored product with the compounds on the plate when the plate is heated. The developed and visualized plate is then ready for analysis of the separation.

10.1a Determination of the R_f

Under a definite set of experimental conditions for a thin-layer chromatographic analysis, a given compound will always travel a fixed distance relative to the distance traveled by the solvent front (Figure 10.2). This ratio of distances is called the R_f. The term R_f stands for *"ratio to the front"* and is expressed as a decimal fraction:

$$R_f = \frac{\text{distance traveled by compound}}{\text{distance traveled by developing solvent front}}$$

The R_f value for a compound depends on its structure and is a physical characteristic of the compound, just as a melting point is a physical characteristic. Whenever a chromatogram is done, the R_f value is calculated for each substance and the experimental conditions recorded. The important data include

1. brand, type of backing, and adsorbent on the thin-layer plate
2. developing solvent
3. method used to visualize the compounds
4. R_f value for each substance

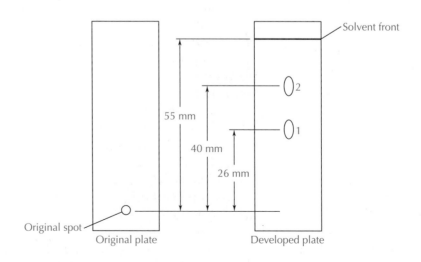

FIGURE 10.2 Calculating the R_f value.

To calculate the R_f value for a given compound, measure the distance that the compound has traveled from where it was originally spotted and the distance that the solvent front has traveled (Figure 10.2). The measurement is made from the *center of a spot.* The best data are obtained from chromatograms in which the spots are less than 5 mm in diameter. If spots show "tailing," measure from the densest point of the spot. The R_f values for the two substances separated in the thin-layer plate in Figure 10.2 are

$$\text{Compound 1:} \quad R_f = \frac{26 \text{ mm}}{55 \text{ mm}} = 0.47$$

$$\text{Compound 2:} \quad R_f = \frac{40 \text{ mm}}{55 \text{ mm}} = 0.73$$

When two samples have identical R_f values, you should not conclude that they are the same compound without doing further analysis. There are perhaps 100 distinct R_f values, whereas there are greater than 10×10^6 organic compounds. Carrying out additional TLC analyses on the two samples, using different solvents or solvent mixtures, would be a way to check on their identities. Further analysis by IR or NMR spectrometry would provide definitive evidence as to whether the compounds were identical or not.

10.2
Plates for Thin-Layer Chromatography

Three solid adsorbents, silica gel ($SiO_2 \cdot xH_2O$), aluminum oxide (Al_2O_3; also called alumina), and cellulose are commonly used for thin-layer chromatography. Many kinds of intermolecular forces cause organic molecules to bind to these solids. Weak van der Waals forces are the only ones that bind nonpolar compounds to the adsorbent. Polar molecules can also adsorb by dipole-dipole interactions, hydrogen bonding, and coordination to the highly polar metal oxide surfaces. The strength of the interaction varies for different compounds, but one generality can be stated: *The more polar the compound, the more strongly it will bind to silica gel or aluminum oxide.* Cellulose is used for the partition chromatography of water-soluble and quite polar organic compounds, such as sugars, amino acids, or nucleic acid derivatives. Cellulose can adsorb up to 20% of its weight in water; the substances being separated partition themselves between the developing solvent and the water molecules that are hydrogen bonded to the cellulose particles.

All solid adsorbents used for TLC are prepared from activated, finely ground powder. Activation usually involves heating the powder to remove adsorbed water. Silica gel is acidic, and it separates acidic and neutral compounds that are not too hydrophilic ("water-loving"). Acidic, basic, or neutral aluminum oxide is available, thus allowing separation of a wide range of nonpolar and polar organic compounds on this adsorbent.

A number of manufacturers sell plates that are precoated with a layer of adsorbent. These plates are available with plastic, glass, and aluminum backing. The plastic-backed silica gel plates are usually the least expensive. They can be cut to any desired size with scissors. The usual size that we suggest is 2.5 × 6.7 cm; this choice allows you to cut a standard 20 × 20 cm sheet into 24 plates. The adsorbent surface is of uniform thickness, usually 0.10 mm. Results are quite reproducible, and sharp separation is normal. The plastic backing is generally a solvent-resistant polyester plastic. The adsorbent is bound to the plastic by solvent-resistant polyvinyl alcohol, which binds tightly to the adsorbent and to the plastic. Precoated plates impregnated with a fluorescent indicator are also available; these plates facilitate the visualization of many colorless compounds [see Technique 10.5].

Thin-layer plates with a glass or aluminum backing are also available in the standard 20 × 20 cm sheets. Both types can be heated without melting; this property is valuable if the plate is to be visualized with a reagent that requires heating [see Technique 10.5]. Aluminum plates can be cut with scissors into convenient sizes for thin-layer plates. We have suggested their use in several experiments.

If the plastic seal on the package containing the precoated sheets has been broken for some time, the thin-layer plates should be activated before use to remove adsorbed water. This practice enhances the reproducibility of R_f values. Activation is done simply by heating the sheets in a clean oven at 100°C for 15–30 min.

Larger thin-layer plates can be used for complex separations. The longer the distance that the developing solvent moves up the plate, the better will be the separation of the compounds being analyzed. Wider plates can also accommodate many more samples. Large plates with an adsorbent layer 1–2 mm thick are used for preparative TLC in which samples of 50–1000 mg are separated. The large sample size allows recovery of the individual components after separation. The sample is applied in a streak across the entire plate near one edge. The references at the end of this chapter describe techniques for making preparative thin-layer plates.

10.3
Sample Application

The sample must be dissolved in a volatile organic solvent; a 1–2% solution works best. The solvent needs a high volatility so that it evaporates almost immediately. Acetone and dichloromethane are commonly used, with chloroform as an alternative, if the sample does not dissolve in these two solvents.

SAFETY PRECAUTION

Work in a hood when you are using CH_2Cl_2 or $CHCl_3$.

If you are analyzing a solid, dissolve 20–40 mg of it in 2 ml of the solvent.

Tiny spots of the sample are carefully applied with a micropipet near one end of the plate. Keeping the spots small makes for the cleanest separation. It is also important not to overload the plate with too much sample, as this leads to large tailing spots and poor separation.

Commercial micropipets are available in 5- and 10-μL sizes, and these work well for applying samples on the plastic-backed plates. The aluminum-backed plates require micropipets with a smaller diameter. A micropipet can be made easily from an open-ended, thin-walled, melting-point capillary tube. The capillary tube is heated at its midpoint with a Bunsen burner.

SAFETY PRECAUTION

Be sure there are no flammable solvents in the vicinity when you are using a Bunsen burner.

A microburner works best because only a small flame is required. While heating the tube, rotate it until it is soft on all sides over a length of 1–2 cm. When the tubing is soft, draw out the heated part until a constricted portion 4–5 cm long is formed (Figure 10.3). After cooling the tube for a minute or so, score it gently at the center with a file and break it into two capillary micropipets. The break must be a clean one, at right angles to the length of the tubing, so that when the tip of the micropipet is touched to the plate, liquid is pulled out by the adsorbent. The diameter at the end of a micropipet should be about $\frac{1}{3}$ mm.

The micropipet is filled by dipping the constricted end into the solution to be analyzed. Only 1–5 μL of the sample solution are needed for most TLC analyses. Hold the micropipet vertically to the plate and apply the sample by touching the micropipet gently

4–5 cm

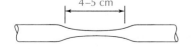

FIGURE 10.3 Constricted capillary tube.

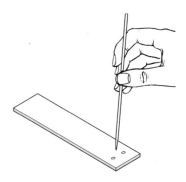

FIGURE 10.4
Spotting the plate.

No type of pen should be used for marking thin-layer plates because the inks will separate during the development and obscure the samples; the graphite and clay used in pencils does not interact with solvents.

and *briefly* to the plate about 1 cm from the bottom edge. Mark the edge of the plate with a pencil at the same height as the center of the spot; this mark indicates the compound's starting point for your R_f calculation. It is important to touch the micropipet to the plate very lightly so that no hole is gouged into the adsorbent. The spot delivered should be no more than *2–3 mm* in diameter to avoid excessive broadening of the spot during the development. If you need to apply more sample, touch the micropipet to the plate a second time at exactly the same place. Let one spot dry before applying the next. The spotting procedure may be repeated numerous times, if necessary.

If you are using the 2.5 × 6.7 cm thin-layer plates, two side-by-side spots can be applied to one plate (Figure 10.4). As a result of diffusion, the spots will become larger during the development step. If the spots get too close to each other or to the edge of the plate, the chromatograms become difficult to interpret.

An authentic standard should be included on the sample plate as a comparison, if it is available. *If two compounds have the same* R_f *value, they may be the same compound; if the* R_f *values differ, they most definitely are not the same compound.* If R_f values are quite close, it is best to run the chromatogram again, using a longer thin-layer plate or a different solvent.

To test for the proper amount of solution to spot on the plate, spot two different amounts on the same slide and see which gives better results. You can determine the best TLC conditions more quickly this way.

10.4
Choices of Developing Solvent

After the spots dry, the thin-layer plate is put in a developing chamber (Figure 10.5). To ensure good chromatographic resolution, the chamber must be saturated with solvent vapors to prevent the evaporation of solvent as it rises up the thin-layer plate. Inserting a piece of filter paper halfway around the inside of the developing bottle saturates the atmosphere with solvent vapor by

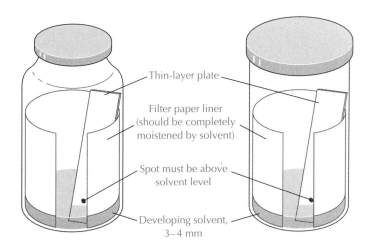

Thin-layer plate

Filter paper liner
(should be completely
moistened by solvent)

Spot must be above
solvent level

Developing solvent,
3–4 mm

FIGURE 10.5 Typical developing chambers.

*The solvent depth in the developing chamber **must** be less than the height of the spots on the thin-layer plate.*

wicking solvent into the upper region of the chamber. Use enough developing solvent to allow a shallow layer (3–4 mm) to remain on the bottom after the closed chamber has been shaken to wet the filter paper. If the solvent level is too high, the spots applied to the TLC plate may be below the solvent level. Under these conditions, the spots dissolve away into the solvent, thereby ruining the chromatogram; all you will see on the developed chromatogram are fuzzy, useless bands that extend across the width of the plate.

Do not touch the adsorbent side of the thin-layer plate with your fingers. Hold the plate by the edges or with a pair of tweezers.

Do not lift or otherwise disturb the chamber while the thin-layer plate is developing in the solvent.

Uncap the developing chamber and place the thin-layer plate inside with a pair of tweezers. Recap the chamber, and let the solvent move up the plate. The adsorbent will become visibly moist. When the solvent front is within 0.5–1 cm of the top of the adsorbent layer, remove the plate from the developing chamber with a pair of tweezers and *immediately* mark the adsorbent at the solvent front with a pencil. To get accurate R_f values, the final position of the front must be marked before any evaporation occurs.

— **SAFETY PRECAUTION** ——

Evaporate the solvent from a developed chromatogram in a fume hood.

The development of a chromatogram is finished within 5–10 min. Let the developing solvent evaporate from the plate before visualizing the results.

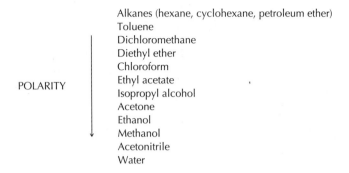

POLARITY

Alkanes (hexane, cyclohexane, petroleum ether)
Toluene
Dichloromethane
Diethyl ether
Chloroform
Ethyl acetate
Isopropyl alcohol
Acetone
Ethanol
Methanol
Acetonitrile
Water

FIGURE 10.6 Common TLC developing solvents.

One vital question remains. What developing solvent should be used? This question has no simple answer, but experience gives a number of helpful leads. In general, you should use a nonpolar developing solvent for nonpolar substances and a polar developing solvent for polar materials (Figures 10.6 and 10.7).

Net chromatographic behavior is the result of competition between the applied compounds and the developing solvent for the active surface sites of the adsorbent. If the solvent is adsorbed too well, the adsorbent is deactivated, R_f values are increased, and separation may be incomplete. Therefore, a solvent that causes all the spotted material to move with the solvent front is too polar, whereas one that does not cause any compounds to move is not polar enough.

Mixing developing solvents yields a solution with a polarity approximately proportional to the ratio of the mixed solvents. The purity of the developing solvent is, therefore, an important factor in the success of your chromatography. For example, chloroform is commonly stabilized with 1% ethanol. This commercial product has a developing power substantially different from that of pure chloroform.

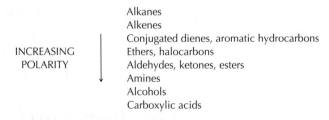

INCREASING
POLARITY

Alkanes
Alkenes
Conjugated dienes, aromatic hydrocarbons
Ethers, halocarbons
Aldehydes, ketones, esters
Amines
Alcohols
Carboxylic acids

FIGURE 10.7 Relative polarity of organic compounds.

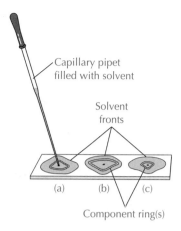

FIGURE 10.8 Rapid method for determining an effective solvent: (a) good development; (b) and (c) poor development.

With a silica gel plate, nonpolar hydrocarbons should be eluted with hydrocarbon solvents, but a mixture containing an alcohol and an ester might be developed with a toluene-dichloromethane mixture. When a very polar solvent is required to move spots on the TLC adsorbent, better results will be obtained by switching to a less active adsorbent and a less polar solvent.

A rapid way to determine the best TLC developing solvent among several possibilities is to spot three or four samples along the length of the same plate (Figure 10.8). Fill a micropipet with the solvent to be tested and gently touch one of the spots. The solvent will diffuse outward in a circle, and the sample will move out with it. Mixtures of compounds will be partially separated and approximate R_f values can be estimated. The components should travel about one-third to one-half as fast as the solvent front.

10.5
Visualization Techniques

The TLC separations of colored compounds can be seen directly. More often than not, however, organic compounds are colorless, so an indirect visualization technique is needed. The simplest of these involves the use of commercial thin-layer plates or adsorbents that contain a fluorescent indicator. A common fluorescent indicator is calcium silicate, activated with lead and manganese. The insoluble inorganic indicator rarely interferes in any way with the chromatographic results and makes visualization straightforward. When the output from a short-wavelength ultraviolet lamp (254 nm) shines on the plate in a darkened room or dark box, the plate fluoresces visible light (Figure 10.9).

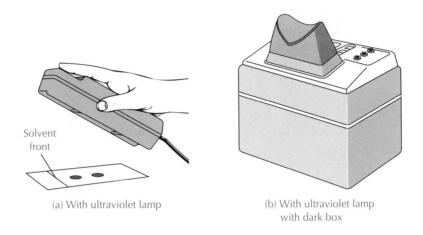

Solvent
front

(a) With ultraviolet lamp

(b) With ultraviolet lamp
with dark box

FIGURE 10.9 Visualization.

The presence of separated compounds is shown by dark spots
on the fluorescent field, because the substances quench the fluorescence (Figure 10.9a). Sometimes the substances are visible by
their own fluorescence. Outline each spot with a pencil while the
plate is under the UV source. This technique gives you a perma-
nent record and allows calculation of R_f values. Not all substances
are visible with fluorescent silica gel, so visualization by the other
methods discussed below should be tried on any sample that may
contain unknown components.

Another way to visualize colorless organic compounds uses
their absorption of iodine (I_2) vapor. The thin-layer plate is put in
a bath of iodine vapor prepared by placing 0.5 g of iodine crystals
in a capped bottle. Colored spots are gradually produced from the
reaction of the separated substances with gaseous iodine. The
spots are dark brown on a white to tan background. After
10–15 min, the plate is removed from the bottle. Sometimes it is
necessary to warm the bottle on a steam bath to increase the
amount of iodine vapor surrounding the thin-layer plate. The colored spots disappear in a short period of time and should be outlined with a pencil *immediately* after the plate is removed from the
iodine bath. The spots will reappear if the plate is again treated
with iodine.

p-Anisaldehyde developing solution: 2 mL of p-anisaldehyde in 2 mL of concentrated sulfuric acid, 36 mL of 95% ethanol, and 5 or 6 drops of acetic acid

Phosphomolybdic acid developing solution: 20% by weight in ethanol

No matter which visualization method is used, the R_f value for each component should be calculated and recorded in your notebook along with the experimental conditions of the chromatogram [see Technique 10.1a].

Visualizing solutions containing reagents that react with the separated substances to form colored compounds can be sprayed on thin-layer plates; alternatively, the thin-layer plates can be dipped in the visualizing solution. Visualization occurs by heating the dipped or sprayed thin-layer plates with a heat gun or on a hot plate. Many of these solutions are specific for certain functional groups. Two common visualizing solutions are *p*-anisaldehyde and phosphomolybdic acid. Consult the references at the end of this chapter for detailed discussions of these and other reagents. When a visualizing solution is used in an experiment in this book, specific instructions are included in that experiment.

The size and intensity of the spots can be used as a rough measure of the relative amounts of the substances. These parameters can be misleading, however, especially with fluorescent visualization. Some organic compounds interact much more intensely with the ultraviolet radiation than do others. It is often possible to obtain quantitative information by running mixtures of known composition alongside the mixture being studied. The spots can also be removed from the plate and each studied spectrometrically. It should be said, however, that quantitative information is not one of the strengths of thin-layer chromatography.

10.6
Summary of TLC Procedure

1. Obtain a precoated thin-layer plate of the proper size for the developing chamber.
2. Spot the plate with a small amount of a 1–2% solution containing the materials to be separated.
3. Develop the chromatogram with a suitable solvent.
4. Mark the solvent front.
5. Visualize the chromatogram and outline the separated spots.
6. Calculate the R_f value for each compound.

References

1. Randerath, K. *Thin-layer Chromatography*; 2nd ed.; Academic Press: New York, 1966.
2. Stahl, E. *TLC: A Laboratory Handbook*; Springer-Verlag: New York, 1969.
3. Touchstone, J. C.; Dobbins, M. F. *Practice of Thin Layer Chromatography*; 2nd ed.; Wiley: New York, 1983.
4. Fried, B.; Sherma, J. *Thin-Layer Chromatography: Techniques and Applications*; 3rd ed.; Chromatographic Science Series, Vol. 66, Marcel Dekker: New York, 1994.
5. Sherma, J.; Fried, B. (Eds.) *Handbook of Thin-Layer Chromatography*; 2nd ed.; Chromatographic Science Series, Vol. 71, Marcel Dekker: New York, 1996.

Questions

1. The R_f value of compound A is 0.34 when developed in hexane and 0.47 when developed in dichloromethane. Compound B has an R_f value of 0.42 in hexane and 0.69 in dichloromethane. Which solvent would be better for separating a mixture of compounds A and B? Explain.

2. A student wishes to use TLC analysis on a mixture containing an alcohol and a ketone. After consulting Figures. 10.6 and 10.7, suggest a possible developing solvent.

3. Two substances moved with the solvent front ($R_f = 1$) during TLC analysis on a silica gel plate, using 2-propanol (isopropyl alcohol) as the developing solvent. Can you conclude that they are identical? If not, what additional experiment(s) would you perform?

4. If the structures of the two compounds in Question 2 are very similar, except for the functional group, which would have the larger R_f (a) on silica gel? (b) on alumina?

Technique 11

GAS-LIQUID CHROMATOGRAPHY

Few techniques have altered the analysis of volatile organic chemicals as much as gas-liquid chromatography (GC). Before GC became widely available, organic chemists usually looked for ways to convert liquid compounds to solids in order to analyze them. Gas-liquid chromatography provides a quick and easy way for both qualitative and quantitative analysis of volatile organic mixtures, in contrast to thin-layer chromatography, which is primarily a qualitative technique. In addition, GC has a truly fantastic ability to separate complex mixtures. However, GC has several limitations. It is useful only for the analysis of compounds that have vapor pressures high enough to allow them to pass through a GC column; and, like TLC, gas-liquid chromatography does not identify compounds unless known samples are available. The current practice of coupling a gas-liquid chromatograph with an IR, NMR, or mass spectrometer combines the superb separation capabilities of GC with the superior identification methods of spectrometry.

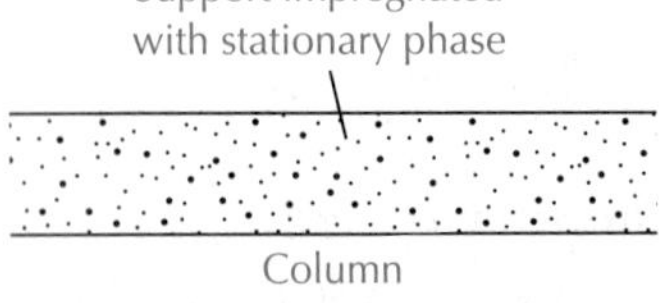

Introduction of practical GC came in 1952, the same year that Archer J. P. Martin and Richard M. L. Synge received the Nobel prize in chemistry for their proposal and study of the GC method. Gas-liquid chromatograph is known by a number of names: *vapor-phase chromatography* (VPC), *gas-liquid partition chromatography* (GLPC), or simply *gas chromatography* (GC). We will use the term GC throughout this book. Gas-liquid chromatography is a tech-

Support impregnated with stationary phase

Column

FIGURE 11.1 Microview of a packed column.

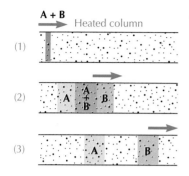

FIGURE 11.2 Stages in the separation of a two-component (A, B) mixture as it moves through the column.

nique that you will use on many occasions as you study organic chemistry. The instrument used for GC is called a *chromatograph*.

In gas-liquid chromatography, the stationary phase consists of a nonvolatile liquid with a high boiling point. For packed column chromatographs, the liquid is coated on a porous, inert solid **support** that is then packed into a tube, which forms the column (Figure 11.1). For capillary column chromatographs, a uniform film of the liquid phase is applied to the interior wall of a capillary tube, leaving a channel through the center for gas passage. In another type of capillary column, the inside of the tube is lined with a thin layer of solid support that is then coated with a film of the liquid phase.

A flow of inert gas, such as helium or nitrogen, serves as the *mobile phase.* When the mixture being separated is injected into the flow of gas at the heated injector port, the components vaporize and pass into the column, where separation occurs. The components of the mixture *partition themselves between the gas and the liquid phases* in an equilibrium that depends on the temperature, the rate of gas flow through the column, and the solubility of the substrates in the liquid phase (Figure 11.2).

11.1
Instrumentation

The basic parts of the chromatographic system are

source of high-pressure pure **carrier gas**
flow controller
heated **sample inlet,** or **injection port**
column
detector
recorder

These components are shown schematically in Figure 11.3.

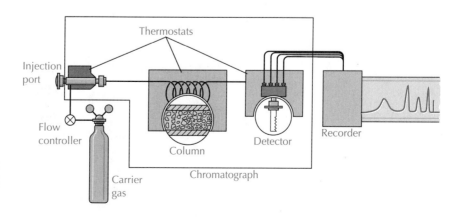

FIGURE 11.3 Schematic diagram of a gas-liquid chromatograph.

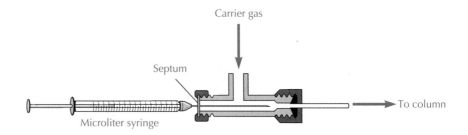

FIGURE 11.4 Injection port during sample injection.

Samples are introduced into the flowing stream of carrier gas in a heated injection port through a sealed rubber septum or gasket (Figure 11.4). A small hypodermic syringe injects the sample through the septum into the stream of carrier gas, where it is immediately vaporized. The injection port connects directly to the head of the column, a metal, glass, or fused-silica tube, that contains the liquid phase (Figure 11.5). The column is enclosed in an oven whose temperature can be carefully regulated, from just above room temperature to 250°C, or even up to 400°C on some instruments. A GC column has thousands of theoretical plates as a result of the huge surface area on which the gas and liquid phases can interact. After the column separates the components in the sample, they pass into a detector, where they produce signals that can be amplified and recorded. Because the three most crucial features of a gas chromatograph are the liquid stationary phase, the type of column, and the detection device, we will discuss these features in more detail.

Packed GC columns   Capillary GC column

FIGURE 11.5 GC columns.

11.2
Columns and Liquid Stationary Phases

A gas chromatograph may have either a packed column or a capillary column; the latter is also called an open tubular column. A packed column usually has an interior diameter of 2 to 4 mm, is 2 to 3 m in length, and contains the packed solid support coated evenly with a liquid phase of 0.05 to 1 μm in thickness. A capillary column is much thinner and has an interior diameter of 0.25 to 0.50 mm, a length of 10 to 100 meters, with coating thicknesses ranging from 0.1 to 5 μm. There are several types of capillary columns available. In a *wall-coated open tubular column* (WCOT), the liquid phase coats the interior surface of the tube, leaving the center open; whereas in a *support-coated open tubular column* (SCOT), the liquid phase coats a thin layer of solid support that is bonded to the capillary wall. *Fused-silica open tubular columns* (FSOT) represent a recent improvement in column design. These columns have better physical strength, lower reactivity toward sample components, and better flexibility for bending into coils than do glass SCOT and WCOT columns. The characteristics of typical GC columns are summarized in Table 11.1.

What happens as a mixture separates on a column? And what factors most influence the efficiency of separation? The answer to the second question involves the surface area of the solid support, the nature of the liquid phase, and the column temperature. We will consider these next.

Mixtures separate because their components interact differently with the liquid phase. The partitioning of different substances between the liquid stationary phase and the vapor phase depends on the relative attraction of the component substances for the liquid phase and on the vapor pressures of the compounds. All gases mix to form solutions, and the solubility of any given compound in the gas phase is a function of its vapor pressure alone. The greater a compound's vapor pressure, the greater will be its tendency to go from the liquid stationary phase to the gas phase. So, in the thousands of liquid-gas equilibria that take place as substances travel through a GC column, a more volatile compound will spend less time in liquid solution and more time in the vapor state, thus traveling through the column faster than a less volatile compound. In general, *lower boiling-point compounds with their higher vapor pressures travel through a GC column faster.*

The solid support in packed columns or support-coated capillary open tubular columns (SCOT columns) consists of a porous, inert material that has a very high surface area. The most commonly used substance is calcined diatomaceous earth, which contains the crushed skeletons of algae, especially diatoms. Its major

Table 11.1. **Characteristics of typical GC columns**

Type of column[a]	Length, m	Inside diameter, mm	Efficiency, plates/m	Sample size, ng
FSOT	10–100	0.1–0.53	2000–4000	10–75
WCOT	10–100	0.25–0.75	1000–4000	10–1000
SCOT	10–100	0.5	600–1200	10–1000
Packed	1–6	2–4	500–1000	$10–10^6$

SOURCE: D. A. Skoog and J. J. Leary, *Principles of Instrumental Analysis,* 4th Ed., 1992, p. 615, by permission of the publishers, Harcourt, Brace and Co., Orlando, FL.
a. FSOT, fused-silica open tubular column; WCOT, wall-coated open tubular column; SCOT, support-coated open tubular column.

component is silica. The efficiency of separation increases with decreasing particle size as a consequence of the expanded surface area available for the liquid coating, although there is a practical lower limit to the particle size imposed by the gas pressure needed to push the mobile phase through the column.

The liquid stationary phase interacts with the materials being separated in many of the ways that have already been discussed for thin-layer chromatography. Dipolar forces, van der Waals forces, hydrogen bonding, and specific compound formation are important intermolecular forces in GC columns. These forces determine the relative volatility of the adsorbed compounds and therefore play dominant roles in the separation process.

As a general rule, the liquid phase is most effective if it is chemically similar to the material being separated. Nonpolar liquid coatings are used to separate nonpolar compounds, whereas polar compounds are best separated with polar liquid phases. In part, this rule is simply a manifestation of the old adage "like dissolves like." Unless the sample dissolves well in the liquid phase, little or no separation occurs. Table 11.2 lists commonly used stationary phases for both packed and capillary columns and the type of compounds that can be separated on them.

The polysiloxanes have the same silicon-oxygen backbone, with some variation in the —R substituent:

Relative	Relative	Chemical	Flexible?
Low	Fast	Best	Yes
Low	Fast	↓	No
Low	Fast		No
High	Slow	Poorest	No

Polydimethyl siloxane has —CH_3 for all —R groups, making it a nonpolar stationary phase. The other polysiloxanes have various numbers of methyl groups replaced by other functional groups such as phenyl (—C_6H_5), 3,3,3-trifluoropropyl (—$C_2H_4CF_3$), and dicyanoallyl (—$C_3H_3(CN)_2$). These functional groups increase the polarity of the liquid and allow the design of a wide variety of stationary phases suited to almost any application.

Polyethylene glycol is commonly used for separating polar compounds and has the structure

$$HO—CH_2—CH_2—(O—CH_2—CH_2)_n—OH$$

Capillary columns provide much better resolution (separation) than packed columns do. A particular liquid phase will separate substances more efficiently on a capillary column than it does on a packed column because more interactions between the liquid phase and the components of the mixture occur per meter of column length in a capillary column. A capillary column is also much longer than a packed column.

The useful temperature range constitutes another important characteristic of the liquid phase. A stationary phase cannot be used under conditions in which it decomposes or in which its vapor pressure is high. Therefore, most columns have an indicated temperature maximum. Table 11.2 includes the temperature limits for the liquid phases listed. All stationary phases will evaporate or "bleed" if they are heated to a high temperature; this vaporized compound then fouls the detector.

The proper choice of liquid phase is a trial-and-error process. We recommend specific columns for the experiments that require GC in this book, although sooner or later you will need to make your own choices. The information in this section should help

Table 11.2. **Common stationary phases for GC**

Stationary phase	Common trade name	Maximum temperature, °C	Common applications
Polydimethyl siloxane	OV-1, SE-30	350	General-purpose nonpolar phase; hydrocarbons; polynuclear aromatics; drugs; steroids; PCBs
Poly(phenylmethyl-dimethyl) siloxane (10% phenyl)	OV-3, SE-52	350	Fatty acid methyl esters; alkaloids; drugs, halogenated compounds
Poly(phenylmethyl) siloxane (50% phenyl)	OV-17	250	Drugs; steroids; pesticides; glycols
Poly(trifluoropropyl-dimethyl) siloxane	OV-210	200	Chlorinated aromatics; nitroaromatics; alkyl-substituted benzenes
Polyethylene glycol	Carbowax 20M	250	Free acids; alcohols; ethers; essential oils; glycols
Poly(dicyanoallyldi-methyl) siloxane	OV-275	240	Polyunsaturated fatty acids; rosin acids; free acids; alcohols

SOURCE: Skoog, D. A.; Leary, J. J.; *Principles of Instrumental Analysis;* 4th ed.; 1992, p. 617, by permission of the publishers, Harcourt, Brace and Co., Orlando, FL.

prepare you to do that. Tables of appropriate liquid phases for specific classes of compounds can be found in the references at the end of the chapter.

11.3 Detectors

Two kinds of detectors are most often used in gas-liquid chromatography: *thermal conductivity detectors* and *flame ionization detectors*. The function of each is to "sense" a material and convert that sensing into an electrical signal.

Thermal Conductivity Detectors

Thermal conductivity detectors operate on the principle that heat can be conducted away from a hot body at a rate that depends on the composition of the gas surrounding the body. In other words,

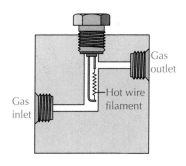

FIGURE 11.6 Thermal conductivity detector.

heat loss is related to gas composition. The electrical component of the detector is a hot wire or filament. Most of the heat loss from the hot wire of the detector occurs by conduction through the gas and depends on the rate at which gas molecules can diffuse to and from the metal surface. Helium, the carrier gas most often used with thermal conductivity detectors, has an extremely high thermal conductivity. However, organic molecules are less efficient heat conductors, because they diffuse more slowly. With only carrier gas flowing, a constant heat loss is maintained and there is constant electrical output. If the gas composition changes, however, the hot filament heats up and its electrical resistance increases. This change creates an imbalance in the electrical circuit that can be recorded.

As the separated materials exit from the column, they enter the detector cell. In practice, the filament of the detector, a tungsten-rhenium or platinum wire, operates at temperatures from 200°C to over 400°C. An enlarged view of the most common thermal conductivity detector is shown in Figure 11.6.

The electrical circuitry of thermal conductivity detectors is a Wheatstone bridge (Figure 11.7). When all resistances are balanced, there is no signal; this balance occurs when only carrier gas is passing through the detector. However, when the resistance of a filament changes as organic molecules pass through the detector, a signal that can be recorded arises. Thermal conductivity detectors have the advantage of stability, simplicity, and recovery of the separated materials, but the disadvantage of low sensitivity relative to that of flame ionization detectors. Because of their low sensitivity, they are unsuitable for use with capillary columns.

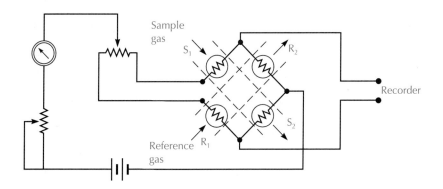

FIGURE 11.7 Thermal conductivity detector, balanced in a Wheatstone bridge.

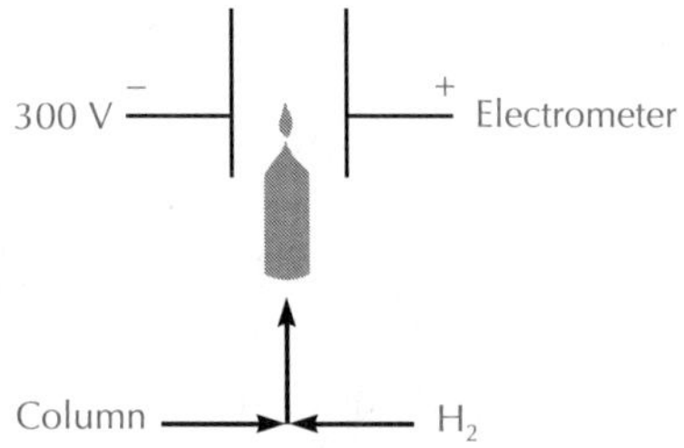

$$300\ V \xrightarrow{\quad -\quad} \quad \xrightarrow{\quad +\quad} \text{Electrometer}$$

Column $\longrightarrow$ $\longleftarrow$ H_2

$$H_2 + O_2 + \text{organic} \xrightarrow{\ \Delta\ } CO_2 + H_2O + 2(\text{ions})^+ + (\text{ions})^- + e^-$$

$$\Sigma(\text{ions})^- + \Sigma e^- \longrightarrow \text{current}$$

FIGURE 11.8 Chemical reactions in the flame ionization detector.

Flame Ionization Detectors   Flame ionization is much more sensitive than thermal conductivity and is used with capillary columns, where the amount of sample reaching the detector is much less than that from a packed column. Noncombustible gases such as water, carbon dioxide, sulfur dioxide, and various nitrogen oxides produce no response with a flame ionization detector, thus allowing the analysis of samples contaminated with water or these other oxides. However, the destruction of the sample is one disadvantage of a flame ion-

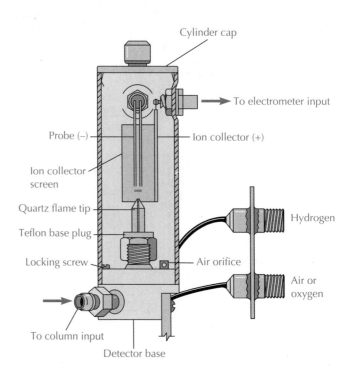

FIGURE 11.9 Flame ionization detector.

ization detector. An instrument equipped with a splitter that detours some of the column effluent before it reaches the detector eliminates this disadvantage.

In a flame ionization detector, the carrier gas leaving the column is mixed with hydrogen and air, and burned. The combustion process produces ions that alter the current output of the detector (Figure 11.8). A typical flame ionization detector is shown in Figure 11.9. In the chromatograph, the electrical output of the flame is fed to an electrometer, whose response can be recorded.

11.4
Interpreting Chromatograms

The recorded response to the electrical signal produced by the detector as the sample passes through it is called a *chromatogram.* A typical chromatogram for a mixture of alcohols is shown in Figure 11.10.

The line at the left shows when the sample was injected into the column. The pen responds to the changes in the electrical signal as each component of the mixture passes through the detector. You will notice that the later peaks are broader. This pattern is typical; the longer a compound remains on the column, the broader its peak will be when it passes through the detector.

Under a definite set of experimental conditions, a compound will always travel through a GC column in a fixed amount of time, called the *retention time.* The retention time for a compound, like the R_f value in thin-layer chromatography, is an important number, and it is reproducible if the same set of experimental conditions is maintained from one analysis to another.

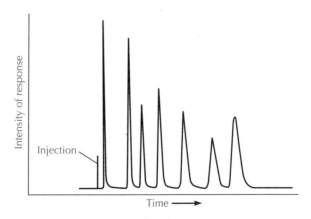

FIGURE 11.10 Typical gas-liquid chromatogram.

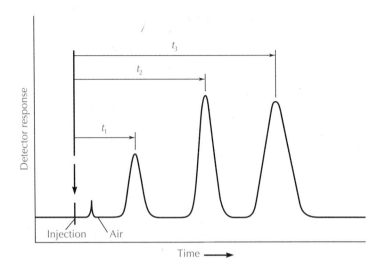

FIGURE 11.11 Measuring retention times.

In Figure 11.11 you can see how retention times are calculated from a GC chart record. The distance from the time of injection to the time that the pen reaches the peak maximum is the retention time for that compound. You can determine the retention time by measuring the distance on the chromatogram with a ruler and dividing the distance by the recorder chart speed. The retention time depends on many factors. Of course, the compound's structure is one of them. Beyond that, the kind and amount of stationary liquid phase used on the column, the length of the column, the carrier gas flow, the column temperature, the solid support, and the column diameter are most important. To some extent, the sample size can also affect the retention time. Whenever you note a retention time in your lab notebook, always record these experimental variables.

Many modern instruments come with digital integrators that determine the peak areas. The printed chromatogram includes a table of relative areas for all the peaks shown and usually gives the retention time in seconds or minutes.

11.5 Practical GC Operating Procedures

The procedure for a GC analysis follows these steps: Decide what column you are going to use. You will usually choose between a polar and a nonpolar column. Make sure that the chromatograph and the recorder are warmed up and ready to go, and that the GC column oven, detector, and injector port are at the correct temper-

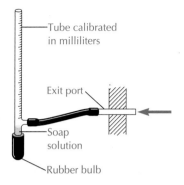

FIGURE 11.12 Bubble flowmeter.

atures. Temperature equilibration usually requires 20 to 30 min for a given set of parameters.

Ascertain that the carrier-gas pressure is properly set. The necessary pressure depends on the instrument and columns you are using, so you should check with your instructor before changing the pressure setting. Smaller diameter columns are normally operated at lower flow rates. For example, with a packed column 6 ft long and $\frac{1}{8}$ inch in diameter, a flow rate of $20-30$ mL·min^{-1} is common. For a $\frac{1}{4}$-inch column of the same length, $60-70$ mL·min^{-1} is usual. Flow rates for capillary columns generally range from 60 to 70 mL·min^{-1}.

A convenient measure of the carrier-gas flow rate in a packed column chromatograph is made at the exit port by using a soap-film (bubble) flowmeter (Figure 11.12). The rate at which a soap film moves up the calibrated tube is measured by using a sweep second hand or stop watch. Capillary column chromatographs have built-in flowmeters.

Turning on the Detector

A thin filament with electricity running through it can oxidize and burn in the presence of oxygen, as happens by design in an incandescent light bulb. Only after the carrier-gas flow has been on for 2 to 3 min should the current be turned on in a thermal conductivity detector. Within a few minutes, the current will stabilize. The flame ionization detector is not ignited until the injector port, column oven, and detector have reached the temperatures needed for the analysis. Before injecting the sample, the detector circuit must be balanced and the proper sensitivity (attenuator) chosen for the analysis. Your instructor will show you how to do these things, because the techniques differ for different instruments.

Sample Size

As soon as the instrument is ready, you can inject your sample. Gas-liquid chromatographs take very small samples; if too much sample is injected, poor separation will occur from overloading of the column. Normally, for a packed column GC, 1 to 3 μL of a volatile mixture are injected through the rubber septum by using a special microliter hypodermic syringe. For capillary columns, the sample must be extremely dilute, usually one or two drops diluted with a milliliter of a volatile solvent such as diethyl ether; diluted samples of 0.5 to 1.0 μL are injected. Consult your instructor about sample preparation and sample size for your particular chromatograph.

Injection Technique

Proper injection technique is important if you want to get well-formed peaks on the chromatogram. The needle should be pushed about 3 cm into the injection port and the injection made immediately with a rapid, smooth motion. This procedure ensures that all the sample reaches the column at the same time. Withdraw the syringe needle quickly after completing the injection.

The time of the injection can be recorded on the chromatogram in several ways. A mark can be made on the recorder base line just after the sample has been injected, but this is hard to do reproducibly. With a packed column, a better way is to include several microliters of air in your syringe. The air is injected at the same time that the sample is, and it comes through the column very quickly. The first tiny peak on Figure 11.11 is due to air. Retention times are then calculated, using this air peak as the injection time. A third way to mark the injection time works well with some packed column instruments. As you inject the sample, turn the base line knob on the recorder back and forth to make a mark on the chart paper. Finally, for an instrument equipped with an automatic digital integrator, simply press the start button at the time of sample injection.

A microliter syringe has a tiny bore that can easily become clogged if it is not rinsed after use. If viscous organic liquids or solutions containing acidic residues are allowed to remain in the syringe, you may find that it is almost impossible to move the plunger in and out. For this reason, a small bottle of acetone is often kept beside each instrument. One or two fillings of the syringe with acetone will normally suffice, if done directly after its use. However, it is unnecessary to rinse the syringe with acetone after each injection in a series of analyses. This practice may even cause confusion, because traces of acetone will show up on the chromatogram. For multiple analyses, it is best to rinse the syringe several times with the sample to be analyzed before filling the syringe with the injection sample.

After injection, wait for the peaks to appear on the moving chart. Sometimes it is difficult to know exactly how long to wait before injecting another sample, because components with unexpectedly long retention times may be in the sample. The analysis of unknown mixtures is a matter of trial and error. However, when you are analyzing mixtures with a known number of components, you need wait only until the last component has come through the column.

When you have finished your analyses, rinse out the hypodermic syringe with acetone and tear off the chart paper record. You will want to firmly attach your GC traces in your notebook, along with a notation of the experimental conditions under which the chromatograms were run.

The experimental parameters to record include injector port temperature, column temperature, detector temperature, gas flow rate, injection sample size, and type of column and nature of its stationary phase.

If the components of your mixture are not well separated, there are a number of factors that can be adjusted. You may have injected too much sample into the column; the column temperature may be too high; the wrong stationary liquid phase may have been used or the packed column may be too short.

11.6
Identification of Components and Quantitative Analysis

Gas-liquid chromatography is useful for both qualitative and quantitative analysis of volatile mixtures. Also, it can quickly help assess the purity of a compound. All these applications will be used in experiments in this book.

Comparison of Retention Times

As with TLC, a compound cannot be identified by GC unless a known sample is available. One method of identification compares the retention time of a known compound with the peaks on the chromatogram of the sample. If the operating conditions of the instrument are unchanged, a match of the known's retention time to one of the sample peaks serves to identify it. This method will not work for a mixture in which the identity of the components is totally unknown, because several compounds could have identical retention times.

Peak Enhancement

Peak enhancement serves as another method for identifying the compounds producing the peaks in a chromatogram. The sample being analyzed is "spiked" with a drop of the known compound and the mixture injected into the gas chromatograph. If the known is identical to one of the compounds in the mixture, only one peak will be observed for that compound, and it will be enhanced relative to the other peaks, because the retention time of the known compound exactly matches that of the unknown compound (Figure 11.13).

Spectrometric Methods

Positive identification of a completely unknown mixture requires the pairing of gas chromatographic methods for separation with

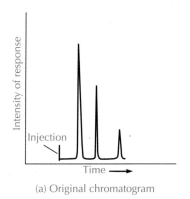

(a) Original chromatogram

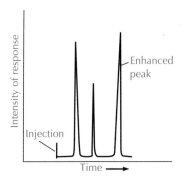

(b) Chromatogram after addition of a known compound identical to a compound in the sample

FIGURE 11.13 Identification by the peak enhancement method.

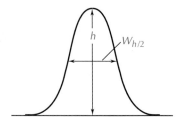

FIGURE 11.14 Determining peak area: h, height; $W_{h/2}$, width at half-height.

spectrometric methods such as IR, NMR, or mass spectrometry. The separated components from the gas chromatographic column may be collected for testing in one of these instruments. More frequently, the instruments are interfaced and the separated components pass directly to the spectrometer from the chromatograph.

Quantification   One of the great advantages of the GC is that approximate quantitative data are almost as easy to obtain as information on the number of components in a mixture. If we assume equal response by the detector to each compound, then *the relative amounts of compounds are proportional to their peak areas.* Each compound has a unique response in the detector, so, for accurate quantitative information, standard mixtures of known concentration must be run for calibration. The response factors nor-mally run from 0.8 to 1.2, so a reasonable estimate can be made simply by measuring the peak area. Most peaks are approxi-mately the shape of either an isosceles or a right triangle, whose areas are simply: $A = \frac{1}{2} \times$ base $\times$ height. The measurement of the base of most GC peaks is difficult, because more abnormalities in the shape occur in this region than in any other. A more accurate estimate of peak area is $A =$ height $\times$ width at half-height (Figure 11.14).

Mechanical or electronic digital integrators, common on most modern chromatographs, determine peak areas. Chromatograms produced by these recorders include a table of data that lists retention times and relative peak areas, making the calculations just described unnecessary. The easiest method for calculating the percentage composition of a mixture is through internal normalization. *The percentage of a compound in a mixture is its peak area divided by the sum of the peak areas.* For accurate quantitative data, careful calibration must be carried out. Detailed discussion of calibration methods can be found in the following references.

References

1. McNair, H. M.; Bonelli, E. J. *Basic Gas Chromatography Varian Aerograph*; 5th ed. Walnut Creek, CA, 1969.

2. Pecsok, R. L. *Principles and Practice of Gas Chromatography;* Wiley-Interscience: New York, 1972.
3. Miller, J. M. *Chromatography: Concepts and Contrasts;* Wiley-Interscience: New York, 1988.
4. Skoog, D. A.; Leary, J. J. *Principles of Instrumental Analysis;* 4th ed.; Saunders: New York, 1992.
5. Ravindrantath, B. *Principles and Practice of Chromatography;* Ellis Harwood Ltd.: Chichester, 1989.
6. Grob, R. L. *Modern Practice of Gas Chromatography;* 2nd ed.; Wiley: New York, 1985.

Questions

1. Why is a GC separation more efficient than a fractional distillation?
2. What characteristics must a carrier gas have to be useful for GC? What must the characteristics of the column packing be?
3. How do (a) the flow rate of the carrier gas and (b) the column temperature affect the retention time of a compound on a GC column?
4. Describe a method whereby a compound can be identified by using GC analysis.
5. Describe two methods by which a compound separated by and collected from a gas chromatograph can be identified.
6. If the resolution of two components in a GC analysis is mediocre but shows some peak separation, what are two adjustments in the operating parameters that can be made to improve the resolution (without changing columns or instruments)?
7. The perfume industry has long prided itself on the very sensitive nature of its testing procedures. Testers were trained over periods of years to smell the various essences when they were present in even the smallest concentrations. Suggest a way that GC could also be useful in perfume testing.
8. Suggest a suitable liquid stationary phase for the separation of (a) ethanol and water; (b) cyclopentanone (bp 130°C) and 2-hexanone (bp 128°C); (c) phenol (bp 182°C) and pentanoic acid (bp 186°C); (d) ethyl acetate (bp 77°C) and carbon tetrachloride (bp 77°C).

Technique 12
LIQUID CHROMATOGRAPHY

The discussion of liquid chromatography (LC), also called column chromatography, and the related methods of flash chromatography and high-performance liquid chromatography (HPLC) completes the triad of chromatographic methods so important in experimental organic chemistry. Thin-layer chromatography [Technique 10] and gas-liquid chromatography [Technique 11] have already been discussed in detail. Like TLC, liquid chromatography can be done under adsorption or partition conditions. Liquid chromatography is generally used to separate mixtures of low volatility. Unlike TLC, liquid chromatography can be carried out with a wide range of sample quantities, ranging from a few nanograms for HPLC up to 100 g for column chromatography.

12.1 Adsorbents

Two adsorbents are commonly used for the stationary phase in liquid chromatography: silica gel ($SiO_2 \cdot xH_2O$) and aluminum oxide (Al_2O_3). Both of these produce a polar stationary phase, with aluminum oxide being more polar; and both are generally used with nonpolar to moderately polar elution solvents as the mobile phase. Most chromatographic separations today use silica gel, because it allows the separation of compounds with a wide range of polarities. Silica gel also has the advantage of being less likely than alumina to cause a chemical reaction with the substances being separated. Aluminum oxide, or alumina, may be used for separations of compounds of low to medium polarity. Polar organic compounds, such as carboxylic acids, amines, phenols, and carbohydrates, adsorb so tightly on an alumina surface that highly polar solvents are needed to dislodge them.

Chromatographic silica gel has 10–20% adsorbed water by weight and acts as the solid support for water under the conditions of partition chromatography. Compounds separate by partitioning themselves between the elution solvent and the water that is strongly adsorbed on the silica surface. The partition equilibria depend on the relative solubilities of the compounds in the two liquid phases. The adsorptive properties of silica gel may vary greatly from one manufacturer to another, or even within different lots of the same grade from one manufacturer. Therefore, the solvent system previously used for a particular analysis may not work for another separation of the same sample mixture, when silica gel from another manufacturer or from a different lot from the same manufacturer is substituted for the stationary phase.

Activated alumina, made explicitly for chromatography, is available commercially as a finely ground powder in neutral (pH 7), basic (pH 10), and acidic (pH 4) grades. Different brands and grades vary enormously in adsorptive properties. Probably the most important factor is the amount of water adsorbed on the surface. Heating the alumina at 400–450°C until no more water is lost produces the most active grade, called Activity I in the classification system of Brockmann. Addition of 3% water gives alumina of Activity II, 6% water content gives Activity III, and 10% water yields Activity IV.

The strength of the adsorption of a substance on aluminum oxide depends on the strength of the bonding forces between the substance and the polar surface of the adsorbent. Dipole-dipole and van der Waals forces, as well as stronger hydrogen-bonding and coordination interactions, are important in the adsorption of compounds to aluminum oxide particles.

The separation of very polar compounds may require an adsorbent of lesser polarity than either silica gel or alumina. For this situation, reverse-phase chromatography may be useful in separating the compounds. ***Reverse-phase chromatography*** uses a stationary phase that is less polar than the mobile phase. These conditions cause elution of the more polar compounds first, whereas the less polar compounds adsorb more tightly to the stationary phase. Figure 12.1 illustrates the separation of two com-

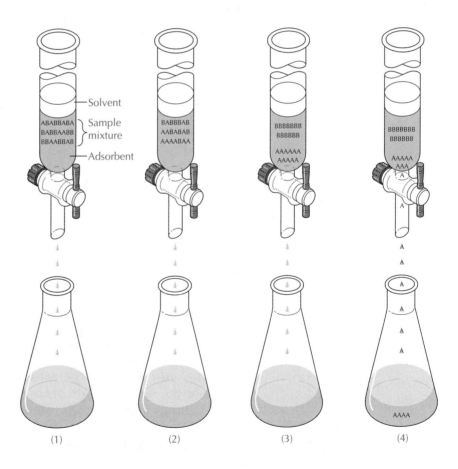

FIGURE 12.1 Stages in the separation of a mixture containing compounds A and B on a chromatographic column. Compound A moves faster and has a higher R_f value than does compound B. Compound B is more strongly adsorbed on the stationary phase.

pounds, A and B, on a chromatographic column. With a polar adsorbent such as alumina or silica gel, the compound represented by A would be less polar than compound B. In reverse-phase chromatography, compound A would be more polar than compound B.

12.2 Elution Solvents

In column chromatography, the solvents used to dislodge the compounds adsorbed on the column *(eluents)* are made progressively more polar. As we discussed in Technique 10 on TLC, nonpolar compounds bind less tightly on the solid adsorbent and dislodge easily with nonpolar solvents. Polar compounds adsorb more tightly to the surfaces of metal oxides. Therefore, polar compounds must be *eluted,* or washed out of the column, with polar solvents.

Consider an activated alumina column with a band of cholesterol adsorbed onto it. Cholesterol is an alcohol with a large surface area and medium polarity.

Cholesterol

Let's compare the action of a nonpolar elution solvent, such as hexane, to that of diethyl ether, a solvent of medium polarity. The primary forces between alumina and hexane molecules are weak van der Waals interactions. Hexane is not bound strongly to the column. The adsorption-desorption equilibria that occur as hexane travels down the column do little to dislodge cholesterol from the surface of the alumina. Diethyl ether, however, binds more strongly with alumina by dipole-dipole and coordination interactions. It can readily displace cholesterol as it travels down the column. In time, a band of cholesterol, dissolved in ether, can be collected from the bottom.

The balance between the activity of the adsorbent and the polarity of the solvent controls the rate at which materials elute from a column. If compounds elute too rapidly and poor separation occurs, either the adsorbent should be stronger or the solvent should be less polar. Conversely, if elution is so slow that only polar solvents are effective, a milder adsorbent is needed. Figure 12.2 gives the usual order of elution, from the least potent (nonpolar) to the most powerful (polar) elution solvent. It should be pointed out that there is no universal series of eluting strengths,

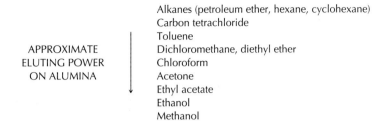

APPROXIMATE
ELUTING POWER
ON ALUMINA

Alkanes (petroleum ether, hexane, cyclohexane)
Carbon tetrachloride
Toluene
Dichloromethane, diethyl ether
Chloroform
Acetone
Ethyl acetate
Ethanol
Methanol

FIGURE 12.2 Relative eluting power of common elution solvents.

because this property depends not only on the activity of the adsorbent but sometimes also on the compounds being separated.

Elution solvents for column chromatography must be rigorously purified and dried for best results. Small quantities of polar impurities can radically enhance the eluting properties of a solvent. The presence of water in a solvent can make it far more powerful. For example, wet acetone may have an eluting power greater than dry ethanol.

12.3 Choosing a Column

After deciding which adsorbent to use for a separation, you need to decide how much adsorbent to use. This choice naturally depends on the amount of material that you wish to separate, but it also depends on the diameter of the cylindrical tube that you use. A long, thin adsorbent column will retain compounds more tenaciously than will a short, fat one. In general, one should use 20 to 30 times as much adsorbent as the material to be separated (Table 12.1). More should be used for a difficult separation, less for an easy one.

An 8:1 or 10:1 ratio of the height of the adsorbent to the column diameter is normal. Alumina has a bulk density of about

Table 12.1. **Representative columns**

Sample, g	Alumina, g	Column dimensions diameter × height in cm
10	250	3.4 × 27
2–3	60	2.0 × 19

FIGURE 12.3 Chromatographic columns.

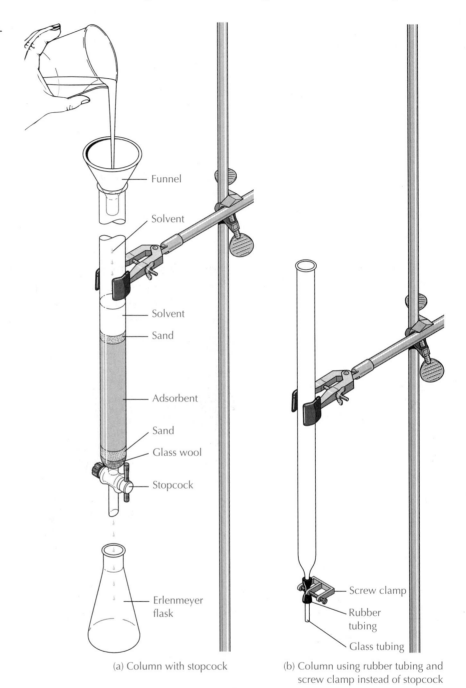

Funnel

Solvent

Solvent

Sand

Adsorbent

Sand

Glass wool

Stopcock

Erlenmeyer flask

Screw clamp

Rubber tubing

Glass tubing

(a) Column with stopcock

(b) Column using rubber tubing and screw clamp instead of stopcock

1 g·cm^{-3}. A calculation using the volume of a cylinder ($\pi r^2 h$) shows, for example, that 20 g of alumina would fill a column having a 1.5 cm diameter to a height of about 11 cm. Therefore, if you are using 20 g of alumina, choose a column with a diameter of about 1.5 cm. Silica gel has a bulk density of about 0.3 g·cm^{-3}, much lower than that of alumina. You would need to use a fatter column for the same amount of silica gel.

Generally we use a chromatography tube with a stopcock near the bottom (Figure 12.3a), although a glass tube with a tapered end, a small piece of rubber tubing closed with a screw clamp and a small piece of glass tubing at the bottom (Figure 12.3b) works just as well.

12.4 Making a Macroscale Column

After you have chosen a chromatography column and weighed out the adsorbent that you need, you can prepare the column for use. A chromatographic column in operation is shown in Figure 12.3a. The construction of such a column is just as crucial to the success of the chromatography as is the choice of adsorbent and elution solvents. If the adsorbent has cracks or channels or if it has a nonhorizontal or irregular surface, you will get poor separation.

Clamp the glass chromatography tube in a vertical position onto a ring stand and fill approximately one-half of it either with the first developing solvent that you plan to use or with a less polar solvent. Add a small piece of glass wool as a plug, and push it to the bottom of the column with a glass rod, making sure all the air bubbles are out of the glass wool. Cover the plug with approximately 6 mm of white sand. The plug and sand serve as a support base to keep the adsorbent in the column and prevent it from clogging the tip.

Pour the adsorbent powder slowly into the solvent-filled column. The stopcock should be closed. Take care that the adsorbent falls *uniformly* to the bottom. The adsorbent column should be firm, but if it is packed too tightly, the flow of elution solvents becomes too slow.

The top of the adsorbent must always be horizontal. Gentle tapping on the side of the column as the adsorbent falls through the solvent prevents the appearance of bubbles in the adsorbent. If large bubbles or channels develop in the column, the adsorbent should be discarded, and the column should be repacked. Any irregularities in the adsorbent may cause poor separation, because part of the advancing material will move faster than the rest. The time consumed in repacking will be much less than the time wasted trying to make a poor column function efficiently.

After all of the adsorbent has been added, carefully pour approximately 4 mm of white sand on top. This layer protects the adsorbent from mechanical disturbances when new solvents are poured into the column later.

If you use a solvent more polar than alkanes or carbon tetrachloride in making an alumina column, you may need to mix, or slurry, the adsorbent in some of the solvent before pouring it into the column. This precaution will prevent the formation of bubbles, which can form when heat is produced by the interaction between polar solvents and the surface of the metal oxide.

Be sure that the adsorbent is covered with solvent at all times during the chromatographic procedure.

The column must *never* be allowed to dry out once it is made; and the solvent level should *never* be allowed to go below the upper level of the adsorbent. Upon drying, the adsorbent may pull away from the walls of the column and channels can form. Once you begin a chromatographic separation, finish it without interruption. When you are finished, the chromatography tube can be emptied by opening the stopcock, pointing the upper end of the column into a beaker, and using gentle air pressure at the tip to push out the adsorbent.

12.5 Elution Techniques

The mixture of compounds being separated is always poured into the column in a solution. The solvent that is the first developing solvent or a solvent very similar in eluting power is the preferred solvent for the sample solution. The solution should be as concentrated as possible, preferably no more than 5 mL in volume. If the sample compounds are not soluble in the first elution solvent, a very small amount of a more polar solvent can be used to dissolve them.

Before the solution of the organic mixture is applied to the column, the solvent used in making the column should be allowed to drip out until the liquid level is just at the top of the upper sand layer. Then close the stopcock and pour the mixture to be separated into the column. After reopening the stopcock and allowing the upper level of the solution to reach the top of the sand, stop the flow, fill the column with elution solvent, and proceed to develop the chromatogram. Follow the same procedure when changing elution solvents.

Elution of the compounds in the mixture being separated is done by using a series of increasingly polar elution solvents. The less polar compounds elute first along with the less polar solvents. Polar compounds usually come out of a column only after a switch to a more polar solvent. As the development of the chromatography proceeds under the proper conditions, the compounds in the mixture put onto the top of the column will separate into distinct bands, or zones (Figure 12.4).

A mixture of two solvents is also commonly used for elution. For example, the development of the chromatography column can begin with hexane and, if nothing is eluted from the end of the column with this solvent, a 10% or 20% solution of ether in hexane can be used next. The ether/hexane solution can be followed by pure diethyl ether. Addition of small amounts of a polar solvent to a less polar one increases the eluting power in a gentle fashion. If the change of solvent is made too abruptly, enough heat may be generated from the alumina/solvent bonding to cause cracking or

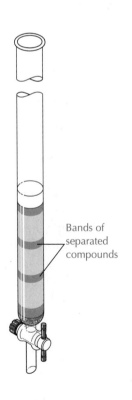

Bands of separated compounds

FIGURE 12.4 Chromatography column during elution.

channeling of the adsorbent column. In some cases, a low-boiling-point elution solvent may actually boil on the column. The bubbles that form will destroy the efficiency of the column.

Choosing an elution solvent is a trial-and-error process.

The proper choice of elution solvents and the amount to use is, in part, a trial-and-error process. We recommend the quantities of specific solvents to use for the experiments that require column chromatography in this book, although sooner or later you will have to make your own decisions. Nevertheless, polar compounds always require more polar elution solvents than do nonpolar compounds. For example, the separation of 1-decene from 2-chlorodecane requires developing solvents of low polarity, such as alkanes or carbon tetrachloride. However, the separation of the alcohol 2-decanol from its oxidation product, 2-decanone, requires more polar solvents, such as benzene or diethyl ether.

Thin-layer chromatography [see Technique 10] is one method that allows you to estimate how good a certain solvent will be as an eluent for a chromatographic separation. You should use the same adsorbent material as you expect to use on the column. The highest R_f value for any of the compounds in the mixture should be about 0.3 for a useful elution solvent.

Flow rate of the solvent depends on the height of the solvent.

A greater height of developing solvent above the adsorbent layer will provide a faster flow through the column. An optimum flow rate is about 2 mL·min^{-1}. Faster rates do not allow enough time for the adsorption equilibria to take place, and incomplete separation may occur. But if the flow is too slow, poor separation may result from the natural diffusion of the bands. A reservoir at the top of a column can be used to maintain a proper height of elution solvent above the adsorbent. A separatory funnel makes a good reservoir. It can be filled with the necessary amount of solvent and clamped onto the ring stand directly above the column. The stopcock of the separatory funnel can be adjusted so that liquid flows into the column as fast as it is flowing out the bottom.

Size of the eluent fractions depends on the experiment.

The size of the eluent fractions collected at the bottom of the column depends on the particular experiment. Common fraction sizes range from 25 to 150 mL. If the separated compounds are colored, it is a simple matter to tell when the different fractions should be collected. However, column chromatography is not limited to colored materials. With an efficient adsorbent column, each compound in the mixture being separated will be eluted separately. After one compound has come through the column, there will be a time lag before the next one appears. Therefore, there will be times when only solvent drips out of the bottom of the column. To ascertain when you should collect a new fraction of eluent, either note the presence of crystals forming on the tip of the column as the solvent evaporates or collect a few drops of liquid on a watch glass and evaporate the solvent. Any of the relatively nonvolatile compounds that are being separated will remain on the watch glass.

Pure components of the sample mixture are recovered by the evaporation of the solvent in the collected fractions. The use of a partial vacuum to aid in solvent evaporation is recommended. The ideal, although expensive, method for removal of solvents uses a rotary vacuum evaporator [see Technique 4.8].

12.6 Microscale Column Chromatography

Column chromatography is an effective way to remove small amounts of impurities from the product of a microscale synthesis. Silica gel usually works as the adsorbent for separating most organic compounds. Thin-layer chromatography on silica gel plates can be used to determine the best solvent system for separating the mixture [see Technique 10.4]; a solvent that moves the desired compound to an R_f of approximately 0.3 can be used for the elution solvent. The separation on a silica gel thin-layer plate with a particular solvent reflects the separation that the mixture will undergo with a silica gel column and the same solvent. A column suitable for separating 50–100 mg of sample can be prepared in a large-volume Pasteur pipet.* Regular-size Pasteur pipets can be used for 10–30 mg of sample.

Assemble all the equipment and reagents that you will need for the entire chromatography procedure before you begin to prepare the column.

Obtain about 50 mL of the elution solvent in a flask fitted with a cork. Place the mixture being separated in a small test tube and add approximately 1 mL of the elution solvent, or some other solvent that is less polar than the elution solvent, to dissolve the sample. Spot a silica gel thin-layer plate with the sample solution; develop and visualize the plate [see Technique 10]. Cork the test tube containing the sample while you analyze the thin-layer plate and prepare the column. Label a series of 10 test tubes (13 × 100 mm) for fraction collection. Pour 4 mL of elution solvent into one test tube and mark the liquid level on the outside of the test tube. Place a corresponding mark on the outside of the other test tubes.

Pack a small plug of glass wool into the stem of a large-volume Pasteur pipet, using a wooden applicator stick or a thin stirring rod (Figure 12.5). Clamp this pipet in a vertical position and place a 25-mL Erlenmeyer flask underneath it to collect the elution solvent that you will be adding to the top of the column. Weigh approximately 3 g of silica gel adsorbent in a tared 50-mL beaker; add enough elution solvent to make a thin slurry. Transfer the adsorbent slurry to the column, using a regular-size Pasteur pipet. Continue adding slurry until the column is two-thirds full of adsorbent. Fill the column four to five times with the elution solvent to pack the adsorbent well. The eluted solvent can be reused for this purpose only.

Be sure that the adsorbent is covered with solvent at all times during the chromatographic procedure.

*Available from Fisher Scientific, catalog no. 13678-8; these pipets have a capacity of 4 mL.

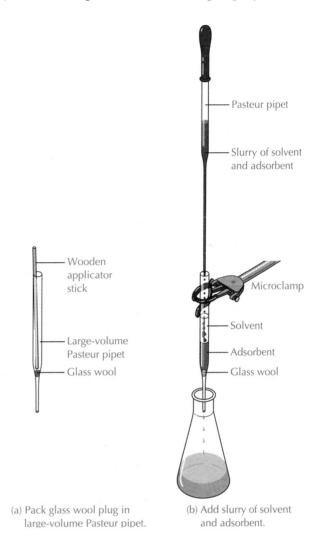

(a) Pack glass wool plug in
large-volume Pasteur pipet.

(b) Add slurry of solvent
and adsorbent.

FIGURE 12.5 Setting up a microscale column.

Allow the solvent level to almost reach the top of the adsorbent and place the test tube labeled Fraction 1 under the column. Begin transferring your sample mixture to the column, using a Pasteur pipet. When all the sample is on the column, begin adding the elution solvent. Collect fractions of approximately 4 mL in your labeled test tubes. Determine the composition of each fraction by TLC. Combine the fractions with the same composition and evaporate the solvent to recover each substance.

12.7
Summary of Column Chromatography Techniques

1. Prepare a properly packed column of adsorbent.
2. Add sample mixture to column in a small volume of solution.

3. Elute the adsorbed compounds with progressively more polar elution solvents.
4. Collect the eluted compounds in fractions from the bottom of the column.
5. Evaporate the solvents to recover the separated compounds.

12.8 Dry Column Chromatography

Dry column chromatography, a variant of column chromatography, is essentially a kind of scaled-up thin-layer chromatography. In wet column chromatography (elution chromatography), solutions of the separated compounds drain from the bottom of the column. In the dry column technique, the mixture separates into bands and the adsorbent is then removed from the column; no solvent is allowed to drip out the bottom. Disposable, plastic, flexible sleeves are often used for columns. The developed column is cut open and the bands of different compounds are cut apart. The compounds are recovered by washing the various sections of adsorbent with a polar solvent.

As in TLC, only one developing solvent is used, and R_f values can be calculated. A preliminary TLC provides a quick way of selecting the developing solvent [see Technique 10.4]. Visualization of the bands is not as simple as in thin-layer chromatography, a problem that limits the usefulness of dry column chromatography. The method is easiest to use with compounds that are colored, but the use of a fluorescent indicator aids in visualizing colorless compounds.

12.9 Flash Chromatography

Simple column chromatography [Techniques 12.1 to 12.6] is extremely time consuming. In the research laboratory, such gravity chromatography has been largely replaced by flash chromatography. The flash technique is not only much faster but also more efficient, because an adsorbent with a very small pore size (23–40 μm, 230–400 mesh) is used. The total time to prepare and completely elute the column is often under 15 min. The small particle size of the stationary phase requires pressures up to 20 pounds per square inch (psi), thus necessitating a special glass apparatus (Figure 12.6) and a source of nitrogen or air pressure. Although it is desirable to have an R_f difference (**DR_f**, component separation) for the compounds being separated of ≥ 0.35, it is possible to separate components with DR_f of ≈ 0.15.

The apparatus consists of a glass column topped by a variable bleed device (Figure 12.6). The bleed device has at its top a Teflon needle valve that controls the pressure applied to the top of the

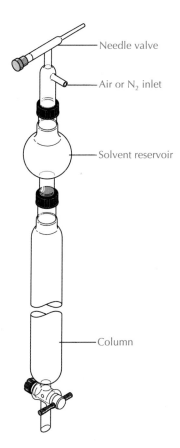

FIGURE 12.6 Apparatus for flash chromatography.

solvent in the column. Table 12.2 provides column and solvent dimensions for preparation of a silica gel column of 5–6 inches in height. We can see from this information that a smaller column diameter requires that the collected fraction sizes be correspondingly smaller. Moreover, as we would expect, the smaller the DR_f value, the smaller the size of the sample that can be placed on the column. The adsorbent is normally not mixed with fluorescent indicator, so every eluent fraction must be monitored by TLC or GC.

Before running the flash column, the TLC characteristics of the sample components should be determined. Ideally a solvent system that gives DR_f of ≥ 0.35 should be used. Systems that have been found useful include petroleum ether (30–60°C) mixed with any one of the following: ethyl acetate, acetone, or dichloromethane (methylene chloride).

Consult Table 12.2 when assembling the column. Either the available column sets the range of sample sizes that can be accommodated, or the size of the sample to be separated indicates the column size needed. To prepare the column, put a glass wool plug at the bottom of the column (a smaller glass tube may be used to insert the plug) and cover this with a thin layer (about one-eighth inch thick) of sand, 50–100 mesh. With the stopcock open, add with tapping, 5–6 inches of 23–40 μm silica gel (Aldrich and other suppliers indicate whether the silica gel is suitable for flash chromatography). Add a second layer of sand (about one-eighth inch thick) and level it with gentle tapping.

Fill the column to the top with the solvent system being used for elution. Insert the connected flow controller and, with the bleed valve open, gently turn on the flow of pressurized gas. Control the pressure by placing your finger over the exit tube (located

Table 12.2. Column dimensions and solvent volumes for flash chromatography

Column diameter, mm	Volume of eluent, mL	Typical sample size, mg		Recommended fraction size, mL
		$\Delta R_f > 0.2$	$\Delta R_f \approx 0.1$	
10	100	100	40	5
20	200	400	160	10
30	400	900	360	20
40	600	1600	600	30
50	1000	2500	1000	50

SOURCE: Still, W. C.; Kahn, M.; Mitra, A. *J. Org. Chem.* **1978,** *43,* 2923–2925.

around the needle valve) and manipulate the pressure such that the column is packed tightly by forcing all entrapped air out the bottom. With practice, you should reach an equilibrium pressure that causes the level of solvent to drop at a rate of 2 inches · min^{-1}. **Never let the column run dry** (the solvent should not go below the level of the top sand layer). When the solvent has just reached the sand, turn off the stopcock and remove the flow controller. Introduce a 20–25% solution of your sample in the elution solvent at the top of the column with a pipette. Reinsert the flow controller and adjust the pressure until the level of solvent at the top of the column drops at about 2 inches · min^{-1}. Collect the proper fraction volume of eluent solution (see Table 12.2) until all the solvent is used or until fraction monitoring indicates that the desired components have been eluted.

If the sample is not very soluble, use a small amount of polar solvent to get the sample on the column.

12.10 High-Performance Liquid Chromatography

Improved techniques and technology for gas chromatography have led to another advance in liquid chromatography, namely, the development of high-performance liquid chromatography (HPLC). High-performance liquid chromatography allows separations and analyses to be completed in minutes instead of hours. The principles of separation are the same, but HPLC gives better separation and detection of compounds present in low concentrations than other liquid chromatographic techniques do. The stationary phase in an HPLC column has a particle size of only 3–10 μm. The enhanced separation and sensitivity of the column comes from the increased surface area provided by these very small particles. However, particles of this small size pack very tightly, a condition that severely restricts the flow of solvent through the column. Pressures of 1000 to 6000 psi are required to force the solvent through the column at a rate of 1 to 2 mL · min^{-1}.

The instrumentation for high-pressure liquid chromatography consists of a column, a sample injection valve, a solvent reservoir, a pump, a detector, and a recorder or computer. Figure 12.7 shows a block diagram of a typical high-pressure liquid chromatograph.

The column can be from 5 to 30 cm long, with an inner diameter of 1–5 mm for analytical HPLC of 0.1–1 mg samples. Preparative HPLC columns range from 5 to 50 cm long with an interior diameter of 2.5–5.0 cm; such columns will separate samples of 0.5 g or more.

Columns used for separating the kinds of mixtures encountered in the organic laboratory commonly employ partition chromatography, in which the stationary phase is a liquid covalently bonded to microporous silica (SiO_2) particles.

Silica particle—Si—OH + (CH$_3$CH$_2$O)$_3$SiR $\longrightarrow$

$$\text{silica particle—Si—O—}\overset{\displaystyle OCH_2CH_3}{\underset{\displaystyle OCH_2CH_3}{\overset{|}{\underset{|}{Si}}}}\text{—R + CH}_3\text{CH}_2\text{OH}$$

Columns suitable for *normal-phase* chromatography have a stationary phase with polar substituents such as —(CH$_2$)$_3$NH$_2$ or —(CH$_2$)$_3$CN for the —R group in the above equation. In normal-phase HPLC, the stationary liquid phase is more polar than the solvent (liquid) phase; therefore, the less polar compounds in a mixture being analyzed elute first. *Reverse-phase* columns with long chain alkyl groups such as —(CH$_2$)$_{17}$CH$_3$ for the —R group are used to separate very polar compounds, for example, amino acids. In this case, the solvent is more polar than the stationary phase, so the more polar compounds elute first.

   The solvent is stored in a reservoir and passes through a filtration system before being pumped through the injector port, the column, and the detector. The solvents used for HPLC must be of very high purity because impurities would degrade the column by irreversible adsorption on the particles of the solid phase. More sophisticated instruments have several solvent reservoirs and a gradient elution system that allows the composition of solvent mixtures to be changed during the separation.

   The detector used for HPLC has to have a high sensitivity, usually in the microgram to nanogram range. The two most common types are ultraviolet (UV) and refractometer detectors. A UV

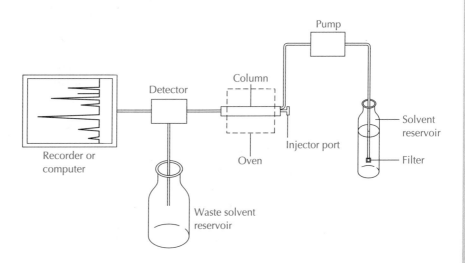

FIGURE 12.7 Block diagram of a typical high-performance liquid chromatograph.

detector is relatively inexpensive and can be used with gradient elution. It detects any organic compound that absorbs in the UV region. The limitations of a UV detector preclude its use with solvents that absorb in the UV region or with samples that do not have a chromophore for UV absorption. Refractometer detectors measure changes in the refractive index of the eluent as solutes move off column in the solvent stream. A refractometer detector cannot be used with gradient elution because the base line would change as the solvent composition changes.

References

1. Bobbitt, J. M.; Schwarting, A. E.; Gritter, R. J. *Introduction to Chromatography*; D. Van Nostrand: New York, 1968.
2. Heftmann, E. (Ed.) *Chromatography*; 2nd ed.; Reinhold: New York, 1967.
3. Miller, J. M. *Chromatography: Concepts and Contrasts*; Wiley-Interscience: New York, 1988.
4. Still, W. C.; Kahn, M.; Mitra, A. *J. Org. Chem.* **1978,** *43*, 2923–2925.
5. Kirkland, J. J.; Snyder, L. R. *Introduction to Modern Liquid Chromatography*; Wiley-Interscience: New York, 1974.
6. Snyder, L. R.; Glajch, J. L.; Kirkland, J. J. *Practical HPLC Method Development*; Wiley: New York, 1988.

Questions

1. Once a column of chromatographic adsorbent is made, it is important not to allow the liquid level to drop below the top of the adsorbent. Why?
2. What precautions must be taken when you introduce a mixture of compounds to be separated onto the adsorbent column?
3. What effect will the following factors have on a chromatographic separation: (a) a very rapid flow rate; (b) too strong an adsorbent; (c) collection of fractions that are too large?
4. Arrange the following compounds in order of decreasing ease of elution from a column of activated alumina: (a) 2-octanol; (b) *meta*-dichlorobenzene; (c) *tert*-butylcyclohexane; (d) benzoic acid.
5. What adsorbent would you choose for the separation of the following compounds by column chromatography: (a) 1,2-diphenylethanol and *trans*-1,2-diphenylethene; (b) benzoic acid and decanoic acid; (c) glucose and sucrose; (d) 2-ethyl-1,3-hexanediol and *N,N-meta*-toluamide?
6. Compare the advantages and limitations of column chromatography with those of thin-layer and gas-liquid chromatography.

POLARIMETRY

Optical activity, the ability of substances to rotate plane-polarized light, played a crucial role in the development of chemistry. It served as the vital link between the molecular structures that chemists write and the real physical world. A major development in the structural theory of chemistry was the concept of the three-dimensional structure of molecules. When van't Hoff and le Bel noted the asymmetry possible in tetrasubstituted carbon compounds, they claimed that their "chemical structures" were identical to the "physical structures" of the molecules. Not only was the fanciful structural theory of the organic chemist useful in explaining the facts of chemistry; it also happened to be "true." Van't Hoff and le Bel could make this claim because their theories of the tetrahedral carbon atom accounted, not only for chemical properties, but also for the optical activity of substances, a physical property.

13.1
Mixtures of Optical Isomers: Separation/Resolution

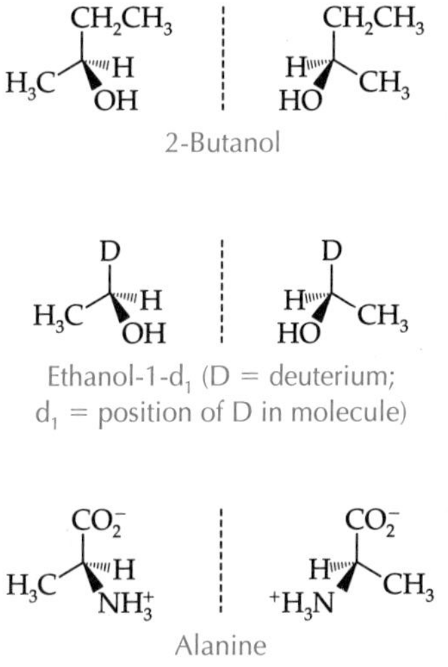

Structures that possess no internal mirror plane of symmetry are said to be chiral, or less accurately, asymmetric. A preliminary suggestion of chirality, a molecular property, can be gained from a *stereocenter* (sometimes called a chiral or asymmetric center), that is, a carbon center bearing four different substituents. More on stereochemistry can be found in Experiment 8 and in all organic textbooks.

A compound whose mirror image is not superimposable on itself is said to be *chiral,* or "handed." Chiral compounds thus possess the property of *enantiomerism.* Enantiomeric pairs are stereoisomers that are nonsuperimposable mirror images. Figure 13.1 depicts the enantiomeric pairs for 2-butanol, 1-deuterioethanol, and the amino acid alanine. Compounds such as these, containing one stereocenter (or "chiral center"), are simple examples of enantiomers. More complex cases of chiral compounds include those with more than one stereocenter in each structure, or those with no stereocenters (for example, helixes and allenes).

The enantiomers of 2-butanol have identical properties (including identical boiling points, IR spectra, NMR spectra, refractive indices, TLC R_f values), except the direction in which they rotate plane-polarized light. One of the enantiomers rotates polarized light in a clockwise direction and is called the *dextrorotatory* isomer, or more simply the *(+)-isomer.* The other enantiomer rotates polarized light counterclockwise and is called

2-Butanol

Ethanol-1-d₁ (D = deuterium; d₁ = position of D in molecule)

Alanine

FIGURE 13.1 Three-dimensional representations of enantiomeric pairs. The dotted line indicates the mirror between each pair.

the *levorotatory* isomer, or (−)-isomer. The rotational power of (+)-2-butanol is exactly the same in the clockwise direction as that of (−)-2-butanol in the counterclockwise direction. Unfortunately, there is no simple theoretical way to predict the direction of the rotation of plane-polarized light on the basis of the absolute configuration of a carbon stereocenter. Thus, in Figure 13.1 it is entirely random as to whether the structure on the left in each pair is dextrorotatory (+) or levorotatory (−).

Usually simple reagents obtained from the stockroom are optically inactive, even when the molecule is chiral. Thus you would normally find that a bottle of 2-butanol is optically inactive. This seeming paradox is explained by the fact that the reduction of 2-butanone with, for example, sodium borohydride, has two ways to proceed. It undergoes the reaction both ways at equal rates, thus giving rise to a 50:50 mixture of the product enantiomers, a product that is optically inactive (Figure 13.2).

An equal mixture of (+)- and (−)-enantiomers is called a *racemic mixture.* Before such a mixture can be separated, the enantiomers must be transformed into stereoisomers having different physical and chemical properties, that is, into *diastereomers.* A mixture of diastereomers is prepared from a racemic mixture by reaction with an optically active substance. The physical property often used for separating or resolving diastereomers is differential solubility.

Resolution with Acids or Bases

The simplest reaction for preparing diastereomers from racemic mixtures is that of an acid with a base to form a salt. For resolution, one of the reactants in the acid-base reaction must be optically active. Neutralization of a racemic amine, for example, by an

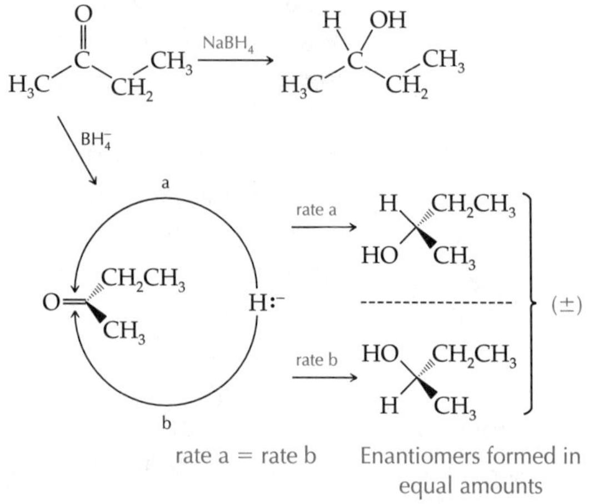

FIGURE 13.2 Reduction of 2-butanone with $NaBH_4$ produces a 50:50 mixture of enantiomers.

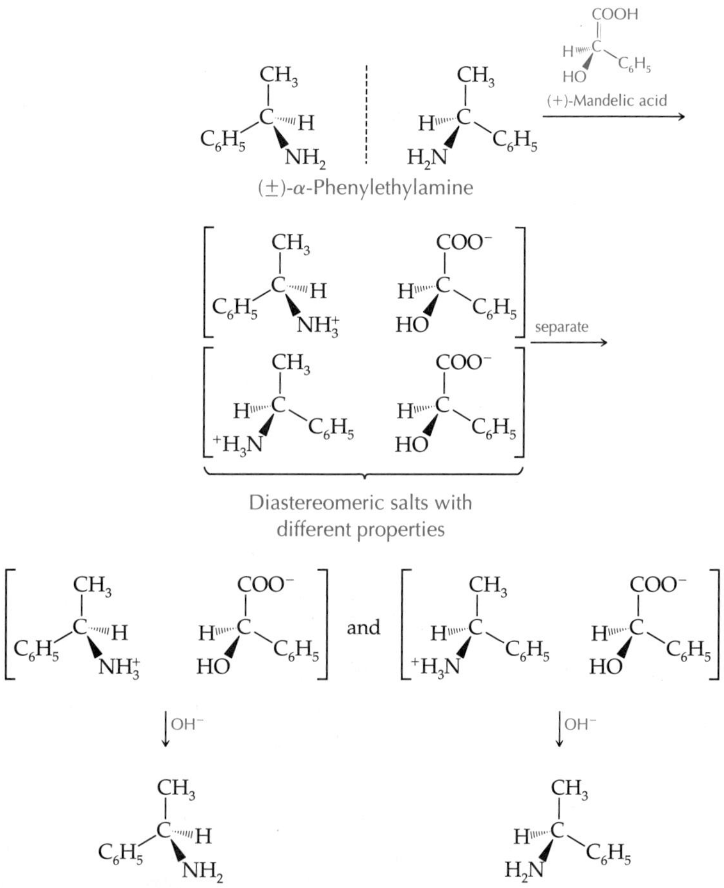

FIGURE 13.3 Resolution of an amine.

optically active carboxylic acid is a method for resolving the amine; neutralization of a carboxylic acid by an optically active amine is a way of resolving the acid. In such neutralizations, two salts are produced. These salts are diastereomers and, as such, differ in their solubilities in various solvents. Therefore, they can be separated by fractional crystallization and the less soluble diastereomeric salt is the more easily obtained. The process for resolution of an amine is represented in Figure 13.3. If we examine the structures of the compounds in Figure 13.3, we find that, in order to achieve a diastereomeric relationship, we need to have a pair of compounds, each of which has at least two chiral centers. When we compare our pair of carboxylate salt structures, we find that in each of the two compounds, the carbon stereocenters bearing the —OH groups have identical configurations, whereas the stereocenters bearing the —NH$_3^+$ groups have opposite configurations. Thus the two salts fit the definition of diastereomers: stereoisomers that are not enantiomers.

Table 13.1. **Optically active acids and bases used for resolutions**

Bases	Acids
Brucine	Tartaric acid
Strychnine	Mandelic acid
Quinine	Malic acid
Cinchonine	Camphor-10-sulfonic acid
α-Phenylethylamine	

Optically active acids and bases frequently used for the resolution of racemic substances are given in Table 13.1. The diastereomers necessary for resolution do not need to be salts. For example, diastereomeric esters may be obtained by reaction of the enantiomers of an alcohol with an optically active acid.

Enzymatic Resolution

An increasingly useful method for the resolution of racemic mixtures is the use of an enzyme that will selectively react with one of the enantiomers. Because all enzymes are chiral molecules, the transition states for the reaction of an enzyme with two enantiomers will be diastereomeric and the energies for forming these two transition states will differ. Thus one of the chiral enantiomers will react faster than the other. In many cases an enzyme will react specifically with only one enantiomer; and this specificity provides an excellent method for resolving a racemic mixture if an enzyme is available that will react with the compound.

Chiral Shift Reagents

Suppose we subject a racemic mixture to some sort of resolution procedure. After resolution has been completed, it is important to determine how successful the process has been. This can be done by one of several procedures in which we quickly and reversibly convert the chiral compound to a more complex one by coordinating the compound to a chiral structure of known stereochemistry. This reaction converts a mixture of enantiomers to a corresponding mixture of diastereomers. The composition of the diastereomers would then correspond to the composition of the original mixture of enantiomers. Below we briefly describe two methods of measuring enantiomeric enrichment: use of lanthanide shift reagents that are chiral and use of chiral chromatography columns.

Derivatives of camphor provide shift reagents that are rich in chiral character. Specifically, Eu(hfc)$_3$, called tris[3-heptafluoropropylhydroxymethylene)-(+)-camphorato] europium is such a compound. This compound undergoes rapid and reversible coordination with Lewis bases (B:). Thus, the following equilibrium is established:

$$Eu(hfc)_3 + B: \rightleftharpoons B:Eu(hfc)_3$$

The product complex B:Eu(hfc)$_3$ brings a paramagnetic ion, Eu^{3+}, into close proximity to the chiral organic compound (B:). This in turn induces changes in the ^{1}H NMR chemical shifts of B: in the complex. The shifts are different for protons (such as the one on the carbon bearing the amino group in α-phenylethylamine) in each of the two coordinated enantiomers because the formation of this diastereomeric complex pair causes those protons to become diastereotopic and thus nonequivalent (Figure 13.4). Therefore, we merely look for the peaks that arise from resonance of the shifted α-protons and measure their areas to determine the diasteromeric composition of the B:Eu(hfc)$_3$ complex, which in turn gives us the enantiomeric composition of the original amine.

Chiral Chromatography

The same general principle is used in chiral chromatography. When a mixture of enantiomers passes through a chiral chro-

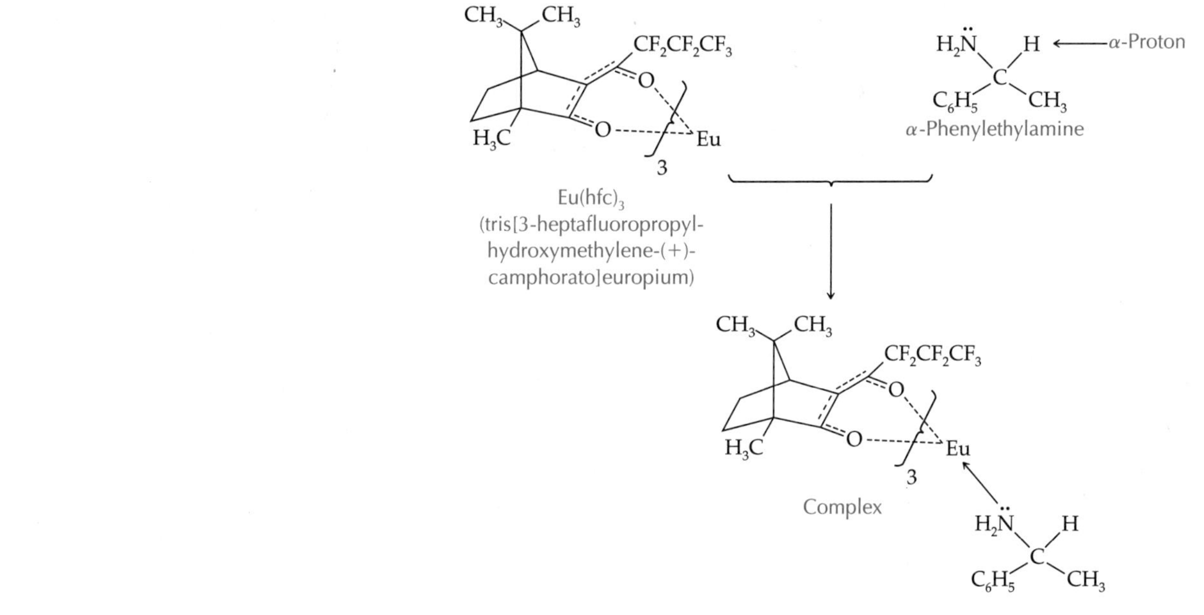

FIGURE 13.4 Structure of the chiral shift complex between Eu(hfc) and α-phenylethylamine.

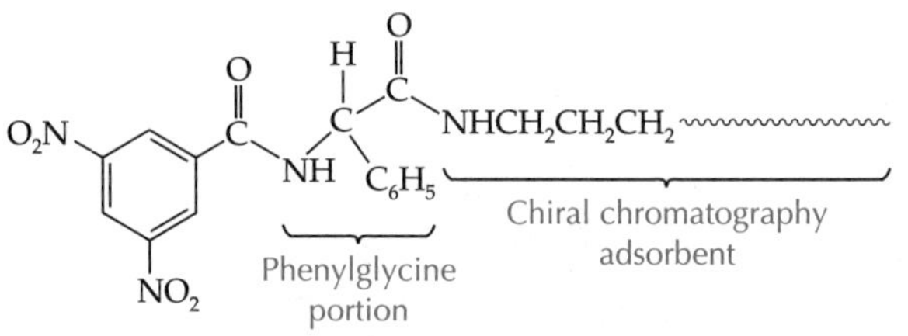

FIGURE 13.5 Bond between the 3,5-dinitrobenzoate derivative of phenylglycine and adsorbent of a chiral chromatography column.

matography column, we observe different coordination strengths between each enantiomer and the stationary phase, differences that lead to separation of the enantiomers. A typical stationary phase that produces this effect is composed of a chiral substance such as albumin or a carbohydrate immobilized by bonding to silica gel. The separation of the enantiomers of the 3,5-dinitrobenzoate derivative of phenylglycine, an amino acid, is an example of chiral chromatography. The less tightly coordinated enantiomer passes through the column first (Figure 13.5).

13.2
Polarimetric Instrumentation

Optical activity is measured with a polarimeter. This instrument, invented by Biot in the early nineteenth century, allows polarized light of a specific wavelength to shine through the sample contained in a long tube. The amount of rotation is measured in the analyzer. A schematic description of a polarimeter is shown in Figure 13.6. All commercially available polarimeters have these same general features. The analyzer of older polarimeters is adjusted

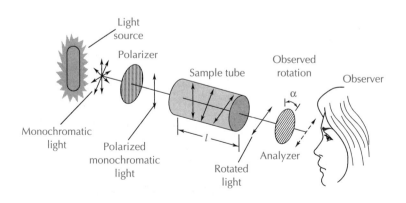

FIGURE 13.6 Schematic description of a polarimeter.

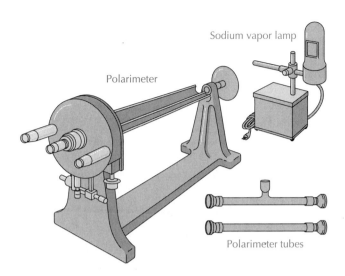

FIGURE 13.7 Polarimetric instrumentation.

manually (Figure 13.7), whereas more recent instruments have all the components housed in a case and are completely automated, with a digital readout of the observed rotation. Both types of instruments use the same kinds of sample tubes (Figure 13.7).

A normal light beam has wave oscillations in all the planes perpendicular to the direction in which the beam is traveling. When the light beam hits certain crystals or a piece of Polaroid plastic, which have ranks and files of molecules arranged in a highly ordered fashion, only the light whose oscillations are in certain planes is transmitted through the solid. The light that gets through is called ***plane-polarized light.*** The remaining waves are refracted away or absorbed by the polarizer. In a rough analogy, the light beam is made to pass through a solid whose atoms are ordered like the slats of wood in a picket fence. Only those light waves whose oscillations are parallel to the slats pass through.

The analyzer is a second polarizer whose files of molecules must also be lined up with the polarized light waves impinging on it in order to transmit the light. If the polarized light has been rotated by an optically active substance in the sample tube, the analyzer must be rotated the same amount to let it through. The rotation is measured in degrees. In practice, it is simpler to calibrate the analyzer to let a minimum of light through, that is, to set the analyzer at zero degrees rotation when its orientation is perpendicular to the polarizing crystal.

Monochromatic light is preferred in polarimetric measurements, because the optical activity or rotatory power of chiral compounds depends on the wavelength of the light used. For example, the rotation at 431 nm (blue light) of a sample of sucrose is 2.8 times greater than it is at 687 nm (red light). The most

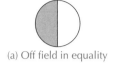 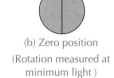 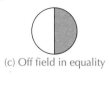

(a) Off field in equality   (b) Zero position   (c) Off field in equality
                                (Rotation measured at
                                  minimum light)

FIGURE 13.8 Representative image in the light field of a polarimeter.

common light source is a sodium lamp, which has two very intense emission lines at 589 and 589.6 nm. This closely spaced doublet is called the **sodium D line.** Another light source that is becoming more popular is a mercury lamp, using the intense 546.1-nm emission. The human eye is more sensitive to the mercury emission in the green region than to the sodium line in the yellow region of the visible spectrum.

A number of techniques are used to detect the rotation of polarized light by an optically active substance. The simplest way would be to rotate the analyzer until no light at all comes through. However, this method depends not only on the sensitivity of our eyes but also on our ability to remember quantitatively the amount of brightness that we have just seen. In practice, this is difficult to do accurately.

Various optical devices can be used to make the measurement of rotation easier. They depend on a sudden change of contrast when the minimum amount of light is transmitted by the analyzer crystal. What you would see in the analyzer of a typical manual polarimeter is shown in Figure 13.8.

13.3
Specific Rotation

The magnitude of the optical rotation depends on the concentration of the solution, the length of the light path through the solution, the wavelength of the light, the nature of the solvent, and the temperature. A typical rotation, that of common table sugar, sucrose, is written in the following manner:

$$[\alpha]_D^{20} = +66.4° \ (H_2O)$$

The symbol $[\alpha]_\lambda^{T°}$ is called the **specific rotation** and is an inherent property of the optically active molecule. $T°$ signifies the temperature of the measurement in degrees Celsius, and λ, the wavelength of light used. In the sucrose example, the sodium D line was used. The specific rotation is calculated from the observed angle of rotation:

$$[\alpha]_{\lambda}^{T°} = \frac{\alpha}{l \cdot c}$$

where α = the observed angle of rotation

l = the length of the light path through the sample (decimeters)

c = the concentration of the sample ($g \cdot mL^{-1}$ of solution)

The cell length is always given in *decimeters* (dm, 10^{-1} m) in the calculation of specific rotation. When a pure, optically active liquid is used as the sample, its concentration is simply the density of the liquid.

Sometimes a rotation of an optically active substance is given as a *molecular rotation:*

$$[M]_{\lambda}^{T°} = \frac{M}{100}[\alpha]$$

where M = molecular weight of the compound.

The value of the specific rotation can change considerably from solvent to solvent. It is even possible for an enantiomer to have a different sign of rotation in two different solvents. These solvent effects are primarily due to specific solvent-solute interactions. The four most popular solvents for polarimetry are water, methanol, ethanol, and chloroform.

The intrinsic specific rotation of a compound is generally considered to be a constant in dilute solutions at a particular temperature and wavelength. However, the specific rotation of a compound may also depend on the concentration of the compound in the solution. If you wish to compare the optical activity of a sample with that obtained by other workers, one should use the same concentration in the same solvent. Sucrose makes an excellent reference compound for polarimetry, because its specific rotation in water is essentially independent of concentration up to 5–10% solutions.

A change in the specific rotation due to temperature variation may be caused by a number of factors, including changes in molecular association, dipole-dipole interactions, conformation, and solvation. When nonpolar solutes are dissolved in nonpolar solvents, variation in the specific rotation with temperature may not be large. But there are cases where the specific rotation varies markedly with temperature. Near room temperature, the rotation of tartaric acid may vary by more than 10% per degree.

In theory, specific rotation should be independent of the concentration of the organic compound. In practice, this is often not the case. Concentration dependence is best avoided by using a dilute solution of the organic compound in a nonassociating solvent, but this is not always practical, especially for polar organic compounds. A record of the specific rotation result should include the solvent and concentration, as well as the temperature and wavelength of light used (D line of sodium is sufficient).

13.4
Using the Polarimeter

The control knobs on various models of polarimeters differ enough that no attempt will be made to discuss the details of operating a polarimeter. Nevertheless, certain techniques of polarimetry are general enough to merit discussion here.

Polarimeter tubes are expensive and must be handled carefully. They come in different lengths, with one-decimeter and two-decimeter tubes being the most common.

When you fill a polarimeter tube with a sample, make sure that the tube has no air bubbles trapped in it, because these will refract the light coming through. Also make sure that there are no suspended particles in a solution whose rotation you wish to measure, or you may get so little transmitted light that measurement of the rotation becomes very difficult. If you have a solution that you suspect is too turbid for polarimetry measurements, filter it by gravity through a small plug of glass wool. If the tube that you are using closes by screwing on an end plate, be careful not to screw it on too tightly, because strain in an end window can produce an apparent optical rotation.

A polarimeter can be standardized by filling a tube with an optically inactive solvent such as distilled water and adjusting the instrument to the zero position (Figure 13.8). Always approach the setting of a manual analyzer from the same direction, so that identical conditions are maintained in your measurements. If you are using a manual instrument, you should use the average of five to seven trials in setting the zero position.

Next you can try your hand at measuring the optical rotation of a sucrose-water solution whose concentration you know accurately. Use a 5.0% or a 10.0% solution. Check on your ability to use the polarimeter by calculating the specific rotation of sucrose, using the equation for [α]. Again, you should calculate the average of five to seven readings of the optical rotation.

To obtain the rotation of an optically active sample, you should first clean out a polarimeter tube with some care. After the tube is clean, it should be rinsed out with the solvent that you plan to use for the solution of your optically active compound. After the tube has been well drained, it should be rinsed with two or three small portions of your solution to ensure that the concentration of the solution in the polarimeter tube is the same as the solution that you made up. You may want to save these optically active rinses because your chiral compound can be recovered from them later.

Fill the tube with your solution, taking care to see that no air bubbles, turbidity, or dirty end windows destroy the validity of your measurements. Again you should use the average of five to seven readings of the rotation to calculate the specific rotation of your optically active substance.

13.5
Enantiomeric Excess (Optical Purity)

The results of polarimetric analysis of a stereoisomer are reported in terms of enantiomeric excess (or, less accurately, optical purity). *Enantiomeric excess* (% ee) is calculated from the expression

$$\% \text{ ee} = \left(\frac{[\alpha]_{\text{observed}}}{[\alpha]_{\text{pure}}} \right) \times 100\%$$

Thus, if we observe a rotation, α, for 2-butanol of $+6.5°$, we can calculate the enantiomeric excess (% ee) of the sample if we know the specific rotation of pure 2-butanol (it is $+13°$):

$$\% \text{ ee} = \left(\frac{6.5}{13} \right) \times 100\% = 50\%$$

It is instructive to examine the molecular composition of 100 molecules of the mixture of % ee = 50%. We have an excess of 50 (+) molecules, which causes the optical activity; and the remaining 50 molecules, because they have no net optical activity, are composed of 25 (+) molecules and 25 (−) molecules. Thus we have a total of 25 (−) molecules, and 50 + 25 = 75 (+) molecules.

References

1. Gordon, A. J.; Ford, R. A. *The Chemist's Companion: A Handbook of Practical Data, Techniques and References;* Wiley-Interscience: New York, 1972.
2. Heller, W. in *Physical Methods of Chemistry;* Vol. I, Part IIIC; Weissberger, A; Rossiter, B. W. (Eds.); Wiley-Interscience: New York, 1971.
3. Morrill, T. C. (Ed.) *Lanthanide Shift Reagents in Stereochemical Analysis;* VCH Publishers: New York, 1986.

Questions

1. A sample of 2-butanol shows an optical rotation of $+3.25°$. Determine the % ee and the molecular composition of this sample. The optical rotation of pure dextrorotatory 2-butanol is $+13°$.

2. A sample of 2-butanol (see Question 1) shows an optical rotation of $-9.75°$. Determine the % ee and the molecular composition of this sample.

3. An optical rotation study gives rise to the result of $\alpha = +140°$. Suggest a dilution experiment to test whether the result is indeed $+140°$ and not $-220°$.

4. The structures of strychnine (R = H) and brucine (R = CH$_3$O) are both common examples of alkaloid bases that can be used for resolutions. These are rich sources of chirality (respectively, $[\alpha]_D = -104°$ and $-85°$ in absolute ethanol). Assume that nitrogen inversion is slow and identify the eight chiral centers in these two nitrogen heterocyclic compounds.

R = H, strychnine
R = CH$_3$O, brucine

5. Only one of the two nitrogens in strychnine and brucine acts as the basic site for the necessary acid-base reaction for a resolution by diastereomeric salts. Which nitrogen, and why?

Tables of Unknowns and Derivatives

The following tables list almost 500 organic compounds and provide representative examples of the 16 functional-group classes that are treated in Part 3, Organic Qualitative Analysis. The functional groups appear in alphabetical order in Tables 1–16. In each table compounds are listed in order of increasing boiling points; solids appear last in each table in order of increasing melting points. We have arbitrarily listed compounds by increasing boiling points unless their melting points are above 40°C. Many pure compounds that melt just above room temperature may be in the liquid state when you encounter them due to depression of their melting points by impurities. Remember also that the melting points we list may be 2–3°C or even higher than the values you measure, since our listed melting points are those of the purified compounds.

When a derivative has been reported to *decompose* upon melting, we place a *d* next to the melting point. When *no solid derivative* was listed in standard reference tables for an alcohol, aldehyde, ketone, amine, carboxylic acid, or phenol, we place a dash in the table in place of the derivative's melting point. In a few cases, two widely different melting points for a compound are listed in the literature; we have included both, the second one in parentheses. The derivatives of multifunctional compounds that can form through derivatization at more than one site are shown by the notation "(mono)," "(di)," or "(tri)" after the melting point.

Table 1. Acyl chlorides

Compound	bp, °C	mp, °C
Acetyl chloride	51	
Propanoyl chloride	80	
Cyclopropanecarboxylic acid chloride	120	
2-Butenoyl chloride (*trans*)	126	
Heptanoyl chloride	176	
Cyclohexanecarboxylic acid chloride	184	
Succinyl dichloride	188	
Benzoyl chloride	197	
Phenylacetyl chloride	210	
4-Chlorobenzoyl chloride	222	
3-Chlorobenzoyl chloride	225	
4-Methylbenzoyl chloride	225	
2-Chlorobenzoyl chloride	233	
4-Methoxybenzoyl chloride	262	26
3-Nitrobenzoyl chloride	278	35
3,5-Dinitrobenzoyl chloride		68
4-Nitrobenzoyl chloride		75

Table 2. Alcohols

Compound	bp, °C	mp, °C	3,5-Dinitro-benzoate	1-Naphthyl-urethane	Phenyl-urethane
Methanol	65		107	124	47
Ethanol	78		93	79	52
2-Propanol	83		122	106	76
2-Methyl-2-propanol	83	25	142	101	136
2-Propen-1-ol	97		49	108	70
1-Propanol	97		74	80	57
2-Butanol	99		75	97	65
2-Methyl-2-butanol	102		116	72	42
2-Methyl-1-propanol	108		86	104	86
1-Butanol	116		64	71	61
3-Pentanol	116		99	95	48
2-Pentanol	119		61	74	——
3-Methyl-3-pentanol	123		96 (62)	83	43
2-Methoxyethanol	124		——	112	——
1-Chloro-2-propanol	127		77	——	——

(continued)

Table 2. **Alcohols**   *(continued)*

Compound	bp, °C	mp, °C	3,5-Dinitro-benzoate	1-Naphthyl-urethane	Phenyl-urethane
2-Methyl-1-butanol	129		70	82	31
3-Methyl-1-butanol	131		61	68	55
2-Chloroethanol	131		95	101	51
4-Methyl-2-pentanol	132		65	88	143
2-Ethoxyethanol	135		75	67	——
3-Hexanol	135		97 (77)	——	——
1-Pentanol	138		46	68	46
2-Hexanol	139		38	60	——
Cyclopentanol	140		115	118	132
3-Hydroxy-2-butanone	145		——	——	——
2,2,2-Trichloroethanol	151	19	142	120	87
1-Hexanol	157		58	60	42
2-Heptanol	159		49	54	——
Cyclohexanol	161	25	112	129	82
1-Amino-2-propanol	163		——	——	——
4-Hydroxy-4-methyl-2-pentanone	166		55	——	——
2-Furylmethanol	171		80	130	45
1-Heptanol	176		47	62	60
2-Octanol	179		32	63	114
1,2-Propanediol	187		——	153 (di)	——
1-Octanol	194		61	67	74
1,2-Ethanediol	197		169 (di)	176 (di)	157 (di)
1-Phenylethanol	202	20	93	106	92
Benzyl alcohol	205		112	134	77
1,3-Propanediol	214		178 (di)	164 (di)	137 (di)
2-Phenylethanol	219		108	119	78
1-Decanol	231		57	73	59
3-Phenyl-1-propanol	236		92	——	45
Cinnamyl alcohol	257	33	121	114	90
1-Dodecanol (lauryl alcohol)	259	24	60	80	74
4-Methoxybenzyl alcohol	259	24	——	——	92
Glycerol	290d		——	191 (tri)	180 (tri)
(−)-Menthol		43	153	119	111
2,2-Dimethyl-1-propanol	113	53	——	100	144
Diphenylmethanol (benzhydrol)	288	69	141	135	139
Benzoin		133 (137)	——	140	165
(−)-Cholesterol		148	——	176	168
(−)-Borneol		208	154	132	138

Table 3. **Aldehydes**

Compound	bp, °C	mp, °C	2,4-Dinitro-phenylhydrazone	Semicar-bazone
Propanal	49		150	89 (154)
2-Butenal (trans)	103		190	199
Hexanal	128		104	106
Heptanal	155		108	109
2-Furancarboxaldehyde	161		229 (212)	202
Cyclohexanecarboxaldehyde	161		172	173
Octanal	171		106	98
Benzaldehyde	179		237	222
Phenylethanal	194	34	121	153
2-Hydroxybenzaldehyde	196		248	231
4-Methylbenzaldehyde	204		232	234 (215)
2-Chlorobenzaldehyde	213		212	225 (146)
3-Phenylpropanal	224		149	127
3-Bromobenzaldehyde	235		——	205
2-Methoxybenzaldehyde	245	38	253	215d
4-Methoxybenzaldehyde	247		254	210
Cinnamaldehyde (trans)	252		255d	215
3,4-Methylenedioxybenzaldehyde	263	37	266d	230
3,4-Dimethoxybenzaldehyde		44	262	177
4-Chlorobenzaldehyde	214	47	254	230
4-Bromobenzaldehyde		56	257	228
3-Nitrobenzaldehyde		58	293d	246
4-Hydroxy-3-methoxybenzaldehyde (Vanillin)		80	271d	229
3-Hydroxybenzaldehyde		104	257d	198
4-Hydroxybenzaldehyde		115	280d	224

Table 4. **Alkanes**

Compound	bp, °C
Cyclopentane	49
2-Methylpentane	60
3-Methylpentane	63
Hexane	69
Cyclohexane	80
Heptane	98
Methylcyclohexane	101
Octane	125
Ethylcyclohexane	131
Cyclooctane	150

Table 5. **Alkenes**

Compound	bp, °C	mp, °C
Cyclopentene	44	
1-Hexene	64	
Cyclohexene	83	
1-Heptene	94	
Cycloheptene	114	
1-Octene	121	
Cyclooctene	144	
1,5-Cyclooctadiene	148	
Camphene	160	51
(+)-Limonene	176	
Indene	182	
Stilbene (*trans*)		124

Table 6. **Alkyl halides**

Compound	bp, °C
2-Chloropropane	36
Dichloromethane	41
1-Chloropropane	46
1-Bromopropene	60
2-Bromopropane	60
2-Chlorobutane	67
1-Chloro-2-methylpropane	68
1-Bromopropane	71
Iodoethane	72
1-Chlorobutane	77
2-Chloro-2-methylbutane	86
2-Bromobutane	90
1-Bromo-2-methylpropane	91
1-Bromobutane	100
1-Chloro-3-methylbutane	100
1-Chloropentane	107
Chlorocyclopentane	114
1-Bromo-3-methylbutane	120
1,3-Dichloropropane	124
1-Bromopentane	130
1-Chlorohexane	134
Bromocyclopentane	137
1-Chloroheptane	160
Bromocyclohexane	166
1-Bromoheptane	174
1-Chlorooctane	180
2-Chloro-1-phenylethane	190
Iodoform	119 (mp)

Table 7. Amides, imides, and lactams

Compound	bp, °C	mp, °C
N,N-Dimethylacetamide	165	
N,N-Diethylformamide	176	
Formamide	195d	
N-Methylacetamide	205	
N-Methylformanilide	244	
Formanilide		48
ε-Caprolactam (hexahydro-2 H-azepin-2-one)		70
Propanamide		79
Acetamide		82
Acrylamide		84
2-Chloroacetanilide		88
Methylurea		101
2-Acetamidotoluene		112
Acetanilide		114
Methacrylamide		116
Succinimide		124
Benzamide		128
Salicylanilide		136
Salicylamide		141
Phenylurea		147
4-Acetamidotoluene		153
Benzanilide		161
4-Acetamidophenol		169
4-Chloroacetanilide		179
Hippuric acid (N-benzoylglycine)		190
Saccharin		220
Caffeine		236
4-Acetamidobenzoic acid		260

Table 8. Amines

Compound	Class*	bp, °C	mp, °C	Benzamide	Benzene-sulfonamide	Picrate
tert-Butylamine	1°	46		134	——	198
Propylamine	1°	49		84	36	135
Diethylamine	2°	55		42	42	155
Allylamine	1°	58		——	39	140

*1°, primary; 2°, secondary; 3°, tertiary.

(continued)

Table 8. **Amines**   *(continued)*

Compound	Class*	bp, °C	mp, °C	Benzamide	Benzene-sulfonamide	Picrate
sec-Butylamine	1°	63		76	70	139
2-Amino-2-methylbutane	1°	77		——	——	183
Butylamine	1°	77		42	——	151
Diisopropylamine	2°	83		——	——	140
Pyrrolidine	2°	89		——	——	112 (163)
Triethylamine	3°	89		——	——	173
1-Aminopentane	1°	104		——	——	139
Piperidine	2°	105		48	93	152
Dipropylamine	2°	110		——	51	75
Ethylenediamine	1°	116		244 (di)	168 (di)	233 (di)
1-Aminohexane	1°	129		40	96	126
Morpholine	2°	130		75	118	146
Diisobutylamine	2°	138		——	55	121
2-Aminoheptane	1°	143		——	——	——
1-Aminoheptane	1°	155		——	——	121
Tripropylamine	3°	156		——	——	116
Dibutylamine	2°	160		——	——	59
1-Amino-2-propanol	1°	163		——	——	142
Indole-3-acetic acid	2°	168		——	——	——
2-Fluoroaniline	1°	176		113	——	——
N,N-Dimethylbenzylamine	3°	182		——	——	93
N-Methylbenzylamine	2°	182		——	——	117
Benzylamine	1°	184		105	88	194
4-Fluoroaniline	1°	186		185	——	——
1-Phenylethylamine	1°	187		120	——	——
2-Phenylethylamine	1°	198		116	69	174
Tributylamine	3°	216		——	——	106
4-Aminomethylpyridine	1° (3°)	230		——	——	——
Aminodiphenylmethane	1°	295		——	——	——
2-Amino-6-methylpyridine	1° (3°)	208	41	90	——	202
2-Benzoylpyridine	3°		42	——	——	130
4-Methylaniline	1°	200	45	158	120	182
2-Aminobiphenyl	1°		49	102	——	——
6-Amino-1-hexanol	1°		57	——	——	——
Methoxyaniline	1°		58	154	95	——
4-Bromoaniline	1°	245d	66	204	134	180
4-Iodoaniline	1°		67	222	——	——
2-Nitroaniline	1°		71	110	104	73
4-Chloroaniline	1°		72	192	122	——

*1°, primary; 2°, secondary; 3°, tertiary.

(continued)

Table 8. **Amines**   *(continued)*

Compound	Class*	bp, °C	mp, °C	Benzamide	Benzene-sulfonamide	Picrate
Tribenzylamine	3°		91	——	——	190
3-Nitroaniline	1°		113	155	136	143
2,4,6-Tribromoaniline	1°		119	198	——	——
3-Aminophenol	1°		122	174 (198)	——	——
3-(Dimethylaminomethyl)-indole	3°		133	——	——	——
2-Aminobenzoic acid	1°		146	182	——	104
4-Nitroaniline	1°		147	199	139	100
3-Aminobenzoic acid	1°		173	248	——	——
2-Aminophenol	1°		174	——	141	——
4-Aminophenol	1°		184	234 (di)	125	——
4-Aminobenzoic acid	1°		189	278	——	——
β-Alanine	1°		200d	120	——	——
Glycine	1°		230d	188	——	202

*1°, primary; 2°, secondary; 3°, tertiary.

Table 9. **Aromatic and polynuclear aromatic hydrocarbons**

Compound	bp, °C	mp, °C
Toluene	111	
Ethylbenzene	136	
1,4-Dimethylbenzene	138	
1,3-Dimethylbenzene	139	
1,2-Dimethylbenzene	144	
Isopropylbenzene (cumene)	152	
Propylbenzene	159	
1,3,5-Trimethylbenzene	163	
tert-Butylbenzene	168	
sec-Butylbenzene	173	
Indane	176	
4-Isopropyltoluene	177	
Indene	182	

(continued)

Table 9. **Aromatic and polynuclear aromatic hydrocarbons** *(continued)*

Compound	bp, °C	mp, °C
1,2,3,4-Tetrahydronaphthalene	207	
2-Methylnaphthalene	241	37
1-Methylnaphthalene	244	
Diphenylmethane	264	
1,4-Di-*tert*-Butylbenzene		76
Naphthalene		80
Triphenylmethane		92
Acenaphthene		96
Fluorene		112
Stilbene (*trans*)		124
Hexamethylbenzene		164
Anthracene		216

Table 10. **Aryl halides**

Compound	bp, °C	mp, °C
Chlorobenzene	132	
Bromobenzene	156	
2-Chlorotoluene	159	
3-Chlorotoluene	162	
4-Chlorotoluene	162	
1,3-Dichlorobenzene	172	
1,2-Dichlorobenzene	179	
4-Bromotoluene	185	28
Iodobenzene	188	
1,2,4-Trichlorobenzene	213	
1-Chloronaphthalene	260	
1-Bromonaphthalene	280	
1,2,3,4-Tetrachlorobenzene		44
1,2,3-Trichlorobenzene		52
1,4-Dichlorobenzene		53
1,3,5-Trichlorobenzene		63
1,4-Bromochlorobenzene		67
1,4-Dibromobenzene		89
Hexachlorobenzene		229

Table 11. **Carboxylic acids**

Compound	bp, °C	mp, °C	MW	Amide	Anilide
Formic acid	100		46	——	50
Acetic acid	118		60	82	114
Propanoic acid	140		74	81	105
Chloroacetic acid	185	61	94	121	135
2-Chloropropanoic acid	186		108	80	92
Dichloroacetic acid	194		128	98	118
2-Bromopropanoic acid	204	24	153	123	99
Bromoacetic acid	208	50	139	91	131
Octanoic acid	236		144	110	57
Decanoic acid	270	30	172	108 (100)	70
3-Chloropropanoic acid		41	107	101	——
Dodecanoic acid (lauric acid)		43	200	99	78
3-Phenylpropanoic acid	280	48 (40)	110	82 (105)	96
Trichloroacetic acid		57	163	141	97
2-Butenoic acid (*trans*)		72	86	160	118
Phenylacetic acid		76	136	156	117
2-Benzoylbenzoic acid (monohydrate)		90	208	165	195
		(anhydrous, 128)	226		
Pentanedioic acid (glutaric acid)		97	132	175 (di)	224 (di)
Phenoxyacetic acid		99	152	101	99
2-Methoxybenzoic acid		100	152	129	——
Citric acid (monohydrate)		100	210	210 (tri)	192 (tri)
		(anhydrous, 153)	192		
Oxalic acid (dihydrate)		104	126	419d (di)	254 (di)
4-Chlorophenylacetic acid		105	170	175	164
2-Methylbenzoic acid		105	136	142	125
3-Methylbenzoic acid		111	136	94	126
(+)- or (−)-Mandelic acid		118	152	133	151
3-Furancarboxylic acid		121	112	169	——
Benzoic acid		122	122	130	160
Maleic acid		130	116	260 (181) (di)	187 (di)
Cinnamic acid (*trans*)		133	148	147	153 (109)
2-Furancarboxylic acid		134	112	143	123
Decanedioic acid (sebacic acid)		134	202	210 (di)	202 (di)
Acetylsalicylic acid		136	180	138	136
2-Chlorobenzoic acid		140	156	142 (202)	114
3-Nitrobenzoic acid		140	167	142	153
2-Nitrobenzoic acid		146	167	174	155
2-Aminobenzoic acid		146	136	109	131
Diphenylacetic acid		148	212	168	180

(continued)

Table 11. **Carboxylic acids**   *(continued)*

Compound	bp, °C	mp, °C	MW	Amide	Anilide
Benzilic acid		150	228	153	174
2-Bromobenzoic acid		150	201	155	141
Hexanedioic acid (adipic acid)		153	146	220 (di)	240 (di)
3-Chlorobenzoic acid		158	156	134	122
2-Hydroxybenzoic acid (salicylic acid)		159	138	139	136
1-Naphthoic acid		160	172	203	163
2-Iodobenzoic acid		162	248	110	141
2,4-Dichlorobenzoic acid		162	191	194	——
(+)-Tartaric acid		169	150	195 (di)	263 (di)
4-Methylbenzoic acid		180	136	160	145
3,4-Dimethoxybenzoic acid		182	182	164	154
4-Methoxybenzoic acid		184	152	167	169
Butandioic acid (succinic acid)		188	118	260 (di)	230 (di)
4-Aminobenzoic acid		189	136		——
Hippuric acid (N-benzoylglycine)		190	179	183	208
β-Alanine		200d	89	——	——
1,2-Benzenedicarboxylic acid (phthalic acid)		206d	166	220 (di)	253 (di)
4-Hydroxybenzoic acid		215	138	162	196
Glycine		230d	75	——	——
4-Nitrobenzoic acid		241	167	201	211
4-Chlorobenzoic acid		242	156	179	194
(E)-2-Butenedioic acid (fumaric acid)		286	116	266 (di)	313 (di)

Table 12. **Esters**

Compound	bp, °C	mp, °C
Ethyl formate	54	
Methyl acetate	57	
Ethyl trifluoroacetate	61	
Ethyl acetate	77	
tert-Butyl acetate	98	
Propyl acetate	101	
Allyl acetate	103	
1-Methylpropyl acetate	112	
2-Methyl-1-propyl acetate	117	

(continued)

Table 12. **Esters** (*continued*)

Compound	bp, °C	mp, °C
Methyl 2-butenoate (*trans*)	119	
1-Butyl acetate	126	
3-Methyl-1-butyl acetate	142	
1-Butyl propanoate	146	
1-Pentyl acetate	149	
Ethyl 2-oxopropanoate	155	
Ethyl hexanoate	168	
Ethyl trichloroacetate	168	
Methyl acetoacetate	170	
2-Furylmethyl acetate	176	
Ethyl acetoacetate	181	
Methyl 2-furoate	181	
Dimethyl butanedioate	196	
Phenyl acetate	197	
Methyl benzoate	199	
4-Hydroxybutanoic acid lactone	204	
Ethyl 4-oxopentanoate	206	
4-Hydroxypentanoic acid lactone	207	
Ethyl benzoate	213	
Benzyl acetate	216	
Diethyl butanedioate	217	
Diethyl (*E*)-2-butenedioate	218	
Diethyl (*Z*)-2-butenedioate	222	
Methyl 2-hydroxybenzoate	223	
(−)-Bornyl acetate	223	29
Ethyl 2-hydroxybenzoate	234	
1-Butyl 4-oxopentanoate	237	
Ethyl decanoate	244	
1-Butyl benzoate	250	
Glyceryl triacetate	258	
Ethyl benzoylacetate	265	
Ethyl dodecanoate	269	
Ethyl cinnamate (*trans*)	271	
Dimethyl phthalate	284	
Dibutyl phthalate	340	
Benzyl benzoate	323	21
Methyl 3-pyridinecarboxylate	209	42
Phenyl 2-hydroxybenzoate		42
1-Naphthyl acetate		48
Methyl 2-hydroxy-2-phenylacetate		52
Phenyl benzoate		70
Diphenyl phthalate		75
Cholesteryl acetate		116 (94)
Ethyl 4-hydroxybenzoate		116
Acetylsalicylic acid (Aspirin)		136

Table 13. **Ethers**

Compound	bp, °C	mp, °C
Tetrahydrofuran	66	
1,2-Dimethoxyethane	85	
Dibutyl ether	143	
Methoxybenzene (anisole)	154	
Ethoxybenzene (phenetole)	168	
4-Methyl-1-methoxybenzene	174	
3-Methyl-1-methoxybenzene	176	
3-Chloro-1-methoxybenzene	193	
2-Chloro-1-methoxybenzene	196	
4-Chloro-1-methoxybenzene	202	
1,2-Dimethoxybenzene	207	
Butyl phenyl ether	210	
2-Bromo-1-methoxybenzene	223	
4-Bromo-1-methoxybenzene	223	
4-Propenyl-1-methoxybenzene (anethole)	237	23
2-Nitro-1-methoxybenzene	273	
4-Nitro-1-methoxybenzene		52
1,4-Dimethoxybenzene	213	60

Table 14. **Ketones**

Compound	bp, °C	mp, °C	2,4-Dinitro-phenylhydrazone	Semicarbazone
Acetone	56		126	190
2-Butanone	80		117	136
2,3-Butanedione	88		314 (di)	235 (mono) 278 (di)
3-Methyl-2-butanone	94		120	113
2-Pentanone	102		144	110
3-Pentanone	102		156	139
3,3-Dimethyl-2-butanone (pinacolone)	106		125	157
4-Methyl-2-pentanone	116		95	133
2,4-Dimethyl-3-pentanone	125		95	160
3-Hexanone	125		130	113
2-Hexanone	129		106 (110)	125

(continued)

Table 14. **Ketones**   *(Continued)*

Compound	bp, °C	mp, °C	2,4-Dinitro-phenylhydrazone	Semicarbazone
Cyclopentanone	131		146	210
2,4-Pentanedione	139		209 (di)	122 (mono)
				209 (di)
4-Heptanone	144		75	133
3-Hydroxy-2-butanone (acetoin)	145		318	185 (202)
3-Heptanone	148		——	103 (152)
2-Heptanone	151		89	127
Ethyl 2-oxopropanoate	155		——	——
Cyclohexanone	156		162	166
4-Hydroxy-4-methyl-2-pentanone	166		203	——
3-Octanone	167		——	——
2,6-Dimethyl-4-heptanone	168		66 (92)	126
3-Methylcyclohexanone	169		155	179 (191)
Methyl acetoacetate	170		——	152
4-Methylcyclohexanone	171		134	199
2-Octanone	172		58	122
Ethyl acetoacetate	181		93	133
Cycloheptanone	181		148	163
5-Nonanone	187		——	90
2,5-Hexanedione	194		257 (di)	185 (mono)
				224 (di)
2,6-Dimethyl-2,5-heptadien-4-one (phorone)	198	27	118	221 (186)
Acetophenone	204	20	240 (250)	198
Ethyl 4-oxopentanoate	206		102	148
1-Phenyl-2-propanone	216	27	156	199
Propiophenone	218	20	191	174
2-Methyl-1-phenyl-1-propanone	222		163	181
1-Phenyl-2-butanone	226		——	135 (146)
4-Methylacetophenone	226	28	260	205
(−)-Carvone	230		191	162
4-Chloroacetophenone	232		231	204 (160)
Butyl 4-oxopentanoate	237		——	——
4-Methoxyacetophenone	258	37	220 (231)	197
Ethyl benzoylacetate	265		——	125
1-Acetylnaphthalene	302		——	289 (233)

(continued)

Table 14. **Ketones**   *(Continued)*

Compound	bp, °C	mp, °C	2,4-Dinitro-phenylhydrazone	Semicarbazone
1,3-Diphenylacetone	330	33	100	146 (126)
2-Benzoylpyridine		42	199	——
Benzoin methyl ether		48	——	——
Benzophenone		48	239 (229)	164
4-Bromoacetophenone		51	230 (237)	208
2-Acetylnaphthalene		53	262	235
1,3-Diphenyl-2-propen-1-one (benzalacetophenone)		58	244	170 (179)
Benzoin ethyl ether		60	——	——
4-Methoxybenzophenone		62	180	——
4-Nitroacetophenone		80	——	——
3-Nitroacetophenone		81	228	257
Fluorenone		83	283	234
Benzil		95	189 (di)	174 (mono) 244 (di)
2-Benzoylbenzoic acid		128	——	——
Benzoin		133 (137)	245	206
4-Hydroxybenzophenone		134	242	194
(±)-Camphor		174	164	247 (237)
(+)-Camphor		180	177	——
Ninhydrin		243	——	——

Table 15. **Nitriles**

Compound	bp, °C	mp, °C
Acetonitrile	82	
3-Methylpropanenitrile	108	
Butanenitrile	117	
2-Butenenitrile (*trans*)	119	
Benzonitrile	190	
Phenylacetonitrile	234	
Butanedinitrile		56
Diphenylacetonitrile		75

Table 16. **Phenols**

Compound	bp, °C	mp, °C	Bromo derivative	1-Naphthyl-urethane	3,5-Dinitro-benzoate
2-Hydroxybenzaldehyde	196		——	——	——
2-Methoxyphenol (guaiacol)	205	32	116 (tri)	118	141
4-Chlorophenol	217	37 (43)	90 (di)	166	186
Methyl 2-hydroxybenzoate	223		——	——	——
2-*tert*-Butylphenol	224		——	——	——
Ethyl 2-hydroxybenzoate	234		——	——	——
Phenyl 2-hydroxybenzoate		42	——	——	——
2-Nitrophenol		45	117 (di)	113	155
5-Methyl-2-isopropyl-phenol (thymol)		50	55 (mono)	160	103
4-Hydroxy-3-methoxy-benzaldehyde (vanillin)		80	160	——	——
1-Naphthol		94	105 (di)	152	217
4-*tert*-Butylphenol		100	65 (di)	110	——
3-Hydroxybenzaldehyde		104	——	——	——
1,2-Dihydroxybenzene (catechol)		106	192 (tetra)	175	152 (di)
1,3-Dihydroxybenzene (resorcinol)		110	112 (di)	206	201 (di)
2-Chloro-4-nitrophenol		111	——		——
4-Nitrophenol		114	142 (di)	151	186
4-Hydroxybenzaldehyde		116	181 (di)	——	——
Ethyl 4-hydroxybenzoate		116	——	——	——
1,3,5-Trihydroxybenzene (phloroglucinol)		117 (dihyd) 219 (anhy)	151 (tri)	——	——
3-Aminophenol		122	——	——	179
2-Naphthol		123	84	156	210
1,2,3-Trihydroxybenzene (pyrogallol)		133	158 (di)	——	205 (tri)
4-Hydroxybenzophenone		134	——	——	——
2-Hydroxybenzanilide		136	——	——	——
2-Hydroxybenzamide		141	——	——	——
2-Hydroxybenzoic acid (salicylic acid)		159	——	——	——
4-Acetamidophenol		169	——	——	——
1,4-Dihydroxybenzene (hydroquinone)		171	186 (di)	——	317 (di)
2-Aminophenol		174	——	——	——
4-Aminophenol		184	——	——	178
4-Hydroxybenzoic acid		215	——	——	——
Phenolphthalein		265	——	——	——

The Literature of Organic Chemistry

Before organic chemists carry out reactions in the laboratory, they need to know what, if anything, is known about the reactions. The success and failures of earlier chemists are all to be found in the chemistry library. The successful completion of any research project requires that you make extensive use of the library to ascertain what is already known about the reaction that you are investigating as well as information about the physical and chemical properties of your compounds.

Four types of volumes are found in all chemistry libraries: textbooks, reference books, chemistry journals, and abstracts and indexes of chemical journals. Textbooks and reference books are frequently called *secondary sources* of information because they compile information from the *primary sources*, the journals in which the original works were published. Abstracts and indexes aid in locating a journal article about a specific topic or compound. In addition to books and journals, libraries now provide access to a vast array of databases through computers connected by phone lines to the computer containing the database. Use of these databases allows a search of the chemistry literature to be completed much more rapidly, and in some cases, more comprehensively, than by a conventional manual search of abstracts and indexes.

Textbooks. Textbooks contain general information. For example, an undergraduate organic chemistry text will tell you that aldehydes can be oxidized with chromic acid to carboxylic acids or that aromatic amines are weak bases. Specific information about compounds, such as the melting point or the ^{1}H NMR spectrum, is not commonly found in undergraduate textbooks. Advanced textbooks provide a more comprehensive treatment of organic reactions and their applications, and they usually include numerous references to the original literature and to review articles. Some representative advanced texts:

1. Carey, F. A.; Sundberg, R. M. *Advanced Organic Chemistry; Part A: Structure and Mechanisms;* and *Part B: Reactions and Synthesis;* 3rd ed.; Plenum: New York, 1990.
2. Lowry, T. H.; Richardson, K. S. *Mechanism and Theory in Organic Chemistry;* 3rd ed.; Harper and Row: New York, 1987.
3. March, J. *Advanced Organic Chemistry: Reactions, Mechanisms and Structures;* 4th ed.; Wiley-Interscience: New York, 1992.

Reference books. Reference books used frequently by organic chemists include the following.

Handbooks   Handbooks provide compilations of physical data, such as the boiling point of bromoethane, the melting point of the 3,5-dinitrobenzoate of 2-butanol, or the specific rotation of R-(+)-limonene. Some handbooks, including *The Merck Index* and the *Aldrich Catalog Handbook of Fine Chemicals,* also list references to the primary literature for preparation methods.

1. Lide, D. R. (Ed.) *The CRC Handbook of Chemistry and Physics;* 77th ed.; CRC Press: Boca Raton, FL, 1996.
2. Lide, D. R.; Milne, G. W. A. (Eds.) *Handbook of Data on Organic Compounds;* 3rd ed.; 7 Vols.; CRC Press: Boca Raton, FL, 1994.
3. Budavari, S.; O'Neil, M. J.; Smith, A.; Heckelman, P. E.; Kinneary, J. F. (Eds.) *The Merck Index: An Encyclopedia of Chemicals, Drugs and Biologicals;* 12th ed.; Merck & Co., Inc.: Whitehouse Station, NJ, 1996.
4. Dean, J. A. (Ed.) *Lange's Handbook of Chemistry;* 14th ed.; McGraw-Hill: New York, 1992.
5. *Aldrich Catalog Handbook of Fine Chemicals;* Aldrich Chemical Co., Milwaukee, WI, published annually.
6. Codagan, J. I. G.; Buckingham, J.; Macdonald, F. (Eds.) *Dictionary of Organic Compounds;* 6th ed.; 9 Vols.; Chapman and Hall: New York, 1996.
7. *Dictionary of Organometallic Compounds;* 2nd ed.; 5 Vols.; Chapman and Hall: London, 1995.
8. *Beilstein's Handbuch der Organischen Chemie;* Springer-Verlag: Berlin. (See discussion under Abstracts and Indexes.)

Spectral Information   The following reference books contain spectra of thousands of organic compounds. Additional compounds are added to the Stadtler collections (refs. 3 and 4) annually.

1. Pouchert, C. J.; Behnke, J. (Eds.) *Aldrich Library of ^{13}C and ^{1}H FT-NMR Spectra;* 3 Vols.; Aldrich Chemical Co., Milwaukee, WI, 1993.

2. Pouchert, C. J. (Ed.) *Aldrich Library of FT-IR Spectra;* 2 Vol.; Aldrich Chemical Co., Milwaukee, WI, 1985.

3. *The Sadtler Collection of High Resolution (NMR) Spectra;* Sadtler Research Laboratories: Philadelphia, 1992.

4. *The Sadtler Reference (IR) Spectra;* Sadtler Research Laboratories: Philadelphia, 1992.

Reactions, Synthetic Procedures, and Techniques   The following list gives representative examples of reference books that contain information about types of organic reactions (refs. 1–6), the wide range of methods used in organic synthesis (refs. 7–17), and descriptions of specific techniques used in organic syntheses (refs. 17–18). They all include extensive references to the primary literature.

1. *Organic Reactions;* Wiley: New York, 1932–present.

2. Coffey, S.; Ansell, M. F. (Eds.) *Rodd's Chemistry of Carbon Compounds;* 2nd ed.; Elsevier: New York, 1964–present.

3. Greene, T. W.; Wuts, P. G. M. *Protective Groups in Organic Synthesis;* 2nd ed.; Wiley: New York, 1991.

4. Katritzky, A. R.; Meth-Cohn, M.; Rees, C. W. (Eds.) *Comprehensive Organic Functional Group Transformations;* 7 Vols.; Pergamon Press: Oxford, UK, 1995.

5. LaRock, R. C. *Comprehensive Organic Transformations: A Guide to Functional Group Preparations;* VCH Publishers: New York, 1989.

6. Mundy, B.; Ellerd, M. G. *Name Reactions and Reagents in Organic Synthesis;* Wiley-Interscience: New York, 1988.

7. *Organic Syntheses;* Wiley: New York, 1932–present. *Collective Volumes I–VI* combine and index ten volumes each through Volume 59, 1988. *Collective Volume VII* compiles Volumes 60–64. Collective Volume VIII (1993) compiles Volumes 65–69. These preparations have been carefully checked in another research laboratory than the one where the procedure was developed.

8. Buehler, C. A.; Pearson, D. F. *Survey of Organic Synthesis;* 2 Vols.; Wiley: New York, 1988.

9. Carruthers, W. *Some Modern Methods of Organic Synthesis;* 3rd ed.; Cambridge University Press: New York, 1992.

10. Fieser, L. F.; Fieser, M. *Reagents for Organic Synthesis;* 17 vols.; Wiley: New York, 1967–1994.

11. Harrison, I. T.; Wade, Jr., L. G.; Smith, M. B. (Eds.) *Compendium of Organic Synthetic Methods;* 8 Vols.; Wiley: New York, 1971–1995.

12. Mackie, R. D.; Smith, D. M. *Guidebook to Organic Synthesis;* Longman: New York, 1982.

13. Paquette, L. A. (Ed.) *Encyclopedia of Reagents for Organic Synthesis;* Wiley: New York, 1995.

14. Sandler, S. R.; Karo, W. *Sourcebook of Advanced Laboratory Preparations;* Academic Press: San Diego, CA, 1992.

15. Trost, B. M.; Fleming, I. (Eds.) *Comprehensive Organic Synthesis;* 9 Vols.; Pergamon Press: Oxford, UK, 1984.

16. Weintraub, P. M. (Ed.) *Annual Reports in Organic Synthesis;* Academic Press: San Diego, CA, 1970–present.

17. Furniss, B. S.; Hannaford, A. J.; Smith, P. W. G.; Tatchell, A. R. (Eds.) *Vogel's Textbook of Practical Organic Chemistry;* 5th ed.; Longman: New York, 1989.

18. Loewenthal, H. J. E. *A Guide for the Perplexed Organic Experimentalist;* 2nd ed.; Wiley: New York, 1990.

Chemistry journals. Primary journals publish original research results and as such, they are the fundamental source of most information about organic chemistry. References to primary chemistry journals are found in the chapters of general reference books such as those listed in the previous section. For example, you might find the boiling point of vinyl benzoate in *The Dictionary of Organic Compounds,* but a detailed preparation method for it will be found in a journal article written by the organic chemist who first prepared the compound. Important current journals that publish original papers in organic chemistry include the following: *Angewandte Chemie* (International Edition in English); *Bulletin of the Chemical Society of Japan* (English); *Canadian Journal of Chemistry; Chemische Berichte* (German); *Journal of the American Chemical Society; Journal of Bioorganic Chemistry; Journal of the Chemical Society* (London), *Chemical Communications; Journal of the Chemical Society* (London), *Perkin Transactions 1; Journal of the Chemical Society* (London), *Perkin Transactions 2; Journal of Heterocyclic Chemistry; Journal of Medicinal Chemistry; Journal of Organic Chemistry; Journal of Physical Organic Chemistry; Organometallics; Nouveau Journal de Chimie* (French); *Synthesis; Synthetic Communications; Tetrahedron; Tetrahedron Letters.*

Abstracts and indexes. Because the literature of chemistry is so vast, finding specific information, such as the preparation of a particular compound or reactions where a specific type of catalyst has been used, would be extremely difficult and time consuming without the existence of a survey of the entire literature of chemistry. *Chemical Abstracts* (CA) provides a survey and is the most complete source of information on chemistry in the world.

Chemical Abstracts condenses the content of journal articles into abstracts and publishes indexes of the abstracts by the following categories: subject, author's name, chemical substance, molecular formula, and patent numbers. To locate a compound in *Chemical Abstracts,* it is necessary to understand the nomenclature systems used both currently and in the past; the *Index Guide* (1991) provides a list of indexing terms, rules, and cross references and is the place to begin a search. Each chemical compound has been assigned a number, called a **registry number,** which greatly facilitates finding references to that compound. Registry numbers can be located in the chemical substance index. In evaluating an abstract you need to keep in mind that abstracts are useful for deciding if an article contains relevant information about the question for which you are seeking information, but because they give only a brief (and sometimes incomplete) summary of an article, you should always consult the original journal article as the final source.

Chemical Abstract Services (CAS), the publishers of *Chemical Abstracts*, also provides a number of databases. One of these databases, called CA, allows access to all information (including the complete abstracts) published in *Chemical Abstracts* since 1967. Companion databases can be used for structures, molecular formulas, chemical names, and papers published before 1967, and for chemicals regulated by the Environmental Protection Agency (EPA) under the Toxic Substances Control Act. Today, most college and university libraries are equipped to search *Chemical Abstracts* using these computerized databases, which are available through several vendors. There is a fee for these services, so it is necessary for you to obtain assistance and training before undertaking an online search. Consult the library at your college or university.

Science Citation Index lists all articles published in the more prominent journals and also lists all the articles which were cited or referred to in current articles. Each year a *Source Index*, *Citation Index*, and *Permuterm Subject Index* are published as part of *Science Citation Index*. *Science Citation Index* is also available in a format that allows online computer searching.

Beilstein's Handbuch der Organischen Chemie (Handbook of Organic Chemistry), while not an abstract journal or index, serves a similar purpose in locating information about organic compounds. This unique compendium covers all known organic compounds on which information was published through 1959. For each compound, the handbook gives the name (or names), the formula, all known physical properties, methods of synthesis, biological properties, derivatives, and any other information reported about it. Organic compounds are organized into three categories: acyclic (Volumes 1–4), isocyclic (carbocyclic; Volumes 5–16), and heterocyclic (Volumes 17–27). The original volumes (Hauptwerk, H) covered the literature through 1909. Four supplements, (Ergänzungswerke, EI–EIV), cover the years 1910–1959 in ten-year increments. A fifth supplement (EV), in English instead of German, is still in preparation and covers the literature from 1960 to 1979.

Every piece of information has a reference to the primary literature and thus data may be checked. Moreover, information is not incorporated into *Beilstein* until it has been evaluated by scientists. The more recent supplements continue to report information on many compounds reported in the original work or earlier supplements. Thus, corrections and updating continue.

Because the system used to classify compounds is complex, several guides to using *Beilstein* have been prepared, including *How to Use Beilstein*, Beilstein Institute, Frankfurt, 1978, and a videocassette by M. M. Vestling, *A Guide to Beilstein's Handbook*, J. Hurley Assoc., Boca Raton, FL, 1988. Even with little or no knowledge of German, it is possible to find physical properties and references to the primary literature; the Beilstein Institute also publishes a booklet entitled *Beilstein Dictionary, German-English*. A search for information on a specific compound is most easily started by use of the formula index *(Formelregister)* rather than the subject index *(Sachregister)*. The second supplement *(Zweites Ergänzungswerk)* has a complete formula index in separate volumes, whereas the fourth supplement has a cumulative formula index and a subject index in each volume *(Band)*. After carbon and hydrogen, elements are listed alphabetically. For example, the compound 4-chlorobromobenzene would be listed at C_6H_4BrCl in the formula index. The boldface number immediately following the name of the compound is the volume number, while the page number in the initial edition follows next. The Roman numeral is the supplement (E) number, and it is followed by the page number in that supplement in the same volume given in boldface after the name of the compound. The entry for an organic compound in the *CRC Handbook of Chemistry and Physics* and the *Aldrich Catalog* also gives the location of the compound in *Beilstein*.

More information about the chemistry library. We urge you to consult the library at your college or university for assistance in conducting a search for information both in the books and journals found in your library as well as online information. The following books and journal articles contain more information about the chemistry library, how to use it, and how to plan and carry out an online search.

1. Bottle, R. T.; Rowland, J. F. B. *Information Sources in Chemistry;* 4th ed.; Bowker-Saur: London, 1993.
2. Loewenthal, H. J. E. *A Guide for the Perplexed Organic Experimentalist;* 2nd ed.; Wiley: New York, 1990, pp 1–81.
3. Maizell, R. E. *How to Find Chemical Information;* 2nd ed.; Wiley: New York, 1987.
4. March, J. *Advanced Organic Chemistry: Reactions, Mechanisms and Structures;* 4th ed.; Wiley-Interscience: New York, 1992, Appendix A.
5. Robinson, H. W.; Buchanan, T. M. *Chemical Information: A Practical Guide to Utilization;* Wiley: New York, 1988.
6. Somerville, A. N. *J. Chem. Educ.* **1991,** *68,* 553–561; 842–853; **1992,** *69,* 379–386.
7. Wienbroer, D. R. *The McGraw-Hill Guide to Electronic Research and Documentation;* McGraw-Hill: New York, 1997.
8. *Using CAS Databases on STN: Student Manual;* American Chemical Society: Washington, 1995.

Index

Note: Page numbers in *italics* indicate illustrations; those followed by t indicate tables.